TABLE B Gestation Table for Cows Based on a 285-Day Gestation Length

Find date of service in upper line. Figure below indicates date due to calve.

Jan	1	2	3	4	5	6	7	8	9	10	11	12	13	14	15	16	17	18	19	20	21	22	23	24	25	26	27	28	29	30	31	
Oct	13	14	15	16	17	18	19	20	21	22	23	24	25	26	27	28	29	30	31	1	2	3	4	5	6	7	8	9	10	11	12	Nov
Feb	1	2	3	4	5	6	7	8	9	10	11	12	13	14	15	16	17	18	19	20	21	22	23	24	25	26	27	28				
Nov	13	14	15	16	17	18	19	20	21	22	23	24	25	26	27	28	29	30	1	2	3	4	5	6	7	8	9	10				Dec
Mar	1	2	3	4	5	6	7	8	9	10	11	12	13	14	15	16	17	18	19	20	21	22	23	24	25	26	27	28	29	30	31	
Dec	11	12	13	14	15	16	17	18	19	20	21	22	23	24	25	26	27	28	29	30	31	1	2	3	4	5	6	7	8	9	10	Jan
Apr	1	2	3	4	5	6	7	8	9	10	11	12	13	14	15	16	17	18	19	20	21	22	23	24	25	26	27	28	29	30		
Jan	11	12	13	14	15	16	17	18	19	20	21	22	23	24	25	26	27	28	29	30	31	1	2	3	4	5	6	7	8	9		Feb
May	1	2	3	4	5	6	7	8	9	10	11	12	13	14	15	16	17	18	19	20	21	22	23	24	25	26	27	28	29	30	31	
Feb	10	11	12	13	14	15	16	17	18	19	20	21	22	23	24	25	26	27	28	1	2	3	4	5	6	7	8	9	10	11	12	Mar
Jun	1	2	3	4	5	6	7	8	9	10	11	12	13	14	15	16	17	18	19	20	21	22	23	24	25	26	27	28	29	30		
Mar	13	14	15	16	17	18	19	20	21	22	23	24	25	26	27	28	29	30	31	1	2	3	4	5	6	7	8	9	10	11		Apr
Jul	1	2	3	4	5	6	7	8	9	10	11	12	13	14	15	16	17	18	19	20	21	22	23	24	25	26	27	28	29	30	31	
Apr	12	13	14	15	16	17	18	19	20	21	22	23	24	25	26	27	28	29	30	1	2	3	4	5	6	7	8	9	10	11	12	May
Aug	1	2	3	4	5	6	7	8	9	10	11	12	13	14	15	16	17	18	19	20	21	22	23	24	25	26	27	28	29	30	31	
May	13	14	15	16	17	18	19	20	21	22	23	24	25	26	27	28	29	30	31	1	2	3	4	5	6	7	8	9	10	11	12	Jun
Sep	1	2	3	4	5	6	7	8	9	10	11	12	13	14	15	16	17	18	19	20	21	22	23	24	25	26	27	28	29	30		
Jun	13	14	15	16	17	18	19	20	21	22	23	24	25	26	27	28	29	30	1	2	3	4	5	6	7	8	9	10	11	12		Jul
Oct	1	2	3	4	5	6	7	8	9	10	11	12	13	14	15	16	17	18	19	20	21	22	23	24	25	26	27	28	29	30	31	
Jul	13	14	15	16	17	18	19	20	21	22	23	24	25	26	27	28	29	30	31	1	2	3	4	5	6	7	8	9	10	11	12	Aug
Nov	1	2	3	4	5	6	7	8	9	10	11	12	13	14	15	16	17	18	19	20	21	22	23	24	25	26	27	28	29	30		
Aug	13	14	15	16	17	18	19	20	21	22	23	24	25	26	27	28	29	30	31	1	2	3	4	5	6	7	8	9	10	11		Sep
Dec	1	2	3	4	5	6	7	8	9	10	11	12	13	14	15	16	17	18	19	20	21	22	23	24	25	26	27	28	29	30	31	
Sep	12	13	14	15	16	17	18	19	20	21	22	23	24	25	26	27	28	29	30	1	2	3	4	5	6	7	8	9	10	11	12	Oct

Beef Production and Management Decisions

Beef Production and Management Decisions

Fourth Edition

THOMAS G. FIELD
Department of Animal Sciences
Colorado State University
Fort Collins, Colorado

ROBERT E. TAYLOR
Department of Animal Sciences
Colorado State University
Fort Collins, Colorado

Upper Saddle River, New Jersey 07458

Library of Congress Cataloging-in-Publication Data

Field, Thomas G. (Thomas Gordon)
Beef production and management decisions / Thomas G. Field,
Robert E. Taylor --4th ed.
p. cm.
Rev. ed. of: Beef production and management decisions / Robert E. Taylor,
Thomas G. Field. 3rd ed. 1999.
Includes bibliographical references and index.
ISBN 0-13-088879-6
1. Beef industry--United States. 2. Beef industry. 3. Beef cattle. 4. Beef.
I. Robert E. Beef production and management decisions. II. Title.

HD9433.U4 T39 2002
636.2'13'068—dc21 2001055151

Editor-in-Chief: *Steve Helba*
Executive Acquisitions Editor: *Debbie Yarnell*
Associate Editor: *Kimberly Yehle*
Editorial Assistant: *Sam Goffinet*
Managing Editor: *Mary Carnis*
Production Management: *Carlisle Communications, Ltd.*
Production Editor: *Bridget Lulay*
Director of Manufacturing and Production: *Bruce Johnson*
Manufacturing Buyer: *Cathleen Petersen*
Marketing Manager: *Jimmy Stephens*
Creative Director: *Cheryl Asherman*
Senior Design Coordinator: *Miguel Ortiz*
Cover Design: *Amy Rosen*
Cover Art: Computer artwork of beef cycle and generational photo, courtesy of Holly Foster.

Pearson Education LTD.
Pearson Education Australia PTY, Limited
Pearson Education Singapore, Pte. Ltd.
Pearson Education North Asia, Ltd.
Pearson Education Canada, Ltd.
Pearson Educaión de Mexico, S. A. de C.V.
Pearson Education—Japan
Pearson Education Malaysia, Pte. Ltd.

10 9 8 7 6 5 4 3 2 1
ISBN 0-13-088879-6

DEDICATION

This book is dedicated to our families for their unending patience and support as we have pursued our interests in the beef cattle industry and our commitment to helping students better understand the management systems philosophy. We have been blessed to grow up in ranching families and to be given the opportunity to serve our community as land grant university faculty.

Parents:	Wallace and Etta Taylor	Fred and Mary Field
Wives:	Carole	Lisa
Children:	Scott, Lori, Mitchell, Alan, and Aaron	Justin, Sean, and Trae

Contents

Preface

The domestication of beef cattle initiated an opportunity for humans to apply their creativity to the formation of the modern beef cattle industry. Beef cattle provide a source of livelihood and fulfillment to producers, feeders, packers, processors, and retailers of beef products and by-products while providing a satisfying source of protein and other goods to the world's consumers. The relationship between humans and cattle provides not only a rich past but also a hopeful future to many societies and cultures throughout the world.

Purpose of the Book

This book is written to serve three primary purposes: 1) to identify the significant biological principles that contribute to the profitable and substainable production of beef cattle, 2) to systematically integrate the biological and economic principles required to make effective management decisions, and 3) to enhance understanding, communication, and cooperation between all segments of the beef industry.

Target Audience

The book is written for four groups—students, beef producers, beef industry organizations and related agribusinesses, and those who influence the decisions that affect the beef cattle industry. The principle goal of the book is to assist beef industry participants in making decisions that contribute to the profitable production of beef cattle that meet the specifications of consumers.

Key Features

The first seven chapters of the book emphasize management principles for the industry and each of the segments of the production and processing chain. Chapters 8, 9, and 10 are focused on structural changes in the industry, the creation of profitable marketing systems, and the impacts of the international trade of beef and beef products. Chapters 11 through 18 deal with the biological and economic principles needed to understand beef cattle productivity. The final two chapters provide an approach to management of information, understanding the past, and preparing for the future of the beef industry.

Changes in This Edition

The primary change is more emphasis on management decisions while incorporating a systems philosophy that allows better integration of the multi-faceted concepts presented. The book provides strong emphasis on improving profitability from both the process of lowering costs and of improving the quality of products. Significant changes in the industry are outlined and a number of useful websites have been provided to allow students to access additional sources of information. The major societal issues facing producers have also been incorporated into many chapters. This edition is also more focused on Total Quality Management and Beef Quality Assurance processes. Changes reflect the input of a number outside reviewers including academicians, cattle producers, and affiliated industry professionals.

Acknowledgments

Appreciation is expressed to the following individuals who have reviewed or contributed material to one or more chapters. Their expertise in the beef industry is well known and their suggestions have given an added dimension to the book.

Dale Blasi, Kansas State University
Randy Blach, Cattle-Fax
Patrick Burns, Colorado State University
Paul Clayton, U.S. Meat Export Federation
Barry Dunn, South Dakota State University
Terry Engle, Colorado State University
Frank Garry, Colorado State University
Temple Grandin, Colorado State Unviersity
David Hawkins, Michigan State University
Andy Herring, Texas Tech University
Doug Hixon, University of Wyoming
Steve Koontz, Colorado State University
Harlan Ritchie, Michigan State University
John Scanga, Colorado State University
Brad Skarr, Iowa State University
Gary Smith, Colorado State University
Tim Stanton, Colorado State University
J. D. Tatum, Colorado State University
Richard Wilham, Iowa State Unversity
Jack Whittier, Colorado State University

Barb Holst has provided a tremendous contribution by typing the manuscript. Her commitment was key to the completion of this revision. A sincere thank you to Bridget Lulay of Carlisle Publishers Services, as well as Debbie Yarnell and Kim Yehle of Prentice Hall whose professional talents helped convert our work into a finished product.

About the Authors

Dr. Taylor was raised on a beef cattle ranch in southern Idaho. He was involved in managing this seedstock, commercial cow-calf and backgrounding operation. He also managed Colorado State University's seedstock herd that is used in teaching, research, and extension activities.

Dr. Taylor received his BS and MS degrees at Utah State University and a PhD from Oklahoma State University. He taught courses in beef production and a graduate-level course in beef cattle management decisions. His research focused on beef cattle management systems. His consulting activities in the U.S. and foreign countries involved all phases of the beef production system.

Dr. Taylor received teaching awards from Iowa State University and Colorado State University, in addition to several regional and national teaching awards. He also received the Distinguished Teaching Award from the American Society of Animal Science recognizing his ability to organize and present materials to students. Many of his concepts for effective teaching are included in this book.

Dr. Field was raised on a Colorado cow-calf and seedstock enterprise. He was involved in the management of the family business after completing his BS degree. He received his BS, MS and PhD from Colorado State University.

Dr. Field teaches courses in introductory animal science, beef production, management decisions, and family ranching. He serves as the director of the Beef Industry Leadership Program and conducts beef cattle management research. He works with beef cattle producers and organizations in the U.S. and internationally.

Dr. Field has received teaching awards from USDA, NACTA, ASAS, and Colorado State University.

In Memorium

Dr. Robert E. Taylor
1934–1998

There is a Chinese proverb that says, "Those who would leave an impression for a year plant rice, those who would leave an impression for 10 years plant a tree, but those who would leave an impression for 100 years educate a human being."

Dr. Robert Taylor was an extraordinary teacher. Those of us who were blessed to study under his guidance found that in addition to all the expected lessons in agriculture and management systems, he was continually teaching us how to live. He delivered the lessons of life by example. Dr. Taylor was a counselor, a mentor, and a role model to thousands of students. First and foremost, he was devoted to his family and his faith.

Dr. Taylor was the embodiment of good stewardship. He was a great stockman who took care of the land, the herds, and the abundant natural resource God has bestowed upon us. He sought truth and understanding with an untarnished sense of inquisitiveness. He was a good steward of the human spirit who without fail carried a sense of hopefulness. He was a source of wisdom, a builder of community, and a visionary leader.

This book is a compilation of Dr. Taylor's life work as a student of agricultural management systems. His wish would be that those who study these lessons might be inspired to best utilize beef cattle for the benefit of humankind.

I hope that his teachings will have the same positive impact that they've had on my life. From the time I first met him as a 19-year-old college student, I have been thankful that he was my teacher and advisor.

Dr. Taylor was a giant and I have been blessed to stand in his shadow.

Tom Field
August 18, 1998

An Overview of the U.S. Beef Industry

Since the time that early humans first painted pictures of cattle on cave walls and took their first taste of beef, the bovine has played a role in the existence of humankind. Whether as a source of wealth, food, clothing, or draft power, cattle have evolved in a symbiotic relationship with people.

The Europeans who first imported cattle to the Americas could not have envisioned the size and scope of the beef industry that would eventually develop in the New World. The cowboys on the trail drives of the late 1800s would not have been able to foresee the changes in the infrastructure and marketing system that allowed the beef industry to move away from a commodity paradigm and toward that of a value-added, consumer-driven business.

The evaluation of agricultural systems has been ongoing for centuries and the emergence of the beef cattle industry resulted from the recognition that domestication of cattle and other livestock would result in a consistent supply of food, fiber, and draft power. The organizational structure of the industry became increasingly complex when the advances of the industrial age allowed rapid increases in production efficiency, permitting people to pursue vocations other than producing their own food supply.

The scientific and information breakthroughs of the last century have heightened agricultural productivity to the point that fewer than 2% of U.S. citizens are directly employed in production agriculture.

However, as fewer people understand or participate in food production, leaders of the beef industry and the agricultural commodities find it important to increase the level of communication between producers, processors, retailers, and consumers. The beef industry has always faced challenges and the present and future are no different. Nonetheless, the relationship between the stockman and his cattle is one with the potential to yield enormous benefits for humans. The thoughtful and diligent study of the beef industry, its associated

FIGURE 1.1 The market steer is the focal point of the U.S. beef industry, although market heifers, cows, and bulls contribute to the beef supply. Time is a major factor in the beef production process from "conception to consumption." Each beef industry segment uses many biological and economic relationships to economically produce a desirable beef product.
Source: Holly Foster.

infrastructure, marketplace, and management offers people the opportunity to apply creativity and energy to a fundamentally important endeavor.

GENERAL OVERVIEW

The beef industry includes breeding, feeding, and marketing cattle with the eventual processing and merchandising of retail products to consumers. The process involves many people and utilizes numerous biological and economic resources. Most important, however, is the time involved: 24 to 36 months are required from breeding time until a food product can be made available to consumers (see Fig. 1.1).

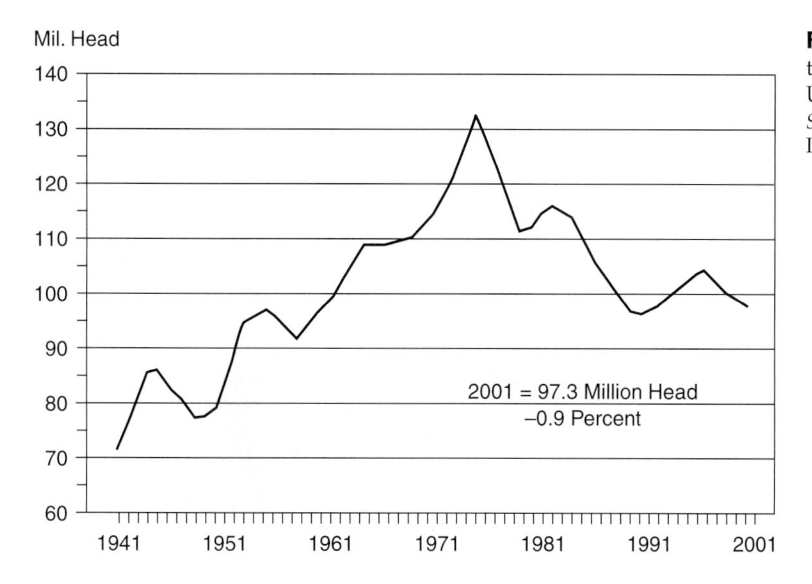

FIGURE 1.2 January 1 total cattle inventory (annual, United States).
Source: Livestock Marketing Information Center.

Commercial beef producers must have a comprehensive knowledge of the beef industry if they are to economically produce desirable breeding, feeder, and fed cattle and the resulting retail products. Since commercial producers are only one segment of the total production process, they must integrate their needs with the demands of other segments if specification beef production is to be a total industry effort.

Numbers, Prices, and Consumption

The beef industry involves people (cattle producers, processors, and consumers), products (number of cattle, pounds produced and consumed), prices, and profitability. Figure 1.2 shows the cattle inventory in the United States over the past 60 years. Beef and dairy numbers are combined because the dairy industry contributes a significant amount of production to the U.S. industry. Cattle numbers increased rapidly from the early 1900s until the mid-1970s, then a dramatic decline occurred. The cattle inventory of approximately 97 million head in 2001 was considerably less than the peak numbers of 132 million head in 1975. There have been significant peaks and valleys in cattle numbers resulting from factors influencing the supply and demand of beef (Fig. 1.2). These cycles reflect the profitable and unprofitable periods of the cattle industry. The influence of cattle cycles in marketing cattle is discussed in detail in Chapter 9.

Table 1.1 highlights some important data about the cattle industry from 1925 to 2001. Changes in population, cattle numbers, product consumption, live cattle prices, and average retail prices can be observed.

Figure 1.3 shows an interesting comparison between cattle numbers and carcass beef production—since 1976, cattle numbers have decreased significantly. Carcass beef production declined initially, remained relatively stable in the 1980s and then increased in the last half of the 1990s, partly due to Canadian imports. The beef industry is currently producing as much beef as it did in 1975 but with about 30 million head fewer cattle for several reasons: (1) most importantly, average carcass weights have increased from 613 lbs in 1970 and 635 lbs in 1980 to 732 lbs in 2001; (2) the feedlot turnover rate has increased from 2 times capacity to 2.4 times capacity, resulting in more cattle available for slaughter; (3) the slaughter age of fed cattle has decreased; (4) the genetic base for heavier cattle at a given age has increased (due to

TABLE 1.1 Cattle Numbers and Prices, Human Population, and Beef Consumption in the United States, 1925–2001

Year	Human Population (mil)	No. Cattle (mil)	No. Beef Cows[a] (mil)	Carcass Beef Produced (bil lbs)	Per Capita Retail Beef Consumption (lbs)	Choice Fed Steer Price ($/cwt)	Retail Choice Beef Price ($/lb)
1925	115.0	63.4	11.2	6.9	44	$10.16	$0.30
1930	122.8	61.0	9.1	5.9	36	10.95	0.35
1935	126.9	68.8	11.1	6.6	39	12.32	0.30
1940	131.8	68.3	10.7	7.2	41	11.86	0.29
1945	139.2	85.6	16.5	10.3	37	17.30	0.33
1950	151.1	78.0	16.7	9.4	47	28.88	0.75
1955	164.0	96.6	25.7	13.2	62	26.93	0.67
1960	179.3	96.2	26.3	14.4	63	25.90	0.80
1965	193.0	109.0	33.4	18.3	75	24.99	0.80
1970	201.9	112.4	36.7	21.5	85	29.45	1.00
1975	213.8	132.0	45.4	23.7	88	45.21	1.52
1980–1984[b]	230.3	114.0	38.2	22.5	78	65.64	2.45
1985–1989[b]	241.5	103.2	33.9	23.5	74	62.99	2.62
1990–1994[b]	250.1	97.9	33.2	23.1	66	74.71	2.86
1995–1999[b]	264.4	101.4	34.5	25.6	67	65.01	2.82
2000	270.0	98.0	33.5	26.7	69	69.00	2.94
2001	276.0	97.7	33.2	26.4	69	NA	NA

[a]For total cows contributing to beef production, add 9.2 million dairy cows to the current beef cow numbers.
[b]Five-year average.
Sources: USDA, NCBA, and Cattle-Fax.

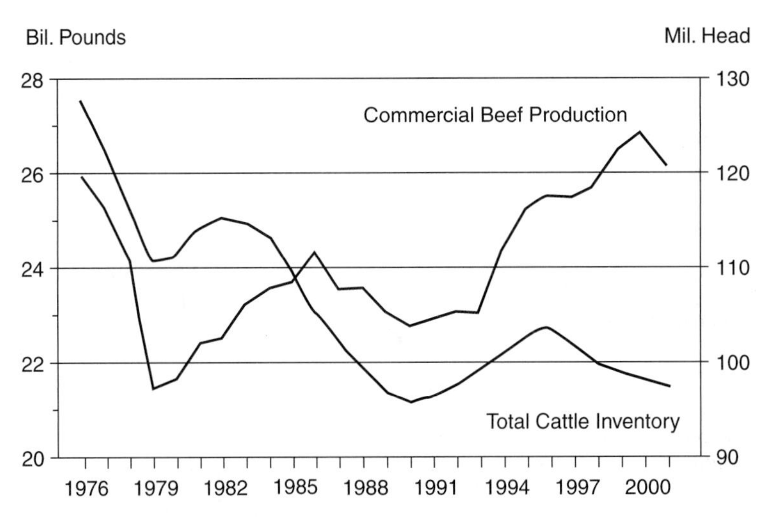

FIGURE 1.3 Beef production versus cattle inventory (January 1, United States).
Source: Livestock Marketing Information Center.

more crossbreeding, increased emphasis on growth in British breeds, and utilization of more Continental breeds by commercial breeders); and (5) increased importations of cattle (primarily from Canada). As a result, there has been an increase in the number of pounds of carcass beef per cow in the breeding herd (Fig. 1.4).

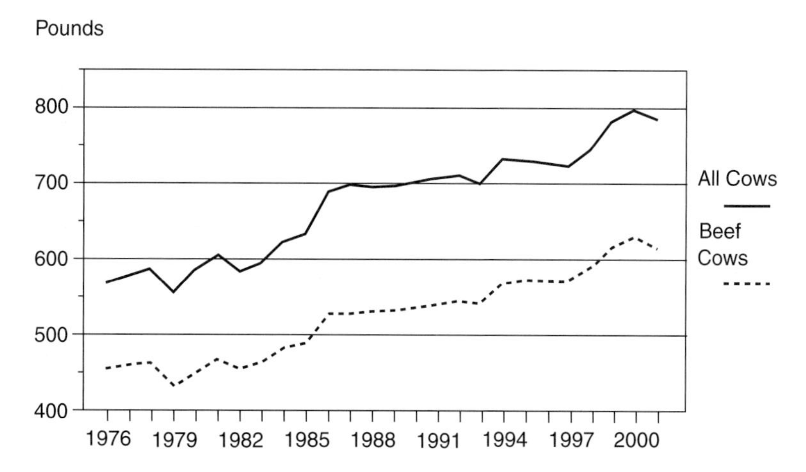

FIGURE 1.4 Carcass weight produced per cow.
Source: Livestock Marketing Information Center.

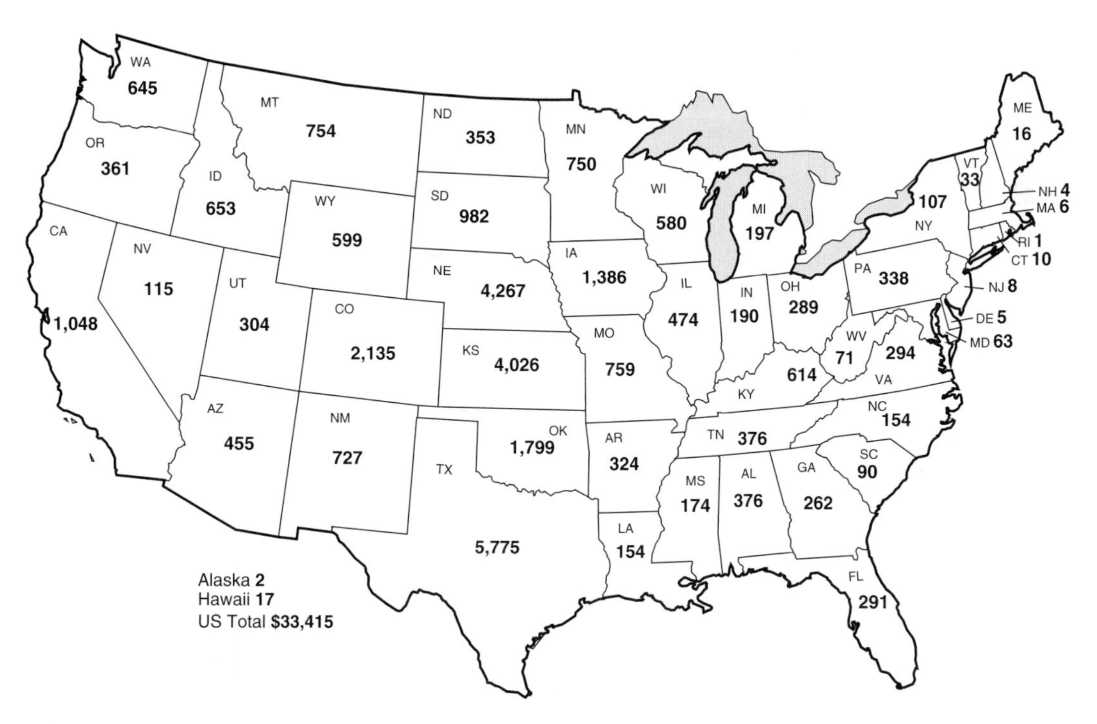

FIGURE 1.5 Cash receipts (million $) from sale of cattle.
Source: USDA (*Agricultural Statistics,* 1999).

CONTRIBUTION TO THE U.S. ECONOMY

One way the beef industry contributes to the U.S. economy is through the sale of live cattle. The average cash receipts received annually from the sale of all agricultural products total approximately $189 billion; $95 billion of that comes from livestock and livestock products, including $33 billion from cattle. Figure 1.5 shows the state totals for cash receipts from the sale of cattle and calves. Texas, Nebraska, Kansas, Colorado, Iowa, Oklahoma, and California generate more than $1 billion each from annual cattle sales. Obviously, the existence of cattle and their production support many other industries that add billions of additional dollars to the U.S. economy. For example, animal health product sales total nearly $2 billion. Other million and multibillion dollar industries—feed, finance, publications, equipment, marketing, AI, and others—are also highly dependent on cattle. The income generated by the beef industry yields a $99 bil to $165 bil. economic multiplier effect. In other words, each dollar generated directly from the sale of cattle and beef generates an additional $3 to $5 in economic impact.

BEEF INDUSTRY SEGMENTS

The term *beef industry* implies that the beef production system is a unified operation subject to an overall management program. However, the beef industry is actually made up of several different segments (Table 1.2) that are linked together through beef animals and products, yet the segments operate somewhat independently from each other. Each segment has different economic parameters and management problems and markets different products. In some cases, segments are in direct competition with one another. In some respects, the various beef industry segments can be considered separate industries because of their distinctly different characteristics.

TABLE 1.2 Overview of the U.S. Beef Industry (Production[a] and Consumption)

Segment	People/Companies	Cattle/Products	Tenderness/Palatability
Seedstock	120,000 breeders 10 AI studs	Approx. 80 cattle breeds (10 breeds are most important, while 5 breeds contribute approx. 60% of the genetics); primarily yearling bulls, semen, and AI certificates	British breeds highest Brahman breed lowest Genetic variation for tenderness exists within a breed
Cow-Calf (yearling stocker)	830,880 producers 105,250 dairy farms (produce 20% of the beef) cow herds < 50 head (80% of all operations); 50% of cow inventory in herds > 100 head	33.4 mil head beef cows 9.2 mil head dairy cows 85% calf crop (SPA) 514 weaning wt. (SPA)	
Feedlot	1,781 feedlots with >1,000 hd capacity in 12 major states	23.2 mil Fed cattle marketed	
Packer	795 plants harvest steers and heifers 97% boxed beef 81% harvested by top four firms 44% purchased on carcass basis	35.7 mil cattle slaughtered 26.2 bil lbs carcass wt *Quality graded (2000)* Prime (3%), Choice (52%), Select (36%) *Yield graded (2000)* 1 (10%), 2 (41%), 3 (36%), 4 (2%), 5 (0.2%) avg. carcass wt (740 lbs, all cattle)	min. fat (0.3 in) prevents cold shortening; electrical stimulation; improves tenderness, aging (14–21 days) increases tenderness

The Seedstock Segment

Seedstock breeders, sometimes referred to as *purebred breeders* or *registered breeders,* are specialized cow-calf producers. Their primary goal is to make genetic improvements that can be utilized by the entire beef industry. Seedstock breeders are predominantly responsible for identification and propagation of genetics that contribute to the profitability of the industry.

Seedstock breeders sell genetic information, breeding animals, semen, and embryos to other breeders and commercial cow-calf producers. Their function is one of service—to provide the genetics that can be economically utilized by the beef industry. The breeders sell breeding animals primarily to commercial cow-calf producers within a 100–150-mile radius of the breeders' operation. Choice of breed—whether one or a combination of breeds—is important in developing a production and marketing program that can best serve the commercial producers in any given area.

The seedstock segment is discussed in further detail in Chapter 4. Chapters 12 and 13 cover the biological relationships for making genetic changes in the economically important traits of cattle.

TABLE 1.2 (Continued)

Segment	People/Companies	Cattle/Products	Tenderness/Palatability
Retailer	128,000 grocery stores	Annual per-capita distribution: hamburger (28 lbs); steaks/roasts (30 lbs); processed (9 lbs)	
Purveyor	More than 360 companies	Center-of-the-plate products	Emphasize high palatability
Consumer	Population: U.S. (276 mil) World (5.9 bil)	*Per capita consumption* (2000) *Retail* / *Boneless* Beef 69 / 66 Pork 54 / 50 Poultry 100 / 70 1.8 oz beef/day (3.4 oz red meat); world per capita carcass beef consumption (23 lbs)	1 in 4 steaks not tender; 30% of beef not satisfactory; 1 carcass affects > 500 people; steaks cooked higher than "medium" tend to be less tender and drier; per capita fat consumption at all time high (fat and taste preferences are highly related)
Exports	Primarily to: (1) Japan (2) Mexico (3) South Korea	2.5 bil lbs (carcass beef) valued at $3.1 bil $1.9 bil (ready-to-cook, poultry)	
Imports	Primarily from: (1) Australia (2) Canada (3) New Zealand	3.0 bil lbs valued at $2.4 bil	Mostly in the form of ground or manufacturing beef

[a]97.3 mil head of cattle, Jan. 1, 2001; world cattle (1.35 bil head); U.S. 2000 agric. cash receipts from agricultural products total ($189 bil); livestock and products ($95 bil); cattle and calves (1st at $33 bil)

(continued)

TABLE 1.2 (Continued) Current U.S. Beef Industry (Financial and Economic)

Segment	Costs	Breakevens	Prices	Profits/Returns
Seedstock ($15 bil investment)		$800–$2,000/ yearling bull	$1,000–$4,000 (to commercial producers); semen $5–$30/unit; AI certificates ($5–$50)	
Cow-Calf ($180 bil investment)	*Annual cow cost* High 1/3 ($490) Avg. ($377) Low 1/3 ($268)	*Weaned calf* High 1/3 ($1.17/lb) Avg. ($0.86/lb) Low 1/3 ($0.61/lb)	*450 lbs* 1993 ($103/cwt) 1995 ($78/cwt) 1997 ($89/cwt) 2000 ($109/cwt)	1986 (–$25/cow) 1991 (+$55/cow) 1996 (–$80/cow) 2000 (+$80/cow)
Yearling/Stocker			*750 lbs* 1993 ($85/cwt) 1995 ($66/cwt) 1997 ($74/cwt) 2000 ($86/cwt)	*Summer Program* 1986 (+$25/hd) 1991 (–$5/hd) 1996 (+$40/hd) 2000 (+$20/hd)
Feedlot ($7.5 bil investment)			*Fed Steer* 1993 ($76/cwt) 1995 ($67/cwt) 1997 ($66/cwt) 2000 ($70/cwt)	1986 (+$26/hd) 1991 (–$40/hd) 1996 (–$10/hd) 2000 (–$5/hd)
Packer ($3.8 bil investment)	Slaughter ($25/hd) Fabrication ($50/hd)	$1.10/lb carcass *(doesn't include* slaughter or fabrication costs or drop credit)	*By-product Value ($/cwt of liveweight)* 1993 ($8.80/cwt) 1995 ($9.60/cwt) 1997 ($10.30/cwt) 2000 ($8.20/cwt)	*Wholesale Value* 1993 ($1.82 lbs) 1995 ($1.64 lbs) 1997 ($1.58 lbs) 1999 ($1.72 lbs)
Retailer ($50 bil investment)			Retail price (Choice; Avg. of all cuts) $2.80/lb; hamburger ($1.45/lb)	*Supermarket Sales* Beef (bil $) *1983* *1993* *1998*: 26.8 21.1 22.1
Purveyor				
Consumer			Market Share % of *meat expenditures* (*1980* *1996* *2000*): Beef 54 42 41; Pork 27 27 28; Poultry 19 31 31	Beef expenditures were $192 per capita. Total consumer spending on beef exceeded $53 bil for the first time.
Exports			9.4% of domestic production is exported; the equivalent of 3.3 mil cattle	Beef value ($3.1 bil) *Beef by-product value* ($1.0 bil)
Imports				Value ($2.4 bil)

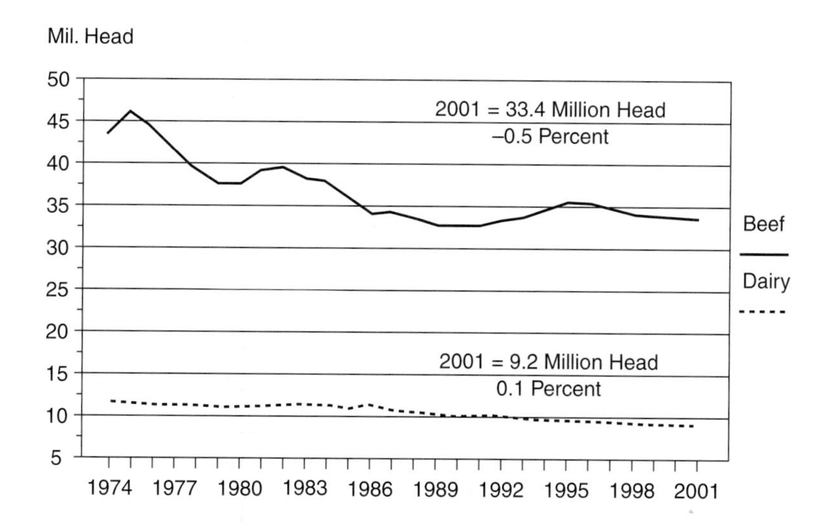

FIGURE 1.6 Beef and dairy cow inventories. *Source:* Livestock Marketing Information Center.

The Commercial Cow-Calf Segment

Commercial cow-calf producers maintain cowherds and raise their calves from birth to weaning. Under ideal conditions each cow is expected to produce one calf each year. The calves are the primary source of revenue for the producer as well as the source of heifers to replace cows that die or are culled.

Changes in cow numbers over the past 15 years reflect an increase in beef cows and a decrease in dairy cows (Fig. 1.6). Total milk production from dairy cows has remained at about the same level due to a marked increase in productivity for each dairy cow. Distribution of beef cows by states is shown in Figure 1.7. Note the concentration of beef cows in the Great Plains. This area, which covers Texas north through North Dakota and the eastern parts of New Mexico, Colorado, Wyoming, and Montana accounts for more than 50% of the total U.S. beef cow population. Some of the Corn Belt and southeastern states also have significant numbers of cows.

Changes in beef cow numbers by states over the past decade are reflected in Figure 1.8. The changes are primarily related to economic conditions and forage supply. The latter can be dramatically influenced by drought, renovation of previously unproductive land, water development, and conversion of cropland to forage production for cattle.

Most of the 33.4 million head of beef cows owned by commercial producers are in herds of 100–1,000 head with a few herds of 15,000–35,000 head under one ownership. Figure 1.9 shows that nearly 80% of beef cow operations have less than 50 head of cows, while controlling only 30% of the cow inventory. The small herd size is not surprising because 45% of the U.S. farms are operated by part-time farmers, and many of their farms are less than 50 acres in size. Approximately one-half of the cows are in herds of greater than 100 head. Yet, fewer than 10% of the enterprises are in this size category.

It takes 300 cows or more to be an economic unit, so there are numerous small beef cow operations that are supplemented with outside income. Increase in cowherd size does not always imply an increased efficiency of production. However, several studies demonstrate that there is a greater return per head as cowherds increase toward 1,000 head. The concentration of 50% of the beef cow inventory in small herds that account for nearly 90% of the total enterprises with the remaining half of the inventory controlled by less than 10% of the cow-calf herds creates challenges in the industry. These challenges include the process of communicating technical and marketing information to managers of small versus large herds, developing supply assurance systems that include both small and large herds, and finding ways for managers of small-sized herds to overcome the cost advantages held by the large enterprises.

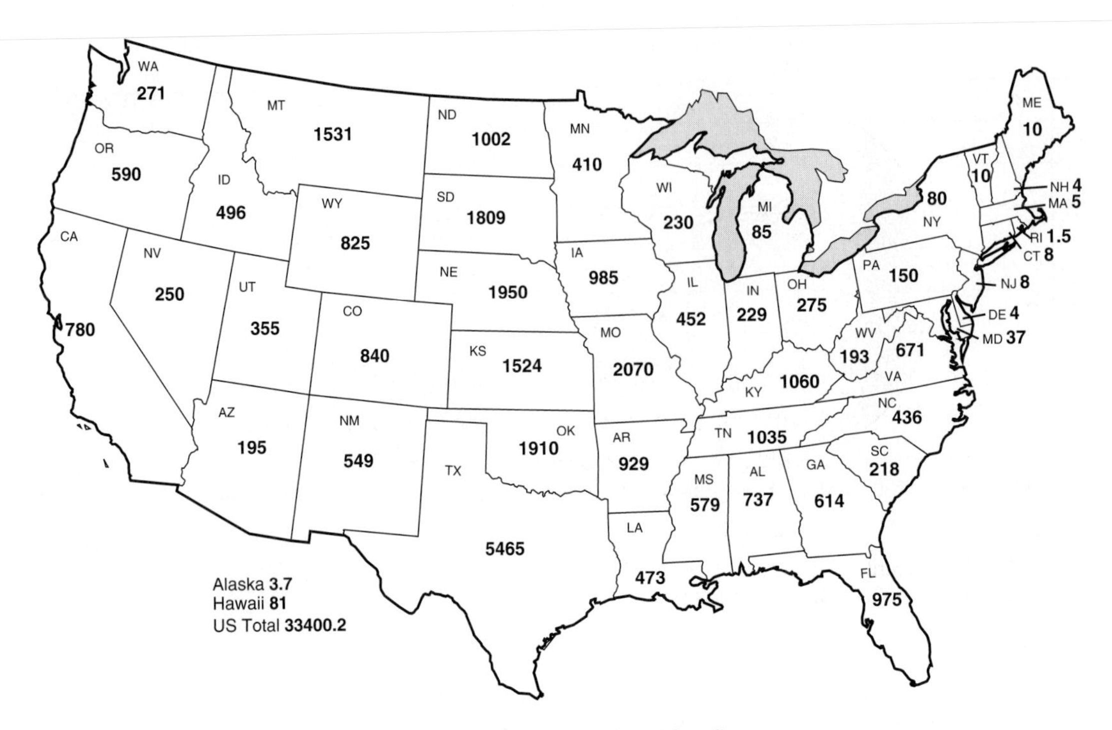

FIGURE 1.7 Beef cows that have calved, January 1, 2001 (1,000 head).
Source: Livestock Marketing Information Center.

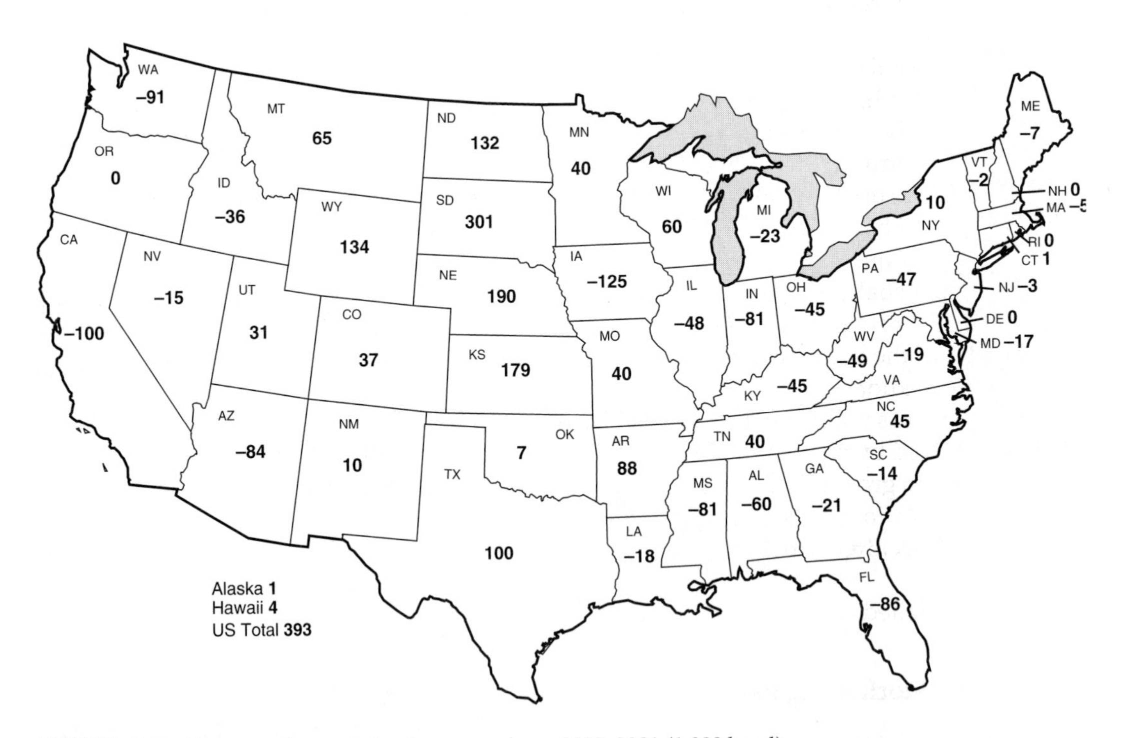

FIGURE 1.8 Ten-year change in beef cow numbers, 1992–2001 (1,000 head).
Source: Livestock Marketing Information Center.

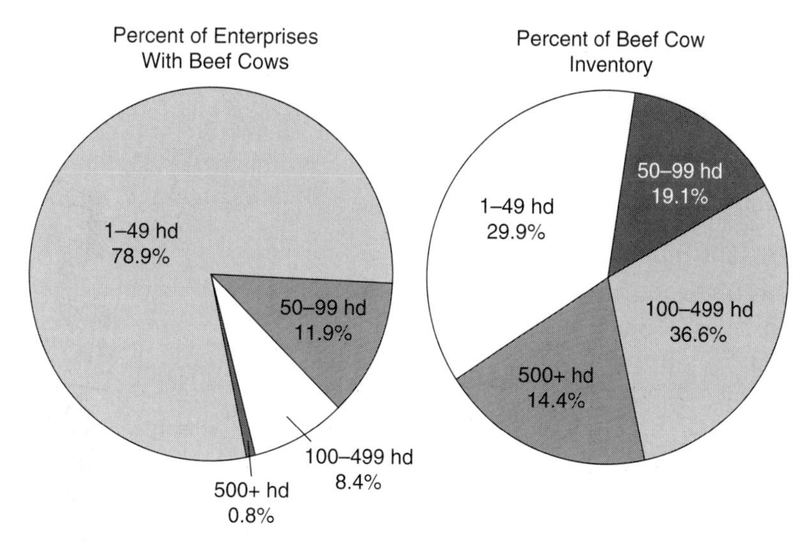

FIGURE 1.9 Percent of beef cattle enterprises and beef cow inventory by size of enterprise.
Source: USDA.

Although there are a few intensively managed cow-calf operations where cows remain in confinement year-round, most are extensively managed operations where cows are maintained on grazed and harvested forage throughout the entire year. Many cow-calf operations are extensively managed in high mountain valleys, plains, and desert areas where 30–100 acres are required per cow, with some supplemental feeds provided. Some cows are maintained on more intensively grazed areas where 1–5 acres per cow are utilized for 5–10 months during the year.

Most cows will calve in late winter and early spring with the majority of calves being born in February, March, and April. Some producers calve their cows in late spring, summer, or fall, primarily to reduce losses from calf scours and to complement their forage production program. Other producers may have both a spring and fall calving program to extend the use of their bulls and to use their labor and forage more efficiently. A few producers continue to calve on a year-round basis; however, critical economic assessments usually do not favor this type of calving program.

Calves are usually weaned at the same time of year, their ages ranging from 5 to 10 months of age. Weaned calves that are heavy (more than 500 lbs) may go directly into the feedlot, but the majority of the lighter calves currently are grown-out on forage for several months before entering the feedlot.

Cow-calf pairs will graze thousands of acres of grasses, legumes, and forbs that can be effectively utilized by ruminants. In many wheat and other small grain-producing areas, cattle will graze green growth in the fall, and early spring, then graze straw aftermath following the harvesting of grain. Cows graze untillable acres and crop aftermath throughout the United States. Cornstalk aftermath for grazing in the fall is very important in the major corn growing regions.

A more detailed discussion of the commercial cow-calf segment is provided in Chapter 5.

The Yearling-Stocker Segment

The *yearling-stocker operator* is responsible for adding weight to weaned calves prior to their shipment to feedlots for additional weight gain prior to harvest. The calves are usually yearlings

(12–20 months of age) by the time they enter the feedlot. Some heavier weaning calves (more than 500 lbs) may go directly to the feedlot, thereby by-passing the yearling-stocker phase.

The yearling-stocker operation usually has available forage—pasture, hay, silage—for feeding during winter months and grazable forage for the spring, summer, and fall months. In spring calving programs, short yearlings (10–14 months of age) may go to feedlots after the winter feeding program, whereas long yearlings (15–20 months of age) will be marketed in the fall following a summer grazing program.

Yearling-stocker operators purchase calves in the fall and/or spring depending on the availability and cost of forage. Some commercial cow-calf operations retain ownership of their calves through the yearling growth stage. Cattle feeders sometimes purchase calves and maintain ownership through both the growing and feedlot phases. These two alternatives are increasing in frequency, making the traditional yearling-stocker operation a minor beef industry segment.

The yearling-stocker segment of the beef industry is discussed in more detail in Chapter 6.

The Feedlot Segment

Feedlots are confinement feeding operations where cattle are fed primarily finishing (high energy) rations prior to harvest. Most feedlot operations feed relatively high grain rations for 90–150 days for economically efficient gains and to improve the palatability of the retail product. Some operations background cattle by feeding them primarily roughage rations prior to the finishing phase.

The number of cattle on feed at a given point in time is shown in Figure 1.10. The annual fed-cattle marketings by state are illustrated in Figure 1.11. Commercial cattle feeders annually feed and market approximately 2.5 times the one-time feedlot capacity, which accounts for the difference in the numbers shown in the two figures.

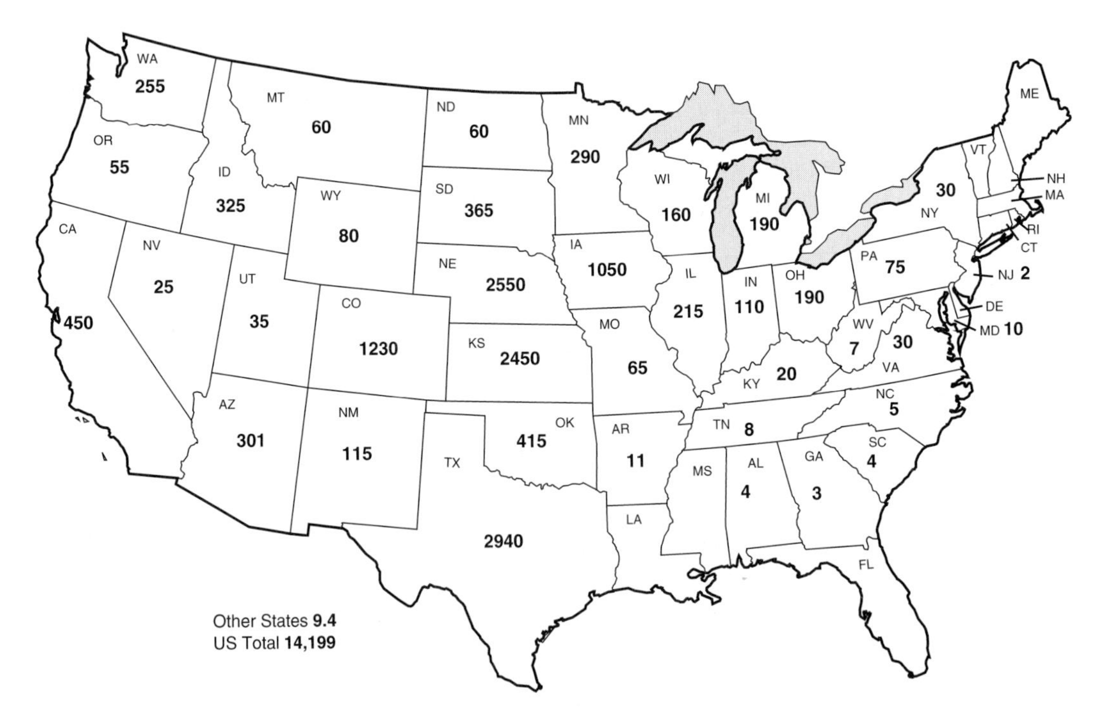

FIGURE 1.10 Cattle on feed January 1, 2001 (1,000 head).
Source: Livestock Marketing Information Center.

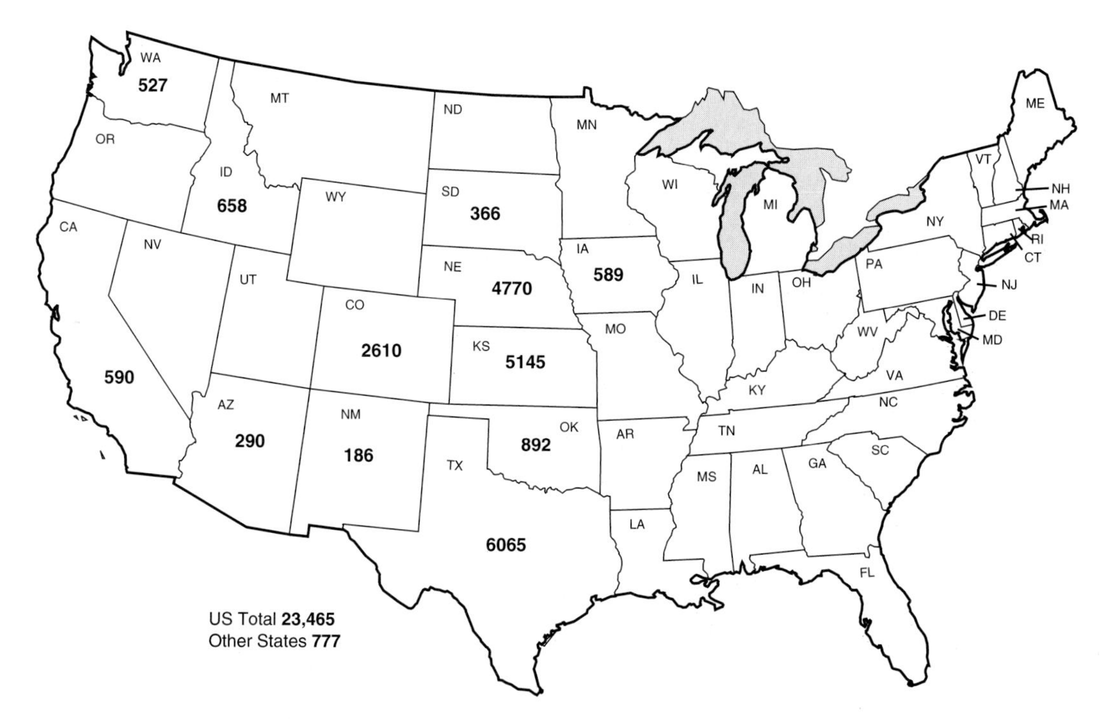

FIGURE 1.11 Fed-cattle marketings by state, 1999 (1,000 head).
Source: Livestock Marketing Information Center.

Cattle-On-Feed reports from the USDA usually give information for only the top 7–12 states. This is because the top 12 states feed > 98% of the cattle, while approximately 88% of the fed cattle are marketed by the top 7 states. The concentration of fed-cattle marketings in the most important cattle feeding states is shown in Figure 1.11. Note the concentration of cattle feeding in the Plains states. The primary reasons for this distribution of fed cattle are the availability of feed grains, the locations of packing plants, and the climatic and geographic conditions that favor cattle feeding.

Figure 1.12 identifies the major cattle feeding states and the average number of cattle on feed per feedlot. Notice that the upper midwestern states of Iowa, South Dakota, and Nebraska tend to have smaller capacities per feedlot as compared with the southern plains and southwestern states. The southern tier of feedlot states has larger feed yards due to more arid and consistent climatic conditions. The larger yards are also typically the primary enterprise of focus by management, whereas, in the Corn Belt region, the feeding enterprise is part of an integrated farming business.

The Packing Segment

The distribution of fed cattle harvested in the various states is shown in Figure 1.13. Cattle are harvested in the same geographical areas where cattle are fed and marketed (compare Figs. 1.10, 1.11, and 1.12). Table 1.3 identifies the major packing companies and their locations. Note the data regarding number of cattle slaughtered and total beef sales for the three largest beef-packing companies.

Packers, purveyors, and retailers slaughter, process, and distribute approximately 26 billion lbs of beef. The magnitude of the beef-packing industry as reflected by number and gender of cattle slaughtered is shown in Figure 1.14. Approximately 29 million head of fed steers

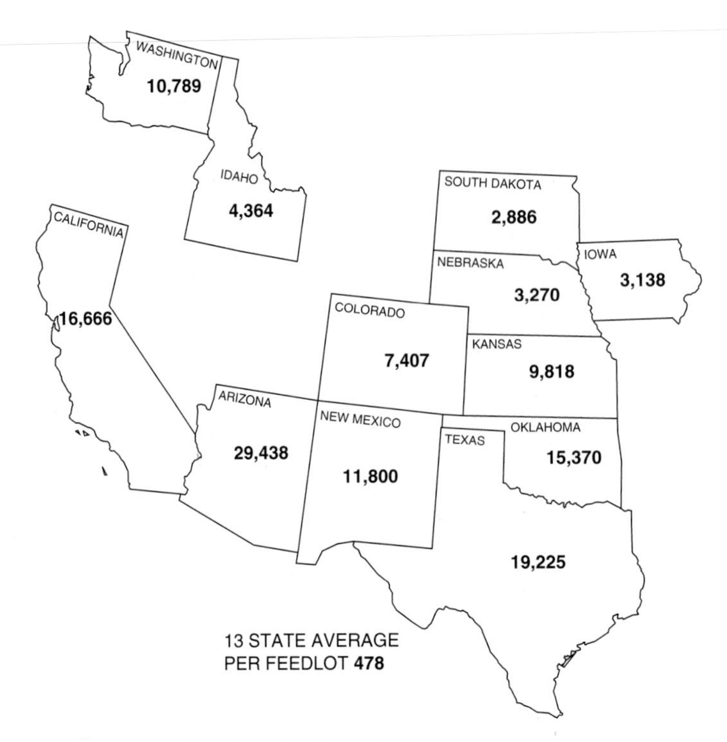

FIGURE 1.12 Average number of cattle on feed per feedlot.
Source: Cattle-Fax.

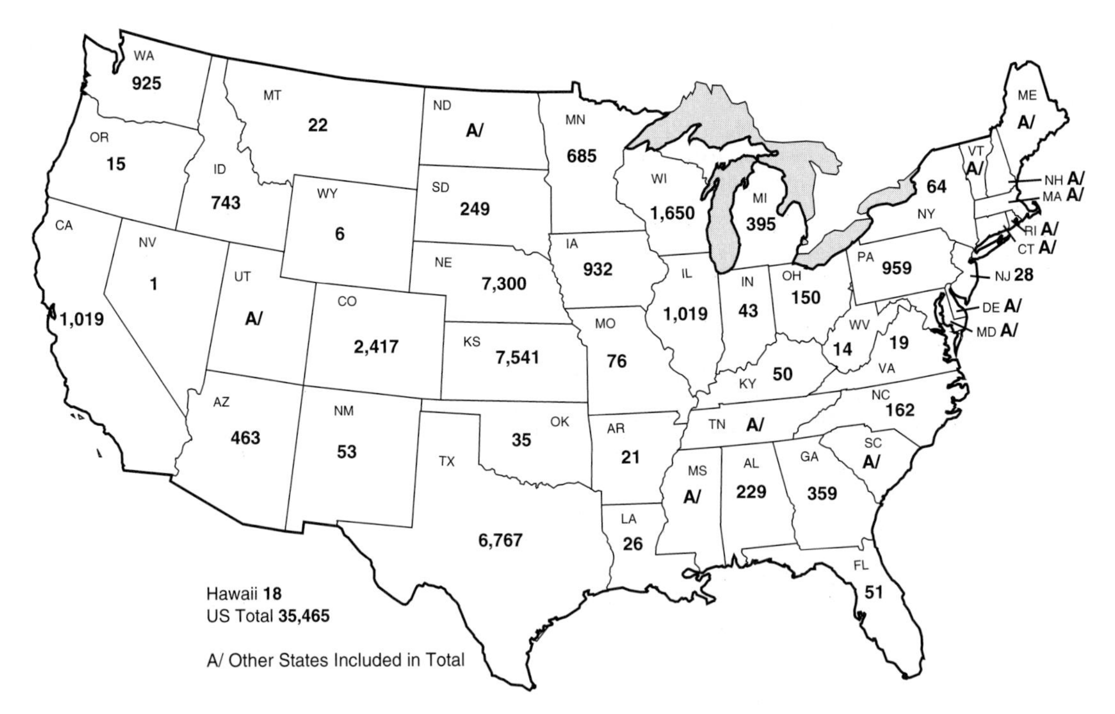

FIGURE 1.13 Commercial cattle harvest, 1999 (1,000 head).
Source: Cattle-Fax.

TABLE 1.3 Top Beef-Packing Companies[a]

Name	2000 Sales ($ bil)	Slaughter Capacity Daily (no. head)	Slaughter Capacity Annual (mil head)	No. Plants	Plant Locations	Head Office
1. IBP, Inc.	12.8	NA	NA	12	TX, ID, NE, IA, KS, IL, WA, Canada	Dakota Dunes, SD
2. Excel Corporation	8.2	NA	NA	6	TX, NE, KS, CO, Canada	Wichita, KS
3. ConAgra Red Meat Company	5.0	22,000	5.8	6	CO, NE, KS, TX, UT, ID	Greeley, CO
4. Farmland National Beef Packing Co. LP	2.7	NA	2.7	2	KS	Kansas City, MO
5. Packerland Packing Co., Inc.	1.4	6,500	1.5	4	WI, NE, AZ, MI	Green Bay, WI

[a]There are approximately 1,250 beef-slaughtering facilities in the United States; however, the top four process 80% of the fed steers and heifers.
Source: NCBA, Directions 2001.

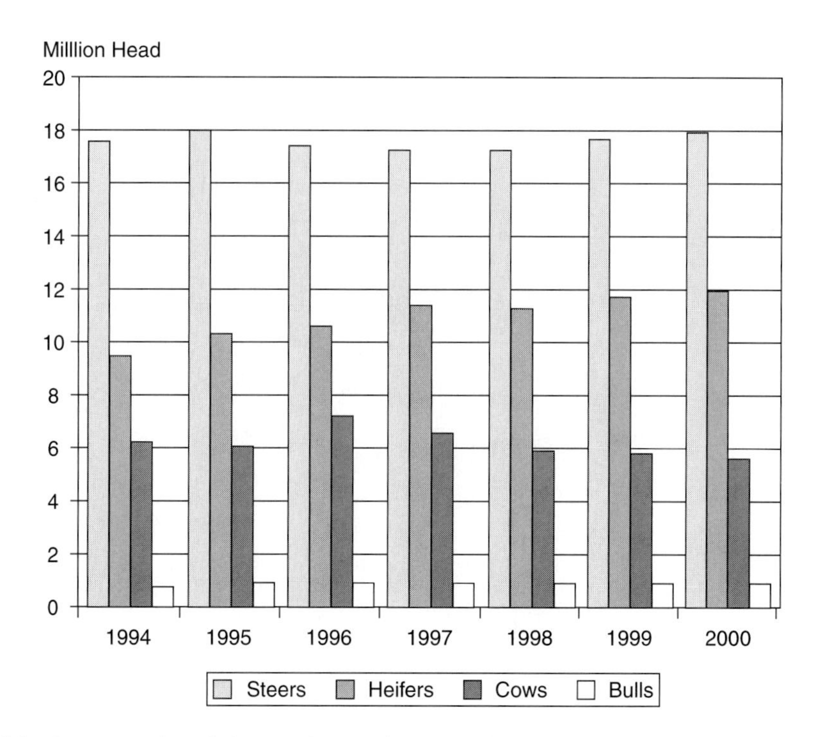

FIGURE 1.14 Commercial cattle harvest by sex class (includes dairy cows).

and heifers comprise the highest percentage of the approximately 35 million head slaughtered annually. The smaller number of nonfed steers and heifers along with cull cows and bulls (that are slaughtered) have had little or no concentrate feeding prior to slaughter. Their rations have been primarily grass and other forages. The number of nonfed steers and heifers slaughtered is approximately 1 million head. This number can increase significantly if fed cattle production is unprofitable for an extended period of time.

Beef sold from packing plants is primarily boxed (> 80% of the beef slaughtered). The boxed beef is primal and subprimal cuts from which much of the bone and excess fat has been removed (Fig. 1.15). The cuts are vacuum-packaged for a longer shelf life. Boxing of beef has proven to be more cost efficient at the packing plant level because (1) labor rates are usually lower at packing plants than retail stores, (2) cutting is usually faster and more efficient as it is done on a moving "disassembly" line by specialized meat cutters, (3) a larger volume of retail product can be handled in less space, (4) more effective use can be made of bone and fat by-products, and (5) transportation costs are reduced, with a more valuable product that can be more easily handled than carcass beef. Retailers have an advantage in buying only those beef cuts that can be more easily merchandised without accepting the entire carcass. In addition, retailers have less spoilage because vacuum-packaged meat has a longer shelf life than carcass beef.

Packers have a preference for carcasses weighing 650–850 lbs. This weight range best fits the preferred size of retail cuts and box size currently used in the boxed beef trade.

Packers are also moving into a case-ready merchandising concept whereby beef is fabricated and packaged either fresh or precooked as a means to capture value. The primary advantages of case-ready products are improved control over food safety, lowered labor costs, improved consistency and yield, enhanced inventory control, and direct delivery of products oriented to consumer preference. The move to case-ready products is a significant development in the industry.

FIGURE 1.15 Boning, trimming, and separation of thin meat cuts from primal cuts is done by workers who perform their specific tasks on stationary platforms on either side of the moving conveyor belt.

The Purveyor Segment

A meat wholesaler, sometimes called a "jobber," is an operator who purchases beef and sells it to a retailer or to another wholesaler. *Purveyors* and *distributors* are two types of beef wholesalers. Purveyors buy beef and perform some fabrication, while distributors buy and sell beef without cutting or changing the product. Purveyors sell almost exclusively to the food service industry (which cooks and sells food for away-from-home or take-out consumption). Purveyors are specialized meat processors who provide highly palatable center-of-the-plate products to food service operators, retail stores, and mail order customers.

Purveyors handle about 5% of the total beef. The number of distributors and purveyors continues to decline, however, so they are becoming less important as a separate beef industry segment. Packers are increasing their sales directly to retailers or through brokers. Fabrication of beef carcasses continues to increase at the packer level.

The Retail Segment

The primary types of *retailers* of beef are shown in Table 1.4, with the approximate sales of groceries handled by each. Supermarkets (retail stores with more than $2 million in sales) constitute more than 77% of all grocery store sales, with the ten largest grocery chains having 60% of the total grocery sales. Chain supermarkets increased sales by nearly 20% from 1989 to 1999, while independent supermarkets and small groceries lost 18.7% and 37.2% of sales over the same time period.

Most of the beef received by grocery stores is as primals, boneless subprimals, or beef for grinding. Historically, almost all retail cuts have been prepared at the store level. The movement to case-ready products is a shift from this approach. Sales by meat, poultry, and fish departments comprise approximately 20% of all grocery store sales, with fresh beef accounting for 30–40% of meat department sales (Table 1.5).

The Consumer Segment

Figure 1.16 shows how beef, the major end-product of cattle production, has been accepted by *consumers*. While per capita retail beef consumption increased markedly from 1960 to 1976, it decreased in 1976 into the early 1990s, after which it remained relatively stable until 1997. Since 1998 beef consumption and sales have begun to show signs of increasing. These increases suggest that opportunities exist to recapture market share. The decline was due primarily to lack of consistency in palatability, smaller supplies of available beef, higher costs relative to other competitive meats, and consumer perceptions of beef's relationship to human health.

TABLE 1.4 Major Beef Retailers

Retailers	Sales ($ bil)
Supermarkets (> $2 mil annual sales)	365.4
Chain supermarkets	292.0
Independent supermarkets	73.4
Other stores (< $1 mil annual sales)	55.3
Convenience stores	29.4
Wholesale club stores	22.6

Sources: FMI; *Progressive Grocer.*

TABLE 1.5 Consumer Expenditures for Meat, Poultry, and Fish in Grocery Stores, 1988 and 1998

Item	1998 ($ bil)	1988 ($ bil)
Fresh meat (beef)	22.1	21.3
Processed meats	14.5	13.3
Frozen meat	0.23	0.20
Canned meat	1.1	1.3
Poultry	17.5	10.6
Fish and seafood	6.9	6.2
Total meat, poultry, and seafood	67.1	58.6
Total grocery sales	414.6	310.8
Food and beverages	303.7	231.3
Nonfood	110.9	79.5

Sources: Supermarket Business Magazine; AMI, 2000 Meat Facts.

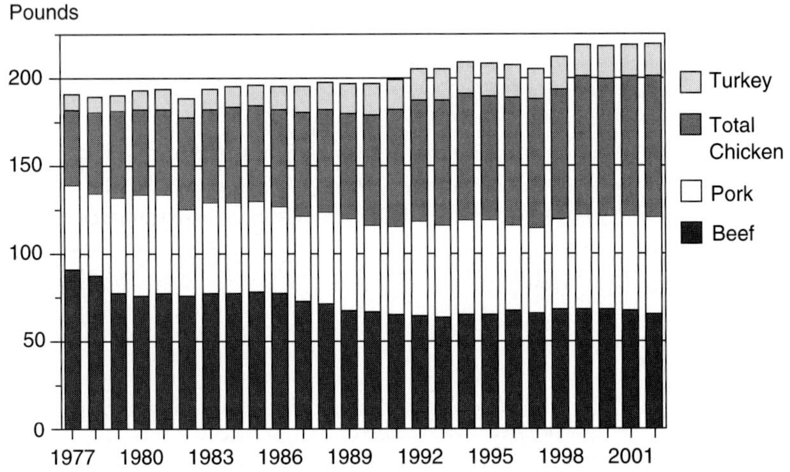

FIGURE 1.16 Annual U.S. per capita red meat and poultry consumption (retail weight).
Source: Livestock Marketing Information Center.

Consumers continue to demand more service and convenience in their food products as their incomes rise or as they have less time to cook, prepare, and eat meals. Time has become a precious commodity as single-parent and two-income families have increased. Time and convenience are reflected in increasing away-from-home meals. Consumers eating at home also want more convenience—they desire products that require minimal preparation time but with a significant amount of choices in regards to flavor. Consumers still want a feeling of having participated in home meal preparation, thus "meal kit" and other meal packaging concepts have gained favor in the supermarket. The active lifestyles of consumers have led to supermarkets accounting for 20% of take-out food sales.

The Beef Belt

Some individuals make reference to the *beef belt* of the United States (Fig. 1.17), which includes the Great Plains states and most of the Corn Belt. These twelve states plus parts of four other

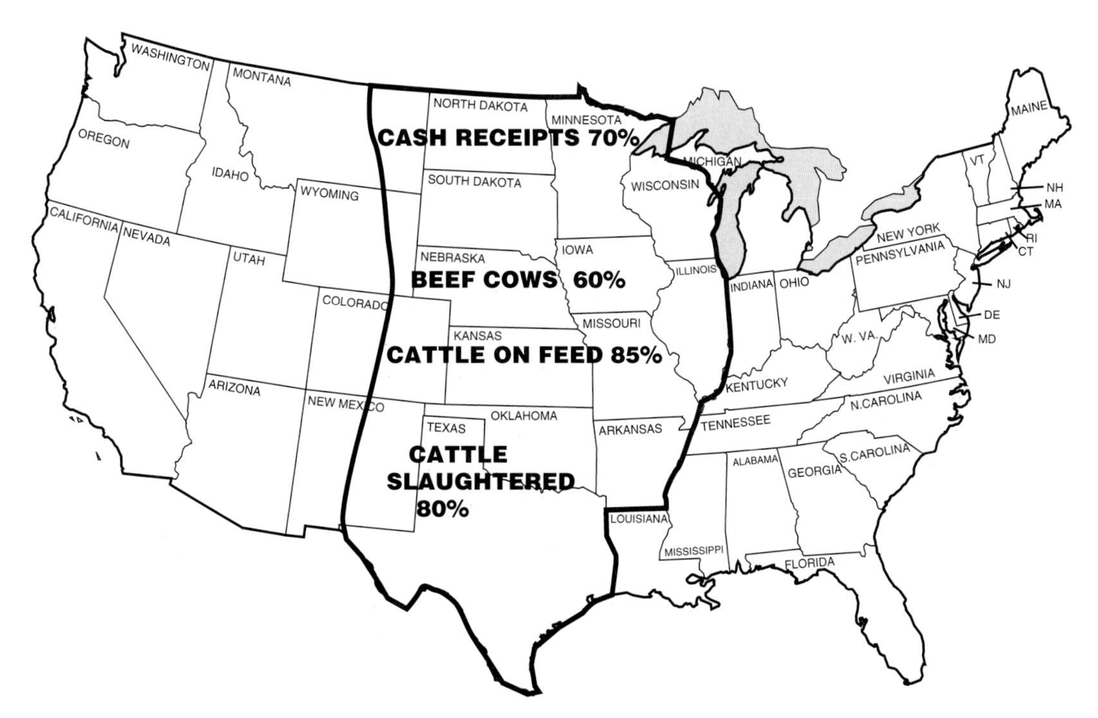

FIGURE 1.17 The beef belt is located in mid-America. Beef production activities are highly concentrated in this area.
Source: Colorado State University.

states account for 70% of the U.S. cash receipts from cattle, 60% of the beef cows, 85% of the fed cattle, and 80% of the cattle harvested.

Beef Industry Goals

A recommended goal for the entire beef industry is to produce a high proportion of highly palatable lean meat (per animal) as efficiently as possible and with a profitable return to all who produce it.

Different industry segments have different end-products for their particular segment; however, lean and palatable beef products are the major end points for all beef industry segments. Efficiency is measured by both biological and economic factors (Chapter 3).

Profitability is the bottom line for each beef segment as well as the total industry. Production practices, management decisions, and marketing choices that affect price per pound and total dollars received for various products are best measured by their contributions to profitability.

Profitability

Costs and returns for cow-calf producers and cattle feeders are shown in Figures 1.18 and 1.19. The net return (income minus cost) has varied dramatically over the years; any one segment might experience a $100 per-head loss to more than a $100 per-head profit at a given point in time. The total costs and returns for production, processing, and merchandising a fed steer are shown in Table 1.6. *Drop-credit* is the value of all products (edible and inedible offal and by-products) other than the carcass (retail cuts).

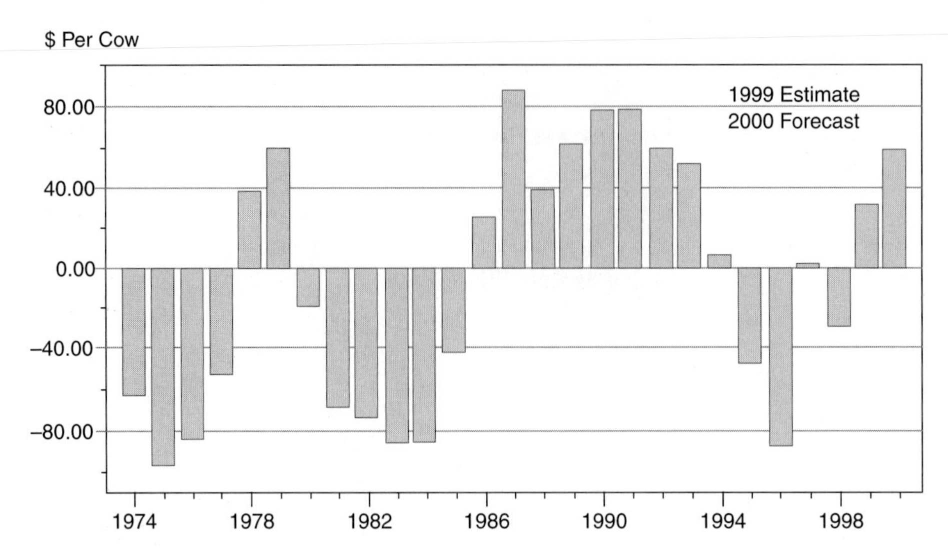

FIGURE 1.18 Estimated annual average cow-calf returns over cash costs.
Source: Livestock Marketing Information Center.

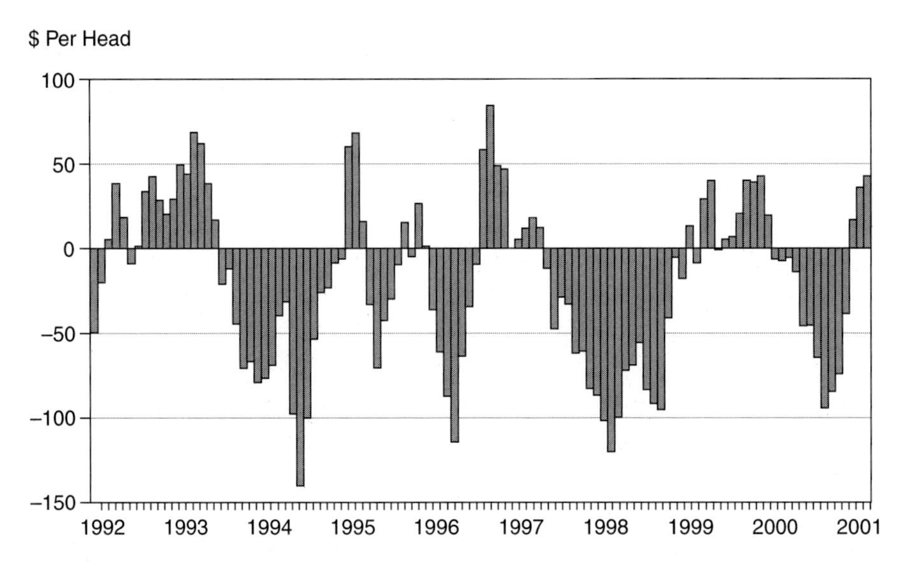

FIGURE 1.19 Average monthly returns to cattle feeders.
Source: Livestock Marketing Information Center.

During the past several decades, there have been few years in which all segments of the beef industry have made a profit during the same year. The late 1980s and early 1990s were profitable years for most segments, but more often profit in one beef industry segment came from a loss in another segment.

There is considerable variation in profitability within a given beef industry segment, both between years and within years. These differences are summarized by the National Cattlemen's Beef Association and Cattle-Fax:

> Clearly, wide cost, performance and price variation exist between individual operations in all segments of the industry. As a result, actual cost and profit/loss differences vary considerably

TABLE 1.6 Total Costs and Returns in Producing a 1,250-lb Choice Slaughter Steer, 1999

Item	Cost	Return
Cost to produce a calf (575-lb steer calf)[a]	$ 445	
Backgrounding yard cost (300-lb gain)[a,b]	148	
Feedlot cost (375-lb gain)[b]	185	
Slaughtering and dressing costs	25	
Fabrication costs	50	
Distribution costs	40	
Additional processing and merchandising	180	
Total cost	$1,073	
500 lbs of retail cuts (@ $2.50)[c]		$1,375
Drop-credit ($8.50/cwt live wt)		106
Total return		$1,481[d]

[a]Does not include land costs.
[b]Includes feed + interest + death loss + overhead + vet.
[c]1250 lbs × .63 dress% = 785 lbs HCW × .70% yield = 550 lbs.
[d] If each segment takes their profits at their point of sale, then this value is reduced.
Source: Adapted from Cattle-Fax data.

between cattle businesses. Even in poor times, some operations may be profitable and expanding, while others are in a loss position and liquidating.

The primary factors affecting profitability for cow-calf, yearling, and feedlot producers are discussed in Chapters 5, 6, and 7, respectively.

BEEF INDUSTRY ORGANIZATIONS

Beef industry organizations have been established to inform, promote, regulate, support research, and exert political influence in areas directly affecting their businesses.

An organizational challenge exists among many of the approximately 850,000 beef cattle producers seeking to operate independently of any organizational structure. Numerous cattle producers are not interested in belonging to beef organizations because they have major sources of income other than that derived from their few head of cattle, while other producers lack trust in industry organizations.

There are many organizations that represent or influence cattle producers. The traditional independent philosophy of cattle producers can serve as a barrier to effective planning and adjustment to change. Those individuals who align with others to direct change are more likely to find success.

The future of beef as a product and of the beef industry is directed by group action. The influence of organizations on the beef industry will be determined by (1) the industry's capacity for linking and coordinating the actions of its organizations and for forming alliances with other groups having common interests, and (2) effective organizational leaders who can effectively represent their members.

All cattle producers must work together for the following reasons:

1. The impact of government on the cattle industry will remain great. This is especially true because politicians represent an urban society with less than 2% of the U.S. population directly involved in production agriculture. Government affairs will continue as a major focus of the beef industry and its representative organizations.

2. To make beef more competitive with other meats, the industry must guide, encourage, and support research and development efforts in production technology and product development and marketing.
3. Cattle producers need a source of information to help them make sound management decisions along with effective educational and training programs for themselves and their employees.
4. The industry needs an effective beef marketing program that includes consumer and market research, product development, product information, promotion, and merchandising. In addition to improving the efficiency of beef production and distribution, the industry needs to support programs that will stabilize or improve beef acceptance.
5. Public information efforts should be expanded to improve public and government acceptance and understanding of beef economics, production, and marketing methods.

Eventually, there may be a unified beef industry organization formed at both state and national levels. Meanwhile, though, beef industry programs should be coordinated to effectively utilize the funding now provided by individual cattle producers to several agricultural organizations.

The major organizations representing individuals or companies within each beef industry segment, and other organizations having an effect on each segment, are shown in Table 1.7. The major organizations involved with the marketing process that moves animals or products from one segment to another are also shown in Table 1.7. Additional organizations that affect the beef industry are noted in the Appendix. Each participant in the beef industry should participate in those organizations that directly affect their business.

TABLE 1.7 Major Organizations Representing or Affecting the Beef Industry[a]

Segments	Major Organizations Influencing Each Segment	Other Organizations Directly Affecting Each Segment[b]
Seedstock producers	American National Cattlewomen Beef Improvement Federation Breed Associations National Cattlemen's Beef Association State Beef Councils State Cattlemen's Associations U.S. Beef Breeds Council	(1) State Department of Agriculture (2) USDA Animal and Plant Health Inspection Service[c] (3) USDA Packers and Stockyards Administration (4) American Association of Bovine Practitioners (5) Occupational Safety and Health Administration
Commercial cow-calf producers	American Farm Bureau American National Cattlewomen Beef Improvement Federation Breed Associations Cattle-Fax County Livestock Growers Organizations National Cattlemen's Beef Association National Grange State Beef Councils State Cattlemen's Association	(6) Environmental Protection Agency (7) American Society of Animal Science (8) USDA-Forest Service (9) USDA-Soil Conservation Service (10) USDA-Agricultural Research Service (11) USDA-Extension Service (12) USDA-Statistical Reporting Service (13) Society for Range Management (14) American Registry of Professional Animal Scientists (15) Livestock Publications Council (16) International Embryo Transfer Society (17) Beef AI Organizations and National Association of Animal Breeders (18) Animal Rights Organizations and Environmental Groups

TABLE 1.7 (Continued)

Segments	Major Organizations Influencing Each Segment	Other Organizations Directly Affecting Each Segment[b]
Feeders	American Farm Bureau American National Cattlewomen Cattle-Fax National Cattlemen's Beef Association State Beef Councils State Cattle Feeder's Association	(1),(2),(3),(4),(5),(6),(10),(11),(12),(14) (15),(18)[d] (19) American Feed Manufacturers Association (20) National Feed Ingredient Association (21) Food and Drug Administration (22) USDA Office of Transportation (23) Animal Health Institute
Packers and processors	American Association of Meat Processors American Meat Institute American Meat Science Association Institute of Food Technologists National Cattlemen's Beef Association National Food Processors Association National Meat Canners Association State Beef Councils State Meat Dealers Association	(1),(2),(3),(4),(5),(6),(12),(14),(15),(18), (21),(22) (24) Federal Trade Commission (25) Labor Unions (26) National Perishable Transport Association (27) USDA Food Safety and Inspection Service (28) USDA Agricultural Marketing Service
Meat retailers/ food service organizations	American Association of Meat Processors American Institute of Food Distributors National Association of Meat Purveyors National Association of Retail Grocers of the United States National Frozen Food Association National Restaurant Association State Meat Dealers Association State Restaurant Association	(1),(2),(3),(5),(6), (18),(21),(22),(24),(25),(26), (27),(28) (29) Food and Drug Law Institute (30) USDA-Food and Nutrition Service (31) Joint Labor Management Commission of the Retail Food Industry (32) National Restaurant Association (33) Food Service and Lodging Institute
Consumer	Community Nutrition Institute Congress Watch Consumer Alert Council Consumer Federation of America National Consumers League	(1),(2),(3),(5),(6),(18),(21),(24),(27),(29),(30) (34) American Public Health Association (35) Food and Energy Council (36) National Academy of Sciences
Marketing points between segments	American Stockyards Association Cattle-Fax National Auctioneers Association National Cattlemen's Beef Association National Livestock Grading and Marketing Association National Livestock Producers Association U.S. Meat Export Federation	(12),(18),(21),(22),(24) (37) Agriculture Trade Council (38) USDA-Economic Management Staff (39) USDA-Marketing and Inspection Management (40) USDA-World Agricultural Outlook Board (41) USDA-Animal Air Transport Association

[a]Addresses and descriptions of these and other organizations are in the Appendix.
[b]Organizations (1) through (18) are applicable to both Seedstock and Commercial Cow-Calf Segments.
[c]USDA organizations can be accessed through <www.usda.gov>.
[d]Repeated numbers refer to the same organizations identified earlier in the table.

TABLE 1.8 Vision Statement and Long-Range Goals of NCBA

Vision statement:	*A dynamic and profitable beef industry that concentrates resources around a unified plan, consistently meets global consumer needs and increases demand.*
Industry priority:	*Increase potential for the profitability of all industry segments.*
Long-range planning goals:	*(1) Increase beef demand by 6 percent by 2004.* *(2) Enhance and protect the business climate for cattle and beef.*

Source: NCBA, 2001.

National Cattlemen's Beef Association (NCBA), 9110 E. Nichols Ave., Centennial, CO 80112 (www.beef.org). The NCBA, with headquarters in Denver, is the national spokesperson for all segments of the nation's beef cattle industry—including cattle breeders, producers, and feeders. The nonprofit trade association was originally formed in 1898. In 1996, the National Cattlemen's Association and the National Live Stock and Meat Board merged to create the NCBA. The NCBA represents approximately 230,000 cattle professionals throughout the country. Membership includes individual members, forty-six affiliated state cattle associations, and twenty-seven affiliated national breed organizations. NCBA programs are financed by funds contributed by individual members as well as by affiliated associations (Table 1.8).

The NCBA prides itself in being a grassroots organization. The organization is structured around a committee and council structure (Consumer Marketing, Communications, Research and Technical Service, Public Policy, Association Services and Administration, Joint Industry Budget, Joint Industry Evaluation, Joint Industry Audit, and Brand-Like Commission, Cow-Calf/Stocker, Feeder, Seedstock, Veal, and Live Cattle Marketing).

The NCBA maintains offices in Washington, DC, in addition to its Denver-based headquarters.

The NCBA provides services cattle producers cannot perform satisfactorily as individuals. Pursuant to this goal, the NCBA performs three basic functions: (1) primarily through its Washington, DC office, it represents the beef cattle industry in the legislative and administrative branches of the federal government; (2) it interprets beef production and beef economics for the public and economic, social, and political developments for the industry; and (3) it provides information to aid members in planning and management decisions.

American National Cattlewomen (ANCW), P.O. Box 3881, Englewood, CO 80155 (www.ancw.org). The ANCW is a commodity group established for participation or the promotion, education, and legislation of beef.

U.S. Meat Export Federation (MEF), 1050 17th Street, Suite 2200, Denver, CO 80265-2073 (www.usmef.org). The MEF is a nonprofit trade association that works with the U.S. meat and livestock industry to identify and develop overseas markets for U.S. beef, veal, pork, lamb, and variety meats. It is based in Denver, with overseas market development offices in Tokyo, Osaka, Singapore, Hong Kong, Beijing, Mexico City, Seoul, Taipei, London, Moscow, and St. Petersburg. USMEF also has representation in the Middle East, Central and South America, and the Caribbean. Through these offices, the MEF coordinates market development programs. Its programs are designed to identify new markets, create widespread product awareness, secure fair market access, provide trade servicing, and assist and educate overseas buyers and U.S. suppliers alike. Established in 1976, the MEF is a cooperator with the Foreign Agricultural Service (FAS) of the U.S. Department of Agriculture. It represents livestock producers and feeders, meat packers, purveyors and exporters, agribusiness and agriservice interests, farm organizations, and grain promotional groups.

The MEF has several sources of funding: its members, overseas private sector interests, beef checkoff money, and the Foreign Agricultural Service (FAS).

State Beef Councils. Most states have a beef council that is funded with checkoff dollars. Their primary objective is to educate consumers about the nutritional aspects of beef and how to best select and prepare beef. The councils communicate with health professionals and food service industry personnel and work to promote beef to consumers.

American Meat Institute (AMI), P.O. Box 3556, Washington, DC 20007 (www.meatami.org). The AMI, founded in 1906, represents meat packers and processors who produce more than 90% of the meat in the United States as well as suppliers of meat equipment, products, and services. Activities include marketing, research, congressional and legislative relationships, improved operating methods and products, conservation, spoilage prevention, and industrial education. Total membership in the organization is approximately 1,100 companies and individuals.

North American Meat Processors Association (NAMP), 1920 Association Drive, Suite 400, Reston, VA 20191-1547 (www.namp.com). NAMP is an international trade association comprised of meat-processing companies and associations who provide the finest center-of-the-plate products and service to food service, retail stores, and mail-order customers. NAMP is also the publisher of the world-renowned publication *The Meat Buyers Guide.*

Food Marketing Institute (FMI), 800 Connecticut Ave., N.W., Washington, DC 20006-2071 (www.fmi.org). Independent grocers, chain stores, and wholesalers are members of the FMI, which maintains a liaison with both the government and consumers. FMI, comprised of 1,500 members, conducts programs in research, education, industry relations, and public affairs.

BEEF INDUSTRY ISSUES

The beef industry has faced numerous issues during past years. Some of these issues are addressed here; others are covered in later chapters. Issues facing the beef industry change frequently, sometimes daily. A current assessment of issues can be made by contacting the NCBA or reading its publications.

The U.S. population continues to increase, but Americans actively involved in agricultural production represent less than 2% of the total population. This disparity in numbers reflects the communication problems between urban people and agricultural producers. The small number of producers have found political representation more difficult than in past years. People in the urban sector have more time, money, and interest to focus on the issues they deem most important.

Some issues are very emotional. Perceptions and personal beliefs get confused with true relationships. Enlightened people identify true relationships and recognize that the media, at times, can be misleading and in some cases, report totally false information.

Cattle producers need an organizational structure like the National Cattlemen's Beef Association to manage issues of national and international scope. Such an organization can objectively project a positive industry image by (1) coordinating a proactive approach to issues affecting the industry; (2) developing sound, technical information to be used by industry leaders; (3) encouraging producers to implement responsible production practices; (4) conducting research among consumers and industry influencers to understand their needs and opinions; and (5) providing credible information to opinion influencers. It is important that the issues facing the beef industry be addressed before they reach a crisis stage.

Environmental Issues

The American public is generally not familiar with the economics of the food production chain. Most people, however, are concerned with food issues such as safety and health and the use and preservation of natural resources. There are influential consumer, nutrition/health, and environmental groups with multimillion dollar budgets that focus on these issues.

Cattle have the unique ability to graze untillable acres and convert plants that humans cannot eat into highly palatable human food. However, while some people know that cattle can effectively use the land, others feel that cattle abuse the land. Cattle producers who implement proper grazing practices prevent overgrazing along streams and rivers (riparian areas). Their grazing management also fosters compatible relationships between livestock and wildlife. The issues of grazing fees on public lands, wetlands, and inferences that overgrazing is the major cause of rangeland desertification are discussed in Chapter 15.

Another environmental issue is the amount of water needed to produce beef. Some environmentalists believe that water should be conserved for humans instead of cattle. Although some argue that 2,500 gallons of water are needed to produce 1 lb of beef, scientific estimates indicate that less than 400 gallons of water are required, most of it being used on irrigated grain.

Animal Rights Issues

Some people view animal welfare and animal rights as the same while others see them as two distinctly different issues. Table 1.9 summarizes several viewpoints of how animals should be perceived and utilized. The extremes of animal exploitation and animal liberation many times involve breaking the law. Beef producers have been concerned with the use and welfare of their animals for centuries. Animals were domesticated to give nomadic people a consistent supply of food and companionship. Draft animals were domesticated for transportation and power. People soon learned that the productive response of animals is greater when they are given proper care.

Today, the nutrition, health, and management needs of farm animals are well known and scientifically based. Evidence suggests that many domesticated animals in the United States receive a more nutritious diet than some humans consume. The veterinary medical profession provides on-farm services, health clinics, and hospital care that are in many ways equal to human health-care services. The members of the NCBA adopted a statement of principles that affirms that cattlemen are united in their philosophy that proper and humane care of the animals they are responsible for is a moral obligation as well as an economic necessity. The tenets of this statement of principle are as follows:

- I believe in the humane treatment of farm animals and in continued stewardship of all natural resources.
- I believe my cattle will be healthier and more productive when good husbandry practices are used.
- I believe that my and future generations will benefit from my ability to sustain and conserve natural resources.
- I will support research efforts directed toward more efficient production of a wholesome food supply.
- I believe it is my responsibility to produce a safe and wholesome product.
- I believe it is the purpose of food animals to serve mankind, and it is the responsibility of all human beings to care for animals in their charge.

The vegetarian influence appears to be a significant factor in the animal rights movement. According to Dr. Michael W. Fox (1986), vice president of the Humane Society of the United States (an animal rights organization):

TABLE 1.9 Animal Welfare and Animal Rights Viewpoints of Several Organizations and Individuals

Category	Viewpoint
Animal exploitation	Animals are here for human use or abuse and they are absolute human property. Groups advocate or conduct activities (e.g., dog fighting, cock fighting, live pigeon shoots) that are illegal in most states. Most activities involve pain or death of the animals, primarily for the entertainment of spectators.
Animal use	Animals exist primarily for human use (e.g., livestock production, hunting, fishing, trapping, rodeos, zoos). Organizations have guidelines for the responsible care of the animals. Harvesting of animals for food should be as painless as possible.
Animal control	These organizations enforce the laws, ordinances, and regulations affecting animals. Animals may be supplied for research. Required to destroy surplus animals. Many advocate spaying or neutering animals.
Animal welfare	National groups, humane societies, and welfare agencies that support the kind treatment of animals. They work within existing laws to accomplish goals. Publicize and document animal abuses to get laws changed. Do not provide animals for research. Require spaying and neutering. Willing to euthanize surplus pets rather than let them suffer.
Animal rights	Believe animals have intrinsic rights that should be guaranteed like human rights. These rights include not being killed, eaten, used for sport or research, or abused in any way. Involves national and local animal rights groups. Some say pets have right to breed. Most would require spaying or neutering.
Animal liberation	Animals should not be forced to work or produce for human benefit. Groups openly call for animal liberation, including pets. Some call themselves "activists" and may condone, encourage, or participate in illegal activities, civil disobedience, or even violence. Believe they have the right to break the law to end animal suffering.

Source: Adapted from K. B. Morgan, *An Overview of Animal-Related Organizations with Some Guidelines for Recognizing Patterns,* Kansas City, MO: Community Animal Control, 1989.

> It may seem offensive to vegetarians that I endorse the concept of treating farm animals humanely because it would seem that I support the livestock and poultry industries. Actually, I don't. . . . [V]egetarianism is . . . an enlightened decision, if not a survival imperative in the long term, and producing and consuming less farm-animal produce, especially meat, are essential steps toward the overall restoration of our culture, agriculture, and environment. (pp. 172, 176)

Of the approximately 7,000 animal protectionist groups in the United States, 400 consider themselves animal rights groups. The animal rights groups number both their membership and budgets in the millions and have grown phenomenally during recent years. Most animal rights activists are female (78%), are college educated (82%), and have annual incomes over $50,000. In addition, many celebrities support the causes of the animal rights movement. The movement has attracted activists from other areas who know how to protest very effectively.

Intensive animal production is one of the primary concerns of animal rights organizations. Some groups—People for the Ethical Treatment of Animals (PETA), the spokes-group for the Animal Liberation Front; the Farm Animal Reform Movement (FARM); and Earth First—urge the reduction and eventual elimination of animals for food. Some groups opt for direct confrontation and the most radical of these has resorted to violence and terrorist activities in pursuit of its goals. Even some of the more traditional organizations are becoming more radical.

For example, The Humane Society of the United States denounces bacon and eggs as "the breakfast of cruelty." According to Regan (1985), some of the more extreme animal rights groups argue that "the rights view will not be satisfied with anything less than the total dissolution of the animal industry as we know it today" (p. 395).

The Animal Industry Foundation (AIF) was established to represent the interests of the livestock and poultry industries. AIF focuses its energies and budget on (1) providing the means for livestock producers to educate the public and (2) fostering solid scientific data in the animal rights debate.

The Farm Animal Welfare Coalition (FAWC), an informal organization of agricultural groups, represents animal producers and animal industry organizations on animal welfare issues. FAWC has cooperative projects with AIF, such as Media Alert, which makes available transcripts of animal rights activities, consumer articles, and related material.

The animal rights groups have great strength in membership, organizational structure, and financing. Their combined budgets are in the tens of millions of dollars, whereas the Animal Industry Foundation has a budget of a few hundred thousand dollars. Animal scientists and the animal industries have not been very effective in delivering information to Congress, state legislators, and the general public.

Animal welfare groups, livestock producers, and others concerned about animal welfare should work together to fund and support bonafide research studies and investigations into the matters of greatest concern. A combination of scientifically based facts and humane treatment of animals may provide the basis for best determining the well-being of both humans and animals.

Health and Food Safety Issues

Consumers have become increasingly more aware of diet and health and the nutritional and safety aspects of food. These issues are discussed in Chapter 2.

Marketing Issues

A major challenge for the beef industry is to generate and maintain high-quality products. This is extremely difficult in a segmented industry where cattle and products are not well identified as they move from one segment to another. Value-based marketing (Chapter 9) may be an unrealistic goal given the current structure of the beef industry that produces a generic commodity of beef.

Quality is an ambiguous word that must be more clearly defined if it is to be effectively communicated and achieved. The production and market specifications presented in Chapter 4 (Table 4.5) are a starting place in putting numerical values on traits that can be used to identify quality. Consumers' definition of quality eventually goes beyond the color of lean meat and the amount of marbling to include products that are reasonably priced, safe, nutritious, consistent, healthful, and that are consistently high in palatability.

Quality is best defined with Total Quality Management (TQM), which is: "Meeting or exceeding your customers' expectation at a cost that represents value to them every time."

Following are some of the major marketing issues related to marketing slaughter cattle:

1. The 2000 Beef Quality audit estimated that $100 per slaughter steer/heifer was lost to the industry because of quality shortfalls. These per-head losses were from waste fat ($51); palatability, such as tenderness and marbling ($24.50); bruises; hide damage; injection sites; and dark cutters ($18); and excessive carcass weights ($6.50). It is estimated that excess fat costs the beef industry $1 billion each year. To remain competitive with other meats and

food products, the beef industry will need to address these costs of production and quality assurance needs.

2. Surveys show that 20% of the cuts from the loin and rib and 60% of retail cuts from the round have unacceptable tenderness problems. Since marbling only accounts for 10–20% of the tenderness differences in beef, quality assurance for tenderness and overall palatability must be addressed more directly. The most frequent consumer complaints of inconsistencies of tenderness and juiciness cannot be ignored without having a devastating effect on the entire beef industry. Traditional marketing and processing methods must be critically evaluated and changed.
3. A beef quality-assurance program is needed so that processors, retailers, and consumers can have confidence in product quality. Cattle producers need to control their health programs in order to maintain consumer confidence that animal drug and medication procedures are well managed. Packers want the assurance that they will encounter a minimum number of injection-site abscesses caused by improper injections of cattle.

The beef industry needs to cooperate to consistently supply high-quality products. The ultimate goal of quality assurance for producers, processors, and retailers is to assure consumers that they are receiving beef products that are safe, healthful, and highly palatable.

MANAGING A MATURE BEEF INDUSTRY

During 1950–1975 the beef industry expanded rapidly with tremendous increases in cattle numbers (Fig. 1.2). From 1975 to 1985, cattle numbers decreased significantly, while during the past decade they have stabilized even though numbers fluctuate as a result of the traditional beef cycles. The beef industry has been called a "mature industry" for the past several years because expansion in cattle numbers or increased beef consumption has not occurred.

Several strategies have been recommended for implementing management systems in a mature beef industry:

1. Know costs and be a low-cost producer.
2. Set up and maintain financial control.
3. Differentiate product for several market niches.
4. Integrate vertically or functionally (via "partnering" and formation of strategic alliances).
5. Master the market tools that are available and create new ones.
6. Be actively involved in organizations that shape rational policy and influence markets for beef products.

The last strategy was discussed in this chapter. The other five strategies are covered primarily in Chapters 3, 5, 6, 7, and 9.

REFERENCES

Publications

Crom, R. J. 1988. *Economics of the U.S. Beef Industry.* Washington, DC: USDA. Agric. Info. Bull. 545.
Directions (Looking Toward 2000). July 1992. Englewood, CO: National Cattlemen.
Fox, M. W. 1986. *Agricide.* New York: Schocken Books.
Helming, B. 1990. Meats: Decade of Danger. *Meat and Poultry* 36: 30.
Hilker, J. H., and Ritchie, H. D. Dec. 1986. *Economics of the Cattle Industry in a Turbulent Decade.* Brookings, SD: South Dakota Cow-Calf Day.

Improving Beef's Consistency and Competitiveness (1991 National Beef Quality Audit). 1992. *American Hereford Journal*, July, p. 99.
Koch, R. M., and Algeo, J. W. 1983. The Beef Industry: Changes and Challenges. *J. Anim. Sci.* 57: 28.
Livestock and Poultry Situation and Outlook Report. Washington, DC: USDA.
McCoy, J. H., and Saehan, M. E. 1988. *Livestock and Meat Marketing.* New York: Van Nostrand.
Meat Industry: A Sampling of the Issues and Controversies Affecting Meat and Poultry All Over the Planet. 1989. *Meat and Poultry* 35: 12–14, 41, 43.
National Cattlemen's Foundation. 1990. *Special Interest Group Profiles.* Washington, DC: D. C. Hill and Knowlton.
Regan, T. 1985. *The Case for Animal Rights.* Berkeley: University of California Press.
Ritchie, H. D. 1982. *Future and Direction of the U.S. Beef Industry.* Proceedings of the Annual Conference on AI and Embryo Transfer in Beef Cattle, Denver, CO.
Rollin, B. E. 1990. Animal Welfare, Animal Rights, and Agriculture. *J. Anim. Sci.* 68: 3,456.
Smith, G. C. 1991. *A Quality Audit of the Beef Industry.* Proceedings of the Range Beef Cow Symposium, Fort Collins, CO.
Ward, C. E. 1988. *Meat Packing Competition and Pricing,* Blacksburg, VA: Research Institute on Livestock Pricing, Virginia Tech University.

Visuals

"Cattlemen Care About the Environment" (1990; 9 min.) and "Issues and Answers for the Nineties" (1990; 7 min.). Communications Department, NCA, Box 3469, Englewood, CO 80111.
"Excel Beef Plant: Slaughter" (1997; 45 min.). CEV, P.O. Box 65265, Lubbock, TX 79464.
"Excel Beef Plant: Fabrication." (1997; 45 min.). CEV, P.O. Box 65265, Lubbock, TX 79464.
Videotapes from the conference on "The Ethics of Humans Using Animals for Food and Fiber" (1991). Univ. of Minnesota, Box 734 Mayo Bldg., Minneapolis, MN 55455.
"What's the Beef" (1991; 15 min.; covers animal welfare, environmental, and human health concerns). CEV, P.O. Box 65265, Lubbock, TX 79464-5265.

Retail Beef Products and Consumers

Beef demand is established by the consumer. The consumer may be in a retail food store purchasing beef for home preparation, in a fast-food outlet where hamburger is usually the focal point, or in a restaurant contemplating steak, prime rib, or some other appealing beef entree. In these situations and others, the consumer's choice of beef in amount and price gives eventual direction and momentum to the needed beef supply.

Beef producers and purveyors must be knowledgeable of the retail beef products they and others eventually produce, how consumers accept these products, and what role they play as producers in creating a positive consumer image for beef. Very simply, if the beef industry cannot stabilize or increase consumer market share, then beef cattle numbers and beef operations will experience a significant downsizing.

RETAIL BEEF PRODUCTS

Consumer products obtained from beef cattle or made from by-products of cattle are shown in Figure 2.1. The carcass and the resulting *retail cuts* are products of primary value, while products of lesser value are classified as *by-products.* By-products are discussed later in the chapter. See Chapter 9 for more detail on yield grades and quality grades.

Retail Cuts

Figure 2.2 identifies the wholesale and retail cuts of beef that are obtained from beef carcasses. Also shown is the carcass location of the retail cuts and the recommended cooking method for each cut. Middle meats come from the rib and loin. They are the higher-valued cuts but they comprise only about 25% of the carcass weight. The round and chuck, known as the end

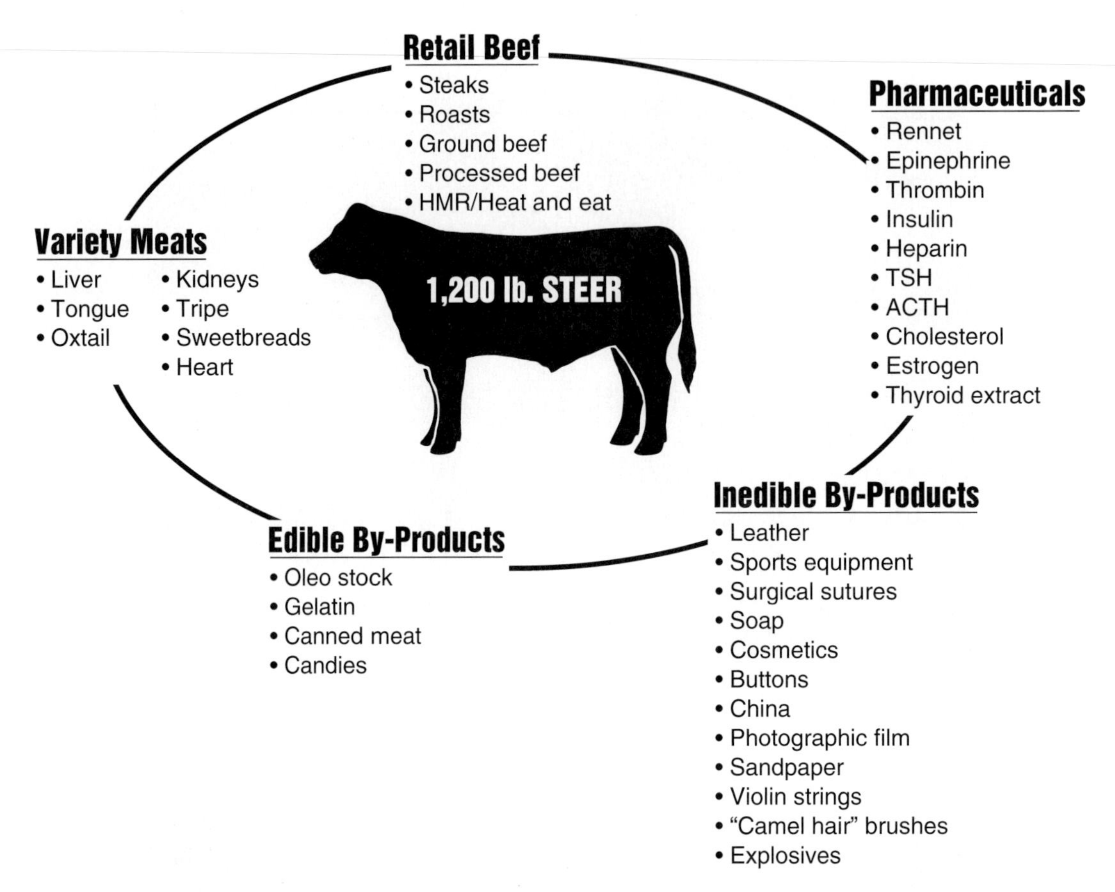

FIGURE 2.1 Products generated by the beef industry include retail beef, pharmaceuticals, variety meats, and by-products (edible and inedible).

meats, represent over 50% of the carcass weight. Cuts from the end meats historically posed problems to the beef industry because they are lower in palatability compared to the middle meats, and because roasts are not considered convenient products by most consumers. As a result, end meats lost considerable value in the market place—as much as 20%. End meats can be ground to make excellent hamburger and processed meats. However, these are lower-valued products and more value-added products are needed that are of intermediate value and that are priced between hamburger and steaks. Thus, the industry has made a concerted effort in product development to improve end meat utilization. Products such as a boneless beef filet from the chuck are a new innovation that competes directly with boneless chicken breast from both a cost and nutritional basis. Another example is a "Rotiss-A-Roast" that is prepared in a manner comparable to rotisserie chicken.

Producers must understand live animal composition and carcass specifications in terms of consumer demand for size of retail cuts, palatability, and consumer lifestyle (convenience and health). Table 2.1 identifies carcass specifications that meet different consumer demands. Shown in this table are the three primary consumer markets—retail store, lean beef, and high palatability beef. If the carcasses identified in Table 2.1 come from cattle that have been selectively bred, fed, and processed, then their properly cooked retail cuts will meet specific consumer preferences identified later in the chapter.

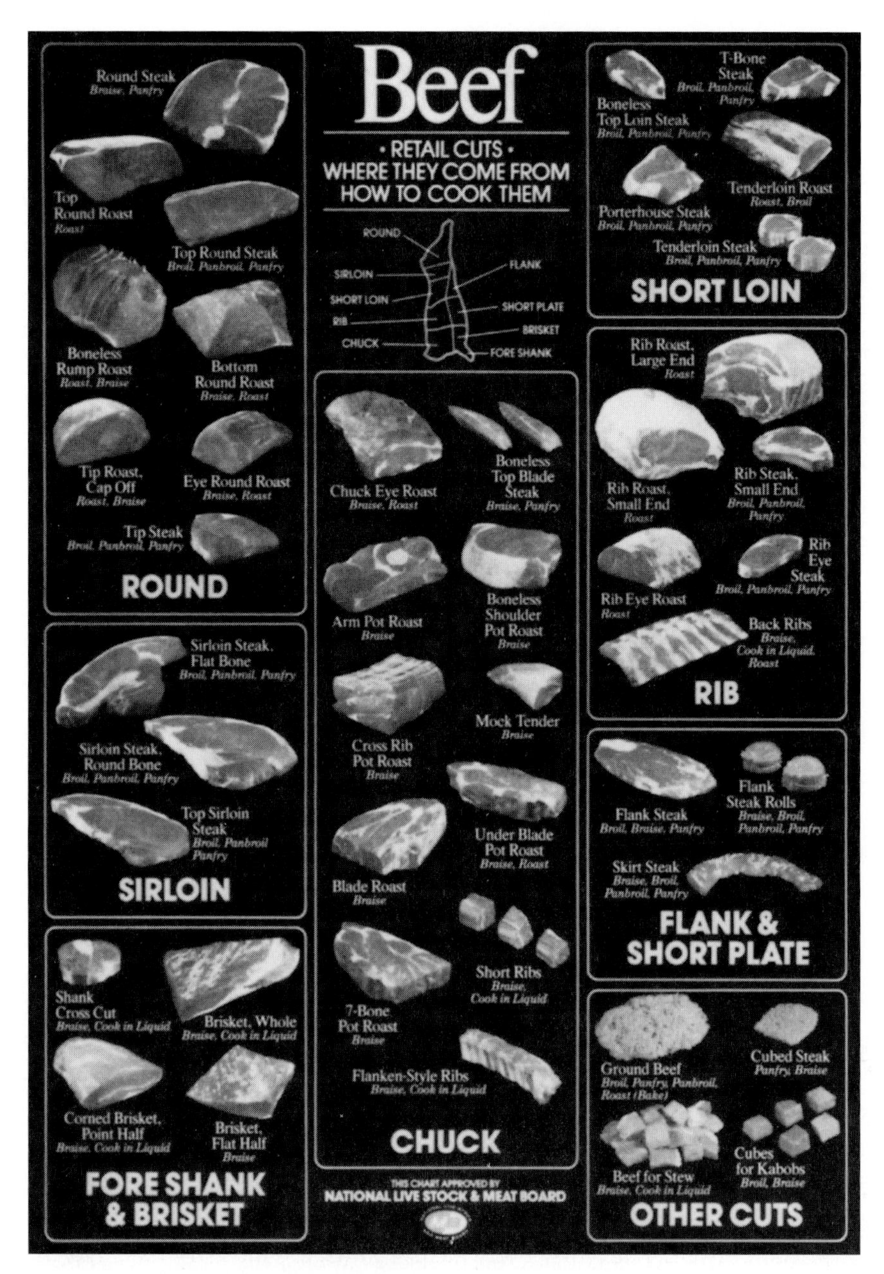

FIGURE 2.2 The wholesale and retail cuts of beef with the recommended type of cooking.
Source: National Cattlemen's Beef Association © 1986.

BEEF CONSUMPTION AND EXPENDITURES

Consumption

Consumption patterns for different foods vary depending on availability, cost, cultural preferences, disposable income for food purchases, technological advances in food processing, and new information on nutrition, health, and food safety—and, to a large extent, on consumer perception about food items and how they fit into their lifestyle. When they are readily available

TABLE 2.1 Carcass Specifications to Meet Consumer Demand

	Specifications by Market Segment[a]		
Trait	Retail Store	Lean	High Palatability
Live weight (lbs)	1,075–1,175	1,150–1,250	1,065–1,325
Carcass weight (lbs)	650–750	700–800	650–850
Fat thickness (in.)	0.2–0.5	Maximum of 0.3	Maximum of 0.8
Ribeye area (sq. in.^{2})	12.0–16.0	12.0–16.0	12.0–16.0
Yield grade	1.5–3.5	1.5–2.9	2.8–3.9
Quality grade	Select/low choice; 100 days on feed	Minimum of low choice	Average choice or higher

[a]There is some overlap between the market segments.
Source: Adapted from Allen (1989).

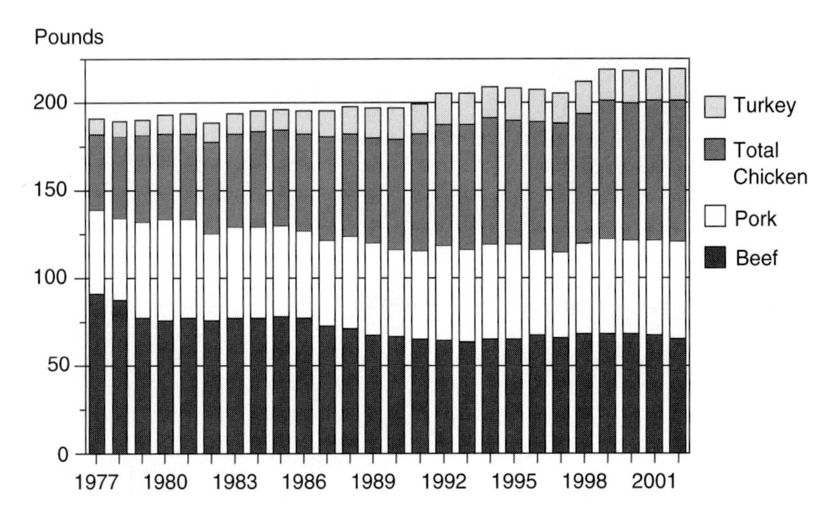

FIGURE 2.3 Per capita meat consumption (retail weight).
Source: Livestock Marketing Information Center.

and the standard of living allows consumers to purchase them, animal food products are preferred over plant food products by most consumers.

For many years, Americans considered beef the *"king of the meats"* by choosing beef in higher amounts over all other meats. However, today's consumers are showing a preference for poultry rather than beef (Fig. 2.3), and poultry is overshadowing beef's leadership in meat consumption. Beef consumption peaked in 1976 and then declined until 1998. However, from early 1999, beef demand increased in seven consecutive quarters. In fact, for the first time, annual consumer spending for beef exceeded $50 billion in 2000, with a final sales value of $52 billion.

Since 1990, per capita poultry consumption has been higher than per capita beef consumption on a retail weight basis, but the two are about equal when compared on a boneless weight basis (Table 2.2). Poultry consumption has increased significantly due primarily to a greater focus on value-added products, consumer perceptions about its healthfulness, and its competitive price. The effect of price cannot be overlooked when comparing the price-quantity relationship of red meat (Fig. 2.4) and poultry (Fig. 2.5). Notice in Figure 2.4 that

TABLE 2.2 Per Capita Annual Meat, Poultry, and Fish Consumption (lbs), 1966–2000

	Red Meat[a]			Poultry[a]			Total (meat, poultry, and fish)[c]
Year	Beef	Pork	Total[b]	Chicken	Turkey	Total	
1966	78 (74)	55 (44)	140 (124)	36 (25)	8 (6)	43 (31)	NA (166)
1976	94 (89)	54 (39)	153 (132)	43 (29)	9 (7)	52 (37)	219 (182)
1986	78 (74)	59 (42)	140 (118)	59 (41)	13 (10)	72 (51)	218 (184)
1996	68 (64)	49 (46)	119 (112)	72 (49)	18 (14)	90 (63)	224 (191)
1997	67 (64)	49 (45)	118 (111)	73 (49)	18 (14)	90 (63)	223 (189)
1998	68 (65)	54 (51)	124 (117)	76 (51)	18 (14)	94 (66)	233 (198)
1999	69 (66)	55 (51)	126 (119)	79 (53)	18 (14)	97 (67)	238 (201)
2000	69 (66)	54 (50)	125 (118)	82 (56)	18 (14)	100 (70)	240 (203)

[a]Retail weight with boneless weight given in parentheses.
[b]Includes veal and lamb for which per capita consumption was approximately 1 lb (boneless) of each annually since 1980.
[c]Per capita consumption of fish and shellfish was 11 lbs in 1966 and 15 lbs in 2000.
Source: USDA (*Food Review*).

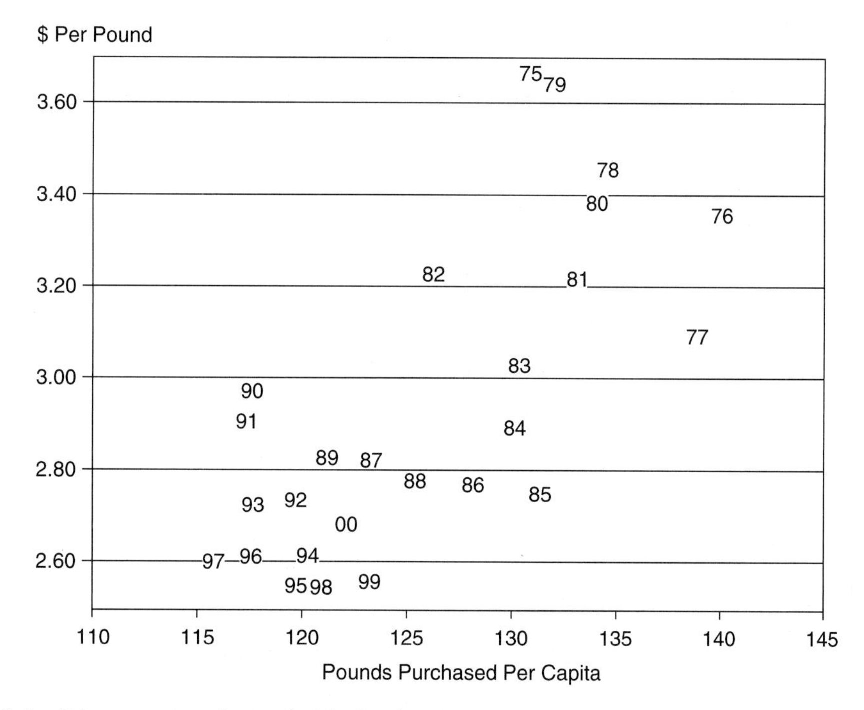

FIGURE 2.4 Price-quantity relationship for beef and pork.
Source: Livestock Marketing Information Center.

declines in price were not necessarily accompanied by increased purchases of red meat. This evidence suggests a real loss in demand. In contrast, Figure 2.5 illustrates that consumers responded to incremental declines in poultry price by increasing their consumption.

Table 2.3 shows the annual per capita consumption of various types of beef products. The decreased beef consumption has been primarily at the expense of beef cuts. Ground beef sales comprise approximately 40% of the current beef consumed. Over the past decade, it would

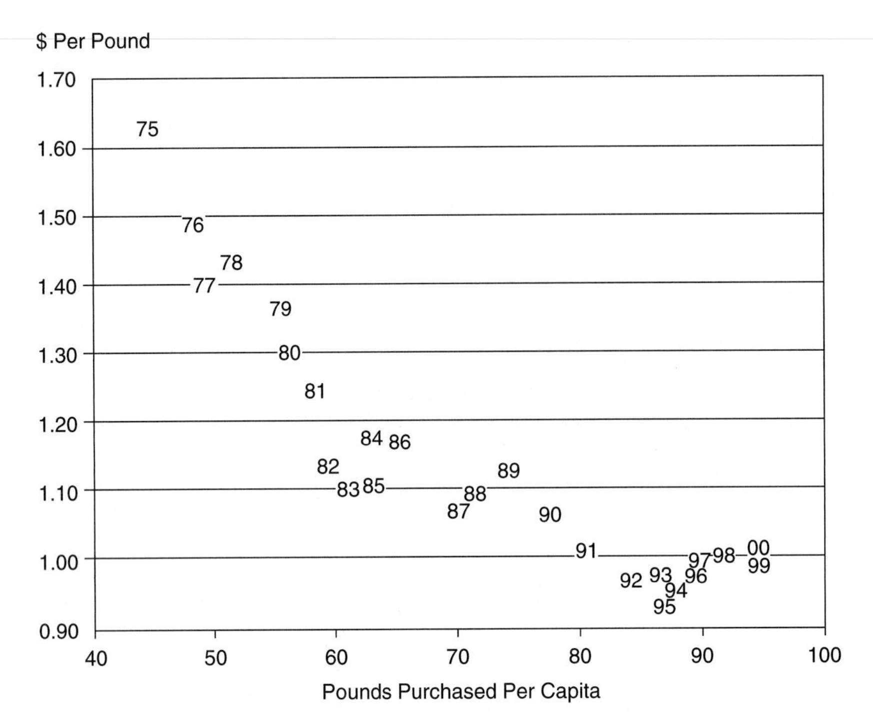

FIGURE 2.5 Price-quantity relationship, for chicken and turkey.
Source: Livestock Marketing Information Center.

TABLE 2.3 Consumption Patterns and Prices of Beef, 1970–2000

Year	Cuts	Ground Beef	Processed Beef	Price Choice Beef	$/lb Ground Beef
1970	51.2	21.0	10.8	1.02	0.66
1980	44.5	24.7	7.4	2.39	1.36
1985	45.7	24.3	5.9	2.29	1.24
1990	31.3	30.6	5.9	2.81	1.59
1995	31.5	26.9	9.0	2.84	1.37
1998	32.5	26.7	8.9	2.77	1.41
2000	28.8	27.9	9.3	2.88	1.45

Sources: NCBA and Cattle-Fax.

appear that demand for beef cuts and ground beef has stabilized. The industry now must determine a means to increase demand for its products.

American consumers have been fortunate to experience continued increases in disposable income while spending proportionally less of that income on food. Today's U.S. consumer spends less than 11% of their income for food. Figure 2.6 illustrates the decline in percent of disposable income spent for meat. However, as the data in Table 2.4 illustrates, in the past 20 years consumers have been willing to pay almost another $150 per capita to purchase meat and poultry. The beef industry has only been able to capture $20 of that amount. Nonetheless, in 2000, consumers spent $53 billion dollars to purchase beef, a new record.

Beef consumption data can be misleading depending on how the information is reported. For example, the data could be based on carcass weight (used in international trade), retail

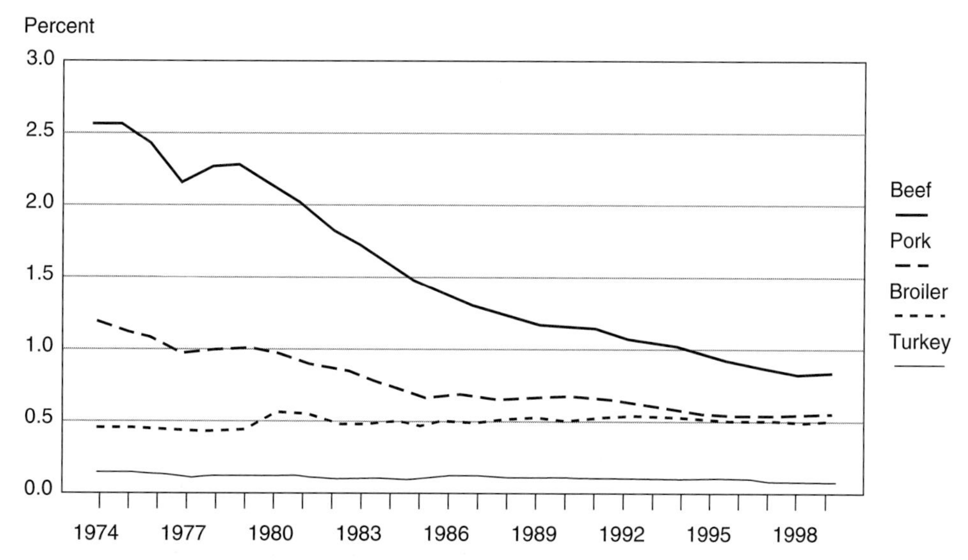

FIGURE 2.6 U. S. Per capita expenditures for meat and poultry (% of disposable income). *Source:* Livestock Marketing Information Center.

TABLE 2.4 Per-Capita Spending on Meat and Beef's Market Share (1980–2000)

Year	Beef ($)	Pork ($)	Broiler ($)	Total Meat ($)	Beef Share (%)
1980	178.82	79.96	44.45	306.07	59.0
1985	180.44	83.78	58.66	326.46	55.9
1990	190.88	105.22	86.92	387.49	49.8
1995	192.94	102.19	99.38	398.33	48.8
1999	199.27	131.98	121.68	452.88	44.0
2000	199.04	126.92	126.94	453.39	43.9

Source: NCBA and Cattle-Fax.

weight, edible weight, or cooked weight basis. For example, per capita consumption for 1999 was carcass weight (94 lbs), retail weight (69 lbs), boneless weight (66 lbs), and estimated amount of beef actually ingested (37 lbs). Beef "disappearance" is a more valid term than beef "consumption" because, even on a retail weight basis, there may be fat trim, bone, cooking, and plate waste losses that represent beef not consumed or ingested. Figure 2.7 shows how the amount of bone in meat dramatically affects the retail versus edible weight disappearance data. Cooking would further reduce the edible weight by 15–30%, depending on the type of meat and cooking method. Americans waste a large amount of food—throwing out 365 lbs a year for every man, woman, and child. Of the nearly 100 billion lbs of food wasted, approximately 16 billion lbs comes from meat, poultry, and fish.

Expenditures

The amount of beef consumed in the United States in earlier years was highly influenced by the per capita disposable income. However, in the late 1970s and 1980s this pattern changed dramatically. The change was influenced by several factors, including health considerations,

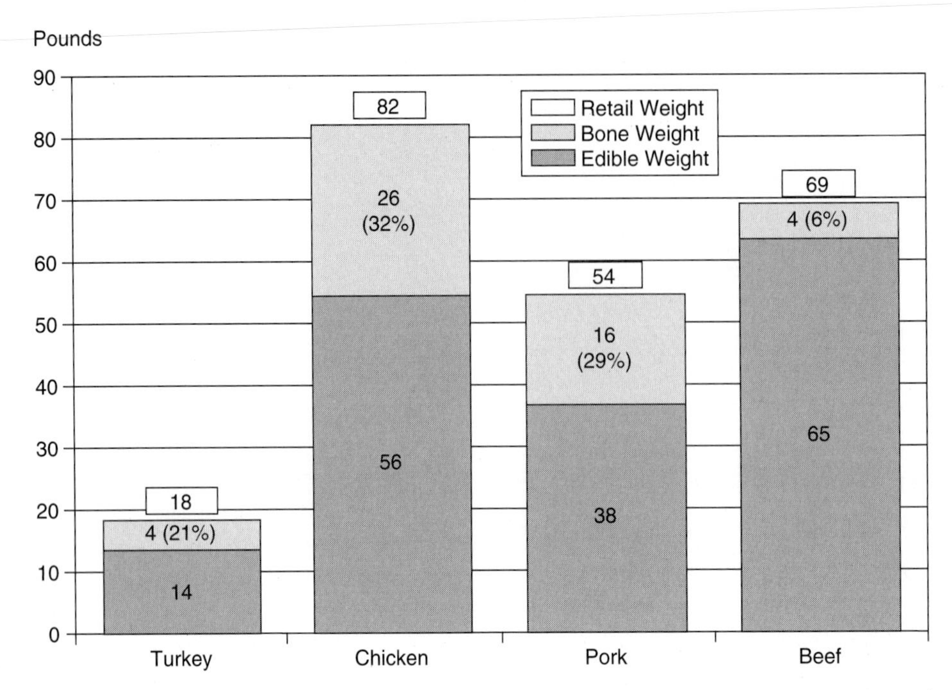

FIGURE 2.7 Per capita disappearance of animal proteins on the basis of retail weight, boned weight, and edible weight.

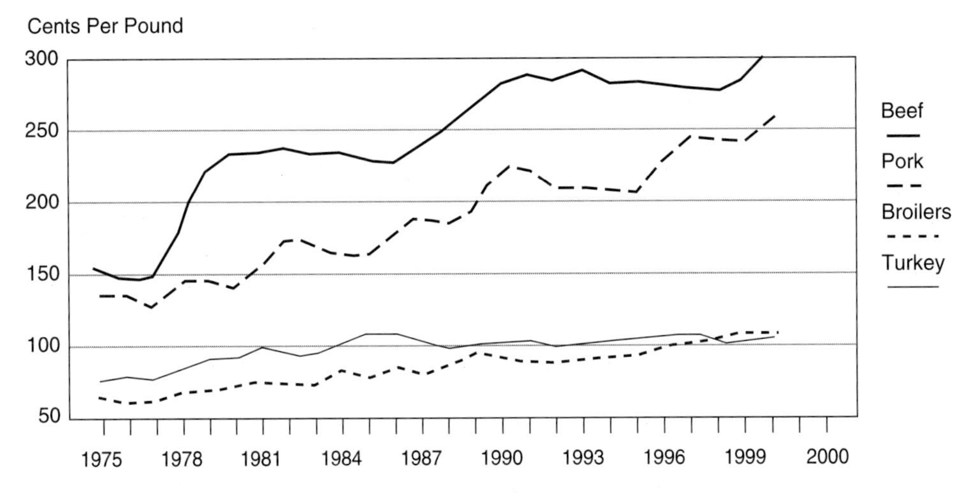

FIGURE 2.8 Retail meat and poultry prices (1975–2000).
Source: Livestock Marketing Information Center.

convenience of retail products, and the competitive price of poultry. The retail price of beef and other competing meats is shown in Figure 2.8. This retail price comparison is somewhat misleading when compared with costs based on a 3-oz boneless serving. Based on a retail weight comparison the price of chicken is approximately 35% of beef, while chicken is 50% the price of beef when compared on edible weight. Since most meat is priced on a retail weight basis, consumers are unaware of price differences of edible portions.

TABLE 2.5 Market Share of Consumer Meat Spending[a] (Percent of Total Meat Expenditures)

Year	Beef (%)	Pork (%)	Chicken (%)	Turkey (%)
1980	53.9	26.9	16.1	3.1
1985	50.3	26.2	19.5	4.0
1990	45.4	27.0	23.1	4.5
1995	44.7	24.8	25.8	4.7
1998	39.8	29.5	26.3	4.4
2000	41.0	27.8	27.2	4.0

[a]The total spent for meat in 2000 was $453 per capita.
Source: USDA.

Consumer market share is determined by the percentage of total meat expenditures that consumers allocate to various products. Table 2.5 illustrates the share of meat spending captured by the beef industry and several of its competitors. The market share loss between 1980 and 1998 cost the beef industry $12.8 billion in consumer expenditures. Calculations by Cattle-Fax showed that if half of this loss could have been regained, the retail beef price would have been $0.36 per lb higher and the fed-steer price would have been $9/cwt higher.

The positive news for beef is that the loss of market share has been halted and since 1999 the industry has begun to win back consumers. Pork and turkey have very stable levels of market share, while chicken has gained share every year in the past two decades. However, in the period from 1998 to 2000, beef's increase in market share of total expenditures outpaced that of chicken.

BEEF PALATABILITY AND CONSUMER PREFERENCES

Consumers obviously eat food for its life-sustaining nutrients, but they typically prefer foods that provide eating satisfaction (taste), convenience, consistency, and a positive contribution to good health. Recent surveys have shown that 91% of U.S. households serve beef, compared with 68%, 80%, and 4% for poultry, pork, and lamb, respectively. Only milk and eggs rank above beef in total household acceptability.

Consumers prefer beef that is tender, flavorful, and juicy, with a high ratio of lean to fat and a high ratio of lean to bone. Producers must understand the primary factors affecting beef palatability so they can assist in designing beef products that have high consumer acceptance.

The 1991 Beef Quality Audit gave the startling observation that one out of four (25%) cooked beef steaks just don't eat right (have inadequate flavor, juiciness, tenderness, and/or overall palatability problems). A Texas A&M study showed that one carcass, poor in palatability characteristics, could negatively affect more than 540 consumers. These and several other studies vividly demonstrate that improvement in beef palatability must occur if consumer market share is to be positively influenced.

Food Service

The market for beef is generally segmented into three categories—at-home preparation, food-service, and export. Consumption of beef from food-service outlets has become a significant market target (Table 2.6). Seventy percent of consumers eat out at least once a week. In 1999, 7.2 billion servings of beef were purchased at food-service establishments (an increase of 13% since 1990). This compares with 5.2 billion servings of poultry, 2 billion of seafood, and 0.5 billion

TABLE 2.6 Percent of Food Expenditures Spent on Food Prepared Away From Home

Year	%
1955	25
2000	45
2010[a]	53

[a]Projected.
Source: Adapted from National Restaurant Association.

TABLE 2.7 Consumption of Beef Offerings at Food-Service Outlets (1999)

Item	% of Sales
Hamburger/cheeseburger	76
Roast beef sandwich	7
Steak entrée	5
Steak sandwich	4
Roast beef/prime rib	3
Other beef entrées	3
Ground beef entrées	2

Source: NCBA.

of pork. Food-service sales typically grow when the economy is strong and consumers continue to demand high degrees of convenience. Beef products preferred by food-service consumers are diverse, but clearly dominated by ground beef offerings (Table 2.7).

Tenderness

Consumers rate tenderness as the most important palatability characteristic of beef. The 1990 National Beef Tenderness Survey showed that 20% of the cuts from the loin and rib and 40% and 50% of the retail cuts from the chuck and round were not satisfactory in tenderness. These tenderness problems can be compounded by improper cooking and serving methods. The beef industry must identify and utilize quality control procedures to assure more consistency in beef tenderness.

Some knowledge of muscle structure (Fig. 2.9) is needed to comprehend tenderness. The structural differences among muscles that help determine the tenderness include:

1. *The amount of connective tissue.* Connective tissue surrounds the myofibrils, while another connective tissue layer covers the muscle fiber. Additional connective tissue layers cover the muscle bundles and also the entire muscle. The more connective tissue, the less tender the beef when it is cooked. Cattle differ in the amount of connective tissue in their total musculature.
2. *Sarcomere length.* The sarcomere is the individual unit of the myofibril that allows the muscle to contract or relax. Longer sarcomeres in myofibrils result in more tender cooked beef. In living muscle, myofibrils shorten during contraction and lengthen during relaxation. Rate of cooling of the carcass during the first few hours postmortem determines sacromere length and thus partially determines of the tenderness of the cooked meat. "Cold shortening" occurs if the muscles stay contracted. The amount of outside fat and marbling help insulate muscle tissue against cold shortening. Research has demonstrated that a minimum of 0.3 in. of fat over the ribeye is sufficient to prevent cold shortening.

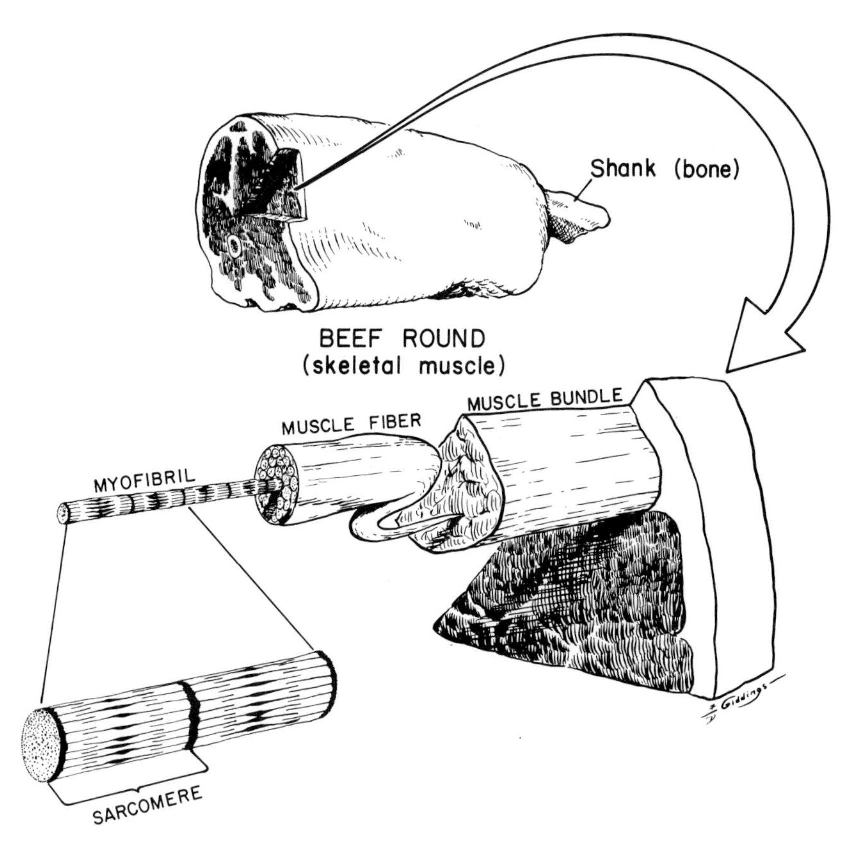

FIGURE 2.9 The fundamental structure of muscle in the beef carcass.
Source: Colorado State University.

3. *Sarcomere degradation.* Beef that is stored (aged) for several days under refrigerated conditions will become more tender. Most of the tenderization will be achieved with 7–14 days of aging. This increase in tenderness due to "aging" is achieved through the activity of proteolytic enzymes (calpains) by causing degradation or a weakened structure of the sarcomeres. *Bos taurus* cattle have less calpastatin (inhibitor of calpain) than *Bos indicus* cattle. Calpastatin prevents calpain from tenderizing the muscle during the aging process.
4. *Marbling.* Marbling makes the cooked beef more tender because fat is less resistant to shear force than muscle fibers or connective tissue. Also, fat lubricates the mouth during chewing causing the consumer to perceive a higher degree of tenderness than in beef containing less marbling. Tenderizing by mechanical means (needling, blade tenderizing) can increase the perceived tenderness of beef. Both the myofibrillar and connective tissue components of tenderness are modified by mechanical tenderization. The mechanical tenderizer physically disrupts the tissue through penetration of small blades into the meat. Mechanical tenderization is used most frequently by institutions that purchase beef that is variable in tenderness. Blade tenderization reduces this variability. There are some food safety concerns with mechanical tenderization because microorganisms on the outside of the muscle could be transferred by the blades to deep within the muscle. The latter increases the risk that cooking temperature would not as easily destroy the microorganisms.

Table 2.8 summarizes the major factors affecting the tenderness of beef. Producing beef that is consistently tender is not easy, however, some in the industry are accomplishing it.

TABLE 2.8 Major Factors Affecting Beef Tenderness

Factor	Effects and Relationships
Breed and biological type within a breed	Brahman and > 50% Brahman cross cattle produce beef that is less tender than other breeds. This difference is partially due to less proteolytic enzyme activity in the muscle of *Bos indicus* cattle. Brahman breeding (when 3/8 or less) combined with other appropriate breeds/biological types (e.g., British breeding) have shown satisfactory evidence of producing beef that is tender. The Senepol, Romosinuano and Tuli breeds are being evaluated as substitutes for the Brahman influence in cross-breeding heat-tolerant cattle. British breeds rank highest in tenderness (Chapter 13). There are differences in tenderness within breeds that show apparent genetic difference between sires in producing tender beef.
Age of animal	As an animal grows older, there is an increase in the cross-linking within and between collagen molecules that makes them less soluble (less susceptible to softening during cooking) and thus less tender. Fed steers and heifers produce the most tender beef between 12–24 mos of age. Young bulls that are fed concentrates can produce reasonably tender beef up to 15–16 mos of age when tenderness decreases.
Feedlot gain	Rapid liveweight gains associated with 100 days of high-concentrate feeding increase tenderness. A high daily rate of gain generates more protein turnover, which in turn is associated with higher myofibril fragmentation and collagen solubility. Cattle gaining 0.5 lbs/day or less have been shown to produce beef that is less tender than that from cattle with more rapid average daily gains.
Rate of carcass cooling,carcass wt, and fat cover	Too-rapid cooling of the carcass results in greater shortening of sarcomeres. Heavier carcasses and those with more fat cover have decreased cold shortening of the sarcomeres.
Aging of carcass or retail cuts	"Aging" is the holding of beef under refrigeration for > 7days to increase tenderness. During this time, there is a proteolytic weakening of the myofibril structures. Storage temperature may be increased during aging to hasten tenderization, as proteolytic enzymes are more active at higher temperatures. Because of food safety concerns, high temperature aging is not recommended.
Electrical stimulation	Administering electrical shocks (high or low voltage) to the carcass during slaughter speeds up rigor mortis, thus reducing cold-shortening. There is evidence that electrical stimulation also causes weakening of the myofibril structure, thus improving tenderness. Strength of voltage and duration of treatment both influence effectiveness of electrical stimulation. Increases in both factors improve tenderness.
Marbling	Higher marbled beef is generally more tender. Marbling accounts for 10% to 20% of the tenderness differences in beef when research results are averaged.
Location of retail cuts	Limb muscles have more collagen to support muscles used for locomotion so they are less tender. Loin muscles have relatively low levels of collagen, thus they are more tender. Some muscles with large amounts of collagen are ground or put through a mechanical tenderizer.
Method of cooking	Collagen softens during moist heat cooking. Under steam cooking, collagen will usually turn to gelatin. Retail cuts with high collagen content are usually cooked with moist heat. Dry heat is used on tender cuts (Fig. 2.2).
Cooking temperature	When internal temperature of beef exceeds 145° F for whole muscle cuts during the cooking process, the meat is less tender because myofibrillar proteins harden with high temperature. USDA recommends 160° F for hamburger, however, this is for food safety reasons and decreases tenderness.
Serving method	Tenderness is greater if beef is served hot (immediately after cooking) rather than allowed to cool to room temperature prior to eating it.

Many more producers, processors, and retailers need to produce and serve more consistently tender and highly palatable beef. Better ways of measuring beef tenderness are needed from production through consumption. Examples of these improvements include the use of DNA markers to identify superior (inferior) gene types, Beef CAM, and other objective carcass evaluation systems. With the implementation of these improved methods, beef tenderness and other beef palatability characteristics could be improved and consumer market share could increase.

Flavor

Beef flavor is determined by specific compounds that are in the intramuscular fat (primarily in the form of marbling fat) of beef muscle.

The primary determinants of flavor desirability of marbling fats as identified by Smith, (1995) are:

1. *Grain-fed versus forage-fed beef.* Most consumers greatly prefer the taste and aroma of the fat from grain-fed, as compared with forage-fed, cattle.
2. *Fed grain for approximately 100 days.* Research has demonstrated that the fat from cattle fed high-concentrate diets for 100 days was of optimal flavor desirability.
3. *Flavor increases as marbling increases.* As the amount of marbling increases from "practically devoid" (characteristic of U.S. Standard) to "slight" (characteristic of U.S. Select) to "small" (characteristic of the lower third of U.S. Choice) to "slightly abundant" (characteristic of U.S. Prime), there are progressive increases in the desirability of flavor in cooked beef.

The characteristic flavor of "aged" beef (beef stored for periods from 7 to 42 days following slaughter) is not attributed to components of fat; rather, the acid-like, tart, nut-like flavor of aged beef arises from "ripening," which generates amines from protein as well as by-products of the breakdown of nucleic acids. Beef can be "wet-aged" (held for long periods of time in vacuum packages) or "dry-aged" (held for long periods of time with no protection or packaging). There is no difference in the extent of tenderization that is achieved in "wet-aged" or "dry-aged" beef that is held, postmortem, for the same period of time. Juiciness is more desirable in "wet-aged" beef while flavor is more intense in "dry-aged" beef because of dehydration and thus concentration of flavor-eliciting compounds and/or growth of molds (Smith, 1995).

Juiciness

Smith (1995) summarized the two major components of juiciness as (1) the release of meat fluids during the first few chews of the meat and (2) the potential stimulating effect of fat on the salivary flow. The second component appears to be more important than the first.

Juiciness is influenced largely by the amount of marbling and degree of doneness. Degree of doneness relates to the amount of intramuscular water (moisture) retained in the meat after it is cooked. More highly marbled beef and beef prepared as "rare" to "medium" would be considered more juicy by consumers; however, if beef cuts are to be cooked to 145° F for safety reasons then "medium" would be considered the minimum degree of doneness. Cooking procedure is very important in influencing the juiciness of beef, as the cooking process that retains the most fat and fluids will yield the juiciest meat. For many consumers, beef that is cooked to the "well-done" state is considered unacceptably dry.

The ability of a muscle to hold its own or added moisture against the forces of heat and pressure is termed "water-holding capacity." Beef that is excessively pale in color usually loses excessive amounts of its moisture during cooking and is dry and powdery when eaten; dark-cutting beef has unusually high water-holding capacity and remains juicy even when cooked to the "well-done" state (Smith, 1995).

Juiciness increases as marbling increases, almost in a linear relationship. Also, beef with higher amounts of marbling can be cooked to higher degrees of doneness (higher temperature end points), yet remain acceptably juicy when compared to beef of lower degrees of marbling.

Tenderness and juiciness are related. The more tender the meat, the more quickly the juices are released by chewing and the more juicy the meat seems to be.

USDA Quality Grade, Fat, and Overall Palatability

Smith et al. (1987) showed that the palatability of loin steaks was rated higher as quality grade increased. Figure 2.10 notes that the undesirable palatability ratings were 59% for Standard, 26% for Select, 11% for Choice, and 5% for Prime. The variability of eating satisfaction among steaks within a quality grade was reduced as quality grades increased from Standard to Prime.

George et al. (1997) evaluated 1,650 steaks from retail outlets in eight U.S. metropolitan centers to characterize a national baseline for beef loin tenderness. Striploin steaks had lower shear force values (were more tender) than top sirloin steaks in all quality grades. Figure 2.11 shows that shear force values significantly improved (were lower) and variation within a quality grade was reduced as quality grades were higher (Prime > upper and low Choice > Select). In a corollary comparison by George et al. (1997) striploin steaks from carcasses grading Select received lower panelist ratings for juiciness, muscle fiber tenderness, amount of connective tissue, and overall tenderness than striploin steaks from higher-quality grades (Choice and Prime).

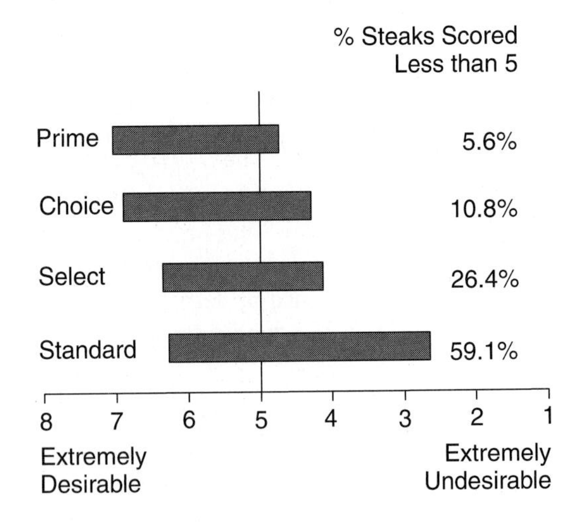

FIGURE 2.10 Percent of loin steaks receiving undesirable overall palatability ratings.
Source: Smith et al., 1987.

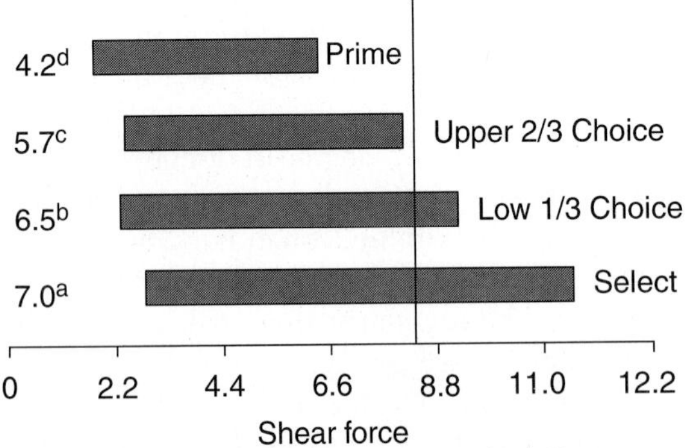

FIGURE 2.11 Striploin steak, by quality grade, means shear force values ∓ 2 standard deviations.
Source: George, 1997.

Numerous surveys have shown that consumers purchase beef and other foods primarily for their taste. The amount of fat in food appears to be highly related to its taste. Research shows that fat consumption may satisfy several physiological and psychological needs. The sensory pleasure response appears to be mediated by the release of endogenous opiate peptides—pleasure-enhancing molecules—manufactured by the human brain (Drewnowski, 1995).

Since opiates mediate the body's response to stress, it may be that preference for fat is nature's way of ensuring an adequate supply of energy. On the other hand, since opiate peptides are also involved in mediating the pleasure response to foods, it may be that fat consumption affects the mental health and well-being of the individual. Foods rich in sugar and fat are sought-after, not for what they do to the body, but for what they may do to the brain. Taste preferences for sugar and fat appear to be under opiate control. These taste preferences for fat are reflected in an increased per capita consumption of fat and oils. Fat and oils intake increased 32% from 1970-1999 even in a society that has been encouraging a reduction in fat consumption.

In the news release for the 1995 National Beef Quality Audit, Dr. Gary C. Smith, Colorado State University, stated "America's beef producers and feedlots—through no fault of their own—have been doing exactly the opposite of what the consumer wants. The 1995 audit revealed consumers want juiciness, flavor and tenderness in their beef cuts, qualities that only can come from having a certain amount of fat marbled into the flesh. Meanwhile, the feedlots have been turning out leaner, more muscular animals with less of the fat that provides these qualities. I expect a shift away from overly muscled cattle, that are too large, back to a medium-sized animal with more marbling."

The 1995 Beef Quality Audit demonstrated that outside fat in cattle had been reduced by selecting lean, more muscular animals and by packers trimming fat. However, quality grade was being reduced as noted in Figure 2.12. The increase in beef of the Select grade at the expense of that in the Choice grade, based on information previously presented, would potentially increase palatability problems for consumers if postmortem technologies such as electrical stimulation and aging were not utilized.

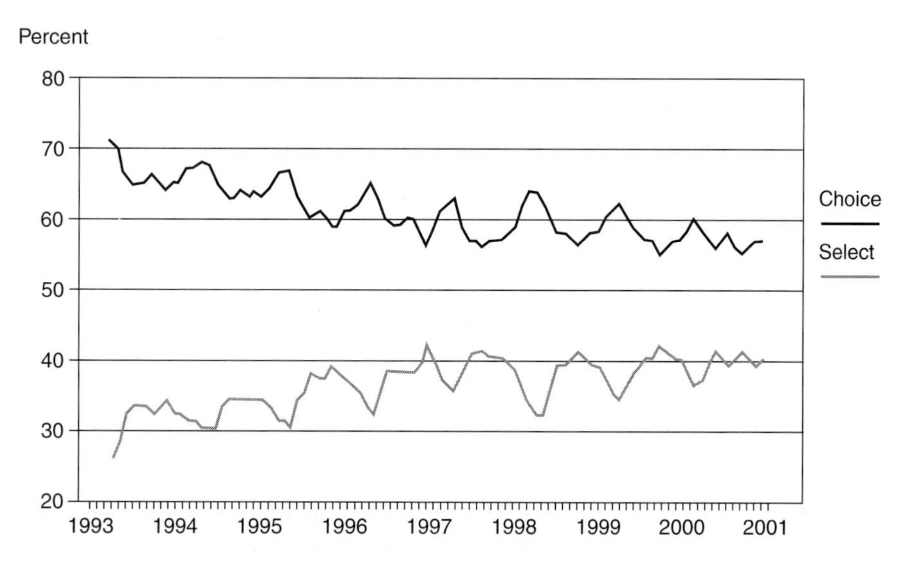

FIGURE 2.12 Beef graded Choice and Select.
Source: Livestock Marketing Information Center.

Research results support the continued use of USDA quality grades as a beef palatability critical control point, with efforts made to add critical control points that can augment, not replace, quality grades—particularly at lower marbling levels. More effective measurements of tenderness are needed to improve prediction of beef palatability. There is reason to believe that if beef is made more consistent in tenderness, juiciness, and flavor that the decline in beef consumption can be corrected. Furthermore, the industry needs to develop and adopt more objective measures of quality and tenderness using technologies such as Beef CAM (used to assess color of lean to predict tenderness) or mechanical probes (objective tenderness/toughness measure).

Lean to Fat

Consumer surveys give evidence that consumers use amount of fat as a selection criterion. The 1988 Market Basket Survey showed that closer trimming of external fat on retail cuts results in a substantial reduction in the amount of fat purchased by consumers. In this survey, more than 42% of the retail beef cuts had no external fat and the overall fat thickness for all retail cuts in the beef case was 1/8 inch. This is a marked change from the 1/4-inch trim in 1986 and the 1/2-inch trim prior to 1986. The beef industry has responded to consumer preference by reducing the fat content presented to consumers' beef products by 25–35% during 1985–1995.

The lean-to-fat and lean-to-bone ratios in beef can be most effectively controlled by proper breeding and feeding programs, with excess fat trimmed and bones removed at the packing plant or retail level. Concentrate feeding, without overfeeding, will keep the lean-to-fat ratio at the desired level, resulting in beef having desired consumer palatability characteristics if fed beef has been processed and cooked correctly. The preferred way to breed and feed cattle is to have a biological type of cattle that when slaughtered will have 0.30–0.45 in. of fat, 700–800 lb carcass weight, and grade Choice 70% of the time. A general selection goal of achieving 70% Choice or better, 70% USDA yield grade of 2 or better, and 0% carcass nonconformation has been suggested.

Beef that is too lean will not be as palatable as beef with at least a minimum level of marbling. Figure 2.13 shows the "window of acceptability" of 3–7% fat content in beef, which is equivalent to beef cuts from the lower part of the Select quality grade to the higher range of the Choice quality grade. Beef cuts with 3% fat or less content will likely not meet palatability expectations, and cuts having more than 7% fat will exceed total fat content for diet and health preferences. The statement "remove the waste fat but keep the taste fat" has considerable merit when optimizing the amount of trimmable fat with the amount of intramuscular fat needed to enhance palatability. The industry must determine the most profitable approach to reducing external or subcutaneous fat—genetics, feeding management, trimming the carcass/cuts, or some combination of these. When high levels of marbling are desired, the use of carcass or retail cut trimming will likely be utilized, which may increase the cost of product to consumer.

Lean to Bone

In the Market Basket Survey discussed earlier, 75% of all beef cuts were boneless, demonstrating that consumers have a preference for lean beef that is free of bone and excess fat. There is a pricing psychology challenge for many consumers, as the price of only lean seems high compared to the same amount of lean with fat and bone included. It is critical that beef marketers clearly differentiate the advantage of boneless beef cuts as compared with bone-in products, such as many poultry products.

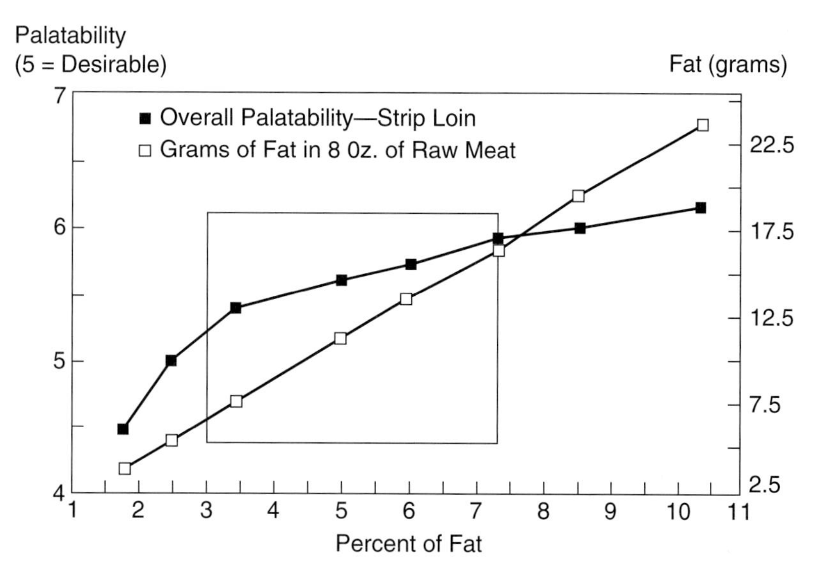

FIGURE 2.13 Window of acceptability for fat content of meat (palatability versus grams of fat, two servings). The window is based on a fat content range of 3.0–7.3%. This is equivalent to meat cuts that grade in the lower range of Select (3.0–4.27% fat content) to those that grade in the high range of Choice (4.28–8.0% fat content).
Source: Savell and Cross (Texas A&M University).

Size of Beef Cuts

The consumer's purchase of beef is significantly influenced by the number of portions needed and the cost per serving. Also, size in relation to thickness of a cut is important and may well dictate the optimum ribeye size and size of other muscles. A T-bone steak, cut an inch and a quarter thick, with an 18-sq.-in. ribeye weighs approximately 32 oz. A T-bone with a 12-sq.-in. ribeye, cut to the same thickness, is a 21-oz steak. There can be palatability problems if steaks are cut too thin because the meat can be easily overcooked and then will be tough and dry. Tatum suggested in the National Beef Quality Audit (1991) that a ribeye size of 11–14 sq. in. could be close to optimum. Size of cut is one of the major reasons why carcass weight specifications should have an upper limit of 800–850 lbs. Unfortunately, the industry has continued to increase carcass size resulting in an average carcass weight of 787 lbs in 2001.

Color

Color of meat in retail cases is a very important selection criterion for most consumers, as they perceive color to be a measure of freshness. Their preference is for a bright, cherry-red color in contrast to a dark, less bright, red color. In past years, this was a useful criterion to evaluate beef because color can be used to separate meat from younger versus older cattle. Beef from older cattle is a darker color and is usually less tender than beef from a younger animal. Currently, however, most meat from older cattle goes into processed meats, where—by mechanical means used in producing ground beef—the tenderness problem is largely eliminated. It should be noted that there is a growing market for whole-muscle cuts from non-fed cattle. Some of the darker-colored beef in retail cases results from meat remaining in the retail case for a period of time. After such meat is cooked, the darker color disappears and eating qualities may be similar to those of brighter-colored meat.

Consumers prefer to buy fresh beef, and attempts to market frozen beef have met with little success. Consumers indicate that the inability to judge the freshness of frozen beef by color is their primary reason for not purchasing it in the frozen condition.

Addition of Vitamin E to feedlot diets increases beef's color retention and extends shelf life in the retail meat case. Feedlots are implementing this recent research finding. Vitamin E costs an average of $1 per head and can return $35 through the extended shelf life.

Identifying Consumer Attitudes and Preferences

The NCBA identified consumer confidence and development of consumer-friendly products as the key factors influencing beef demand. Consumer confidence is the result of perceptions about food safety, nutrition, and the overall enjoyment of an eating experience. Delivery of consumer-friendly products is dependent on meeting customer expectations in regards to convenience and value.

Beginning in the 1980s, when beef demand losses were recognized as significant, the industry began to transform itself into a consumer-driven entity. The first step of this process was to quantify and understand the wants and needs of customers. A variety of factors influence consumer demand—the amount of disposable income, age, ethnic background, and family status, for example.

Lifestyles of consumers have changed with 70% of women in the workforce, an increase in the number of hours in the average workweek, a rise in the number of single-parent households, and the generally busier lives of children and young adults. NCBA-funded studies have shown that as a result of these time pressures, consumers spend less time on meal planning and preparation. For example, approximately 65% of all dinner decisions are made on the same day as consumption, with about 75% of those not having made a choice for dinner by 4:30 P.M. Preparation of meals from scratch is nearly a lost art as approximately 75% of households spend less than 45 minutes on meal preparation. The demand for simplicity and convenience presents a dilemma to the traditional merchandising of beef as fresh whole-muscle cuts.

Furthermore, consumer cooking skill is generally declining with 1 in 3 customers admitting that their lack of cooking knowledge prevents them from buying certain beef cuts. In fact, most consumers are confused by the use of anatomical names for retail beef cuts, the lack of cooking information on packages of fresh beef, and the poor user-friendliness of the self-serve meat case. When surveyed, nearly all consumers indicated that they would use cooking instructions if they were available. Furthermore, 78% of respondents believed these instructions would encourage increased beef sales.

Table 2.9 illustrates the ranking of factors that are very important when food purchases are made. It is interesting to note that taste is the most important factor over the past decade, with nutrition and product safety ranked as very important by approximately 70% of consumers.

TABLE 2.9 Percentage of Consumers Ranking Various Factors as "Very Important" in Making Food Purchases

Factor	1990	1992	1994	1996	1998	1999
Taste	88	89	90	88	89	92
Nutrition	75	77	76	78	76	70
Product safety	71	71	69	75	75	70
Price	66	75	70	66	64	63
Ease of preparation	33	36	34	36	37	35
Time to prepare	36	41	36	38	36	35

Source: Adapted from Trends, FMI, 2000.

The consumer market is highly differentiated and, thus, a multitude of opportunities are available for producers to develop specialty beef products designed to meet the unique requirements of a specific niche market. Examples include products such as *Certified Angus Beef* that focuses on the consumer segment that desires the flavor and palatability characteristics associated with relatively high levels of marbling. Another example is *Laura's Lean,* a product that is produced to meet the needs of consumers who desire beef that has minimal fat content.

The natural/organic market receives much attention but still comprises only about one percent of U.S. grocery sales. Still, this product category accounts for about $25 billion in annual sales. Beef products such as *Coleman Natural Beef* have focused on capturing a share of the organic/natural market. Specific details on marketing branded products are presented in Chapters 8 and 9.

SIX PRIMARY BEEF CONSUMER PREFERENCES

The NCBA and others have recently identified "six drivers" or consumer preferences that must be addressed if consumer market share is to be recaptured. They are (1) reinforce the consumer's love of beef, (2) deliver on the consumer's demand for consistent quality (high palatability), (3) deliver on the consumer's demand for convenient meat solutions, (4) enhance beef as an integral part of a healthy diet, (5) enhance the consumer's perceived value of beef, (6) deliver on the consumer's demand for a safe product. Each of these "drivers" or preferences will be discussed in this chapter.

Beef Lovers

Consumers are attracted to beef because of its palatability characteristics—tenderness, juiciness, and flavor. If the beef is highly palatable and highly repeatable in palatability, certain beef consumers, who are categorized as "beef lovers," frequently return to their preferred eating experiences. Consumers in this category are willing to pay relatively high prices for steak or prime rib and they will be frequent, repeat customers if they have a highly desirable beef eating experience each time. In fact, consumer demand for middle-meat cuts such as steak and prime rib has been growing rapidly in the past decade. The fastest-growing restaurant category has been the "steakhouse."

Beef lovers are high-volume beef consumers who comprise one-third of all beef consumers, yet they consume nearly 70% of all beef products sold. Their annual per capita beef consumption is approximately 174 lbs, which is about 10 times greater than that of the light beef users.

The beef industry is challenged in meeting the needs of beef lovers. A beef palatability study (Smith, 1995) concluded that consumers encounter what they consider to be eating problems with beef approximately one-third of the time. This occurs even in white tablecloth restaurants. There are indications that beef's market share can be lost even within the beef lovers' group. Beef that is consistently high in palatability is needed to retain beef lovers' loyalty.

Consistent Quality (highly palatable beef products)

Beef is not consistent in the palatability characteristics of tenderness, juiciness, and flavor. This is true for the same beef cut whether selected in a restaurant or from the retail meat case. Consistency and uniformity are a must if beef's market share is to be stabilized or increased.

Consistency can be significantly improved by implementing management practices at critical control points in the total system, both pre-harvest and post-harvest. Factors affecting tenderness, the most important palatability characteristic of beef, were previously identified in Table 2.8. These become the critical control points for improving the consistency of tenderness.

A recent research study by Tatum et al. (1997), in cooperation with the NCBA, demonstrated how tenderness could be significantly improved. Nonconformance was identified in both top sirloin and top loin steaks with Warner-Bratzler shear force values ≥ 10 lbs. Baseline comparisons were with beef from carcasses that had not been electrically stimulated, where one group of carcasses was aged 3 days (worst-case scenario where nonconformance was 64%) and a second carcass group was aged for 21 days (normal scenario with a 28% nonconformance rate). Cattle in the research project were sired by 31 bulls representing 8 breeds. Sire differences, electrical stimulation, aging, and calcium-activated tenderization were evaluated. The most effective critical control points to improve tenderness were:

- Selecting the top 25% of sires based on progeny group means for 14-day top-loin shear force, and
- High-voltage electrical stimulation of all carcasses followed by 14 to 21 days of post-mortem aging.

Together, these intervention strategies reduced the expected rate of nonconformance to about 6% (one in 17) for top sirloin steaks and 1% (one in 100) for top loin steaks.

While tenderness is extremely important, it is doubtful that beef can compete for market share with only a lean tender product. Broilers do not have a tenderness problem and the meat has a bland taste and frequently is dry. However, they enhance the palatability of their product by adding flavor and juiciness components through further processing and cooking. Beef's market share can be enhanced by consistently producing tender products that are also consistently high in beef's natural flavor and juiciness. However, opportunities to add flavor and juiciness components to increase the variety of convenient beef products should not be overlooked. The emergence of fully cooked, ready-to-eat beef products in the retail market illustrates the potential to increase sales. IBP, the largest U.S. packer, introduced a branded product line—*Thomas E. Wilson*—in 2000 that is expected to generate sales of more than $1 billion within the first three years.

Convenient Beef Products

Consumers live in a fast-paced society that demands food that can be prepared in a short period of time. The growing number of two-income families has increased the demand for convenience foods. This demand is the factor that motivated IBP to move into the sale of branded case-ready fresh beef and fully cooked, easy to prepare beef items.

Some individuals feel that lack of "convenience" has had a significant influence on decline in beef demand—in some cases, greater than health concerns. They cite evidence that a shift in demand for health concerns would have led to increased purchases of whole broilers and a decline in hamburger consumption, which has not been observed. They indicate that the major shift has been from beef table cuts to processed chicken parts, which reflects convenience fostered by growth in food outlets.

In the late 1980s and early 1990s, the best-selling grocery products were convenience items that featured easy, rapid preparation and little cleanup. The leading products were (1) microwaveable items, (2) just-add-water and just-add-meat items, and (3) bottled juices. The signal is for packers, processors, and retailers to develop and merchandise more convenient beef products. Beef has not been competitive with other products in the deli, the frozen entrée section, or the shelf-stable microwave section of the supermarket.

Today, more food is being purchased away from home, primarily in some 33,000 fast-food establishments. Higher-income families purchase more of their food away from home and have a higher per capita consumption of beef. This is further evidence that beef consumption is closely associated with standard of living. Beef consumption patterns continue to be greatly influenced by the economic well-being of the country and the lifestyle of consumers. Food purchased away from home accounts for more than 47.5 % of the $789 billion Americans spend on food. In most cases, cost of food, particularly meat items, is higher than for the same food prepared at home. About one-third of the total hamburger is consumed away from home, primarily in fast-food outlets.

A growing market outlet for grocery sales utilizes online ordering. Sales from this source are expected to grow to more than $10 billion by 2003.

Healthy Beef Products

Recent consumer surveys indicate that Americans are sensitive to health and nutrition relationships in the foods they purchase. Some people changed their diets during the 1980s and 1990s, primarily for health or nutritional reasons, although consumer surveys continually rank "taste" above "nutrition" as most important in food selection.

Most nutritionists agree that a healthy diet should contain all the required nutrients and enough calories to balance energy expenditure. Unfortunately, nutrient deficiencies or excesses exist in much of the world. Therefore, food choices or lack of choices, determines the nutritional status of most people. Obesity caused by caloric excesses has become a leading nutritional problem in the United States with approximately one-third of the population classified as obese.

The nutritional advantages of beef simply cannot be ignored (Table 2.10). The most compelling reason to include beef as a component of the diet is its nutrient density. Nutrient-dense foods are defined as those that offer a higher proportion of several recommended nutrients than calories. Beef is a particularly good source of zinc, iron, protein, and the B vitamins (Table 2.11). A 3-oz serving of beef provides less than 10 % of the calories in a 2,000-calorie diet while contributing more than 10% of the recommended daily allowances for protein, iron, zinc, niacin, and vitamins B_6 and B_{12}. Table 2.12 illustrates comparative amounts of several foods to acquire the same nutrients available in a 3-oz serving of beef.

Ease of nutrient absorption is also of concern. For example, iron is present in two forms—heme and non-heme. Heme iron is the most easily absorbed and is the predominant form of iron in red meats.

Unfortunately, many American citizens are deficient in iron (40%) and zinc (73%). These deficiencies are particularly noticeable in the female population. Approximately 75% of females between the ages of 12 and 49 are iron and zinc deficient. This group of consumers, particularly teenagers, tend to be low consumers of beef despite the fact that beef provides 67 to 58 percent of the zinc and iron, respectively, available in the food supply.

TABLE 2.10 Key Nutrients Found in Beef and Their Associated Functions

Nutrient	Function
Zinc	Enhance immune function, promote wound healing, essential for normal growth and cognitive abilities through childhood
Iron	Oxygen carrier, enhance body energy
Protein (beef is a complete protein that provides all the essential amino acids)	Building block for muscle development, required for repair of cells and regulation of metabolic processes
B Vitamins (riboflavin, niacin, B_{12})	Assists in energy metabolism, promotes skin health, aids digestion fosters normal appetite, promotes normal nerve function

TABLE 2.11 Nutrient Content of 3 oz of Cooked Beef and its Contribution to RDA[a]

Nutrient	Amount	% RDA
Calories	189	10
Protein	26 gm	56
Fat[b]	8.7 gm	14
Cholesterol[b]	76 mg	24
Sodium	55 mg	2
Iron	2.7 mg	14
Zinc	6.0 mg	39
Niacin	3.6 mg	17
B_{12}	2.3 mcg	39

[a]Recommended Daily Allowance based on a 2,000-calorie/day diet and the RDA for women age 23–51.
[b]The American Heart Association recommends that fat contribute not more than 30% of total daily calories and that cholesterol be limited to 300 mg/day.
Source: USDA.

TABLE 2.12 Equivalent Servings Required to Equal the Amount of Nutrient Found in a 3-oz Serving of Beef

Nutrient	Equivalent Serving
Zinc	12 (3.25 oz) cans of tuna
Vitamin B_{12}	7 chicken breasts
Iron	3 cups of spinach
Riboflavin	2 1/3 chicken breasts
Thiamin	2 chicken breasts

Source: USDA.

In regards to meeting their protein requirement, only 45% of teenage males and 32% of teenage females actually consume sufficient protein. At the same time, teenagers consume 6 to 10 daily servings from the sugar, fat and oil category with 40 percent of their energy requirements being met by foodstuffs considered to be low in total nutritional value.

Despite the evident dietary advantages of beef, concerns have been raised in regards to the role of meat in the incidence of some human diseases.

The relationship between diet and human health is a controversial and complex topic. The consumption of beef has been linked to two of the most dreaded human diseases—coronary heart disease (CHD) and cancer. Consumer perceptions have been influenced to accept these alleged relationships, and consumers have reduced their consumption of some animal products accordingly.

Before examining the root causes of human disease, it is worthwhile to understand the concept of risk analysis. American citizens are enjoying an ever-increasing life span, better health, and a higher standard of living than their ancestors could have imagined. Despite this evidence of well-being, widespread public worry about diet-health relationships persists. These worries are perpetrated when reports fail to account for the following:

- the wholesale extrapolation of results obtained from lab animal models to humans,
- the fact that natural compounds may contribute significantly more risk than man-made compounds, and
- the effect of dose rate on disease incidence.

In a media-dominated environment where sensationalistic headlines are all too common, consumers are advised to be wary of "junk science." Warning signs that results of a study are being incorrectly represented are as follows:

1. The results recommend changes that offer quick-fix promises.
2. Foods are described as "good" versus "bad."
3. Simplistic conclusions are offered from a complex study.
4. The study was not peer reviewed.
5. Recommendations ignore differences between individuals or among groups.
6. Results are interpreted to offer significant negative consequences from a specific food item or diet selection.

Sound, well-based research work often appears to be lost in an emotionally charged issue. Many organizations and individuals have occasionally based judgments and decisions on emotion rather than on the best accumulated research facts. Some individuals feel that consumer perceptions are directed by some self-proclaimed "diet and health experts" and a communications system that builds part of its readership on sensationalism. Scientific principles should be identified so that decisions are based on true relationships.

The major known risk factors associated with coronary heart disease are genetics (a family history of CHD), high blood cholesterol, smoking, hypertension, physical inactivity, and obesity. Obesity caused by excess caloric intake is a major nutritional problem in the United States. Of the leading ten causes of death in the United States, obesity is considered a risk factor in five (coronary heart disease, stroke, hypertension, type II diabetes, and some forms of cancer). More than 50% of U.S. adults exceed their recommended weights. Interestingly enough, survey results suggest that consumers are less concerned about caloric and fat intake than they were in 1990. In a ten-year span, the percentage of consumers who reported that they were always conscious of caloric intake fell from 40 to 25%. Consumers reporting that they were always cautious about their fat intake fell from 51 to 33% over the same time period. It is generally accepted that consumption of animal fat by humans causes an increase in the level of cholesterol in the blood, while consumption of vegetable oils (polyunsaturated fats) causes a decrease in blood cholesterol concentration. These relationships led to the theory that there is a relationship between consumption of animal fats and the incidence of atherosclerosis ("plugging" of the arteries with fatty tissue), which in turn results in an increased likelihood of death from coronary heart disease.

Evidence supporting the proposed blood cholesterol—heart disease relationship is still theoretical. Studies in which dietary fat intake has been modified, either in kind or amount, did not show significantly reduced mortality rates. Changing diets from animal fats to vegetable fats has not improved the heart disease record. There are data that suggest that poor health conditions can also result from people eating diets high in polyunsaturated fats.

Most consumers do not know the difference between saturated and polyunsaturated fats. Through many margarine and vegetable oil commercials, however, they have been informed that saturated is "bad" and unsaturated is "good." In a review of the diet and heart disease relationship, some medical doctors argue that the dietary-heart hypothesis became popular because a combination of the urgent pressure of special interest groups or health agencies, oil-food companies, and ambitious scientists had transformed that fragile hypothesis into treatment dogma. Diet-heart enthusiasts then began using a show of hands or mail polls to influence national food policy.

Senator George McGovern's committee on Nutrition and Human Needs in the late 1960s and early 1970s shifted its emphasis from developing policy aimed at eliminating malnutrition to dealing with the issue of consuming too many calories. Largely influenced by a self-proclaimed diet expert, McGovern's committee released a document entitled "Dietary

Goals for the United States" based on two days of testimony. Written by a journalist with absolutely no background in science, nutrition, or health, the report would become the basis for a national policy focused on dietary fat. Twenty-five years later there is still not compelling and clear evidence as to the effect of dietary cholesterol and fat intake on human longevity.

The relationship between cholesterol intake and death from coronary heart disease is minimal. All saturated fatty acids are not equal in terms of their effect on serum cholesterol. It is important to recognize that less than half of the fatty acids found in beef fat are saturated. The 18-carbon length fatty acid (stearic acid) has either a neutral effect on serum cholesterol or may actually lower serum cholesterol concentrations when substituted for other saturated fatty acids. Stearic acid comprises almost one-third of the saturated fatty acids in beef.

For example, a Select grade steak will have about 50% of its total fat in monounsaturated form of which nearly 90% is oleic acid (the beneficial fat found in olive oil). The remaining one-half of the total fat is saturated but one-third of that is stearic acid, which is potentially beneficial, but at the very worst neutral in its effect. In total, more than one-half, and as much as three-quarters, of the fat in the steak will lower cholesterol levels (Taubes, 2001).

Cholesterol is a naturally occurring substance in the human body. Every cell manufactures cholesterol on a daily basis. The average human turns over (uses and replenishes) 2,000 mg of cholesterol daily. Average dietary consumption of cholesterol is approximately 600 mg daily. Therefore, the body makes 1,400 mg each day to meet its needs.

Cholesterol cannot be used by the body unless it is joined with a water-soluble protein, creating complexes known as lipoproteins. There are several different types of lipoproteins. Research workers have identified two of these lipoproteins: HDL (high-density lipoprotein) and LDL (low-density lipoprotein). High blood levels of the LDLs have been generally associated with increased cardiovascular problems, while some research data show that higher blood levels of HDLs may reduce heart attacks by 20%. Additional research with laboratory animals has shown that those fed beef had HDL levels 33% higher than animals fed soybean diets.

There is evidence of genetic differences in the proportion of HDLs and LDLs in individuals. People who are overweight, nonexercisers, and cigarette smokers have higher proportions of LDLs than those who are lean, exercisers, and nonsmokers.

Although the exact roles of dietary cholesterol and blood levels in the development of coronary heart disease are not known, it would appear logical from the existing information to use prudence in implementing drastic changes in dietary habits. Certainly, those individuals with high health risks primarily due to genetic background should take the greatest precautions.

Figure 2.14 shows that on an average per capita basis, red meat and poultry supply less than 16 grams of fat per day. This level of fat (15.6 g) is less than 24% of the 67 grams of fat per day recommended by the American Heart Association for a 2,000-calorie-per-day diet. The 15.6 grams of fat represent 30% of the calories from fat where each gram of fat contains 9 calories.

Some individuals argue that consumers eat too much red meat and that excessive red meat consumption causes cancer. Evidence to support this statement is questionable, and excessive consumption must be defined. Previous data have shown that the average daily per capita beef consumption is approximately 1.8 oz. Figure 2.15 shows the per capita red meat consumption at approximately 3.4 oz/day while a more recent report gives 2.6 oz/day. The 3.4 oz provides approximately 25 g of protein, which is less than half the average recommended daily allowance (RDA) for protein. The American Heart Association recommends 3.5 oz of cooked meat per person on a daily basis. Based on this recommendation, the average U.S. per capita consumption of beef and other red meat is not excessive (Fig. 2.15).

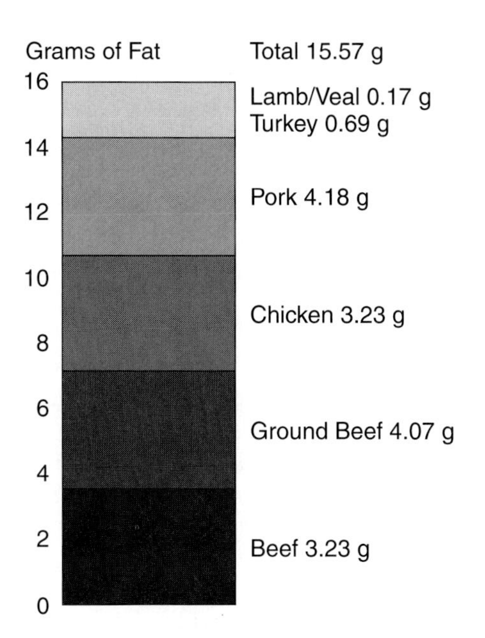

FIGURE 2.14 Average daily per capita fat consumption from cooked meat.
Source: USDA.

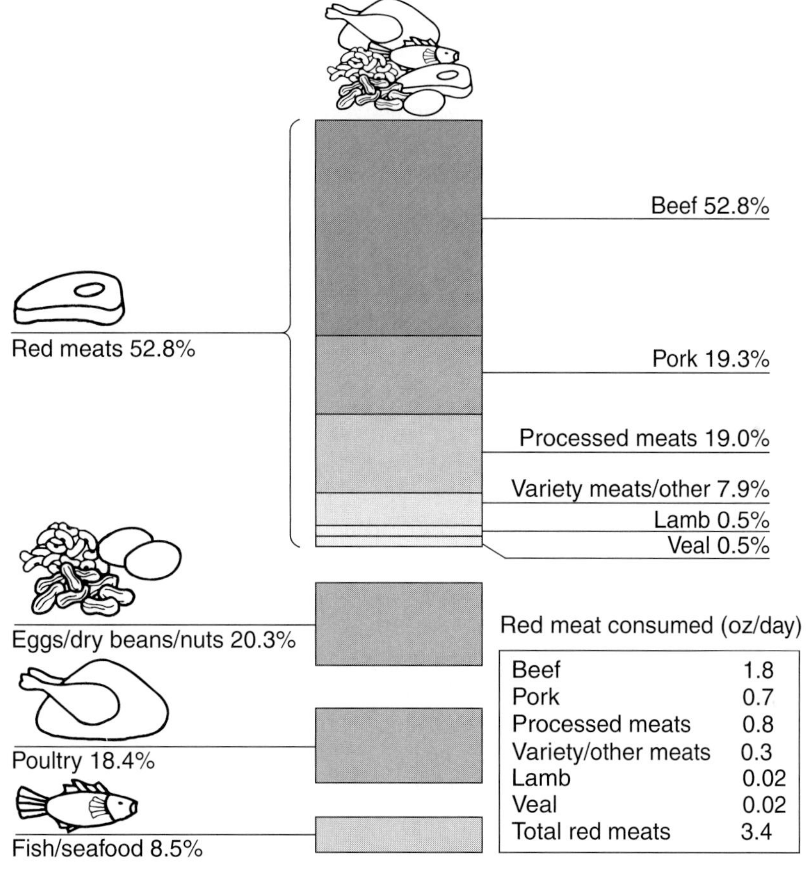

Red meat consumed (oz/day)	
Beef	1.8
Pork	0.7
Processed meats	0.8
Variety/other meats	0.3
Lamb	0.02
Veal	0.02
Total red meats	3.4

Figure 2.15 Daily per capita meat consumption as part of meat, poultry, fish, dry beans, eggs, and nuts group (Food Guide Pyramid).
Source: National Live Stock and Meat Board.

Based on a review of several types of epidemiological studies, Austin et al. (1997) concluded that red meat consumption has not been clearly shown to be a risk factor for cancer. If an association is shown between red meat and cancer, one of three explanations is possible: (1) the association is due to chance or some bias in the study method; (2) the association is due to confounding (i.e., meat and cancer are associated only because they both are related to some common underlying condition such as a low intake of cancer-protective fruits/vegetables); (3) the association is real. Whether or not it can be concluded that red meat is a risk factor for cancer depends on several criteria: consistency of the association, strength of the association (relative risk), specificity of the association, and congruence with existing knowledge (e.g., is there an explanation or biological mechanism).

The criteria of "consistency" requires that the association be repeated under different circumstances, such as among various population groups and among individuals within a population. In the case of red meat and cancer, the relationship is not consistently demonstrated within the population. When a positive association between red meat and a specific cancer is demonstrated, it generally is weak. Moreover, the association between consumption of red meat of different types (beef, pork, lamb, processed meats) and specific cancers is inconsistent. Although it has been suggested that meat components such as fat, protein, or iron, or chemicals formed during the cooking of meat might be carcinogenic, these hypotheses remain unproven. On the contrary, meat contains some components such as conjugated linoleic acid, a fatty acid, which may protect against cancer.

Inferring a relationship from epidemiological studies of diet and chronic disease is particularly difficult due to several characteristics of both diet and chronic disease. First, there are several problems with accurately quantifying dietary intake. This was a major problem in most of the epidemiological studies reviewed by Austin et al. (1997). Second, diseases such as cancer are caused by a variety of genetic and environmental factors. Diet (and red meat intake) is only one of many lifestyle factors that may influence risk of developing a disease. Also, chronic diseases such as cancer tend to have a long latency period during which time changes in many factors may occur. For these reasons, it is difficult to determine if an association, such as one between red meat and cancer, is real.

Epidemiological investigations can identify risk factors, not a cause-and-effect relationship. Only when epidemiological findings are supported by information from other types of scientific studies, such as experimental animal studies and human clinical trials, can a decision regarding a causal relationship be made on firmer ground.

The U.S. government has released several reports that outline dietary guidelines and goals for American citizens. Some argue that it is the government's responsibility to provide people with information about diet; others support scientifically based guidelines, but believe Americans should have freedom of choice.

The *Dietary Guidelines* published by the USDA and Department of Health and Human Services (HHS) include broad recommendations and are a reasonable attempt to initiate sound nutritional practices among individuals. The revised 1990 *Dietary Guidelines,* which draw heavily on two diet and health reports by the National Academy of Sciences and the U.S. Surgeon General, recommended that not more than 30% of daily calories come from fat and that less than 10% come from saturated fat. Some of the earlier versions of the *Guidelines* implied that red meat was a health risk by recommending that Americans eat less meat. The revised guidelines, however, convey a more positive recommendation for lean red meat consumption. In 1992, the USDA began use of the Food Guide Pyramid (Fig. 2.16) in conjunction with the nutritional education programs. The pyramid replaced the wheel graphic used to display the four basic food groups that has been used in nutritional educational programs since the 1950s.

In the period 1970-99, red meat consumption declined by 11% while per capita fat and oil intake rose 31%. Soft drink consumption over the same time period jumped 109%. Moderating

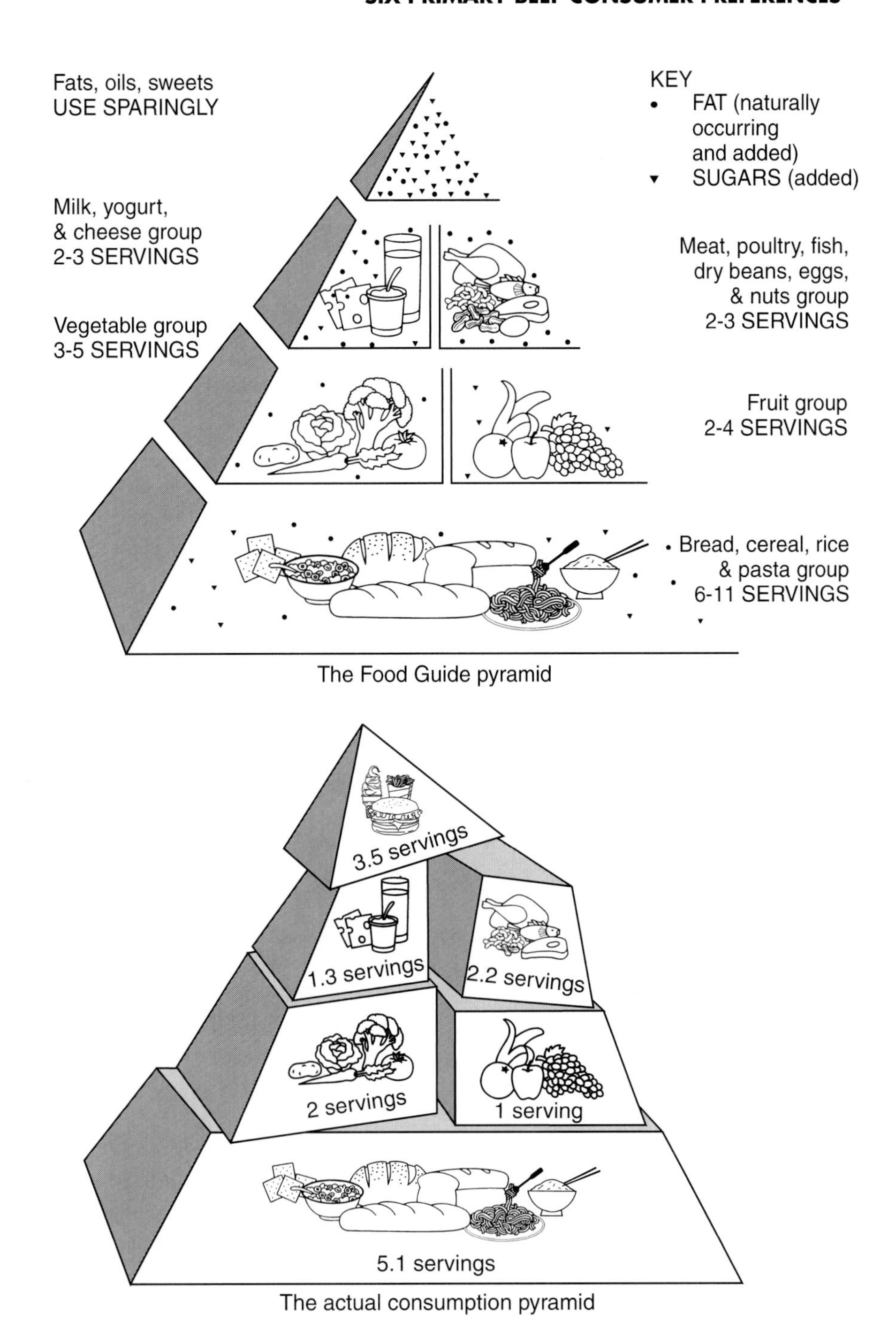

FIGURE 2.16 A comparison between the Food Guide Pyramid and actual consumption.
Source: USDA and National Live Stock and Meat Board.

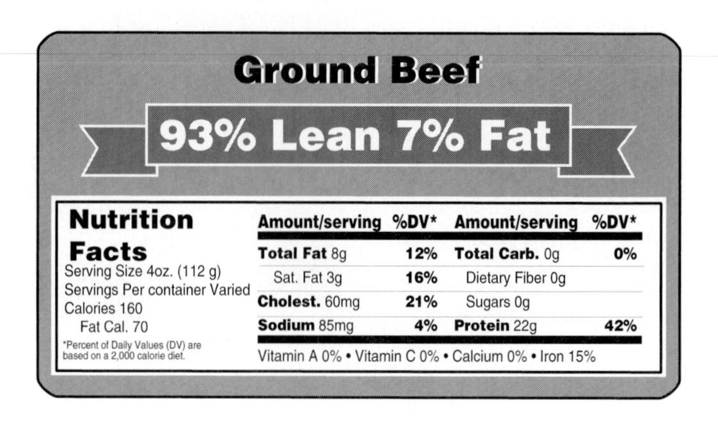

FIGURE 2.17 Examples of Nutrition Facts label for ground beef. *Source:* USDA.

fat consumption is a challenge for many consumers because taste is associated with the fat content of foods. Many consumers rank taste as the most important criteria in their food selection. Because of this preference for fat in foods, the 'actual consumption' pyramid is out of balance compared to the "recommended" Food Guide Pyramid (Fig. 2.16). The meat, poultry, fish, dry beans, eggs, and nuts group is in balance.

In recent years, the beef industry has met consumer demands by providing beef trimmed to a fat level of 1/8 inch or less and has encouraged the publication of the 1990 version of the USDA's *Handbook* 8-13. That handbook contains nutritional information on beef trimmed to 1/4 inch of fat. Some retail beef cuts are labeled with nutrition facts (Fig. 2.17) although they are not required by law to be labeled.

In many respects, the dietary guidelines, consumption goals, and nutritional information have not changed all that much over the past 20 years. Adherence to the suggested *Dietary Guidelines* can be easier by following these suggestions of the Dietary Guidelines Alliance:

1. **Be realistic**—make small, incremental changes in eating and exercise habits.
2. **Be adventurous**—expand your tastes to enjoy a variety of foods.
3. **Be flexible**—balance food consumption and exercise regime over several days instead of focusing on one meal or one day.
4. **Be sensible**—enjoy all foods—just don't overdo it. Choose sensible portion sizes.
5. **Be active**—exercise is a key to maintaining appropriate weight.

Lean, palatable beef not only fits into a healthy diet but also satisfies the taste preferences of many people.

Growth Promotants and Antibiotics

Estrogens are naturally occurring substances essential for life's processes, especially for reproduction. Small amounts of estrogen are beneficial, but large amounts of estrogen may be harmful. Most substances produce harmful effects when given in excessive amounts; this is true of essential nutrients such as water and minerals (salt) as well as for hormones.

Currently, many feeder calves and feedlot cattle receive ear implants containing estrogen to stimulate weight gain by increasing muscle growth. Some consumers believe that eating growth-promotant-implanted beef is harmful to human health. However, as Table 2.13 shows, beef from implanted steers contains low levels of estrogen when compared with other foods and to the amount produced daily by the human body.

TABLE 2.13 Estrogen Production in Humans and Content in Selected Foods

Human/Food	Nanograms[a] of Estrogen
Daily human estrogen production	
Child (before puberty)	50,000
Adult male	135,000
Adult female	480,000
Pregnant female	20,000,000
Estrogen in one serving	
Steak from implanted steer	1.9
Steak from nonimplanted steer	1.2
Coleslaw	2,724
Split pea soup	908
Chocolate ice cream	1,387

[a]One nanogram equals one-billionth of a gram.

Consistent with good management practices, cattle producers use government-approved antibiotics to prevent and treat diseases. There is no evidence that beef originating from cattle correctly treated with antibiotics causes human illness due to the generation of antibiotic-resistant bacteria. In fact, many of the antibiotics used with cattle are not used in human medicine. The cattle industry's Beef Quality Assurance program involves veterinarians and cattle managers to ensure proper withdrawal times are followed when antibiotics are administered. Monitoring by USDA shows that there are almost no (< .0001) violative residues of antibiotics in beef from fed steers and heifers. For non-fed cows and bulls the incidence rate of violative residues was higher (< .003) but still extremely low. Violations are mostly traced back to dairy cows. Most credible experts in the area of bacterial resistance focus attention on the overprescription of antibiotics in human medicine.

Safe Beef Products

There are health risks with most foods as evidenced by frequent communications in the media and in the scientific literature. Some of these apparent problems are sensationalized but later prove to be of little concern or they are easily resolved; other problems need resolution even though they involve a small percentage of the population. The U.S. food supply is also more safe than at any other time in history, and longevity of the human population continues to increase.

Although the food industry does not have a perfect record with regard to food safety, U.S. consumers enjoy a plentiful food supply that can arguably be called the safest in the world. Life is full of risks and relative risk must be carefully evaluated when considering food safety issues (Table 2.14).

Recent surveys show that consumers are completely or mostly confident that supermarket food (including beef) is safe and wholesome, although more than 80% view food safety as an important public issue. Despite these findings, some consumers continue to voice concerns about bacterial contamination of meat.

Two of the most harmful bacteria associated with meat are *Salmonella* and *E. coli* 0157:H7. These bacteria may be found on meat as a result of contamination during processing, handling, or preparation. Compared to poultry, the incidence of *Salmonella* organisms on fresh beef is quite low and there are few reported illnesses resulting from *Salmonella* in beef.

E. coli 0157:H7 has caused severe illness in several people (most frequently the very young, very old, and immunosuppressed) primarily through their consumption of contaminated

TABLE 2.14 Relative Risk of Death from a Variety of Circumstances

Relative Risk of Death (per million)	Circumstance
220	Auto accident
38	Drowning
29	Fire-related
0.6	Lightning
0.2	Venom
0.02	Botulism
0.01	Salmonellosis

Source: Mossell (1988).

ground beef. Ground beef is more susceptible to contamination because as beef is ground, the surface area is greatly increased and handling is extensive. In addition, any bacteria on the exterior of meat prior to grinding is distributed throughout the product as it is ground and blended; so, ground beef is more likely to have pathogens in its interior than are steaks and roasts. By cooking whole-muscle cuts (e.g., steaks and roasts) to a safe internal temperature of at least 145° F, the bacteria that are present on the meat's surface will be destroyed by cooking. It is recommended to cook hamburgers to an internal temperature of 160° F to destroy bacteria embedded in the meat.

The USDA's Food Safety and Inspection Service (FSIS) has modified its meat inspection system to reduce harmful bacterial contamination of meat. The final rules on pathogen reduction and use of HACCP (Hazard Analysis and Critical Control Points) were issued in 1996. Each meat plant must develop and follow HACCP plans to identify and prevent specific food safety hazards associated with each product or production process. Also, meat plants must regularly test carcasses for the generic form of *E. coli* which is the most reliable indicator of fecal contamination of carcasses. Fecal material is the primary pathway for contamination with intestinal bacteria that include *Salmonella, Campylobacter,* and *E. coli* 0157:H7.

Consumers need to know that they can exert a great deal of control in preventing food-borne illnesses from meat and other foods by following these safe handling instructions: Keep meat refrigerated or frozen, thaw in refrigerator or microwave, keep raw meat separate from other foods, wash working surfaces, including cutting boards, utensils, and hands after touching raw meat, cook thoroughly, keep hot foods hot and refrigerate leftovers immediately or discard.

Consumers frequently misunderstand the source of food-borne illness outbreaks. In one survey, 18% of consumers identified food processing plants as the most likely source of food safety problems. Yet, the vast majority of illnesses are not due to mistakes made at processing but failures in food service establishments or homes relative to storage, handling, and preparation of foods.

Bovine Spongiform Encephalopathy (BSE), commonly referred to as "mad cow disease," is a fatal brain disease of cattle. The disease agent, a rogue protein referred to as a prion, is found only in brain tissue, the spinal cord, and the retina of infected cattle. Its prevention or eradication is important to protect cattle herds and the human beef supply.

Transmissible BSE was first diagnosed in Great Britain in 1986 and the highest incidence has been in that country with other cases confirmed primarily in other European countries. No case of BSE has been confirmed in the United States. A surveillance program has been in place for more than a decade.

BSE has been linked to a human disease variant—Creutzfeldt-Jakob disease (vCJD). There is evidence that some people may be genetically predisposed to infection by the BSE agent as all tested vCJD patients have been homozygous at a particular codon.

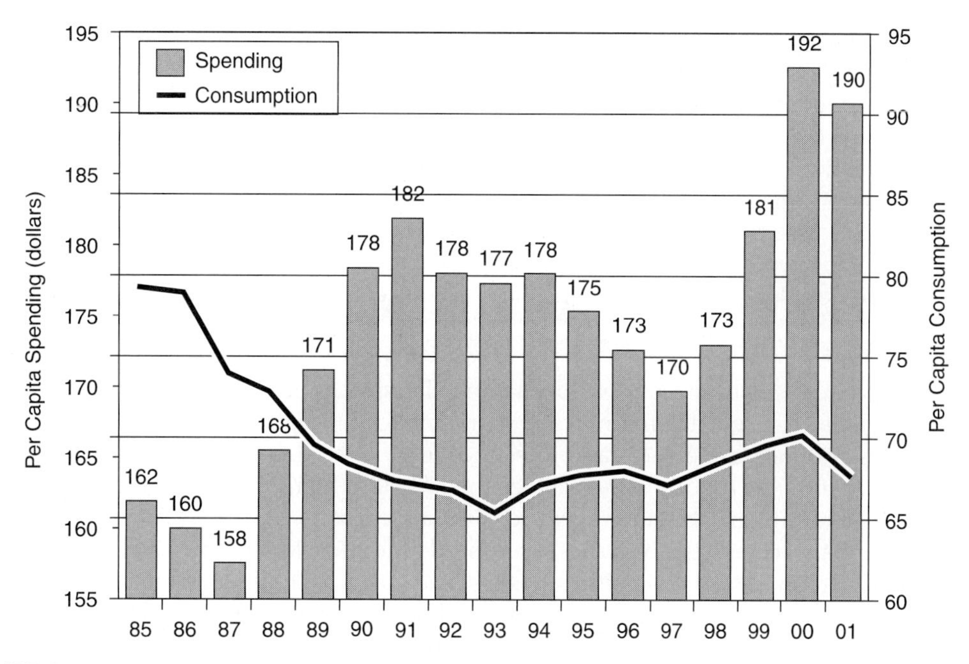

FIGURE 2.18 Per capita beef spending and consumption.
Source: USDA and Cattle-Fax.

To prevent BSE from entering the United States, the USDA has restricted the importation of live ruminants and ruminant products from countries where BSE is known to exist. No beef has been imported from the United Kingdom since 1985, prior to the BSE outbreak in 1986. Also banned are ruminant protein by-products (with the exception of blood and milk products and a few others produced in the United States). These bans have been in place for more than a decade.

Beef's Value and Perceived Value

Beans, peanut butter, eggs, milk, and chicken are typically cheaper sources of protein than beef on a per-unit cost basis. Most consumers, however, give preference to the palatability characteristics of beef and will pay the higher price unless extreme economic pressures dictate otherwise.

Perceived value to some consumers is based on the palatability of beef—how it tastes and the entire eating experience. These consumers are willing to pay relatively higher prices for this perceived value of beef when compared with other competing meats. Because of their willingness to pay a premium price for beef, per capita beef spending has generally grown despite the losses in per capita consumption on a weight basis (Fig. 2.18).

BY-PRODUCTS

The hide is the best-known by-product and usually the highest valued. The hide provides three types of leather (latigo, suede, tooling) used in sports equipment, luggage, boots, and shoes. Leather also provides felt, certain textiles, a base for many ointments and insulation, and as a binder for plaster and asphalt. The hair from the hide is used to produce insulation and rug pads. Fine hair from the ear is used to make artist's brushes, so-called camel hair brushes. Gelatin from hides is used in foods, film, and glues. Artificial skin for severely burned humans has been made from cowhide, shark cartilage, and plastic.

The primary edible by-products are called *variety meats* and include the liver, heart, kidney, brain, tripe (walls of the stomach), sweetbread (thymus), and tongue. These products have long been known for their high nutritive value and are considered gourmet items by some consumers. An average 1,100-lb slaughter steer produces approximately 34 lbs of variety meats. Because the U.S. per capita consumption (disappearance) of variety meats is only about 9 lbs, surplus variety meats are exported to countries that have a preference for them. A more detailed discussion of the export markets for variety meats is presented in Chapter 10.

Other edible by-products come from fats (e.g., oleo stock and oil for margarine and baker's shortening, while oleo sterine is used for making chewing gum and certain candies); bones, horns, and skins (e.g., gelatin for making marshmallows, yogurt, ice cream, mayonnaise, canned meats and gelatin dessert); and intestines (e.g., natural sausage casings and surgical sutures). Other inedible by-products, besides the hide and hair, come from inedible fats and fatty acids (e.g., antifreeze, binding agent for asphalt in roads, stearic acid to produce tires, candles, cellophane, ceramics, cosmetics, crayons, deodorants, detergent, insecticides, insulation, linoleum, freon, perfumes, paints, plastics, shoe cream, shaving cream, soaps, textiles, pet foods, and floor wax); from bones, horns, and hooves (e.g., animal feeds, buttons, bone china, combs, piano keys, and bone charcoal, which is used in production of high-grade steel ball bearings); from collagen-based adhesives (e.g., glues, adhesives, bandages, wallpaper, sheet rock, emery boards); and from nonedible gelatin for photographic film.

Table 2.15 shows the source of the primary pharmaceuticals from beef cattle and their value to humans. To obtain 1 lb of dry insulin, processors must obtain pancreas glands from approximately 60,000 cattle. One cow's pancreas can supply a diabetic patient with a two-day supply of insulin. Synthetic insulin and other synthetic pharmaceuticals may reduce the demand for certain cattle by-products.

Iron (from the blood), vitamin B_{12}, and liver extract (from the liver), and calcium and phosphorus (from bone meal) are nutrients that are used in human and livestock nutrition.

This summary of beef by-products is not complete. Research has identified, and no doubt will continue to identify, useful by-products from beef cattle.

Value of By-Products

By-product value is first determined at the packer's segment and is quoted on dollars per hundredweight of liveweight of the slaughter animal. For example, the average by-product value in March 2001 was $9.32 per hundredweight.

MANAGEMENT SYSTEMS HIGHLIGHTS

- W. Edwards Deming, a great contributor to industrial management systems and TQM (Total Quality Management), taught that the focus must be on the consumer—that consumer products should be continually improved and at reduced costs. This chapter emphasizes improving beef products to achieve higher levels of palatability and consumer acceptance. Ways of reducing costs will be covered in later chapters.
- Management decisions must be made in ways intended to stabilize or increase beef's percentage of consumer meat expenditures (market share).
- Profitability, in part, is determined by prices consumers pay for retail cuts of beef and beef by-products that comprise a large number of consumer products.

TABLE 2.15 The Pharmaceuticals from Beef Cattle—Their Source and Utilization

Pharmaceutical	Source	Uses
Epinephrine	Adrenal gland	Relief of hay fever, asthma, and other allergies; heart stimulation
Thrombin	Blood	Assists in blood coagulation; treatment of wounds; skin grafting
Fibrinolysin	Blood	Dead tissue removal; wound cleansing agent; healing of skin from ulcers or burns
Desoxycholic acid	Bile	Used in synthesis of cortisone for asthma and arthritis
Liver extract	Liver	Treatment of anemia
Ox bile extract	Liver	Treatment of indigestion, constipation, and bile tract disorders
Heparin	Lungs	Anti-coagulant
Insulin	Pancreas	Treatment of diabetes
Chymotrypsin	Pancreas	Remove dead tissue; treatment of localized inflammation and swelling
Glucagon	Pancreas	Counteracts insulin shock; treatment of some psychiatric disorders
Trypsin	Pancreas	Cleansing of wounds
Rennet	Stomach	Assists infants in digesting milk; cheese making
Ovarian hormone	Ovary	To treat painful menstruation and prevent abortion
Parathyroid hormone	Parathyroid gland	Treatment of human parathyroid deficiency
Corticotrophin (ACTH)	Pituitary gland	Diagnostic assessment of adrenal gland function; treatment of psoriasis, allergies, mononucleosis, and leukemia
Hyaluronidase	Testicle	Enzyme that aids drug penetration into cells.
Thyrotropin (TSH)	Pituitary gland	Stimulates functions of thyroid gland
Vasopressin	Pituitary gland	Control of renal function
Cholesterol	Nervous tissue	Male sex hormone synthesis
Thyroid extract	Thyroid gland	Treatment of cretinisin
Amfetin (trade name)	Amniotic fluid	Reduces postoperative pain and nausea and enhances intestinal peristalsis

- Focusing on an optimum slaughter steer with the following specifications would solve part of the palatability and acceptability problems of consumers: 700–800 lb carcass weight, 0.30–0.45 in. of fat over ribeye, and groups of high percentage British breed cattle that produce carcasses that are 70% USDA Choice or better, and with a minimum of non-conformance for weight, maturity, dark cutters, and other carcass defects.
- Beef cattle industry participants have as their highest obligation the responsibility for producing the most satisfying, nutritious, and safe food products possible. The incorporation of quality assurance plans, HACCP protocols, and other safeguards are of critical importance to producers and consumers.

REFERENCES

Publications

Aberle, E. D., Reeves, E. S., Judge, M. D., Hunsley, R. E., and Perry, T. W. 1981. Palatability and Muscle Characteristics of Cattle with Controlled Weight Gain: Time on a High Energy Diet. *J. Anim. Sci.* 52: 757.

Allen, D. 1989. Carcass specifications to meet consumer demand. Proc. Beef Improvement Federation Research Symposium. Wichita, KS.

Anderson, B. A., and Hoke, I. M. 1990. *Composition of Foods: Beef Products.* Washington, DC: USDA. Agric. Handbook No. 8-13.

Austin, H., and McBean, L. D. 1997. *Red Meat and Cancer: A Review of Current Epidemiological Findings.* Chicago, IL: National Cattlemen's Beef Association.

Beef Customer Satisfaction. 1995. Chicago, IL: National Live Stock and Meat Board.

Belk, K. E., Sofos, J. N., Scanga, J. A., and Smith, G. C. 2001. *U.S. Red Meat: A Pledge to Minimize Risk to Public Health.* Proceedings paper for United States Meat Export Federation.

Brown, P., Will, R. G., Bradley, R., Asher, D. M., and Detwiler, L. 2001. *Bovine Spongiform Encephalopathy and Variant Creutzfeldt-Jakob Disease: Background, Evolution, and Current Concerns.* Centers for Disease Control (www.cdc.gov).

Cannell, R. C., Tatum, J. D., Belk, K. E., Wise, J. W., Clayton, R. P., and Smith, G. C. 1999. Dual-Component Video Image Analysis System As a Predictor of Beef Carcass Red Meat Yield Percentage and for Augmenting Application of USDA Yield Grades. *J. Anim. Sci.* 77: 2942.

Cattle and Beef Handbook. 1999. Nutrition and Health. Englewood, CO: National Cattlemen's Beef Association.

Contributions of Animal Products to Healthful Diets. 1997. Ames, IA: Council for Agricultural Science and Technology.

Demand Strategies. The Meat Consumer. 1992. Chicago, IL: National Live Stock and Meat Board.

Dietary Guidelines for Americans. 2000. Washington, DC: USDA and U.S. Dept. of Health and Human Services.

Dietary Guidelines for Americans. 1995. Washington, DC: USDA and U.S. Dept. of Health and Human Services.

Drewnowski, A. 1995. Impact of Taste Preferences on Dietary Choices and Food Consumption Patterns. *Food and Nutrition News* 67:15.

Eating in America Today: A Dietary Pattern and Intake Report (Edition II). 1994. Chicago, IL: National Live Stock and Meat Board.

Field, T. F., Garcia, J., Ahola, J. 1996. Quantification of the utilization of edible and inedible beef by-products. Final Report to NCBA. Colorado State University. Fort Collins, CO.

Food Consumption, Prices, and Expenditures, 1970–1999. 2000. Washington, DC: USDA.

Food Facts and Health. 1991. Ames, IA: Council for Agricultural Science and Technology. Report No. 118.

Food and Nutrition News. 1998. Vol. 70, No. 1. Young, M. K. and R. Korpolinski, eds.

Food and Nutrition News. 1998. Vol. 70, No. 3. Young, M. K. and R. Korpolinski, eds.

Food Marketing Institute. 2000. *Trends in the United States: Consumer Attitudes and the Supermarket.* Washington, DC.

Food Review (periodical). Washington, DC: USDA.

George, M. H. 1997. Retail Loin Palatability Survey. Chap. 2. Ph.D. thesis. Fort Collins, CO: Colorado State University.

Gutherie, J. F., and Roper, N. 1992. Animal Products: Their Contribution to a Balanced Diet. *Food Review* 15: 29.

Hays, V. W., and Black, C. A. 1989. *Antibiotics for Animals: The Antibiotic Resistance Issue.* Ames, IA: Council for Agricultural Science and Technology. Comments from Report No. 1989–2.

Hedrick, H. B., Aberle, E. D., Forrest, J. C., and Judge, M. D. 1994. *Principles of Meat Science.* San Francisco: W. H. Freeman.

Hodge, S. G. 2000. Consumer Attitudes Toward Natural Beef Products. M.S. thesis. Fort Collins, CO Colorado State University.
Livestock and Poultry Situation and Outlook Report. Washington, DC: USDA.
Morgan, J. B., et al. 1990. *National Beef Tenderness Survey. Beef Cattle Research in Texas.* College Station, TX: Texas Agric. Expt. Sta. PR 4819–4865.
Mossell. 1988. Monograph. Auburn University, Auburn, AL.
National Research Council, Committee on Diet and Health, Food and Nutrition Board 1989. *Diet and Health. Implications for Reducing Chronic Disease Risk.* Washington, DC: National Academy Press.
NCBA. 2001. Beef Industry Long Range Plan. Englewood, CO, NCBA.
Niyo, K.A. (ed.). 1997. *Contribution of Animal Products to Healthful Diets.* CAST Report 131. Ames, IA: Council for Agricultural Science and Technology.
Nutrition and Your Health: Dietary Guidelines for Americans. 1985. Washington, DC: USDA; U.S. Dept. of Health, Education, and Welfare.
Pyramid Servings Data. 1999. Results from USDA's 1994–96 continuing survey of food intake by individuals. Agricultural Research Service–USDA.
Roeber, D. L., Cannell, R. C., Belk, K. E., Tatum, J. D., and Smith, G. E. 2000. *Effects of a Unique Application of Electrical Stimulation on Tenderness, Color, and Quality Attributes of the Beef Longissimus Muscle. J. Anim. Sci.* 78:1504.
Romans, J. R., Jones, K. W., Costello, W. J., Carlson, C. W., and Ziegler, P. T. 1977. *The Meat We Eat.* Danville, IL: Interstate Printers and Publishers.
Savell, J. W., Cross, H. R., Hale, D. S., and Beasley, L. 1988. *National Beef Market Basket Survey.* College Station, TX: Texas Agric. Expt. Sta. Tech. Rpt. 88–1.
Smith, G. C., *Meat Science—The Palatability Piece.* 1995. Certified Angus Beef Program, South Sioux City, IA.
Smith, G. C., et al. 1987. Relationships of USDA quality grades to cooked beef palatability. *J. Food Qual.* 10:269.
Surgeon General's Report on Nutrition and Health. 1988. Washington, DC: Department of Health and Human Services. Publication No. 88–50210.
Tatum, J. D., George, M. H., Belk, K. E., and Smith, G. C. 1997. *An overview of a TQM approach for improving beef tenderness.* Fort Collins, CO: Colorado State University.
Taubes, G. 2001. *The Soft science of dietary fat.* Science (Vol. 90, 2536).
The National Beef Quality Audit. 1991. Englewood, CO: National Cattlemen's Beef Association.
The National Beef Quality Audit. 1995. Englewood, CO: National Cattlemen's Beef Association.

Visuals

"Issues and Answers for the Nineties" (1990; 9.5 min.); "Cattlemen Care About the Environment" (1990; 7 min.); "Cattlemen Care about Beef Safety" (1991; 10 min.). National Cattlemen's Beef Assoc., Communications Dept., Box 3469, Englewood, CO 80155.
"Animal Agriculture: Myths and Facts" (1990; 18 min.). Animal Industry Foundation, P.O. Box 9522, Arlington, VA 22209–0522.
"Muscle Profiling and Bovine Myology. (2001; CD-Rom). National Cattlemen's Beef Association, Box 3469, Englewood, CO 80155.

Management Systems: Integrated and Holistic Resource Management

Management systems or integrated resource management (IRM) and holistic resource management (HRM) involve (1) understanding the available resources and the principles associated with each resource and (2) knowing how the resources are interrelated so that integrated and holistic management decisions can be effectively implemented.

RESOURCES AND PRINCIPLES

A ranch, farm, or feedlot can be divided into individual resource components. A *resource* is anything that a manager can obtain from the physical environment to meet the needs of the beef cattle operation. Within each resource area there are biological and economic principles that will affect productivity and profitability.

Figure 3.1 identifies the major resources that are part of the total management program. Some resources are perpetual (e.g., sun, wind, and water); others are nonrenewable (e.g., fossil fuels, phosphates); and still others are renewable (e.g., air, water, fertile soil, plants, and animals). A manager must be knowledgeable of each type of resource if the goal in Figure 3.1 is to be accomplished. Furthermore, maximum net profit may not always be the ultimate goal because the sacrifice of quality of life may be too great (e.g., working 16–20 hours a day versus 8–10 hours). However, a goal that approaches maximum net profit is often realistic for many beef operations.

Successful beef producers and beef industry organizations identify and apply important principles that assure their individual and collective success. Principles are truths or natural laws that can be applied universally. A truth is a fact, law, or verified hypothesis. Truth is knowledge of things as they are, as they were, and as they are to come. Thus, principles

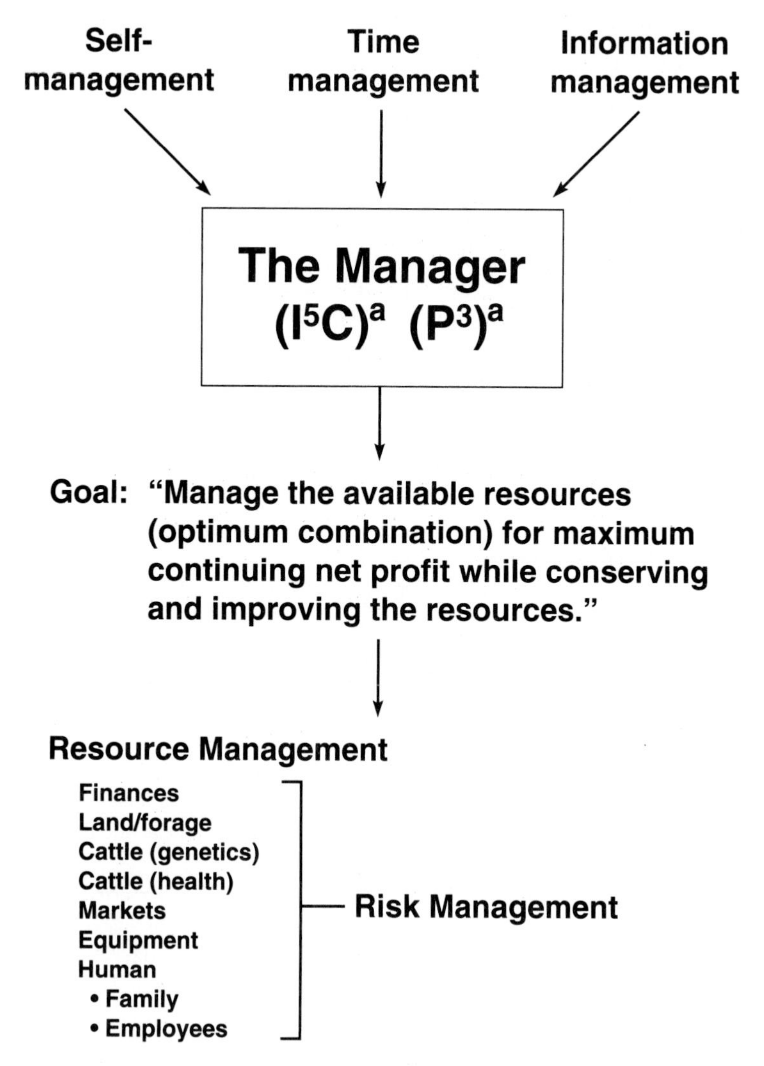

FIGURE 3.1 An overview of a manager's challenge: continued self-development, focusing on a central goal and effectively managing all the resources.

transcend time and humans are continually motivated to identify new principles and effectively utilize time-proven principles that will improve their well-being.

Principles are the same across cultures, however, the application of these truths, through production practices, can vary from one place to another. For example, the reproduction principles involve hormones and target organs that function to create estrous cycles and sex cells that result in a new calf crop. For the reproduction principles to be manifested, nutrition principles must also be utilized. Energy, protein, and other nutrients in essential amounts must be available at specific time periods or some of the reproduction principles will not be expressed. While the reproduction and nutrition principles apply universally, how the nutrition (forage or feed) is provided—what plant species and amounts—can vary widely from enterprise to enterprise. Therefore, while the reproduction and nutrition principles remain the same over wide geographical areas, the application of these principles through production practices can vary even in enterprises only a few miles apart.

THE HUMAN RESOURCE

Without question the human resource is of the greatest importance. The foundation of the industry is the strength of human creativity and willpower (Fig. 3.2).

Self-management, discussed later in the chapter, is an integral part of human resource management. It is human beings who develop mission statements, set goals, understand resource relationships, make decisions, and provide the labor that implements the written management plan. Successful managers understand that they lead (not manage) people and manage the other resources (Fig. 3.3). People are persuaded more by leadership example than by words.

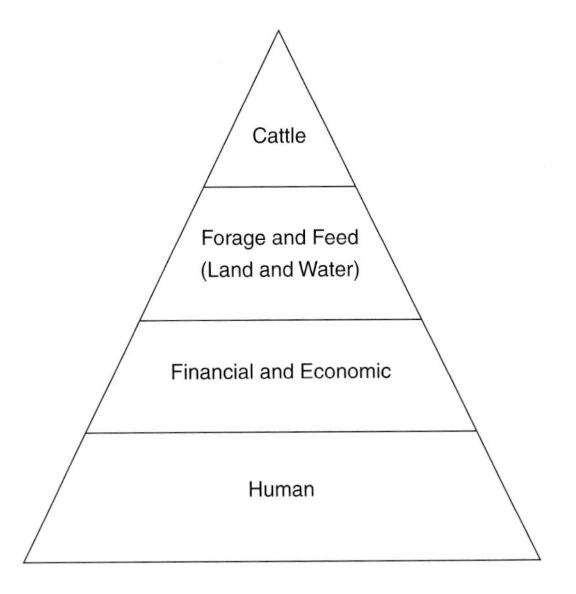

FIGURE 3.2 The foundation of the beef industry is the human resource. Application of human creativity to the other resources is the basis of profitability.
Source: Colorado State University.

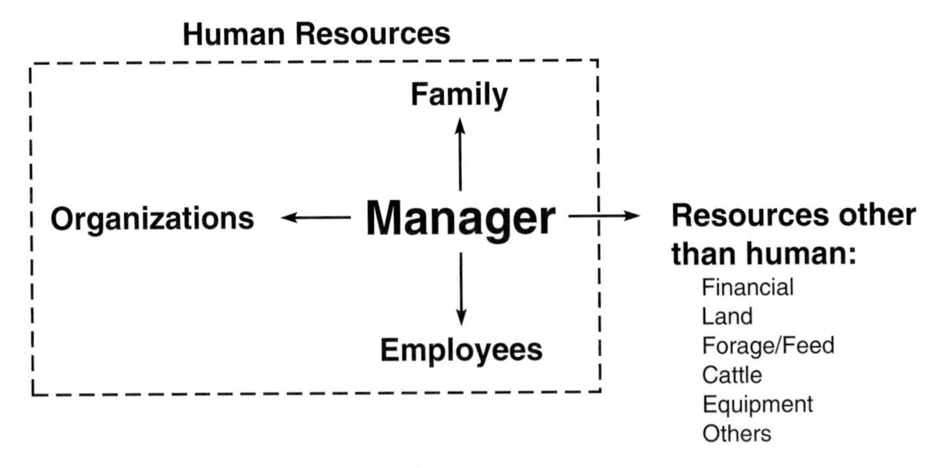

FIGURE 3.3 Effective managers are involved in numerous resource areas, giving priority to human resources (themselves, employees, family, and people in organizations).

TABLE 3.1 Comparing Leadership and Management

Leadership	Management
Creates new paradigm/vision	Works within the paradigm/vision
Works on the system	Works within the system
Leads people	Manages things
Focuses on effectiveness	Focuses on efficiency
Does the right thing	Does things right
Works strategically	Works tactically

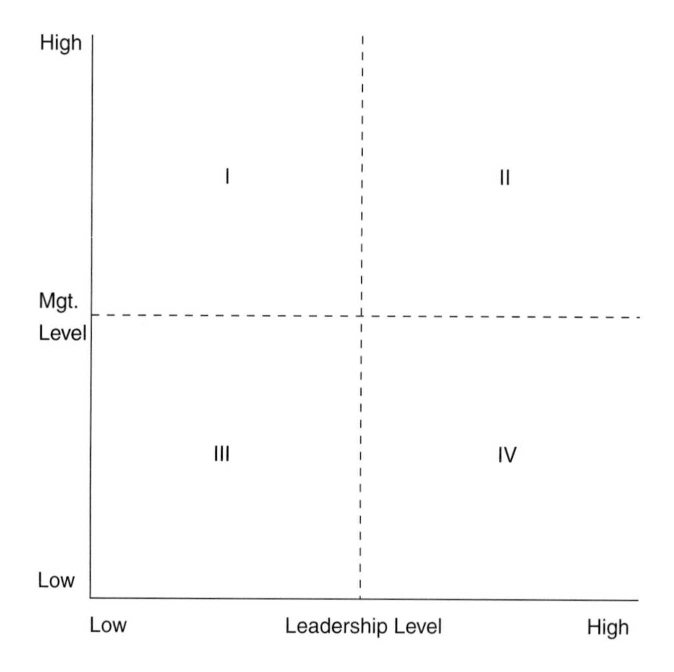

FIGURE 3.4 Four combinations of leadership and management levels. Quadrant II is preferred.
Source: Colorado State University.

Leadership and Management

Leadership and management are somewhat different yet their interrelation must be understood and implemented to assure continuing success. Management is the process of taking an organization along an established route as smoothly and efficiently as is possible. Leadership, on the other hand, is the process of moving an organization into uncharted waters by effectively understanding and implementing change. Table 3.1 shows the differences between leadership and management. Figure 3.4 demonstrates the importance of combining leadership and management so that high levels of performance can be obtained.

It is interesting that in Table 3.1 the comparison of "doing the right things" before "doing things right" reinforces the long-time proven 80–20% rule of business (Fig. 3.5). Another way of approaching this issue is to consider that a leader's primary task is to make doing the right thing easy and the wrong thing hard.

Leadership and management identify the need to create and implement paradigms. A paradigm is the map of our mind's perceptions, how a person sees the world or a particular situation. One person may have a paradigm that the current beef industry is self-destructing, while another person has a paradigm that the current beef industry has numerous opportunities to be highly successful.

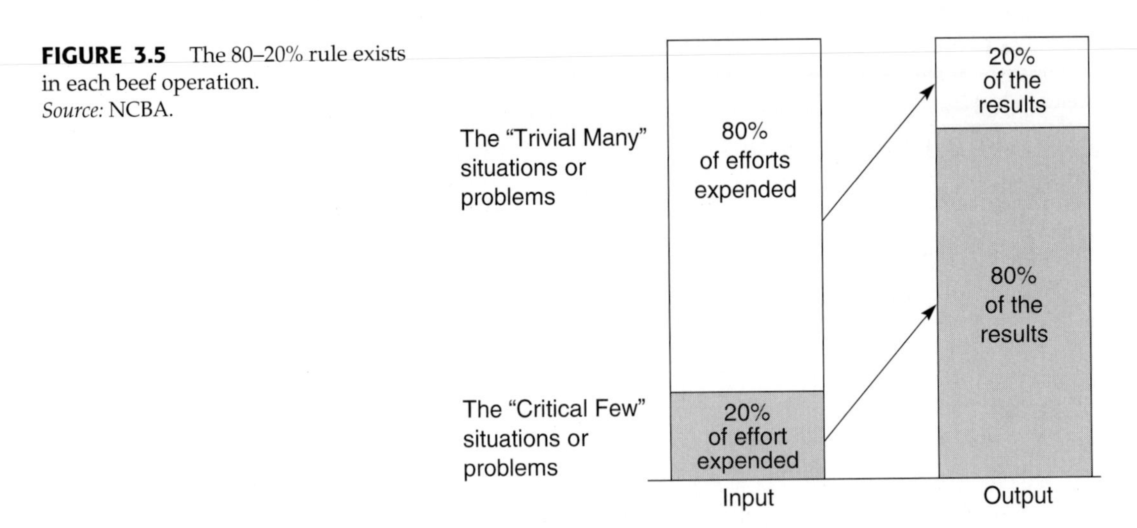

FIGURE 3.5 The 80–20% rule exists in each beef operation.
Source: NCBA.

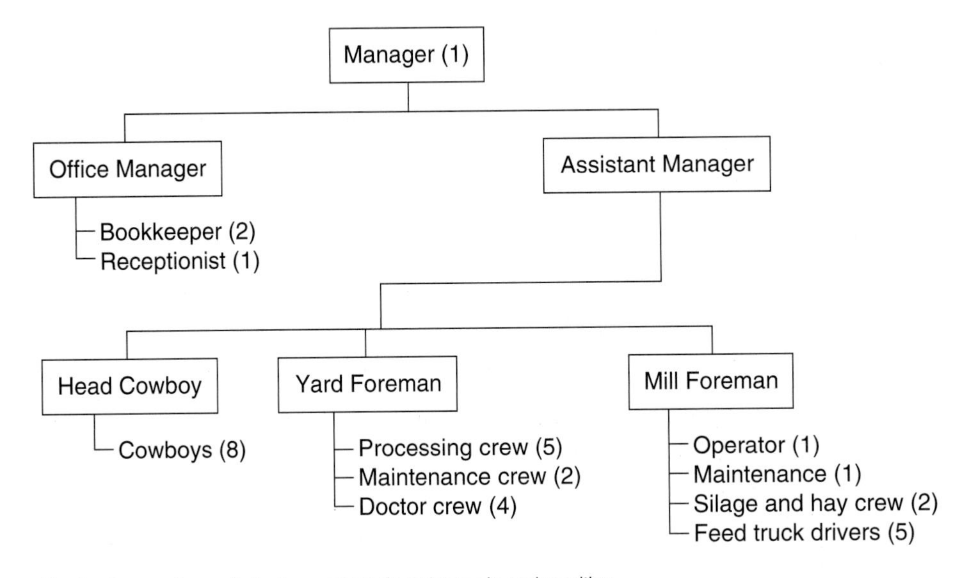

FIGURE 3.6 Organizational structure of a large commercial cattle feeding operation.
Source: Colorado State University.

The manager of a beef cattle operation may be an owner-operator with only a minimum of additional labor or someone in charge of a complex organizational structure involving several employees (see Fig. 3.6). An effective manager, whether involved in a one-person operation or a large complex one, needs to:

1. Develop a written mission statement; identify and implement short-, intermediate-, and long-range goals as part of a written management plan.
2. Set priorities and allocate resources accordingly.
3. Know what needs to be done and at what point in time.
4. Keep abreast of current knowledge related to the enterprise and the beef industry.

5. Know how to use time effectively.
6. Be self-motivated.
7. Manage oneself so that others can learn from a good role model.
8. Communicate responsibilities effectively to all employees and encourage a team approach.
9. Attend to the physical, emotional, and financial needs of oneself and those involved in the operation (employees and family).
10. Motivate employees to perform at optimal capacity.
11. Conduct honest business dealings.
12. Remove or reduce high risks affecting profit.
13. Provide a management information system that gives timely and accurate feedback on decisions.
14. Be profit-oriented.

Beef producers must manage each operation as a business to survive economically and to sustain high net returns. Economic pressures associated with competition encourage producers to manage their operations as business enterprises. Beef producers invest time and money not only in their businesses but also in organizations that help to shape government policy and influence the marketplace for beef products.

Mission/Vision Statements

Mission/vision statements put paradigms into words. Leadership provides new paradigms through creation of mission/vision statements. One person should not do this alone, unless it is only for them personally. All people in the organization should participate in the development of the mission statement so a sense of ownership of the vision is widespread. The organization may be a family team or a larger team involving several employees.

A mission/vision statement is a carefully prepared document that captures the organization's or team's purpose and value. It is the first critical step in making vision a reality for every team member. A mission/vision statement is just as realistic for an individual as a group while the material is written for a team, an individual can be easily substituted. A vision statement has three components.

1. The mission statement is a written statement of purpose designed to inspire team members to commit to the team's vision.
2. The glossary defines key words and phrases in the mission statement; this prevents differing interpretations of the mission.
3. The guiding principles are the crucial values that guide relationships with clients and one another. Examples of core values are teamwork, communication, honesty, accountability, and satisfaction.

An effective mission statement will answer four questions that capture the team reason for being: (1) Who are we? (2) What do we do? (3) For whom do we do it? (4) Why do we do it?

In addition, the mission statement must be:

1. Memorable (easily memorized—it must be enduring).
2. Compelling—it must inspire and motivate people—move them into action.
3. Focused on customer service—the purpose and mission of the team must be to serve its clients.
4. Able to create a future for the team.
5. Able to provide a core of values that all team members believe and to which they are committed.

An example of a mission statement applicable to a beef operation or the total beef industry follows:

> Produce low-cost/high-profit cattle that consistently yield competitively priced, highly palatable and consistently uniform retail products.

Additional examples of mission statements from ranches and industry organizations:

> We are continually striving to improve the efficiency of converting God's forage into healthy, nutritious and great tasting beef to better feed His people.
>
> R. A. Brown Ranch, Throckmorton, TX

> We are proactive and we achieve success by creating opportunities and progressively realizing our goals in a fun, challenging, and encouraging environment, where we continually improve our people, our ranch, our community, and the ecosystem in which we sustain a net profit.
>
> Paint Rock Canyon Enterprises, Hyattville, WY

> Our mission is to provide our members with excellence and innovation in our programs and services to further advance the quality, value, and reliability of Red Angus cattle. We pursue this mission to promote the economic well-being and satisfaction of our members and their customers and to equip them to be progressive cattle breeders.
>
> Red Angus Association of America, Denton, TX

Goals

Goals or strategies are needed to more specifically define the action to achieve the mission/vision statement. They assure success based on the following definition: Success is the progressive realization of worthwhile goals. Goals should be few, specific, measurable, and with a reasonable time frame to accomplish them. A vague goal would be "to improve the quality of the cattle." "Quality" has several definitions and means different things to different producers. A more specific, measurable goal for a cow-calf producer might be: the breakeven price on weaned calves will be lowered from $.86 per pound to $.73 per pound in three years.

It is interesting to review the Beef Industry Long Range Plan (January 2001) as written by the National Cattlemen's Beef Association (NCBA) to evaluate their vision statement and goals.

Vision Statement

> *A dynamic and profitable beef industry, which concentrates resources around a unified plan, consistently meets global consumer needs and increases demand.*

The two long-range plan goals for NCBA are (1) to increase beef demand 6% by 2004 and (2) enhance and protect the business climate for cattle and beef.

The American Angus Association listed ten goals in its 1997 mission statement. Four of these were to (1) achieve Certified Angus Beef acceptance of 30% in ten years, (2) achieve 80% Angus-based beef cattle population in ten years, (3) report carcass trait EPD on 60% of the bulls in the main Sire Evaluation Report, and (4) strive to incorporate ultrasound data into carcass EPD within three years.

The following are two additional outcomes selected from NCBA's "Quality and Consistency" and "Production Efficiency" areas of focus: (1) Reduce consumer dissatisfaction due to variability in eating quality (especially tenderness) by 50% by 2000; (2) Average costs will be reduced by 15% in the total beef production system—from farm to table—by the year 2000. Increased production of USDA Choice and Prime carcasses while reducing B-maturity carcasses provides evidence of progress towards goal #1. There is less evidence to support achievement of goal #2.

Human Resources

Human resources may include the manager, hired labor (seasonal or full-time), and labor supplied by family members. Successful managers know how to accomplish the following:

1. Assess the optimum labor needs for a business.
2. Identify prospective employees who can effectively contribute to the enterprise.
3. Motivate and adequately reward employees, not only monetarily but also through increased responsibility and opportunity for them to manage their own time.
4. Communicate goals, as well as the objectives and plans to accomplish the goals, so each employee or family member understands his or her role in achieving the desired level of productivity and profitability.
5. Create a unified team approach by implementing a group-designed mission statement that effectively directs all activities of the business.

Effective communication skills (including listening, writing, and speaking) are required of managers in order to utilize the human resource to the fullest extent. The importance of effective communication cannot be overemphasized in writing and communicating a plan and in understanding and motivating people in the desired direction. Listening to understand employees and other people is an important communication skill that is often overlooked by managers but is well understood and practiced by leaders.

Family Relationships

Family relationships in a family-owned-and-operated beef operation can be enhanced or destroyed by effective or ineffective communication (Fig. 3.7). The latter is tragic and usually is not perceived as a potential outcome when family members start working together. In addition to the five points mentioned earlier regarding successful management of the labor resource, the following items are pertinent to successful family operations:

1. Apply sound business principles rather than assuming things will work out simply because people are within the family. Involve all family members in financial decisions. At the same time, incorporate family values and goals into the strategic plan.
2. Evaluate other successful family operations. Determine why they are successful and how they resolve difficulties. Include all family members in the written plan of responsibilities (e.g., who will make the decisions, how each family member will be paid, how vacation and other time away from the business will be handled). Develop a systematic approach to deal with potential conflict. Also, assure that compensation to family members aligns with contributions to the business.
3. Hold weekly family councils for additional planning, evaluating, and problem solving. Create an environment that encourages open communication.
4. Recognize that family relationships have a higher priority than programs and profitability but that all can be compatible.
5. Have patience and tolerance with age differences in the family. Provide roles for family members and involve everyone (e.g., spouses, in-laws) in development of plans and goals.
6. Recognize that management changes can occur too fast or too slow in how they affect family relationships and profitability of the operation.
7. Assure that family members have the skills and abilities required to fulfill their job roles. Maintain the same performance standards for both family and nonfamily employees.

Family businesses will only be successful in the long-term when the appropriate balance of planning, productivity, communication, trust, and respect are obtained.

FIGURE 3.7 Family ranches typically involve multiple generations of people.

Zimmerman and Fetsch (1994) offer the following steps to building a model that allows for consensus decision making in an environment that is both open and supportive.

1. Establish family rules and a shared vision.
2. Improve family communication and hold regular family meetings to enhance communication, delegation, and business effectiveness.
3. Create departments and appoint managers to spread responsibility among family members and to allow people a chance to develop their own levels of expertise. This process promotes shared responsibility, accountability, and training of people.
4. Develop job lists to allow the prioritization of tasks and the allocation of resources to assure their timely completion.
5. Establish monthly calendars to open communication between the departments of the business while helping family members share both their family and business needs. This step is critical in minimizing the surprises that originate from poor communication.
6. Resolve equality issues if they exist. Common equality issues include deskwork versus physical labor, generational/gender pay equity, and on- versus off-farm employment.

Employees

The NCBA (2001) identified the need for accessing competent people to manage and work in the beef industry as one of its most critical business climate drivers. Perhaps no other issue is more critical to the success of the beef industry than to attract and retain high-quality employees.

Successful employer programs revolve around communication of expectations, ongoing training and education, goal setting, and effective performance reviews. Just as is the case when dealing with family members, respect and trust are of critical importance.

The keys to excellent people management are as follows:

1. Communicate expectations clearly.
2. Provide frequent feedback to employees with a goal of improvement. Seek input from employees.
3. Say thanks for a job well done and share credit.
4. Match a person's abilities to the job assigned.
5. Seek ways to help people grow and learn.
6. Avoid micromanagement.

The Planning Process

The manager of a beef cattle enterprise is responsible for planning and decision-making. In general, the management process involves (1) developing a written plan, (2) taking action, (3) evaluation, then (4) repeating the process. More detailed steps of the process are shown in Figure 3.8. The management process should not result in unproductive cycles but in meaningful progress toward written goals within required time limits (Fig. 3.9).

Managers should develop an effective management plan, *in writing*. A written plan makes it easier for everyone involved to identify the mission statement and goals, makes it easier to implement a plan of action of the operation, and makes periodic evaluations more effective.

A written plan describes who will do what and when. The plan provides direction and motivation for the decision makers. Written goals and objective evaluations allow producers to

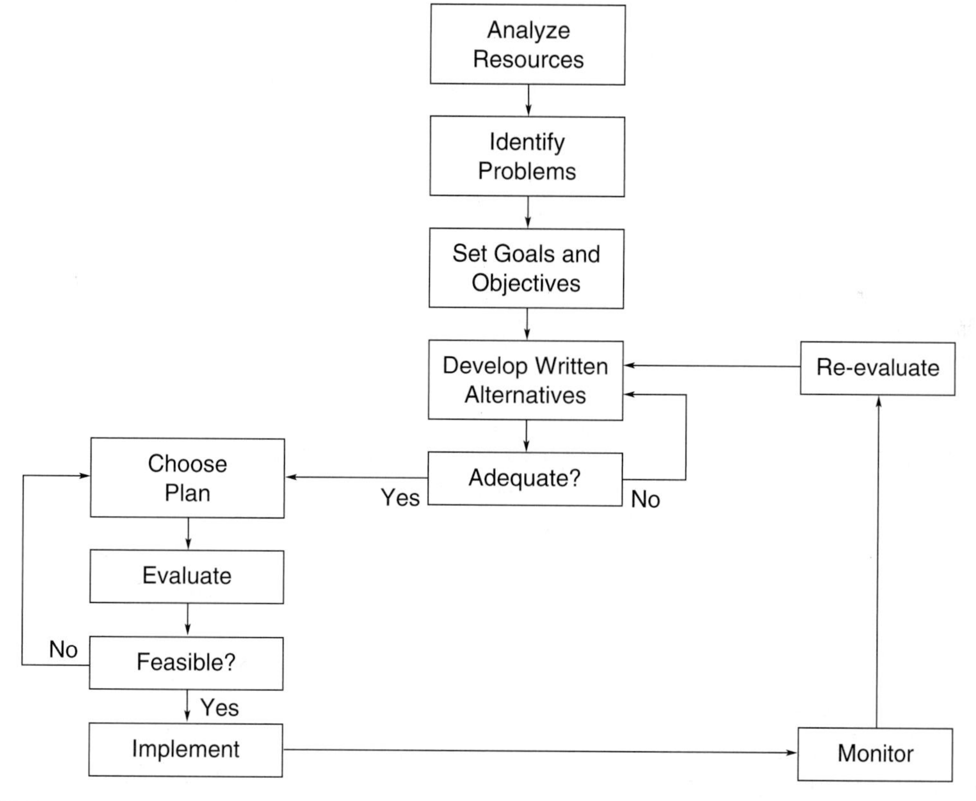

FIGURE 3.8 Major component parts of the planning process.
Source: Colorado State University.

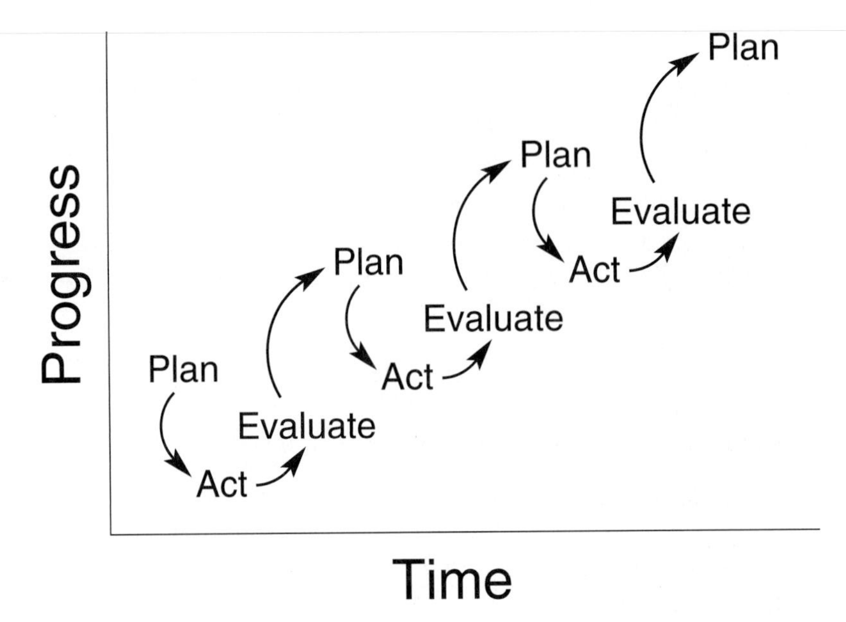

FIGURE 3.9 Progress is intended to be accomplished in the management decisions process.
Source: Colorado State University.

measure improvements and weaknesses in the operation and prioritize needed management changes. An unwritten goal is not likely to be realized.

A written plan for a cattle operation can be determined in part by answering the following questions:

1. What do I want my cattle to do for me and my customers?
2. What are my cattle presently doing for me and my customers (e.g., what are the current levels of productivity and profitability)?
3. Are my cattle best matched to their environment? Is the biological type of cattle, emphasizing reproductive efficiency, well integrated with the most economical combination of forage/feed, markets, labor, and other resources?

A producer's written plan should also be consistent with beef industry goals and should take into account the biological and economical constraints imposed by resource variability of individual operations, yet be consistent with the two major beef industry goals—(1) achieving low-cost production and (2) increasing consumer market share.

Family-owned businesses should extend the planning process into two additional areas—succession and estate transfer. A succession plan prepares the business for the transfer of power by identifying and training the individual designated as the "heir apparent." The estate plan anticipates the passing of the present owners and details the process of transferring assets from one generation to the next.

Time Management

Effective managers understand that end results are most important. To achieve the desired end results, priority must be given to time utilization. Every manager has the same resource—a fixed amount of time—but how effectively and efficiently that resource is used is the difference between a good manager and a poor manager (Fig. 3.1).

All beef cattle operations have the "trivial many" and the "critical few" situations or problems. Managers should know the difference between these two categories and how each may affect the end results. An understanding of the 80–20% rule is important to the success of a manager (see Fig. 3.5). Priorities can be established if managers are asking, "Am I doing the right things?" rather than, "Am I doing things right?" Once the right things are identified, then it is important to do them right. This gives direction to effective time management.

Some managers construct their own management calendars, listing important activities that must be accomplished during certain times of the month. A management calendar used by one cow-calf producer is shown in Table 3.2. This producer reviews the major activities that must be planned and accomplished each month, then uses a weekly calendar to establish priorities for these and other activities.

TABLE 3.2 Management Calendar for a Selected Cow-Calf Operation with Spring and Fall Calving Herds[a]

Major Activities	Jan	Feb	Mar	Apr	May	June	July	Aug	Sept	Oct	Nov	Dec
Planning, assessing goals, record analysis	xxxx	xxxx	xxxx	xxxx	xxxx	xxxx	xxxx	xxxx	xxxx	xxxx	xxxx	xxxx
Calving		ss	ssss	ss				ff	ffff	f		
Breeding preparation (semen, equipment, bulls)			s						f			
Vaccinate cows (Leptospirosis & Vibriosis)	s						f					
Vaccinate heifers (Clostridium C, IBR, Lepto, BVD)				s						f		
Synchronize heifers				s						f		
Synchronize cows				s						f		
Breed heifers					sss	ssss					ffff	ff
Breed cows					ss	ssss	ss				fff	ff
Brand, castrate, dehorn calves				s						f		
Vaccinate calves (Blackleg, Malignant Edema)				s						f		
Pregnancy test		f							s			
Fertilize pastures				x								
Control pinkeye						xxxx	xxxx	xxxx				
Fly control tags						x						
Wean calves			f							s		
Process weaning weights			f							s		
Vaccinate at weaning (IBR, PI_3, Leptospirosis)			f							s		

(continued)

TABLE 3.2 (Continued)

Major Activities	Jan	Feb	Mar	Apr	May	June	July	Aug	Sept	Oct	Nov	Dec
Retag cows		f	f						s	s		
Grub control (cows & bulls)									x			
Grub control (calves)									s			
Vaccinate heifers for brucellosis				f							s	
Cull cows (sell or feed)					f					s		
Vaccinate cows (Clostridum C)	s						f					
Select replacement heifers (weaning, yearling, and pregnancy test)		f	s	f					sf	s		
Increase precalving level of feed	s											
Keep facilities and equipment repaired	x	x	x	x	x	x	x	x	x	x	x	x
Income tax preparation and filing	x											
Income tax evaluation and adjustment									x			
Cash flow update and evaluation	x	x	x	x	x	x	x	x	x	x	x	x
Check salt and mineral supply	x	x	x	x	x	x	x	x	x	x	x	x
Evaluation target weights for replacement heifers		s				f		f				s
Evaluate management plan	x	x	x	x	x	x	x	x	x	x	x	x
Spend time planning, both short- and long-term	x	x	x	x	x	x	x	x	x	x	x	x
Check water supply	xxxx	xxxx	xxxx	xxxx	xxxx	xxxx	xxxx	xxxx	xxxx	xxxx	xxxx	xxxx
Attend seminar, workshop, or educational conference	x	x									x	x
Plan next year's budget and cash flow statement										x		

[a]s = spring calving herd, f = fall calving herd, x = applies to both herds. Placement of letter within the month shows the week designation. Type of activities and timing of activities may be different for other operations.

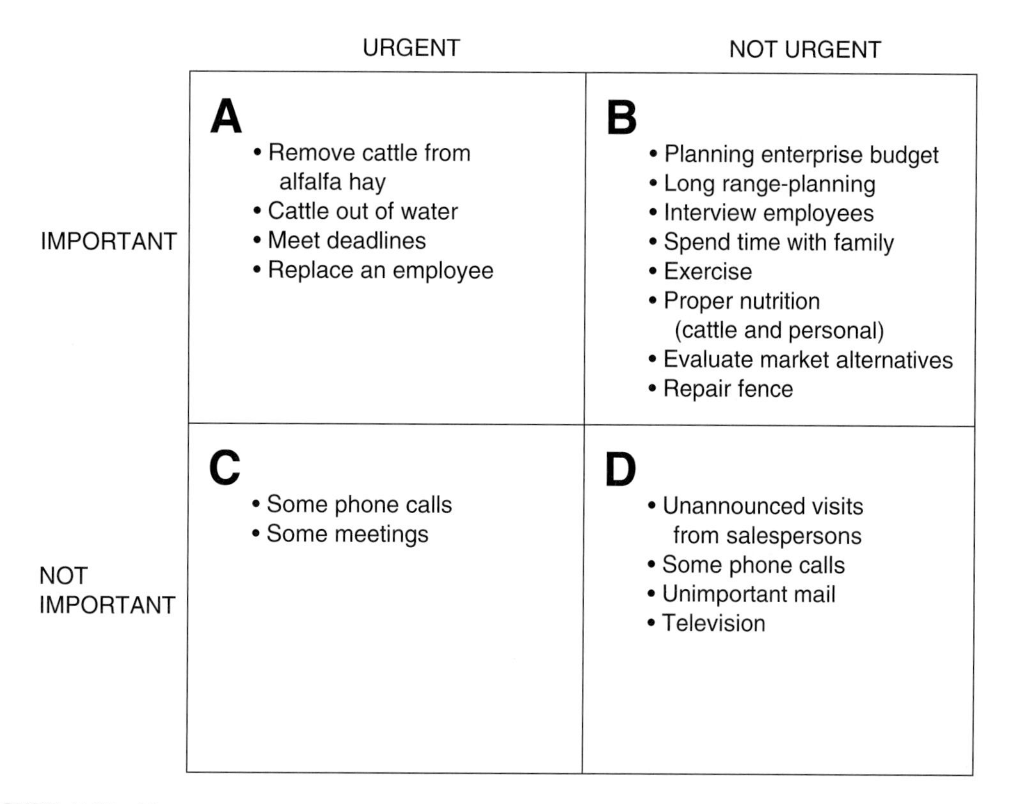

FIGURE 3.10 Time management priorities A, B, C, and D identify the four quadrants where activities can be categorized. Some example activities are shown.
Source: Adapted from Covey.

Establishing time priorities will help identify the 20% of the activities that will produce the 80% of the desired results. Figure 3.10 shows a method for managing time effectively. The demands on a producer's time can be separated into one of the four quadrants: A (Important and Urgent), B (Important and Not Urgent), C (Not Important and Urgent), and D (Not Important and Not Urgent). Obviously, the 20% time activities that count the most will be located in quadrants A and B. As management improves, increasingly more time will be spent in quadrant B.

To get into quadrant B, management should ask three questions: (1) Is it important? (2) Can I significantly influence it? (3) Is it measurable? If the activity passes all three tests, then management should direct attention to it. Finally, management should deal with no more than three important items at any one time.

There is often not enough time to achieve all the activities that a manager wants to accomplish. So how can a manager make sure that the top 20% of priorities are accomplished? Available time comes from quadrant D by saying no to these activities. An example would be saying no to the time expected by a salesperson that calls or stops by. That salesperson will be available if the manager identifies the need to evaluate the product as part of quadrant A or B.

Most managers spend their time in quadrants A and C with urgent activities. Some time can be gained by spending less time in quadrant C. Most of the crisis management comes in quadrant A. Crisis management is critical in most cases but often can be prevented by spending more time in quadrant B. For example, water and forage development in quadrant B can alleviate much of the crisis drought management that will appear in quadrant A.

Information Management

Valid information is an essential part of making intelligent management decisions (Fig. 3.1), for any decision is only as good as the information on which it is based. The manager should ask the right questions about the operation, then ask several basic questions about the available information: (1) What is needed? (2) Where can valid information be obtained? (3) What will the information cost—both in time and money? (4) Will the information help to achieve the goals of the business (e.g., profit)?

The importance of valid information in the decision-making process cannot be overemphasized. Obtaining useful information is an ongoing process, and successful managers devote part of their time to obtaining and assessing information. Obtaining valid information involves identifying true principles or natural laws or understanding things as they actually exist. The application of true principles brings stability, survival, and profitability to an operation.

There are two primary sources of useful information: (1) the information generated within the business and (2) that obtained from outside the operation. The information obtained within the business is most useful in making daily, weekly, and monthly management decisions, while outside information is more useful for developing future management plans. Care should be exercised in collecting data from within the operation that can be effectively translated into useful information. The process of collecting, recording, summarizing, and utilizing data should focus on enhancing profitability and sustainability. Production, financial, environmental, and other records are needed to manage available resources.

The amount of outside information is voluminous. One of the greatest challenges of this generation is to manage the so-called information explosion. Choosing what to read and who to listen to and how to evaluate the information obtained are crucial management decisions. Choosing what new technology to implement and evaluating its cost effectiveness are also important to managing information effectively. Equally important is determining the most useful information to obtain within the operation.

Chapter 19 discusses in more detail how to identify and manage information.

Self-Management

Most individuals are born with leadership and management talents and abilities that may or may not be developed later on in life. It is useful to reflect back on five-year blocks of one's life to identify the personal characteristics and management skills that have been realized and acquired. This process provides motivation to develop other management abilities in the future.

Many outstanding managers continue to improve their personal management skills by following the examples of role models, attending self-improvement seminars, listening to motivational tapes, and reading material on personal and human development. Weaknesses can be overcome and management skills can be enhanced by building self-esteem and consistently practicing the desired skills. Achieving self-improvement goals will in turn enhance the accomplishment of goals identified for the beef operation.

Someone has stated, "That which we persist in doing becomes easiest for us to do, not that the nature of the thing itself has changed but that our power to do it increases." Personal commitment changes a promise into reality.

Figure 3.1 identifies the major challenges facing a manager. Possibly, time management and information management could be combined into self-management. Also, the manager is included in the human resource management category. However, it is important to separate time, information, and self-management in order to see their importance and interrelationships. The realization of the goal identified in Figure 3.1 will measure how well all resources are managed.

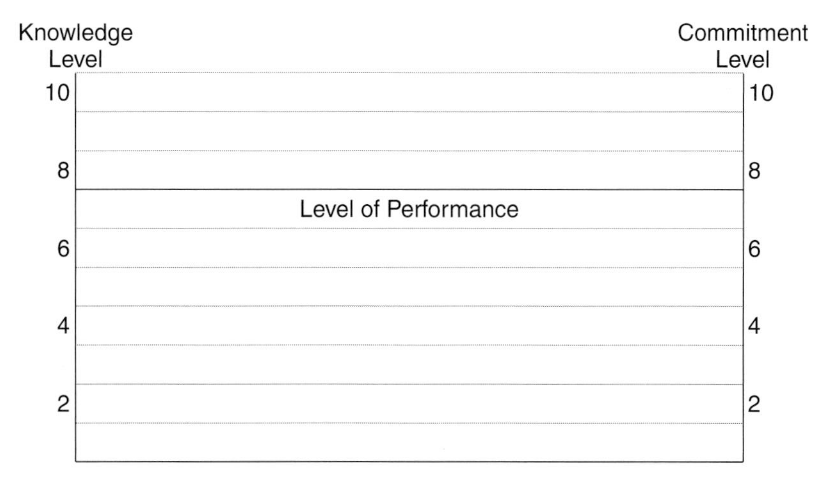

FIGURE 3.11 Level of performance, through self-management, can be raised by simultaneously increasing knowledge and commitment.

Effective leaders and managers improve their personal performance level over time. They focus primarily on increasing their knowledge and making greater commitment to implementing this knowledge. An example is shown in Figure 3.11 where the performance level of an individual is a 7 out of a possible 10. It is important to note that improving knowledge and commitment must be done in concert, otherwise the performance will be at the lower of the two components.

Included in self-management is knowing how to manage or prevent stress. The inability to manage long-term stress can damage a manager's health and be destructive to family and employee relationships. Eventually, continued severe stress can cause the emotional and financial collapse of the human and physical resources in an operation. A manager's emotional responses to continued stress will appear in one or more of the following ways: denial of problems, depression, withdrawal, blaming others, or blaming oneself for all the problems.

Even the best managers will have to make difficult decisions that may ultimately fail. Although stress cannot be totally eliminated, it can be reduced to manageable levels by doing the following:

1. Knowing one's personal limitations, both physical and financial.
2. Developing risk management plans consistent with one's physical, financial, and emotional limitations.
3. Working as a team with one's family and/or employees. Sharing disappointments and successes.
4. Improving one's problem-solving skills and increasing the accuracy of one's decisions.
5. Accepting the reality of this statement: "God grant me the serenity to accept the things I cannot change; courage to change the things I can; and the wisdom to know the difference."
6. Taking time to relax each day by leaving behind the pressures of work and the challenging decisions yet to be made. Relaxation activities should be included in quadrant B activities (Fig. 3.10).
7. Maintaining one's physical well-being through a well-balanced diet and adequate exercise.
8. Developing a sense of humor. The ability to laugh at oneself or a situation can relieve tension and enhance effective communication with others involved.

THE FINANCIAL RESOURCE

As fulfilling as the agricultural lifestyle can be, only by effective business management can the desired lifestyle be assured. Thus, it is of critical importance that beef and cattle producers develop their business management skills.

The costs, revenues, profitability, and net worth of a beef enterprise can only be critically assessed with a useful set of production and financial records. A record system is needed to (1) monitor cash flow and maintain financial control of the operation, (2) analyze the business so that effective management decisions can be made, (3) make loan applications, and (4) report and manage taxes. Managers should choose the record system that they can most effectively implement to make management decisions within the operation and to communicate financial information needed by themselves and other people such as bankers or farm advisors. Although good records do not ensure success, successful managers usually have access to good records.

Financial and Economic Records

Table 3.3 identifies financial records used by most successful cattle managers. Examples of several of these records are shown in the Appendix and in later chapters. Figure 3.12 shows a record systems flow chart that correlates with most of the financial reports described in Table 3.3. Production records that reflect cattle numbers and pounds are needed to generate the financial records. These production records are discussed in detail in Chapters 4 through 7.

Credit and money management become most crucial during periods of inflation, high interest rates, land depreciation, and relatively low cattle prices. Prudent use of credit can enable a cattle operation to grow more rapidly than it could through the use of reinvested earnings and savings, so long as borrowed funds return more over time than they cost. Thus, beef producers have to look to credit as a financial tool and learn to use it effectively.

Financial records are used to understand operating expenses (cash and noncash), interest for working capital, and term debt. Financial records do not evaluate opportunity costs or equity capital invested in the enterprise. *Economic records* deal with the opportunity costs associated with the enterprise as well as the expenses listed in the financial analyses.

Economic evaluations compare the potential returns from alternative enterprises to the beef operation. Economic analysis is typically used to evaluate longer-term decisions.

Cash Versus Accrual Accounting

The *cash accounting* method recognizes income and expense items only at the time actual cash is received or paid. *Accrual accounting* recognizes income and expenses when they are effectively earned or incurred, rather than when cash changes hands.

The cash flow statement is an important part of the financial picture of a beef operation, however, it is only one part of the financial picture. The record system of a beef operation should be maintained on the accrual accounting system or converted to the accrual system through inventory adjustments. The cash flow statement only measures cash inflow and cash outflow and it says nothing about the profitability of the business. The accrual accounting method associated with the cash flow statement, income statement, and balance sheet will show profitability and overall financial strength and position.

Profit can be defined several ways (Fig. 3.13). Usually profit is referred to as total revenue (gross proceeds) less total expenses or costs. Profit is an economic term and should be evaluated from a strict accounting point of view using generally accepted accounting practices (GAAP). The only true measure of profit comes from accrual or accrual-adjusted income statements.

TABLE 3.3 Financial Records for Beef Cattle Operations

Financial Record	Description and Purpose
Cash transactions[a]	The recording of all cash receipts and expenditures is the simplest, yet most time-consuming of all financial records. This provides most of the information needed for completing cash flow statements, filing income tax returns, and making loan applications.
Balance sheet[a]	Provides a financial picture of the operation at one point in time—usually on the last day of the year. It reflects the net worth of the operation. Net worth = assets (what is owned) minus liabilities (what is owed).
Income statement[a]	A moving financial picture that describes most of the changes in net worth from one balance sheet to the next. Net income is calculated by subtracting the expenditures (cash, decrease in inventory, and depreciation) from income (cash receipts and increases in inventory).
Cash flow statement[a]	Shows cash generated and cash needed on a periodic basis (usually monthly) throughout the year. It assesses times when money must be borrowed and times when money is available for additional purchases, investment, or retiring existing debts. A cash flow budget can be used to plan for the next calendar year. An active cash flow statement tracks what is actually happening and evaluates the accuracy of the cash flow budget.
Enterprise budget[b]	Identifies costs and returns associated with a specific product or enterprise. It can aid in making financial decisions by identifying specific problem areas where management changes can be made. Enterprise budgets are also useful where operations have more than one enterprise or where additional enterprises are being considered. Components are production and marketing assumptions, operating receipts, direct costs, net receipts, and breakeven analysis. Estimates of market weight and price can be used in assessing risk in production decisions.
Partial budget[c]	Involves only those income and expense items that would change when implementing a proposed management decision.
Income tax	Form 1040 Schedule F is the primary income tax form for sole proprietors and individual partners in a partnership. Producers who file Schedule F have basically completed a cash basis income statement. There are numerous other forms and 1040 schedules (e.g., Asset Sales, Asset Purchases, Self-employment Tax, Farm Rental Income and Expenses, Tax Withholding, Depreciation) that may be completed, depending on the individual operation and circumstances.

[a]See the Appendix for examples of forms.
[b]See Chapter 5 for an example.
[c]An example is shown in Table 3.10.

Income Tax

Frequently, managers feel that they are ineffective if income taxes must be paid. However, paying little or no income tax should not be a major goal of the operation. A common attitude is reflected in the comment of some cattle producers, "I've been ranching 20 years and haven't made any money, but I haven't paid any taxes either." Well-managed beef operations will pay income taxes if maximizing after-tax income is a goal. Therefore, it is not poor management to pay income tax, but paying more than is owed is bad business.

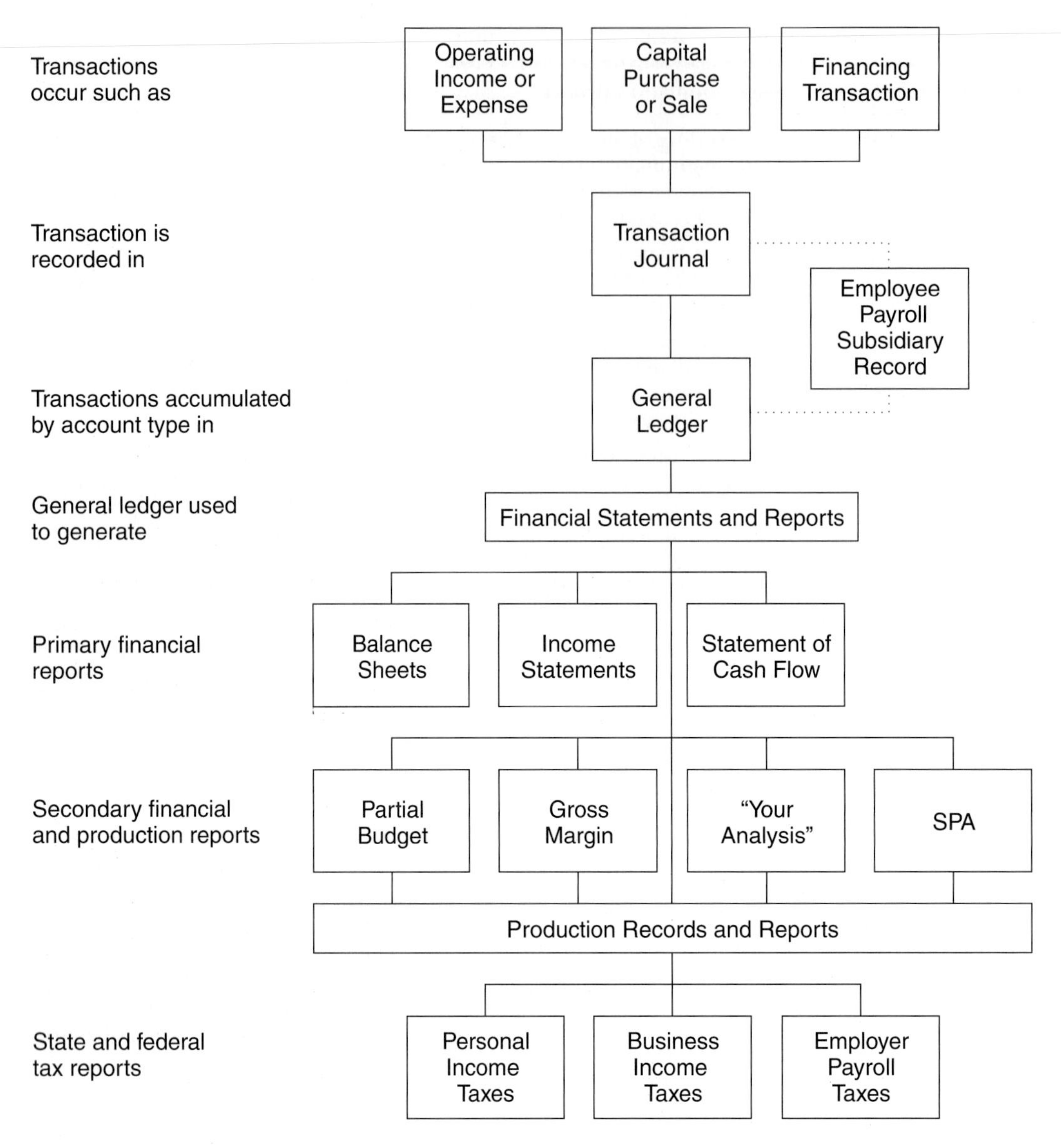

FIGURE 3.12 Record system flow chart.
Source: Adapted from Sheep Production Handbook.

Tax laws are complex and constantly changing. Most cattle producers should consult a qualified tax advisor for both tax reporting and longer-term tax management.

Beef producers who know the basic structure of federal income taxes are in a better position to manage the financial resources of the operation. They may estimate their tax position before the tax year is over for better short-term tax planning. Longer-term tax planning may well involve an appropriate estate plan.

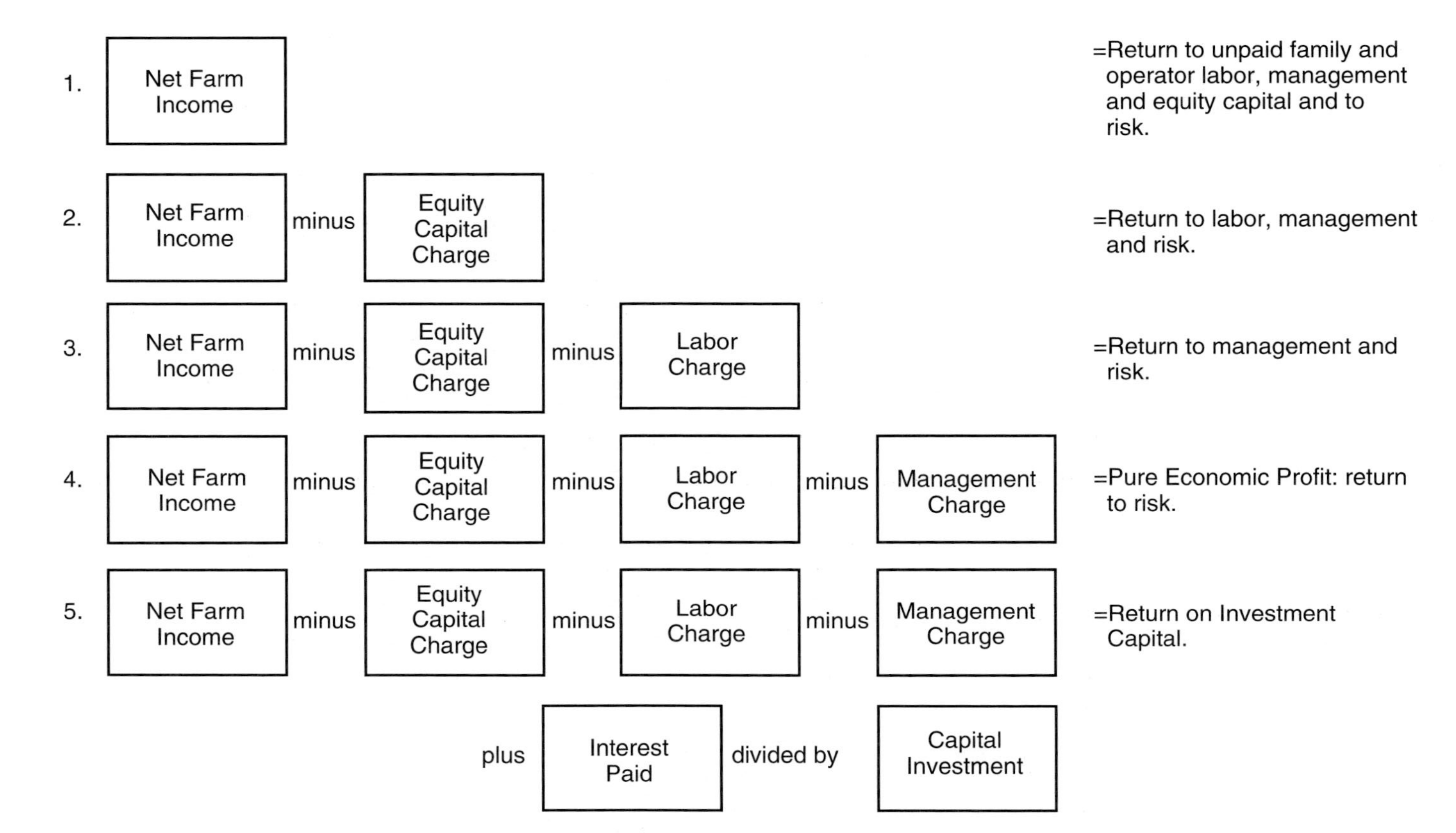

FIGURE 3.13 Alternative bottom-line definitions used to summarize beef cow herd profits.
Source: Hughes, Fargo, ND: North Dakota State Univ.

Estate Planning

Estate and gift taxes can be minimized if adequate records are kept. Estate planning must be done to pass assets from generation to generation without also passing on major financial problems.

Under a 1997 law, the unified credit exemption will be phased up to $1 million by the year 2006 for all estates. In addition to the increased unified credit, family farms and ranches can utilize a new family business exemption. Effective in 1998, family businesses now have a combined unified credit and exclusion of $1.3 million per entity, which would change slightly each year due to the interplay between the unified credit and family business exclusion.

Producers can make $10,000 annual cash gifts to children and grandchildren without paying a federal gift tax. It is recommended that the services of a lawyer and tax advisor be employed for appropriate estate and gift planning.

LAND AND FEED RESOURCES

Land and feed are two important resources in beef cattle operations. They are obviously interrelated, as land (soil) is needed to produce the feed resource for cattle. In cow-calf operations, it is the largest component of investment cost on a per-cow basis. Feed costs are typically the highest (usually over 50%) of the annual operating costs in feedlot and cow-calf operations.

Much of the land cattle graze is not suitable for producing crops for human consumption. Even the higher-valued land produces crop by-products that cattle can utilize.

The last part of the goal stated in Figure 3.1— "...while conserving and improving the resources"—has a particular reference in land and feed (forage). Advocates of holistic resource management argue that the interactions of animals with soil and plants have a significant effect on conserving and improving those resources. Their primary point is that it is the responsibility of cattle to adapt to what the land provides. The land, soil, and other renewable resources should not be over utilized and damaged in an attempt to meet the needs of the wrong biological type of cattle.

One of the most significant challenges in cow-calf management is to match the biological types of cattle (puberty, mature size, growth, and milk production) to the most economical feed resource. The "matching" implies economically producing the optimum number of pounds while sustaining or improving the feed resource. "Most economical" usually implies utilizing grazed forage rather than large amount of higher-priced harvested or supplemental feeds. This topic is covered in more detail in Chapters 5 and 15.

THE CATTLE RESOURCE

The typical business has production and market specifications that describe how it produces highly preferred consumer products. The beef business is no exception. The industry has struggled over the years not only to define such specifications but also to identify and communicate the general direction of the industry and the specific goals of each segment. The reasons for its struggles include the following:

1. Until recently, the beef industry has not had a well-defined concept of consumers and their preferences. Even today, much more refinement is needed in this area. Market-driven businesses serve very specific consumer markets where the goal is to cater to the specific and well-defined needs of targeted customers.
2. The industry is comprised of several segments (seedstock, commercial cow-calf, yearling-stocker, feeder, purveyor, packer, retailer, and consumer) where animal and product

identity is typically lost as it passes from one segment to another. Also, communication between segments is difficult and in some cases almost impossible, especially between segments that are widely separated (e.g., seedstock and packer). Each segment is driven by different economic and value determining parameters.

3. With a loss of animal and product identity between segments and a lack of useful production and financial records within certain segments, the needed specifications are often replaced with general descriptions and opinions. Some opinions are valid, while others are grossly incorrect. Even generalized opinions are not as useful in improving productivity and profitability as cost-effective records.
4. Both seedstock and commercial cow-calf producers produce cattle in a wide variety of environmental conditions. These variable conditions (such as climate and feed) dictate different production specifications if cattle are to be both productive and profitable.
5. Many, if not the majority of, individuals involved in cow-calf production (both seedstock and commercial cow-calf) do not solely derive their livelihood from their cattle operations. Their reasons for having cattle are oriented toward recreation, diversion from city life, or the availability of several acres with grass. Typically, the cattle are highly supplemented, sometimes with high priced feed and usually with income from outside the cattle operation. Some "part-time" cattle producers do manage costs and generate profits on their operations.

Cattle producers who are profit-oriented and earn their livelihood from their cattle want cattle that are both productive and profitable. The major components of productivity and profitability are shown in Table 3.4. Usually, cattle that are profitable are also productive. However, some highly productive cattle are not profitable

Productivity and profitability, as presented in Table 3.4, need to be examined in more detail. Table 3.5 shows beef cattle productivity and market specifications that are applicable to the entire beef industry. The three broad classifications of these traits are reproduction,

TABLE 3.4 Factors Affecting Cattle Productivity and Profitability

Productivity	Profitability
Numbers	Production costs
Cows	
Calves	Price and total value of products
Pounds	
Weaning weight	Weaned calves; yearlings
Yearling weight	
Carcass weight	Fed cattle; carcasses
Product composition	
Fat	By-products; boxed beef
Lean	
Bone	Retail products
Product palatability	
Tenderness	Economic efficiency
Flavor	Breakeven prices
Juiciness	Cost-benefit ratios
Biological efficiency	Sustained net profit

Source: Colorado State University.

TABLE 3.5 Production and Market Specifications for Beef Cattle

Trait	Optimum Range[a]	Industry Target[b]
Reproduction		
Age at puberty (months)	12–16	14
Scrotal circumference (cm), yearling	32–40	36
Reproductive tract score at 14 mos of age	4–5	5
Weight at puberty (lb)		
Heifers	600–800	700
Bulls	900–1,100	1,000
Age at first calving (mos)	23–25	24
Birth weight		
Calves from cows (lb)	75–95	85
Calves from heifers (lb)	60–80	70
Body condition score (BCS) at calving	4–6	5
Postpartum interval (days)	55–95	75
Calving interval (days)	365–390	365
Calving season (days)	45–90	65
Calf crop weaned (% of cows exposed)	80–95	85
Cow longevity (years of age)	9–15	12
Growth		
Mature cow weight (lb) at BCS 5	900–1,300	1,100
Weaning weight (steer; lb at 7 mos)	450–600	525
Yearling weight (steer; lb at 365 days)		
Grazed and/or backgrounded	600–800	700
Weaning to feedlot	900–1,100	1,000
Feedlot gain (lb per day)	2.5–3.5	3.0
Feedlot feed efficiency (steers; lb feed/lb gain)	5–7[c]	6[c]
Days on feed (high-energy feedlot ration)	60–120	90
Carcass		
Carcass weight (lb)	650–850	750
Quality grade	Select–Choice	Choice
Yield grade	1.5–3.5	2.5
Fat thickness (in.)	0.10–0.60	0.30
Ribeye area (sq. in.)	11–15	13
Palatability (% fat in retail cuts)	3–7	5
Muscle to bone (lb muscle to lb bone)	3.5–4.5	4.0
Lean yield per day of age (lb)		
Weaned steer to feedlot	0.75–0.95	0.85
Grazed yearling steer to feedlot	0.45–0.65	0.55
Frame score		
Steers	4–6	5
Cows	4–6	5
Bulls		
Maternal cross	4–6	5
Terminal cross	5–7	6

[a]Range will include most commercial beef operations where an optimum combination of productivity and profitability is desired.
[b]Target gives a central focus applicable to many commercial beef operations. Deviation from this target and optimum range is dependent on market, economic, and environmental conditions in specific commercial beef operations.
[c]High-energy ration.
Source: Colorado State University.

growth, and carcass. These areas identify the production part of the industry in producing numbers and pounds—and eventually producing pounds of consumer products, which are primarily proportions of palatable lean, fat, and bone. The specifications of retail beef products preferred by consumers were discussed in Chapter 2.

Table 3.4 implies the need for records that measure the productivity of cattle and how closely the cattle meet the specifications needed by the industry and individual operations. A more detailed discussion of cost-effective production records is presented in Chapters 4 through 7.

Cattle Identification

Cattle productivity and profitability can be measured on an individual animal, herd level, or per-acre basis. As cattle move through the production, marketing, and processing stages of the industry, the reasons for individual animal identification may vary. However, when viewed from a total industry perspective, there are increasing benefits for the development of a coordinated identification system.

A total quality management approach to meeting the needs of consumers calls for the implementation of a unified identification system because of the following reasons:

1. Identification systems may be required to gain access to international markets.
2. To enhance consumer confidence in the production and processing of beef.
3. For facilitation of health monitoring and disease surveillance.
4. To provide a means of trace-back in the case of food safety failures or disease outbreaks. Furthermore, animal identification systems may be desirable from a production perspective as a means to (1) assist measurement of production parameters, (2) track financial performance, (3) facilitate monitoring of quality assurance efforts, and (4) to track herd health performance.

Individual animal identification is most common in seedstock herds. Historically, hot iron brands have been used to mark the hides of cattle with the distinctive marks of individual owners. Branding has been considered an effective and affordable way to deter theft while identifying herd of origin.

Tags, tattoos, earmarks, and ear clips are also frequently used in the industry. Electronic identification using microchip technology is being adopted by some in the industry. Convenience and cost effectiveness will ultimately determine the degree of industry acceptance.

As alliances and other coordinated production systems become more widespread, the development of more sophisticated systems of identification will be developed. Identification systems will be discussed in several of the following chapters (4, 5, and 6, in particular).

THE EQUIPMENT RESOURCE

The equipment resource is discussed separately because some cattle operations are equipment intensive and this can be very costly. Machinery costs are high relative to other costs and are very high relative to cattle prices (Table 3.6). Repair costs are approximately 50 cents for each $1 spent on machinery purchases.

Most commercial cow-calf operations must have low production costs to be profitable. Sometimes it is difficult for the manager to differentiate the "needs" from the "wants" in a high-tech world of mechanization. Although equipment that can provide a life style of convenience and labor saving devices is appealing, a careful economic analysis of its cost should be made.

TABLE 3.6 Indexes[a] of Prices Received and Prices Paid by Beef Producers

Item	Year			
	1993	1995	1998	2000
Prices received				
Beef	100	77	85	104
Prices paid				
Fuel	–	–	84	136
Fertilizer	97	122	112	110
Ag chemicals	107	115	122	120
Supplies and repairs	107	112	119	124
Trucks	109	121	119	119
Machinery	106	121	132	137
Building materials	105	114	129	121

[a]Index based on 1990–1992 = 100. If index is on 1910–1914 = 100, then the 1995 prices-received index is 646, while the prices-paid index is 1,443.
Source: ERS/USDA.

Ownership costs (sometimes called *fixed* or *overhead costs*) of equipment include depreciation, interest, taxes, insurance, and storage. Repairs, fuel, oil, and operating labor are the primary variable (operating) costs. Some of the equipment options that should be considered in terms of economics are new versus used purchases, leasing, custom hiring, and joint ownership.

Some managers of low-cost cattle operations utilize a minimum of equipment because of the dependency on fossil fuel. Some of these operations show a marked increase in profitability by reducing high fuel costs and equipment repair bills and the high costs of replacement machinery. They also exhibit more stable operations in the long term, because fossil fuel (as a nonrenewable resource) may become scarce or cost prohibitive.

The dilemma of the cost-price squeeze (Table 3.6) that characterizes the beef commodity market continues to challenge the management ability of most beef producers. In Table 3.6, beef prices in 1993 and 2000 were toward the high part of the beef cycle, while in 1995 and 1998 the prices were at the low end of the beef cycle.

THE MARKET RESOURCE

Managing the market resource requires knowing consumer demand (discussed in Chapter 2) and the carcass specifications presented in Table 3.5. Producers are then concerned with the choice of various markets that will return the highest level of profitability to the enterprise. Too often producers chose market highs without recognizing the costs associated with obtaining these premiums.

In addition, marketing involves the cost of goods and services needed to produce the cattle. Profitability, then, is the difference between costs and income. While this is simplified, it is important to recognize that beef producers are price takers and not price makers. In other words, beef producers cannot pass on their higher production costs by increasing the price of their saleable products. Individual producers can only make minor changes in the market price for the cattle they sell. Good beef managers recognize that profitability can be enhanced more effectively by managing the costs of production, and then utilizing a value-based marketing program such as retained ownership (marketing on a grid), or selling a branded product, or being part of an alliance, rather than attempting to increase market price of a commodity product.

More information on the marketing principles that affect management decisions is presented in Chapter 9.

MANAGEMENT SYSTEMS

Because the resource combinations for each operation are different, no single, fixed management approach for successful cattle production is applicable to all operations. The differences among beef cattle enterprises can include such variables as levels of forage production, marketing alternatives, energy costs, debt structure, biological type of cattle, environmental conditions (e.g., weather), costs of feed nutrients and labor, and management competence levels. These variables and their interactions pose challenges to the producer, who must combine them into sound management decisions for a specific operation. There are, however, principles that apply to all producers. The definition of principle was discussed at the beginning of this chapter. Principles, in addition to management practices, are discussed throughout this book. Detailed information on sources of information for the individual resources and management systems are in Chapter 19.

A management systems approach provides a method of systematically organizing the information needed to make valid management decisions. It permits many variables to be critically assessed and analyzed in terms of their contributions to the desired end point. Management systems involve *systems thinking* rather than *linear thinking*. An example of *linear thinking* would be a producer implementing several management practices to increase weaning weight because the more pounds sold the higher the revenue and the assumption that profit per cow is also higher. *Systems thinking* would consider how increasing weaning weight might impact costs by increasing milk production, increasing mature cow weight, increasing forage consumption, and how body condition score and reproductive performance might be changed especially under a low-cost environment. The *systems thinking* example is consistent with actual studies that show herd weaning weight and net profit per cow to be independent of one another.

Systems thinking considers circles of influence and how each circle influences or interacts with other circles that are important in the system. These circles of influence are noted for a cow-calf focus (Fig. 3.14), a beef industry focus (Fig. 3.15), and a biological system for cattle (Fig. 3.16). *Integrated Resource Management* (IRM) is one of several management systems approaches being utilized to integrate the various resources into the management goal identified in Figure 3.1. The manager is challenged to make management decisions from a broad perspective, yet must also be aware of how each resource interacts with the others. These other management systems that take an integrated approach *include Holistic Resource Management* (HRM), *Ranching for Profit*, and others. These management systems approaches are used by successful cow-calf managers today. However, some successful managers have used IRM/HRM and other management systems concepts for decades.

All management programs should eventually be evaluated with the profitability formula:

$$\text{Profit (or loss)} = (\text{production} \times \text{price}) - \text{costs}$$

Production is usually reflected in numbers and pounds. High and sustained profitability will meet the goal identified in Figure 3.1 and result in an optimum combination of production, price and costs. Profitability in the short-term must be balanced with the need to conserve and improve the resources associated with the enterprise.

It has been a common practice to increase or maximize production of cattle by applying known biological relationships. Cattle production has typically been maximized without careful

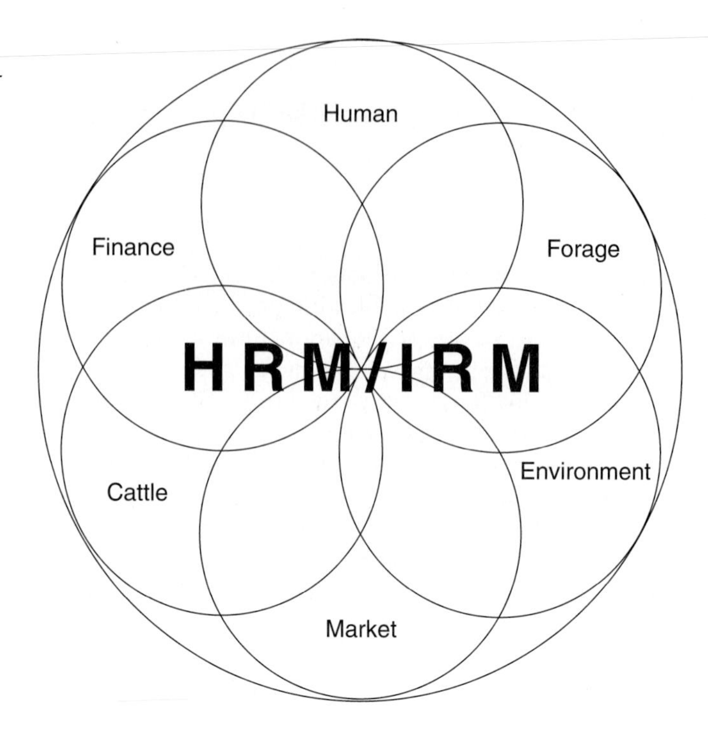

FIGURE 3.14 Circles of influence involving systems thinking for a cow-calf management system.
Source: Colorado State University.

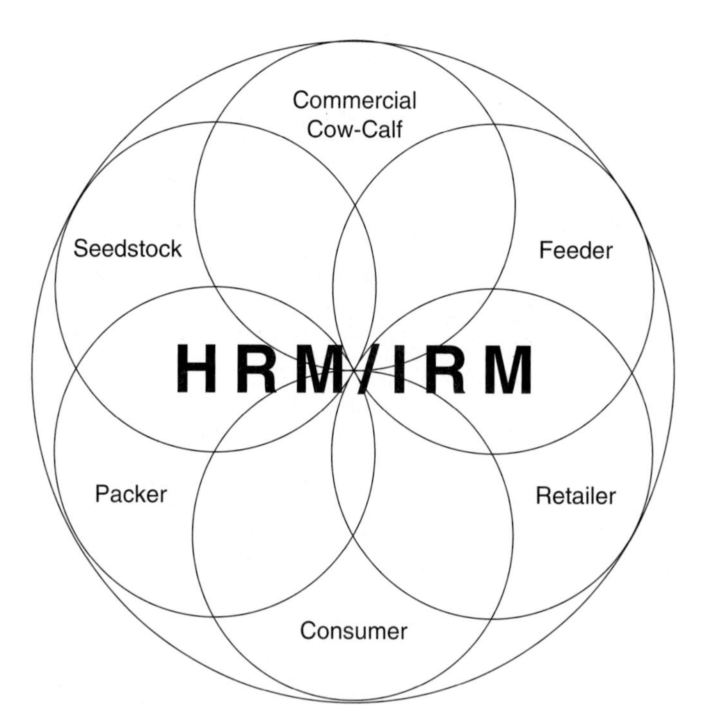

FIGURE 3.15 Circles of influence involving systems thinking for a beef industry management system.
Source: Colorado State University.

consideration of costs and how increased productivity relates to land, feed, and management resources. The need to optimize rather than maximize cattle productivity continues to receive more and more attention. Thus, there is an increased interest in a management systems approach that can optimize production and maximize net profit.

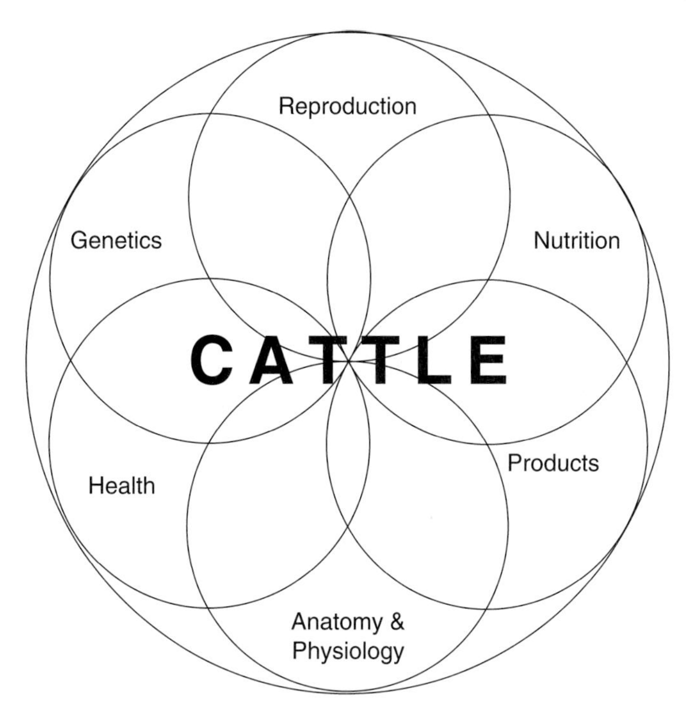

FIGURE 3.16 Major components (principles) of the biological system of cattle.
Source: Colorado State University.

The Principle of Optimums

What type of beef cattle (e.g., traits such as size and composition of products) poses the questions of how much fat, how little fat, how big in size, or how small in size. Should we focus on maximums, minimums, or optimums? In the beef industry, there are numerous references to maximizing performance—for example, maximizing milk production and weaning weight, maximizing heterosis, maximizing reproductive efficiency, and maximizing profit. Most of the maximizing refers to increasing production without considering costs. In most cases for these traits, maximum profitability will be reached before production is maximized (Figs. 3.17 and 3.18). Caution should be exercised in maximizing profit because on a short-term basis working additional hours or expecting employees to expend more effort may result in a loss of health and/or a loss of spouse and family members. Approaching the maximization of profit may be more realistic if it is stated "Maximizing continuing net profits" where this level of profit can be sustained over a long period of time. This approach gives a balance (or an optimum combination) of production and production capability under low-cost production.

Maximums and minimums are rather easily defined, but defining optimums is more of a challenge. Several definitions follow:

Optimum:	Most favorable or most conducive to a given end especially under fixed conditions; best possible under a restriction expressed or implied.
Optimize:	To make as perfect, effective, or functional as possible.
Optimal:	Most desirable or satisfactory.
Optimization:	Process of orchestrating the effects of all components (resources) toward the achievement of the stated aim.

Some of the key points in these definitions are "to a given end," "under fixed conditions," "under a restriction," and "achievement of a stated aim." These become defined with a mission/vision statement and goals.

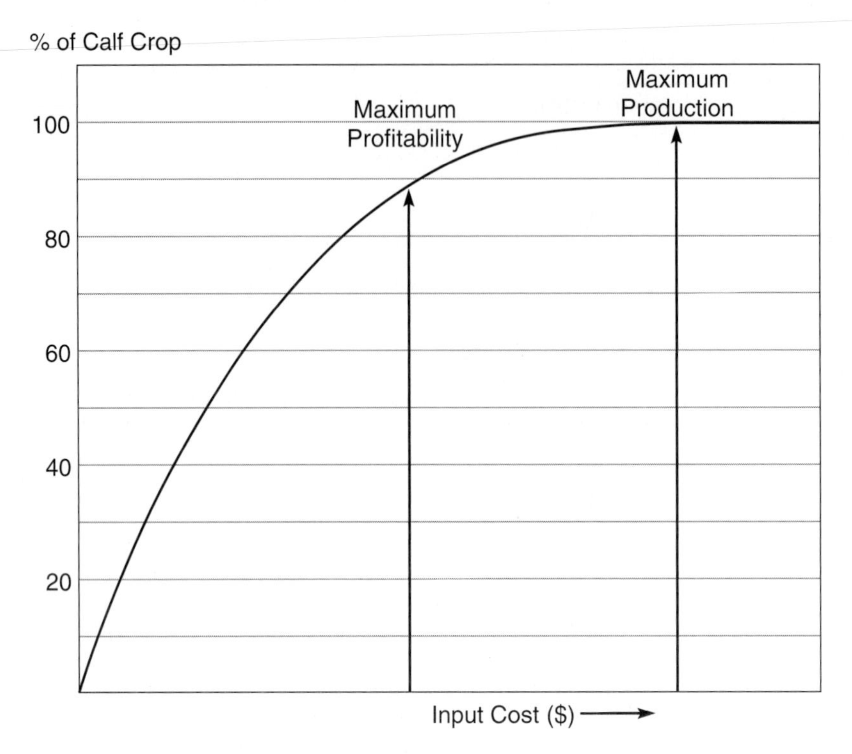

FIGURE 3.17 Maximum productivity versus maximum profitability.
Source: Colorado State University.

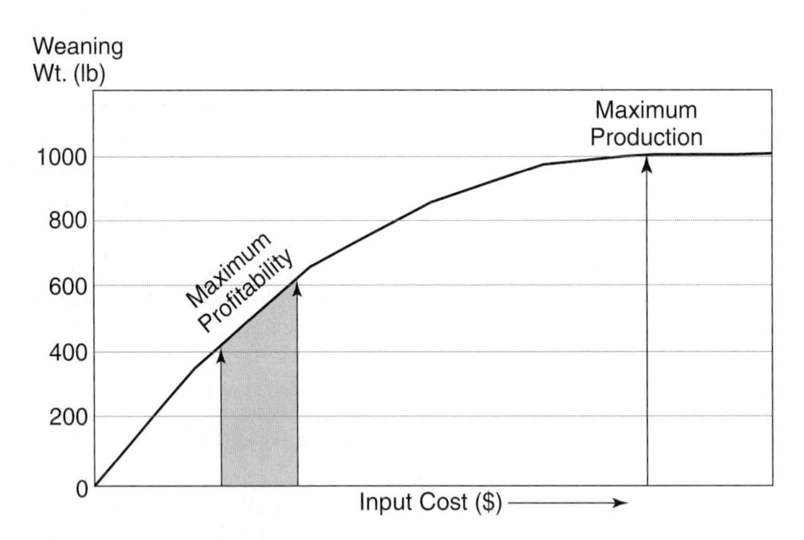

FIGURE 3.18 Maximum profitability and maximum production for average weaning weights of calves (5–9 months of age).
Source: Colorado State University.

TABLE 3.7 Achieving Optimum Levels of Productivity by Considering Profitability

Production Level	Profitability
High	Lower (costs higher than returns)
↓	↓
Optimum	Highest (best combination of costs and returns)
↑	↑
Low	Lower (returns lower than costs)

TABLE 3.8 Management Challenges with Minimum and Maximum Pounds

Too Few Pounds	Problem	Too Many Pounds	Problem
Low sale weight	Fewer dollars	Heavy birth weight	Higher calf losses; longer rebreeding
More days to slaughter weight	Higher gain cost, poor feed efficiency	Heavy mature cow weight	Higher maintenance feed cost; lose maternal traits[a]
Light carcass weight	Overhead cost per lb of carcass too high for packer	Too high milk production (weaning weight)	High maintenance feed cost; lose maternal traits
Light weight replacement heifers	Won't calve at 24 months of age	Heavy carcass weight	Size of cuts too big; lose consumer acceptability

[a]Maternal are primarily reproduction traits expressed by the cow.

Table 3.7 shows how productivity levels and profitability levels can be combined to achieve optimums in productivity. The arrows in Table 3.7 show that as high and low levels of productivity move towards an optimum level, then profitability increases. With profitability one would want to approach maximum levels as long as they were sustainable.

Producers sell pounds as weaned calves, yearlings, fed cattle, and carcasses. Maximums, minimums, and optimums can be considered for the traits that measure pounds (Table 3.8).

Consumers prefer highly palatable, lean beef. Maximums, minimums, and optimums can be considered for those traits that measure fatness in beef cattle (Table 3.9).

Table 3.5 showed optimum levels of pounds (birth weight, weaning weight, slaughter weight, carcass weight, mature cow weight) and optimum levels of fatness (fat thickness, quality grade, % fat in retail cuts, and body condition score) in breeding cattle.

It should not be surprising that some of the basic biological truths will not be useful or applicable to a specific cattle operation because an economic analysis will not permit them to be included in sound management decisions. For example, one well-known biological relationship is that calves born early in the calving season will have heavier weaning weights. Because of this relationship, some producers continue to move the calving season to an earlier time of the year. However, the focus on maximizing weaning weight may increase costs and reduce fertility. An evaluation of many beef cattle herds found that changing the calving season to later in the year better matched forage availability and increased profits significantly. On one ranch, changing calving season alone was estimated to increase the ranch's carrying capacity by 30–40% in terms of animal units. In addition, annual cow cost was reduced 18% per cow because of the reduction in winter feed requirements.

TABLE 3.9 Management Challenges with Minimum and Maximum Amounts of Fat

Too Little Fat	Problem	Too Much Fat	Problem
Less than 0.30 in. on carcass	Cold shortening (tenderness problem); difficult to grade choice	Yield grade 3.5 or higher	Low retail cutout carcass discounts; poor consumer product; uneconomical feed use
Less than 3% intramuscular fat	Unacceptable taste to many consumers	Greater than 7% intramuscular fat	Fat getting too high to meet health guidelines
Body condition score (BCS) 3 and less at breeding	Cows—longer postpartum interval	BCS 7, 8, and 9 at breading	Poor reproduction; uneconomical feed use
	Heifers—may not calve at 24 months of age		
	Higher production costs		

Biological Efficiency versus Economic Efficiency

Tess (1995) summarized this key topic at the 1995 NCA Cattlemen's College:

- Biological efficiency does not predict economic rankings very well.
- End-users want economic comparisons.
- Production/management systems are ranked confidently on (1) breakeven prices and (2) net profit.

Optimums can be effectively measured by having sustained low breakeven prices and high profits that can be measured at endpoints such as weaning, stocker/yearling, feedlot, carcass, and retail cuts. Breakeven prices and consumer preferences determine the optimum production and market specifications applicable to each beef industry trait (Table 3.5).

The difference between biological efficiency and economic efficiency must be understood. *Biological efficiency* is measured in biological units—for example, the number of pounds of beef produced relative to the number of pounds of feed consumed. *Economic efficiency* is measured in economic units (dollars)—for example, the number of dollars returned for each dollar spent. While the two types of efficiency are interrelated, they can be very different. A management practice that is biologically efficient may not be economically efficient or feasible. In fact, many measures of biological efficiency are of limited value in measuring the performance of the total integrated system. For example, producers may understand that increasing the feed supply to their cows would increase calf crop percent from 85% to 95%. However, in some cases, this practice may not be economically efficient because the income from an increased number of calves would not cover the increased feed and other costs.

Another example of economic versus biological efficiency is feeding cattle to change from one carcass quality grade to another. It may be economically efficient to feed cattle a longer period of time in an attempt to move cattle from Select to Choice because of the price spread between the two grades. However, biological efficiency (pound of feed per pound of gain) is unfavorable during this period. Furthermore, once 100–120 days on feed have been surpassed, it is difficult to gain significant improvement in USDA Quality Grade.

Rather than contrast and compare biological efficiency and economic efficiency, it is best to combine them. The term *bioeconomic efficiency* reflects that combination. Management systems emphasize the best combination of biological and economic efficiencies.

TABLE 3.10 Partial Budget Outline

Added revenue—expected increase in revenue from products sold and services rendered as a result of the proposed change.	$______	
Reduced expenses—estimated annual expenses that will be eliminated or decreased if the change is made.	$______	
Added revenue plus reduced expenses (total credits).		$______
Added expenses—new expenses and increases in current expenses directly associated with the proposed change (e.g., depreciation, interest, repairs, taxes, insurance, labor, materials, and marketing expenses).	$______	
Reduced revenue—expected value of revenue that will no longer be received if the change is made.	$______	
Added expenses plus reduced revenue (total debits).		$______
Difference (change in net income).		$______

Source: E. N. Castle et al., *Farm Business Management* (New York: Macmillan, 1987).

A decision to implement a proposed biologically efficient production practice can be economically assessed by using the partial budget outline (Table 3.10). It is a simple method for evaluating bioeconomic efficiency.

Risk Management

Beef production involves making many decisions with uncertain outcomes. Risk management is a decision-making process that evaluates the chance or probability of adverse outcomes. Some of the risks that bring uncertainty into the management decision process are weather, changes in cattle prices, changes in input costs, equipment breakdown, changes in government regulations, variability in animal and crop performance, disease, and labor and human issues (e.g., loss of an employee or personal health).

Weather can reduce calf crop percent, such as by causing high death losses of young calves when they are exposed to wet and cold conditions. This risk can be reduced by providing additional shelter and avoiding a large concentration of calves when chances of severe storms are highest.

Drought can significantly affect feed supply and cause an unnecessary reduction in herd size (Fig. 3.19). Producers in drought-prone areas manage drought risk by keeping herd size below the numbers that can be grazed in the best forage years. Also, some producers maintain a year's supply of stored forage to compensate for frequent years of drought.

Drought can reduce feed supply at critical times and thereby reduce next year's calf crop, particularly when cows are thin at calving time or when the feed supply is low from calving through rebreeding. The risk can be managed by providing supplemental feed to cows or calves. Calves can be creep-fed during periods of low forage supply, which allows the cows to maintain a higher body condition and thus have good reproduction. Calves can also be weaned early to help assure a higher calf crop next year.

Market prices have a history of wide fluctuations on a yearly, a monthly, and even a weekly basis. Producers can study price trends, know current market prices, forward-contract their cattle, and hedge their cattle using the futures market (see Chapter 9) to manage risks of price changes.

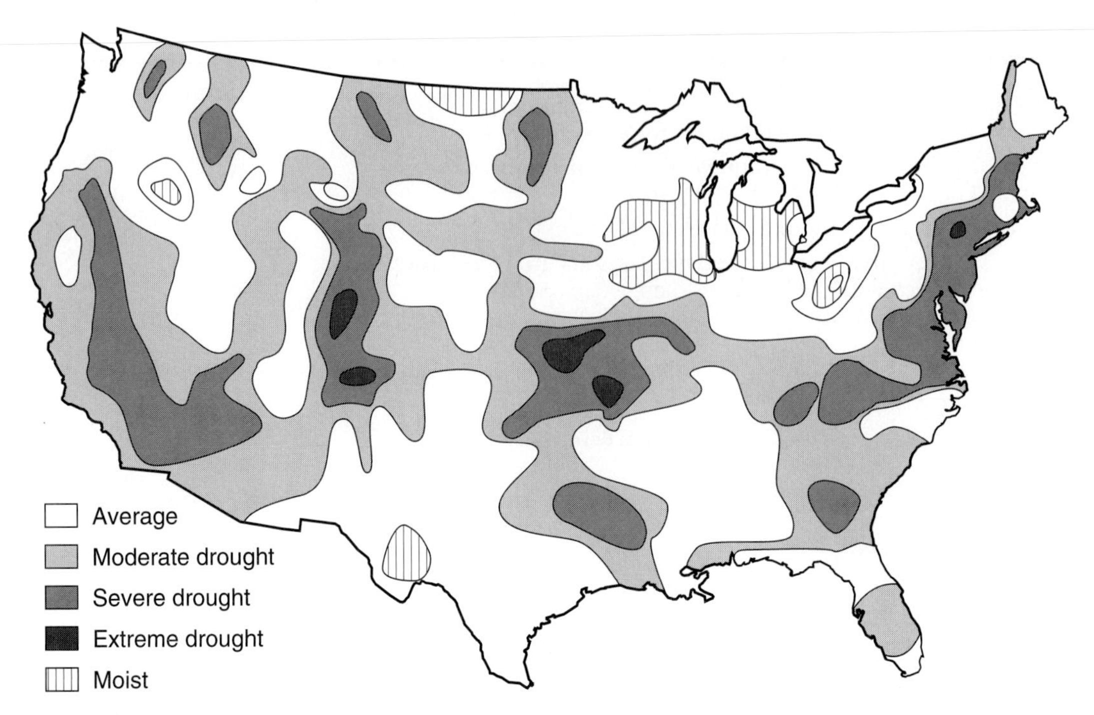

FIGURE 3.19 Drought status in the United States at one point in time. Droughts result in large annual fluctuation in feed supply. Drought occurs with greater frequency in certain areas of the United States and has to be managed accordingly.
Source: Western Livestock Round-Up.

MANAGEMENT SYSTEMS HIGHLIGHTS

- Application of leadership principles creates the need for a management systems approach.
- A written plan is developed that includes a vision/mission statement, strategies/goals, and values.
- "Systems thinking" should have a higher priority than "linear thinking." "Systems thinking" involves the interaction of "circles of influence" and their interactions that affect the desired outcome.
- Management systems involve a holistic and integrative approach to producing highly preferred consumer products at a reduced cost while developing, conserving, and improving resources.
- The management plan should integrate resources but realize that in most cases resource priorities will be human > financial > forage > cattle.
- Optimums will be reached before maximums. Economic efficiency is combined with biological efficiency.
- Optimums appear to be best measured by sustaining low breakeven prices and maintaining high levels of profitability while conserving and improving the resources.

REFERENCES

Publications

Bourdon, R. 1986. *The Systems Concept of Beef Production.* Beef Improvement Federation. (BIF) Fact Sheet FS8.

Castle, E. N., Becker, M. H., and Nelson, A. G. 1987. *Farm Business Management. The Decision-making Process.* New York: Macmillan.

Covey, S. R. 1989. *The Seven Habits of Highly Effective People.* New York: Simon & Schuster.

Dolezal, S. L., and Pollert, H. 1998. Cow-calf production record software. Oklahoma Cooperative Extension Service.

Field, T. G., and Taylor, R. E. 1997. Emotion, Tradition and Business. Proc. of BIF Research Symposium, Dickinson, ND.

Fowler, J. M., and Torell, L. A. 1987. Economic Conditions Affecting Ranch Profitability. Rangelands 9: 55.

Gersick, K. E., Davis, J. A., Hampton, M. M., and Lansberg, I. 1997. Generation to generation: Life cycles of the family business. Harvard Business School Press. Boston, MA.

Gutierrez, P. H. 1993. Cost control using economic analysis and SPA. Proc., The Range Beef Cow Symposium, Gering, NE.

Heifetz, R. A. and Laurie, D. L. 1997. The Work of Leadership. Harvard Business Review, Jan.-Feb.

Irvine, D. 1997. Succeeding with succession—from parenting to partnership. Proc. Alberta Hereford Association 106th Annual Meeting and Conference. Calgary, Alberta.

Kotter, J. P. 1996. Leading change. Harvard Business School Press, Boston, MA.

Maddux, J. 1981. *The Man in Management.* Proceedings of the Range Beef Cow: A Symposium on Production, VII. Rapid City, SD.

Miller, W. C., Brinks, J. S., and Greathouse, G. A. 1985, A Systems Analysis Model for Cattle Ranch Management. *Proc. West. Sec., Amer. Soc. Anim. Sci.* 36: 90.

NCA-IRM-SPA™ Guidelines for Production and Performance Analysis for the Cow-Calf Enterprise. Details the computation, interpretation, and limitation of each SPA performance measure. Call (303) 694-0305, or write to Michael Smith, National Cattlemen's Association, 5420 South Quebec Street, P.O. Box 3469, Englewood, CO 80155. Cost is $10 per copy.

NCA-IRM-SPA™ Workbook for the Cow-Calf Enterprise. Facilitates the recording of the cattle and feed use inventory information and schedules for the preparation of the financial statements. The workbook includes the NCA-IRM-SPA Guidelines. Cost is $25. To order call (409) 845-8012, or write to James McGrann, Department of Agricultural Economics, Texas A&M University, College Station, TX 77843-2124.

Parsons, S. D. 1992. *Ranching for Profit (Self-Study Guide).* Albuquerque, NM: Ranch Management Consultants.

Savory, A. 1999. *Holistic Resource Management: A new framework for decision making.* Washington, DC: Island Press.

Swigert, S. 1996. A new top hand: your computer. Proceedings, Beef Improvement Federation.

Tess, M. W. Production Systems and Profit. 1995 NCA Cattlemen's College.

White, L. D., Troxel, T. R., Pena, J. G., and Guynn, D. E. 1988. Total Ranch Management: Meet Ranch Goals. *Rangelands* 10: 3.

White, R. S., and Short, R. E. (eds.). 1987. *Achieving Efficient Use of Rangeland Resources.* Miles City, MT: Montana Agric. Expt. Sta, and USDA-ARS.

Zimmerman, T. S., and Fetsch, R. J. 1994. Family Ranching and Farming—A Consensus Management Model to Improve Family Functioning and Decrease Work Stress. *Family Relations* 43.

Visuals/Audiotapes

Covey, S. R. 1991. *The Seven Habits of Highly Effective People* (audiocassette seminar). Covey Leadership Center, 3507 N. University Ave., Suite 100, Provo, UT 84604-4479.

Parsons, S. D. (1991; Videotape). "Grazing for Profit" (section 3—Time Management). Ranch Management Consultants, 7719 Rio Grande Blvd. N.W., Albuquerque, NM 87107.
Sutherland, David. (1998; Videotape). "The Farmer's Wife." Distributed by PBS Home Video (www.pbs.com).
Videotapes on financial statement preparation and analysis following the Farm Financial Standards. For catalog phone (617) 282-8358 or write Farm Credit Services, Education Training Department, LW 77, 375 Jackson Street, St. Paul, MN 55101.

Computer Programs

CHAPS III. Integrated production and financial management software. Contact North Dakota State University, 1089 State Avenue, Dickinson, ND 58601, phone (701) 227-2348.

FINYEAR (Farm/Ranch Financial Statement Preparation Package). For producers without financial statements this software will generate the balance sheets, accrual income statement, and statement of cash flows. Contact James McGrann's office at (409) 845-8012.

Integrated Farm Financial Statements (IFFS). This program facilitates preparation of financial statements following the Farm Financial Standards as well as projected statements and enterprise budgets. Call Oklahoma State University at (405) 744-6081 to order.

NCA-IRM-SPA™ for Financial Calculations (SPAF). This program integrates financial and production information to generate the cow-calf enterprise financial performance report. Obtained from James McGrann's office at (409) 845-8012.

NCA-IRM-SPA™ for Production Calculations (SPAP). This software generates the first three SPA reports including (1) description, (2) reproduction and production performance, and (3) grazing performance measures. SPAP is programmed by AGRO Systems in cooperation with Texas A&M and is available from AGRO-Systems, Brighton, IL 62012-9900. Call (618) 372-3000 or James McGrann's office at (409) 845-8012.

PC-COWCARD. Production data management software. Contact University of Nebraska, P.O. Box 830918, Lincoln, NE 69583-0918, phone (402) 472-5571.

Management Decisions for Seedstock Breeders

A brief description of the seedstock segment of the beef industry was given in Chapter 1. Since the primary purpose of seedstock breeders is to provide genetic resources to commercial cow-calf producers, Chapters 12 and 13 should be carefully reviewed and understood before proceeding with Chapter 4. Chapters 2 and 5 are also applicable to breeders, since they are specialized cow-calf producers who make genetic inputs to low production costs and contribute to the production of highly palatable beef products. Parts of Chapter 17 cover growth and development, which is important to the selection and management of breeding cattle.

Breeders provide genetic inputs and related services to the entire beef industry. They should understand how their decisions integrate with the management decisions made by other beef industry segments in producing cattle that meet production, market, and consumer specifications. Breeders must effectively communicate with cow-calf producers to help fulfill the needs of these commercial producers. Cow-calf producers sometimes confuse their needs and wants. For example, in the 1980s, many cow-calf producers wanted bulls that were extreme in frame size and milk production only to find out in later years that moderate frame and moderate milk production were needed to maximize profits.

BREEDING PROGRAM GOALS AND OBJECTIVES

Current breeding programs are diverse for seedstock breeders. Some breeders focus on producing grand champions at one or more major shows, other breeders want the sale average of their commercial bulls to be a minimum of $2,500 per head, while other breeders want to breed the highest gaining bulls at a major bull testing station or maximize other traits such as milk production, marbling, or yearling weight. While these breeding programs are noteworthy, they may be antagonistic to the management system goals and objectives when cost of production, producing beef products consistent with consumer demand, and total beef industry profitability are the main focal areas. Seedstock producers should focus on providing the genetics that contribute to low breakeven prices, consistently high profits, and increased consumer market share.

Animal Identification

Breeders must have individual animal identification for pedigree registration and for recording and assessing performance. Ear tattoos and plastic ear tags are the most common methods of identification (Fig. 4.1). An ear tattoo is usually required by the breed association as permanent identification, and the same tattoo number is recommended for both ears because a single tattoo may be difficult to read.

Ear tags are a more convenient method of identification than tattoos because the tags can be read at a distance and when cattle are in a chute. Ear tag losses are a problem (usually 2–10% per year are lost). When the ear tag is lost, the tattoo serves as the backup system for maintaining individual animal identity.

Establishing unique herd identities for each animal is a key to maintaining appropriate and convenient production and pedigree records. Some producers use the last digit of the year as the first number for the ear tag. For example, the first calf born in 1999 would be given tag number 901 (9 for 1999 and 01 for being the first calf). However, this system may yield problems. For example, after 10 years, some females may have duplicate numbers.

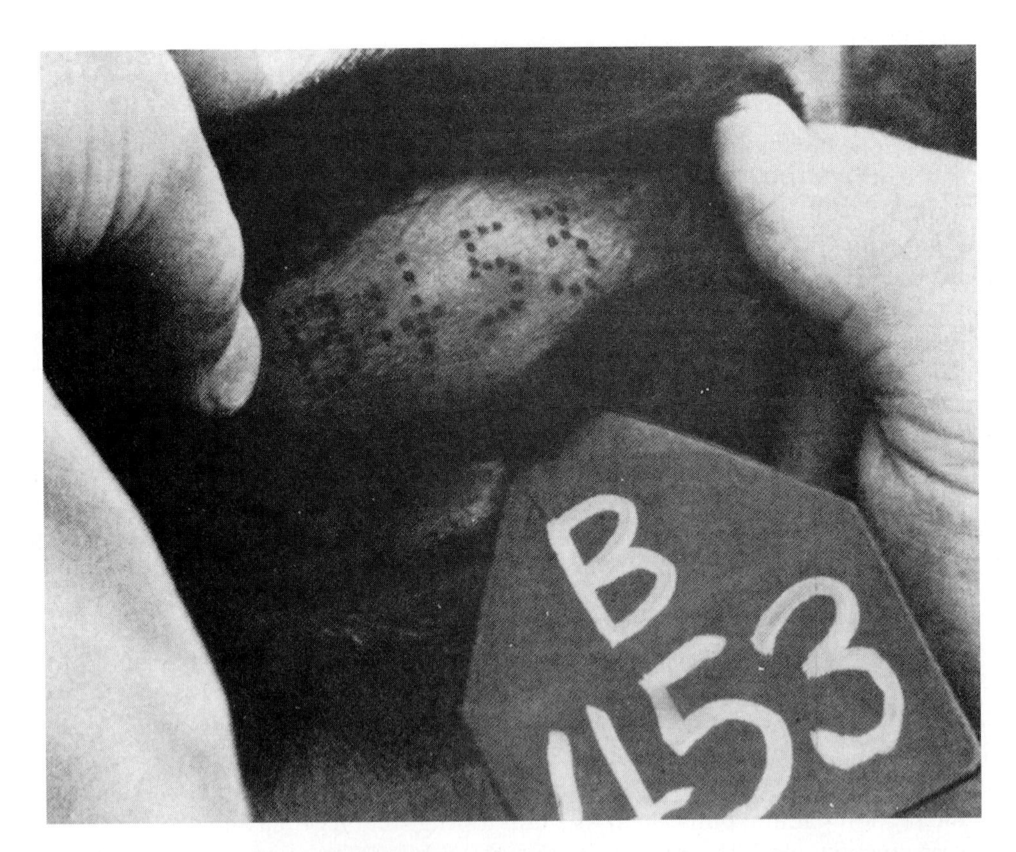

FIGURE 4.1 Double identification in the calf's ear using a tattoo and plastic ear tag. Note the same identification is used with both methods.
Source: American Angus Association.

This is particularly challenging to computer records, which cannot distinguish between 1989 cow numbered 901 and 1999 cow numbered 901. Some breeders can solve this problem by changing ear tag colors every 10 years and including the tag color as part of the identification system.

Some producers use letters to represent the years in order to prevent the 10-year duplication of numbers; for example, A = 1991, B = 1992, C = 1993. . .M = 2002, N = 2003, and P = 2004. The difficulty with this system is that unless all producers use the same letter-year system it is difficult for them to communicate the ages of the cattle to other producers. A standardized year-letter system can be found in the Appendix. Note that I, O, Q, and V are not used.

Number brands can be applied to the hide as either hot-iron brands or freeze brands. Hot-iron brands are usually effective, although the time needed to apply several numbers and to clip long hair to make the numbers readable are definite disadvantages. Furthermore, hot-iron brands as well as freeze brands cause significant damage to the hide and thus reduce its potential value as a by-product.

Freeze brands are applied to an area where the hair has been clipped. The brands are cooled in dry ice and alcohol or liquid nitrogen. The hair, where the brand was applied, comes in white after a few weeks. There are advantages of freeze brands over hot-iron branding: less trauma to the hide and larger cattle accept the process more readily. The disadvantages of freeze brands are the additional costs, increased time required for application, and too frequently some numbers are difficult to read. The best results are achieved on black cattle, whereas poor results occur on white or light-colored cattle.

PRODUCTION RECORDS

Some breeding cattle currently are evaluated and selected only on visual appraisal, even though it is well known that visual appraisal alone is a poor indicator of most economically important traits measured at the cow-calf level. Using scales and other measuring instrumentation gives breeders the objective measurements needed to identify beef cattle superiority or inferiority for several traits affecting commercial profitability.

Most breed associations make performance record systems available to their members for a nominal processing fee. The breed performance programs are variously named according to the breed; for example, Angus Herd Improvement Records (AHIR), Total Performance Records (TPR, for Hereford), Performance Registry System (for Simmental), and Charolais Herd Improvement Program (CHIP).

A genetic evaluation system is only as good as the accuracy and volume of data submitted to the national association. In an attempt to increase participation in the breed performance program and as a result enhance the reliability of genetic estimates generated, many breed organizations are moving to whole herd reporting. *Whole herd reporting* systems combine the fees for registry, pedigree transfer, and performance records calculation into one fee based on the cow herd inventory. Such an approach increases the amount of data submitted and allows for the development of new expected progeny differences (EPDs) for a variety of reproductive and convenience traits.

The traits traditionally measured by most breed performance record programs are birth weight, weaning weight, yearling weight, and milk production (indirectly). Ratios of these weights are usually calculated for individual animals within the herd. However, the ratios are meaningful and valid only when the animals are compared with a contemporary group of animals exposed to a similar environment. Expected progeny differences provide an accurate means to compare animals within a breed. Table 4.1 outlines a basic breed performance program showing both the input records and the final analysis records.

A *contemporary group* is a set of calves that have been raised together and managed in a similar manner. Typically, contemporary groups are composed of animals of the same gender.

TABLE 4.1 Outline of a Breeding and Performance Program

I. INVENTORY
  A. Record input
    1. Add cows and bulls to inventory and identify cows leaving the herd with regard to reasons for disposal
  B. Record analysis output
    1. Current inventory; frequency of, and reasons for, involuntary culling
  C. Selection decisions
    1. Evaluate whether the genetic composition of the cattle is consistent with the management system and long-term goals
    2. Look for specific problems in certain sire lines

II. BREEDING
  A. Record input
    1. Sire
    2. Dates bred or time exposed
    3. Date and result of pregnancy exam
  B. Record analysis output
    1. Expected calving dates
    2. Percentage bred of those exposed for each sire and for the entire group

III. CALVING
  A. Record input
    1. Birth date
    2. Birth weight
    3. Calving ease score
    4. Calf ID (identification)
    5. Sire and dam ID
  B. Record analysis output
    1. Adjusted birth weight and ratio
    2. Calving ease score (separately for first-calf heifers and for older cows)
    3. Gestation length (if breeding dates are known) adjusted for cow age and sex of calf
    4. Proportion calving in each of 21 days of the calving season
    5. Percent calf crop of those exposed
    6. Birth weight and calving ease EPDs
    7. Calving interval

IV. WEANING (CALF AGE RANGE OF 160–250 DAYS)
  A. Record input
    1. Date weighed
    2. Calf weight
    3. Management code
    4. Contemporary group code
    5. Cow weight, hip height, and condition score
    6. Cow pregnancy status
  B. Record analysis output
    1. Adjusted 205-day weaning weight and ratio
    2. Repeat of calving information
    3. Weaning-direct and weaning-milk EPDs for each calf
    4. Weaning-direct and weaning-milk EPDs for each cow and bull
    5. Mature size EPDs for cows and their sires
  C. Selection decisions
    1. Make initial selection/culling of heifers based on EPDs and structural problems
    2. Cull cows based on soundness and EPDs
    3. Consider culling open cows

V. YEARLING (AGE RANGE OF 330–390 DAYS)
  A. Record input
    1. Dates weighed and measured
    2. Weight, hip height, and scrotal circumference
    3. Management code
    4. Contemporary group code
  B. Record analysis output
    1. Adjusted yearling weight ratio, adjusted linear measures and frame score
    2. Repeat of calving and weaning information
    3. EPDs for weight and appropriate linear measures for each calf, dam, and sire
  C. Selection decisions
    1. Select bulls by comparing the breeding values for each trait of relevance, both for sires in use and the yearlings
    2. Cull any heifers that are structurally unsound and too extreme in frame (large or small)
    3. Select replacement heifers based on EPDs

VI. CONTRIBUTION TO COMMERCIAL COW-CALF, FEEDLOT, AND CARCASS (RETAIL PRODUCT) PRODUCTION
  A. Record input from sire progeny groups of herd sample from commercial herds
    1. Production costs for weaned calves, feedlot cattle, and carcass weight
    2. Carcass weight, quality grade, and yield grade
    3. Warner-Bratzler Shear or other measures of tenderness and total palatability
  B. Record output
    1. Breakeven prices for weaned calves, feedlot cattle, and carcasses
    2. EPDs for carcass weight, marbling, and percent retail product
    3. EPDs for tenderness and other measures of palatability
  C. Selection decisions
    1. Select bulls whose progeny excel in low breakeven prices
    2. Select bulls whose progeny excel in optimum carcass weights, marbling, percent retail product tenderness, and overall palatability

Source: Adapted from BIF Guidelines for Uniform Beef Improvement Programs.

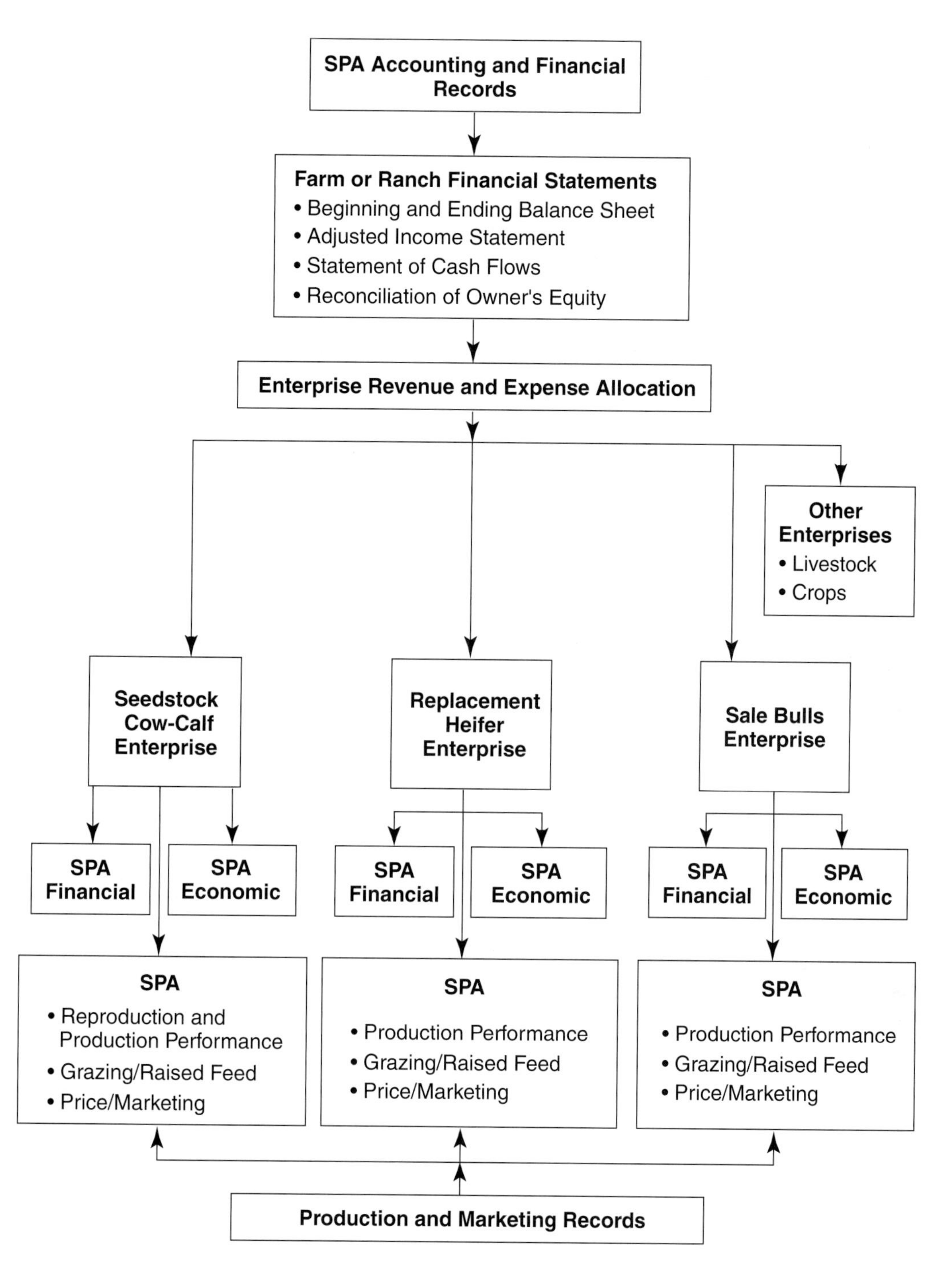

FIGURE 4.2 Farm or ranch financial statements and seedstock cattle Standardized Performance Analysis (SPA) linkages.
Source: National Cattlemen's Beef Association.

FINANCIAL RECORDS

The most effective record system for seedstock breeders combines financial and production records (Fig. 4.2). More detail for the Standardized Performance Analysis (SPA) is given in Chapter 5. It is recommended that a separate record analysis be done for cow-calf, replacement heifers, and sale bull enterprises. Breakeven prices are then available for each of the three enterprises. Sale bull breakeven prices will typically fall between $1,000 and $2,000.

Once performance records are generated, it is important to put them into an appropriate format so they can be most effectively utilized. Some seedstock and commercial producers hand calculate their performance record data and maintain their own record forms. Most breeders and producers, however, submit their records to a breed association where the records are computerized and returned in a standardized and user-friendly form. Performance records that are used to calculate EPDs are especially useful to breeders.

Useful Performance Records

Useful performance records have the following characteristics:

1. They are objectively, accurately, and honestly measured.
2. They are adjusted for environmental effects (e.g., age of calf, age of dam) so that they are comparable.
3. They permit differences in performance records to be compared (*not* the absolute value of the records).
4. They can be compared within a contemporary group to minimize the environmental effects and improve the accuracy of estimating genetic differences.

BULL SELECTION

Importance of the Bull

Effective bull selection should focus on two primary areas: (1) producing live calves and (2) making genetic improvements in economically important traits. Our focus here is on genetic improvement; producing live calves is discussed in detail in Chapter 11.

Even though a bull contributes 50% of its genetic material to each calf, the magnitude of the bull's contribution is greater because of the increased number of offspring he sires. The genetic importance of successive bulls used in a herd is shown in Figure 4.3. As observed in the figure, the three bulls (A, B, and C) contribute nearly 90% of the genetic material to the

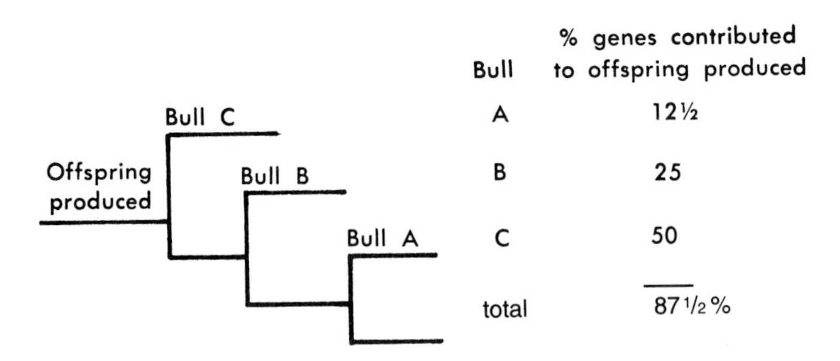

FIGURE 4.3 Genetic contribution of three bulls to the offspring currently produced in the herd. *Source:* Colorado State University.

TABLE 4.2 Genetic Contribution of Bull Selection in Cattle Selected for Weaning or Yearling Weight

	Selection Differential[a] (selection practiced)	
	Weaning Wt (lbs)	Yearling Wt (lbs)
Bulls	78	140
Heifers	19	18
Total	97	158
% from bull	78/97 = 80%	140/158 = 89%

[a]Calculation of selection differential is shown in Chapter 12.
Source: U.S. Meat Animal Research Center, Clay Center, Nebraska.

FIGURE 4.4 Purebred breeders must produce genetically superior bulls that will complement well-planned crossbreeding programs.
Source: American Brahman Breeders Association.

calves currently in the herd. Studies conducted by the U.S. Meat Animal Research Center, as shown in Table 4.2, clearly demonstrate that 80–90% of genetic improvement in a herd comes from bull selection.

Emphasis on bull selection does not diminish the importance of good beef females because genetically superior bulls typically do not have poor dams and superior bulls sire productive and profitable daughters (Fig. 4.4). Most genetic superiority or inferiority of cows, however, depends on the bulls previously used in the herd. Optimum genetic input coming from effective sire selection dictates where priorities should be placed.

Sire Summaries

The development of sire summaries has made sire selection more effective. Most breed associations publish a sire summary that is updated annually or twice annually. Sire summaries can be requested from the breed associations listed in the Appendix either as hard copies or obtained from the association web site.

Expected progeny difference (EPD) and *accuracy* (ACC) are the important terms used in understanding sire summaries. An EPD combines into one number a measurement of genetic potential based on the individual's performance and that of related animals (such as the sire, dam, siblings, progeny, and other relatives). EPD is expressed as a plus or minus value, reflecting the genetic transmitting ability of a sire for a particular trait. The most common EPDs reported for bulls, heifers, and cows are birth weight EPD, milk EPD, weaning growth EPD, maternal EPD (which includes milk and preweaning growth), and yearling weight EPD (Fig. 4.5).

Table 4.3 shows sire summary data for several Angus bulls. If Bull B and Bull C were used in the same herd (each on an equal group of cows), the expected performance of their calves would be as follows:

Bull B's calves would be 16 lbs heavier at birth.
Bull B's calves would weigh 59 lbs more at weaning.
Bull B's calves would weigh 78 lbs more as yearlings.
Bull B's daughters would wean 7 lbs more calf.

As shown in the table, Bull A and Bull D have an optimum combination of all traits. Bull E is a promising young sire with an excellent combination of EPDs. However, the accuracy (ACC) is relatively low and could significantly change with more progeny and as daughters start producing.

Accuracy is a measure of expected change in the EPD as additional progeny data become available. EPDs with ACC of 0.90 and higher would be expected to change very little, whereas EPDs with ACC below 0.70 might change dramatically with additional progeny data. Figure 4.6 shows how a higher ACC reduces the expected change in the EPD. For Bull B with a 0.40 accuracy, a 15.4 lb change (in either direction) in yearling weight EPD is just as likely as a 5.1 lb change for Bull A with a 0.80 accuracy. The likelihood of change is the same for both bulls, however, it is the magnitude of the possible change that differs.

Some breeds have EPDs for traits other than birth weight, weaning weight, and yearling weight (see Table 4.4). EPDs are not available for several economically important traits, especially maternal traits related to total reproductive performance.

TABLE 4.3 Selected Data from an Angus Sire Evaluation Report

	Birth Weight		Weaning Weight (direct)		Weaning Weight Maternal				Yearling Weight	
					Milk			Combined Value		
Sire	EPD	ACC	EPD	ACC	EPD	ACC	DTS[a]		EPD	ACC
A	−1	0.95	+20	0.95	+1	0.93	331	+12	+39	0.94
B	+9	0.95	+47	0.95	−17	0.89	88	+6	+74	0.90
C	−7	0.83	−12	0.85	−10	0.79	29	−16	−4	0.85
D	+1	0.95	+22	0.95	+3	0.92	230	+14	+54	0.94
E	−1	0.73	+15	0.70	+12	0.42	0	+20	+45	0.66

[a]DTS = daughters.

Each bull listed in this report is compared to every other bull. The analysis takes into account only the differences expressed in each herd in which the bulls were used. For example, Bull A has a weaning EPD of +30 lbs and Bull B has a waning EPD of +20 lbs. If you randomly mate these bulls in your herd, you could expect Bull A's calves to weigh 10 lbs more at weaning than Bull B's progeny (30 – 20 = 10).

EXPECTED PROGENY DIFFERENCES

PRODUCTION									CARCASS						ULTRASOUND BODY COMPOSITION					
BW AC	WW AC	Milk AC	Herds Daus	YW AC	YH AC	MW AC	MH AC	SC AC	CW AC	Marb AC	RE AC	Fat AC	%RP AC	Grp Pg	%IMF AC	RE AC	Fat AC	Rump Fat AC	%RP AC	Grp Pg
+5.3	+31	+28	751	+59	+.7	+29	+.7	+1.38	+13	+.08	+.36	+.003	+.2	+19	−.12	+.31	−.003	−.030	+.60	55
.99	.99	.99	2836	.99	.99	.95	.97	.99	.94	.95	.93	.93	.93	197	.71	.71	.72	.75	.71	109

GROWTH & MATERNAL EVALUATION

Expected progeny difference (EPD) is the estimate of how future progeny of each sire are expected to perform in each of the traits listed. EPD is expressed in pounds, either plus or minus. Interim EPD may appear in cases wherein there is insufficient information to allow the calculation of an EPD through National Cattle Evaluation (NCE) procedures. This EPD will be preceded by an "I" and is calculated from pedigree information by the AHIR Department of the Association.

Accuracy (AC) is the reliability that can be placed on the EPD. An accuracy of close to 1.0 indicates higher reliability.

Weaning weight direct (WW) expressed in pounds, is a predictor of a sire's ability to transmit weaning growth to his progeny compared to that of an average sire.

Weaning weight maternal (Milk) is a predictor of a sire's genetic merit for milk and mothering ability as expressed in his daughters compared to daughters of an average sire. In other words, it is that part of weaning weight attributed to milk and mothering ability.

Daus reflects the number of daughters that have progeny weaning weight records.

Herds indicates the number of herds from which daughters are reported.

Yearling weight (YW), expressed in pounds, is a predictor of a sire's ability to transmit yearling growth to his progeny compared to that of an average sire.

Yearling height (YH) is a predictor of a sire's ability to transmit yearling height, expressed in inches, compared to that of an average sire.

Mature daughter weight EPD (MW), expressed in pounds, is a predictor of the difference in mature size of daughters of a sire compared to daughters of an average sire.

Mature daughter height EPD (MH), expressed in inches, is a predictor of the difference in mature height of daughters of a sire compared to daughters of an average sire.

Scrotal circumference EPD (SC), expressed in centimeters, is a predictor of the difference in transmitting ability for scrotal size compared to that of an average sire.

CARCASS EVALUATION

Carcass weight EPD (CW), expressed in pounds, is a predictor of the differences in hot carcass weight of a sire's progeny at a given end point compared to progeny of an average sire.

Marbling EPD (Marb) is an expression, as a fraction, of the difference in a subjective USDA marbling score of a sire's progeny at a given end point compared to progeny of an average sire.

Ribeye area EPD (RE), expressed in square inches, is a predictor of the difference in ribeye area of a sire's progeny at a given end point.

Fat thickness EPD (Fat), expressed in inches, is a predictor of the difference in external fat thickness at the 12th–13th ribs of a sire's progeny at a given end point compared to progeny of an average sire.

Percent Retail Product (%RP) is a predictor of the difference in pounds of salable retail product of a given sire's progeny compared to the progeny of an average sire in the Angus breed.

Group/Progeny (Grp/Pg) reflects the number of contemporary groups and the number of carcasses for each sire included in the analysis.

ULTRASOUND BODY COMPOSITION EPD

Intramuscular Fat (%IMF) is a predictor of the difference in a sire's progeny for percent intramuscular fat in the ribeye muscle compared to an average sire.

Ribeye Area (RE) is a predictor of the difference in square inches of ribeye area of a sire's progeny compared to the progeny of an average sire.

Fat Thickness (Fat), expressed in inches, is a predictor of the difference in external fat thickness at the 12th–13th ribs of a sire's progeny compared to the progeny of an average sire.

Rump Fat (RMPFT), expressed in inches, is a predictor of the difference in external fat thickness as measured on the rump at a point between hooks and pins of a sire's progeny compared to the progeny of an average sire.

Percent Retail Product (%RP) is a predictor of the difference in pounds of salable retail product of a sire's progeny compared to the progeny of an average sire.

Group/Progeny (Grp/Pg) is the number of contemporary groups and number of progeny considered in this analysis. Contemporary groups are defined by scan data and member code.

FIGURE 4.5 Explanation of a sire's EPDs in a sire evaluation report.
Source: Angus Sire Evaluation Report, Fall 2001.

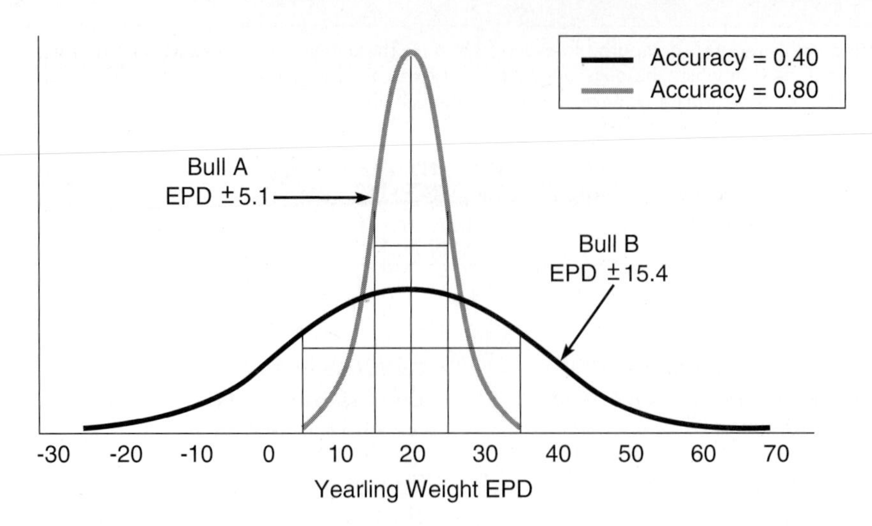

FIGURE 4.6 Possible change in EPDs where two bulls have the same yearling weight EPD (+20) but different accuracies (0.40 vs. 0.80).
Source: American Simmental Association.

TABLE 4.4 Sire Summary Data (EPDs or Ratios) for Selected Breeds

Traits	Angus[c]	Brahman[c]	Charolais[c]	Gelbvieh[c]	Hereford/ Polled Hereford[c]	Limousin[c]	Red Angus[c]	Simmental[c]
Calving ease[a]								
–first calf[b]	o	o	o	+	o	o	o	+
–maternal[c]	o	o	o	+	o	o	o	+
Scrotal circumference	+	o	o	+	+	+	+	o
Birth weight	+	+	+	+	+	+	+	+
Gestation length	o	o	o	+	o	+	o	o
Weaning weight	+	+	+	+	+	+	+	+
Yearling weight	+	+	+	+	+	+	+	+
Milk[d]	+	+	+	+	+	+	+	+
Maternal (milk + ½ weaning weight)[d]	+	o	+	+	+	+	+	+
Disposition (docility)	o	o	o	o	o	+	o	o
Stayability	o	o	o	o	o	+	+	o
Heifer pregnancy	o	o	o	o	o	o	+	o
Yearling hip height	+	o	o	o	o	o	o	o
Mature cow weight	+	o	o	o	o	o	o	o
Mature daughter height	+	o	o	o	o	o	o	o
Carcass weight	+	o	+	+	o	+	+	+
Marbling	+[f]	o	+	+	+[g]	+[f]	+	+
Ribeye area	+[f]	o	+	+	+[g]	+[f]	o	o
Fat thickness	+[f]	o	+	+	+[g]	+[f]	+	o
Percent retail product	+[f]	o	o	o	o	o	o	+
Ultrasound rump fat	+	o	o	o	o	o	o	o
Grid value	o	o	o	+	o	o	o	o
Warner-Bratzler Shear Force	o	o	o	o	o	o	o	+

[a]Computed from birth weights.
[b]Expressed as a ratio, not an EPD. Measures calving ease of bulls' own calves.
[c]Expressed as a ratio, not an EPD. Measures calving ease of a bull's daughters.
[d]Computed from weaning weights.
[e]+ = trait in sire summary; o = trait not in sire summary.
[f]Computed from both progeny and ultrasound data.
[g]Computed from ultrasound data.
Source: Breed sire summaries.

EPDs are needed that could be used to improve cost-effective calf crop percentage in commercial cow-calf herds. The reproductive complex is difficult to measure and most performance programs do not have adequate information available today to use for percent calf crop. It is imperative that measurements be researched and used. Total reproductive performance is a function of age and weight at puberty, conception rate, gestation length, calving ease, post-partum interval, and longevity. These components of reproductive efficiency are measured directly or indirectly through correlated traits (e.g., birth weight, milk production, mature weight, scrotal circumference, and body condition score). These traits interact with available resources (e.g., forage, low-cost production) compounding the evaluation problems. Clearly, research needs support and resulting measures need to be utilized by breed programs. The use of the specific traits measured should help improve the reproductive performance, especially where genetic selection can be combined with optimum levels of heterosis in commercial production.

Bull selection is complex because the desired genetic improvement is for a combination of several traits. Continuous selection for only one trait may result in problems in other traits. A good example is selecting for yearling weight alone, which results in increased birth weight because the two traits are genetically correlated. Birth weight is associated with calving difficulty, so increased birth weight might be a problem. Large increases in birth EPD can occur and calving difficulty can be serious.

The biggest challenge in bull selection is selecting bulls that will improve maternal traits. Frequently, too much emphasis is placed on yearling growth and frame size that often results in excessive mature size. These traits can be antagonistic to maternal traits, such as birth weight, early puberty, and maintaining a cow size consistent with an economical feed supply. Figure 4.7 shows an example of stacking a pedigree for maternal traits for yearling bull selection where the bulls are used natural service on heifers. If an older bull can be used AI, then the emphasis would be on accuracies higher than are available on young bulls. Examples of older bulls that could be used AI are the sires and maternal grandsires of the yearling bulls shown in Figure 4.7.

Table 4.5 identifies the bull selection criteria for commercial beef producers. These criteria should also help seedstock breeders focus on their selection criteria. First, the breeding program goals and the traits are identified. Second, the bull selection criteria are identified, emphasizing maternal traits, then relatively high growth, and carcass traits without causing a serious antagonism to maternal traits. Maternal, growth, and carcass traits would be balanced in selecting bulls of one biological type if a commercial producer was using only one breed, a rotational crossbreeding program, producing the cows to be used in a terminal crossbreeding program, or selecting bulls for a composite breed. Finally, Table 4.5 identifies the traits emphasized in the selection of terminal cross bulls. In selecting terminal cross sires, little emphasis is placed on maternal traits because replacement heifers from this cross are not kept in the herd.

Figure 4.8 shows a performance pedigree that includes the bull's pedigree and performance information on the bull, his sire, and his dam. Also included are performance data on the bull, his sire and dam, and their progeny. Figure 4.9 identifies a bull that combines maternal, growth, and carcass traits into an excellent package. The EPDs shown are balanced, with high accuracies for most of the traits.

Birth Weight

Birth weight should be objectively measured and recorded on all calves (live and dead) within 24 hours of birth. Estimates of birth weights are inaccurate and are of limited value.

Age of cow and sex of calf influence the birth weight of the calf. Table 4.6 shows the BIF age of dam adjustments for birth weights so that birth-weight records can be standardized for

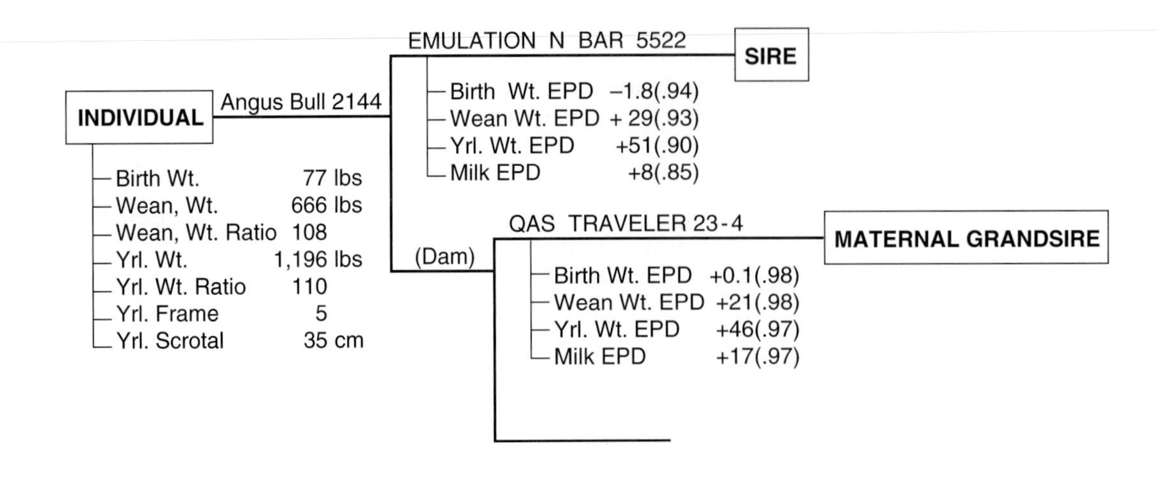

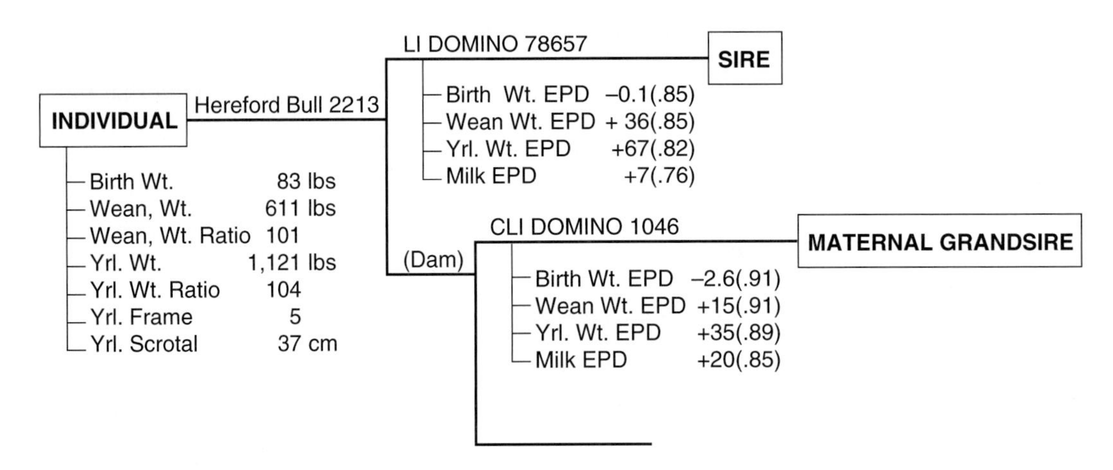

FIGURE 4.7 Yearing bull selection with a balanced trait emphasis. Bulls can be used naturally on replacement heifers.
Source: Colorado State University.

comparison. Some breeds use BIF adjustments in their record analysis while other breeds have their own specific adjustment factors. The birth weights of calves from 2-year-old heifers of different breeds are 4–8 lbs lighter than the birth weights of mature cows (5–10 years of age). Bull calves are usually 5–8 lbs heavier than heifer calves. A sex adjustment of birth weight is usually not done because performance records are compared within sex of calf categories.

Birth weight evaluations are more critical when calving 2-year-old heifers. Actual birth weights reflect calving difficulty; however, adjusted birth weights and birth weight EPDs are better measures of genetic differences.

Calving Ease

Birth weight is the best measure of calving ease or calving difficulty (dystocia). Although calving ease scores can be recorded and utilized, they are more subjective than birth weight. For example, a heifer may be manually assisted to give birth earlier than needed, but with

TABLE 4.5 Bull Selection Criteria for Commercial Beef Producers and Breeders

Goals and Traits	Bull Selection Criteria for Maternal Traits[a]	Selection Criteria For Terminal Cross Bulls[c]
BREEDING PROGRAM GOALS Selecting for an **optimum** combination of **maternal, growth, and carcass traits** to **maximize profitability** (Avoiding genetic antagonisms and environmental conflicts that come with maximum production or single trait selection) Provide genetic input so **cows** can be **matched with their environment**—cows that wean more lifetime pounds of calf without overtaxing the forage, labor, or financial resources Stack pedigrees for maternal traits MATERNAL TRAITS **Early puberty/high conception:** Calve by 24 months of age **Calving ease:** Moderate birth weights (65–80 lb hfrs; 75–90 lb cows). Calf shape (head, shoulders, hips) which relates to unassisted births **Early rebreeding and longevity:** ("fleshing ability"): "5" body condition score (BCS) at calving without high cost feeding **Mature weight:** Medium sized cows (1,000–1,250 lbs; BCS 5) **Milk production:** Moderate (wean 500–550 lb calves under average feed supply) GROWTH AND CARCASS TRAITS—EARLY GROWTH AND COMPOSITION Rapid gains—relatively heavy weaning and yearling weights within medium (4–6) frame size. Yield grade 2, grading $\geq$ 70% Choice (steers harvested at 1,200 lbs) FUNCTIONAL TRAITS (LONGEVITY) Udders (shape, teats, pigment) Eye pigment Disposition Structural soundness	**Birth weight EPD:** Preferably under +1.0 lb. Evaluate calving ease of daughters. **Scrotal circumference:** 32–40 cm at 365 days of age. EPD above +0.3 cm. Passed breeding soundness exam **Milk EPD:** under +15[b] **Maternal EPD:** 20–40 (prefer +25) **Weaning weight EPD:** $\geq$ + 20 lbs **Yearling frame size:** 4.0–6.0 (smaller frame size will adapt better to harsher environments—e.g., less feed, more severe weather, less intensive care) **Mature weight:** Under 2,000 lbs with BCS of 5 (preference for future). Currently bulls under 2,500 lbs will have to be considered **Body condition:** Backfat of approx. 0.20 inches at yearling weights of 1,100–1,250 lbs. Monitor reproduction of daughter's BCS **Yearling weight EPD:** $\geq$ + 40 lbs **Accuracies:** All EPDs of 0.90 and higher (older, progeny tested sires) for AI bulls. Young bulls without EPDs or with accuracies below 0.90—evaluate trait ratios and pedigree EPDs. Select sons of bulls that meet EPDs listed above **Growth and carcass traits as noted in goals** **Functional traits:** Visual evaluation of bull and his daughters **Visual:** Functional traits as noted in goals sufficiently attractive to sire calves, which would not be economically discriminated against in the marketplace. Preference for "adequate middle" as medium frame size cattle need middle for feed capacity	**Birth weight EPD:** Preferably no higher than +5.0; want calf birth weights in 80–100 lb range (calves only from cows) **Scrotal circumference:** 32–40 cm at 365 days of age. Passed breeding soundness exam **Milk EPD:** Not considered **Maternal EPD:** Not considered **Weaning weight EPD:** +40 lbs or higher **Yearling weight EPD:** +60 lbs or higher **Yearling frame size:** 6.0–8.0 (should be evaluated with frame size of cows so slaughter progeny will average 5.0–6.0) **Mature weight:** No upper limit as long as birth weight, frame size, and carcass weights are kept in desired range **Body condition:** Evaluate with body condition of cows so that carcass fat of 0.4 in. progeny at 1,100–1,300 lb slaughter weights **Accuracies:** All EPDs of 0.90 and higher (older, progeny tested sires) for AI bulls. Young bulls without EPDs or with accuracies below 0.90—evaluate trait ratios and pedigree EPDs. Select sons of bulls that meet EPDs listed above **Growth and carcass traits as noted in goals** **Visual:** Functional traits as noted in goals sufficiently attractive to sire calves that would not be economically discriminated against in the marketplace

[a]EPDs are Angus based.
[b]Assumes extensive management under a grazed forage scenario.
[c]EPDs are for Continental breeds; assumes no breeding replacement female will be kept from these bulls.

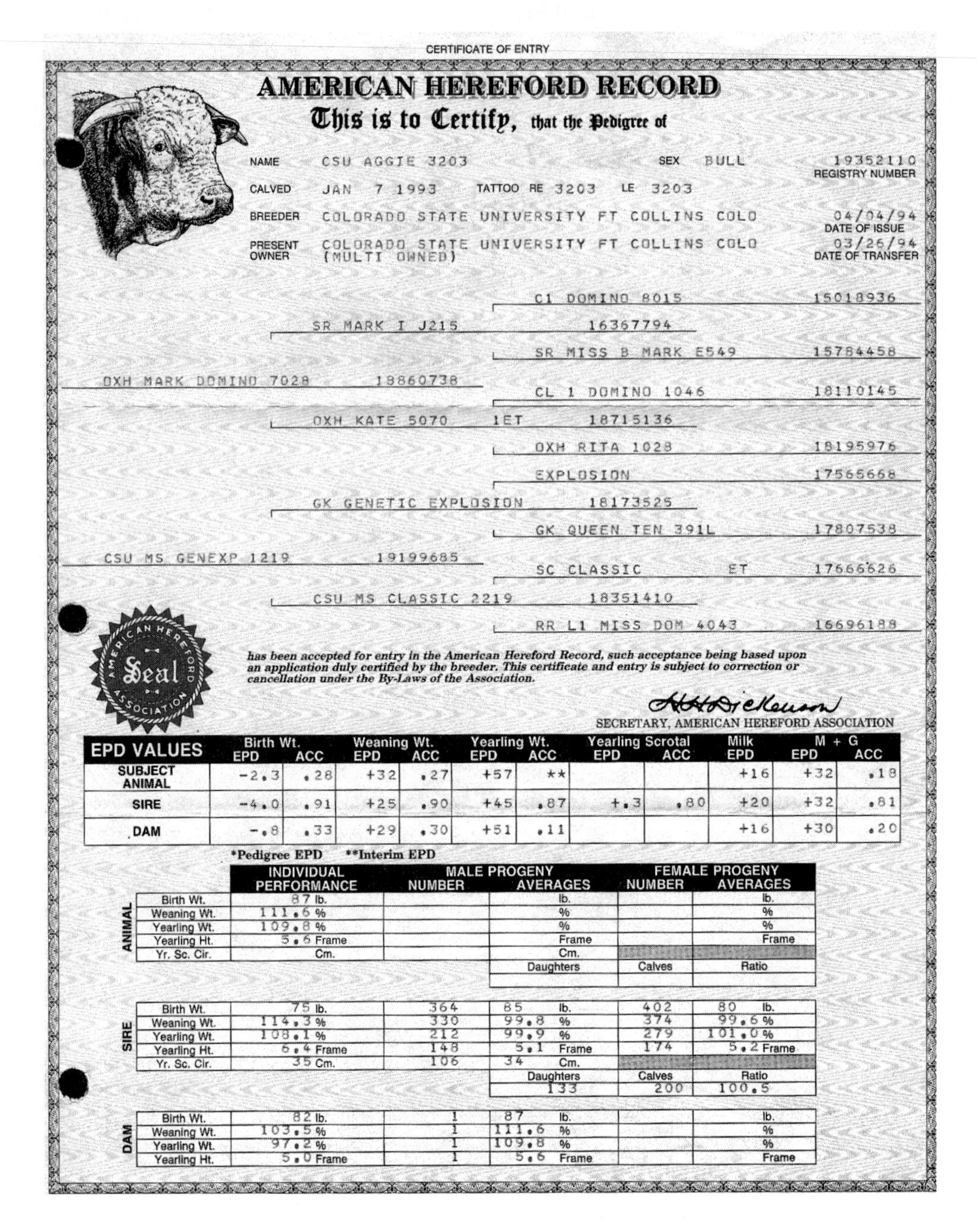

CERTIFICATE OF ENTRY

AMERICAN HEREFORD RECORD

This is to Certify, that the Pedigree of

NAME CSU AGGIE 3203 SEX BULL 19352110 REGISTRY NUMBER

CALVED JAN 7 1993 TATTOO RE 3203 LE 3203

BREEDER COLORADO STATE UNIVERSITY FT COLLINS COLO 04/04/94 DATE OF ISSUE

PRESENT OWNER COLORADO STATE UNIVERSITY FT COLLINS COLO (MULTI OWNED) 03/26/94 DATE OF TRANSFER

C1 DOMINO 8015 15018936
SR MARK I J215 16367794
SR MISS B MARK E549 15784458
OXH MARK DOMINO 7028 18860738
CL 1 DOMINO 1046 18110145
OXH KATE 5070 1ET 18715136
OXH RITA 1028 18195976
EXPLOSION 17565668
GK GENETIC EXPLOSION 18173525
GK QUEEN TEN 391L 17807538
CSU MS GENEXP 1219 19199685
SC CLASSIC ET 17666626
CSU MS CLASSIC 2219 18351410
RR L1 MISS DOM 4043 16696188

has been accepted for entry in the American Hereford Record, such acceptance being based upon an application duly certified by the breeder. This certificate and entry is subject to correction or cancellation under the By-Laws of the Association.

SECRETARY, AMERICAN HEREFORD ASSOCIATION

EPD VALUES	Birth Wt. EPD	Birth Wt. ACC	Weaning Wt. EPD	Weaning Wt. ACC	Yearling Wt. EPD	Yearling Wt. ACC	Yearling Scrotal EPD	Yearling Scrotal ACC	Milk EPD	M + G EPD	M + G ACC
SUBJECT ANIMAL	-2.3	.28	+32	.27	+57	**			+16	+32	.18
SIRE	-4.0	.91	+25	.90	+45	.87	+.3	.80	+20	+32	.81
DAM	-.8	.33	+29	.30	+51	.11			+16	+30	.20

*Pedigree EPD **Interim EPD

ANIMAL	INDIVIDUAL PERFORMANCE	MALE PROGENY NUMBER	MALE PROGENY AVERAGES	FEMALE PROGENY NUMBER	FEMALE PROGENY AVERAGES
Birth Wt.	87 lb.		lb.		lb.
Weaning Wt.	111.6 %		%		%
Yearling Wt.	109.8 %		%		%
Yearling Ht.	5.6 Frame		Frame		Frame
Yr. Sc. Cir.	Cm.		Cm.		
			Daughters	Calves	Ratio

SIRE	INDIVIDUAL PERFORMANCE	MALE PROGENY NUMBER	MALE PROGENY AVERAGES	FEMALE PROGENY NUMBER	FEMALE PROGENY AVERAGES
Birth Wt.	75 lb.	364	85 lb.	402	80 lb.
Weaning Wt.	114.3 %	330	99.8 %	374	99.6 %
Yearling Wt.	108.1 %	212	99.9 %	279	101.0 %
Yearling Ht.	6.4 Frame	148	5.1 Frame	174	5.2 Frame
Yr. Sc. Cir.	35 Cm.	106	34 Cm.		
			Daughters	Calves	Ratio
			133	200	100.5

DAM	INDIVIDUAL PERFORMANCE	MALE PROGENY NUMBER	MALE PROGENY AVERAGES	FEMALE PROGENY NUMBER	FEMALE PROGENY AVERAGES
Birth Wt.	82 lb.	1	87 lb.		lb.
Weaning Wt.	103.5 %	1	111.6 %		%
Yearling Wt.	97.2 %	1	109.8 %		%
Yearling Ht.	5.0 Frame	1	5.6 Frame		Frame

FIGURE 4.8 Performance pedigree of a Hereford bull.
Source: American Hereford Association.

FIGURE 4.9 An Angus bull (QAS Traveler 23-4) that has an excellent combination of maternal, growth, and carcass traits as noted in the following EPDs. His accuracies (ACC) are high because he has produced more than 10,000 calves with birth, weaning, and yearling weight records in more than 1,000 herds. He has sired more than 2,400 daughters that have weaned calves in more than 500 herds.

	EPD	ACC		EPD	ACC
Birth weight	+0.4	0.99	Mature daughter weight	−3	0.97
Weaning weight	+22	0.99	Carcass weight	+1	0.97
Milk	+19	0.99	Marbling	+0.15	0.98
Yearling weight	+49	0.99	Ribeye area	−0.01	0.97

Source: American Breeders Service and the American Angus Association.

TABLE 4.6 Age of Dam Adjustment Factors for Birth Weight and Weaning Weight

Age of Dam (years)	Birth Weight[a] (lbs)	Weaning Weight (lbs)	
		Male	Female
2	8	60	54
3	5	40	36
4	2	20	18
5–10	0	0	0
11 and older	3	20	18

[a]Standard birth weight is male (75 lbs) and female (70 lbs). Standard birth weights for male calves are as high as 90–92 lbs for some breeds.

Source: Beef Improvement Federation (BIF) Guidelines for Uniform Beef Improvement Programs.

TABLE 4.7 Calving Ease Scores

Score	Description
1	No difficulty; no assistance
2	Minor difficulty; some assistance
3	Major difficulty; usually mechanical assistance
4	Caesarean section or other surgery
5	Abnormal presentation

Source: BIF Guidelines for Uniform Beef Improvement Programs.

additional time she could have calved without assistance. Furthermore, an easy pull to one producer may be a difficult pull to someone else. The recommended calving ease scores are shown in Table 4.7.

Weaning Weight

Preweaning growth is best measured as the adjusted 205-day weight. Calves should be weighed as near to 205 days of age as possible, with 160–250 days being the recommended acceptable range. Calves weighed outside this range will result in information that cannot be meaningfully compared to other performance data without risk of error.

Weaning weights are adjusted for age of calf and age of dam by using the following formula:

$$\text{Adjusted 205-day wt (lbs)} = \frac{\text{Actual weaning weight} - \text{birth weight}}{\text{Age in days}} \times 205 + \text{birth weight} + \text{age of dam adjustment}$$

Weaning weights need to be adjusted for age of dam due to the variation in milk production by different aged females. For example, a 7-year-old cow typically produces more milk than a 2-year-old. The appropriate number of pounds must be added to the 205-day weight, previously adjusted for age of calf (see Table 4.6). For example, if the 205-day weight of a bull calf from a 2-year-old heifer is 465 lbs, then 60 lbs would be added (for a total of 525 lbs) to adjust to the 5- to 10-year-old cow basis. Many breeds have their own weaning weight adjustments that differ from the BIF adjustment.

Weaning weight ratio is then computed by dividing each calf's adjusted 205-day weaning weight by the average of its contemporary group. The ratio is expressed as a percentage of the contemporary group. For example, a heifer calf with an adjusted weaning weight of 500 lbs, where the contemporary group average was 550 lbs, would have a weaning weight ratio of:

$$\frac{500}{550} \times 100 = 90.9$$

Thus, the example calf is 9.1% (100 − 90.9) lighter than the average of her contemporary group.

Yearling Weight

After obtaining data on postweaning gain over a 100–150-day feeding period, the adjusted yearling weight can be calculated as follows:

$$\text{Adjusted 365-day wt} = 160 \times \frac{\text{Actual final weight} - \text{actual weaning weight}}{\text{Number of days between weights}} + \text{Adjusted 205-day wt}$$

Yearling weight ratios are computed separately for each contemporary group by dividing the individual animal's record by the contemporary group average and multiplying by 100.

Mature Cow Weight

Mature cow weight is an economically important trait because it is part of a biological-type description. It is related to reproductive performance, when cows are matched to their economical feed environment, and to the carcass weight of a cow's progeny.

Mature cow weights, however, are not recorded by most breeders. In 1988, the American Angus Association began to encourage breeders to record mature weights, so in its *Sire Evaluation Report* mature cow weight EPDs are reported. Other breeds have added or are considering inclusion of mature cow weight in their sire evaluations. Table 4.8 shows the effects of age on mature cow weight as well as the adjustments used for different body condition scores for the Angus breed.

Milk

Milk EPDs are used to compare genetic differences in weaning weight due to genes for milk production passed from parents to daughters. In essence, milk EPDs help breeders make comparisons between potential parents for their ability to contribute to weaning weight as a result of lactational differences. These estimates also help breeders match the desired levels of milk production to the available forage resources in both their own and their customer's production environment.

Total maternal EPDs combine the effects of weaning weight (direct) EPD and milk EPD to allow comparisons of weaning weight differences of calves from daughters. Total maternal EPDs are calculated by adding milk EPD to one-half of the weaning EPD.

Carcass Traits

The traits most commonly of interest are carcass weight, ribeye area, marbling or intramuscular fat score, percent retail product, backfat, and tenderness. Carcass weight and tenderness (estimated by Warner-Bratzler shear force) are measured directly on progeny. Ribeye area, marbling score, percent retail product, and backfat can be measured via direct progeny test-

TABLE 4.8 Mature Angus Cow Weights, by Cow Age and Body Condition Score Adjustments

Cow Age (years)	Mature Cow Weights (lbs)	Body Condition Score (BCS)	Weight Adjustment (lbs) for BCS
2	1,085	2	+256
3	1,128	3	+201
4	1,232	4	+152
5	1,299	5	+86
6	1,332	6	0
7	1,340	7	−88
8	1,340	8	−190
9	1,326	–	–
10	1,319	–	–
11	1,294	–	–

Source: American Angus Association, *Sire Evaluation Report,* Spring 2001.

ing or estimated via ultrasound measures on live animals. Furthermore, the American Gelbvieh Association released a grid merit EPD in 2001 to allow for an economic comparison between sires for their progeny's ability to perform under the Gelbvieh Alliance grid pricing system.

Other Traits

Scrotal circumference, stayability, and docility are other traits included in some sire summaries. Scrotal circumference EPDs are utilized to ascertain differences in age of puberty. The desired level of performance in this trait will vary based on the breeds being used and their intended purpose.

Stayability is a measure of longevity that assesses the likelihood of a female remaining in the herd to at least six years of age. EPDs for this trait are expressed in percentages to allow comparisons of the percentage of a sire's daughters expressing longevity.

Docility is an important trait in some breeds where concerns have been expressed about the temperament of certain sire groups. Docility EPDs are calculated from subjective scoring of temperament as a 1 to 6 scale with 1 being docile and 6 being very excitable.

Locating Genetically Superior Bulls

Commercial and seedstock producers should identify breeders who keep honest, comparative records on their cattle. Most performance-minded seedstock producers have useful weaning, yearling, and birth weight records on their bulls. Increasingly, seedstock producers obtain feedlot and carcass data on progeny from their own bulls or use bulls that have carcass EPDs. Seedstock breeders should provide meaningful performance data on their bulls and commercial producers should demand information that impacts the profitability formula. Excellent performance records can be obtained and made available from the farm or ranch. Ratios should be evaluated. However, they are meaningful only when the bulls have been treated alike in a similar environment.

The best approach for selecting outstanding bulls involves evaluating the records first and then selecting several bulls whose records reflect the desired level of performance. From that group, bulls can be selected that meet the buyer's goals for skeletal size, freedom from predisposition to fat, scrotal size, and other soundness traits.

Bull testing stations provide a place where bulls from several sources can be compared. Comparisons are valid for rate of gain, skeletal size, and scrotal circumference but not for birth weights, weaning weights, and most maternal traits. Weight-per-day-of-age comparisons have some validity; however, caution should be exercised because too much emphasis can be placed on growth traits when maternal traits should be emphasized. Some testing stations test maternal bulls (e.g., calving ease based on birth weights) in a separate contemporary group.

Full-Service Genetic Providers

Because of the dramatic impact from sire selection, it is important for cow-calf producers to carefully select not only their sires but also the breeders from whom they purchase genetics. The interface between seedstock and cow-calf producers is of critical importance. The communication of goals, needs, and the ability to meet those needs is largely responsible for any progress made by individual producers to become more competitive and for the industry to regain lost market share. A generally accepted management theorem is that 70% of customer loss can be attributed to the lack of a personal relationship with the supplier.

An analogy that describes the importance of this relationship is provided in the following statement from Bob Reed of the Llano Estacano Vineyard and Winery, "Once the grapes hit the dock, it's too late for us to correct the fruit. You can make good or bad wine from good grapes, but you cannot make good wine from bad grapes. So, we try to help our farmers grow premium grapes for our wine."

Given this philosophy, the seedstock sector is changing from being a supplier of bulls to becoming a supplier of value-added genetics supported by a full array of information and services. Some of the services being provided include:

1. Information packaged as a decision-support tool.
2. Providing market outlets for calves from customers' herds.
3. Facilitating the transfer of knowledge via customer seminars, newsletters, and discussion groups.
4. Offering more than one breed or composite to customers.
5. Assisting customers with the development of quality assurance programs.
6. Facilitating involvement in alliances and branded product initiatives.
7. Facilitating the collection, summarization, and transfer of useful information within the beef industry.

The personal characteristics of a customer-oriented seedstock supplier include (1) a desire to actively listen to customers and the willingness to seek frequent opportunities for interactions with past and potential clients, (2) makes on-site visits to customers during key times when the performance of the genetics is on the line (calving, pregnancy checking, weaning), (3) offers customers significant input into the sire selection process for the seedstock herd, and (4) has a passion for the quality of the product and the long-term well-being of the client.

Bulls Versus Steers

Seedstock and commercial producers should be aware of how sex differences affect the productivity and performance of economically important traits. Sexual maturity, as reflected in bulls, steers, and heifers, has a marked effect on growth and carcass characteristics (Chapter 17). Most feedlot and carcass goals are identified with steers since most cattle going to slaughter are steers.

Table 4.9 compares the expected productivity differences between steers and bulls. The data in the table demonstrate that selected bulls need to be higher in performance than their

TABLE 4.9 Comparative Average Productivity of Bulls and Steers[a]

Trait	Bulls	Steers
205-day weight (lbs)	525	500
Feedlot gain (lbs/day)	3.25	3.00
365-day weight (lbs)	1,050	980
Yearling hip height (in)	51	49
Feed/lb gain	6.8	7.8
Backfat (in)	0.2	0.4
Ribeye area (sq in)	13.0	11.0
Cutability (%)	51%	48%
Quality grade	Select	Choice

[a]The growth data are compared at same age, while carcass data are compared at same slaughter weight.
Source: Adapted from several different sources.

counterpart steers. Producers that select bulls at the same level of productivity expected in steers will generally be disappointed in the siring ability of the bulls.

What Is a Bull Worth?

This question is typically posed by producers evaluating bulls with high performance records. The base price for a bull is his salvage or slaughter value. Any value of the bull, above salvage price, is based on the bull's ability to sire live calves and the genetic superiority of his calves.

The value of high-performing versus lower-performing bulls should be evaluated. Some commercial producers use the guideline that the cost of a new bull should be approximately the value of three weaning steers or two yearling steers. From this example, the value of commercial bulls can be in the $1,500–$2,500 range. Table 4.10 illustrates the price trends for Hereford and Limousin bulls in 2000. A more detailed evaluation of pricing commercial bulls is given in Chapter 5 where bull cost per cow or calf and "using the right genetics" are presented. Seedstock production of bulls for an alliance or a commercial cow-calf operation raising their own bulls can produce yearling bulls at approximately $1,000 per head breakeven.

Seedstock breeders who market most of their bulls to other seedstock breeders can justify prices above $10,000, especially if they are selling semen and AI certificates. Some of these breeders will sell two units of semen and an AI certificate for $25 or more. Some bulls have generated more than $1 million in semen and certificate sales. It is more difficult to justify extremely high prices for individual females (Table 4.11).

Seedstock breeders can maximize the progress they make by utilizing reproductive technologies such as artificial insemination, embryo transfer, or the use of sexed semen. By virtue of these procedures, a breeder can access the very best genetics from a national, if not an international, gene pool. These techniques also make it possible for selection of proven parents with high associated accuracies.

SELECTING REPLACEMENT HEIFERS

Heifers, as replacement breeding females, can be selected for several traits at different stages of their productive life. The objective is to identify heifers that will conceive early in the breeding season, calve easily, give a flow of milk consistent with the most economical feed supply, wean a relatively heavy calf, and make a desirable genetic contribution to the calf's postweaning growth and carcass merit while remaining free of structural defects to allow a high degree of longevity.

TABLE 4.10 Average and Top Lot Prices for Hereford and Limousin Breeding Cattle, 2000

	Hereford		Limousin	
	N	$/hd	N	$/hd
Bulls				
Average	5,349	2,263	3,715	2,131
Top Lot	1	100,000	1	47,000
Females				
Average	5,580	1,667	5,613	1,803
Top Lot	NA	NA	1	71,000

Source: American Hereford Association and North American Limousin Foundation, 2001.

Beef producers have found it challenging to determine which of the young heifers will make the most productive cows. When a cowherd begins to reach an optimum level of performance, then producers should cull heifers at weaning and yearling ages that are in the low and high categories for the various traits. Heifers in the middle one-half of the group are then kept to evaluate their pregnancy status after breeding. Such an approach helps to hold performance of the cowherd at or near the current level.

Table 4.11 shows the process that some producers use to select the most productive replacement heifers. This selection process assumes that more heifers will be retained at each stage of production than the actual number of cows to be replaced in the herd. The number of replacement heifers that producers keep is based primarily on cost of production and market value at the various stages of production. More heifers than the number needed should be kept through pregnancy check time. Shortly after the end of the breeding season, a final selection of heifers is based on early pregnancy.

TABLE 4.11 Replacement Heifer Selection Guidelines at Different Productive Stages

Stage of Heifer's Productive Life	Emphasis on Productive Trait	
	Primary	Secondary
Weaning (7–10 months of age)	Cull only the heifers whose actual weight is too light to prevent them from showing estrus by 14 mos of age. Also consider the economics of the weight gains needed to have puberty expressed. Also cull heifers that are too large in frame and birth weight	Weaning weight ratio Weaning EPD Milk EPD Predispostion to fatness Adequate skeletal frame Skeletal soundness Cull for bad disposition Dam's performance summary
Yearling (12–15 months of age)	Cull heifers that have not reached the desired target breeding weight (e.g., minimum of 675–750 lbs for medium-sized breeds or cross; minimum of 750–850 lbs for large-sized breeds and crosses). Also, cull heifers that are extreme in size and weight	Weaning weight EPD Milk EPD Yearling weight ratio Yearling EPD Predisposition to fatness Adequate skeletal frame Skeletal soundness
After breeding (19–21 months of age)	Cull heifers that are not pregnant and those that will calve in the latter one-third of the calving season	Weaning weight EPD Milk EPD Yearling EPD Predisposition to fatness Adequate skeletal frame Skeletal soundness
After weaning first calf (31–34 months of age)	Cull to the number of first-calf heifers actually needed in the cowherd based on early pregnancy and weaning weight performance of the first calf. Preferably all the calves from these heifers have been sired by the same bull	

SELECTING COWS

Unless a producer is involved in a herd size reduction program, relatively few cows can be culled to enhance genetic progress in the herd. Most cows basically cull themselves because of poor reproductive performance, death loss, and age. Producers can practice some culling on performance records if the death loss is low, if cows over 9–10 years of age can sustain high levels of productivity for a few more years, or if a higher percentage of heifers are kept as replacements. The guideline, "select heifers on EPDs and cull cows on actual performance," is a good approach to follow.

Cows can be culled on pregnancy status and calving interval. This basis for culling appears to be more economic than selection for genetic progress. Open cows can rebreed and produce calves regularly during the remainder of their productive life. The economic decision to cull open cows should be based more on replacement cost and the time value of money. It may appear obvious that the commercial producer should more frequently cull the open females; however, the decision is not as obvious or easy for a seedstock breeder to cull open females with desirable genetic merit. This decision is controversial in the minds of many breeders. The decision, however, should be made on the basis of known biological and economic relationships and not on opinions. Complicating the decision is the use of AI. Failures in heat detection, semen thawing, and insemination technique can result in lowered rates of fertility that should not be attributed to the female.

In most herds, more than 50% of culling is due to unsatisfactory reproductive performance. Cow longevity and profitability, for both seedstock and commercial producers, can be increased by matching the biological type of cow (mature weight, level of milk production, and body condition score) to the most economical feed conditions.

Culling cows to assist in making genetic change in a herd is relatively simple if the cows performed together under similar environmental conditions. One of the biggest problems in comparing cow records is that the cows likely have produced calves from different bulls. Relatively large genetic differences in the bulls used in the herd would make the cows' records less comparable than when all cows are bred to the same bull. This is usually not a serious problem when producers are culling the bottom 10–20% of their cows on performance records. Table 4.12 compares the performance records of high- and low-producing cows of the same herd. The lower weight ratios identify the low-producing cows.

Caution should be exercised in culling pregnant cows that are below herd average in performance. Heifers that replace these cows are usually high cost inputs and the weaning weights of their calves can be less than the calves from culled cows. The latter is due to age-of-dam effect on weaning weight. Some cow-calf producers can market their less-productive cows to other producers who have better feed resources. These cows can be sold above slaughter price and still be profitable for the new owners. These cows can then be economically replaced with heifers that will improve profitable herd performance.

TABLE 4.12 Performance Data on High- and Low-Producing Cows in the Same Herd

Cow	No. Calves	Weaning Wt (lbs)	Weaning Wt Ratio	Yearling Wt Ratio
1	9	583	112	108
2	9[a]	464	89	94

[a]One calf died before weaning (not computed in averages).

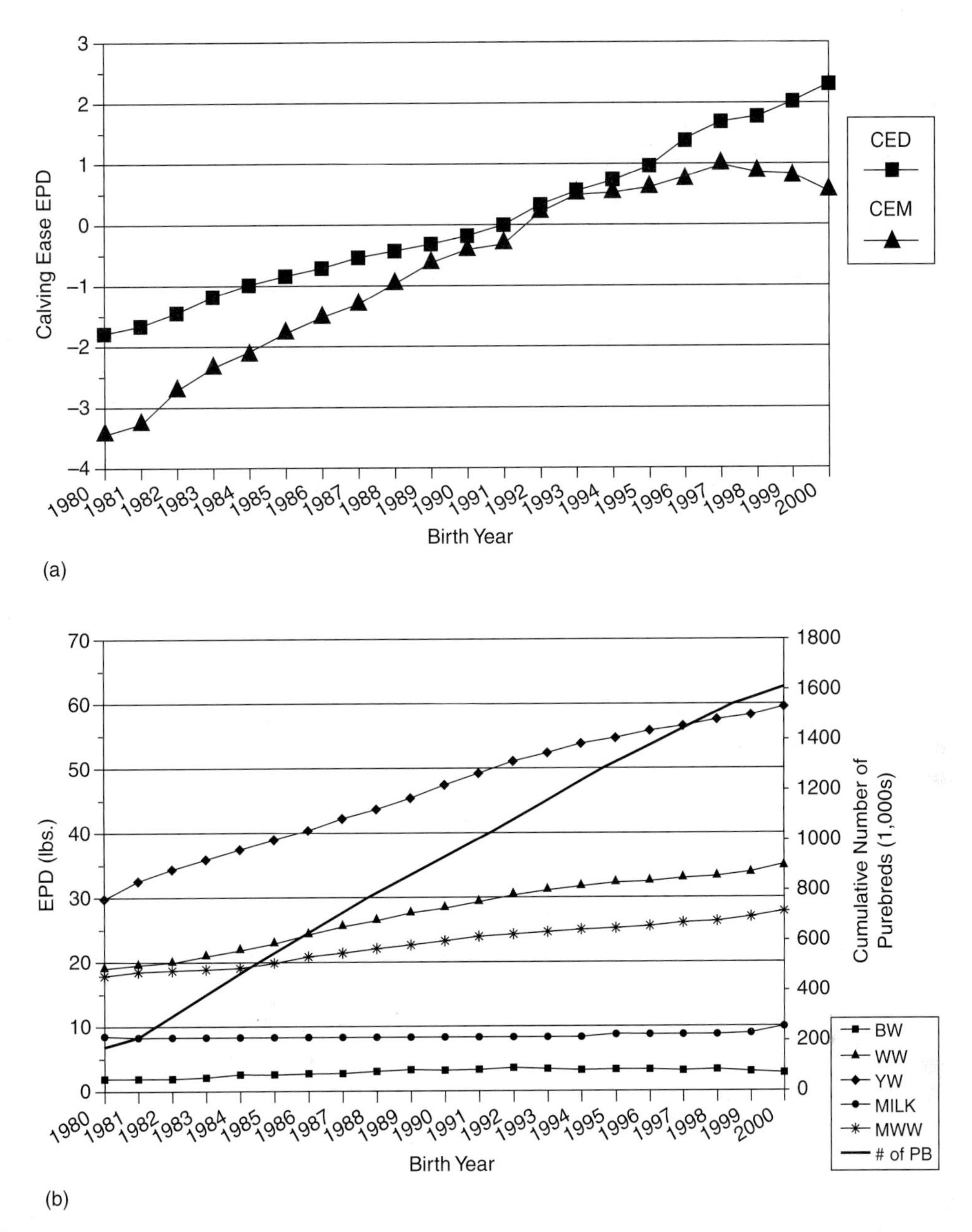

FIGURE 4.10 Genetic trends in Simmental cattle for (a) calving ease and (b) growth and maternal traits. *Source:* American Simmental Association.

The Genetic Trend

The genetic trend for a seedstock herd or breed can be monitored by evaluating the change in EPDs over the years. Figure 4.10 shows the genetic trend for several traits in the Simmental breed. Comparing a herd trend to the breed trend is a useful benchmark. Table 4.13 shows the genetic trend for a Hereford herd and for the Hereford breed. In the Hereford herd example, the objective was to emphasize maternal traits (e.g., keeping birth weight EPD below 0, while increasing weaning weight EPD and yearling weight EPD). The breeder wants to keep the

TABLE 4.13 The Genetic Trend in Progeny EPDs for a Hereford Herd Compared to Breed Trend[a]

Year	Birth Wt	Weaning Wt	Milk	Yearling Wt	Scrotal Circumference
1980	+2.1 (+0.3)	+21 (+4)	+6 (+1)	+29 (+7)	+0.2 (+0)
1983	+1.9 (+1)	+23 (+10)	+7 (+2)	+36 (+17)	+0.5 (+0.1)
1987	+1.4 (+1.9)	+28 (+17)	+11 (+3)	+47 (+28)	+0.5 (+0.1)
1990	−0.9 (+2.8)	+24 (+22)	+12 (+4)	+43 (+37)	+0.9 (+0.2)
1995	−0.4 (+3.6)	+24 (+29)	+16 (+8)	+47 (+48)	+0.7 (+0.4)
1998	−0.7 (+3.7)	+28 (+32)	+19 (+10)	+56 (+53)	+0.8 (+0.4)
2000	−0.2 (+3.9)	+30 (+34)	+12 (+11)	+57 (+58)	+1.0 (+0.5)

[a]Breed trend in parentheses.
Source: Colorado State University.

milk EPD at about +10 lbs. In 1995, this herd was approaching its production targets of 70–80 lb birth weights for calves from heifers, 80–90-lb birth weights for calves from cows, 500–550 lb weaning weights, yearling frame score of 5, and 1,100-lb yearling weights for bulls having 34–38 cm scrotal circumference. Since 1995, the goal of the herd has been to stabilize performance in these ranges.

MARKETING DECISIONS

The marketable products for seedstock breeders include breeding animals, semen, and embryos. These products can have a commercial value (e.g., the slaughter price or the value of semen or breeding stock to commercial cow-calf producers), or they can be valued at several times the commercial base when used by other seedstock breeders. In either case, the true value of the products should be determined—that is, how they contribute to the animal and product specifications (Table 3.5). There is also value associated with the service provided to customers (e.g., convenience; confidence that the products will make economic contributions to the customer's products).

Marketing involves sales, advertising and promotion, pricing, service, and other factors that fulfill the customer's needs. The breeder must know the value and cost of the products to be sold, the available customers, and the marketing plan for communicating the value of the products to customers. In addition, the cost-effectiveness of the various advertising and promotion programs should be known.

The Marketing Plan

A written marketing plan should include goals, objectives, and strategies. It should also identify customers, assess the competition, point out the uniqueness of the product, and evaluate resources, including production costs and marketing budgets. Breeders should produce what they plan to market, and then market the products profitably. The marketing plan should also be goal-driven. Goals might include identifying the number of customer contacts to be made either by phone or in person, reassessing the advertising copy and budget, assisting customers in the collection and use of feedlot and carcass data, or developing a new direct mail instrument.

Breeders marketing bulls to commercial cow-calf producers using bulls for natural service know that the geographical area for their customers is typically within a 100–200-mile radius of their operation. Other breeders may sell semen to commercial producers or semen and breeding cattle to other breeders on a national or international scale. Thus, the merchandising area can be broad or restricted depending on the customers being targeted by the breeder.

An essential part of the marketing plan is to know production costs. Enterprise budgets should be kept so that breakeven prices can be calculated for weaned calves, yearling bulls and heifers, and semen (if it is produced for sale). The marketing plan should be developed based on a realistic assessment of the strengths and weaknesses of the genetic product being offered for sale. The following questions should be asked to accomplish this assessment:

1. Who are my customers?
2. What do my customers want and what constraints do they face?
3. How well does my program match the goals of my customers?
4. Upon what basis do my customers assess cattle (performance records, visual appraisal, etc.)?
5. Who are my competitors and how does my product/service compare with theirs?
6. Who are potential partners among my competitors with whom I might form an alliance?
7. How much are my cattle worth and how should they be valued?

Marketing Alternatives

The three primary marketing alternatives available to breeders are *private treaty sales, consignment sales,* and *production sales.* A breeder could use one or all of these marketing options, though the decision should be based on anticipated net returns and the method that will best fit the needs and personality of the breeder.

Private treaty sales involve selling cattle on the farm or ranch to one buyer at a time, whereas consignment sales involve cattle from several breeders' herds. In production sales, breeders host their own sale. Cattle in consignment and production sales are usually sold at auction (Fig. 4.11). Table 4.14 shows the advantages and disadvantages of each marketing alternative.

FIGURE 4.11 The selection decisions made by cattle breeders at bull sales provide the foundation of genetic merit for commercial cow-calf herds.
Source: Holly Foster.

TABLE 4.14 Advantages and Disadvantages of Private Treaty Sales, Consignment Sales, and Production Sales

Type of Sale	Advantages	Disadvantages
Private treaty	Less risk because breeder sets the price Buyers see total program Lower sale costs	Cattle could be overvalued or undervalued; difficult to set price Breeder must be an effective communicator and salesperson Usually only one buyer at a time More advertising intensive over time Determining order in which buyers get access
Consignment	Several potential customers come together Sale costs are divided among consignors Could increase private treaty sales Helps establish value of private treaty cattle	Cattle compared to other breeder's cattle Sale management may not be professional Cattle must be well displayed to be competitive Consignor may not select the right cattle or plan far enough in advance More price risk and higher costs
Production	Opportunity to expand market area Sale arranged by professionals Buyers see total program Breeder controls sale arrangements Cattle not competing with those of other breeders Encourages competitive bidding	Need 40–50 lots to have a good sale and reduce sale costs Encouraged to sell inferior cattle May not attract enough buyers to meet expected sale average More price risk and higher costs (may be as high as 20% of the sale gross)

Source: Adapted from various sources.

Another marketing alternative for seedstock breeders is to participate in alliances. The alliance can be a simple agreement between two or more breeders or a more involved alliance where genetics are provided and integrated into a total industry plan.

One example of a cooperative effort between two seedstock breeders is combining their two production sales into one sale and in having one advertising program instead of two. This effort reduces costs and provides buyers with larger numbers of genetics products.

In an industry alliance involving several production and processing segments, a seedstock breeder understands the needed genetic inputs. Genetics are supplied by contracts that meet biological and economic specifications. Costs are reduced because there are minimum advertising and sale costs. The genetic inputs are evaluated in the entire system with information feedback returning to the breeder. The breeders then adjust their breeding programs based on this input to improve future genetic inputs.

Advertising and Promotion

Promotion is the act of furthering the acceptance and sale of merchandise through advertising and publicity. Promotion is not solely advertising, for breeders can promote their products without spending large amounts of money. Promotion communicates to customers where breeders are located and what they have to sell.

Some forms of cost-effective herd promotion include the following:

1. A farm or ranch sign; a sign on a truck or trailer.
2. A neat, well-kept operation that reflects attention to management.
3. Brochures that are handed out or mailed.

TABLE 4.15 Advertising Checklist

OVERVIEW
Be bold and aggressive
Advertise frequently
Be patient—results usually do not occur overnight
Start planning early
Set aside planning time
Map target area
Budget
Draw up your calendar
Choose your media
Plan your positioning (go for top-of-mind awareness)
BASIC INGREDIENTS OF AN AD
Attention-grabbing headline that offers a benefit
Dominant illustration
Persuasive body copy that is believable, concise, consistent
Consistent logo/signature
Distinct look
A SUCCESSFUL AD WILL . . .
Stop the reader and arouse interest
Create desire for the product
Encourage a decision
Ask for the order and sell cattle

Source: Polled Hereford World, June–July, 1992.

4. Sponsoring youth activities; breed association sponsored activities (e.g., field days, animal evaluation clinics, tours, and picnics).
5. One-on-one communication by phone or visits.

Advertising typically involves placing ads in the media (publications, radio, personal videotapes) and direct mail. The latter can be a cost-effective and profitable approach when the mailing lists are complete and updated regularly.

Ads can be placed in local newspapers and in regional or national publications. Table 4.15 shows an advertising checklist.

Marketing Effectiveness

The effectiveness of a marketing plan is best measured by profitability and customer satisfaction. Profitability should be achieved for both the buyer and seller. Breeders must sell higher than their breakeven price to realize profits. Commercial producers will realize profits when breeders' products help meet production and market targets that include matching biological types of cows to their most cost-effective feed environment.

In the end, customers want more than just a bull. They are seeking seedstock backed by excellent genetic performance data, economic information, and the sense of confidence that comes from dealing with breeders who guarantee their products.

MANAGEMENT SYSTEMS HIGHLIGHTS

1. Seedstock breeders are the genetic engineers for the beef industry.
2. Breeders should know how their genetic products integrate and contribute to the costs, profits, and production of highly palatable consumer products.

3. Breeders should have records to calculate breakeven prices for weaned calves, sale bulls/semen, and heifer enterprises.
4. Seedstock producers can reduce production costs for their own operation and the total industry by forming alliances with other breeders and with other beef industry segments.
5. Successful seedstock breeders should focus both on the process of generating superior genetics and offering services that add value to those genetics.

REFERENCES

Publications

Boggs, D. 1992. *Understanding and Using Sire Summaries.* Stillwater, OK: Beef Improvement Federation. BIF-FS3.

Bourdon, R. M. 1997. *Understanding Animal Breeding.* Upper Saddle River, NJ: Prentice-Hall.

Brinks, J. S., and Bourdon, R. M. 1989. Replacement Heifer Selection. *BEEF* 25: 28.

Bull and Heifer Replacement Workshops (Proc.). 1990. Fort Collins, CO: Colorado State University.

Cartwright, T. C. 1979. Size as a Component of Beef Production Efficiency: Cow-Calf Production. *J. Anim. Sci.* 48: 974.

Field, T. G. 1995. *Creating a Revolution to Enhance the Competitive Position of the Beef Industry.* Proceedings of the Beef Improvement Association Conference. Armidale, NSW, Australia.

Field, T. G. 1999. *Meanwhile Down on the Farm.* Proceedings of the Australian Hereford Society Cattlemen's Conference. Brisbane, Australia.

Gelbvieh Seedstock Marketing Handbook. 1996. American Gelbvieh Association, Westminister, CO.

Gibb, J., Boggess, M. V., and Wagner, W. 1992. *Understanding Performance Pedigrees.* Stillwater, OK: Beef Improvement Federation. BIF-FS2.

Guidelines for Uniform Beef Improvement Programs. Beef Improvement Federation, 1996. Ron Bolze, Northwest Research Extension Center, 105 Experiment Farm Road, Colby, KS 67701.

Legates, J. E. 1990. *Breeding and Improvement of Farm Animals.* New York: McGraw-Hill.

Performance Record Programs and National Sire Evaluation Summaries (available from most beef breed associations).

Silcox, R., and McGraw, R. 1992. *Commercial Beef Sire Selection.* Stillwater, OK: Beef Improvement Federation. BIF-FS9.

Welcome to the Marketplace. 1992. *Polled Hereford World,* June–July.

Visuals

"Like Begats Like" (1989; 8.5 min.). American Angus Association, 3201 Frederick Blvd., St. Joseph, MO 64501.

"Using EPDs in Sire Selection" (1988; 11 min.). American Simmental Association, 1 Simmental Way, Bozeman, MT 59715.

Computer Programs

Most major breed associations have programs where data in the sire summaries can be accessed. Breeders can easily identify bulls that meet their EPD and accuracy specifications. Some associations also have programs for planned mating options. Breeders enter the EPDs on a cow and have it matched against a sire the breeder has selected. The program gives the projected EPDs for the calves produced from the mating. Addresses and web sites of the breed associations are in the Appendix. Sire summaries for certain breeds can be accessed through their web site.

Commercial Cow-Calf Management Decisions

A task force committee of the National Cattlemen's Association sounded a warning in 1989 with the following statement:

> Low-cost producers (in all segments of the production chain) will survive in this system of competitive markets. Others [high-cost producers] will eventually be unable to compete and will exit the business.

Initially, this pointed summary statement fell on deaf ears as cattle prices were relatively good and many producers said they couldn't reduce costs as each year their costs were increasing. When cattle prices drop during the downside of the cattle cycle, producers are pressured to seriously evaluate their management options, including production costs. Table 5.1 illustrates that the percentage of profitable cow-calf producers can decrease rapidly during 3–4 years of transition in the beef cycle. The price drop of nearly 40 cents/lb for 500 lb calves is the reason for the dramatic change in profitability.

High-cost producers continue to exit the cattle business or deplete their assets. Other cow-calf producers have positioned or are positioning themselves as low-cost producers. An increased number of producers are finding that low-cost production is not only possible but exciting and essential in assuring survivability and improving profitability.

It is important to remember that profitable, low-cost producers are not successful only by virtue of their ability to manage cost of production. Instead, they have the ability to optimize productivity, the flexibility to take advantage of opportunities in the marketplace, and they are effective in both the planning and implementation processes of their business. Profitable cow-calf enterprises are the result of the application of creativity, hard work, and excellent risk management.

TABLE 5.1 Cow-Calf Producers' Profitability in Different Years of the Beef Cycle

	1993		1996		2000	
Cow-Calf Category	% of Producers	Ave. Price of 500-lb Calves	% of Producers	Ave. Price of 500-lb Calves	% of Producers	Ave. Price of 500-lb Calves
Profitable	72	$1.00/cwt	15	$62/cwt	90	$106/cwt
Near breakeven	22	$1.00/cwt	42	$62/cwt	2	$106/cwt
Nonprofitable	6	$1.00/cwt	43	$62/cwt	8	$106/cwt

Source: Adapted from Cattle-Fax data.

CREATING THE VISION

Most important in profitable cow-calf management is the ability to see the enterprise as a system of interacting components and understanding the relationship between and among the components. Successful managers understand the impact of the change in one component on other factors or resources. The system can be evaluated as being composed of six primary components: natural resources, the family, economic and finance factors, the production system, sociopolitical influences, and spiritual and cultural expectations (Dunn, 2000). It is important to review Chapter 3 prior to continuing this chapter.

Many low-cost producers have a written management plan that identifies their mission statement with specific goals to accomplish the plan. Two examples of mission statements are noted:

(1) Produce low-cost/high-profit cattle that yield competitively priced, highly palatable, beef products.
(2) Manage the available resources (optimum, low-cost combination) for maximum continuing net profit, while conserving and improving the resources.

Specific, measurable goals might be: (1) Reduce annual cow cost $10 per cow per year for the next 5 years, or (2) Reduce the weaned calf breakeven price from $0.80/lb to $0.60/lb in two years.

A big challenge for most cow-calf producers in meeting these mission statements and goals is developing a cost-effective financial and production records system. Nationally, it is estimated that only 5–10% of cow-calf producers actually calculate cost of production. It is difficult to manage what is not measured.

PROFIT-ORIENTED MANAGEMENT DECISIONS (FINANCIAL RESOURCE)

The Profitability Formula

Simply stated,

Profit or <loss> = income − costs

Profit-oriented management decisions are directed toward increasing income, decreasing costs, or both.

The profitability formula in slightly more detail is:

Profit or <loss> = (pounds × price) − costs

To increase profit (or minimize loss), cow-calf producers focus on three areas: (1) increase pounds, (2) increase price, and (3) decrease costs. Eventually, producers want to achieve an optimum combination of pounds, price, and costs so that profitability can be maximized.

Producers have the least management influence over prices because they are price takers not price makers. Marketing management, however, should not be ignored because good managers can influence price by a few cents a pound or several dollars a head depending on the timing of their decisions. Furthermore, by taking opportunity of retained ownership options, additional profits may be generated. Well-planned production and marketing through retained ownership and/or alliances can increase returns by several hundred dollars per head.

The profitability formula in more detail is:

$$\text{Profit or } \langle\text{loss}\rangle = [(\%\text{ calf crop} \times \text{weaning wt}) \times \text{price}] + [(\text{lb market cows and bulls}) \times \text{price}] - \text{costs}$$

There are more "pounds sold" in cow-calf operations than only "pounds of calves" (e.g., pounds of market cows and market bulls).

Breakeven Price Analysis

The profitability formula also can be expressed as a breakeven price analysis. Basically, it determines the price the calves must bring to cover the costs of production. Simply stated:

$$\text{Breakeven price} = \frac{\text{Annual cow cost}}{\text{Average weaning wt} \times \%\text{ calf crop}}$$

An example for an average producer would be:

$$\text{Breakeven price} = \frac{\$300}{500\text{ lbs} \times 85\%} \text{ or } \frac{300}{425\text{ lbs}}$$

$$\text{Breakeven price} = \$70.59/\text{cwt}$$

Obviously, if the sale price for this producer is above $70.59/cwt, a profit is realized; if the calves sell for less than $70.59, a loss is incurred.

Table 5.2 shows several breakeven prices for varying annual cow costs, weaning weights, and calf crop percentages. The circled breakeven prices reflect widely different management levels; for example, breakeven prices under low management ($116.67/cwt); average management ($70.59/cwt); and high management ($43.85). These breakeven prices do not consider pounds of cull cows and cull bulls sold per cow, so the actual breakeven prices would be lower than those shown in the table. However, the breakeven price analysis is a simple way of looking at the profitability of a cow-calf operation. A producer could take the approach of breaking even on calf sales, then having sale of cull cows and bulls as a profit.

The breakeven price formula that accounts for all cattle sales is as follows:

$$\text{Breakeven price} = \frac{(\text{annual cow cost} - \text{value of market cows and bulls sold})}{\text{Average weaning weight} \times \%\text{ calf crop}}$$

Breakeven price analysis can be used to evaluate the economics of different management alternatives.

Table 5.3 shows cow productivity, annual cow costs, and calf breakeven prices for average, low-cost, and high-cost producers in the various regions of the United States.

It is evident that low-cost producers can be found in each region of the country and that the difference between low-cost and high-cost enterprises in net income per cow is significant. Table 5.4 shows that many of the factors contributing to these differences are largely due to cost per cow. While pounds weaned per cow is important, its contribution to net income is far outshadowed by cost per cow. This difference shows producers where management decisions should first be focused.

TABLE 5.2 Breakeven Price[a] with Varying Calf Crop Percentages, Annual Cow Costs, and Weaning Weights

Calf Crop (weaned, %)	Annual Cow Cost ($)	Average Calf Weight at Weaning		
		400 lbs	500 lbs	600 lbs
		Dollars per Hundredweight		
95	$350	$ 92.10	$73.68	$61.40
95	300	78.95	63.16	52.64
95	250	65.79	52.63	43.85
85	350	$102.94	$82.35	$68.63
85	300	88.24	70.59	58.82
85	250	73.53	58.25	49.02
75	350	$116.67	$93.33	$77.78
75	300	100.00	80.00	66.67
75	250	83.83	66.67	55.56

[a]Value of cull cows and bulls are not calculated in these breakeven prices.

TABLE 5.3 Cow-Calf Producers' Profits by Cost Group

	Weaning %	× WW (lbs)	Lb/Wean Cow Exposed	Cost of Production ($/cow)	Cost of Production ($/cwt)	Net Income ($/cow)
Southwest (TX, OK, NM) 1991–1999						
Top 25%	85	538	455	315	56	141
2nd 25%	84	530	441	358	72	35
3rd 25%	81	518	419	392	89	(39)
Low 25%	81	502	410	544	133	(222)
Difference between high versus low profit	+4	+36	+45	−229	−77	+363
North Dakota IRM—1997						
Top ⅓	88	560	470	313	62	68
Middle ⅓	88	556	468	389	100	(23)
Low ⅓ (high cost)	89	536	441	465	157	(145)
Difference between high versus low profit	−1	+24	+29	−152	−95	+213
Iowa State IRM—1994–1999						
Top ¼	83	490	407	335	63	96
Middle ¼	83	482	400	413	91	(13)
Low ¼	80	454	363	545	138	(195)
Difference between high versus low profit	+3	+36	+44	−110	−75	+291
Michigan IRM—1999						
High profit 50%—Michigan	87	523	440	298	52.0	NA[a]
Average Michigan herd	85	499	417	379	76.5	NA[a]
Average Illinois herd	83	462	384	314	67.7	NA[a]

[a]Return to capital, labor, and management was $17,490 for high-profit Michigan herds, $7,325 for average Michigan herd, and $3,599 for average Illinois herd.
Source: Multiple sources.

TABLE 5.4 Comparison of Primary Factors Affecting Net Income—High Versus Low Profit

Region	Data	Value lb Weaned/Cow Exposed[a] ($)	Cost of Production per Cow ($)	Percent of Profit Total Advantages Due to Lower Costs
Southwest	1991–1999	+45	−229	83
North Dakota IRM	1997	+29	−152	84
Iowa State IRM	1994–1999	+44	−110	71

[a]Assumes calf price at $1.00/lb.
Source: Multiple sources.

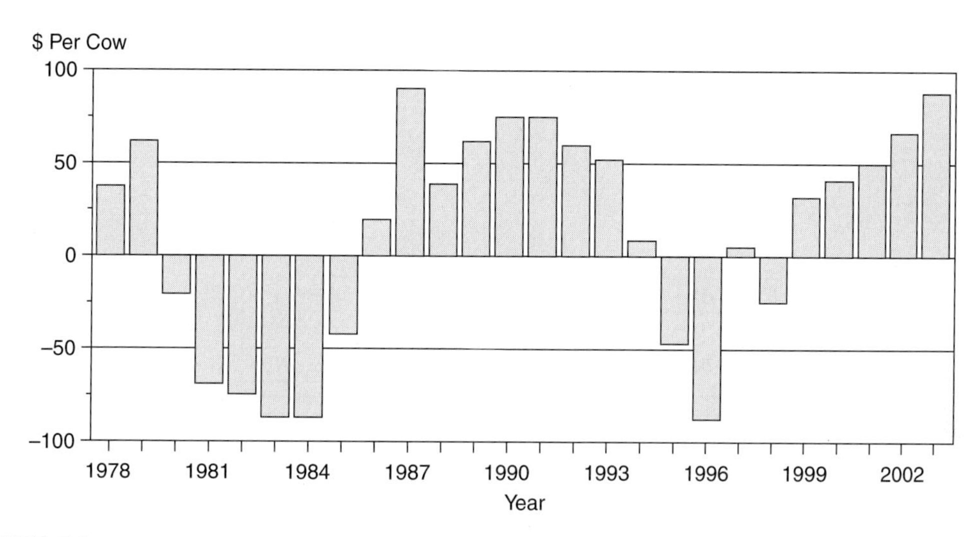

FIGURE 5.1 Estimated average cow-calf returns (over cash costs—including pasture rent).
Source: Livestock Marketing Information Center.

The beef cow is a factory that converts resources (input costs) into products (output value) that eventually result in highly palatable protein products (beef) to meet consumer preferences. The expectation is that the cow-calf operation will be profitable by the output value per cow exceeding the input costs. Figure 5.1 shows average cow-calf profitability has fluctuated over the past several years. This figure follows the beef cattle cycle quite closely—years of average losses and years of profitability. Average returns per cow can fluctuate widely within a year when individual operations are compared. However, individual producers can deviate from these average returns and have consistently profitable cows even in a "down market."

FACTORS AFFECTING POUNDS OF CALF WEANED

The number of pounds of calf weaned per cow in the breeding herd reflects both weaning weight and percent calf crop weaned. Table 5.5 demonstrates that both factors are important in calculating pounds of calf weaned per cow exposed. Producers should thus attempt to improve both traits when economically feasible. Further analysis of the table shows that a 1% change in percent calf crop is approximately equivalent to a 5 lb change in weaning weight.

Figure 5.2 shows the major factors that a cow-calf producer should critically evaluate in order to improve pounds of calf weaned per cow exposed. The two component parts (e.g., percent calf crop and weaning weight) are described in detail in this chapter. Each operation will differ in levels of productivity in each area. This also implies that priorities for improvement

TABLE 5.5 Calculation of Pounds of Calf Weaned per Cow Exposed

	Average Weaning Wt of All Calves (lbs)				
% Calf Crop	400	450	500	550	600
	Pounds of Calf Weaned per Cow Exposed[a]				
75	300	338	375	412	450
80	320	360	400	440	480
85	340	382	425	467	510
90	360	405	450	495	540
95	380	427	475	522	570
100	400	450	500	550	600

[a]Calculated by multiplying % calf crop by average weaning weight.

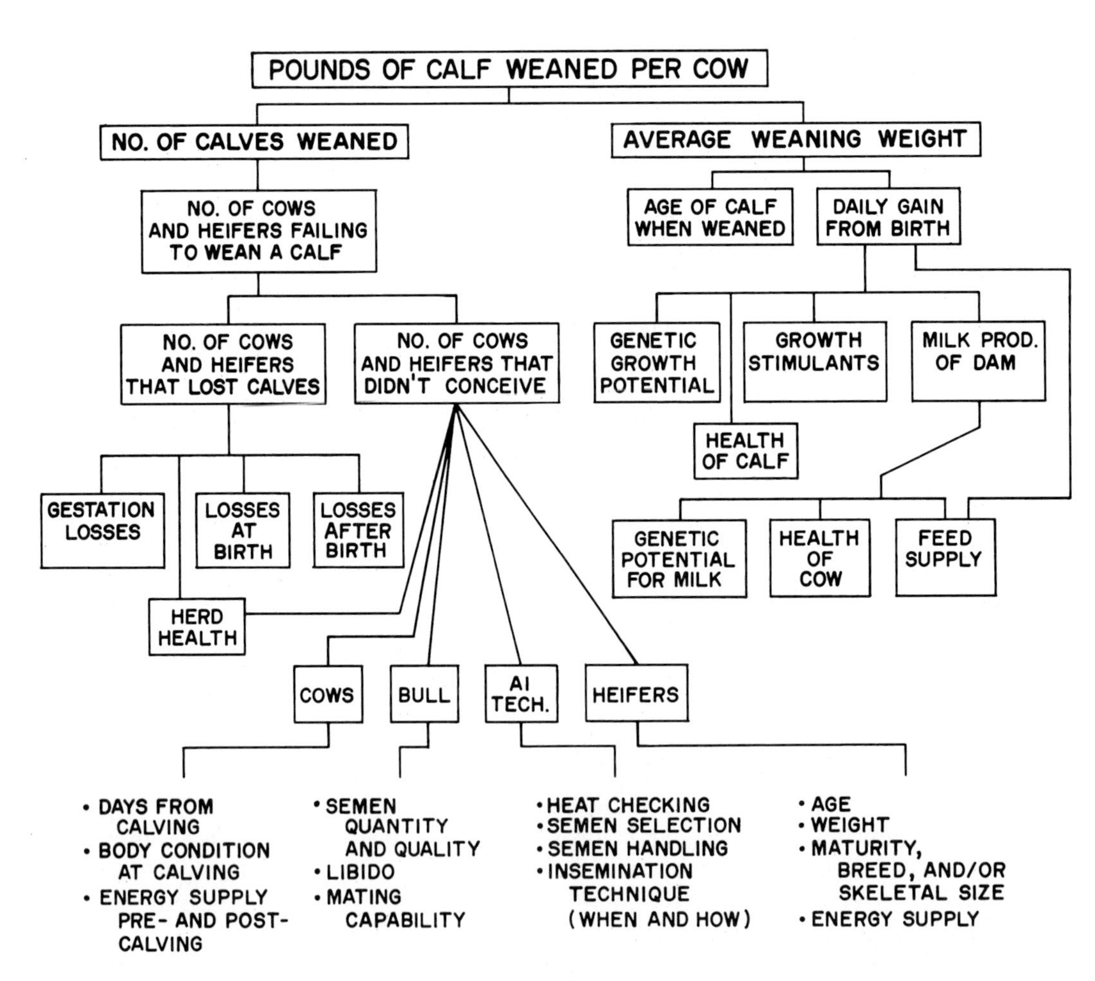

Figure 5.2 The major factors affecting pounds of calf weaned per cow bred.
Source: Colorado State University.

will usually differ for different operations. In past years, when cow-calf producers wanted to produce more beef, they typically purchased additional land and cows. Today, the emphasis has shifted increasing the productivity and dollar return on productive units—the cow and the acre of land.

The remainder of this chapter emphasizes improving productivity and profitability for a cow, acre, and total operation. The various factors affecting productivity should be critically evaluated as to how they impact both costs and returns in the profitability formula.

MANAGING PERCENT CALF CROP

Calf crop percentages can be calculated several ways. The calculation that best reflects production efficiency is as follows:

$$\%\text{ net calf crop} = \frac{\text{No. calves weaned}}{\text{No. cows in breeding herd (previous year)}} \times 100$$

This measure of percent calf crop, based on the number of cows exposed, is used throughout this book. If other measures of percent calf crop are used, they should account for all losses in reproductive efficiency.

Calf crop percentages for most cowherds in the United States will be between 60% and 100%. Extreme environmental conditions of drought, severe spring storms, and major disease problems may cause the calf crop percentage of individual herds to fall below 60%. Individual herds can document calf crop percentages over 95% for over 10 consecutive years; however, these are exceptional herds and there is evidence that this may not be optimum for many herds. The estimated average percent calf crop weaned in the United States, as previously calculated, is between 70% and 85%.

Economics of Percent Calf Crop

Most cow-calf producers expect a cow to produce a live calf every year and to do so profitably (Figure 5.3). While this is a desired goal for an individual cow, it usually is not a realistic economic goal because beef cows are managed in herds or groups and typically under extensive conditions. However, forage-based management systems provide opportunities for beef cows to be managed under low-cost scenarios. As Table 5.2 illustrates, percent calf crop tends to be optimized at levels between 80–85%.

FIGURE 5.3 A live calf born and weaned is a primary goal of the commercial cow-calf producer.
Source: American Hereford Association.

TABLE 5.6 Effect of Various Calf Crop Percentages on Breakeven Prices (assumes a $300 annual cow cost and 500 lb weaning weights)

Calf Crop %	Lb Calf Weaned (per cow exposed)	Breakeven Price (per cwt)	Change in Breakeven Price (per cwt) for Each 5% Change in Calf Crop %
100	500	$60.00	
			$3.16
95	475	63.16	
			3.51
90	450	66.67	
			3.92
85	425	70.59	
			4.41
80	400	75.00	
			5.00
75	375	80.00	
			5.71
70	350	85.71	

Reproductive performance, as reflected in calf crop percentages, is the most economically important production trait. It cannot be neglected or poorly managed without severe economic effects. Table 5.6 shows how changes in percent calf crop can affect the breakeven price. Note in the table that as calf crop percentage improves, its effect on breakeven price is diminished.

Improving percent calf crop should be based on understanding the effect of management alternatives on costs and returns. Figure 3.17 (Chapter 3) shows that maximum profitability is reached before a 100% calf crop is obtained. The following evaluates the cost effectiveness of changing percent calf crop: A producer ($70.59/cwt breakeven with 500 lb weaning weights, 85% calf crop and $300 annual cow costs) wants to raise calf crop from 85% to 90%. If weaning weight remains at 500 lbs, what is the largest increase in annual cow cost that can be accepted and still remain at a $70.59/cwt breakeven? We can solve for x if 425 lbs would increase to 450 lbs (500×0.90) as a result of increasing percent calf crop:

$$\frac{\$300}{425 \text{ lbs}} = \frac{x}{450 \text{ lbs}}$$

$$425x = 135{,}000$$

$$x = \$317.65$$

If the producer could increase calf crop percentage from 85% to 90% at a cost per cow of less than $17.65, this management practice appears to be economically feasible.

Factors Affecting Calf Crop Percentage

Table 5.7 shows the relative importance of factors affecting the net calf crop in a herd where detailed records have been kept over several years. The most important factors affecting net calf crop in this example were failure of the female to become pregnant during the breeding season and calf losses at or shortly after birth. These two factors accounted for 82% of the reduction in net calf crop. Therefore, management to improve calf crop percentages must reduce losses from all four loss categories but especially from these two major areas. Although the

TABLE 5.7 Factors Affecting Net Calf Crop in a Disease-Free Beef Herd Bred by Natural Service (14-year summary)[a]

Factor	Number	% Reduction in Net Calf Crop
Females not pregnant at end of breeding season	2,232	17.4
Perinatal calf deaths	821	6.4
Calf deaths, birth to weaning	372	2.9
Calf deaths during gestation	295	2.3
Net calf crop weaned (%)	9,107	71.0
Total	12,827	100.0

[a]Includes females 14 mos to 10 years of age during breeding seasons of 45 or 60 days' duration.
Source: U.S. Livestock and Range Research Station, Miles City, Montana.

FIGURE 5.4 Percent calf crop is a critical trait to profitable cow-calf production. An effective management plan can significantly improve performance in the number of live calves.
Source: Holly Foster.

losses identified in Table 5.7 pertained to a herd of cattle in a research study, they are similar to the types of losses that occur in producers' herds.

Factors affecting calf losses will vary in individual herds. Producers should maintain adequate records so they will know where the greatest losses are occurring. A management program to improve calf crop percentage can be applied by producers being knowledgeable of losses and the cause-and-effect relationships affecting each loss area (Fig. 5.4). Table 5.8 illustrates a nine-year trend in calf survival and weaning percentage on a large western ranch.

In this herd (Table 5.8), the average death loss from branding to weaning in years 1–6 was 4.5%. In the three years following the initiation of the vaccination protocol, the death loss was reduced to 2.7%. On average, 16 more calves were weaned following the management change.

TABLE 5.8 Trend of Calf Survival and Weaning Percent on a Large Western Ranch

Year	Cows Exposed[a] (N)	Calves Branded (N)	Calf Survival to One Month of Age (%)	Calves Died Branding/ Weaning (N)	Calves Weaned (N)	Calf Crop Weaned (%)
1	914	855	94	48	807	88
2	918	888	97	50	838	91
3	881	829	94	27	802	91
4	877	848	97	33	815	92
5	882	840	95	42	798	89
6	898	862	96	33	829	92
7[b]	917	851	93	24	827	89
8	908	850	94	17	833	91
9	900	876	97	28	848	94

[a]Only open cows sold.
[b]A new vaccination program for respiratory disease was initiated at branding.

If an average weaning weight of 450 lbs and an average price of $89/cwt are assumed, then an additional $6,408 (16 × 450 × $.89) were generated. If the cost of the vaccine was $0.20/hd, and the average number of calves at branding was 859 head, then the total cost was $1,718 (859 × $.20). Thus, the management change generated a net return of $4,690 ($6,408 − $1,718) per year.

Recent research indicates that approximately two-thirds of the 20% reduction in calf crop percentage is due to fertilization failure and embryonic death. Part of this loss can be attributed to hormone imbalances, poor uterine environments, and sex cells lacking viability. It is not clear how to prevent these losses through management programs. Excellent herd health, nutrition, selection, and other current management practices should help prevent some fertilization failures and embryonic losses.

Some of the major biological principles affecting calf crop percentage are summarized in Table 5.9.

MANAGING WEANING WEIGHTS

Calves are typically born in the spring and weaned in the fall. Most calves weaned are between 6 and 10 months of age. Some calves are born in the fall; in some operations, cows calve year-round.

Weaning weights in individual herds will vary from approximately 300 lbs to more than 700 lbs. Significant increases in weaning weights have been made in recent years. Average weaning weights at 220 days of age are approximately 500 lbs in herds surveyed by USDA (NAHMS, 1997).

Economic Considerations

As noted earlier in the chapter, increasing the weaning weights of calves can lower the breakeven price required to cover production costs. Table 5.10 shows that at a given calf crop percentage and an annual cow cost, the increments in weaning weight have a decreasing economic advantage. Weaning weight, then, should be increased in a herd as long as it is cost effective. Average herd weaning weights appear to be approaching an optimum at around 500 lbs to 550 lbs (Table 5.2) for most cow-calf enterprises. The marketing goals of the enterprise will also have a significant impact on the targeted weaning weight.

TABLE 5.9 Major Management Principles Affecting Percent Calf Crop

1. **Age and Weight at Puberty**
   a. Heifers calving at 2 years of age should reach puberty prior to 14 months of age and have one or two estrous cycles prior to breeding. Fertility of the first estrus is usually lower than those of subsequent estrous periods.
   b. Heifers need adequate nutrition to reach a target weight at breeding that is consistent with biological type (target weight at breeding should be 60–65% of mature cow weight based on a body condition score of 5).
   c. Body condition score (BCS) of replacement heifers at breeding should be a minimum of 5 and preferably 6.
   d. Crossbred heifers will reach puberty earlier and at lighter weights when compared to their straightbred counterparts.
2. **Breeding Heifers**
   a. Synchronizing estrus of replacement heifers and breeding them 2 weeks prior to cow herd, if cost effective, are excellent management tools to prevent culling these females at 3–4 years of age because they are late calvers.
3. **Feeding Heifers**
   a. Pregnant heifers should not be underfed or overfed during gestation. The best guide is to have a BCS of 5 or 6 at calving time. Heifers that are underfed will have a low BCS and a long postpartum interval. They also will produce calves that have a higher death loss and reduction in calf weight resulting from scours and other diseases.
4. **Calving Ease**
   a. Dystocia (calving difficulty) is a primary cause of calf deaths at birth. It costs $850 million annually. Additional losses can occur later due to the poor immune system of a stressed calf and a longer rebreeding period for the stressed dam.
   b. Dystocia is influenced primarily by birth weight of the calf, and birth weight is controlled most effectively through sire selection (see Chapter 4).
   c. The dam's pelvic area has a minor influence on dystocia; however, caution must be exercised in selecting for larger pelvic area. Larger-framed heifers typically have larger pelvic areas but in turn produce calves with larger birth weights. Selection of pelvic area within an optimum frame size may have merit. Growth implants will increase pelvic area at breeding, but there is little difference in pelvic area at calving time between implanted and nonimplanted heifers.
   d. Two-year-old heifers experience more calving difficulty than older females. Therefore, bull selection and management at calving time is more crucial for these young females. Mature females can experience no calving difficulty with extremely large calves (over 100 lbs). However, if heifers with heavy birth weights are saved as replacements they may experience calving difficulty, have heavier mature weight, and experience rebreeding problems, which reduces percent calf crop in later generations.
5. **Post-Partum Interval**
   a. Cost-effective BCS at calving, influenced primarily by nutrition, milk production level, and mature weight, is the best guide in managing the postpartum interval. Early weaning (5–6 months of age) of calves that is consistent with forage supply can help cows gain in BCS prior to the winter season. The cow has a maximum postpartum interval of 70–90 days if she produces a calf on a yearly basis. Postpartum intervals are most challenging on females being bred for a second calf. Calving seasons of approximately 70 days have been shown to be most profitable.

The following example shows one method of economically assessing a proposed management practice: A producer ($70.59/cwt breakeven with 500 lb weaning weights, 85% calf crop, and $300 annual cow costs) determines that by using a growth implant, pounds of calf weaned per cow exposed could be increased 10 lbs with an increased cost of $2 per cow. What happens to the breakeven price?

$$\frac{\$302}{435 \text{ lbs}} = \$69.42$$

TABLE 5.10 Changes in Breakeven Prices with 50 lb Increments in Weaning Weight (assumes an 80% calf crop and a $300 annual cow cost)

Weaning Wt (lbs)	Lbs of Calf Weaned	Breakeven Price (cwt)	Change in Breakeven Price (cwt)
350	280	$107.14	
			$13.39
400	320	93.75	
			10.42
450	360	83.33	
			8.33
500	400	75.00	
			6.82
550	440	68.18	
			5.68
600	480	62.50	

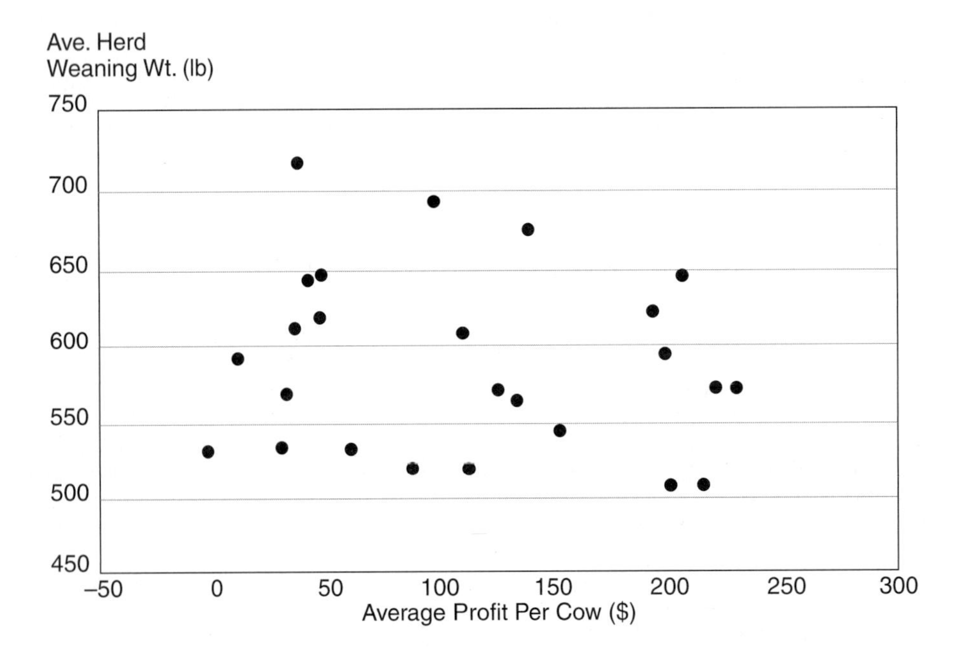

FIGURE 5.5 Average weaning weight and profit per cow. Each dot represents one herd.
Source: North Dakota IRM herds. North Dakota Extension Service.

Breakeven price is lowered from $70.59/cwt to $69.42, so this management decision could be profitable.

Figure 5.5 shows that herd average weaning weights between 500 lbs and 700 lbs are not related to profit per cow. Other studies evaluating the profitability of cowherds in the Great Plains have shown that weaning weight compared with cost of production and reproductive rate is a distant third in factors affecting profitability. These studies suggest that weaning weight has an optimum level within each cow-calf herd and that deviation from this optimum could decrease profits.

Factors Affecting Weaning Weight

Table 5.11 shows the major management principles affecting weaning weights. The application of these principles can be used to reach an optimum weaning weight for a herd. Commercial producers combining excellent cattle genetics with well-managed feed and health environments will have relatively heavy calves at weaning (Fig. 5.6).

Several management practices that result in a high percentage of calves being born early in the calving season were identified earlier in the chapter. These practices increase weaning weight because the calves are older when weaned.

TABLE 5.11 Major Management Principles Affecting the Weaning Weights of Calves

1. **Age Of Calf When Weaned**
   a. Increasing calf age increases weaning weight. Age at weaning should be determined by the economical costs, especially feed costs.
2. **Calving Season**
   a. Calves born in summer and fall usually weigh less than calves born in spring. Exceptions to this occur in some geographical areas of the United States, where forage quantity and quality are high in the fall.
3. **Growth Implants**
   a. Implanting suckling calves increases weight 10–25 lbs. Consideration should be given to the market endpoint when establishing an implant program.
   b. Caution should be exercised in implanting replacement heifers, as their reproduction may be affected.
4. **Feed Supply**
   a. Cost-effective management practices should focus on increasing the amount of grazed forage and its nutrient quality during the time when cows are milking the heaviest and calves are growing rapidly from dam's milk and the direct consumption of forage.
   b. Weaning weights can be increased by feeding more harvested and purchased feed to cows and calves; however, this is usually not cost effective, especially over the long term.
5. **Selection**
   a. Genetic selection, primarily through bulls, can improve weaning weights 3–5 lbs per year. Weaning weight EPDs and milk EPDs should be emphasized if weaning weights need to be improved.
   b. Selection for yearling weight EPD will increase weaning weight nearly as much as selecting only for weaning weight EPD. Selection for yearling weight EPD not only increases weaning weight but yearling weight and feedlot gain as well.
   c. Caution should be exercised so that genetic selection for growth and milk does not exceed the most economical grazed forage program. Otherwise, BCS and reproduction will decline or a higher-cost feeding program (more harvested and purchased feeds) must be implemented to maintain reproduction at a relatively high level.
6. **Heterosis**
   a. Utilization of heterosis (hybrid vigor) through well-planned crossbreeding can increase weaning weight by 10% and pounds of calf weaned per cow exposed by 20% over the average of the breeds in the crosses.
   b. See section 5c in this table.
7. **Health**
   a. Calves that do not experience calving difficulty and that receive adequate levels of colostrum during the first 24 hours after birth usually weigh heavier at weaning time.
   b. An excellent, cost-effective cow-calf health program assures an adequate milk flow from the cow and the ability of the calf to express its weaning weight potential.

FIGURE 5.6 A good crossbreeding system with excellent forage will usually produce heavy calves at weaning.

Genetic Factors Influencing Calf Gains

When an adequate nutrient supply is maintained, the most limiting factor influencing gain is the genetic makeup of the healthy calf. A major genetic influence on nutrient supply comes in the genetic ability of the cow to produce an optimum flow of milk. Growth potential of the calf and milk production of the cow are improved genetically through selection and crossbreeding.

The gene combination affecting growth can be changed through sire, heifer, and cow selection. It has been well demonstrated that effective selection can result in a genetic increase of 3–5 lbs per year in the weaning weights of calves. Most of this change will come through bull selection based on evidence of genetic superiority in weaning and yearling weights. Some genetic improvement can be made through cow and heifer selection, but the amount of progress is limited due to the number of females that must be kept.

There is a breed effect on weaning weight where larger, faster-gaining breeds will have heavier weaning weights at the same age. This relationship is not only true when comparing breed averages but is also noted when comparing biological types within a breed. Crossbreeding for the average cow-calf producer can produce a 20% increase in pounds of calf weaned per cow exposed as compared to a straight breeding system. Most of this increase occurs from improved reproductive performance; however, 25–40% of the increase is the effect of heterosis on (1) growth potential of the crossbred calf and (2) increased milk production of the crossbred cow. An economic evaluation of crossbreeding showed that a 16–20% increase in net income after taxes could result through effective use of crossbreeding. However, implementing high levels of heterosis and selection could yield weaning weights beyond the herd optimum thus lowering profitability.

A cow-calf producer with an average weaning weight of 400 lbs and an 80% calf crop in a herd of straightbred cows produces 320 lbs of calf weaned per cow exposed. If this producer implemented an effective selection and crossbreeding program, the pounds of calf weaned per cow exposed would be 414 lbs in approximately 6 years. Assuming the annual cow cost remained constant at $300, this would change the breakeven price from $0.94/lb to $0.72/lb.

Many commercial producers want to know what breeds to use in a crossbreeding program. Certain breeds do complement one another more satisfactorily, however, producers need to pursue the genetic resource in more depth. It is extremely important to understand the variations that exist within each breed (see Chapter 13). In many cases, the biological types of cattle within the breeds have more economic significance than the heterosis obtained by crossing two or more breeds (see Table 5.12). However, it should not become a question of genetic superiority versus heterosis for commercial producers. They should take advantage of both to optimize productivity and maximize profitability.

Growth Stimulants

The most common growth stimulants for calves are ear implants of Compudose®, Synovex®, and Ralgro®. Implants can be given to young calves when they are a few days or weeks old. (The average growth responses to various implants are shown in Chapter 7.) Implants should not be used on replacement heifers because of possible detrimental effects on reproduction.

The cost of a Ralgro® or a Synovex® implant is approximately $1. The approximate 10–20 pound increase in weaning weight well justifies the implant and labor cost. Proper implanting procedures as noted in Chapter 7 must be followed. The growth-promoting response of Compudose® lasts over a 200-day period. The implant costs approximately $2.50 and the expected response is a 5% increase in preweaning gain.

Feed Supply

Environment and the genetic potential of the calf determine the calf's weaning weight. Environmental effects such as health of cows and calves and feed supply are extremely important. It is estimated that approximately 60–70% of weaning weight is accounted for by milk production of the cow; the remaining 30–40% comes from grass and other forage that the calf consumes directly. Therefore, available forage to both cows and calves will significantly affect the weights of the calves at weaning time. Feed supply can be significantly affected by weather conditions, particularly precipitation. Amount and distribution of precipitation affects weaning weights through the amount and nutrient content of the forage.

TABLE 5.12 Crossbreeding and Within-Breed Selection

Breed and Level of Performance	Weaning Wt (lbs)	Average Weaning Wt (lbs)	Weaning Wt after Crossbreeding (lbs)	Breakeven[a] Price
Low A	450			
		430	452[b]	$83.87
Low B	410			
Medium A	485			
		475	499[b]	$75.19
Medium B	465			
High A	540			
		520	546[b]	$68.65
High B	500			

[a]Assumes 80% calf crop and $300 annual cow cost.
[b]Assumes a 5% heterosis above the average of the two breeds.

Month of birth has some effect on weaning weights of calves. Highest daily gains appear to be realized when calves are born 2–3 months prior to green forage production. Calves born earlier and particularly calves born later (summer) usually have slower daily gains. Most cows have a peak in milk production approximately two months following calving. Milk production appears to be sustained longer if cows have lush, green pasture about the time they are peaking in milk production. Calf gains are more dependent on milk early in the calf's life as compared with the last couple of months prior to weaning.

The nursing calf usually has an adequate supply of nutrients when receiving cow's milk and an ample amount of forage. Salt and a phosphorus supplement should be provided to cows on a year-round basis. Most mature forages are low in phosphorus, so phosphorus supplementation should be considered

More important than weaning weight per calf or pounds of calf weaned per cow exposed, however, are pounds of beef produced per unit of land area. This measurement reflects not only the improvement made in cattle but also the range and pasture improvements accumulated over a period of several years. An excellent demonstration of this is shown in Table 5.13, where range reseeding and grazing intensity, along with improvements in the productivity of cattle, are reflected in calf production at the Eastern Colorado Research Center. Records showed little improvement in average weaning weight or calf crop percent over the 20-year period. Total animal units maintained on the same land area changed from 144 to 360, and salable beef increased from 78,750 lbs to 154,915 lbs during the 20-year forage and cattle improvement program.

Creep Feeding of Calves

Creep feeding can add an additional 20–50 lbs to the weaning weights of calves. Caution should be exercised with creep feeding, however, because the added weight is often not economical and the gaining ability of the calves after weaning may be affected.

Replacement heifer calves should not be creep fed because future milk production is reduced. Apparently fat accumulation in the udder decreases the amount of secretory tissue that develops. A producer should compute the expected cost of gain from creep feeding and assess the future production and utilization of the weaned calves before deciding to creep feed. Additional information on creep feeding is given in Chapter 14.

MANAGING ANNUAL COW COSTS AND RETURNS

Obviously, the breakeven price of calves is reduced if a producer can lower annual cow costs while maintaining percent calf crop and weaning weight at the same level (Table 5.14).

TABLE 5.13 Calf Production at the Eastern Colorado Range Research Center from Cows 3 Years Old and Older, 1953–1972

Years	No. Cows	Weaning Wt (lbs)	% Weaned	Lbs Weaned per Cow	Acres per Animal Unit	Lbs of Calf per Acre
1st five years	75	430	90%	387	23	17
2nd five years	81	434	95	412	21	20
3rd five years	96	431	94	405	19	22
4th five years	118	431	93	401	15	27
20-Year average	–	432	93	402	19	21

Source: Colorado State University Experiment Station Progress Report, PR73-80.

TABLE 5.14 Effect of Changing Annual Cow Cost on Breakeven Prices (assumes an 80% calf crop and 500-lb weaning weights)

Annual Cow Cost	Pounds of Calf Weaned	Breakeven Price (cwt)	Change in Breakeven Price (cwt)
$450	400	$1.12	
			$0.12
400	400	1.00	
			0.12
350	400	0.88	
			0.13
300	400	0.75	
			0.13
250	400	0.62	
			0.12
200	400	0.50	

Sometimes, it is profitable to lower annual cow costs even if weaning weight or percent calf crop is decreased.

Consider the following example: A producer ($70.59/cwt breakeven with 500 lb weaning weights, 85% calf crop, and $300 annual cow costs) determines that feed costs could be lowered $50 per cow but that it would lower the calf crop from 85% to 80%. Would this be a feasible management decision if weaning weight remained the same?

$$\frac{\$250}{500 \times 0.80} = \frac{\$250}{400 \text{ lbs}} = \$62.50$$

This would be a profitable management decision, even though it may not seem logical to lower percent calf crop. The breakeven price is lowered from $70.59 per cwt to $62.50 per cwt and the profit per cow would be higher at the $250 annual cow cost even though fewer calves would be sold.

Assessments of the costs of production are the most neglected area in many commercial cow-calf operations. Adequate expense records must be maintained so that cost areas can be carefully analyzed. Each producer should know where major costs occur and how the costs can be lowered or held in line with the level of productivity of the cattle (Fig. 5.7).

Table 5.15 shows a regional summary of the costs and returns for 55 commercial herds with 59,217 exposed females in 12 states where standardized performance analysis (SPA) has been utilized. The average difference in annual cow cost, calculated on a financial basis, between the high-return and low-return producers across all regions was $114/hd. While regional differences exist, it is clear that irrespective of location, cow-calf producers have the ability to maintain effective control over annual cow costs. SPA is a standardized cow-calf enterprise production and financial performance analysis system defined by the Cow-Calf Financial Analysis Subcommittee of the National Integrated Resource Management (IRM) Coordinating Committee of the National Cattlemen's Beef Association (NCBA). SPA facilitates comparisons of an operation's performance between years, producers, production regions, and production systems. The analysis is based on fiscal year production and financial data.

SPA includes performance measures for reproduction, production, grazing and raised feed, marketing, and financial and economic performance. Figure 5.8 shows the different components for an economic analysis and a financial analysis. SPA is an integrated analysis that links both financial and production performance. Cow-calf producers should conduct a complete SPA analysis or a similar type of analysis annually.

FIGURE 5.7 Cow-call producers must keep accurate income and expense records to assess annual cow costs and the profitability of their operations.

TABLE 5.15 Summary of Annual Cow Costs and Breakeven Prices by Region and Profitability Level

Region	Annual Cow Cost ($/cwt)	Calf Breakeven Price ($/cwt)
Northwest		
High return 1/3	275	53
Low return 1/3	399	87
Southwest		
High return 1/3	254	51
Low return 1/3	358	83
Midwest		
High return 1/3	239	51
Low return 1/3	340	83
South Plains		
High return 1/3	244	48
Low return 1/3	365	81
Southeast		
High return 1/3	224	48
Low return 1/3	342	84

Source: NCBA and Cattle-Fax.

FIGURE 5.8 Comparison of an economic analysis and a financial (accounting) analysis for a beef cow herd.
Source: Hughes, Fargo, ND: North Dakota State University.

SPA worksheets and software have been developed to accommodate the producer's annual production and accounting data and to calculate the selected performance measures. Information is generated in a manner that clearly focuses on measures that can be used to identify opportunities to reduce costs of production. More detail about SPA can be obtained from the sources listed in the References section at the end of the chapter. For producers who feel SPA is too complex, they may want to consider SPA-EZ, which is described in the Appendix.

The detailed cost summary from SPA helps producers identify major cost items so that attention can be directed to the management of those costs. SPA can help producers evaluate how well they are doing by comparing their results with those of other producers. The information generated by SPA can be used to implement labor and management incentives aimed at reducing costs. SPA provides a useful system of benchmarks for cow-calf producers. However, benchmark comparisons need to be made with caution. The most useful comparisons are internal benchmarks that allow managers to study and evaluate the performance of their enterprise over time within the company. External benchmarks that allow comparisons to other herds should be utilized only when it will contribute to valuable and sustained improvement.

Managers are advised to use benchmarking comparisons to achieve the following objectives:

1. Identify superior performance.
2. Define best management practices and understand those factors that contribute to or impact superior performance.
3. Adopt and adapt best management practices that are suitable to the enterprise and the human resource.

Production and financial records are needed to make effective management decisions. An enterprise budget is a logical place to start. Observe the major headings in the cow-calf enterprise budget shown in Table 5.16: "Operating Receipts," "Direct Costs," and "Property and

TABLE 5.16 Cow-Calf Enterprise Budget (440 head of spring calving cows)—2000

Production and Marketing Assumptions

Livestock Description	Unit	Market Wt	Market Month(s)	No. Head Marketed	% of Cow Herd
1. Steer calf	lb	500	Nov.	198	45
2. Heifer calf	lb	470	Nov.	172	39.09
3. Cull cows	lb	1,000		56	12.73
4. Cull bull	lb	1,600		3	0.68
1.3% Cow death loss; 13% replacement ratio			91% weaned calf crop		

Operating receipts, Livestock sales

Description	Unit	Market Price ($/lb)	Production per Cow	Value per Cow ($)	Total ($)
1. Steer Calf	lb	1.11	225.00	249.75	109,890.00
2. Heifer Calf	lb	1.04	183.73	191.08	84,075.00
3. Cull Cows	lb	.38	127.27	48.36	21,279.00
4. Cull Bull	lb	.42	10.91	4.58	2,016.00
5. Total Receipts				$493.77	$217,260.00

Direct Costs-Feed, Purchased and Raised

Description	Unit	Price	Production per Cow	Value per Cow ($)	Total ($)
6. Pasture rent	aum	$14.00	3.00	$ 42.00	$18,480.00
7. Aftermath	aum	10.00	1.00	10.00	4,400.00
8. Hay	ton	95.00	0.45	42.75	18,810.00
9. State lease	aum	4.00	3.00	12.00	5,280.00
10. Salt and mineral	lb	0.29	100.00	29.00	12,760.00
11. Total feed expenses				$135.75	$59,730.00
Other Operating Costs:					
12. Hired labor				$35.00	$15,400.00
13. Repairs: machinery, building, fences				9.45	4,158.00
14. Machine hire/truck				1.24	547.75
15. Supplies				1.81	796.40
16. Veterinary medicine				7.50	3,300.00
17. Fuel, oil, lubricants				21.00	9,240.00
18. Breeding fees, pregnancy check, etc.				1.50	660.00
19. Miscellaneous				3.00	1,320.00
20. Utilities				2.35	1,034.00
21. Transportation				5.00	2,200.00
22. Marketing expenses				8.55	3,762.00
23. Herd bulls				8.52	3,748.80
24. Interest on operating capital			8.0%	8.41	3,700.40
25. Total other operating costs				$113.33	$49,867.40

Livestock Purchased (or raised) for Resale

Description	Unit	Price	Quantity	Per Cow	Total
28. Total cash operating expenses (line 11 + line 25)				$249.08	$109,597.40
Property and Ownership Costs					
29. Depreciation					
30. Taxes				20.59	9,060.00
31. Insurance				−6.26	2,754.00
32. General overhead (5% of operating costs)				2.50	1,100.00
34. Total property and ownership costs				12.45	5,480.00
35. Total direct costs (line 28 + line 34)				41.80	18,394.00
36. Net receipts (line 5 − line 35)				290.88	127,991.40
(Returns to capital, land, management, risk)				$202.89	$89,268.60

[a]Lines 26–27 omitted because no purchase-resale costs occurred; line 33 is Other.

Source: Adapted from Gutierrez et al., 1990 (2000 data included). DARE Information Report IR:90-2.

Ownership Costs." In this budget, receipts per cow are $493.77, with cash operating expenses $249.08 and property and ownership costs $41.80 totaling $290.88. This results in net receipts per cow of $202.89.

The enterprise budget, an SPA evaluation, or another financial analysis is needed to determine when production (percent calf crop and weaning weights) and costs may be out of line or on target. These comparisons in similar geographical areas throughout the United States show that feed costs can be twice as high and breakeven prices on calves half the amount in one herd compared with another herd. Commercial cow-calf producers need to ask the questions, "Am I a low-cost producer?"—"If not, can I be one?"—"If I can be a low-cost producer, how can I do it?"

Factors Affecting Costs and Returns

Low-cost producers in the NCA-IRM-SPA Database were requested to identify the primary factors that determined their low breakeven prices (Table 5.17).

Even though these "top 5" are ranked as most important, the other, or miscellaneous costs, should not be overlooked. These smaller expense items (i.e., fuel, machinery, repair, supplies, utilities, taxes, etc.) account for less than 20% of an average budget. Yet, they make up over 40% of the difference in cow cost between high-cost and low-cost producers. These smaller expense items should be compiled and evaluated on a regular basis.

The impact of a 10% change on breakeven prices and returns is evaluated in Table 5.18. Note that it would be nearly impossible to attain the full return from all the factors. However, the information in the table points to areas where good decisions can have a beneficial impact on profits.

TABLE 5.17 Top Five Ways Low-Cost Producers Reduce Costs

1. Reduce harvested and supplemental feed costs (40%*)
2. Use rotational grazing (better pasture management) (30%*)
3. Use the right genetics (27%*)
4. Reduce labor costs (25%*)
5. Implement a strong herd health program (19%*)

*Percent of respondents.
Source: NCA-IRM-SPA Database (1995 NCA Cattlemen's College).

TABLE 5.18 The Impact of a 10% Change in Key Factors on Cow-Calf Breakeven Prices and Returns

Factor	Change (%)	Decrease in Breakeven Price ($/cwt)	Increase in Return ($/cow)
Total feed cost	−10	4.27	21.58
Weaned calf crop	+10	7.87	39.82
Weaning weight[a]	+10	6.62	33.47
Interest cost	−10	0.18	0.93
Cull cow weight[b]	+10	1.00	5.08
All combined		19.94	100.88

[a]Equivalent to a 10% increase in calf price.
[b]Equivalent to a 10% increase in market cow price.
Source: Adapted from Cattle-Fax.

TABLE 5.19 High-Return Versus Low-Return Producers

	High 1/3	Low 1/3	Difference
Calf breakeven	$53/cwt	$86/cwt	$151/cow
Annual cow cost	$259/hd	$357/hd	$98/cow
Raised and purchased feed	$58/hd	$137/hd	$79/cow
Supplemental feed/hd	2,129 lbs	2,867 lbs	738 lb/cow
Feed cost per ton	$54	$96	$42/ton
Debt per cow	$186	$427	$241/cow
Financing cost per cow	$11	$41	$30/cow
Percent weaned	90%	88%	2%
Weaning weight (lb)	569	528	41 lb/calf
Lb weaned per cow exposed	509	454	55 lb/cow
Bull purchase price/hd	$1,996	$1,874	$122/bull
Herd health cost/hd	$15.36	$15.87	$0.51/hd

Source: National SPA Data Base.

A comparison of SPA data from high-return versus low-return producers also allows the identification of key areas requiring careful planning and effective plan implementation to achieve profitability (Table 5.19). In general, high-return enterprises had lower cow costs, lower feed costs, were less reliant on supplemental feed, had lower debt, and higher production per cow.

Reducing Feed Costs

Feed costs (purchased feed, harvested feed, and grazed forage) comprise approximately 60% of the annual cow costs in many cow-calf operations. Since feed cost is the largest component cost and different feeds have varying costs, this area must receive high priority in assessing production costs. Chapter 14 (Nutrition) and Chapter 15 (Managing Forage Resources) include principles that are applicable to assessing feed costs.

Most low-cost producers have reduced their feed costs by increasing the number of days cows graze while decreasing the amount of harvested and purchased feed fed per cow. Extending the grazing season and feeding less hay or harvested feed has been accomplished primarily by:

1. Matching calving season with green forage production. For the west and northern Great Plains, several low-cost producers are calving in April and May rather than February and March. Summer calving (June–August) shows substantial reductions in feed and labor costs (Lardy, 1998). High nutrient demands of the cow and calf are less costly if forage is grazed rather than feeding hay or other harvested or purchased feed.
2. Extending green grazing through intensive and rotational grazing systems. This also can increase carrying capacity (more cows for the same land area), which reduces the fixed cost per cow.
3. Improving range/pasture forage production and utilization through burning, reseeding, interseeding, and grazing a mixture of grasses and legumes where possible.
4. Stockpiling forage for late fall, winter, and early spring grazing.
5. Weaning calves earlier (i.e., 5 months of age) so cows can increase in body condition score (BCS 6 or 7) prior to the winter months. Less feeding of harvested feed occurs because cows in this condition can lose 75–150 lbs prior to calving and still maintain excellent reproductive performance.

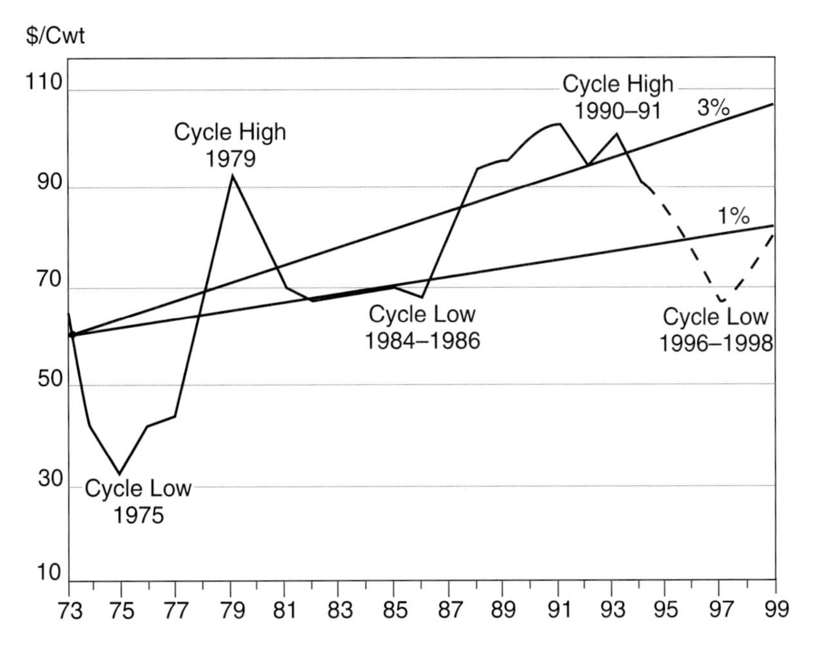

FIGURE 5.9 How 1% versus 3% inflation affects breakeven prices.
Source: Simonds (1995); adapted from NCBA 1995 Cattlemen's College.

Feed costs can be reduced up to 15% by feeding replacement heifers to have low to moderate gains during most of the growing phase (Lynch, 1997). The heifers' growth is accelerated about 60 days prior to the breeding season with target weights achieved at breeding. The feeding of a high-concentrate diet following a period of feed restriction does not alter milk production like that experienced in creep feeding heifers prior to weaning.

Equipment and fuel costs are significantly reduced when cow-calf producers develop forage programs where cows graze more days during the year and less harvested feed is fed. The cost of on-farm fuel increased by 100% from 1999 to 2000. While producers have no control over fuel prices, they can exert some control over substantial overusage. Fossil fuel, machinery, and some supplies increase in cost due to inflation, while renewable resources such as water and sunlight needed to grow grazable forage are less inflationary. Figure 5.9 shows how different rates of inflation (1% to 3%) affect breakeven prices with time. The 1% inflation is more profitable than the 3% inflation during cattle cycle price fluctuations.

Utilizing the "Right Genetics"

Low-cost producers understand the importance of matching the most functional type of cow to a low-cost environment consisting primarily of grazed forage. These "right genetics" imply effective sire selection for a cost-effective biological type of cow while utilizing heterosis by crossing two or more breeds or using a composite breeding program. Most cow-calf producers describe the biological type of cow that best fits this low-cost environment as noted in Table 5.20.

Table 5.21 identifies the bull selection criteria that produces the biological type of female described in Table 5.20. It also identifies bulls that will sire low-cost progeny in the feedlot and superior carcasses in addition to low-cost cows.

TABLE 5.20 Biological Type of Cow for Low-Cost Production

Trait	Description
Mature cow size	Moderate (1,000–1,200 lbs; BCS 5)
Milk	Moderate (can wean 475–550 lb calf [7 mos of age] with an average forage supply)
Muscling	Moderate (average thickness through the stifle area; progeny of same biological type have 11–14 sq. in. ribeye)
Fatness	Fleshing ability (can increase BCS during late summer and early fall grazing); progeny of same biological type have yield grades from 2.0–3.4 at 700–750-lb carcass weights
Longevity/Stayability	Cow produces a calf each year (365 days) beyond 9 years of age. This trait appears to be influenced primarily by cost-effective fleshing ability (BCS). Most cows leave the herd early because of poor reproduction. They are culled primarily as two-to-four-year-olds because they are open or late breeders—their cost-effective BCS is usually below 5.

TABLE 5.21 Bull Selection Criteria Contributing to Low-Cost Production of Superior Carcasses

Evidence that this bull and other similar breeding:

1. Performed in one or more low-cost commercial cow-calf operations where the calves have been evaluated in the feedlot and carcass. Performance specifications in these low-cost herds:
   a. Weaning breakeven price below $0.65/lb (includes all costs).
   b. Feedlot performance of calves
      - Rate of gain of minimum 3.2 lbs/day from 500–600 lbs to slaughter weight (1,100–1,300 lbs); slaughter age 13–15 months.
      - Feedlot cost of gain (<$0.60/lb with $3.00/bu corn).
   c. Carcass performance of calves
      - >70% choice
      - Yield grade 2.0–3.5
      - Satisfactory tenderness (e.g., WBS <8 lbs/in.)
2. Longevity and stayability of daughters—they remain in the herd beyond 10 years of age (highly related to fleshing ability and early rebreeding of 2- and 3-year-old cows under low-cost production).
3. Performance specifications for individual bulls (prefer to also see half-brothers to this bull having similar specifications)—the following EPDs are Angus based (see across-breed EPDs for comparison):
   a. Birth Wt. EPD: Maximum of +1.5 (<0 for heifers); bull needs to sire actual birth weights as follows: heifers' calves (65–80 lbs) and cows' calves (75–90 lbs).
   b. Weaning Wt. EPD: +15 to +25 lbs.
   c. Milk EPD: Under + 15 lbs.
   d. Yearling Wt. EPD: +30 to 50 lbs.
   e. Scrotal Circumference EPD: Above +0.3 cm and/or minimum yearling scrotal circumference of 34 cm.
   f. Yearling Frame Score: Maximum 5.9 with most of the bull's calves in the 4.0–5.5 range (prefer bull's mature wt near 2,000 lbs and cow's mature wt below 1,200 lbs based on body condition score of 5).
   g. Pedigree (parents and grandparents) stacked for specifications noted in a. through f. and many of those noted in h.
   h. Additional evaluations needed: (1) disposition, (2) pulmonary arterial pressure (adaptability to high altitude), (3) adaptability to varying weather and feed conditions, (4) reduction of variation in a given trait (e.g., smaller standard deviation in birth weight of heifers' calves), (5) other reproduction EPDs (e.g., fertility, calving ease, body condition score, etc.).

Source: Colorado State University.

TABLE 5.22 TDN Requirements for Cows of Various Weights and with Different Milk Production Levels

Cow Wt (lbs)	Annual TDN for Maintenance (lbs)	Daily lbs of TDN Required for Maintenance and Milk at Various Milk Production Levels		
		8 lbs	16 lbs	24 lbs
800	2,315	9.4	12.0	14.7
1,000	2,709	10.5	13.1	15.8
1,200	3,098	11.5	14.2	16.8
1,400	3,472	12.6	15.2	17.9
1,600	3,832	13.6	16.3	18.9

Source: Adapted from NRC, *Nutrient Requirements of Beef Cattle,* 1984.

Mature weight of cow and level of milk production are important factors affecting annual cow cost. Table 5.22 and Figure 5.10 show how cow weight and level of milk production affect energy requirements. These requirements demonstrate that more feed is required when weight or milk production increases. A cow's stage of production has a significant impact on her feed requirements. For example, a 1,000 lb cow has higher requirements for energy and crude protein of 19% and 23% in late gestation as compared with the second trimester of pregnancy. The same sized cow requires 29% more TDN and 48% more crude protein when she is milking as compared with her second trimester. Anticipating these differences and adapting nutritional management accordingly is a key to profitability.

There is disagreement about the size of cow needed for commercial production. Many producers feel that cow size will and should vary according to environmental conditions, particularly feed supply. There are observations that large cows that wean heavier calves will not become pregnant in areas where the most economical feed supply comes primarily from grazed forage. There is increasing evidence that cows with moderate mature weights have lower annual costs than larger cows (Fig. 5.11).

When cow size is increased, the amount and nutrient quality of feed given to heifers are increased so they will reach puberty in sufficient time to calve at two years of age. The appropriate cow size is best determined by the amount and nutrient quality of available feed, cost of feed, realizing that grazed forage is usually lowest in cost, and the desired carcass weight and grades of the resulting progeny. In the latter category, carcass weights of 700–800 lbs, Yield Grade 2, and Choice Quality Grade are reasonable targets for the majority of market steers and heifers. These carcass specifications strongly suggest a mature cow weight range of 1,000–1,200 lbs (BCS 5). There are numerous examples of moderate-sized cows (1,000–1,200 lbs in average body condition) that can reproduce regularly and wean 500–550 lb calves with excellent feedlot and carcass performance. Recent surveys show that the average mature weight of commercial cows is 1,000–1,150 lbs in many herds (NAHMS). Cow weights appear to be heavier in many seedstock herds. For example, the average registered 5-year-old Angus cow weighs 1,300 lbs (Angus Sire Evaluation Report, Spring 2001).

A terminal crossbreeding program can be used to achieve economic efficiency of smaller-sized cows that produce calves with high feedlot and carcass performance. Cow lines should be selected for moderate size and high maternal performance. Sire lines should receive selection emphasis for growth and cutability. Table 5.23 gives economic support for this type of crossbreeding program.

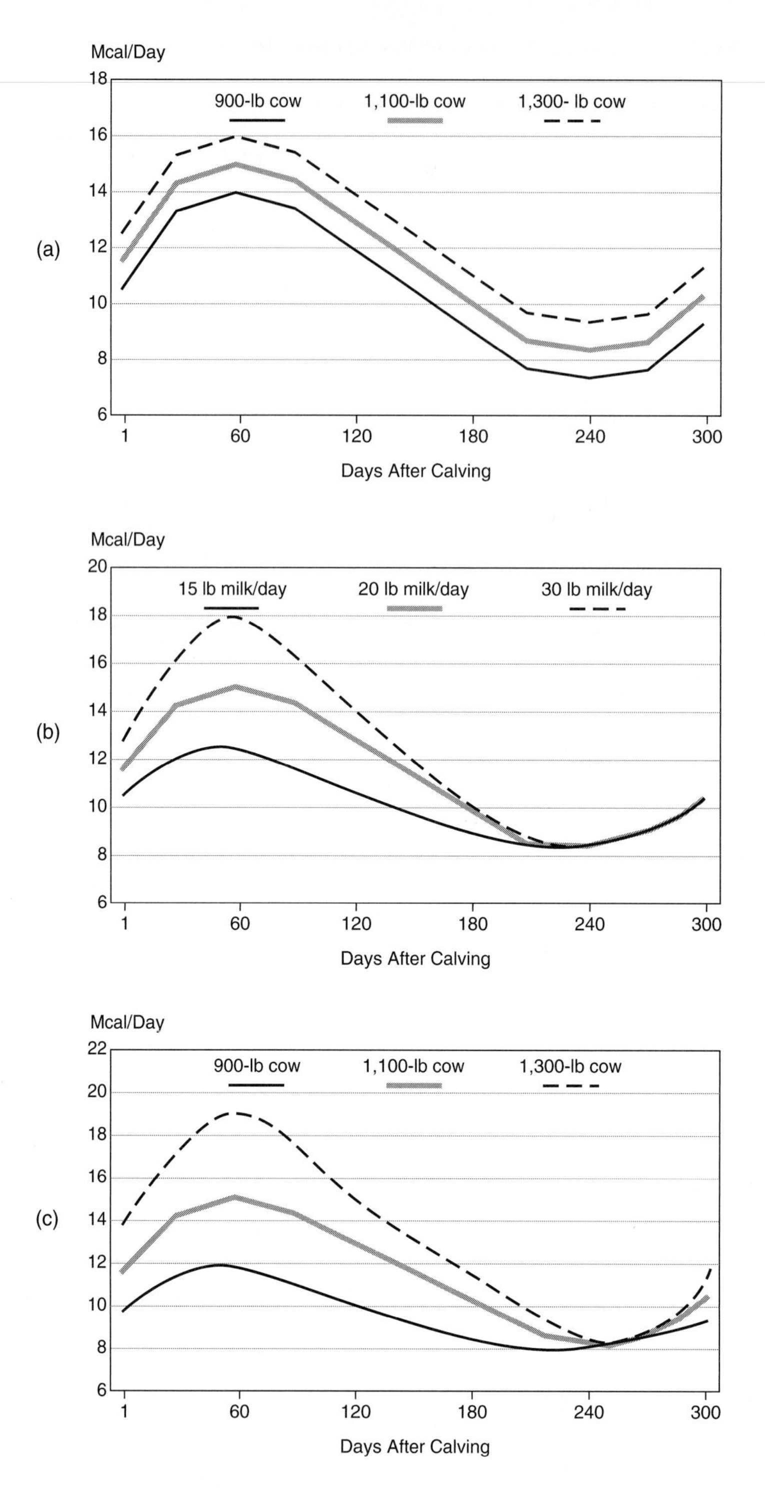

FIGURE 5.10 Effects of cow size and milk production on net energy requirements.
Source: Kansas State University. (a) Effect of cow size on net energy requirements for 20 lbs of milk production. (b) Effect of milk production on net energy requirements for a 1,000 lb cow. (c) Combined effects of cow size and milk production on net energy requirements.

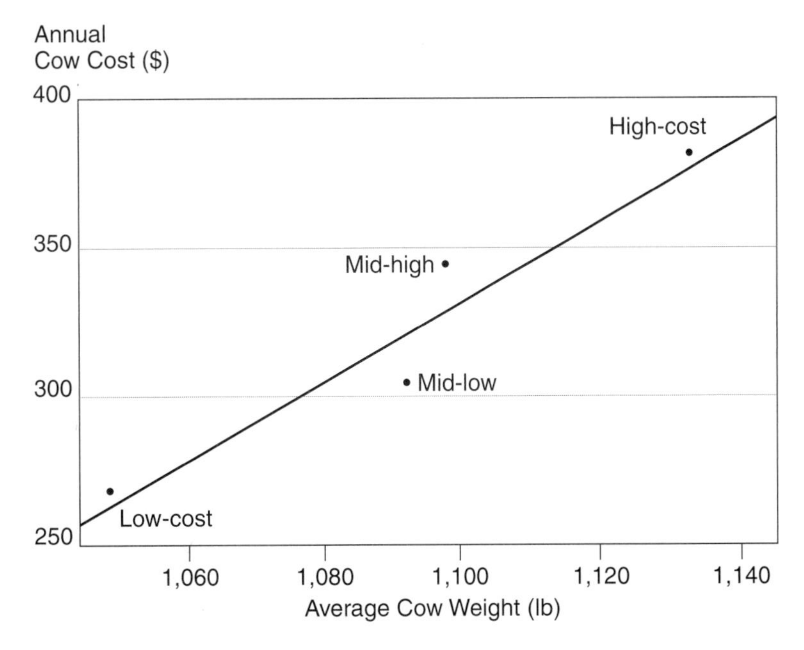

FIGURE 5.11 Relationship of cow weight to annual cow cost.
Source: Cattle-Fax.

TABLE 5.23 Percent Return on Investment[a] from Combinations of Mature Size for Two Breeding Systems Under a Pasture Program

	Size of Cow		
Breeding System and Size of Sire	**Small**	**Medium**	**Large**
Crossbreeding			
Small	15.9%	15.9%	15.0%
Medium	16.7	16.6	15.4
Large	17.7	17.4	16.0
Straightbreeding	14.8	14.9	14.9

[a]Return on investment is defined as total income less expenses, other than interest it is expressed as a percentage of total expenditures.
Source: J. Anim. Sci. 41: 1238.

The breed-cross of the cow has been shown to affect the amount of feed required for maintenance when mature cow weights are similar. Angus-Hereford crossbred cows required 19%, 14%, and 16% less feed than Jersey, Charolais, and Simmental cross cows, respectively (Ferrell and Jenkins, 1984). All cows were compared on a similar mature weight. These differences in maintenance requirements are related primarily to level of milk production. Similar differences in maintenance requirements are observed when comparing cows of high versus low milk production within the same breed. These differences appear to persist, even during the dry period, thus changing the energy requirements previously shown in Figure 5.10 b and c.

A research study (Van Oijen, et al., 1993) compared the economic efficiency of three groups of cows of the same mature size but differing in level of milk production—high (> 20 lbs milk/day), medium (18 lbs milk/day), and low (14 lbs milk/day). The low milk group was more economically efficient (dollar value of output per $100 of total input cost) than the other higher milk groups. The low milk group would not be considered low by industry standards as that amount of milk would produce 500 lb calves with good forage availability.

The emphasis on increased frame size, growth rate, and mature size during the past 15 years has significantly increased dystocia (calving difficulty). It is estimated that dystocia, primarily due to birth weight, costs the beef industry $850 million each year. This large cost can be reduced by selecting bulls that sire lighter birth weights and by saving replacement heifers with moderate birth weights (< 90 lbs) and from bulls with birth weight EPDs less than 0 (Angus base). Calves with moderate birth weights can have excellent growth (weaning weights, feedlot gain, and yearling weights) by emphasizing the appropriate selection criteria. Attempting to maximize heterosis may increase production costs above optimal cost levels of birth weight, milk production, weaning weight, and mature cow weight. The primary value of heterosis in low-cost cow herds is (1) increased calf survival and (2) increased cow longevity—best measured by lifetime performance of cows. One research study demonstrated that by 12 years of age crossbred cows produced 1.6 more calves and 875 lbs more calf weaning weight than straightbred cows (Sacco et al., 1989).

Bull costs can range from $10–$85 per cow as determined primarily by initial purchase price and bull-to-cow ratios (Table 5.24). There is ample evidence that the "right genetics" can be purchased for less than $2,000 per bull and that bull-to-cow ratios can be above 1 to 30 in many operations. However, genetic selection is not a decision that can be short-changed. One of the most important control points in the beef industry is the choice of both sires and mating systems.

Alliances between seedstock breeders, cow-calf producers, and cattle feeders could reduce the breakeven price for yearling bulls to $1,000–$1,500 per head. This cost reduction of several hundred dollars per bull could be shared so that profitability is enhanced for all three groups in the alliance. What is needed to implement this cost reduction is vision and commitment. Alliances will encourage appropriate priorities to reproduction, growth, and carcass traits so that the antagonisms between these traits can be avoided and effectively managed.

Reducing Labor Costs

It is common to find cow-calf operations of similar size with varying numbers of employees. For example, a 300 cow-calf operation may have 1–2 employees while another of the same size has 3–4 employees. In this example, labor cost per cow in one operation is two to four times higher than that of the other operation.

Some large, low-cost operations have reduced labor cost by having one employee responsible for 800–1,000 cows with some labor-intensive activities (e.g., branding and weaning)

TABLE 5.24 Bull Cost Per Cow Based on Varying Bull Purchase Prices and Bull-to-Cow Ratios

Bull Purchase Price	Total Bull Cost (4 yrs)[a]	Bull-to-Cow Ratio[b] 1:20	1:30	1:40
		Bull Cost per Cow		
$1,000	$1,220	$16.94	$11.30	$ 8.47
2,000	2,640	36.66	24.44	18.33
3,000	4,060	56.39	37.59	28.19
4,000	5,480	76.11	50.74	38.05

[a]Assumes $1,000 salvage value, $200/yr bull cost (feed, maintenance, etc.), 10% risk (based on purchase price), 8% interest for 4 years.
[b]Assumes bull is used for 4 years; 90% conception rate each year.
Source: Colorado State University.

shared by several employees. Some smaller operations can improve their forage management and increase the number of cows by 20–30% on the same land area. This increased carrying capacity can decrease labor cost per cow.

The well-proven 80–20% rule of business tells labor/management to do the "right things" first before doing "things right." Work activities should receive priority that answers the question, "How can my time be spent most effectively in lowering cost of production by concentrating on $100/hour activities before $5/hour activities?" Generally this means more time spent on financial, forage, and human resource management prior to cattle management. Cattle management should not be overlooked, however, as cattle that are more problem-free (i.e., ease of calving, good dispositions, good mothering ability, etc.) require less labor—thus being more cost effective.

Less labor is usually required by improving grazing management and decreasing dependence on mechanical harvesting of forage. Some operations have reduced labor and equipment costs through contracting these services. At the same time, once the appropriate number of employees has been determined, it is critical to attract and retain good talent. Employee development is typically an area of weakness in most agricultural production settings.

Maintaining a Strong Herd Health Program

Low-cost producers do not cut corners on herd health programs, yet they manage herd health costs and returns precisely. Most low-cost producers realize that cattle death loss is typically a small part of total health-related losses. In many herds, sickness of cows and calves is very costly because of lower reproduction rates, reduced weight gain, poor feed conversion, and high health costs associated with clinical cases of sickness. Subclinical illnesses in cattle are difficult to identify, however, they are costly since animal performance is reduced while input costs remain high.

Most cost-effective herd health programs focus on preventing (1) reproductive diseases in the cows, (2) calf scours, and (3) respiratory diseases in the calves. Specific herd health programs and their associated costs depend on risk factors such as (1) health history of the herd and other herds in the immediate area, (2) nutritional program, (3) exposure to other animals, (4) stress factors associated with the environment, (5) marketing program, and (6) mismanaged cattle (vaccinations and treating animals are not substitutes for poor management).

Strong herd health programs can reduce feedlot costs (lower costs of gain, less death loss, and lower health treatment costs) and reduce carcass losses through proper administration of health products. These health strategies can enhance profits, especially in retained ownership programs.

Herd health is usually less costly when preventative plans are developed and implemented with a veterinarian rather than the higher costs of treatment when "crisis" health problems occur.

MATCHING COWS TO THEIR ECONOMICAL ENVIRONMENT

Table 5.25 outlines the major management factors affecting annual cow costs. Producers who understand these factors and manage them effectively are likely to be high-profit commercial producers.

The management challenge for cow-calf producers is to match the biological type of cows to their most cost-effective environment (Fig. 5.12). This is not an easy task, but a logical sequence can be found in the decision-making process:

1. Costs must be measured and recorded.
2. Feed costs should be carefully assessed and compared with those of other operations in the same area.

TABLE 5.25 Major Management Principles Affecting Annual Cow Costs

1. ENTERPRISE BUDGET
   a. Annual cow costs can be determined and monitored through an enterprise budget. Producers must know annual cow costs to calculate breakeven prices and assess net returns.
   b. The individual cost components of the enterprise budget should be accurately recorded, then compared with other operations with similar environmental conditions. A specific cow-calf producer can then determine if his or her annual cow costs are high, low, or about average. Particular attention should be paid to feed, machinery, labor, interest, and health costs. Comparisons within the enterprise across time are also useful.
   c. Feed costs should be most critically assessed because they comprise from 40–70% of the annual cow cost. Producers should determine their most economical feed environment, then decide what biological types of cows are best matched to this feed environment. Grazed forage is usually the most economical feed environment, as harvested and purchased feeds are more expensive in most cases.
2. BIOLOGICAL TYPES OF COWS
   a. *Puberty.* Heifers that express early puberty and calve at 24 months of age under a low-cost feeding program will contribute to lower annual cow costs. This implies optimum weaning weights and frame size and using breeds (biological types within breeds) that are known for early puberty. An economical feeding program from weaning to a target breeding weight is essential.
   b. *Calving ease.* Calving ease, especially needed in 2-year-old and 3-year-old females, reduces annual cow costs as less labor is required, health costs for cow and calf decrease, females will breed earlier and thus require less feed to produce a calf per cow per year.
   c. *Milk production.* If milk production is too low, the weaning weight of the calf is poor and the cow puts the nutrients "on her back" instead of "in her calf." Milk production that is too high results in a heavy calf but an open cow or late bred cow, which will increase annual cow cost. Research studies show that moderate levels of milk are more cost effective and profitable.
   d. *Mature weight.* Research studies demonstrate that moderate mature cow weights (1,000–1,250 lbs under most cow-calf environments) are most profitable. Cows larger than these optimum weights can be biologically efficient but may not be economically efficient.
   e. *Rebreeding.* Rebreeding performance is best measured by body condition score (BCS) at calving time. Managing the BCS by utilizing primarily grazed forage with minimum amounts of harvested and purchased feeds will keep annual cow cost low.
   f. *Longevity.* Cows that reproduce regularly under a low-cost feed environment will remain in the herd for a long period of time. This situation requires a lower heifer replacement rate which usually lowers annual cow cost. Replacement heifer costs are high in most herds and high rates of heifer replacement are costly.
3. MANAGING FEED COSTS
   a. *Calving season.* The calving season for many operations should be approximately 30–45 days prior to grazed green forage. Earlier calving seasons will require large amounts of more expensive harvested and purchased feeds. Later calving seasons do not effectively utilize the peak of high-quality nutrients from grazed forage.
   b. See sections 1c, 2a, 2c, 2d, and 2e in this table. When BCS of cows is low at calving and breeding, producers should determine the most economical management decision—e.g., increase the supplemental feed to increase BCS or change biological type of cows to increase BCS on the original feed supply.
   c. *Heifer development.* Delaying the majority of weight gain until late in heifer development may decrease costs (up to 15%) without detrimental effects on reproductive performance.
4. INCREASING COW NUMBERS ON SAME LAND AREA
   a. Overhead costs (e.g., land payments, insurance, taxes, vehicle costs, and others) are expenses that remain approximately the same regardless of the number of beef cows in the operation. As cow numbers increase on the same land area, total overhead costs change little if any; however, overhead costs per cow will decrease.
   b. Cow numbers can be increased on the same land area primarily by improving forage production and grazing management. This reduces the number of acres required per cow (see Table 5.13).

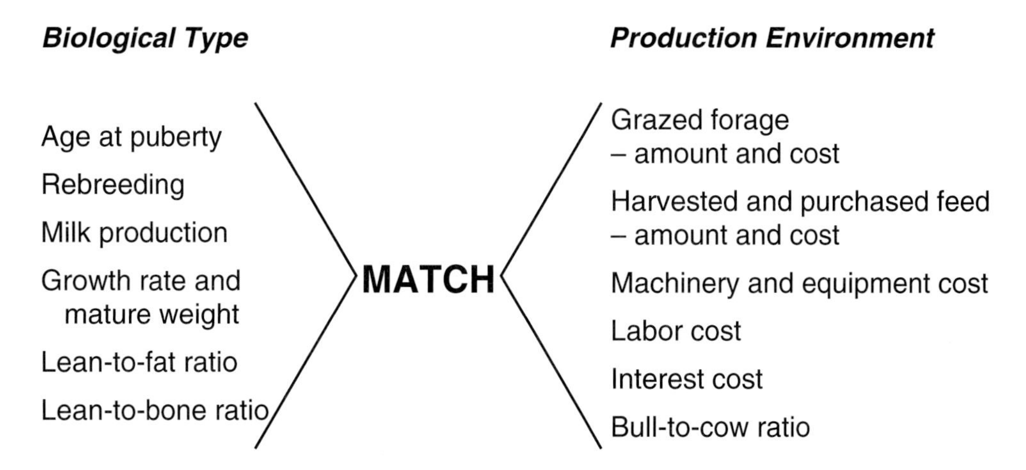

FIGURE 5.12 Matching biological types of cows to the most economical production environment is a management challenge that significantly affects profitability.

3. The most economical feed supply should be determined by answering the following questions:
   a. Can the grazing of forage be extended for the cows during the year and still meet nutrient requirements?
   b. Does the calving season best match the available green forage (usually calving a month prior to adequate levels of green forage)?
   c. Could the fall grazing season be extended by weaning calves earlier and allowing cows to add body condition prior to winter weather?
4. After identifying the most economical feed supply, determine the biological type of cattle that will be most profitable.
5. Integrate all management decisions by calculating the breakeven price on calves.

Figure 5.12 identifies the primary traits used in defining biological types. Breeds and breed types have been used to reflect biological types (refer to Chapters 4 and 13). Breeds can be used in a general sense but for individual operations biological types must be described more specifically because several biological types exist within a breed.

Tables 5.26 and 5.27 give the results of an excellent research study that focused on biological and economic assessments of different biological types of cows in a typical northern Great Plains environment. While the results may not apply to all other cow-calf environments, the importance of evaluating the economics of different biological types is emphasized. An analysis of biological types in several environments, however, indicates that moderate birth weight, moderate milk, moderate mature weight, and moderate frame size are realistic targets for many cow-calf producers.

The data in Tables 5.26 and 5.27 demonstrate that moderate performance in certain traits will result in maximum profits. However, it should be noted that the poor performance of the Herefords occurred because only 59% of the original group of heifers reached puberty at 14 months of age. There are different biological types of Herefords that perform differently for the traits shown. In addition, when Simmental bulls were crossed on Hereford cows to produce the S × H crossbreds, this mating group experienced the largest birth weight, most calving difficulty, and the lowest calf survival rate.

Managing cow-calf resources implies matching the cows profitably to the available resources. Therefore, in most of the western and Great Plains states and possibly other states as

TABLE 5.26 Performance of Different Breed Types (Biological Types) over Several Calf Crops

	Biological Type			Milk Production per Day (lbs)		Weaning Wt (lbs) per Female Exposed[a]			
Breed Type of Dam[b]	Mature Size	Milk	Puberty/ Rebreeding	2-Year-Old Dams[c]	5–8-Year-Old Dams	2-Year-Old Dams[d]	5–8-Year-Old Dams	Mature Cow Wt (lbs)	% Cows in Herd after 5 Years
HH	Medium	Low	Late	16.5	15.8	213	345	1,225	41
AH	Medium	Medium	Early	18.5	20.0	359	359	1,270	52
1S3H	Medium	Low+	Medium	18.4	18.7	260	394	1,240	53
SH	Medium+	Medium	Early	20.9	22.4	365	416	1,290	67
3S1H	Large	High	Late+	22.1	24.4	315	372	1,330	42

[a]Weaning weight at 180 days of age.
[b]Hereford (H), Angus (A), and Simmental (S).
[c]Taken 40 days into lactation.
[d]Taken 130 days into lactation.
Source: Montana State University (*J. Anim. Sci.* 68:54, 1910); video, "Cows That Fit Montana."

TABLE 5.27 Simulated Measures of Lifetime Performance for Different Biological Types of Cows[a]

Breed Type of Dam	Biological Type	Number of Matings	Lb Calf Produced per Cow Exposed	Ratio of Calf Wt to Cow Wt Sold	Breakeven Steer Price ($/lb)	Net Return per Cow Exposed ($/cow)
HH	Table 5.26	199	283	1.12	$1.02	$−6.44
AH	Table 5.26	243	339	1.56	0.87	61.50
1S3H	Table 5.26	229	332	1.55	0.90	45.57
SH	Table 5.26	246	370	1.70	0.82	81.78
3S1H	Table 5.26	199	338	1.27	0.89	48.40

[a]Five replications were simulated for each breed type of dam × sire breed combination. Each replication simulated the life-cycle inputs and outputs for sixty replacement heifers in each biological type.
Source: Montana State University (Proceedings, Western Section American Society of Animal Science 43: 43).

well, the cows should be of a medium-frame size, have medium milk, have early sexual maturity, and have a feed supply that originates from low-cost grazed forage.

Replacement Heifer Development

Generally, replacement heifers are weaned in the fall, are bred the following spring or summer, and calve the following spring at two years of age. From an economic standpoint, this is the best management scheme for many operations. It is not recommended that replacement heifers be creep fed because it leads to deposition of fat into the mammary system that can be detrimental to future milk production.

Puberty is primarily a function of age and weight. Most heifers are old enough to breed at 14–16 months of age. The critical limit is weight. A general management guideline has been to feed heifers to attain 65% of their mature weight by the start of the breeding season. To ac-

TABLE 5.28 Performance of Heifers Raised on Three Different Nutritional Development Programs

					Summary of Female Culling (N)						
							Death Loss in First Calves				
Treatment	N	Weight at Breeding (lbs)	Total ME Intake (med/high)	Pregnancy (%)	Open/ Aborted Yearling	Open as 2-Year Olds	Birth	<72 hrs	>72 hrs	Total
High[a]	94	865	3,072	90.4	10	6	4	3	9	16
Medium[b]	94	845	2,854	92.6	9	7	2	1	6	9
Low-High[c]	94	843	2,652	91.4	10	3	5	6	9	20

[a]263 kcal ME/$(BW_{kg})^{0.75}$ for 205 days.
[b]238 kcal ME/$(BW_{kg})^{0.75}$ for 205 days.
[c]257 kcal ME/$(BW_{kg})^{0.75}$ for 83 days followed by 277 kcal ME/$(BW_{kg})^{0.75}$ for 122 days.
Source: Freetly, Ferrell, and Jenkins, (2001).

complish this, most heifers need to gain 0.5 to 1.5 lbs per day depending on their weaning weight. There are two primary approaches to achieving target breeding weights—feed heifers to gain at a consistent rate over the full development period or feed heifers to grow slowly with a ration increase to boost gains in the two months prior to mating. Research results suggest that heifers fed to gain 0.50 lb/day early with an accelerated gain of 2.5 lbs/day two months prior to breeding are equal to, if not superior, in reproductive performance as compared with females fed to gain at a constant rate. Furthermore, under most feed price conditions, the slow-fast program was cheaper to implement. As Table 5.28 demonstrates, pregnancy rates are comparable for high, medium, and low-high dietary protocols. The same study found that these dietary regimes had no effect on birth weight, calving difficulty, or first-calf weaning weight. Furthermore, heifers consumed the least on the low-high treatment. However, there was an increase in calf death loss for the low-high protocol particularly when compared with the medium feeding level.

Determining the number of females to be retained each year is a critical decision in terms of its impact on both expenses and revenues. Four heifer retention strategies were evaluated by Iowa State researchers. Steady size (SS) to return the same number of heifers each year, cash flow (CF) to maintain income at a steady level (differing numbers of heifers sold to attain the goal), dollar-cost averaging (DCA) in which the same value of heifers is retained annually, and rolling average value (RAV) in which the 10-year average value of heifers is retained.

Table 5.29 shows that DCA and RAV strategies were the most profitable. This results from the effect of having lower-cost heifers entering production during the price peaks of the cattle cycle, and thus, more calves are sold at higher prices. Also, when heifer prices are high, more females are sold.

However, there is a downside. Producers must accept more variation in year-to-year cash flow and producers must have flexibility in terms of access to land as the herd size is variable over time.

Evaluating Management Alternatives

Management alternatives for cow-calf producers can be evaluated by determining how they affect productivity and costs. Table 5.30 shows how several common management decisions can affect the breakeven price for an operation that has base production management practices. Base production implies that none of the management practices listed in Table 5.30 have been implemented.

TABLE 5.29 Heifer Retention Strategy/Impact on Revenue, Return Over Economic Cost, and Return Over Cash Cost (30-year period)

Annual Revenue	Average ($1,000)	Minimum ($1,000)	Maximum ($1,000)	Ending Total Revenue ($1,000)
SS	43.6	26.9	64.7	39.6
CF	36.4	14.0	65.9	14.0
DCA	47.4	24.7	96.2	41.8
RAV	43.8	22.5	75.1	49.2
Return Over Total Economic Cost				
SS	−1.8	−16.3	19.4	0.5
CF	−0.9	−11.2	2.9	2.7
DCA	0.1	−21.1	37.5	1.7
RAV	−0.4	−17.6	27.8	3.1
Return Over Cash Cost				
SS	4.8	−7.9	27.2	5.9
CF	4.1	2.9	6.4	4.7
DCA	6.5	−14.9	48.0	7.8
RAV	5.6	−12.4	35.9	8.4

SS = Steady Size, CF = Cash Flow, DCA = Dollar-Cost Averaging, RAV = Rolling Average Value.
Source: Iowa State University, (2000).

TABLE 5.30 Evaluating Management Decisions Via Breakeven Prices

Management Practice	% Calf Crop	Weaning Wt (lbs)	Lb Calf Weaned per Cow Exposed	Annual Cow Cost ($)	Breakeven Prices ($/cwt of calf)	Implement
Base production	80	440	352	325	92.33	
Growth implant	80	450	360	327	90.80	Yes
Creep feed	80	480	384	360	93.75	No
Crossbreeding[a]	82	460	377	330	87.53	Yes
Reduce supplemental feed	78	430	335	300	89.55	Yes

[a]Progeny of a 2-breed cross.
Source: Colorado State University.

Partial budgeting (covered in Chapter 3) is utilized in Table 5.30 to determine how each management practice increases production (percent calf crop, weaning weight, or both) and to estimate the cost of implementing the management practice. For example, for growth implants the weaning weight is estimated to increase 10 lbs per calf with a projected cost of $2 per calf. This would lower the breakeven price from $92.33 (base production) to $90.80.

Note that even those decisions that lower productivity may yield economic benefits to the enterprise. Once again, it is important to point out that decision-making in a diverse system requires an understanding of the dynamics in the system (Chapter 3). Barry Dunn of South Dakota State University (2000) describes five key characteristics of complex systems: (1) They are tightly linked such that an impact on one component yields multiple responses. (2) Because they are dynamic, change occurs in multiple time scales (short versus intermediate versus long-term). (3) They are resistant to the obvious solution. (4) Complex systems are often counterintuitive as the response to a decision may be quite distant in time and space and, therefore, the simple solution may yield very undesirable consequences. (5) Trade-offs between long- and short-term benefits are often demonstrated. Decision-making must deal with these characteristics of the system.

Low breakeven prices are achieved when percent calf crop and weaning weight are relatively high and costs are kept low. Low-cost producers will have more years of profitability than high-cost producers during typical periods of fluctuating calf prices (Fig. 5.13).

Table 5.31 shows the actual costs from a very low-cost cow-calf operation. Besides demonstrating how low-cost some operations can be, it also justifies how low-cost operations can be profitable year after year. Prices of 500 lb calves have not dropped below 50 cents per pound during the past 20 years.

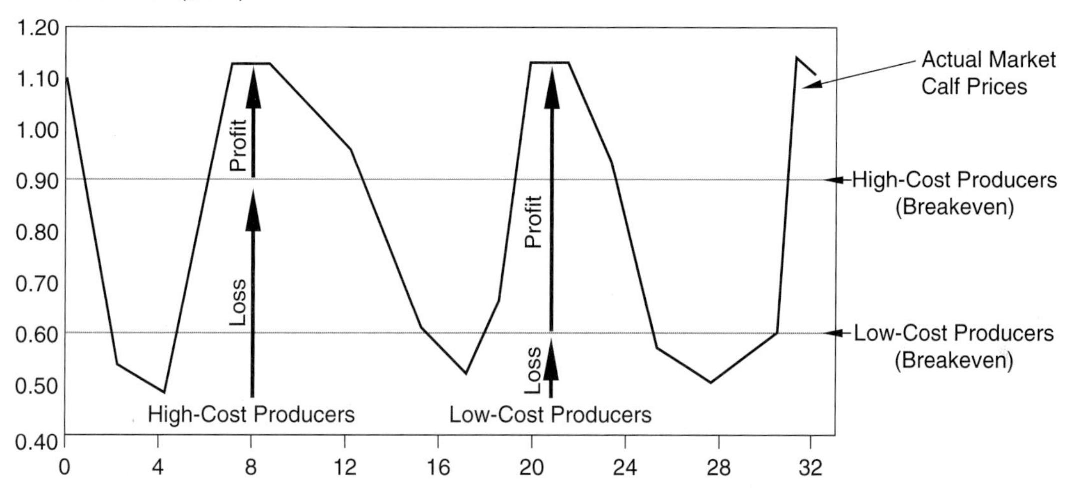

FIGURE 5.13 Returns for high- and low-cost producers as calf prices change.
Source: Colorado State University.

TABLE 5.31 Annual Costs for a Cow-Calf Operation That Is Considered Low-Cost (1996)

	Cost		
Cost Category	$ per Cow	$ per Calf	Cents per lb of Calf
Labor and fringe benefits	$31	$33.94	$.0879
Feed (purchased or produced that are fed on the ranch)	17	18.91	.0490
Pasture (grazing, crop aftermath; all overhead is included including fences, water development, equipment, etc.)	64	70.25	.1820
Cow depreciation	3	3.39	.0088
Other direct costs (vet., medicine, taxes, etc); all capital costs not covered in pasture category including land[a] and other capital items	20	22.42	.0581
Office and administration	16	17.62	.0456
Total	$151	$166.53	$.4313[b]

[a]Costs associated with initial costs of land are not included.
[b]Cost per pound for 1995 was $0.4386
Source: An actual cow-calf operation involving a herd size of several hundred cows.

MARKETING DECISIONS

Marketing begins with the decision about what will be produced. Thus, for cow-calf producers marketing actually starts at breeding time. However, too many producers do not begin thinking about marketing until a few weeks or days prior to weaning the calves.

Marketing decisions should be based on the profitability formula:

$$\text{Profitability} = (\text{product} \times \text{price}) - \text{costs}$$

Product is usually reflected in numbers and pounds and may include breeding bulls, weaned calves, background yearlings, slaughter cattle, carcass, or retail products. Price of product and cost of production are also important components of the marketing decision process. If cattle producers are to make intelligent and profitable marketing decisions, they must:

1. Know the cost of production and breakeven prices.
2. Know the value of the product they have produced, including how it will perform and be valued by other segments of the industry.
3. Understand prices (e.g., factors affecting prices, how to obtain current price information).
4. Know how to integrate the value of product, cost of production, and prices in evaluating different market alternatives.

Marketing Alternatives

Cow-calf producers have several marketing alternatives. They may (1) sell calves at weaning, (2) sell bred heifers, (3) sell cull cows for slaughter or for breeding purposes to other producers who can make the cows profitable for a few more years, or (4) retain ownership via backgrounding, or grazing yearlings, or through the feedlot, or retail stage. Any combination of these alternatives may also be used.

Selling calves at weaning is the most common marketing alternative because (1) it is a traditional marketing method; (2) the loan agency offer requires immediate payment rather than extending credit over a longer period of time; and (3) income tax considerations arising from shifting income from one year to another may restrict changes in the marketing strategy.

Table 5.32 gives some price guidelines that can be used to help make a decision about retained ownership. For example, assume the fall weaned calves would be expected to bring \$65 per cwt next May and the projected cost of gain was 55 cents a pound. If weaned calf prices were above \$70 per cwt, producers should consider selling the weaned calves or retaining ownership if calf prices were below \$70.

More cow-calf producers are retaining ownership of their calves through the growing and feedlot stages. Producers who know the cost of production and the value of the calves in

TABLE 5.32 Comparing Retained Ownership Versus Selling Weaned Calves

Projected Fed Price ($/cwt)	Cost of Feedlot Gain ($/cwt)		
	50	55	60
	Weaned Calf Sale Price ($/cwt)		
60	65	60	55
65	75	70	65
70	85	80	75

Source: Cattle-Fax Cow-Calf Focus.

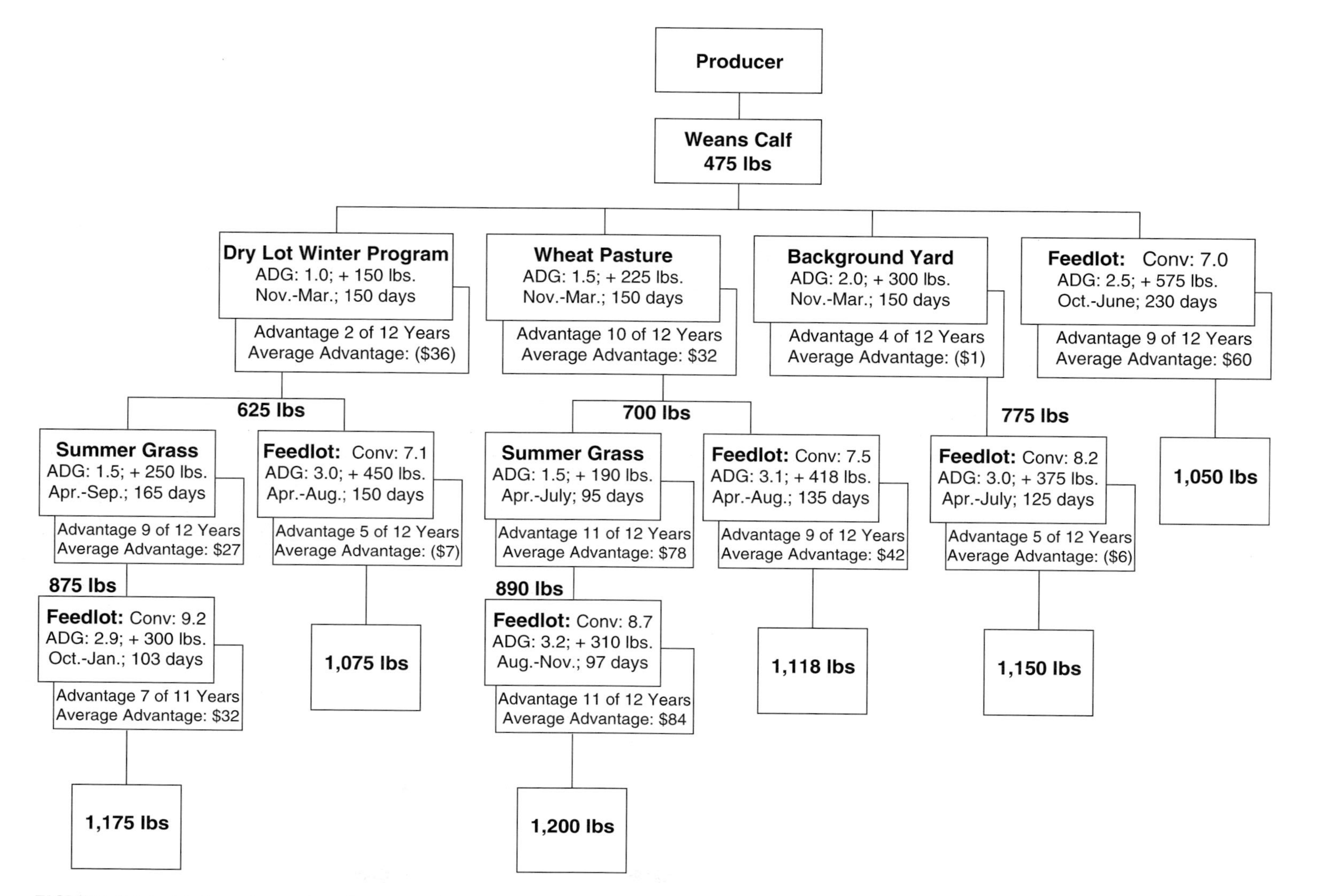

FIGURE 5.14 Various alternatives for retained ownership of weaned calves (for weaned calf of 475 lbs).
Source: Cattle-Fax.

and out of the feedlot are using retained ownership to a profitable advantage. In some cases producers enter an alliance with the cattle feeder rather than totally own the cattle themselves. The alliance may share the value of the cattle and the profit or loss in the feeding enterprise.

Figure 5.14 shows various alternatives for cow-calf producers retaining ownership of weaned calves. Market weights, performance of the calves, months involved, and economic advantage of retained ownership compared with selling the calves at weaning are shown. Table 5.33 shows the economic comparison of selling the weaned calf versus retaining ownership through several other production and marketing programs. These comparisons are based on average costs of production.

Cow-calf producers considering retaining ownership of their calves should first evaluate the daily gain, cost of gain, health, and net return on a sample of these calves through the feedlot and carcass production phases. This reduces the risk of making an incorrect management decision on retained ownership. Being aware of market and price trends is also helpful.

ESTABLISHING A COMMERCIAL COW-CALF OPERATION

Many cow-calf operations have existed for decades, being inherited over several successive generations. During the past 10–20 years, land has become a major investment commodity primarily because of an expanding population with large available financial resources. Thus, the price of land has exceeded its ability to be profitable under many phases of agricultural production, including the production of beef cattle.

Cost of a commercial cow-calf operation can be evaluated on a per-cow basis. This would include the costs of land, cow, buildings, equipment, and other minor expenses. It is not uncommon to have investment costs of $1,500–$4,000 on a per-cow basis. With an interest cost of 10%, this amounts to an annual interest cost of $150–$400. Thus with interest cost and a $300 annual cow cost, the annual production cost would be approximately $450–$700 per cow. Even when calf prices are high, they will not cover all of these costs. For example, if 450 lb calves sold for $1 per pound, a $1,500 investment cost per cow would require approximately $450 to break even ($300 annual cow cost plus $150 interest). Producers cannot plan on high calf prices to continue for long periods of time if beef continues as a commodity product subject to the typical beef cattle cycles.

Total capital investments for an economical cattle operation are extremely high. Several studies show that an economical unit is approximately 300 head. At $3,000 per cow, the total investment would be $900,000. It is evident that high total capital investment plus high annual interest costs prevent many potential beef producers from purchasing new beef operations. Some established cow-calf producers may be encountering a financial crisis because of an excessive debt load. They may solve this financial problem by selling some assets (e.g., land) or generating additional income from other resources (e.g., hunting, fishing, etc.). These solutions may allow them to keep an economic cow-calf unit intact.

Nearly half of the beef producers in the United States are 55 years of age or older. The primary reason for aging cow-calf producers is that younger people do not have the capital or management experience. Some individuals are successfully leasing cow-calf operations rather than trying to purchase the land.

Land prices and interest costs do fluctuate over time (see Table 5.34). There are certain times when cattle operations can be purchased on an economically sound basis. These purchases may be made by new producers or by established producers seeking to add to an existing operation. It is important to also evaluate the regional differences in land values (Fig. 5.15). Note that cow-calf production is generally concentrated in those states with the lowest pasture land values.

TABLE 5.33 Spring Calving Comparison of Retained Ownership Options to Sale of Weaned Calf 475 lb Weaned Steer Calf ($/hd), 1980–1998

Sold as a Weaned Calf In Fall of Same Year					Advantage of Programs Compared to Selling at Weaning								
Year Calf Born	Average Cost of Producing Calf	Average Price Received	Profit or (Loss)	Year Sold	Dry Lot Winter Program	Dry Lot Winter Program and Summer Grass	Dry Lot Winter Program, Summer Grass, and Feedlot	Wheat Pasture	Wheat Pasture and Summer Grass	Wheat Pasture, Summer Grass, and Feedlot	Wheat Pasture and Feedlot	Back-ground Yard	Back-ground Yard and Feedlot
1980	318	384	66	1981	(50)	(24)	(35)	(8)	(39)	(2)	17	(58)	(22)
1981	333	323	(10)	1982	(18)	43	40	31	69	70	92	16	77
1982	332	319	(13)	1983	10	(8)	123	74	35	37	41	61	39
1983	363	311	(52)	1984	8	41	105	80	69	126	71	37	38
1984	369	331	(38)	1985	(14)	(29)	50	53	21	144	(47)	32	(76)
1985	359	327	(32)	1986	(58)	36	99	4	21	156	62	(10)	28
1986	340	334	(6)	1987	26	139	157	64	110	182	152	97	184
1987	323	425	102	1988	16	67	107	67	34	114	94	96	79
1988	368	457	89	1989	(47)	45	137	26	86	121	67	24	27
1989	385	466	81	1990	(15)	81	145	20	107	184	98	29	67
1990	390	480	90	1991	20	41	40	53	95	48	(21)	69	39
1991	378	471	93	1992	(51)	32	160	(9)	54	119	56	13	51
1992	382	444	62	1993	32	95	62	62	123	103	118	101	132
1993	389	477	88	1994	(36)	(63)	62	7	(18)	32	(28)	16	(55)
1994	384	400	16	1995	(61)	(46)	(41)	(23)	(24)	46	(9)	0	4
1995	382	332	(50)	1996	(106)	3	71	(44)	5	97	(17)	(68)	(60)
1996	399	313	(86)	1997	60	135	68	50	182	137	86	65	70
1997	391	436	45	1998	(21)	(100)	(66)	(8)	(43)	(18)	(86)	0	(78)
1998	405	373	(32)	1999	(4)			26				50	
			Average Advantage		(17)	35	79	30	55	101	49	31	36
			Highest Return		60	139	160	80	182	184	152	101	184
			Lowest Return		(106)	(63)	(41)	(44)	(39)	(2)	(47)	(68)	(76)

*Cattle are sold in January of the next year.
Source: Cattle-Fax.

TABLE 5.34 Land and Interest Costs, 1970–2000

Year	Prime Interest Rate (%)[a]	Land Value ($/acre)[b]
1970	7.9	196
1975	7.9	340
1980	15.3	737
1985	9.9	713
1990	10.0	668
1995	8.8	832
1997	8.5	926
1999	8.0	1,020
2000	9.2	1,050

[a]From 1970–2000, the lowest prime rate was 4.75% for Feb. of 1972; the highest was 20.5% for August of 1981.
[b]These are average farm real estate values in the United States. Land prices for pasture would be lower. Average pasture values per acre for 1997, 1998, 1999, and 2000 were $466, $489, $503, and $517, respectively.
Source: USDA, Federal Reserve Bank.

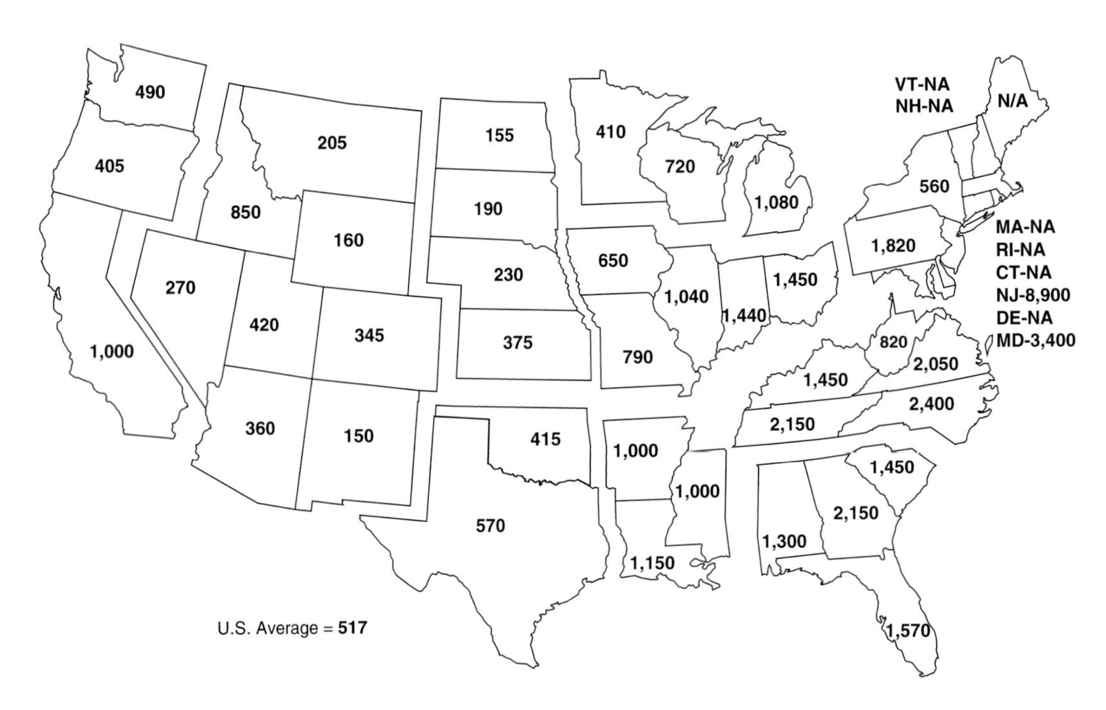

FIGURE 5.15 Pasture Average Value per Acre, January 1, 2000.
Source: NASS, USDA.

As noted earlier in the chapter an assessment of economical cow-calf operations shows that a large number of cow-calf producers do not depend primarily or even substantially on beef cattle for their incomes. A 1997 NAHMS survey showed only 380 out of 2,713 cow-calf operations have their primary source of income from the beef herd. A cowherd size of 300 cows generated approximately 80% of producers' annual income (NAHMS). The producers having 100 or fewer cattle have other, more important sources of agricultural income or they have off-farm jobs. They are not producing cattle for significant income purposes but as a means of maintaining a lifestyle outside the city or of using marginal cropland. These producers are not as likely to make production and marketing decisions based on the same economic considerations directly affecting other beef cattle producers.

Integrating and Sustaining the Resources

The challenge for a producer's leadership and management decisions process is to integrate and sustain the human, financial, forage, and cattle resources. Integration of resources reflects how the resources interact with one another. For example, if a producer decides to increase calf crop percentage, how much more forage or supplemental feed will it require, or how much less forage will it take if mature weight and milk production are decreased, and how will these decisions affect the breakeven price of the weaned calves?

The focus of the managers should be—"How low can I get the breakeven price and still sustain it over time?" If a resource cannot be sustained or improved it is likely being depleted. Short-term economics reflects resource depletion, however, long-term economics with sustained breakeven and sustained profits should be the goal.

Cow-calf producers with a vision of long-term profitability will have an optimum combination of cattle productivity (reproduction, growth, and end product) and low-cost production. Even with retained ownership programs, economic priorities for the most profitable cow-calf producers should be:

1. Cost of production (keeping breakeven prices of calves near $.60/lb or lower). This assumes all costs are measured and effectively managed while achieving optimal levels of percent calf crop and weaning weight.
2. Reproduction (optimum levels of percent calf crop—typically in the mid 80% to low 90% for most cow-calf operations—based on calves weaned per 100 cows exposed). Optimal levels of reproduction are achieved under low-cost, extended grazing programs.
3. Growth (moderate average weaning weights in the 475–550 lb range for 7-month-old calves, assuming an average forage availability; feedlot gain for calves in the 3.0–4.0 lbs/day range).
4. End product (700–800 lb carcass weights; 2.0–3.4 yield grades, while avoiding discounts on yield grade 4s; quality grade of >70% choice). Choices of breeds and biological type for tenderness and palatability to improve consumer market share are presented in Chapter 13.

Cost of production and maternal traits (primarily reproduction) are far more important than growth and end product in determining cow-calf profitability. Therefore, cow-calf producers should have vision and motivation to keep a focus on "first things first." Alliances between seedstock, cow-calf, and feedlot producers will encourage optimal combinations of cow, feedlot, and carcass traits for low-cost/high-profit cattle production.

MANAGEMENT SYSTEMS HIGHLIGHTS

1. Producers should provide the leadership to change paradigms and to have a written plan to accomplish the vision/mission and goals to lower costs of production and increase consumer market share.
2. Major focus is reducing costs of production by:
   - Evaluating calving season, utilizing grazing systems to extend the green forage, stockpiling forage for fall, winter, and early spring, lease feeding aftermath, and minimizing the use of harvested and purchased supplemental feed.

- Matching breed and biological type of cattle to the forage management system.
- Reducing land costs associated with cow-calf operation.
- Reducing equipment costs by harvesting less hay and other supplemental feed.
- Optimizing improvements in reproduction efficiency, genetic selection, and heterosis.
- Evaluating bull-to-cow ratio and bull purchase costs (assess sources of bulls costing < $2,000) and consider bull-to-cow ratios of greater than 1 to 30.
- Using cost-effective health programs.
- Evaluating alternatives in heifer development.

3. Making holistic and integrative management decisions that will sustain or improve all the resources (e.g., land, financial, cattle, and people) while increasing or sustaining high profits is the goal.

REFERENCES

Publications

Adams, D. C., et al. 1996. Matching the Cow with Forage Resources. *Rangelands* 18: 57.

American Angus Association. 2001. *Sire Evaluation.* Saint Joseph, MO.

Azzam, S. M., et al. 1993. Environmental Effects on Neonatal Mortality of Beef Calves. *J. Anim. Sci.* 71: 282.

Bellows, R. A., Staigmiller, R. B., and Short, R. E. 1990. *Studies on Calving Difficulty.* Research for Rangeland Based Beef Production. Bozeman, MT: Montana Agric. Expt. Station.

Beverly, J. R., and Spitzer, J. C. 1980. *Management of Replacement Heifers for a High Reproductive and Calving Rate.* Texas A and M Ext. Public. B-1213.

Blasi, D., and Corah, L. R. 1993. *Assessment of Forage Resources to Determine the Ideal Cow.* Proceedings of the Cow Calf Conference III. Manhattan, KS: Kansas State University.

Cattle Industry Reference Guide. 2001. Cattle-Fax, Greenwood Village, CO 80111.

Corah, L. R., and Blanding, M. R. 1992. Use All Four Keys to Higher Profits in the Cow Herd. *BEEF* (Spring).

Cow-Calf Focus (periodical). Cattle-Fax, P.O. Box 3947, Englewood, CO 80155.

Cow-Calf Management (Reducing Production Costs). 1991. *BEEF.* 27, no. 7A (Spring).

Cowley, J., Buskirk, D. D., and Black, J. R. 2001. *MSU-IRM-SPA Analysis for Cow-Calf Operations—1999 Summary.* East Lansing, MI. Research and Demonstration Report. Michigan State University.

Davis, K. C., Tess, M. W., Kress, D. D., Boornbos, D. D., Anderson, D. C., and Greer, R. C. 1992. *Live Cycle Evaluation of Five Biological Types of Beef Cattle in a Range Production System.* Proc., WSASAS 43: 43.

Dunn, B. March 2000. Interview with *Beef Today.*

Ferrell, C. L., and Jenkins, T. G. 1984. Energy utilization by mature, nonpregnant, nonlactating cows of different types. *J. Anim. Sci.* 58: 234.

Fields, M. J., and Sand, R. S. 1995. *Factors Affecting Calf Crop.* Boca Raton, FL: CRC Press.

Freetly, H. C., Ferrell, C. L., and Jenkins, T. G. 2001. Production Performance of Beef Cows Raised on Three Different Nutritionally Controlled Heifer Development Programs. *J. Anim. Sci.* 79: 819.

Gutierrez, P. H., and Dalsted, N. L. 1990. *Buying Versus Raising Replacement Heifers: Economic Issues.* Bull and Heifer Replacement Workshops. Fort Collins, CO: Colorado State University.

Hughes, H. 1998. *A Comparative Analysis of a North Dakota Demonstration Herd to the North Dakota Database.* Fargo, ND: North Dakota State University.

Hughes, H. March 2000. *My Top Four Recommendations for Improving the Profitability of a Beef Cow Herd. Hereford World.*

Iowa Beef Center. 2000. *Summary of Iowa Beef Cow Business Record.* Ames, IA: Iowa State University.

Lardy, G., et al. 1998. *Spring Versus Summer Calving for the Nebraska Sandhills: Production Characteristics.* Nebraska Beef Report.

Lawrence, J. D. 2000. *Profiting from the Cattle Cycle: Alternative Cow Herd Investment Strategies.* Presentation to the Southern Plains Beef Symposium, Ardmore, OK.

Lynch, J. M., et al. 1997. *Influence of Timing of Gain on Growth and Reproductive Performance of Beef Replacement Heifers. J. Anim. Sci.* 75: 1715.
McGrann, J. NCA-IRM-SPA. 1992. *Workbook for the Cow-Calf Enterprise.* College Station, TX: Dept. of Agric. Economics, Texas A&M University.
McGrann, J. M. 2000. *Cost Effective Decisions Needed: Cow-Calf SPA Results for Texas—1991–99.* College Station, TX: Texas A&M University.
McGrann, J., and Walter, S. 1995. PE102, *Reducing Costs with IRM-SPA Data.* Cattlemen's College (NCA). Nashville, TN.
National Animal Health Monitoring System (NAHMS). Part I. Reference of 1997 Beef Cow-Calf Management Practices. USDA, APHIS, VS.
National Animal Health Monitoring System (NAHMS). Part II. Reference of 1997 Beef Cow-Calf Management Practices. USDA, APHIS, VS.
National Health Monitoring System (NAHMS). Part III. Reference of 1997 Beef Cow-Calf Production Management and Disease Control. Jan. 1998. USDA, APHIS, VS.
NCA-IRM-SPAO. 1992. *Guidelines for Production and Performance Analysis for the Cow-Calf Enterprise.* Englewood, CO: National Cattlemen's Association.
Outlook and Strategies: 2001 and Beyond. 2001. Greenwood Village, CO: Cattle-Fax.
Retained Ownership. 1999. 7th ed. Greenwood Village, CO: Cattle-Fax.
Ritchie, H. D. The Optimum Cow: What Criteria Must She Meet? *Feedstuffs,* Aug. 21, 1995.
Sacco, R. E., and Baker, J. F., et al. 1989. Lifetime productivity of straightbred and F_1 cows of a five-breed diallel. *J. Anim. Sci.* 67: 1964.
Simonds, G. 1995. *Matching Cattle Nutrient Requirements to a Ranch's Forage Resource.* Intermountain Cow Symposium (Jan. 4–5), Twin Falls, ID.
Staigmiller, R. B., Short, R. E., and Bellows, R. A. 1990. *Developing Replacement Heifers to Enhance Lifetime Productivity.* Research for Rangeland Based Beef Production. Bozeman, MT: Montana Agric. Expt. Station.
Taylor, R. E., and Field, T. G. 1995. *Achieving Cow-Calf Profitability Through Low-Cost Production.* Proceedings of the Range Beef Cow Symposium XIII, Gearing, NE.
USDA. 2000. *Agricultural Land Values.* Washington, DC: National Agricultural Statistics Service.
Van Oijen, M., Montano-Bermudez, M., and Nielsen, M. K. 1993. Economical and Biological Efficiencies of Beef Cattle Differing in Level of Milk Production. *J. Anim. Sci.* 71: 44.

Visuals

"Cows that Fit Montana" (1985; 12 min.). Montana State University, Dept. of Animal Science, Bozeman, MT 59717.

Computer Programs

NCA-IRM-SPA for Financial Calculations (SPAF). This program integrates financial and production information to generate the cow-calf enterprise financial performance report. Cost is $25 for Texas residents and $35 for out-of-state residents, obtained from James McGrann's office at (409) 845-8012.

NCA-IRM-SPA for Production Calculations (SPAP). This software generates the first three SPA reports, including (1) description, (2) reproduction and production performance, and (3) grazing performance measures. SPAP is programmed by AGRO Systems in cooperation with Texas A&M and is available from AGRO-Systems, Brighton, IL 62012-9900. Call (618) 372-3000 or James McGrann's office at (409) 845-8012. Cost is $52.

FINYEAR (Farm/Ranch Financial Statement Preparation Package). For producers without financial statements this software will generate the balance sheets, accrual income statement, and statement of cash flows. Cost is $75 for Texas residents and $100 for out-of-state residents. Contact James McGrann's office at (409) 845-8012.

Integrated Farm Financial Statements (IFFS). This facilitates preparation of financial statements following the Farm Financial Standards as well as projected statements and enterprise budgets. Call Oklahoma State University at (405) 744-6081 for this software package. Cost is $150.

Replacement Heifer Breakeven Analysis Template. SCM-9013. Fort Collins, CO: Colorado State University.

COWSHARE—Program analyzes the profitability of share arrangements for cow-calf or cow-yearling operations. Department of Animal Science, University of Nebraska, Lincoln, NE 68503.

Yearling-Stocker Management Decisions

Yearling-stocker operations handle cattle that are fed and managed for growth prior to going to a feedlot for finishing. These operations typically involve steers and heifers after they have been weaned. Some stocker operations utilize thin cows that can make rapid and efficient gains prior to slaughter. Replacement heifers, intended to go into the breeding herd, can be included in the stocker category. This chapter, however, focuses primarily on steers and heifers being grown prior to the finishing phase in the feedlot. Parts of Chapters 12, 14, and 17 should be integrated with this chapter. Development of replacement heifers is covered in Chapters 5, 9, and 14.

The profitability of a stocker program depends on cost and availability of grass or forage, calves or yearlings, and health management. There are several alternative yearling-stocker programs (Fig. 6.1) located primarily in areas of the United States where there is an abundance of available forage for grazing and proximity to feedyards (Table 6.1). The top three states for stocker cattle production are Texas, Kansas, and Oklahoma. The choice of program depends mainly on the available feed, projected cost of gain, and weight and gain potential of the weaned calves. Calves that weigh more than 500 lbs, are moderate to heavy muscled, and

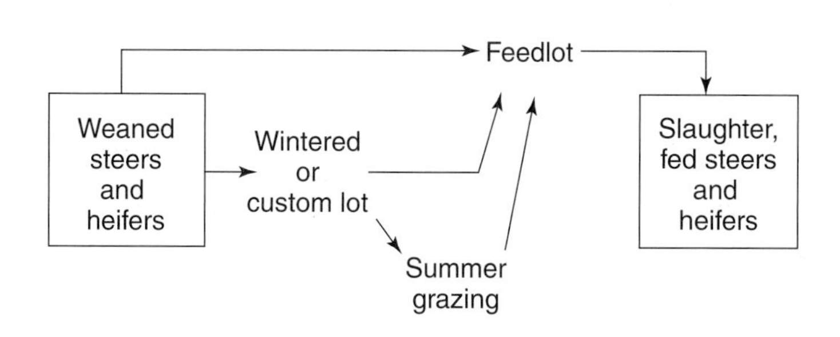

FIGURE 6.1 Alternative production pathways for weaned steers and heifers.
Source: Colorado Sate University.

TABLE 6.1 Areas of Grazed Yearling-Stocker Production in the United States

Geographic Area/States	Description
Northwest	
Oregon and Washington	Grazing season is mid-April to early October; calves originate within the state and from surrounding states (ID, NV, and CA). Turnout weight is 500–650 lbs with off-weights of 750–950 lbs. Destination is feedlots within the states and to ID, CA, CO, NE, and KS.
West	
California	Primarily 3 regions: (1) Sacramento north—resembles OR, NV, and UT (see those states). (2) Sacramento to Bakersfield—the winter/spring grazing season begins Dec. 15–Jan. 15. Turnout weight is 400–450 lbs on calves originated within the states, NV, AZ, UT, and Mexico. Off-weights are 750–900 lbs in May and June. Cattle go primarily to feed yards in CA, WA, ID, and CO. (3) Bakersfield south to Mexico and east to AZ is comparable to the winter-spring grazing pattern of AZ.
Colorado, Idaho, Montana, Nevada, Utah, Wyoming	Grazing can be mid-April through September. Turnout weights are typically 500–700 lbs with market weights of 750–950 lbs. Cattle go to feedlots in their own and bordering states and to CA, NE, and KS.
Southwest	
Arizona	Arizona has two main grazing seasons: (1) April to October and (2) mid-December to April 1. During summer grazing a few stockers run in the higher elevations with turnout weights of 500–600 lbs with off-weights of 600–800 lbs. The winter grazing period in the desert utilizes light calves weighing 350–550 lbs with off-weights of 600–750 lbs. These lightweight calves originate in AZ, TX, Gulf Coast states, and Mexico. They are marketed to feedlots primarily in AZ, CA, TX, CO, and KS.
New Mexico	New Mexico is similar to summer grazing in Arizona. They are marketed to feedlots similar to Arizona with the Oklahoma Panhandle being an additional marketing outlet.
Great Plains	
North and South Dakota	Grazing primarily west of Missouri River during May to mid-October. Turnout weights are 525–600 lbs; off-grass weights are 750-850 lbs. Cattle go primarily to NE, KS, and CO feedlots.
Nebraska	Grazing season is mid-May to mid-October. Turnout weights are 550–650 lbs; off-grass weights are 800–900 lbs. Majority of yearlings will go to Nebraska feedlots.
Kansas	Summer grazing from mid-April to mid-October. Intensively grazed cattle (70% of total) may be marketed mid-July to mid-August. Turnout weights are 500–600 lbs; off-grass weights are 700–900 lbs. Winter wheat grazing is mid-November to mid-March with some limited graze-out until mid-May. Most wheat-grazed cattle are marketed in March. Most cattle go to feedlots in the Great Plains area.
Oklahoma	Grazing is from mid-April to mid-September which occurs primarily in Western Oklahoma Panhandle and extreme Northeast corner of the state (Osage county). Cattle originate from within the state, MO, TX, or Southeast. Turnout weights are 575–625 lbs; off-grass weights are 800–900 lbs. Fall wheat grazing is mid-November to mid-March involving 400–450 lbs calves. Five month gains of 250–270 lbs are expected. Yearlings are marketed primarily to Texas/Oklahoma Panhandle and Kansas feedlots.

TABLE 6.1 (Continued)

Geographic Area/States	Description
Great Plains (Cont'd)	
Texas	Summer grazing season is mid-March to mid-August. Light calves (450–550 lbs) are grazed on native grasses in western part of the state and a combination of native grasses and improved grasses (e.g., Bermuda grass) in the eastern and southern regions of the state. Cattle are sold to feedlots weighing 675–775 lbs. Fall grazing programs involve wheat, oats, and some rye grass during mid-October through March. In-weights and out-weights are similar to those on summer grazing program.
Midwest	
Iowa, Minnesota, Illinois, Wisconsin, Indiana, Ohio, Michigan	Southern Iowa produces a limited number of yearlings, but the Midwest is typically not a stocker-yearling area.
Missouri	Missouri is a diverse cattle state with numerous grazing seasons and various combinations of cow-calf and stocker operations. The summer grazing season generally begins in late April and runs through September. Turnout weights are 500–600 lbs; off-grass weights are 650–850 lbs. Most of the stockers will be sold to feedlots in MO, KS, TX, NE, IA, and IL.
Northeast	There are limited numbers of stocker-yearling cattle produced in the Northeast area of the United States.
Southeast	
Kentucky and Tennessee	Due to widespread spring calving in these states, grazing programs are somewhat limited to fall through early spring. Improved pastures consist of rye grass, blue grass, and clover. The calves are raised to 700–850 lbs, then sold to feedlots primarily in the Central and Southern Plains and Corn belt states.
Louisiana	Summer stocker programs are limited due to heat and humidity. Winter grazing operations using primarily native calves consist of rye grass, oats, and some wheat in the north. Yearling cattle weighing 650–850 lbs will be sold primarily to Texas Panhandle feedlots in early spring.
Florida and Georgia	Florida produces few stocker-yearling cattle due to weather conditions and grazing programs that focus primarily on cow-calf. Grazing programs are more indicative of typical Gulf Coast stocker operations. Calves in these states and areas are marketed in April and June. This marketing time coincides with KS, TX, and OK stocker demand because of the available grass at this time.

Source: Adapted from Cattle-Fax Resources, Inc.®

have a high gain potential are likely to go directly to the feedlot or to be backgrounded for a few weeks and then fed a finishing ration. Calves lighter than 500 lbs with a low to moderate gain potential are more likely to be grown through the winter, then pastured the next summer before entering the feedlot.

Some yearling-stocker operations involve the same farm or ranch ownership of calves from birth through the feedlot finishing phase. Other operations involve cattle being purchased and sold several times between weaning and slaughter. The number of yearling-stocker operations has decreased because more cow-calf producers are retaining ownership through the yearling-stocker phase and more feedlot operators are maintaining ownership of calves through the backgrounding and feeding phases. Breeding improvement programs of cow-calf producers also have resulted in calves with heavier weaning weights and the ability to gain rapidly if they are placed directly into the feedlot.

Regardless of who owns the weaned calves the yearling-stocker operation is still a separate enterprise that is differentiated from cow-calf production and the feedlot phase. Thus, a separate enterprise budget is an important part of the yearling-stocker operation if effective management decisions are to be made.

The primary function of the yearling-stocker operation is to market available forage and high roughage feeds such as grass, crop residues (corn stalks, grain stubble, beet tops), wheat pasture, and silage. Other concerns are using labor efficiently and using regions that are best suited for summer grazing.

A distinction is sometimes made between *wintering* and *backgrounding.* A wintering program usually emphasizes slower gains on high roughage diets with the intent that the calves go to a pasture program in the following spring and summer. Backgrounding emphasizes a faster rate of gain, using more grain and less roughage to prepare the calf for a more immediate feedlot placement. Approximately 40% of backgrounded calves are retained by cow-calf producers, 40% are retained by background operations, and 20% are retained by feedlots. Purchase of stocker cattle is greatest in the fall months (Oct.—30%, Nov.—15%, and Sept.—10%) with another peak of activity in the spring (March, April, and May—6% each). A vast majority of stocker cattle weigh between 400–600 lbs at the time of purchase.

COMPUTING BREAKEVEN PRICES

Yearling-stocker operators who do not maintain ownership of their calves need to decide whether they want to buy stocker cattle for their available forage, sell the harvested forage, or rent the grazable forage to someone else. Individuals operating backgrounding lots do not have this type of flexibility in choosing among the alternatives of cattle ownership, though they can provide custom backgrounding. Owners of specialized backgrounding lots have high overhead costs that must be absorbed if profitability is maintained. Thus, during periods of poor economic returns, cattle backgrounders must decide how many cattle to purchase or to contract with other owners in an attempt to minimize losses. The basic question asked is, "Can a profit be calculated using the best available information?" If the calculations do not indicate a profit, then the decision would likely be to sell or rent the forage rather than take the risk of the selling price increasing several months later. Yearling-stocker operators consider some of the following calculations in making such decisions:

Estimating a Breakeven Price on Gain

Estimated cost:

400 lbs (purchase wt)
× $1.00 (current price)
$400 (dollars paid)

Estimated gain:

780 lbs (sale wt)
− 400 lbs (purchase wt)
380 lbs
380 lbs ÷ 360 days = 1.05 lb/day (gain)

Estimated sale value:

780 lbs (sale wt)
× $0.80 (sale price)
$624 (projected sale value)

Breakeven price on gain:

$624 (projected sale value)
− 400 (dollars paid)
$224
$224 ÷ 380 lbs = $58.95/cwt

If the cost of gain is less than $58.95 per 100 lbs, the assessment in the preceding example appears profitable (assuming the projected gain and projected sale price are valid estimates).

Estimating the Breakeven Sale Price

This is used where past costs of gain are known—for example, average of $.40/lb—and a producer expects to purchase 425 lb calves and sell 700 lb yearlings.

Total cost of gain:

700 lbs (sale wt)
− 425 lbs (purchase wt)
275 (total gain)
275 lbs × $0.40 = $110 (total cost of gain)

Estimated purchase cost:

425 lbs (purchase wt)
× $1.00 (price per lb)
$425 (dollars paid)

Necessary selling price:

$425 (dollars paid)
+ 110 (cost of gain)
$535 (total cost)
$535 ÷ 700 lbs = $76 per cwt

A selling price over $76 per cwt would be profitable, assuming the estimated cost of gain is valid. Calves need to gain approximately 2 lbs per day to have a $0.40 cost per lb of gain.

After calculating the projected cost of gain and the breakeven selling price, the producer can prepare budgets based on different rates of gain in the wintering or backgrounding program. This assumes that a cow-calf producer might be evaluating these alternatives, rather than selling the calves in the fall, or using a custom backgrounding operation. Producers with only a forage-type operation have little flexibility in altering the growth rate of the cattle grazed. They prepare budgets to decide whether cattle should be purchased or whether their forage should be leased to another producer.

THE BUDGETING PROCESS

Examples of the budgeting process for wintering and backgrounding operations are shown in Table 6.2. The faster-gain programs show a higher return potential than the slower-gain programs. Even though the total costs are higher for the faster-gain programs, their cost per pound of gain is less. This occurs because a higher percentage of feed is used for growth rather than for maintaining body weight.

In some cases, slower-gain programs may provide a useful alternative to selling weaned calves in the fall because there is a market for feed that might be difficult to sell. Also, if calves are retained through the winter, they could command a higher price in the spring, or pasture gains may compensate for the reduced returns of the wintering program. In this case, producers with a wintering program and a summer pasture program should compute a separate budget for each program.

Table 6.2 is only an example of the budgeting process. Producers should prepare their own individual budgets and use their specific costs. If valid cost and return data are available from past years, they can increase the accuracy of budget estimates for the current year. Table 6.3 lists the primary factors contained in a yearling-stocker budget and shows how management decisions influencing these factors can affect profitability. In this example,

TABLE 6.2 The Budgeting Process for Alternative Management Programs for Weaned Calves

	Wintering Rate of Gain		Backgrounding Rate of Gain	
	Low	High	Low	High
Initial wt (lbs)	425	425	425	425
Average daily gain	0.9	1.25	1.5	1.75
No. days feeding	150	150	150	150
Sale wt (after 2% shrink)	549	600	637	674
Purchase cost (425 lbs at $1)	$425	$425	$425	$425
Feed costs	63	77	82	90
Other variable costs	26	26	26	26
Total costs	$514	$528	$533	$541
Breakeven sales price (total cost/market wt) × 100 =	$93.62	$88.00	$83.67	$80.27
Estimated selling price/cwt	$90.00	$88.00	$88.00	$87.00
Estimated returns/cwt	(−$ 3.62)	$ 0.00	$ 4.32	$ 4.91
Estimated returns/head	(−$19.87)	$ 0.00	$27.58	$45.36

Source: Adapted from *Western Livestock Round-Up* (USDA).

TABLE 6.3 Effect of a 10% Change in Breakeven Price on a Yearling-Stocker Budget[a]

Factor	Change (%)	Decrease in Breakeven Price ($/cwt)	Increase in Profit ($/head)
Sale price	+10%	$ 0.00	$42.01
Purchase price	−10	6.57	37.26
Average daily gain	+10	2.27	12.61
Pasture cost	−10	0.68	5.08
Interest cost	−10	0.27	1.25
Death loss	−10	0.19	1.11
All factors combined	+10	$9.02	$99.32

[a]Taken from a Nov. 1999–Mar. 2000 budget that included a 525-lb calf ($466), winter pasture and feed ($64), other operating costs ($20), and interest ($14)—a total cost of $564 and a profit of $47/head.
Source: Cattle-Fax.

purchase price, sale price, and the gaining ability of the cattle have the greatest influence on profit. Figure 6.2 shows how average profits have varied over the years for yearling-stocker operations.

Table 6.4 shows an enterprise budget for summer grazing yearling cattle. The budget assumes 150 days on pasture for season-long grazing and 75 days on pasture for early-intensive, a purchase weight of 550 lbs, a pasture lease rate of $51.60/hd and $44.46/hd for season-long and early-intensive systems, respectively. Average daily gains of 1.5, 1.2, 1.9, and 1.5 lbs/day for season-long levels 1 and 2 as well as early-intensive levels 1 and 2, respectively, were also assumed.

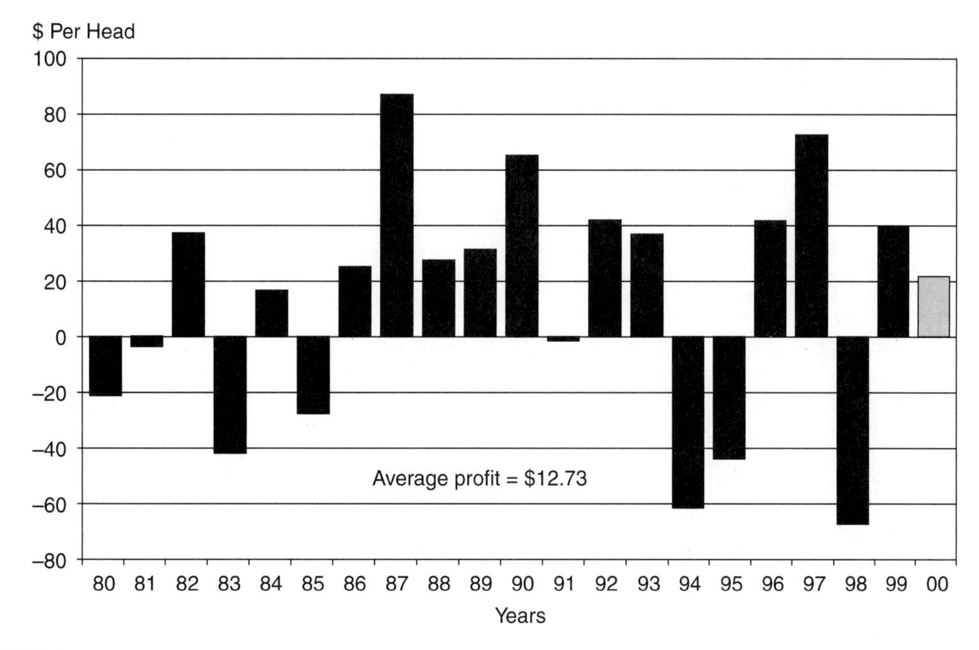

FIGURE 6.2 Average yearling-stocker profit or loss (summer grazing program; purchase Feb.-Mar. at 500 lbs; sell Sept.-Oct. at 750 lbs).
Source: Cattle-Fax.

MANAGEMENT CONSIDERATIONS

Market Prices

The marketing decisions that affect market prices often have the greatest effect on the profit potential of yearling operations. Stocker operators typically deal with both sides of the marketing equation—purchase and sale. Purchase weight and price, along with sale weight and price, have a major influence on total dollars returned to the operation (Table 6.2). Cattle prices may be discounted or enhanced based on a number of factors (Table 6.5).

The buy-sell margin—that is, the difference between calf purchase price and selling price—should be carefully evaluated by a breakeven price analysis. High buy-sell margins limit profitability. As a rule of thumb, the calf-buying price should not exceed the selling price by more than 15%.

Stocker-yearling producers prefer to purchase shrunk cattle at average prices and cattle that are low to moderate in condition, healthy, relatively light for their age, and moderate or better in both muscle and frame. Although the cow-calf producer might identify these animals as somewhat "mismanaged cattle," they are potentially profitable to the yearling operator. Calves heavy at weaning and that possess the ability for rapid gains need to be utilized in the finishing phase of the feedlot, not in the yearling operation. An exception to this might occur when yearling feed costs and interest rates are very low.

Feed prices, forage supplies, and projected calf and yearling prices are needed to determine the best time for marketing. Seasonal price patterns suggest that fall calf prices are usually the lowest for the marketing year. Even though spring prices (and total returns) are often higher than in the fall, potential net returns using a simple budget should be the deciding factor. Producers can obtain some flexibility by considering the purchase of lighter weight calves or heifers when market conditions are favorable.

TABLE 6.4 Cost-Return Projection—Summer Grazing Steers in Western Kansas

	Season-Long		Early-Intensive		
	Level 1	Level 2	Level 1	Level 2	Your Farm
VARIABLE COSTS PER HEAD[1]					
1. Pasture	$ 51.60	$ 51.60	$ 44.46	$ 44.46	$
2. Silage					
3. Hay					
4. Grain					
5. Supplement					
6. Mineral—salt	4.80	4.80	2.40	2.40	
7. Feed processing					
8. Labor	6.48	6.48	4.32	4.32	
9. Veterinary, drugs, supplies	9.00	9.00	7.50	7.50	
10. Marketing costs	3.00	3.00	3.00	3.00	
11. Hauling/yardage					
12. Utilities, fuel, oil	4.00	4.00	3.00	3.00	
13. Facilities and equipment repairs	5.00	5.00	4.00	4.00	
14. Miscellaneous	5.00	5.00	4.00	4.00	
15. Interest on purchased livestock + ½ variable costs	24.36	24.36	12.01	12.01	
A. TOTAL VARIABLE COSTS	$ 113.24	$ 113.24	$ 84.69	$ 84.69	
FIXED COSTS PER HEAD					
16. Depreciation on facilities and equipment	$ 1.30	$ 1.30	$ 0.65	$ 0.00	
17. Interest on facilities and equipment[1]	2.03	2.03	1.01	1.01	
18. Insurance and taxes on facilities and equipment	0.53	0.53	0.26	0.26	
B. TOTAL FIXED COSTS	$ 3.85	$ 3.85	$ 1.93	$ 1.28	
C. TOTAL COSTS PER HEAD (A+B)	$ 117.09	$ 117.09	$ 86.61	$ 85.96	
RETURNS PER HEAD					
19. Market steer	$ 660.59	$ 633.17	$ 611.16	$ 592.52	
20. Less cost of steer	540.21	540.21	540.21	540.21	
21. Less Death Loss: (1.5% of line 19)	9.91	9.50	9.17	8.89	
D. GROSS RETURN PER HEAD	$ 110.47	$ 83.47	$ 61.78	$ 43.42	
E. RETURNS OVER VARIABLE COSTS (D−A)	$ −2.76	$ −29.77	$ −22.91	$ −41.27	
F. RETURNS OVER TOTAL COSTS (D−C)	$ −6.61	$ −33.62	$ −24.83	$ −42.54	
G. AVERAGE SELLING PRICE NEEDED PER CWT:					
22. To cover variable cost and feeder (A + 20) ÷ (net selling weight)[2]	$ 85.60	$ 90.88	$$ 91.55	$ 95.69	
23. To cover total cost and feeder (C + 20) ÷ (net selling weight)[2]	$ 86.10	$ 91.41	$ 91.83	$ 95.88	
H. TOTAL FEED COST (lines 1 through 7)	$ 56.40	$ 56.40	$ 46.86	$ 46.86	
24. Hundredweight produced	2.21	1.65	1.29	1.00	
25. Feed cost per hundredweight (H ÷ 24)	$ 25.50	$ 34.10	$ 36.28	$ 46.98	
I. ASSET TURNOVER (D ÷ INVESTMENT)[3]	19.37%	14.64%	11.13%	7.82%	
J. NET RETURN ON INVESTMENT [(F + 15 + 17) ÷ INVESTMENT][3]	3.47%	−1.27%	−2.13%	−5.32%	

[1]Original cost of facilities and equipment plus salvage value divided by 2, times an interest rate of 10%.
[2]Net selling weight = selling weight−(death loss% × selling weight).
[3]Investment equals total cost of purchased animal and value of facilities and equipment.
Source: Adapted from Dumler, Jones, and O'Brien (2000).

TABLE 6.5 Premiums and Discounts Paid for Feeder Cattle

Factor	Impact on Price
Breed	Herefords sell at premium to Angus (1980s) Angus sell at premium to Herefords (1990s) Black hides sell at a premium >25% Brahman discounted 2–3 times those with < 25% Brahman
Weight, condition, muscle	Cattle < 600 lbs and light muscled discounted $18–26;/cwt compared to heavy muscled Lighter weight cattle that are healthy, ≥ mod. framed, ≥ mod. muscle sell at a premium to heavier weight calves (premium declines as feed prices rise) Thin cattle more likely to be discounted in fall compared to spring Overconditioned feeders discounted $1–3/cwt
Fill	Overfilled cattle heavily discounted
Health status	Sick calves discounted $10–25/cwt
Lot size	Cattle sold in truck-sized lots bring a premium compared to small lots ($7.50/cwt for feeders; $4.50/cwt for yearlings)

Source: Adapted from multiple sources.

TABLE 6.6 Full-season Flint Hills (KS) Lease Rates for Stocker Cattle

	Average$/hd/Season			
Year	500–700 lbs	> 700 lbs	Avg $/acre	Avg (range) $/lb of gain
1997	62.7	72.1	15.9	.32 (.31–.38)
1998	63.6	74.6	16.9	.32 (.30–.38)
1999	58.8	63.8	16.0	.31 (.24–.36)
2000	63.7	62.3	16.3	.33 (.25–.40)
2001	65.5	75.4	16.3	.31 (.22–.40)

Source: Adapted from Morgan, (2001).

Conditions can change, especially prices, from the time a budget is constructed. The conditions that affect costs and returns should be periodically evaluated to determine the best marketing alternatives. The method used in the previous year may not be the best approach for the current year. In addition, when large numbers of cattle are held over the winter, heavy marketings of yearlings in the spring can cause reduced market prices to be depressed. Prudent producers may want to take a hard look at forward-pricing opportunities to lock in profitable returns.

Pasture Leases

Leasing or renting agreements for pasture are influenced by a variety of factors including the costs associated with land ownership, the tenants' expected gross or net income, demand for use of the land, federal farm programs, and land productivity. Rental rates are typically determined by a fixed rate per head per season, a per acre rate, or a rate per pound of gain (Table 6.6).

TABLE 6.7 Assigned Responsibilities in Pasture Lease Agreements (%)

	Small Grain Winter Grazing			Pasture Grazing		
Task	Landowner	Cattle owner	Both	Landowner	Cattle owner	Both
Checking livestock	25	66	9	47	49	3
Salt and minerals	20	78	2	46	52	2
Fencing materials	58	35	7	44	45	12
Fencing labor	47	49	4	45	45	9
Fertilizer	59	22	19	49	45	6
Supplemental feeding	19	80	2	48	50	2
Water	54	41	6	45	45	9

Source: Adapted from Doye et al. (2000).

The more productive the land and the lower the likelihood of drought results in higher dryland pasture lease fees. The average pasture lease fees ($/acre) from 1993 to 1998 for Oklahoma, Texas, Kansas, and Missouri were $8.65, $5.73, $12.20, and $19.08 per acre, respectively.

The pasture lease agreement typically allocates responsibility to either the land owner, the stocker operator, or both for activities such as checking cattle, maintaining fences and water sources, and providing salt and mineral supplementation (Table 6.7).

Cattle Health

While the costs associated with cattle that die are obvious, cattle operators often do not recognize all of the costs associated with sick cattle. The average sick animal shrinks 10–20%. Considerable additional labor is required per sick animal. Medicine, treatment programs, and death losses are expensive. An average three-day medication cost for a 400-lb calf can cost approximately $10. A 5% death loss increases the cost of each calf $10–$20, or 5–8%. Since the cost of "one time through the chute" is considered equivalent to a seven-day-feeding period, it is cheaper to prevent disease than it is to treat it. An example of cash health expenses is shown in Table 6.8.

The best way to prevent disease in a yearling-stocker operation is to know the origin of the cattle and their previous health program as well as working to minimize stress levels. Purchasing "fresh cattle" involves less risk than buying feeder calves of unknown origin that have likely been transported long distances and exposed to infected cattle. Shipping stress can significantly decrease the profitability and production efficiency of native-range-based stocker cattle enterprises. A Texas shipping stress study showed the morbid heifers returned $20–$52 per head less than healthy heifers. Morbid replacement heifers had a conception rate that was 9% lower and a one-month later pregnancy date than healthy heifers (Pinchak, 1995). The average cost for health products, excluding labor and implants, in a 2001 survey was $7.28/hd.

Producers must also have an excellent health and nutrition program upon receiving the cattle. How the cattle are handled and fed during the first five weeks will have a significant effect on their health and profitability.

A *Drovers'* survey in 2001 found that stocker operators reported respiratory disease as the greatest health problem (40% of respondents), followed by internal and external parasites (15%), pinkeye and other ocular problems (11%), and lameness and foot rot (7%). A Kansas State study showed that approximately 9% of all stocker cattle exhibited signs of morbidity in the first 30 days following arrival with two-thirds of the cases resulting from respiratory disease.

TABLE 6.8 Cash Health Expenses for Yearling-Stocker Cattle

Production Period and Treatment		Cost per Head ($/hd)	Your Value[a]
Newly arrived or retained cattle (routine process of cattle)			
1. Ear implant for growth[b]		$1.00	______
2. Ear tags for external parasites		1.00	______
3. Blackleg and malignant edema		0.12	______
4. IBR, PI_3, BVD, BRSV		0.50	______
5. *Lepto Pomona*		0.15	______
6. Vitamins A and D (for dry range condition)		0.10	______
7. Treatment of lice and grubs		0.30	______
8. Worming		1.50	______
Subtotal		$4.67	
Sick pen cattle (example treatments)			
Treatment 1: Oxytetracycline			
Treatment 2: Sulfamethazine boluses			
Treatment 3: Procaine Penicillin G			
Treatment 4: Erythromycin			
Treatment 5: Tylosin			
Treatment 6: Procaine Penicillin G			
Total sick pen cost[c]	$12.00 × 0.25 =	$3.00	______
Additional veterinary and medical expenses			
Routine vet calls[d]	$30.00 × 0.044 =	$1.33	______
Total		$9.00	______

[a]Prices change so current costs should be determined.
[b] Implants last about 120 days. Exclude implants on wheat grazeout and heifers kept for breeding.
[c]$12 per head × portion of cattle treated (assumes 25% of cattle are treated).
[d] $30 per call with 4.4% of the cattle treated.
Source: Adapted from the Oklahoma Agricultural Experiment Station Publication, *Beef and Pasture Systems for Oklahoma—A Business Management Manual.*

Bovine respiratory disease (BRD) is seldom the result of a single factor. It is usually caused by a combination of stress, virus infection, and invasion of the lungs by pathogenic bacteria such as *Pasteurella* and *Hemophilus.* Stress undermines the natural defenses built into the linings of the trachea and bronchi. Respiratory viruses (such as IBR, PI_3, BVD, BRSV, and others) further damage these natural defenses. Ultimately, pathogenic bacteria find a wide-open road into the lungs, where they localize, multiply, and cause the severe damage called BRD, or pneumonia or shipping fever. More detailed health programs are discussed in Chapter 16.

Nutrition

Successful nutritional programs for stocker-yearling cattle must be carefully coordinated with a well-planned health program. Many newly received cattle are stressed; they have been sold through several markets, exposed to diseases, and shipped long distances through varying kinds of weather without feed and water. How these cattle adjust to their new environment will be largely determined by complementary nutritional and health programs.

Low feed intake and water consumption are the two biggest nutritional challenges for newly arrived stocker-yearling cattle. These challenges exist for the first 3–4 weeks. It is important that the cattle consume feed before filling up with water. The physical environment for receiving calves is of importance with dust minimization being the most critical factor.

TABLE 6.9 General Ration Guidelines for Receiving Stocker Calves (450–600 lbs)

Ration Component	Recommendation
Dry matter	70–85%
Crude protein	14%[a]
Crude fiber	25%
Potassium	1.4%[b]
Copper	20 ppm[c]
Zinc	100 ppm[c]

[a]Highly stressed lightweight cattle should receive 16%.
[b]Electrolyte responsible for fluid replenishment of tissue.
[c]Enhance immune response.

If possible, newly received stockers should be managed in a grass lot instead of a drylot. Morbidity rates may be reduced by as much as 40% as a result of this practice. Calves should receive two or more lbs of high-quality hay (brome, Bermuda grass, or native grass hays) in addition to a starter ration (chopped feeds). The starter ration should minimize dust to prevent respiratory stress, minimize fermented feeds (silage, high-moisture grain) to enhance intake, and avoid excessive starch to prevent acidosis.

Table 6.9 shows other nutritional recommendations for 450–600 lb stocker-yearling cattle that can increase feed intake and contribute to the nutritional well-being of cattle during the initial adjustment period. Protein should be supplemented on a pounds-per-head basis rather than a percent-of-ration basis because of the variability in feed consumption. Natural protein is recommended in the ration, with only minimal amounts of nonprotein nitrogen. Bypass protein may also be beneficial. Potassium additions tend to improve the performance and health of cattle.

Water should be provided 3–4 hours after the cattle's arrival. Water tanks should be clean and the cattle (especially small cattle) should be able to reach the water without being frightened. Water that is allowed to run continually into the tank or trough may attract the cattle to the water.

Calves may be fed roughage on the ground or in feedbunks (Figs. 6.3 and 6.4). Feeding cattle in feedbunks allows more flexibility in moving from a wintering to a backgrounding program in that the grain can be more conveniently added to the feeding program.

Some yearling-stocker cattle are not fed harvested feeds through most of the wintering program, particularly when the climate, in combination with certain forage systems, permits green forage to be grazed nearly on a year-round basis. In certain areas of the Southeast, for example, winter annuals such as rye, oats, ryegrass, wheat, and barley provide a forage base for winter grazing (Fig. 6.5). Also, some types of clover planted in the fall will allow grazing by late January in certain areas. Table 6.10 shows several rations for wintering and backgrounding calves at different daily gains.

Rate of Gain

Rate of gain is affected by the health, body condition, and nutrition of cattle. Most producers prefer the daily rate of gain of backgrounded and grazed cattle to be 1.5 lbs per day or higher. Cattle that gain faster usually are more profitable. In some cases, cattle with a high rate of gain lose money, but they lose less than slower-gaining cattle (Fig. 6.6). Fast-gaining cattle may not be profitable if production costs are too high or when the buy-sell price margin is too low. Slow-gaining cattle usually do not produce enough pounds to cover maintenance feed and

FIGURE 6.3 Most calves are grown on high roughage feeds during the winter.
Source: Holly Foster.

FIGURE 6.4 High roughage can be fed in feedbunks and grain can also be fed if deemed desirable. These cattle receive alfalfa hay during the winter, then they are on pasture during the spring and summer.

TABLE 6.10 Rations for Different Rates of Gain for Growing Calves

	300–400 lbs			400–500 lbs			500–700 lbs		
Ration	.5–1 lb Day Gain	1–1.5 lbs Day Gain	1.5–2 lbs Day Gain	.5–1 lb Day Gain	1–1.5 lbs Day Gain	1.5–2 lbs Day Gain	0.5–1 lb Day Gain	1–1.5 lbs Day Gain	1.5–2 lbs Day Gain
1. Corn Silage	6.5	11.5	—	6.5	21.0	11.0	—	19.0	21.0
Alfalfa hay, midbloom	7.5	4.0	5.0	10.5	5.5	5.0	17.0	9.5	5.5
Corn	—	1.5	5.0	—	—	4.5	—	—	3.0
Dicalcium phosphate	0.02	0.04	0.04	0.03	0.03	0.04	0.04	0.04	0.02
2. Corn Silage	—	—	17.5	—	—	30.5	—	—	38.5
Corn	—	—	1.5	—	—	—	—	—	—
32% protein supplement	—	—	1.7	—	—	1.6	—	—	1.8
Dicalcium phosphate	—	—	0.01	—	—	0.01	—	—	—
3. Sorgo Silage	12.0	11.5	—	12.5	20.5	11.0	5.5	34.0	21.5
Alfalfa hay, midbloom	6.0	4.0	5.0	9.0	4.5	4.5	15.0	5.0	5.0
Corn	—	2.5	5.0	—	2.0	5.0	—	—	4.5
Dicalcium phosphate	0.02	0.03	0.04	0.03	0.02	0.03	0.04	0.04	0.02
4. Prairie hay, early bloom	8.0	6.5	4.5	11.5	9.0	7.0	15.5	13.5	10.0
Corn	—	3.0	4.5	—	3.5	5.5	—	2.5	6.0
32% plant protein suppl.	2.0	1.0	1.5	1.5	1.0	1.0	1.0	1.0	1.0
Dicalcium phosphate	—	0.01	0.08	—	—	—	0.02	—	—
5. Alfalfa hay, midbloom	10.0	7.0	5.0	13.5	10.0	7.5	17.0	14.0	10.5
Corn	—	3.0	5.0	—	4.0	6.0	—	3.0	6.5
Dicalcium phosphate	0.02	0.03	0.04	0.03	0.01	0.03	0.03	0.02	—
6. Alfalfa hay, early bloom	10.0	7.5	5.5	13.5	11.0	8.0	16.5	15.5	11.5
Corn	—	2.5	4.5	—	2.5	5.5	—	1.5	5.0
Dicalcium phosphate	0.01	0.02	0.04	0.01	—	0.02	0.01	0.01	—

FIGURE 6.5 Steers grazing a winter annual in Alabama.
Source: Alabama Agric. Expt. Sta.

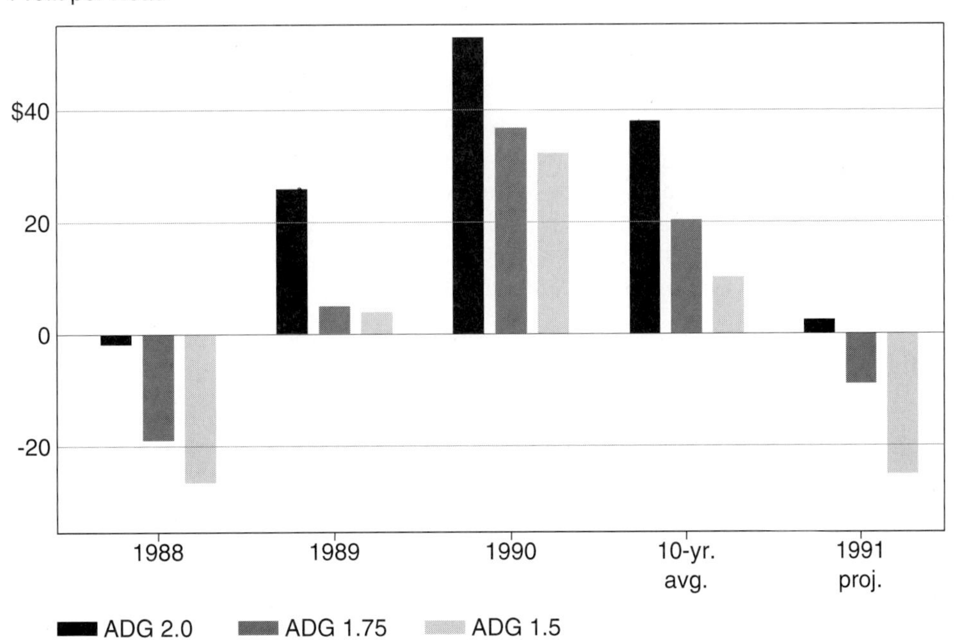

FIGURE 6.6 Cattle gains on wheat pasture as related to profitability.
Source: Kansas State University.

TABLE 6.11 Evaluating Feed Cost per Pound of Gain with Varying Rates of Gain

Daily Gain	Feed (lbs)			Cost of Feed ($)			Total	Overhead Cost per	Total Daily	Cost/lb Gain
(lbs)	Hay	Silage	Corn	Hay	Silage	Corn	($)	Day ($)	Cost ($)	($)
0.74	14.2	—	—	$0.35	—	—	$0.38	$0.39	$0.77	$1.05
1.22	5.5	20.0	—	0.14	$0.19	—	0.35	0.39	0.74	0.61
1.52	4.0	16.4	3.0	0.10	0.16	$0.12	0.40	0.39	0.79	0.52

Source: Backgrounding Enterprise Budgets for Colorado, Information Report (IR) 87-7.

TABLE 6.12 Effects of Ionophores on Different Types of Pastures

	Pasture Types			
	Native Growing	Native Dormant	Improved	Small Grains
Number of trials	54	9	10	7
Number of cattle	3,880	649	504	470
Average daily gain (lbs/day)	1.41	1.02	1.58	1.58
Ionophore response (lb/day)	0.18	0.12	0.14	0.11

Source: Michigan State University (summary of eighty research trials).

interest costs. Table 6.11 shows how the cost of gain can be projected by combining feed costs with various rates of gain.

Growth stimulants, such as ear implants or ionophores, are included in the management programs of many yearling-stocker programs. Ear implants are commonly used in wintering, backgrounding, and grazing programs with a similar gain response noted regardless of the implant type used (Chapter 7 identifies the available implants). Monensin (which has the trade name Rumensin) and lasalocid (which has the trade name Bovatec) are ionophores—antibiotics that alter rumen metabolism—used for wintering, backgrounding, and grazing programs. The response of grazing cattle to the two types of ionophores is similar (Table 6.12).

Producers identify gaining ability in healthy cattle by associating gain with a relatively low body condition score. Genetics also affect gaining ability, however, it is difficult to assess unless the source and breeding of the cattle are known.

Production Systems

Numerous alternatives are available for managing weaned calves to harvest weights. Efficiency of production includes the growing of calves prior to receiving finishing diets in the feedlot. Frequently, the economics of production only consider a single phase of the total production system. As a consequence, one segment of the beef industry may make management decisions based on maximum profit while they manage the animals that may adversely affect the profit of the next owner. Economic efficiency of the total production system could be reduced. For example, cost per pound of gain is usually lower when calves are wintered at a relatively fast rate of gain and feedlot operators also want high daily gains so cost of gain will be relatively low. However, high feedlot gains may not be realized if calves wintered at high rates of gain are grazed during the summer months.

The University of Nebraska has made excellent economic assessments of several production systems evaluating weaned calves through the feedlot. The following highlights are given for some of these systems:

- Cattle, grazing corn stalks, that had a relatively low winter gain (0.79 lb/day) compensated during the summer and experienced faster summer gains than those wintered at a higher rate (2.0 lbs/day). Most of the compensatory gain was achieved during the first 62 days of grazing. Compensatory growth is a phenomenon of rapid growth by cattle, which usually follows a period of growth restriction.
- Cattle that were on grass for the short grazing period (62 days) had faster finishing gains and tended to be more efficient. The cattle that were wintered at a fast rate (2.0 lbs/day) and pastured for the full summer period (120 days) had a higher breakeven. Cattle wintered at a fast rate of gain (2.0 lbs/day) should only be grazed on high-quality forage in the spring and early summer to be competitive with systems that have lower winter input costs (Morris, 1996).
- Systems where grazing ended in September compared with November, using native Sandhills range or grazing red clover interseeded in bromegrass had the lowest slaughter breakeven costs. Maximizing grazed forage gain, while cost of gain is low, reduces overall breakeven costs of forage systems. Grazing combinations of warm- and cool-season forages allow for optimizing forage quality (Shain, 1996).
- Medium frame size cattle can be grown in forage grazing systems before finishing, and these systems provide opportunity for reducing carcass fat by shortening the finishing days on feed (Vieselmeyer, 1995).

Wheat grazing offers three unique enterprise options to producers: (1) harvest grain only, (2) graze and harvest grain, or (3) graze only. If option 2 is chosen, then grazing is terminated prior to wheat attaining the jointing stage. Wheat grazing in the Midwest is typically initiated in the late fall months. Supplementation is provided while the plant is dormant. As the pasture becomes lush it is critical to manage cattle to avoid losses associated with bloat.

A cost-return budget for winter wheat grazing is provided in Table 6.13. The budget assumes 120 days on pasture at a rate of \$1.75/cwt/month, 240 lbs/hd of grain sorghum, 350 lbs/hd of hay, and 30 lbs/hd of salt and mineral as supplement, and purchase weight of 425 lbs and 450 lbs for heifers and steers, respectively. An average daily gain of 2.2 and 1.8 lbs/day for steers in levels 1 and 2, and 2.0 and 1.6 lbs/day for heifers in levels 1 and 2 is also assumed.

Yearling-stocker production systems can be evaluated on how they contribute to a carcass breakeven price. For example:

$$\underset{(\$/\text{lb})}{\text{Carcass breakeven}} = \frac{\text{Weaned calf cost} + \text{yearling-stocker cost} + \text{feedlot}}{\text{Carcass weight}} = \frac{\text{Total cost}}{\text{Carcass weight}}$$

Since cost of gain in each production phase affects the other two phases, it is important to achieve an optimum combination of all three phases of production.

MANAGEMENT SYSTEMS HIGHLIGHTS

1. Initial purchase price (cost of weaned calf) and sale price of yearling-stocker cattle have the largest effect on profitability. Management decision priorities should focus on these two prices.
2. Cost per pound of gain is influenced by forage/feed, health, labor, and marketing costs.
3. Obtaining large liveweight gains from grazed forage, while cost of gain is low, reduces overall breakeven costs.

TABLE 6.13 Cost-Return Projection—Winter Wheat Grazing

	Steers		Heifers		
	Level 1	Level 2	Level 1	Level 2	Your Farm
VARIABLE COSTS PER HEAD					
1. Pasture	$ 40.74	$ 39.06	$ 38.15	$ 36.47	$
2. Silage					
3. Hay	8.75	8.75	8.75	8.75	
4. Grain	8.47	8.47	8.47	8.47	
5. Supplement					
6. Mineral and salt	7.20	7.20	7.20	7.20	
7. Feed processing					
8. Labor	4.86	4.86	4.86	4.86	
9. Veterinary, drugs, supplies	9.00	9.00	9.00	9.00	
10. Marketing costs	3.00	3.00	3.00	3.00	
11. Hauling/yardage	10.00	10.00	10.00	10.00	
12. Utilities, fuel, oil					
13. Equipment repairs	5.00	5.00	5.00	5.00	
14. Miscellaneous	4.00	4.00	4.00	4.00	
15. Interest on purchased livestock + ½ variable costs	16.51	16.48	14.20	14.17	
A. TOTAL VARIABLE COSTS	$ 117.53	$ 115.82	$ 112.63	$ 110.93	
FIXED COSTS PER HEAD					
16. Depreciation on facilities and equipment	$ 4.42	$ 4.42	$ 4.42	$ 4.42	
17. Interest on facilities and equipment[1]	4.39	4.39	4.39	4.39	
18. Insurance and taxes on fence and equipment	1.14	1.14	1.14	1.14	
B. TOTAL FIXED COSTS	$ 9.95	$ 9.95	$ 9.95	$ 9.95	
C. TOTAL COSTS PER HEAD (A+B)	$ 127.47	$ 125.77	$ 122.58	$ 120.87	
RETURNS PER HEAD					
19. Market animal (see Table 1)	$ 594.41	$ 594.91	$ 553.17	$ 516.50	
20. Less cost of animal:					
Steers: (see Table 1)	444 69	444.69			
Heifers: (see Table 1)			376.85	376.85	
21. Less death loss: (2% of line 19)	12.48	11.89	11.06	10.33	
D. GROSS RETURN PER HEAD	$ 166.62	$ 137.83	$ 165.26	$ 129.33	
E. RETURNS OVER VARIABLE COSTS (D−A)	$ 49.09	$ 22.01	$ 52.63	$ 18.40	
F. RETURNS OVER TOTAL COSTS (D−C)	$ 39.15	$ 12.07	$ 42.68	$ 8.46	
G. AVERAGE SELLING PRICE NEEDED:					
22. To cover variable cost and feeder (A + 20) ÷ (net selling weight)[2]	$ 80.35	$ 85.88	$ 75.11	$ 80.67	
23. To cover total cost and feeder (C + 20) ÷ (net selling weight)[2]	$ 81.77	$ 87.40	$ 76.63	$ 82.31	
H. TOTAL FEED COST (lines 1 through 7)	$ 65.16	$ 63.48	$ 62.57	$ 60.89	
24. Hundredweight produced:	2.50	2.03	2.27	1.80	
25. Feed cost per hundredweight (H ÷ 24)	$ 26.09	$ 31.32	$ 27.60	$ 33.89	
I. ASSET TURNOVER (D ÷ INVESTMENT)[3]	32.69%	27.04%	37.40%	29.27%	
J. NET RETURN ON INVESTMENT [(F + 15 + 17) ÷ INVESTMENT][3]	11.78%	6.46%	13.87%	6.11%	

[1]Original cost of fence and equipment plus salvage value divided by 2, times an interest rate of 10%.
[2]Net selling weight = selling weight − (death loss% × selling weight).
[3]Investment equals total cost of purchased animal and value of fence and equipment.
Source: Adapted from O'Brien, Jones, and Dumler (2000).

4. Carcass breakeven prices are an excellent way to measure the contribution of yearling-stocker cost of gain to the total production system from the weaned calf through the carcass.

REFERENCES

Publications

Bock, B. J., Hannah, S. M., Brazle, F. K., Corah, L. R., and Kuhl, G. L. 1991. *Stocker Cattle Management and Nutrition.* AES-CE Report C723. Kansas State University, Manhattan, KS.

Doye, D. E., Klettke, D., and Fisher, B. L. 2000. *Oklahoma Pasture Rental Rates: 1998–99.* Extension Report CR-216-0499. Oklahoma State University, Stillwater, OK.

Drovers 2000 Stocker Survey. *Drovers* 129, no. 4, p. 55.

Dumler, T. J., Jones, R., and O'Brien, D. M. 2000. *Summer Grazing of Steers in Western Kansas.* KSU Farm Mgmt. Guide MF-1007. Kansas State University, Manhattan, KS.

McReynolds, K. L. 1986. *Grazing Yearling Beef.* Kansas Farm Mgmt. Guide MF-591. Kansas State University, Manhattan, KS.

Miller, B. 1991. Get the Most from Your Wheat. *BEEF Today* (October).

Morgan, T. H. 2001. Flint Hills Pasture Lease Rates Show Mixed Trends. *Feedstuffs,* May 14, p. 8.

Morris, C., et al. 1996. *Beef Production Systems from Weaning to Slaughter in Western Nebraska.* Nebraska Beef Report, University of Nebraska, Lincoln, NE.

O'Brien, D. M., Jones, R., and Dumler, T. J. 2000. *Winter Wheat Grazing.* KSU Farm Mgmt. Guide MF-1009. Kansas State University, Manhattan, KS.

Pinchak, W. E., et al. 1995. *Shipping Stress Impacts on Production Efficiency and Profitability of Stocker Cattle.* Beef Cattle Research in Texas. Texas A&M University.

Proceedings of the Light Cattle Management Seminars. 1982. Colorado State University, Fort Collins, CO.

Roth, L., Klopfenstein, T., and Sahs, W. 1988. *Corn Stalklage for Growth Calves—A Review.* Beef Cattle Report MP-53: 51. University of Nebraska, Lincoln, NE.

Rust, S. R. 1988. *Ionophores for Grazing Cattle.* Ext. Bull. E-2100. Michigan State University, East Lansing, MI.

Shain, D., et al. 1996. *Grazing Systems Utilizing Forage Combinations.* Nebraska Beef Report, Lincoln, NE.

Sindt, M., Klopfenstein, T., and Stock, R. 1990. *Production Systems to Increase Summer Gain.* Nebraska Beef Cattle Report, University of Nebraska, Lincoln, NE.

Stocker-Feeder Management Guide. 1985. University of Idaho Cooperative Extension Service, Moscow, ID.

Vieselmeyer, B., Klopfenstein, T., et al. 1995. *Physiological and Economic Changes of Beef Cattle During Finishing.* Nebraska Beef Report, University of Nebraska, Lincoln, NE.

Computer Programs

Cattle-Fax. 1995. *Retained Ownership Analysis Software.* 5720 South Quebec Street, Englewood, CO 80111.

Feedlot Management Decisions

The cattle feeding industry is dependent on three primary factors: (1) source of feeder cattle, (2) market outlet for fed cattle, and (3) source of high-energy feed. Since feed cost comprises most of the expense in a feedlot operation (excluding the cost of feeder cattle), the location of feeding operations is generally concentrated where the major portion of feed grains is produced.

The term *fed cattle* refers to cattle that have been fed concentrates (grains). Typically, for several months prior to slaughter the cattle are fed in confinement areas or feedlots where distance to feed, water, and possibly shelter is minimized.

The major cattle feeding areas in the United States are shown in Figure 7.1. These areas are feed-producing regions where grains and roughages are grown. The locations are determined by soil type, growing season, and rainfall or irrigation water. The heaviest concentration of fed cattle in the United States is in eastern Colorado, Nebraska (along the Platte River), the high plains of Texas, southwestern Kansas, and the western half of Iowa.

An overview of the cattle feeding segment is given in Chapter 1. A review of Chapters 1, 8, 9, 13, 14, and 17 is recommended before proceeding with this chapter.

HISTORY OF THE FED CATTLE INDUSTRY

The major growth of the fed cattle industry occurred between 1945 and 1972, followed by a leveling off of fed cattle numbers. Some cattle had been grain-fed during the early development of the United States; however, the large commercial cattle feeding industry began to emerge after World War II. Most of the growth during the following two decades was due to an oversupply of cheap feed grains produced in the United States and relatively cheap fossil fuels. The development of the cattle feeding industry has closely followed corn production, which is the major feed grain in the United States.

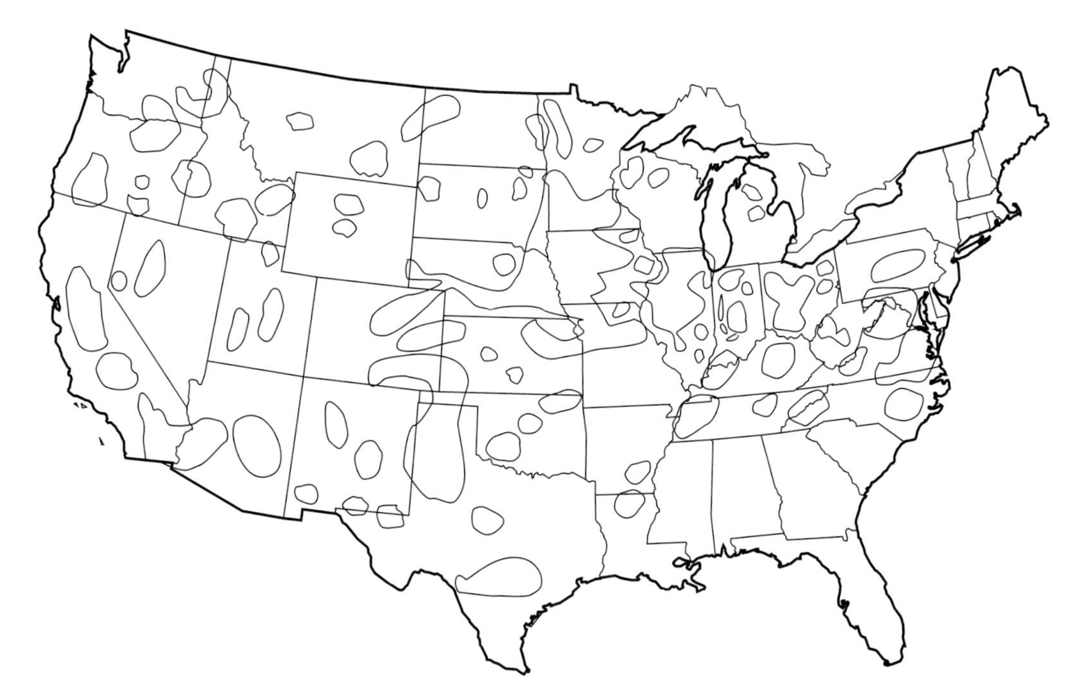

FIGURE 7.1 Major cattle feeding areas in the United States. The circled areas represent location but not volume of cattle fed or backgrounded.
Source: USDA.

The availability and price of feed grains dictate the number of cattle that are grain-fed as well as where changes in number of cattle fed occur in various parts of the United States. Certain Great Plains states (Texas, Kansas, and Nebraska) and Corn Belt states (Iowa, Illinois, and Indiana) continue to have a surplus of feed grain production over that which is currently fed. The increase in fed cattle numbers during the past ten years occurred primarily in the states with surplus feed grains. The efficiency of production is important as cattle feeders in California in the 1960s demonstrated that they could import both feeder cattle and grain and still make cattle feeding a profitable enterprise. Development of additional irrigation water in several of the Plains states has greatly enhanced feed grain production, which, in turn, has increased the number of cattle fed in that area, while cattle feeding in California has decreased.

TYPES OF CATTLE FEEDING OPERATIONS

There are two basic types of cattle feeding operations: the commercial feeder and the farmer-feeder. The two types are generally distinguished by type of ownership and size of feedlot. Commercial feedlots are usually defined as having more than 1,000-head capacity and farmer-feeder feedlots as having less than 1,000-head one-time capacity.

The farmer-feeder operation is usually owned and operated by an individual or family (see Fig. 7.2). The commercial feedlot may be owned by an individual, a partnership, or a corporation—the last type is most common, especially as feedlot size increases (Figs. 7.3 and 7.4). Commercial feedlots may own the cattle, feed the cattle owned by someone else (called *custom cattle feeding* or *custom feedlots*), or engage in some combination of the two. Custom-fed cattle are owned by other cattle feeders, investors, cattle producers, or packers and are fed on a contractual basis.

FIGURE 7.2 A farmer-feeder operation is typically family-owned and is one enterprise in a integrated farm.
Source: Holly Foster.

FIGURE 7.3 Aerial view of a 35,000-head capacity commercial feedlot. Some of the more important component parts of the feedlot are: (1) feedmill (grain and supplement storage, feed mixing facility), (2) collection area for feedlot runoff—primarily for pollution control, (3) hay storage, (4) open pit storage of feed—usually silage or ground, high-moisture grain, (5) hospital area—for isolation and treatment of sick cattle, (6) alley-way for movement of cattle from unloading and processing area, to loading area for marketing, and movement of cattle to hospital pens, and (7) feed alleys—each side has a fence-line feedbunk into which feed is augured from feed trucks. Other important areas not shown are unloading–loading areas and handling facilities for processing recently acquired feeder cattle.
Source: Successful Farming Magazine.

FIGURE 7.4 A commercial feed yard depends on the use of mechanized feed delivery systems where thousands of cattle are feed in concrete banks.
Source: Holly Foster.

Even though the number of cattle fed in feedlots having more than 1,000-head capacity is increasing while feedlots less than 1,000 head are decreasing, there are some advantages to a farmer-feeder operation. Consider the following advantages and disadvantages:

1. The farmer-feeder can utilize cattle as a market for homegrown feeds.
2. The farmer-feeder can effectively utilize high roughage feeds in a backgrounding or warm-up operation.
3. The farmer-feeder can distribute available labor over several different enterprises with cattle feeding being only one of those enterprises.
4. The farmer-feeder may have advantages in flexibility. When cattle feeding is unprofitable, the farmer-feeder can totally close out the cattle feeding and divert time and dollars into other phases of the farming operation. The large commercial feedlot has much higher overhead costs, which are fixed regardless of whether the pens are filled with cattle or are empty.
5. Commercial feedlots usually obtain and analyze more records and information and make more effective management decisions. Commercial feedlots typically utilize more professional expertise (consultants) in managing nutrition, health, and marketing.
6. Custom cattle feeding greatly reduces the operating capital requirement for the commercial feedlot and shifts some of the risk to the customer. However, one of the greatest risks for commercial feedlots is keeping the lots full of cattle. Customers will stop putting cattle into custom lots when financial losses occur over a period of time, creating the problem identified in item 4. Farmer-feeders usually feed just one group of cattle per year, which requires leaving the facilities vacant for several months. Commercial feedlots attempt to keep their lots full year-round, so they have a cattle turnover rate of 2–2.5 times per year.

Feeding Dairy Calves

Most of the bull calves from the 9.1 million dairy cows become feedlot steers with a predominance of them being Holstein. They are usually fed by farmer-feeders, although some commercial feedlots will feed several pens of Holsteins each year.

These young dairy calves are managed differently than calves from the beef breeds because they are removed from the cows shortly after calving. This is because milk production is emphasized, thus dairy beef is a by-product of the dairy industry. Veal is produced from some 1–3-month-old calves yielding carcasses of less than 150 lbs. Other dairy calves are grown on milk replacer and other feeds then placed in the feedlot for finishing.

Holstein cattle consume relatively large amounts of feed and gain rapidly and efficiently up to 1,100 lbs; they become quite inefficient at heavier weights primarily due to their increased requirements for maintenance. Net energy requirements for maintenance are about 12% higher than those established for beef cattle. From a carcass standpoint, Holsteins have a lower dressing percentage and lower muscle-to-bone ratio. However, Holstein cattle marble quite well and have high cutability carcasses. They are more uniform in carcass characteristics than most other beef breeds.

MANAGING A FEEDLOT OPERATION

The five primary factors needed to assess a feedlot operation are shown in Figure 7.5, where the major component parts of each factor are listed under each heading.

Simply stated, profitability in a cattle feeding operation is determined by this formula:

$$\text{Profitability} = \text{income} - \text{costs}$$

The factors listed in the last column of Figure 7.5 determine the income; the factors listed in the other four columns determine expenses. All five factors are included in the enterprise budget shown in Table 7.1. In 1996 a feeder calf was worth $420 and the cost of feed was $251.60, which accounted for 57% and 34% of the total cost, respectively. By comparison, the low cost of feed and the high cost of feeder cattle in 2000 created a different outlook. By comparing the yearly variations in cost of feeder steer, feed costs, and selling price, the primary reason for the difference in profit or loss can be determined. Some of these costs and selling prices can vary tremendously within the same year.

Figure 7.6 shows the highly variable average returns per head over the past several years, reflecting the high financial risk in cattle feeding. However, well-managed cattle feeding operations will have fewer years of losses and more years of profit than are shown in Figure 7.6. The detailed discussion of the major factors and their component parts shown in Figure 7.5 will identify some of the management decisions implemented by successful cattle feeders.

Cost Facilities Investment	Nonfeed Costs of Feeder Cattle	Nonfeed Per Pound of Gain	Total Costs Per Pound of Gain	Dollars Received
Land Pens Equipment Feed Mill Office	Grade Weight Shrink Transportation Gain Potential Yardage Association Fees Brand Inspection	Ration and Markup Performance of the Cattle —Gain —Efficiency Length of Feeding	Death Loss Labor Taxes Interest Insurance Utilities Veterinary Repairs	Market Option Weight Dressing % Quality Grade Yield Grade Manure/ Compost

FIGURE 7.5 Major factors utilized in analyzing a feedlot operation.
Source: Colorado State University.

TABLE 7.1 Budget Showing Costs and Return for a 2000 Feedlot

Budget Assumptions

In-Weight	750 lbs	Conversion Rate (AF)	7.9 lbs
Out-Weight	1,175 lbs	In-Price ($/cwt)	$84.14
Net Gain	450 lbs	Out-Price ($/cwt)	$68.00
Days on Feed	133 days	Operating Interest Rate	10.0%
Average Daily Gain	3.3 lbs	Death Loss	1.0%

Operation Budget, March–July 2000

	Total Cost ($/head)	Percent of Total
Feeder Cost	$638.55	74.3%
Feed	170.00	19.8%
Interest	26.93	3.1%
Death Loss	6.39	0.7%
Vet-Med	6.00	0.7%
Overhead	12.00	1.4%
Total Cost/Head Sold	$859.87	100.0%
Total Income/Head	$804.24	
Net Profit (Loss)/Head	$ (55.62)	
Breakeven ($/cwt)	$ 71.66	
Total Cost of Gain ($/cwt)	$ 42.54	

Source: Catttle-Fax.

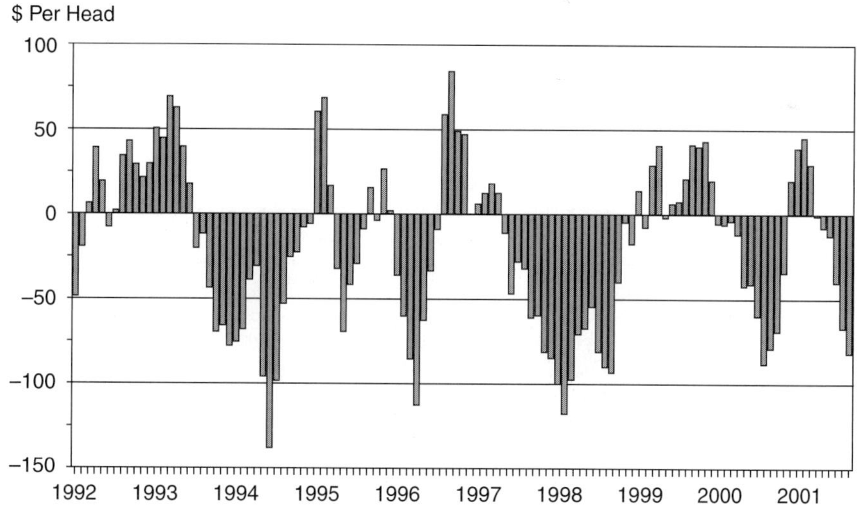

FIGURE 7.6 Average returns to cattle feeders feeding 725-lb steers in the southern plains.
Source: Livestock Marketing Information Center.

Facilities Investment

Facility planning efforts should take into account the following objectives:

1. Minimize animal and employee stress while assuring safety for each.
2. Contribute to profitability and the production of safe, wholesome products.
3. Protect the surrounding environment.
4. Allow for the effective delivery of feed and care to cattle.

Selecting a feedlot site must be carefully accomplished. A number of factors must be considered to assure that the aforementioned objectives are met (Table 7.2).

Most commercial feedlot facilities throughout the United States are similar in terms of layout and equipment. The typical feedlot has an open lot with dirt pens that can hold 100–500 head of cattle. The pens are usually mounded in the center so that the cattle have a dry resting area. The fences are made of pole, cable, or pipe. There are fence-lined feedbunks with concrete aprons inside the pens, where the cattle stand while eating. A feedmill to process grains and other feeds is usually a part of the facility. Feed is distributed to the fence-lined bunks with specialized feed trucks that mix the feed while traveling to the feed bunks. Bunker trench silos hold corn silage and other roughages. Grains may also be stored in these silos; however, they are often stored in steel bins located above the ground. In 1960 the cost of a typical feedlot facility was $30 to $60 per head; today the cost of building a new feedlot is approximately $100–$130 per head while an established feedlot can be purchased for about $75 per head.

Facilities for farmer-feeder operations vary from unpaved, wood-fenced pens to paved lots with shelter (sheds) to partial confinement or total confinement feedlots. The latter may have manure collection pits under the cattle with feed stored in airtight facilities. Most feeds are stored in upright silos and grain bins. Feeds are typically processed on the farm and distributed to feedbunks (located inside or outside the pens) with tractor-powered equipment. The costs of such facilities are shown in Table 7.3. The costs are computed for farmer-feeders under Midwest feeding conditions, which is where the majority of farmer-feeders are located. However, the costs are similar for farmer-feeders in other areas as well.

Most total confinement feedlots are farmer-feeder feedlots located in the Midwest with fewer than a 1,000-head capacity. The world's largest confinement feedlot—a commercial feedlot owned by J. R. Simplot Company and located in eastern Oregon—has a capacity of 32,000 head and was built at an estimated cost of $6.2 million.

TABLE 7.2 Site Specifications for Feedyards

Factor	Specification
Land slope	2–5%
Soil type	≥25% clay
Land mass	1 ac/100 hd for pens, alleys, and feed roads
Pen size	300 ft^2/hd (250 ft^2/hd in drier climates)
Feedbunk type	Fence-line feeders with 12-ft concrete aprons
Bunk space[a]	Arriving cattle—24 in./hd Backgrounding cattle (500–700 lbs)—18 in./hd Finishing cattle—9–12 in./hd
Other considerations	Minimum distance from bunk to water channels of 400 ft; 175–250 ft from bunk to back of pen; minimum of 200 ft from back of pen to nearest water channel

[a]Young cattle tend to eat in groups, thus the need for more space. One time a day feeding systems require more space than two times a day.
Source: Adapted from Iowa State University (2000).

TABLE 7.3 Construction and Annual Maintenance Costs of Various Feedlot Types and Capacities

	750 Hd ($/hd)		1,500 Hd ($/hd)		5,000 Hd ($/hd)	
Type	Construction	Maintenance	Construction	Maintenance	Construction	Maintence
Earthen lot with windbreak	138	17	178	20	150	17
Earthen lot with shed	283	33	320	37	292	33
Concrete lot with shed	311	34	326	36	298	32
Confinement with solid floor	426	47	433	48	410	45
Confinement with slotted floor	532	59	538	60	515	57

Source: Iowa State University (2001).

Environmental Management

The four primary environmental management issues for feedyards are dust, odor, flies, and water quality. Dust management is best accomplished via regular pen maintenance with timely manure removal in late spring. The regular use of box scrapers reduces manure accumulation without cratering the pen hardpan. The use of overhead sprinklers or water truck sprayers can be utilized to maintain the recommended moisture rate of 25–35% in a loose soil layer of 1 in. or less. Water delivery systems may be cost prohibitive. For example, to increase moisture from 10% to 35% in 1 in. of manure requires approximately 14 gal of water per head of pen capacity. Achieving the same moisture increase in 2 in. of manure would require roughly 28 gal per head of pen capacity.

Control of odor requires regular pen maintenance, the use of correctly constructed and maintained runoff holding ponds, and proper nutritional management. The use of a more precisely balanced ration to avoid overfeeding of phosphorus is a key strategy.

Research is being conducted to use plant-oil extracts or fat extracts as a treatment for pen surfaces to control both odor and dust. Plant-oil extracts may be useful in inhibition of manure fermentation as a means to preserve nutrient value and suppress odor. These extracts may also have antimicrobial properties that help to inhibit the growth of organisms that contribute to food-borne illness.

Another strategy is to compost manure. The benefits of composting are to reduce volume and weight, kill weed seeds, concentrate nutrients, reduce odor, and to reduce fly populations. The disadvantages are labor and equipment costs, storage space, and nutrient loss. The nutrients in manure vary by type of lot construction (Table 7.4).

TABLE 7.4 Nutrients in Manure from Four Sources (lb/hd/yr)

Source	N	P_2O_5	K_2O
Solid manure from open lots	65	41	65
Solid manure from deep-bedded building	132	66	132
Liquid runoff from open lots	5	2	11
Liquid manure from deep pit	89	55	79

Source: Iowa State University (2001).

Fly control can be achieved via chemical, biological, or combination strategies. Chemical controls used alone are typically costly, short term in effect, and increase risk of human or environmental chemical exposure. The use of biological controls via fly-parasites has been successfully adopted in the industry.

Fly-parasites, also known as parasitic wasps, lay their eggs inside fly pupae. The fly pupae then become a food source for the fly-parasites. The use of early releases of fly-parasites before fly season followed by scheduled weekly releases is recommended. The cost of biological control ranges from $.20–$1.00 per head of cattle.

Water quality issues are likely to have a significant impact on feedlots and other intensive animal management facilities. Waste management is a primary concern as the decomposition of manure can negatively affect water quality via pathogens, nitrate, ammonia, phosphorus, salts, and organic salts. Incorrect handling, storage, or land application of manure can result in contamination of groundwater or surface water.

In an active feedyard, a layer of soil and manure becomes sufficiently compacted to create a seal that serves as a barrier to seepage. If the integrity of this layer is ensured, then water infiltration can be kept to less than 0.05 in. per day. It is critical that this layer be left undisturbed during pen cleaning.

Applying manure or wastewater directly to agricultural lands requires thorough knowledge about the following factors associated with a particular site—soil type, slope, irrigation practices, precipitation levels, crop nutrient requirements, nutrient levels of manure, and proximity to waterways or wells. The goal of manure management is essentially to collect, store, and apply wastes to lands at appropriate agronomic rates with the objectives of optimizing crop growth rates, economic returns, and protecting water quality.

Runoff problems can be minimized via employment of Best Management Practices (BMP) such as building up-gradient ditches, dams, grass filter strips, filter fences, and other appropriate controlled drainage systems. Lagoons or ponds may be required to contain wastewater and runoff. BMP for manure handling are provided in Table 7.5.

TABLE 7.5 Summary of Best Management Practices for Manure Handling, Storage, and Application

Analyze for nutrient content
Account for available N from the total system
Apply to land areas large enough to accommodate manure volume
Calculate long-term manure loading rates
Maintain records of manure and soil analysis; application volume, timing, and methodology; additional fertilizer applications; and plant yields
Base manure application rates upon site-specific nutrient plans
Incorporate manure soon after application to avoid runoff
Determine application protocol based on soil composition and risk of aquifer contamination
Apply manure uniformly utilizing correct calibration
Utilize buffer zones to prevent water contamination
Use grass strips to catch and filter nutrients and sediments from runoff
Use rotational application schemes when planting high N use crops or forages
Locate manure stockpiles away from wells
Divert run-off via ditches, terraces, etc.
Maintain integrity of manure-soil seal when cleaning feedlot pens

Source: Adapted from Colorado State University (1994).

Breakeven Prices

Computation of breakeven prices is the central focus of a feedlot risk assessment and management program. Breakeven prices can be calculated on the price that can be paid for feeder cattle, cost of gain, and (or) sale price on fed cattle. Breakeven prices are usually calculated on an individual lot of cattle that are fed as a unit. Costs are estimated based on current and future market prices and gain costs that have been previously experienced.

The breakeven price for final liveweight of the fed cattle is calculated as follows:

$$\begin{array}{c}\text{Breakeven price}\\ \text{(per cwt. of live weight)}\end{array} = \frac{\text{Cost of feeder cattle} + \text{feed cost} + \text{nonfeed costs (Fig. 7.5)}}{\text{Estimated final weight (lb) of fed cattle}}$$

An example is:

$$\begin{array}{l}\text{Break even price}\\ \text{(cost/cwt of}\\ \text{live weight)}\end{array} = \frac{(750 \times \$0.70) + (450 \text{ lb} \times \$0.50) + (450 + \$0.10)}{1{,}200 \text{ lb}} = \frac{\$795}{1{,}200 \text{ lb}} = \$66.25$$

Breakeven price calculations for cost of feeder cattle and cost per pound of gain are presented later in the chapter.

As alliances and other retained ownership programs are implemented, a total production breakeven price should be calculated and utilized. A carcass weight breakeven price combines the costs of the cow-calf, yearling-stocker, and feedlot production phases. The calculations are:

$$\begin{array}{c}\text{Breakeven price}\\ \text{(cost/lb of carcass)}\end{array} = \frac{\text{Cost of weaned calf} + \text{yearling-stocker cost} + \text{feedlot cost}}{\text{Carcass weight (lb)}}$$

Using an example of current industry costs and weights for steers:

$$\begin{array}{c}\text{Break even price}\\ \text{(cost/lb carcass)}\end{array} = \frac{\$404 + \$113 + \$256}{770 \text{ lb}} = \frac{\$773}{770 \text{ lb}} = \$1.00/\text{lb}$$

A variety of factors can change the profitability of a pen of cattle. Poor health management resulting in high levels of morbidity, mortality, and realizers; poor rates of gain; and carcass discounts are likely contributors to financial losses. Table 7.6 summarizes data from the Texas Ranch to Rail Program as to the causes of economic losses. For the cattle fed in 1999–2000, it took the gains of 335 cattle (26%) with the lowest profits to offset the losses incurred by the 10% of the cattle that lost money.

TABLE 7.6 Causes of Economic Losses at the Feedyard

Cause	N	Average loss ($/hd)
Death	18	608.54
Realizers	17	281.95
Dark cutters	23	86.53
Poor gain[a]	47	47.09
Poor gain and grade discount[b]	18	89.54
Yield Grade 4 and 5 discount	11	26.93

[a]Gain was 24% below average; medicine cost 3 times greater than average.
[b]Gain was 27% below average; medicine cost 5 times greater than average.
Source: Texas A&M University (2001).

Cost of Feeder Cattle

Cattle feeders usually calculate what they can pay for feeder cattle based on projected feeding costs (feed and yardage) and anticipated selling price of slaughter cattle (Table 7.7). The price spread between feeders and fed steers is an important consideration for cattle feeders; however, it should be noted the cattle feeder is projecting the price of fed cattle 4–6 months from the time the feeder cattle are purchased. The prices of both feeder cattle and fed cattle can vary considerably during this short time period.

Projected fed steer price and price of corn are used to estimate breakeven purchase price for feeder cattle. Tables 7.8 and 7.9 show the breakeven purchase price for 550 lbs and 750 lbs feeder steers.

Feeder cattle are typically acquired from cow-calf producers or stocker operators via direct purchase or by virtue of a custom-feeding arrangement. The sources and ownership of feeder cattle are listed in Table 7.10.

TABLE 7.7 Computation for Projected Price for Feeder Cattle

Item	Price (per cwt)	Total Dollars
Sell 1,152 lb steer (1,200 lbs with 4% shrink after 120 days)	$65	$749
Cost of 450 lbs gain feed cost (0.45/lb) + nonfeed cost ($0.10/lb)	$54	$243
Value of feeder steer (sale price − gain cost)		$506
Price feeder can pay for 750 lb steer (breakeven price per cwt)	$68	

TABLE 7.8 Breakeven Purchase Price for 550 lb Feeder Steers

Fed-Steer Price	Corn Price ($/bu)						
	2.00	2.40	2.80	3.20	3.60	4.00	4.40
60.00	73.29	67.23	61.18	55.14	49.10	43.06	37.02
62.00	77.04	70.98	64.93	58.89	52.85	46.81	40.77
64.00	80.79	74.73	68.68	62.64	56.60	50.56	44.52
66.00	84.54	78.48	72.43	66.39	60.35	54.31	48.27
68.00	88.29	82.23	76.18	70.14	64.10	58.06	52.02
70.00	92.04	85.98	79.93	73.89	67.85	61.81	55.77

Source: Cattle-Fax.

TABLE 7.9 Breakeven Purchase Price for 750 lbs Feeder Steers

Fed-Steer Price	Corn Price ($/bu)						
	2.00	2.40	2.80	3.20	3.60	4.00	4.40
60.00	66.45	63.09	59.73	56.43	53.13	49.83	46.53
62.00	69.40	66.04	62.68	59.38	56.08	52.78	49.48
64.00	72.35	68.99	65.63	62.33	59.03	55.73	52.43
66.00	75.30	71.94	68.58	65.28	61.98	58.68	55.38
68.00	78.25	74.89	71.53	68.23	64.93	61.63	58.33
70.00	81.20	77.84	74.48	71.18	67.88	64.58	61.28

Source: Cattle-Fax.

TABLE 7.10 Source and Ownership of Cattle in Two Sizes of U.S. Feedyards

	< 8,000 Hd Capacity	> 8,000 Hd Capacity
Source of purchased cattle:		
At auction	47	31
Via direct sale	24	23
Custom fed or joint ownership	25	44
Born in herd owned by feedyard	3	1
Other	1	1
Type of ownership:		
By feedyard	52	34
Joint ownership	9	8
Custom fed	39	58

Source: NAHMS (2000).

TABLE 7.11 Feedlot Performance of Preconditioned versus Nonpreconditioned Cattle

	Preconditioned Direct from Ranch	Not Preconditioned Auction Market
Head (N)	71	66
In-weight (lbs)	612	619
Out-weight (lbs)	1,263	1,308
Days on feed	258	278
Average daily gain (lbs/hd)	2.52	2.48
Health cost ($/hd)	1.48	6.13
Pulls (%)	2	47
Retreats (%)	0	25
Castration (%)	0	2
Dehorn (%)	1	23

Source: Adapted from Iowa State University (2001).

The condition of newly received cattle is of paramount importance to the feedyard manager. *Preconditioned* calves are those that have been weaned, trained to acquire feed and water from bunks and tanks, have been immunized, and are generally experiencing low stress at the time of sale. Preconditioned cattle are typically better able to remain healthy and productive in the first 30 days of arrival at a feedyard. A comparison of preconditioned versus not preconditioned cattle is provided in Table 7.11.

Nonweaned calves arriving at the feedyard are 3–4 times more likely to contract bovine respiratory disease (BRD) than weaned calves independent of age or vaccine status. Steers treated for BRD have net profits that are $57.48 per head lower than nontreated contemporaries. Forty percent of BRD occurs in the first 14 days of arrival and 81% of the high morbidity cases expressed BRD in the first 42 days in the feedyard. Research consistently demonstrates that as the number of times cattle are pulled due to morbidity average daily gain declines, percent USDA Select and Standards increases, and profits decline.

Cost of Gain

Once feeder cattle are placed in a feedlot, cost of gain becomes a major economic concern for the cattle feeder. Total cost of gain typically is divided into two parts: (1) feed cost per pound (or cwt) gain and (2) nonfeed cost per pound (or cwt) gain.

Breakeven cost per pound of gain is calculated as follows:

$$\frac{\text{Breakeven cost}}{\text{(price per lb of gain)}} = \frac{\text{Value (\$) of fed cattle} - \text{cost (\$) of feeder cattle}}{\text{Total pounds of gain}}$$

$$\frac{\text{Breakeven cost}}{\text{(price per lb of gain)}} = \frac{(1{,}200\text{ lb} \times \$0.65) - (700\text{ lb} \times \$0.70) = \$290}{500\text{ lb}} \quad \frac{\$290}{500\text{ lb}} = \$0.58\text{ lb}$$

Grain prices have a tremendous effect on cost of gains for the cattle feeder. The price of feed comprises 70% to 80% of the total gain cost. Other feed grains (e.g., sorghum, barley, and oats) are important grains in certain cattle feeding areas, however, their total production is only about 20% of corn production and their price follows very closely the price of corn. Figure 7.7 shows the price of corn during 1995–2001.

Corn prices increased rapidly in 1996 and peaked during midyear. This escalation in corn price resulted from a sizable drop in corn supplies. There was an increased demand for corn primarily from the U.S. livestock and poultry industries. Corn usage jumped to a record high and if this rate had continued the existing domestic supplies would have been depleted.

Supply and demand for corn are both important in affecting corn price. Besides feed grains, corn is used for human food, seed, industrial use (e.g., fuel production), and approximately 20% is exported. Corn production in the United States is about 9.5 billion lbs per year. In 1996, carryover corn supplies, or inventory carried from one year to the next, was the lowest in nearly two decades. As a result, prices nearly doubled to $4.75 per bushel. This precipitated a dramatic decline in feeder cattle prices. However, an excellent U.S. corn crop during the fall of 1996 increased the corn supply and corn prices dropped rapidly. Corn prices in 2000–2001 have been at or below $2.00 per bushel. These rapid changes in corn prices have a significant effect on the cost of gains and the profitability of feeding cattle.

Factors Affecting Feed Cost per Pound of Gain

Most factors affecting feed cost per pound of gain are reflected in the animals' daily rate of gain and pounds of feed required per pound of gain. There are many factors affecting rate and

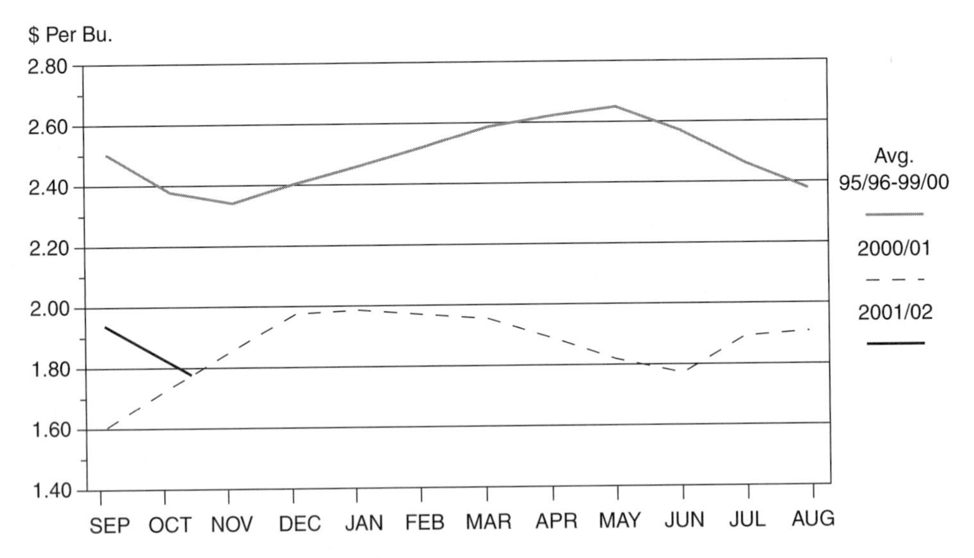

FIGURE 7.7 Monthly average price of U.S. corn.
Source: Livestock Marketing Information Center.

TABLE 7.12 Evaluating the Cost-and-Return Comparison of a Factor Influencing Rate and Efficiency of Gain for a 100-Day Feeding Period

Item	Control Group		Added Factor Having Positive Influence on Rate and Efficiency of Gain
Feed intake (lbs)	2,200		2,052
Total gain (lbs)	280		320
Feed saved (lbs)		148	
Value of feed saved @ $0.06/lb		$ 8.88	
Additional gain (lbs)		40	
Value of gain @ $0.64/lb		$ 25.60	
Advantage of gain and feed saved		$ 34.48	
Additional cost (assume added factor—$0.04/head/day)		−$ 4.00	
Total return per head		$ 30.48	

efficiency of gain, the value of which can be determined by a simple cost-return comparison as shown in Table 7.12.

Feed efficiency is an economically important trait for the cattle feeder. It has been determined that a 5% improvement in feed efficiency is equivalent to any one of the following:

1. Reducing ration costs on a dry matter basis by $8 per ton.
2. Reducing purchase costs of the feeder animal by $1.75/cwt.
3. Increasing daily gain by 0.6 lbs per day.
4. Reducing the interest rate on operating capital from 15% to 9.5%.

Daily gain and feed efficiency have a relatively high relationship (more rapid gains result in less feed required per pound of gain) where cattle are fed from the same in-weights to the same slaughter weights or to the same carcass compositional endpoint. The primary reason for this relationship is that cattle that gain more rapidly require less feed to maintain their body weight as they are fed for shorter periods of time.

Note in Table 7.13 that daily gain approaches 3 lbs per day before a higher percentage of the feed goes to gain rather than to maintenance. In this example, a difference of 1 lb of gain per day, within the same initial weight of cattle, accounts for a $25–$30 difference in maintenance cost. Maintenance feed is all cost with no pounds gained; however, it is needed to keep the animal alive, functioning, and ready to gain weight.

Since feed efficiency has such a significant relationship to cost of gain, a discussion of major factors affecting it is warranted. The three most important variables affecting feed efficiency are (1) energy density of the feed, (2) amount of dry matter consumed, and (3) rate of gain. Rations high in roughage fill the rumen with bulk, thus limiting energy intake and gain. Physical distention of the rumen is one of two main factors controlling dry matter intake. Adding grain to a high-roughage ration reduces fill, thereby increasing energy intake and improving gain.

Feed intake with high-energy rations is controlled by the products of digestion curtailing the animals' appetite (called *chemostatic regulation*) rather than rumen fill. Thus, increasing the energy density in an already high-energy ration usually has little effect on gain. Simply stated, improvements in feed efficiency are reflected mainly as increases in gain in low-energy rations and as reductions in feed intake in high-energy rations.

TABLE 7.13 Relationship of Gain-to-Maintenance Requirements for Cattle Fed to a 1,150-lb Market Weight

Initial Weight	Average Daily Gain	Days to 1,150 lbs	Lb Corn[a] for Maintenance	Value of Corn for Maintenance	Lb Corn for Growth after Maintenance	% Feed for Maintenance
400	2.0	375	2,310	$92.40	1,586	59%
	2.5	300	1,848	73.92	1,629	53
	3.0	250	1,540	61.60	1,670	48
600	2.5	220	1,421	56.84	1,309	52
	3.0	183	1,182	47.28	1,339	47
	3.5	157	1,014	40.56	1,375	42

[a]Total feedlot ration converted to corn equivalent to simplify example.
Source: Ideal Beef Memo (January 1982).

There are several ways the cattle feeder can improve feed efficiency in a feedlot operation:

1. Increasing the energy density of the feed by feeding less roughage. The highest practical energy level for a finishing ration is in the range of 0.64–0.68 Mcal/lb net energy for gain. Rations containing higher energy levels increase the risk of serious problems such as acidosis. Some farmer-feeders, raising their own silage, may feed high roughage levels to optimize the amount of feed produced per acre with cost of gain.
2. Processing grain is another method of improving feed efficiency. Steam flaking and reconstituting milo improves feed efficiency 8–17% depending on the processing method used and the type of grain; however, cost of feed is also increased.
3. Feeding additives and growth stimulants can improve feed efficiency from 3% to 12%. Several of these stimulate rate of gain but also indirectly improve feed efficiency due to the relationship between gain and feed efficiency. A more detailed discussion follows later in the chapter.
4. Feeding cattle to the proper composition of gain. It takes more feed to deposit a pound of fat than a pound of lean. Cattle that mature too early and cattle fed beyond their logical slaughter potential will be very inefficient and gain will be costly (see Fig. 7.8). Cattle that are capable of gaining rapidly and producing a yield grade 2 carcass at 1,100–1,250 lbs liveweight will most likely be highly efficient in feed utilization.
5. Keeping cattle free of climatic stresses such as muddy pens and minimizing their exposure to extremely cold or hot temperatures will contribute marked changes in feed efficiency. Cattle exposed to these stresses usually have slower gains through reduced feed consumption (particularly in times of extreme heat stress) and higher maintenance requirements with resulting poor feed conversions. Mounds in the feed pens or well-drained pens will help keep cattle dry and free of muddy conditions. Cattle fed in pens with 4 in. of mud will have feed efficiency lowered by approximately 10%. Table 7.14 provides management approaches for stress reduction.
6. Cattle that have the genetic ability to gain rapidly will produce efficient gains. These cattle have been genetically selected for appetite, which significantly increases their consumption of feed. Most cattle feeders have not tapped this genetic resource because compensatory gain has been available. Cattle with compensatory gain are typically light for their age, which means they have been given somewhat restricted nutrition during part or all of their life. When they are fed finishing rations in the feedlot, they compensate for the earlier restriction of feed by gaining very rapidly. Cattle with compensatory gain are not as

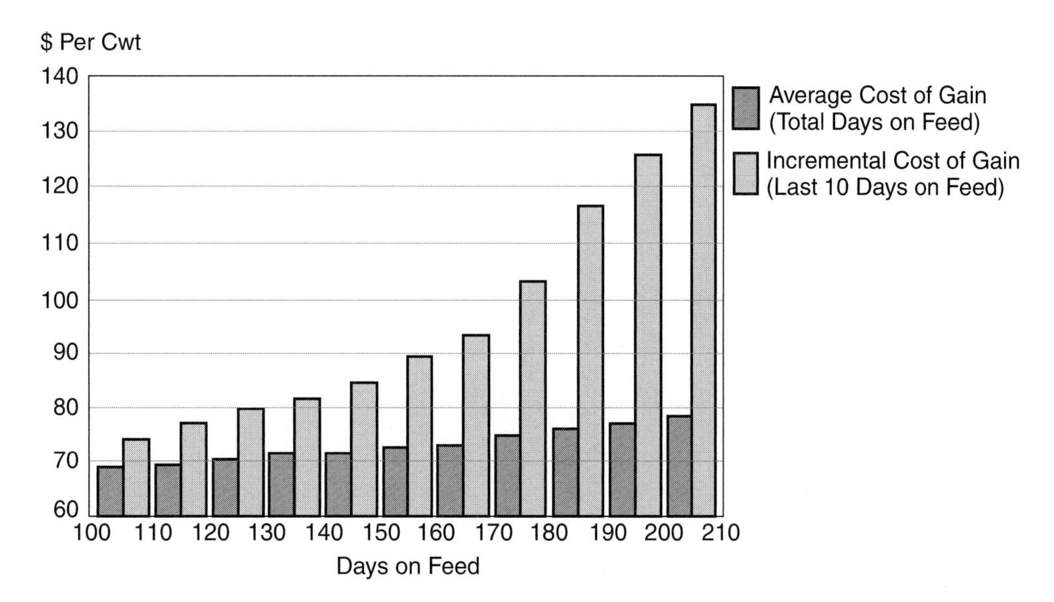

FIGURE 7.8 Average and incremental cost of feedlot gain with increasing days of feed. Gain costs computed using 1981 cost data.
Source: Beef Business Bulletin (NCBA).

TABLE 7.14 Factors Initiating Stress and Management Protocols to Reduce the Impact of Stress

Factor	Effect of stress	Method(s) to reduce stress
Behavior		
Novel experiences[a]	Reduced gain, increased dark cutters	Standardize handling, processing, and transport to minimize stress
Cattle handling	Reduced gain, increased dark cutters	Minimize noise, hotshots, animal discomfort during processing and transport, don't overcrowd tubs and alleyways
Environment		
Pen condition	Reduced gain, increased morbidity	Keep pens dry, well drained, and mounded; minimize dust in pens and feed
Shade and windbreaks	Increased maintenance cost, poorer feed efficiency and gain	Provide shade and sprinklers to minimize effects of extreme heat, and windbreaks to minimize wind chill; don't ship or process in the hottest part of the day
Insects	Increased eye problems, disease, loss of performance	Control flies via biological or chemical control
Management		
Feed and water	Reduced performance	Make water available at arrival, during feeding periods, and prior to shipment; feed consistently to allow cattle to settle into a routine of habit
Bullers	Reduced gain plus increased injury and bruising	Relocate bullers immediately
Processing and transport	Reduced gain, increased injury; increased bruises and dark cutters	Wait 48 hours following arrival for processing

[a]New pens, handling by people and on horseback versus on foot, etc.
Source: Adapted from multiple sources.

numerous as in past years, primarily because those producing them have found it unprofitable. Also, feeders have been more competitive in the marketplace for these cattle, so their economic advantage has been reduced compared with several years ago. Feeder cattle with the genetic potential to reach 1,100–1,250 lbs at 14–16 months of age and that have desirable carcasses are becoming more numerous. Cattle feeders who can identify and purchase these genetically superior cattle can usually expect low gain costs because of excellent feed conversion. It has been estimated that cost of gain can be lowered $4–$8 per hundredweight by feeding cattle with high genetic gaining ability.

7. Sex of feedlot cattle also affects feed conversion, though differences in feed efficiency vary depending on the slaughter endpoint. It is generally accepted that heifers are less efficient than steers and bulls and that bulls are more efficient than steers. These relationships are valid when the different sexes are slaughtered at similar slaughter weights. Heifers typically have 10–15% lower average daily gains, 6–8% poorer feed efficiency, and slightly higher death losses than steers.

Factors Affecting Nonfeed Costs

Among nonfeed costs per pound of gain, gaining ability, health, yardage, and interest rates are the most important. Cattle with more rapid daily gains have less maintenance feed cost and less overhead cost because of the fewer days required to accumulate a given amount of weight gain. Table 7.13 shows that 600-lb cattle gaining 2.5 lbs per day require 63 more days to reach 1,150 lbs than cattle gaining 3.5 lbs per day. At a nonfeed cost of 35 cents per day, slow-gaining cattle require approximately $22 more per head than fast-gaining steers to break even at the same harvest weight.

Healthy cattle are less costly because they do not require additional labor for treatment and the cost of medication is eliminated. Also, healthy cattle will have a lower death loss and a higher rate of gain because of a consistent high-feed consumption pattern (Table 7.15). Some surveys have shown that a 1% death loss costs $5 to $10 per head marketed. And, a 10% morbidity rate costs about $2 per head marketed for medication costs alone. Another study showed that feedlot cattle that did not get sick had $90 more net return than sick cattle in the same feedlot.

Respiratory diseases (e.g., BRD or bovine respiratory disease) are consistently the major health problem, usually accounting for approximately 75% of the morbidity and 50% of the mortality.

TABLE 7.15 Comparison of Sick versus Healthy Cattle for Feedlot, Quality Grade, and Profitability Performance

Trait	Sick	Healthy
Herd	218	1,080
Death loss (%)	5.5	0.7
Ave. daily gain (lbs/day)	2.65	3.08
Total cost of gain ($/cwt)	62.32	49.03
Medicine cost ($/hd)	26.78	0.00
Net return ($/hd)	23.31	146.17
Quality grade (%)		
Choice	37	54
Select	53	43
Standard	10	3

Source: Texas A&M Ranch to Rail, 1999–2000.

Feeder cattle from one source, arriving from backgrounding operations, and low-stressed, weaned calves are usually lower risk cattle for respiratory and other disease problems.

Eighty percent of feedyards ride or walk pens two or more times a day for cattle received within 14 days. For cattle that have been at the yard for at least one month, pen observation is typically reduced to once daily (70% of feedyards) with 25% of feedyards conducting pen checking activities twice or more per day. About one-third of cattle are processed within 12 hours of arrival while the remaining two-thirds are processed between 13 and 72 hours of arrival. Standard processing protocols vary somewhat by feedlot size (Table 7.16). The inclusion of nonlactating, mature cows with newly received calves may reduce morbidity, improve average daily gain, and increase the speed by which calves learn to eat at a bunk.

Table 7.17 shows the influence of interest rates on the daily costs. Cost per pound of gain can be affected by several cents when the interest rate varies a few percentage points.

Cattle feeding is a high-risk business because of fluctuating market prices, the biological variability of the cattle being fed, and the complexity of the many other factors that require careful management. The variation in cattle performance in the feedyard and on the rail can have a dramatic effect on feedyard profitability (Tables 7.18 and 7.19).

TABLE 7.16 Percent of Processed Cattle Receiving Various Protocols

	Feedyard Capacity	
Feedyard Processing Protocol	<8,000 Hd	>8,000 Hd
Vaccinated against respiratory diseases	98	98
Vaccinated against clostridial diseases	81	77
Receiving an injectable antibiotic	17	19
Receiving an implant	89	97
Treated for parasites	94	98

Source: NAHMS (2000).

TABLE 7.17 Daily Cost per Head at Various Interest Rates (assumes an $825 total cost of animal and feed)

Annual Interest Rate	Daily Interest Cost per Head ($)
15%	$0.34
12	0.27
9	0.20

TABLE 7.18 Variation in Feedlot Performance of 1,311 Cattle from 111 Ranches

A.D.G. (lb/day)	Cattle (%)	Cost of Gain Feed ($/cwt)	Cattle (%)	Medicine Cost ($/hd)	Cattle (%)
≤ 2.5	7	≤ 40	15	0	30
2.51–2.75	15	41–45	30	≤ 5	37
2.76–3.00	23	46–50	23	5.01–10	18
>3.0	55	51–55	22	10.01–15	6
		>55	10	>15	9

Source: Adapted from Texas A&M Ranch to Rail, 1999–2000.

TABLE 7.19 Variation in Carcass Performance and Profitability of 1,311 Cattle from 111 Ranches

Carcass Weight (lbs)	Cattle (%)	Fat Thickness (in.)	Cattle (%)	Ribeye Area (in.)	Cattle (%)	Yield Grade	Cattle (%)	Quality Grade	Cattle (%)	Profit ($/hd)	Cattle (%)
≤ 650	2	< .20	22	≤ 11	2	1	15	Prime	0	< 0	6
651–750	22	.20–.29	14	11.1–13	31	2	48	Choice	51	1–50	9
751–850	50	.30–.39	20	13.1–15	41	3	32	Select	45	51–100	17
> 850	26	.40–.49	13	15.1–17	21	4	10	Standard	4	101–150	37
		.50–.59	13	>17	5					>150	31
		.60–.69	9								
		>.70	9								

Source: Texas A&M Ranch to Rail, 1999–2000.

Total Dollars Received

Sale price of the market steers and heifers has the most significant effect on total dollars received in the cattle feeding operation. Market prices can fluctuate widely between years and within a year with both having a major influence on profits (Fig. 7.9).

Most fed cattle are sold on a liveweight basis where packers' buyers bid on pens of cattle. There is an increasing number of feedlot cattle that are marketed on a carcass weight and grade criteria. Premiums and discounts are paid primarily for carcass weight, quality grade, and yield grade differences. These carcass characteristics predict some consumer preferences—for example, carcass weight (size of retail cuts), yield grade (percent fat to lean), quality grade (juiciness, flavor, and tenderness), and others. A more detailed discussion on how prices and other marketing factors (e.g., sex, liveweight, shrink, grades, and grid markets) can be evaluated is covered in Chapter 9.

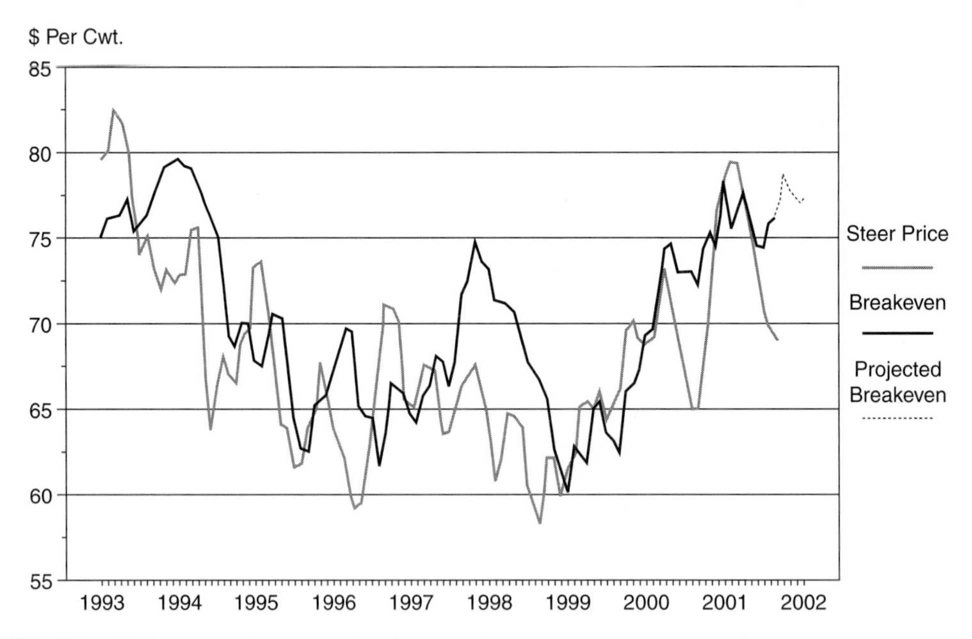

FIGURE 7.9 Choice steer price versus breakeven price.
Source: Livestock Marketing Information Center.

TABLE 7.20 Effect of a 10% Change on the Breakeven Price in a 1989 Feedlot Budget[a]

Factor	Change (%)	Decrease in Breakeven Price ($/cwt)	Increase in Profit ($/head)
Sale price	+10%	$0.00	$ 80.42
Purchase price	−10	5.57	66.84
Feed cost[b]	−10	1.44	17.31
Average daily gain	+10	2.74	32.84
Interest cost	−10	0.22	2.69
Death loss	−10	0.06	0.66
All factors		$10.03	200.76

[a]Based on a 2000 budget—assumes a finished weight of 1,200 lbs at $68/cwt.
[b]Equivalent of a 10% improvement in feed efficiency.
Source: Cattle-Fax.

Manure, as a by-product of the feedlot, can be a minor source of income. Some feedlots will compost the manure that is later sold primarily for home lawns and gardens. Other feedlots will sell or give the manure to farmers or to contract haulers who transport the manure from the feedlot. Some manure is not acceptable to farmers if the salt content is high.

MANAGEMENT PRIORITIES

Table 7.20 shows the impact of changes in the major factors affecting breakeven price and profit. Obviously, changes in sale price, purchase price, and feed costs have major effects. An economic analysis of 200 feedlots in the Midwest (Lawrence, 1997) showed that fed cattle prices and feeder cattle prices explained over 70% of the profitability.

Because cattle prices vary considerably with time, cattle feeders must implement risk management techniques to protect their investments and ensure profitability. The futures market, with the use of hedging and options, are risk management instruments currently used by many cattle feeders. These instruments are discussed in detail in Chapter 9.

Implants and Additives

Rate of gain and feed efficiency can be improved by using growth promotant implants and feed additives. Some commonly used *implants* and *feed additives* for calves, yearlings-stockers, and growing-finishing cattle are shown in Tables 7.21 and 7.22.

Implants can be categorized into several classifications—medium potency estrogens, lower potency estrogens, androgens, higher potency combinations, and lower potency combinations. When choosing an implant program, consider cost of implant, expected response, type of cattle, and convenience. The proper combination of implants can yield improvements in daily gain, feed efficiency, and lowered cost of gain of up to 20%, 15%, and 10%, respectively.

Implants containing trenbelone acetate are typically considered terminal implants or the last implant given. The general recommendation is that these products should be administered in the last 100 to 140 days on feed. Early implants tend to be estrogenic in nature.

For implants to be effective, they must be implanted properly. First, the animal must be restrained so that movement of the head is minimal during the implanting process. Determine where the implant will be located (Fig. 7.10), then insert the needle the proper distance away from that point, push it to the hilt, and back the needle off slightly before commencing to squeeze the trigger and withdraw the needle. This prevents possible crushing of the implant,

TABLE 7.21 Approved Growth Promotant Implants

Implant[a]	Ingredient, Level (mg), and Use	Class[b]
Calfoid	10 estradiol benzoate/100 progesterone Calves < 400 lbs	LPE
Component E-C	10 estradiol benzoate/100 progesterone Calves < 400 lbs	LPE
Component E-S	20 estradiol benzoate/200 progesterone Steers >400 lbs	MPE
Component E-H	20 estradiol benzoate/200 testosterone Heifers > 400 lbs	MPE
Component T-S	140 trenbelone acetate Finishing steers	A
Component T-H	200 trenbelone acetate Finishing heifers	A
Component TE-S	24 estradiol 17 beta/120 trenbelone acetate Finishing steers	CHP
Compudose 200	24 estradiol 17 beta All classes	MPE
Encore	48 estradiol 17 beta All classes	MPE
Finaplix-S	140 trenbelone acetate Finishing steers	A
Finaplix-H	200 trenbelone acetate Finishing heifers	A
Implus-S	20 estradiol benzoate/200 progesterone Steers >400 lbs	MPE
Implus-H	20 estradiol benzoate/200 testosterone Heifers >400 lbs	MPE
Ralgro	36 zeranol All classes	LPE
Ralgro Magnum	72 zeranol Finishing steers	MPE
Revalor-S	24 estradiol 17 beta/120 trenbelone acetate Finishing steers	CHP
Revalor-H	14 estradiol 17 beta/140 trenbelone acetate Finishing heifers	CHP
Revalor-G	8 estradiol 17 beta/40 trenbelone acetate Pasture steers and heifers	CLP
Revalor 200	20 estradiol 17 beta/200 trenbelone acetate Finishing steers	CHP
Synovex-C	10 estradiol benzoate/100 progesterone Calves <400 lbs	LPE
Synovex-S	20 estradiol benzoate/200 progesterone Steers >400 lbs	MPE
Synovex-H	20 estradiol benzoate/200 testosterone Heifers >400 lbs	MPE
Synovex-Plus	28 estradiol benzoate/200 trenbelone acetate Finishing steers	CHP

[a]Follow manufacturer's directions for implanting method and withdrawal time.
[b]A = Androgen, CHP = combination high potency, CLP = combination low potency, LPE = low potency estrogen, MPE = medium potency estrogen.

TABLE 7.22 Selected Feed Additives for Growing-Finishing Cattle[a]

Common or Trade Name (Additive)	Claims	Recommended Level	% Improvement	
			Gain	Efficiency
Aureomycin (Chlortetracycline)	Improves gain and feed efficiency; reduces liver abcesses	70 mg/hd/day	3–5	3–5
Bovatec (Lasalocid Sodium)	An ionophore that increases the proportion of propionic acid in the rumen and decreases occurrences of acidosis and bloat	10–30 g per ton of feed	2	8
MGA (Melengestrol Acetate)	A synthetic progesterone that inhibits estrus	0.25–0.50 mg per day	5–11	5
Rumensin (Monensin)	An ionophore that increases the proportion of propionic acid in the rumen and decreases occurrences of acidosis	5–30 g/ton of complete feed; do not feed more than 360/mg/hd/day	0	5–12
Terramycin (Oxytetracycline)	Improves gain and feed efficiency and reduces liver abcesses	75 mg/hd/day	3–5	3–5
Tylan (Tylosin)	Same as Terramycin	60–90 mg per head per day	3–5	3–5

[a]Follow manufacturer's directions for proper feeding and withdrawal times.

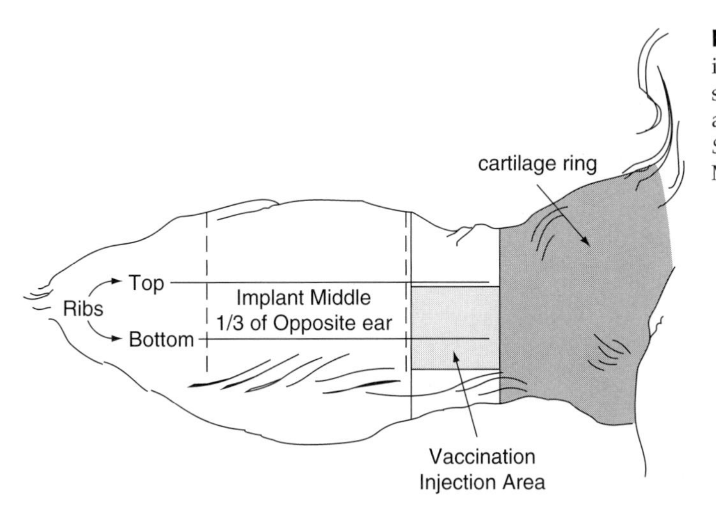

FIGURE 7.10 Proper location of implant and subcutaneous vaccination site. It is not recommended to implant and vaccinate in the same ear. *Source:* Colorado Beef Quality Assurance Manual.

which might lead to decreased effectiveness or possible problems due to too-rapid absorption. The needle must be sharp; if a bigger hole is put into the skin, it invites bacteria into the wound and may cause infection and poor absorption of the pellets.

Once the needle has been removed from the ear, it is a good practice to press the injection opening between the thumb and finger so the opening tends to seal. This practice is not done to prevent the implants from coming out but to prevent infection of an open wound.

Assuring proper implanting protocols is critical. The ear should be cleaned with a brush and then disinfected with a chlorohexadine solution prior to placement of the implant. This procedure reduces the possibility of an abscessed ear. Cattle with abscessed implant sites will experience as much as a 0.2 lbs per day reduction in gain. The net economic effect of an infected ear can approach $20.00 per head.

According to an Oklahoma State University study the economic returns from implanting were $18.32 per implant investment if cattle were sold live and $13.53 per implant investment if cattle were sold grade and yield (Duckett et al., 1996). The difference in profit was due to the decreased percent of prime and choice carcasses from the implanted cattle. Producers should be cautious in the use of aggressive implant protocols due to potential negative effects in quality.

The trade-off is between the increased growth of fed cattle versus the impact on product quality. A comparison of cattle receiving trenbelone acetate (TBA) implants versus those not receiving TBA showed a decline of 11% in USDA Prime and Choice grades, but a gain of more than 40 lbs in carcass weight in the treated animals. The advantage in weight is typically more profitable except in those cases when the Choice-Select spread is exceptionally high.

However, from a consumer perspective, the aggressive use of high-potency implants may yield undesirable effects in terms of tenderness, marbling, and overall acceptability. Finding the correct balance among these factors is an important management consideration. Producers attempting to hit high-quality branded beef targets may be best served by using conservative implant strategies. Table 7.23 illustrates the implant practices of feedyards surveyed by the National Animal Health Monitoring System of the USDA.

Certain feed additives are known as ionophores. Ionophores are antibiotics obtained from streptomyces microorganisms that alter rumen metabolism and improve animal performance. Currently two ionophores—monensin (trade name Rumensin) and lasalocid (trade name Bovatec)—are approved by the FDA for commercial use in cattle (see Table 7.22).

Total Quality Management (TQM)

W. Edwards Deming has given the world a Total Quality Management (TQM) process of quality production rather than relying on an inspection process to eliminate the creation of mistakes. Since quality has many different meanings to different people, Deming defines

TABLE 7.23 Percent of Feedlot Cattle Receiving Implants by Feedlot Size and Cattle Weight Category

	Size of Feedyard	
Number of Times Implanted	< 8,000 Hd (%)	>8,000 Hd (%)
	Steers and Heifers <700 lbs at Placement	
0	6	1
1	24	17
2	67	75
≥ 3	3	7
	Steers and Heifers >700 lbs at Placement	
0	8	2
1	66	67
2	26	31
≥ 3	0	0

Source: NAHMS (2000).

quality as meeting or exceeding your customers' expectations at a cost that represents value to them every time.

Mies (1993) reviews how Deming's 14-point TQM plan applies to feedyard management. The 14 points are:

1. Create a constancy of purpose toward improvement of the product and service, with a plan to become competitive, stay in business, and provide jobs.
2. We can no longer live with commonly accepted levels of delays, mistakes, defective materials, and defective workmanship.
3. Cease dependence on inspection to achieve quality. Eliminate the need for inspection on a mass basis by building quality into the production in the first place.
4. End the practice of awarding business on the basis of price. Instead, depend on meaningful measures of quality, along with price. Move toward a single supplier for any one item in a long-term relationship based on loyalty and trust.
5. Improve constantly and forever the system of production and service to increase quality of the product and constantly decrease costs.
6. Institute modern methods of training for all employees.
7. Institute leadership. Focus supervisors on helping people do a better job. Ensure that immediate action is taken on conditions detrimental to quality.
8. Drive out fear so everyone may work efficiently for the company.
9. Break down barriers between departments and encourage problem solving through teamwork.
10. Eliminate arbitrary goals, posters, and slogans for the work force that seek new levels of productivity without providing methods.
11. Use statistical methods for continuing improvement of quality and productivity. Eliminate work standards that prescribe numerical goals.
12. Remove barriers that rob workers of their pride of workmanship.
13. Institute a vigorous program of education and training.
14. Take action to accomplish transformation. The transformation is everybody's job.

MANAGEMENT SYSTEMS HIGHLIGHTS

1. W. Edwards Deming helped numerous businesses become successful by implementing a Total Quality Management (TQM) process of quality production.
2. Deming defined quality in TQM as meeting or exceeding your customers' expectations at a cost that represents value to them every time.
3. Mies (1993) applies Deming's 14-point TQM plan to feedyard management. These 14 points are applicable to all production segments of the beef industry—both individually and collectively.
4. The carcass, as an endpoint of feedlot production, is an excellent place to evaluate cost of production and an initial assessment of value to the consumer.
5. Breakeven price of carcasses measures production costs from two or more cost centers. For example:

$$\text{Cost/lb of carcass} = \frac{\text{Cost of weaned calf} + \text{yearling-stocker cost} + \text{feedlot cost}}{\text{Carcass weight}} = \frac{\text{Total cost (\$)}}{\text{Lb of carcass}}$$

6. The following carcass measurements give an initial assessment of value to the consumer:
   - Carcass weight—size of retail cuts
   - Yield grade—fat to lean composition
   - Quality grade—taste (juiciness, flavor, and tenderness)

- Color of lean—acceptable appearance in retail meat case
- Hump—tenderness

REFERENCES

Publications

Albin, R. C., and Thompson, G. B. 1996. *Cattle Feeding: A Guide to Management.* Amarillo, TX: Trafton Printing.

Brink, T. 2000. The Impact of TBA Implants on Gelbvieh-Sired Cattle. *Gelbvieh World* (Feb.).

Colorado State University and Texas A&M University. 1999. *Impact of Feedlot Growth Promotant Implant Strategies on Carcass Grade Characteristics and Subsequent Cooked Beef Palatability Traits When Applied to Small/Medium Framed, 3-Way, British Crossbred Steers.* Final report to NCBA.

Duckett, S. K., Wagner, D. G., Owens, F. N., Dolezal, H. G., and Gill, D. R., 1996. Effects of Estrogenic and Androgenic Implants on Performance, Carcass Traits and Meat Tenderness in Feedlot Steers: A Review. *The Professional Animal Scientist.* 12: 205–214.

Great Plains Beef Cattle Feeding Handbook. Cooperative Extension Services of CO, KS, MT, NE, NM, ND, OK, SD, TX, and WY.

Harner, J. P., and Murphy, J. P. 1998. *Planning Cattle Feedlots.* Kansas State University Ext. Publ. MF-2316. Manhattan, KS.

Iowa State University. 2001. *Beef Feedlot Systems Manual.* ISU Ext. Publ. 1867. Ames, IA.

Lawrence, J. D. 2001. *Preconditioning: Does It Pay?* BEEF. Spring special Cow-Calf Newsletter.

Lay, D. 2000. *Growth Promotants in Cattle.* Iowa State University Fall Extension Newsletter.

Livestock and Poultry Situation and Outlook Report. 1970–2001. Washington, DC: USDA.

McNeill, J. W. 2000. *Where the Profits Came From . . . and What Caused the Losses.* Ranch to Rail Summary. Texas A&M University, College Station, TX.

Mies, W. L. 1993. *Adapting Total Quality Management to Feedyard Management.* Englewood, CO: The National Cattlemen's Association.

Morse, D. 1995. Environmental Considerations of Livestock Producers. *J. Anim. Sci.* 73: 2733.

Owens, F. N., et al. 1997. The effect of grain source and grain processing on performance of feedlot cattle: A review. *J. Anim. Sci.* 75: 868.

Roeber, D. L., Cannell, R. C., Belk, K. E., Miller, R. K., Tatum, J. D., and Smith, G. C. 2000. Implant Strategies During Feeding and Impact on Carcass Grades and Consumer Acceptability. *J. Anim. Sci.* 78: 1867–1874.

Schaefer, D., Hirschinger, C., and Klemme, R. 1986. *Wisconsin Farm Enterprise Budgets—Holstein Steers,* Report number A3360. Madison, WI: University of Wisconsin.

Smith, R. A. 1998. Impact of Disease on Feedlot Performance: A Review. *J. Anim. Sci.* 76: 272.

Texas A&M University. 2000. 1999–2000 Ranch to Rail Summary Report. College Station, TX.

USDA. 2000. *Baseline References of Feedlot Management Practices.* National Animal Health Monitoring System. APHIS,VS.

Vieselmeyer, B., et al. 1995. *Physiological and Economic Changes of Beef Cattle During Finishing.* Nebraska Beef Report, Lincoln, NE.

Visuals

"The Calf Path" (1996; 10 min.). Texas Cattle Feeders Assn., Amarillo, TX.

"Cattle Handling Principles to Reduce Stress" (1998; 54 min.). Grandin Livestock Handling System, Inc., Fort Collins, CO.

"The Feedyard" (1993; 27 min.). Creative Educational Video, Lubbock, TX.

"Holstein Beef Production: Birth to 350 Pounds" (1990; 24 min.); "Holstein Beef Production: Finishing Holstein Steers" (1990; 25 min.). University of Wisconsin, Madison, WI.

"The New Beef Quality Era" (2000; 10 min.). Microbeef Technologies, Inc., Amarillo, TX.

Computer Programs

Hi-Plains Systems, Inc., P.O. Box 8152, Amarillo, TX 79109.
Taurus (see Chapter 14).

Websites

Entomology site focused on beneficial insects: www.kunafin.com
Biocontrol information network: www.biconet.com
Clean water action plan: www.cleanwater.gov

Concentration, Integration, and Alliances in the Beef Industry

The following statements aptly describe the focus of this chapter:

> Industry structure will be shaped by individual decisions in response to economic forces in the free enterprise system. Low-cost producers (in all segments of the production chain) will survive in this system of competitive markets; others will eventually be unable to compete, and will exit the business. Some of the production assets of those exiting will be accumulated by remaining producers. Consequently, the size of the "average" operation will increase. Increased acquisitions, and expanding size of operations, can be expected in all sectors of the beef industry. (NCA 1989 Task Force Report, p. 4)
>
> If the industry is not able to form alliances and working arrangements between cow-calf, feedlot, and packing operations, it will not make much improvement in its current marketing system. (Topper Thorpe, Cattle-Fax)
>
> Vertical coordination from farms forward will become absolutely necessary. And not too many of us will be selected for participation in supply chains unless individuals are cost-efficient, innovative, and partner-willing. (Ray Goldberg, Ph.D., Harvard Business School)
>
> The power of a partnership can elevate your business and your partners' to levels never thought possible. By sharing information, risks, and rewards, partners can do more together than acting independently. (Sheldon Lavin, OSI Group, Meat Processor)

Beef industry structural changes, as influenced by concentration, integration, and alliances affect costs of production, production practices, consumer demand, market prices (cattle and retail product), and profitability.

CONCENTRATION AND INTEGRATION

Concentration level is one measure used to determine the degree of competitiveness in an industry or market. It measures the combined sales of the leading firms as a share of total market or industry sales. Concentration measures are used in antitrust law to determine the competitive impact of mergers and acquisitions within the same market.

Concentration occurs in the beef industry when fewer businesses own or control more of the total cow-calf, feeder, or packer supply. The most noted change in beef industry concentration has occurred in the packing segment since there are currently less than half the number of packing plants as in the 1980s. Among the major beef packing plants identified in Chapter 1, four companies—IBP, ConAgra, Excel, and Farmland National Beef—slaughter approximately 80% of the steers and heifers and handle approximately 85% of the boxed beef. In 1980, the biggest four packers controlled only about one-third of the fed cattle harvest (Table 8.1).

Figure 8.1 shows that most of the largest packing plants in the United States are located in the Plains states. Approximately two-thirds of these plants are owned by the four big packers. More than 4,000 head of cattle per day are harvested at some of these plants.

Many producers and some producer organizations were vocal in expressing belief that the packer concentration had increased market power in too few hands and resulted in lower prices paid for fed cattle. Research into those accusations generally shows small negative impacts from increased concentration. Concerns about packer concentration tend to be voiced most ardently when prices are low. For example, the greatest cause of the depressed prices during 1996–1997 was due to the oversupply of cattle that was characteristic of that phase of the cattle cycle.

Concentration in meatpacking has resulted primarily due to the pressure for packing plants to become more cost-competitive. Economy of size that results in reduced per-herd harvest and fabrication costs has led to increased concentration. Estimates are that a $5.00 reduction in processing cost per head could translate into an increase of $0.25 to $0.50 per cwt higher price for fed cattle.

Another market concentration issue is captive supply—the acquisition of cattle via non-cash market channels or by direct ownership of fed cattle by packers. In the past 20 years, the percent of total harvest accounted for by packer-fed steers and heifers has declined from 6% in 1975 to 3.5% in 1998 (Table 8.2). This level of activity does not suggest a loss of competitiveness in the market.

However, since 1990, the number of cattle procured by the top four packers via forward contracts and marketing agreements has been approximately 20% (Table 8.2). While economic

TABLE 8.1 Steer and Heifer Slaughter Concentration by the Top Four Firms

Year	Federally Inspected Plants (N)	Top Four Plants[1] (N)	Head (mil)	Concentration (% of total slaughter)
1980	1,411	23	9.5	35.7
1985	1,451	20	14.1	50.2
1990	1,105	26	19.1	71.6
1995	836	27	22.8	79.3
1997	822	27	23.1	79.5
1999	795	24	24.2	81.4

[1] Individual plants served by the top four packers.
Source: Packers and Stockyard (2000).

FIGURE 8.1 Location of packing plants harvesting > 1,000 head per day per plant.

TABLE 8.2 Number of Steers and Heifers Fed by Packers

Year	Number of Packer-Fed Steers and Heifers (1,000)	Percent of Total Slaughter (%)	Percent of Cattle from Forward Contracts and Marketing Agreements[1]
1975	1,284	6.0	NA
1980	874	3.6	NA
1985	861	3.3	NA
1990	1,257	4.9	20.1
1995	973	3.6	21.1
1996	896	3.2	19.2
1998	1,022	3.5	18.9

[1] Four largest beef packers.
Source: Packers and Stockyards (2000).

research studies have failed to show this is an unfair trade practice, industry watchdogs continue to monitor the level of price competitiveness in the fed cattle market.

The Justice Department is solely responsible for investigating possible price fixing, bid rigging, and other illegal activities sometimes associated with concentration. It determines whether competitors (remaining firms) in a given region have the ability to increase prices (for products) or decrease prices (for inputs, e.g., cattle) for a substantial period of time (at least a year). If the Justice Department finds that a certain merger was not in the interest of an industry, the judicial process is utilized. In recent years, the Justice Department has reviewed and allowed a number of mergers in the packing segment. In June of 2001, following months of legal wrangling, the single largest merger in the packing industry's history was allowed by a federal court. The merger of IBP and Tyson joins the largest beef packer, the second-largest pork packer, and the largest poultry processor.

Concentration has also occurred in the cattle feeding segment but less dramatically than in the packing segment. The number of feedlots has decreased from 122,000 in 1970 to 41,000 in the mid-1990s. The projected number of feedlots in the year 2010 is 21,000. In 2000, the top 25 cattle feeding companies operated 106 feedlots and marketed 38% of the total fed cattle marketings. Cactus Feeders, Inc., in Amarillo, Texas, had a one-time capacity of 460,000 cattle in its nine feedyards. Assuming a 2.5 times turnover, Cactus feeds more than 1 million cattle a year.

Feedlots located in the 12 major cattle feeding states market nearly 95% of the nation's fed cattle. Table 8.3 shows that these feedlots are decreasing in number with the larger feedlots increasing in the number of head marketed. Feedlots with a one-time capacity of 16,000 head and higher, marketed 11.2, 13.7, and 15.7 mil head in 1985, 1995, and 2000, respectively.

Table 8.4 illustrates the trend in the feedyard industry whereby the largest companies account for the vast majority of the total marketings. Feedyards with a one-time capacity of greater than 16,000 head comprise approximately 11% of feedlots with at least 1,000-head capacity. Yet, these large producers account for almost 80% of all fed cattle marketings.

Concentration can also be viewed from the perspective of where feedlot cattle are located in the United States (Fig. 8.2). The larger packing plants are found in areas where cattle feeding is concentrated. An abundance of high-energy feeds (grain and silage), desirable climatic conditions (dry and relatively mild temperatures), and a sparse human population are the economic stimuli for location of the feedlots. The cost of transporting live cattle is the primary incentive for packing plants locating close to feedlots.

TABLE 8.3 Number of Feedlots and Number of Head Marketed in the 12 Major Cattle Feeding States

			< 1,000 Head Feedlots		>1,000 Head Feedlots	
Year	Total Feedlots (thousand)	Total Marketed (mil head)	No. Feedlots (thousand)	No. Marketed (mil head)	No. Feedlots (thousand)	No. Marketed (mil head)
1985[a]	50.9	22.9	49.2	4.6	1.7	18.3
1995[a]	41.4	23.4	39.4	2.3	1.9	21.1
2000	NA[a]	24.9	NA[b]	1.7	1.8	23.3

[a] Top 13 states.
[b] Feedlots with < 1,000-head capacity not reported.
Source: USDA.

TABLE 8.4 Number of Feedlots and Cattle Marketed by Size of Feedlot (>1,000 hd), 1999

	1,000–8,000	8,000–16,000	16,000–32,000	> 32,000	Total
Number of lots	1,674	193	141	111	2,119
Percent	79.0	9.1	6.2	5.2	100.0
Number of cattle marketed (1,000 hd)	4,258	3,093	5,197	10,982	23,530
Percent	18.1	13.1	22.1	46.7	100.0

Source: USDA.

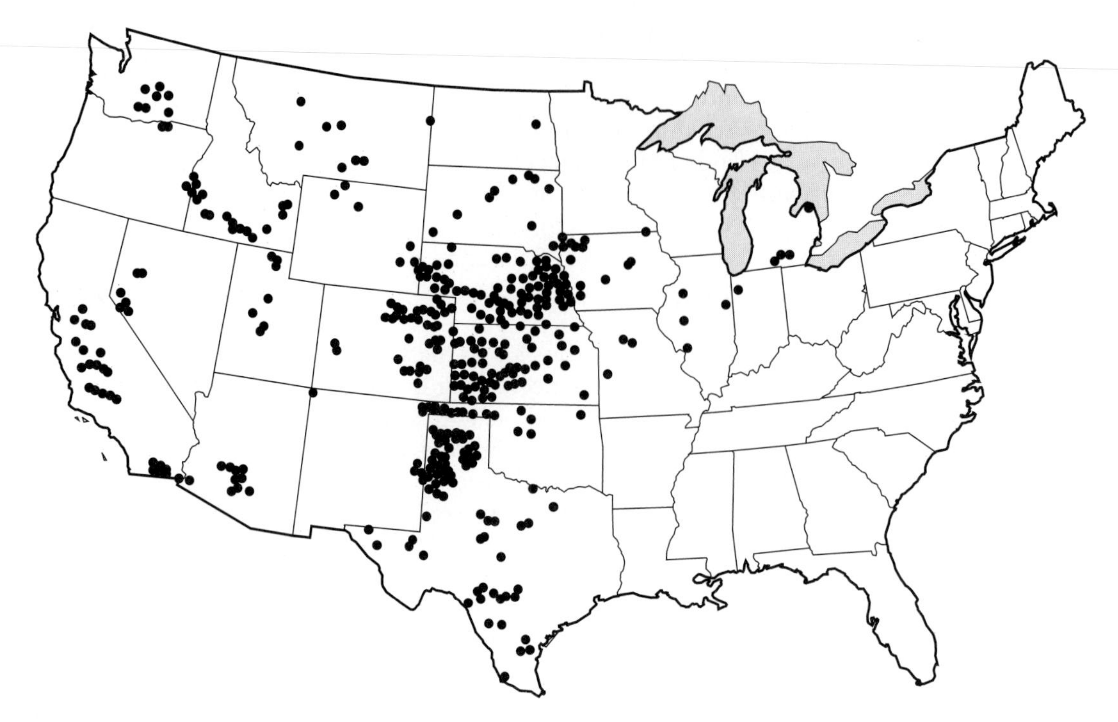

FIGURE 8.2 Location of feedyard with > 4,000 head one-time capacity.
Source: USDA and Cattle-Fax.

TABLE 8.5 Changes in the Number of Cow-Calf Operations in the United States

	Year			
	1970	1990	2000	2010
No. of Operations	1,200,000	920,000	830,880	740,000

Source: Adapted from USDA data.

Concentration is the least apparent in the cow-calf segment compared with other segments because land costs are usually high relative to the return on investment from cattle revenues. However, the number of cow-calf operations has declined by almost 400,000 since 1970 and it is projected that this decline will continue in future years (Table 8.5). Only about 10% of cow-calf operations have more than 100 head of breeding cows on inventory. Yet, these larger enterprises account for more than 50% of all beef cows in the United States (Table 8.6). Many of the enterprises with fewer than 100 head of beef cows depend on outside income sources to support family living. Whether these smaller herds are maintained as part of a diversified agricultural business or to sustain a desired lifestyle, they will likely continue to remain an important component of the industry.

Figure 8.3 shows the location of cow-calf operations with at least 500 cows. These larger operations are widely distributed across the United States. It is estimated that in the year 2010 there will be 7,000 of these larger operations that will control 20% of the total cow herd. There are approximately 30 cow-calf owners each with a minimum of 5,000 cows. The largest operation is approximately 35,000 cows.

TABLE 8.6 Cow-Calf Producers and Production by Various Herd Size

Number of Operations	Total (N)	Percent by Size Group			
		1–49	50–99	100–499	500+
1995	897,660	79.8	11.7	7.8	0.6
1996	885,980	79.4	12.0	7.9	0.6
1997	872,840	79.3	11.9	8.0	0.6
1998	855,460	79.3	11.8	8.2	0.6
1999	844,170	78.9	12.0	8.4	0.6
2000	830,880	78.6	12.1	8.6	0.7
January 1 Inventory—1,000 hd					
1995	35,190	31.2	19.2	35.3	14.3
1996	35,319	30.8	19.6	35.4	14.2
1997	34,458	30.4	19.4	35.9	14.3
1998	33,885	30.4	18.9	36.1	14.6
1999	33,745	29.9	19.1	36.6	14.4
2000	33,569	29.3	19.2	36.8	14.7

Source: USDA (2000).

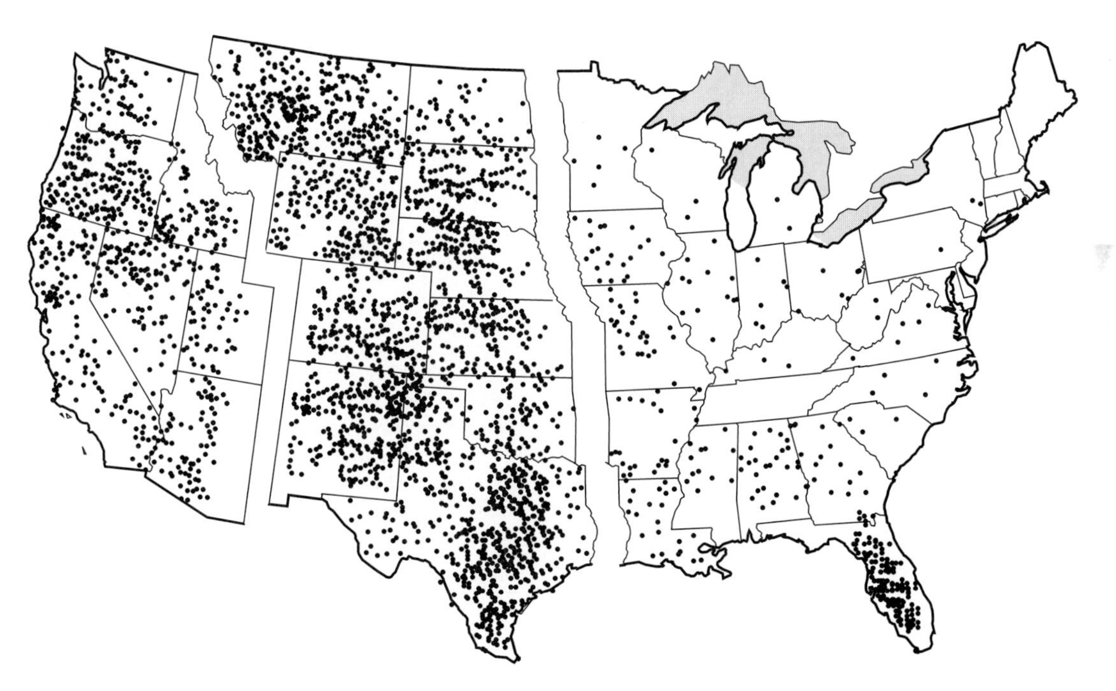

FIGURE 8.3 Location of Cow-calf operations having at least 500 cows per operation. One dot represents one cow-calf operation.
Source: U.S. Census data.

Simultaneously, the food processing and distribution system that extends beyond the packer has also undergone rapid consolidation. The 30 largest food wholesalers control 75% of the wholesale food market while only five companies control 40% of the retail food market (Table 8.7). Again, one of the major drivers of consolidation is the need for economics of scale by companies operating on thin margins. The net profit as a percent of sales in supermarkets ranges from .012 to .0103.

It is important to note that consolidation has been most significant closest to the consumer with the least amount of concentration occurring at the cow-calf level (Table 8.8).

TABLE 8.7 Top 10 Grocery Stores (sales)

Company	Stores Owned	Sales ($ Bil)	% of U.S. Sales
Wal-Mart Supercenters (AR)	862	57.2	11.1
Kroger, Co. (OH)	2,359	49.2	9.6
Albertsons (ID)	2,514	36.4	7.1
Safeway (CA)	1,726	33.2	6.4
Ahold USA (VA)	1,208	27.5	5.3
Supervalue (MN)	457	23.3	4.5
Fleming (TX)	164	14.7	2.9
Publix Super Markets (FL)	645	14.0	2.7
Winn-Dixie Stores (FL)	1,160	13.8	2.7
Delhaize America (NC)	1,425	12.7	2.5

Source: Supermarket News (www.supermarketnews.com).

TABLE 8.8 Largest Five Seedstock, Cow-Calf, Feeder, Packer, and Grocery Companies in the United States

Seedstock	Calves Born	Cow-Calf	Cows
Leachman Cattle Co. (MT)	3,340	Deseret Cattle and Citrus (FL)	40,000
Stevinson/Basin, Inc. (MT)	2,900	J. R. Simplot Co. (ID)	32,500
Summitcrest Farm (OH)	1,939	King Ranch, Inc. (TX)	25,000
DeBruycker Charolais (MT)	1,600	Lykes Bros., Inc. (FL)	20,550
Express Ranches (OK)	1,575	Parker Ranch, Inc. (HI)	18,000
Top 5 share (%)	NA	Top 5 share (%)	.04%
Feedyard	**One-time Capacity**	**Packers**	**2000 Sales ($ Bil)**
Cactus Feeders, Inc. (TX)	460,000	IBP, Inc. (SD)	12.8
Conti Beef, LLC (CO)	425,000	Excel Corp. (KS)	8.2
ConAgra Cattle Feeding (CO)	420,000	ConAgra Beef (CO)	5.0
Caprock Industries (TX)	296,000	Farmland National (MO)	2.7
National Farms (MO)	270,000	Packerland Parking (WI)	1.4
Top 5 share (%)	13%	Top 5 share (%)	84%
Grocery Company	**2000 Sales ($ Bil)**		
Wal-Mart Supercenters	57.2		
The Kroger Co.	49.2		
Albertsons, Inc.	36.4		
Safeway, Inc.	33.2		
Ahold USA, Inc.	27.5		
Top 5 share (%)	40%		

Source: NCBA and FMI, 2001.

The structural changes that have occurred in the beef industry have forced managers and industry leaders to look for unique and innovative approaches to improve profitability in the new landscape. One option that exists in the industry is to realign ownership or contractual arrangements via integration.

Vertical integration occurs when two or more stages of production are controlled by the same owner. *Horizontal integration* takes place when supporting or competitive goods and services are controlled or owned by the same company. Some examples of horizontal integration include a feedlot that owns or controls the grain supply or a packing company that has multispecies plants (e.g., pork and poultry).

Vertical integration or coordination occurs in the beef industry through ownership, contractual arrangements, and formula pricing involving two or more segments. For example, the big three packers own outright or purchase under a forward contract or formula pricing relationship about 20% of their slaughter needs. More contracting would allow vertical coordination rather than vertical integration. Through this process, large packers would not own the cattle but they would have some control over numbers, quality, and prices. Feeders could have similar contracts with cow-calf producers.

In the latter half of the 1990s, the major packing companies began to build or acquire case-ready plants that would enable them to deliver prepackaged beef products direct to retail stores that would require no additional handling other than to stock the products into the sales case. IBP launched its own brand, Thomas E. Wilson®, in an effort to provide higher value to its customers. Harris Ranch launched its own line of precooked and fresh products, as did the alliance group Ranchers' Renaissance.

Future Beef Operations (FBO) was launched in 2000 with the intent to build state-of-the-art case-ready processing plants supplied via contract-grower arrangements. FBO has a contractual agreement with Safeway to become the sole supplier of beef to their retail divisions. FBO also plans to market its own line of pet foods, blueside leathers, and other by-products as a means to capture value.

The most significant integration in the beef industry has been in the feeding and packing segments. Other examples of integration have not impacted the total industry as much as integration of feeding and packing, but they have been important in reducing production costs and in creating "niche marketing."

Integration of the Broiler Industry

The broiler industry is more highly integrated, both vertically and horizontally, than the beef industry. However, the broiler industry is less concentrated than the beef industry at the packing and processing level. One broiler company may own or control most of the production and marketing operations identified in Figure 8.4. In 1960, the 20 largest broiler firms controlled 52 slaughter plants and 32% of total United States slaughter. In 2000, the top 10 firms controlled 70% of the broiler production in the United States.

Broiler production is concentrated in the states of Arkansas, Georgia, Alabama, North Carolina, Mississippi, Texas, Maryland, Delaware, California, and Virginia. (California is the only state with significant production located outside the "broiler belt.") These states produce nearly 85% of U.S. broilers.

Broiler production has increased tremendously over the past several decades, from 3.7 billion lbs of ready-to-cook broilers in 1960 to more than 30 billion lbs in 2000. The broiler industry is concentrating growout operations into fewer but larger farms. Almost 85% of all broiler production is conducted under a contract grower arrangement with the remaining 15% almost totally owned by the packer-processor. The five largest firms control almost half of all current broiler production and slaughter. These poultry giants are bringing integration to the pork industry as well.

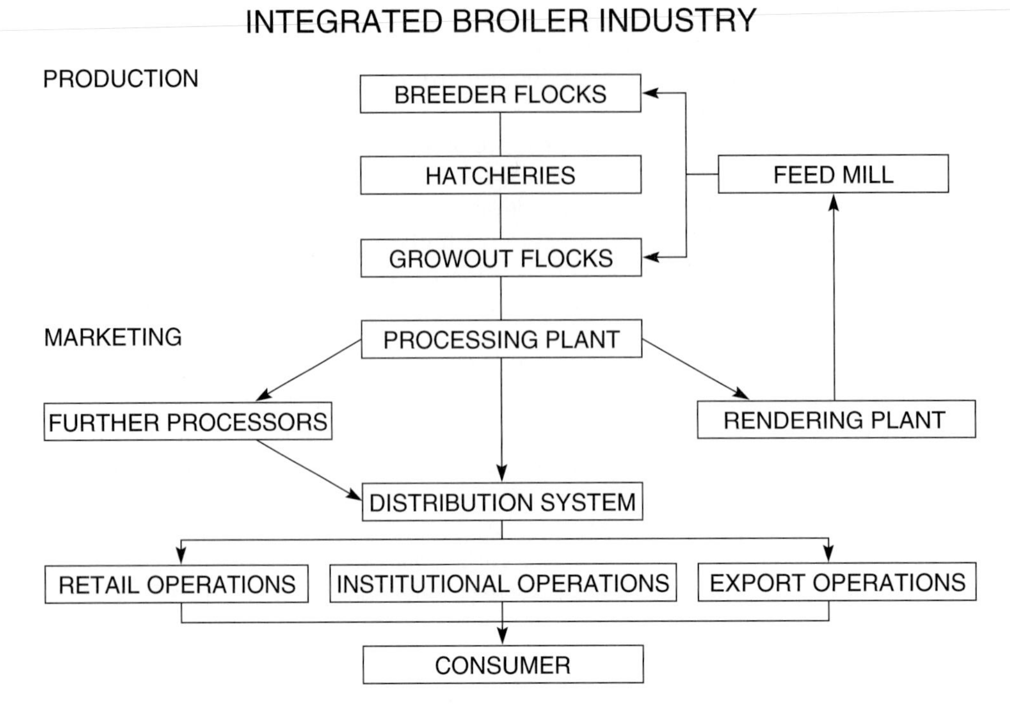

FIGURE 8.4 Integration in the broiler industry.

Topper Thorpe of Cattle-Fax compared the structure of the beef industry to the poultry industry by stating, "In the final analysis, the beef industry structure will continue to change. It is making some progress in spite of its structure and the variability of its product. But more progress is needed. While the beef industry will never look like the poultry industry, some valuable lessons can be learned through comparison."

The poultry industry is evaluating the next level of integration called "Supply Chain Optimization" where there is a single measure of success—net margin (Seagraves and MacDonald, 1997). Supply chain optimization decisions consider and optimize the interactions between costs and revenue throughout the total production system. This concept is not new, nor is it unique to the poultry industry. W. Edwards Deming, the father of total quality management, referred to the lack of coordination between business units as suboptimization. He described the practice of suboptimization as one in which managers pursue the goals of lowest possible cost without receiving feedback on how their decisions affect the results of the operation as a whole. An example currently used is that companies make use of strains of birds that are not the easiest or cheapest to grow but a final product is created that is more in line with the needs of the company's customers.

Vertical integration by ownership is unlikely in the beef industry for the following reasons. First of all, the cow-calf and stocker segments are land-based and low-margin. Thus, players farther down the supply chain are unlikely to be interested in assuming more risk and debt. Second, the size and scope of the beef industry is too cumbersome to facilitate total ownership. For example, a contract for the delivery of beef to a major retail customer would require production from 200,000 head of fed cattle and a capital outlay of $1.7 billion. The production requirement would include: three million acres of forage, 10–15 backgrounding enterprises, five feedlots, 125,000 acres of cropland, one packing/fabrication plant, two case-ready processing plants, and a $250 million line of operating credit (Garrett, Future Beef Operations, 2000).

However, contractual arrangements, development of supply chains, and other partnerships are likely to become predominant.

FORCES DRIVING STRUCTURAL CHANGE

Concentration and integration in the industry result from a variety of causative factors. Among these are the advantages of economies of scale. The benefits that arise from capturing economies of scale include the opportunity to lower costs via the advantages of volume purchasing power that lowers per unit input cost, capability to access useful new technology, the ability to spread fixed costs over higher levels of production, and gaining access to markets as a result of scale.

In a commodity industry, where by definition the average producer breaks even over time, one of the best opportunities to capture profits is via the process of lowering costs. In the beef industry, cow-calf herds of greater than 500 head have about one-half of the production costs in producing a weaned calf as compared with herds with less than 50 head (Table 8.9). The four largest beef packers have a 3% advantage in cost of production—a significant advantage in a low-margin industry.

Consolidation in the agricultural sector has also been driven by consumer expectations for plentiful and relatively cheap foods. The net result for U.S. consumers has been that instead of paying 21 cents of each earned dollar for food, as was the case 50 years ago, they now pay 11 cents. The corporate superstores, be it Wal-Mart or Safeway, are part of the economic landscape because they meet a demand. They are there because consumers want to do business in an environment where access to a variety of brand names is readily available. Commodity production systems of the past are not capable of meeting consumer demands for a plethora of value-added products that offer an assurance of quality and value.

The consequences of industry structural change have been summarized by Boehlje (1995) and Field (2000) as follows:

1. The deadwood gets trimmed as the low-profit, inflexible players are removed from the system.
2. As the focus shifts from segment to system, total cost reduction in the system becomes important. To assure critical communication of information within the system, the attention given to data capture, transfer, and utilization is magnified.

TABLE 8.9 Cost of Production and Profitability Comparisons of Various Sized Cow-Calf Herds (Southwest: TX, OK, NM), 1991–1999

Size of Herd	Weaning (%)	Weaning Weight (lbs)	Lb Weaned/ Exposed	Capital Investment Per Cow ($)	Cost of Production Per Cow ($)	Cost of Production ($/cwt)	Net Income/ Cow
1–49	85	512	441	4,751	496	114	(114)
50–99	83	502	414	4,001	384	81	10
100–199	82	507	418	3,321	421	92	(55)
200–299	83	533	439	4,007	411	88	(24)
300–499	83	533	441	2,994	403	87	(21)
500–999	82	531	437	2,913	370	77	23
1,000+	80	538	428	3,371	317	69	42
Average	83	522	431	3,605	402	87	(21)

Source: SPA Summary.

3. Inputs (health program, genetics, nutrition) become a contractual obligation instead of a free-agent decision. This is necessary to assure that customers receive a very specialized product that meets detailed specifications.
4. Value-added, specification-oriented decisions replace mass quantity production goals.
5. As product specifications become more clearly defined, then the need for standardization at critical control points becomes very important.
6. Spot markets are replaced by contracts, formulas, and performance-oriented pricing systems.
7. Risk management is expanded to include pricing, environmental impact, animal welfare, and partnerships with suppliers and customers.
8. Information becomes the cake not the frosting. Access, sharing, and integration of data become the center of attention.
9. Control measures at the interface with consumers and the point of genetic decision-making become increasingly critical.
10. Partnerships and alliances take the place of isolated, free-agent decision-making.

Consolidation results from consumer demand, public policy, survival of the fittest in an environment of increasing competition, and the loss of producers who choose to retire rather than adapt. The benefits of consolidation include lower production costs at the farm and ranch level, lower food prices for consumers, and a more competitive position for the United States in global markets. The drawbacks include the loss of some people and firms from the agricultural sector and dramatic changes in rural communities both in terms of culture and infrastructure.

ALLIANCES/BUSINESS RELATIONSHIPS

The term "alliance" has been a growing concept in the beef industry since the mid-to-late 1990s. A 2000 survey by BEEF magazine showed that approximately 40% of beef producers were involved or planned to be involved in some type of marketing alliance. Cow-calf producers had most of their alliances with seedstock breeders, breed associations, and feedlots. About 40% of the total supply of fed cattle are marketed via an alliance, contract, or formula pricing system.

Part of the stimulus in developing alliances came from a strategic alliance pilot project coordinated by the National Cattlemen's Association (now NCBA) in 1993. The project was sponsored by several beef organizations and involved a partnership of producers, feedlots, packers, and retailers as a group. The strategic alliance cattle were superior to other groups of typical feedlot cattle in both cost of production and retail product. Among the strategic alliance cattle there were large individual differences in productivity and profitability. This demonstrated that future alliances could select and manage cattle that would be superior in cost of production and retail product.

Alliances usually involve a contractual arrangement between different segments of the beef industry. These alliances are a type of integration where different individuals and companies can operate somewhat independently of one another but still share in risks and profits when cattle and beef products meet certain specifications (Fig. 8.5). Success of alliances is highly dependent on participants having common paradigms, effective communication, trust, and "win-win" attitudes. Alliances can be simple by involving only two or more producers or organizations or they can be very complex where all segments of the beef industry are in an alliance. These alliances can be organized and driven by producers, packers, breed associations, or the total industry. The net result of each alliance is that bulls, feeder cattle, carcasses, retail products, or all of these are produced for targeted markets.

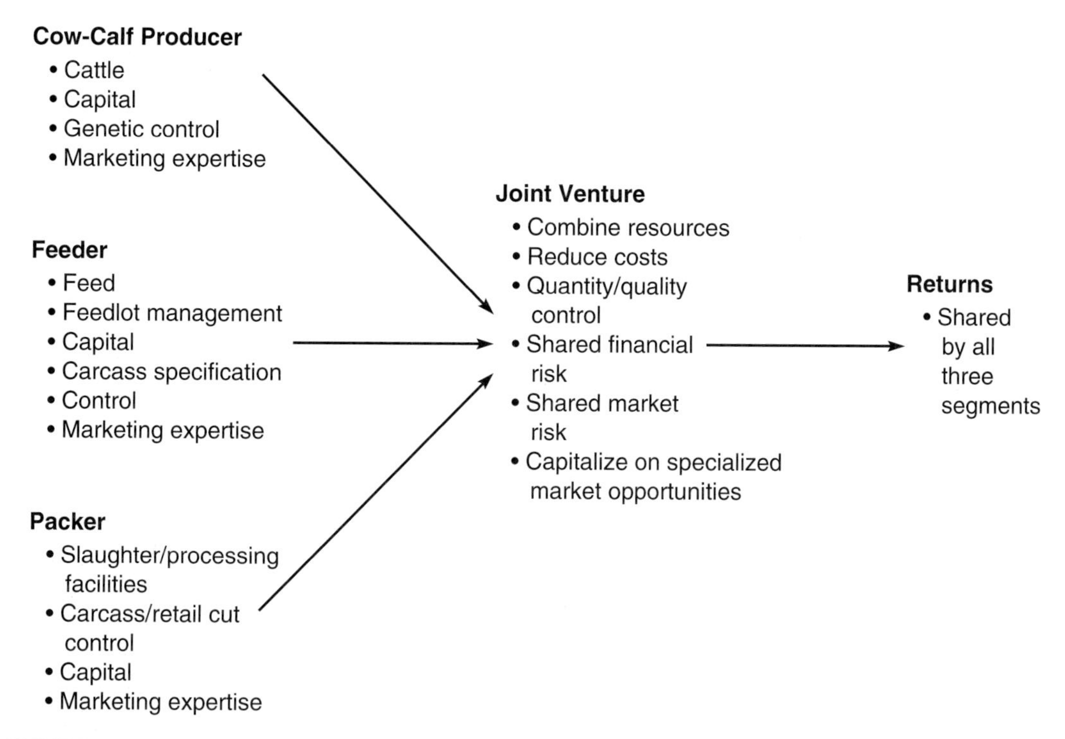

FIGURE 8.5 An example of an alliance where three segments make contractual arrangements to reduce costs and risks, improve retail products, and enhance profit.

While alliances can differ widely from one another, most current alliances are marketing alliances, breed alliances, and closed cooperatives. Breed alliances represent efforts to increase markets for specific breeds. Closed cooperatives are member-owned organizations that attempt to control the product from cows to consumers. They sell to participate in all income-generating segments of the beef industry by combining producers, processors, and retailers into one coordinated business. Table 8.10 shows some of the major alliances, their volume of production, and the specifications. Most of the specifications focus on live cattle and carcasses.

Several of the alliances have branded beef products while some organizations have branded products without the name of an alliance attached to it. Belk and Smith (1997) noted that the intent for branded beef programs, as they are presently evolving, is to "use source verification and (or) cattle/carcass sorting to either lessen variability in—or guarantee consistency in quality, palatability, composition and (or) safety (real or perceived) of beef. Branding creates opportunities to implement quality management techniques across all sectors of the marketing chain."

Many of the branded beef products are USDA "Certified" and/or "Verified." USDA meat graders certify carcasses as meeting or exceeding live requirements, quality factors, yield factors, and other carcass characteristics. Verified implies that the USDA graders verify the production process of the carcasses so the contract requirements are met. Cattle and carcasses, or the processes used to generate them, that are found by USDA graders to be in compliance with the specifications or contractual obligation, receive one of the two USDA shields: (1) the "Accepted as Specified" shield or (2) "Process Verified" shield (Belk and Smith, 1997).

The market for "*Natural*" and "*Organic*" beef in the United States is presently less than 2% of beef sales—expected to grow but conservatively. *Natural* implies that the cattle have received no growth promotants (hormones) and/or antibiotics.

TABLE 8.10 Beef Alliances and Branded Products

Alliance Web Address	Year Established	Cattle in 1999 (N)	Ave. Premium Pd./Hd. ($/hd)	Cost ($/hd)	Minimum Cattle (N)	Required Genetics	Production Practices Required[a]	Associated w/Branded Product
Angus America *www.angusamerica.com*	1996	90,000	17.5	3 to 6	35	>50 % Angus	None	Y
Angus Gene Net *www.angusgene.net*	1998	50,000	17	3	30	None	None	N
B3R Country Meats	1986	20,000	60	0	40	None	S, W, P, N	Y
Beef Advantage Project, L.P.	1996	NA	NA	NA	NA	50% British & Cont. <25% *Bos indicus*	S, W, P	Y
MLE/Southern States Coop. *www.southernstates-coop.com*	1995	NA	NA	NA	Load lots	75–100% British <25% *Continental*	S, W	Y
Certified Angus Beef *www.cabprogram.com*	1978	1.9 million	0 to 5	0	1	>50% Angus	None	Y
Certified Hereford Beef *www.hereford.org*	1995	45,804	NA	0	1	<50% Hereford 100% British	Packer dependent	Y
Coleman Natural Products, Inc. *www.colemannatural.com*	1979	NA	NA	NA	1	British and British cross preferred	S, P, N	Y
ConAgra Better Beef, L.L.C. *www.conagra.com*	1995	175,000	NA	NA	250	100% British or 50% British 50% Cont.	S, P	Y
Decatur Beef Alliance	1994	40,000	NA	12	Load lot	<25% *Bos indicus*	S	N
Farmland Supreme Beef *www.agribeef.com*	1995	80,000	15.57	2.50 to 4	70	75% British 25% Continental	None	Y
Gelbvieh Alliance *www.gelbvieh.com*	1995	40,000	13.37	1 to 5	1	None	None	Y
Hitch Enterprises *www.hitchok.com*	1992	22,218	NA	4	Load lot	None	W, P	N
Iowa Quality Beef Supply Net.	1999	10,000	NA	Up to 6	Load lot	None	None	N
Laura's Lean Beef Company *www.laurasleanbeef.com*	1985	45,000	65	None	20	75% Continental	S, W, P, N	Y
Lean Limousin Beef Co.	1988	7,000	3/cwt	None	1	>50% Limousin	None	Y
Maverick Ranch Beef Assn.	1985	50,000	NA	None	Load lot	Prefer Continental	None	Y

TABLE 8.10 Beef Alliances and Branded Products—(*continued*)

Alliance Web Address	Year Established	Cattle in 1999 (N)	Ave. Premium Pd./Hd. ($/hd)	Cost ($/hd)	Minimum Cattle (N)	Required Genetics	Production Practices Required[a]	Associated w/Branded Product
Nebraska Corn-Fed Beef *www.necornfedbeef.com*	1997	28,000	14.88	4 to 5	1	50–75% British, rest Continental	S	Y
Nolan Ryan's Tender Aged Beef	2000	NA	NA	None	1	Prefer % *Bos indicus*	N	Y
Oregon Country Beef *www.oregoncountrybeef.com*	1986	8,000	NA	None	1	None Continental	S, W, P, N	Y
Painted Hills Natural Beef *www.natural-beef.com*	1997	1,980	.04/lb over ave	None	1	100% British, prefer Angus	S, N	Y
Performance Plus–Retained Ownership	1993	NA	32	7.5	1	50–75% British, 25–50% Continental	S, W, P	N
Perfomance Plus–Sale Barn	1993	NA	27.5	8	1	50–75% British, 25–50% Continental	S, W, P	N
ProBeef Producer Network *www.probeef.com*	1997	20,000	52	5	100	>50% Continental	S, W, P	Y
Ranchers Renaissance	1997	100,000	NA	NA	NA	<25% *Bos indicus*	S, W, P	Y
Red Angus Feeder Calf Certification *www.redangus1.org*	1995	63,250	NA	1.25	1	One registered Red Angus parent	S	Y
Red Oak Farms Premium Hereford Beef *www.redoakfarms.com*	1996	NA	NA	None	1	>50% Hereford, 100% British	S	Y
Seedstock Plus *www.seedstockplus.com*	2000	NA	NA	None	1	Gelbvieh, Angus, Red Angus	S, W, P	Y
U.S. Premium Beef, Ltd. *www.uspremiumbeef.com*	1986	620,000	15.25 grid 17.99 dividend	10	20	Low % *Bos indicus*	None	Y
Western Beef Alliance	1994	20,000	31.5	5 to 8	40	50% British 50% Continental	S	Y
Western Ranchers Beef Cooperative	1997	7,600	52	3	1	None	S, W, P, N	Y

(*continued*)

TABLE 8.10 Beef Alliances and Branded Products—(*continued*)

Alliance Web Address	Year Established	Cattle in 1999 (N)	Ave. Premium Pd./Hd. ($/hd)	Cost ($/hd)	Minimum Cattle (N)	Required Genetics	Production Practices Required[a]	Associated w/Branded Product
CALF-BASED PROGRAMS								
Certified Angus Beef Feedlot Licensing Program	1999	23,116	NA	0.25	1	>50% Angus	None	Y
Farmland Beef Connection–Black Angus Link	1999	14,000	11.5	3 enroll 4 data	20	51% Angus, 49% British	S, W, P	Y
Hi-Pro Producer's Edge	1993	35,000	NA	NA	1	<25% *Bos indicus*	S, W, P	N
Kent Beef Alliance	1997	NA	NA	NA	1	None	S	N
Land O' Lakes Strategic Alliance *www.beeflinks.com*	1994	15,000	NA	3 to 10	40	No *Bos indicus*	None	N
Montana Beef Network *www.mtbeef.org*	1999	18,600	NA	1 to 6	1	None	S, W, P	N
MorrMan's Value Trac	1996	6,863	15.56	2 to 6	5	None	S	N

[a] S = source verification, W = weaned, P = preconditioned, N = natural (no hormones, implants).
Source: Adapted from *Drover's* and *BEEF.*

USDA released their Organic Production and Handling Standards in December 2000. The rules state that animals raised for harvest must be raised under organic management from the last one-third of gestation. Feed must be 100% organic although vitamin and mineral supplements are allowed. Growth promotant hormones and antibiotics are disallowed. Furthermore, these animals must have access to the outdoors and cattle must have access to pasture. A USDA organic seal may be used as part of a certification program starting in October 2002.

Several alliances use value-based formulas and grids in their marketing programs. A more detailed discussion of grid and value-based formulas is in Chapter 9 (The Marketing System).

The most successful alliances will have a combination of cost of production (low breakevens), improved retail products and high net profits. The majority of current alliances emphasize only feedlot costs, and carcass traits (weight, quality grade, and yield grade). It should be recognized that consumer market share is not likely to be sustained or improved with only a lean product (e.g., yield grades 1 and 2) or high-quality grades (e.g., Choice). More effective measures of tenderness and high palatability must be identified and utilized.

Alliances are typically formed to better meet consumer expectations by improving the process of production and processing via the application of Total Quality Management (TQM), Hazard Analysis Critical Control Point (HACCP), or Best Manufacturing Practices (BMP) principles to improve profitability via a combination of cost-control and value-added efforts, and to better manage total risk in the system.

However, several factors stand in the way of creating successful alliances. These include lack of trust, availability of capital, poor planning, lack of management expertise, inflexibility, poor decision-making, and lack of commitment to the alliance when economic conditions shift.

Producers who desire the opportunity to participate in alliances should carefully research the following questions:

1. Which management practices will I have to change?
2. Do I prefer collaborative or independent decision-making?
3. How much financial risk can I undertake in both the short and long term?
4. Do I have sufficient borrowing power?
5. Do the goals of the alliance align with those of my enterprise?
6. What are the costs and benefits?
7. How much experience does the management of the alliance have in critical areas?

Alliances function as part of a free enterprise system and, as such, do not have total assurance of success. However, when the partners and conditions are aligned within an alliance structure, partners can achieve profitability while experiencing the meaningful sense of fulfillment that comes from successful innovation.

Even if highly palatable retail products can be consistently produced and effectively marketed, the beef industry could have large numbers of producers exiting the business and cow numbers decreasing because of high production costs. It is possible to make $50 per head with superior feedlot and carcass performance, yet lose $150 per head in the cow-calf segment. If both cost of production and retail product is emphasized (Table 8.11), there are several hundred dollars profit that can be captured in both the live cattle production and processing (carcass/retail cut) phases. The combined total is greater than emphasizing only one of the segments. There are current examples of the same cattle that can make more than $100 per head in the cow-calf production and more than $100 per head in the feedlot and carcass phases. This higher return is compared with average cattle and carcasses in the current beef industry. Targets and specifications for cost of production, breakevens, and retail products are identified in Chapters 2 through Chapter 7.

TABLE 8.11 Components of a Total Beef Industry Alliance

Low Cost of Production and Higher Net Profit	Retail Products and Improved Market Share
• Low weaned calf breakeven price ($/cwt) • Low yearling breakeven ($/cwt) • Low feedlot breakeven ($/cwt) • Low carcass weight breakeven, e.g., $\frac{\text{Weaning cost + yearling cost + feedlot cost}}{\text{Carcass weight}} = \text{Cost per lb of carcass}$ • Resources (e.g., forage) sustained or improved to optimum levels • Continuing high net margin (profit) each year	• Consistently uniform and highly palatable retail products (e.g., tenderness, flavor, and juiciness) • Highly tender —breed and sire selection within breed —low Warner-Bratzler Shear —carcasses electrically stimulated —carcasses aged 14–21 days • Carcass quality grade • Carcass yield grade • Optimum carcass weights • Carcasses free of defects (e.g., dark cutters, injection-site blemishes, bruises, etc.) • Free of microorganism contamination and harmful residues • Product has an environmentally safe image • Cost-competitive with other meats • Regain market share to 50% or higher • Value-added by-products (e.g., hides free of brands or butt branded, large choice of convenient, and highly palatable retail products)

Source: Adapted from several sources.

Table 8.11 shows how a carcass breakeven price can be calculated for beef. This emphasizes the need for an optimum combination of costs rather than low costs in one of the production phases that may result in high cost in another production phase. For example, a group of yearling cattle may have high gains with a low breakeven price, but these high yearling gains may cause lower feedlot gains and a higher breakeven price.

Source Verification, Traceback, and Identification Systems

As consumers, federal regulatory agencies, and international markets continue to increase demand for source verification and traceback systems as part of a total quality assurance effort, the pressure to develop supply chain or national identification systems grows. There are a variety of reasons for adoption of a source verification system. Perhaps the most important of these is the ability to better capture and utilize data from conception to consumption on a host of factors influencing profitability. Furthermore, the ability to source verify may provide sufficient market differentiation to enhance profitability, at least in the short term. Finally, the development of a beef industry identification system would provide some degree of ability to track a food safety problem to its root cause. Unfortunately, traceback systems in regards to food safety issues are limited due to the many opportunities for cross-contamination.

The challenge in the design of source verification and identification systems is finding an approach that is both functional at each industry segment and cost-effective. For these reasons, producers are understandably hesitant to accept federally mandated schemes. The most likely scenario is the development of voluntary systems tied to alliance on branded beef production systems. It is absolutely critical to note that identification systems will not prevent problems and, as such, should be considered as one piece of a broader quality program.

An ideal identification system would be able to provide unique identification of individual animals, their family origin, and transaction tracking to allow for recording where an animal goes (specific stocker, feeder, etc.) prior to consumption. Furthermore, this system would be tied to a highly user-friendly database system to record meaningful data on an animal's management and performance.

There are a variety of identification methods that can be utilized individually or in combination. These include visual identification (ear tags—plastic, metal clips; back tags; brands, tattoos; ear notches; and color patterns) and nonvisual identification (electronic and biometric). Electronic identification systems include bar codes, two-dimensional symbols, and optical character recognition (scanners to record information from an existing tag). Biometric approaches include retinal or iris scanning, retinal imaging, DNA sequencing, and antibody fingerprinting.

The most likely technologies to be of value to the beef industry are RFID devices in the form of ear tags ($3 to $4 per device), subcutaneous injectable transponders ($4 to $5 per device), or rumen boluses ($8 to $9 per device). Retinal imaging offers potential but is not yet field-ready. The use of injectable or bolus systems are problematic in terms of assuring that these devices are not permitted to adulterate the food supply.

Certainly technology will bring new alternatives and improvements. However, it is no longer a question of if source verification will occur but when. The integration of identification systems with data management, quality assurance, and branded-product initiatives will clearly bring benefits to producers, processors, and consumers. These benefits will not be derived without the challenges associated with implementation of a new approach.

MANAGEMENT SYSTEMS HIGHLIGHTS

1. Effective industry integration (contractual or functional) utilizes systems thinking and management systems planning.
2. W. Edwards Deming described the practice of suboptimization (opposite of optimization) as the lack of coordination between business units. Managers of these business units practice suboptimization when they pursue goals of lowest possible cost without receiving feedback on how their decisions affect the results of the operation (system) as a whole.
3. Total beef industry alliances should involve an integrated and optimized management system from genetics (seedstock) to product (consumer).
4. The most profitable alliances will focus on a total management system to combine low cost of production, value-based marketing, and high net margin in consistently supplying highly palatable consumer products.
5. Alliances that utilize management systems principles are more likely to succeed. Human resource management principles of trust, synergism, effective communication, and "win-win" attitudes are among the most important principles that must be implemented first.
6. If the beef industry does not effectively implement the management systems approach, consumer market share will continue to decline (Fig. 8.6). Cattle numbers will decrease and more producers, processors, and retailers will exit the beef business.

FIGURE 8.6 If the beef industry segments cannot develop cooperative working relationships, pork and poultry will continue to capture more consumer market share.
Source: Beef Bottom Line, North American Limousin Foundation.

REFERENCES

Publications

Alliance Resource Guide. Sept. 2000. Shawnee Mission, KS: Drovers.

Alliance Yellow Pages. Aug. 2000. Overland Park, KS: BEEF.

Alternative Marketing Programs. 1999. 7th ed. Greenwood Village, CO: Cattle-Fax.

Barkema, A., and Drabenstott, M. 1996. *Consolidation and Change in Heartland Agriculture.* Federal Reserve Bank of Kansas City. Kansas City, MO.

Belk, K. E., and Smith, G. C. Nov. 1997. *Branded Beef—Where, When, What and Why?* Pharacia—Upjohn Cattle Feeders Seminar, Phoenix, AZ.

Boehlje, K. 1995. *Industrialization of Agriculture: What Are the Consequences?* Proceedings of the Industrialization of Heartland Agriculture Conference, Minneapolis, MN.

Cattle-Fax. 2001. *Outlook and Strategies.* Greenwood Village, CO.

Corah, L. 1997. *An Overview of Alliances and Packers Interest in Genetic Control.* Proceedings of the Virginia Beef Cow-Calf Conference, Blacksburg, VA: Virginia Cooperative Extension.

Council for Agricultural Science and Technology. 2001. *Vertical Coordination of Agriculture.* Ames, IA.

Drabenstoh, M. 1999. Consolidation in U.S. Agriculture: The New Rural Landscape and Rural Policy. *Economic Review* (1st quarter), p. 63.

Field, T. G. 2000. *Balancing the Economical and Social Importance of Ruminants with Their Environmental Impact.* Proceedings of the World Buiatrics Congress, Punte del Estee, Uruguay.

Garrett, R. 2000. Personal correspondence.
Ishmael, W. 1997. Pick a Target, Then Aim. *BEEF Today* (Jan.).
Lamm, R. L., and Beshear, M. 1998. From the Plains to the Plate: Can the Beef Industry Regain Market Share? *Economic Review* (3rd quarter), p. 1.
National Cattlemen. 2000. The 15 Largest. (June/July).
NCA. *Beef in a Competitive World.* NCA Beef Industry Concentration/Integration Task Force Report, Oct. 25, 1989. Englewood, CO: NCA.
Perez, A., and Christensen, L. A. 1990. *Recent Developments in the Location and Size of Broiler Growout Operations.* Livestock and Poultry Situation and Outlook Report. Washington, DC: USDA, LPS-41.
Report and Implementation Recommendations of the Joint Brand-like Initiative Task Force. National Cattlemen's Beef Association, January 23, 1998.
Richie, H. D. 2001. *Understanding Alliances : What They Are, How They Function, and Questions to Ask.* Research and Demonstration Report, Michigan State University.
Seagraves, M., and MacDonald, B. Jan. 1997. Poultry Industry Must Look to Next Level of Integration. *Feedstuffs* 69: 20.
Smith, G. C. 2001. *Increasing Value in the Supply Chain.* Proceedings of the Conference of the Canadian Meat Council, Vancouver, BC.
Thorpe, T. 1991. A Question of Structure. Is the Beef Industry too Segmented for its Own Good? *National Cattlemen* 6: 34–36.
USDA. 2000. *Grain Inspection.* Packers and Stockyard Statistical Report. Washington, DC: Packers and Stockyard Administration.
Ward, C. E. 2000. *Characteristics and Dynamics of Alliances in the Beef Industry.* Stillwater, OK: Oklahoma State University.
Ward, C. E. 1997. *Key Findings from the USDA Packers Concentration Study.* Stillwater, OK: Oklahoma State University.

Visuals

"Strategic Alliances" (1993; 18 min.). National Cattlemen's Association, Englewood, CO.

Websites

BEEF Magazine: www.beef.mag.com
National Cattlemen's Beef Association: www.beef.org
Drovers: www.drovers.com
Federal Reserve Bank of Kansas City: www.kc.frb.org
See Table 8.10 for alliance web sites.

The Marketing System

The process of producing breeding cattle, feeder cattle, fed and non-fed cattle (35 million head), carcass beef (26.1 billion lbs), and retail cuts for the consumer in the United States is accomplished through a vast and complex marketing system. *Marketing* is the physical movement, transformation, and pricing of goods and services, with numerous buyers and sellers working to move cattle and beef products from the point of production to the point of consumption. Price determination is the result of interactions of supply and demand to establish market price levels. Market price levels together with access to market information, the size and structure of the market, and other related factors are utilized by buyers and sellers to find a price for a particular product. This process is referred to as price discovery.

Producers need to understand marketing to produce products preferred by other beef industry segments, including the consumer. They also need to decide intelligently among various marketing alternatives and understand how animals and products are priced if profitable cattle are produced and effectively marketed. This chapter, which covers marketing principles, should be integrated with the marketing information discussed in Chapters 1–8 and 10.

MARKET CLASSES AND GRADES

Market classes and grades have been established to segregate cattle, carcasses, and products into uniform groups based on the preferences of buyers and sellers. The USDA has established market classes and grades to make the marketing process simpler and more easily communicated. Use of *USDA grades* is voluntary. Some packers and other organizations, including alliances, have their own private grades and brand names that they use in combination with USDA grades. An understanding of market classes and grades helps producers recognize the quantity and quality of products they supply to consumers.

Several national surveys have shown that consumers are not familiar with most USDA grades for beef (e.g., Prime, Choice, Select, and so on). Consumers typically confuse these grades with the intent of the inspection mark. They often assume that all beef is graded and frequently report certain grades when no such grade exists for beef (e.g., "Grade A," "First

Cut," and "Grade AA"). Many consumers are familiar with Choice beef, associating it with a desirable product, but usually they do not understand the grading criteria.

Markets for Slaughter Cattle

Slaughter cattle are separated into classes based primarily on age and sex. Age of the animal has a significant effect on tenderness, with younger animals typically producing more tender meat than older animals. Age classifications for meat from cattle are *veal, calf,* and *beef.* Veal is from young calves, 1–3 months of age, with carcasses weighing less than 150 lbs. Calf is from animals ranging in age from 3 months to 10 months and with carcass weights between 150 lbs and 300 lbs. Beef comes from more mature cattle, over 12 months of age, having carcass weights higher than 300 lbs. The classes and grades established by the USDA for cattle are based on sex, quality grade, and yield grade, all of which are used in the classification of both live cattle and their carcasses (see Table 9.1).

The sex classes for cattle are *heifer, cow, steer, bull,* and *bullock.* Sex classes separate cattle and carcasses into more uniform carcass weights and tenderness groups and identify how carcasses are processed. Occasionally, the sex class of *stag* is used by the livestock industry to refer to males that have been castrated after their secondary sex characteristics have developed. *Heiferette* is a sex class between heifer and cow that identifies a first-calf heifer (approximately 23–26 months old) that loses her calf or has her calf weaned early. The female is put into the feedlot and typically produces a youthful carcass that has the potential to grade Choice.

Quality Grades

Quality grades are intended to measure certain consumer palatability characteristics, whereas yield grades measure amount of fat, lean, and bone in the carcass. Fed steers representing some of the eight quality grades and five yield grades are shown in Figures 9.1 and 9.2. To a large degree, slaughter cattle of different quality and yield grades reflect visual differences primarily in fatness and to a lesser degree in muscling.

The factors used to determine quality grades are (1) bone maturity, (2) marbling, (3) lean color, and (4) firmness and texture of lean tissue. Marbling and bone maturity are the two most important factors. Marbling—flecks of fat within the ribeye muscle or intramuscular fat—is visually evaluated at the twelfth and thirteenth rib interface. The nine degrees of marbling vary from abundant to practically devoid (Fig. 9.3).

TABLE 9.1 Official USDA Grade Standards for Live Slaughter Cattle and Their Carcasses

Class or Kind	Quality Grades (highest to lowest)	Yield Grades (highest to lowest)
Beef Steer and heifer	Prime, Choice, Select, Standard, Commercial, Utility, Cutter, Canner	1, 2, 3, 4, 5
Cow	Choice, Select, Standard, Commercial, Utility, Cutter, Canner	1, 2, 3, 4, 5
Bullock	Prime, Choice, Select, Standard, Utility	1, 2, 3, 4, 5
Bull	No designated quality grades	1, 2, 3, 4, 5
Veal	Prime, Choice, Select, Standard, Utility	NA
Calf	Prime, Choice, Select, Standard, Utility	NA

Source: USDA

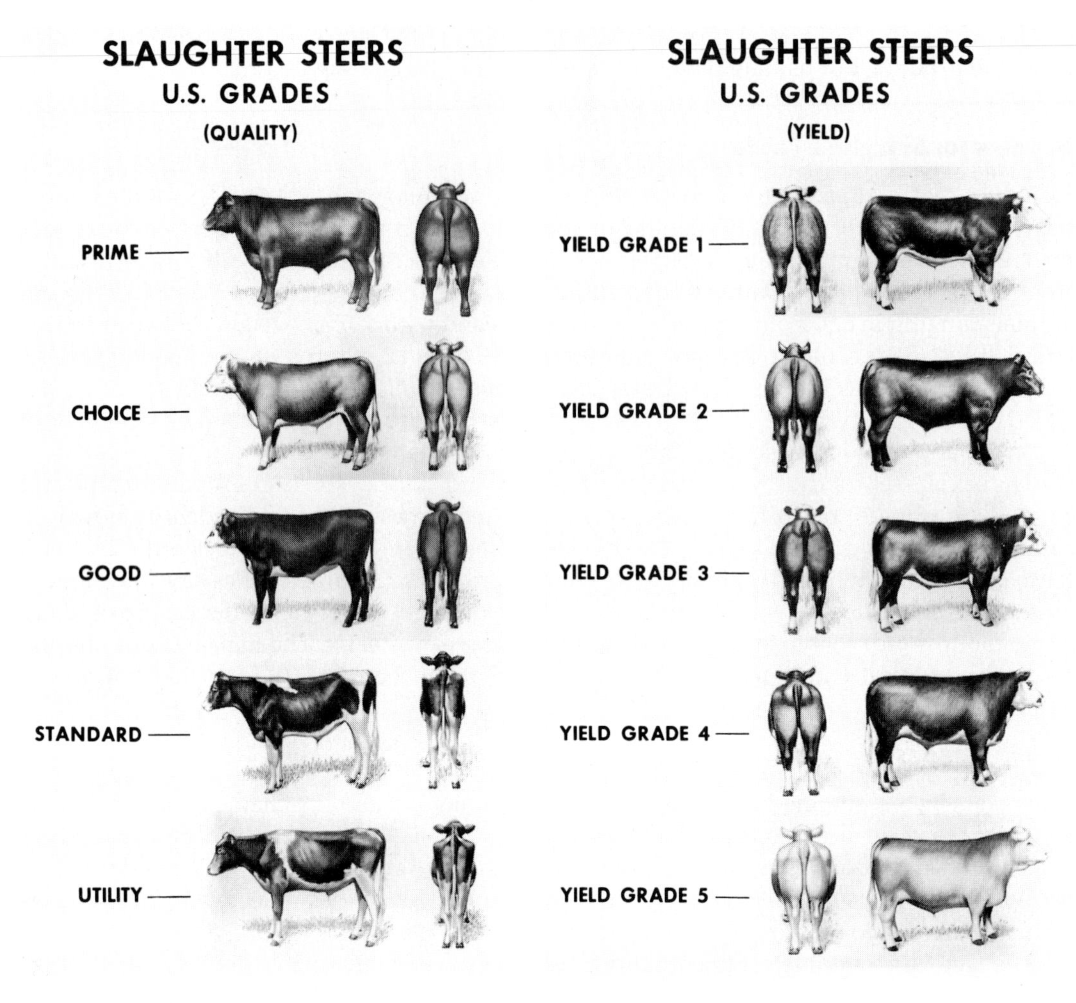

FIGURE 9.1 USDA quality grades not shown are Commercial, Cutter, and Canner, which are reserved primarily for cows.
Source: USDA.

FIGURE 9.2 USDA yield grades for slaughter cattle.
Source: USDA.

Maturity classifications are based on the relationship that as animals increase in age, meat tenderness decreases. Bone maturity estimates the physiological age of the animal and maturity is determined primarily by bone ossification. After the carcass has been split into two halves, the dorsal processes of the vertebrae are exposed and then evaluated for amount of ossification. Carcasses with A maturity have the largest amount of cartilage on the tips of the vertebrae processes, while complete ossification has occurred in carcasses with E maturity. Carcasses with C, D, or E maturity are eligible for only the Commercial, Utility, Cutter, and Canner grades.

Figure 9.4 shows how the final carcass quality grade is determined by combining maturity and marbling; Figure 9.5 visually depicts several quality grades. Note in Figure 9.4 that as maturity increases, a carcass usually has to have more marbling to stay in the same quality grade. Most fed steers and heifers are harvested before reaching 30 months of age. B maturity carcasses are only graded Choice or higher if they have marbling scores of modest or higher. Note

VERY ABUNDANT
ABUNDANT
MODERATELY ABUNDANT
SLIGHTLY ABUNDANT
MODERATE
MODEST
SMALL
SLIGHT
TRACES

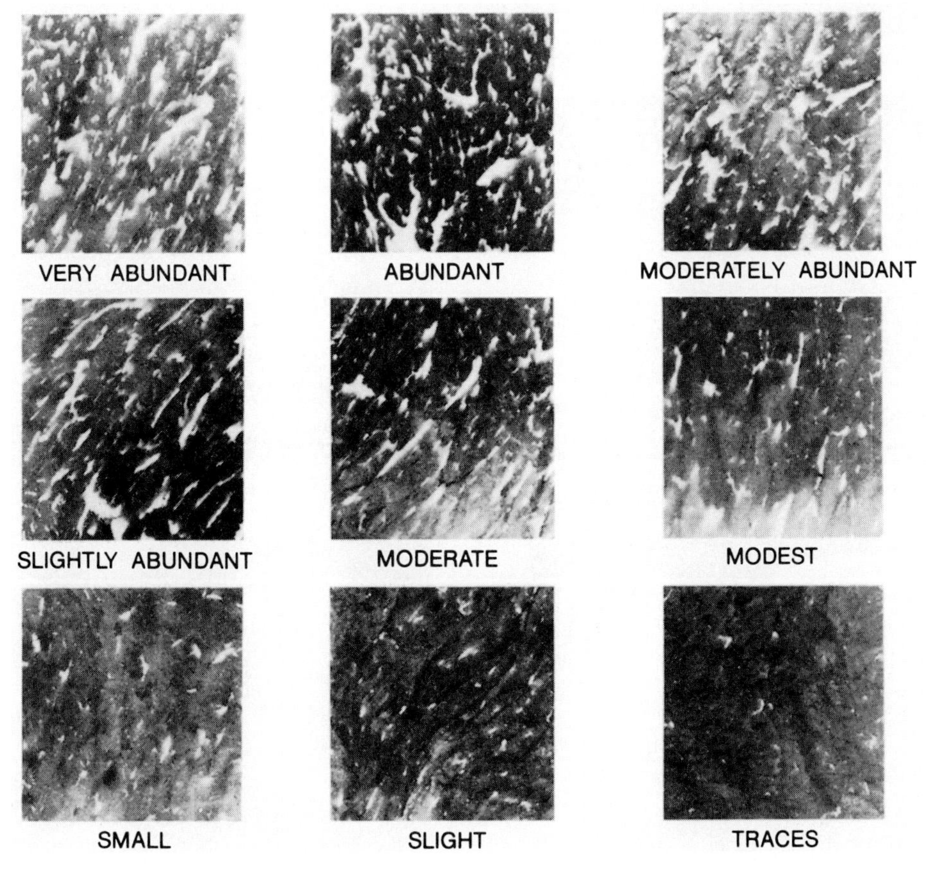

FIGURE 9.3 Typical degrees of marbling used to quality grade beef carcasses. Practically devoid not shown.
Source: USDA.

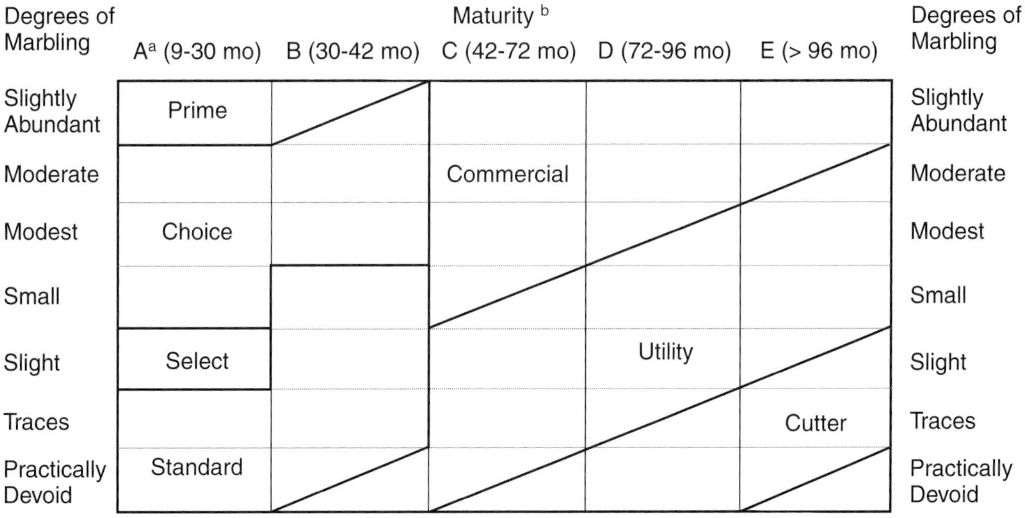

Degrees of Marbling
Maturity b
Aa (9-30 mo)
B (30-42 mo)
C (42-72 mo)
D (72-96 mo)
E (> 96 mo)
Degrees of Marbling
Slightly Abundant
Moderate
Modest
Small
Slight
Traces
Practically Devoid
Prime
Choice
Select
Standard
Commercial
Utility
Cutter

[a] Assumes that firmness of lean is comparably developed with the degree of marbling and that the carcass is not a "dark cutter."

[b] Maturity increases from left to right (A through E).

[c] The A maturity portion of the Figure is the only portion applicable to bullock carcasses.

FIGURE 9.4 Relationship between marbling, maturity, and carcass quality grade.[a]
Source: USDA.

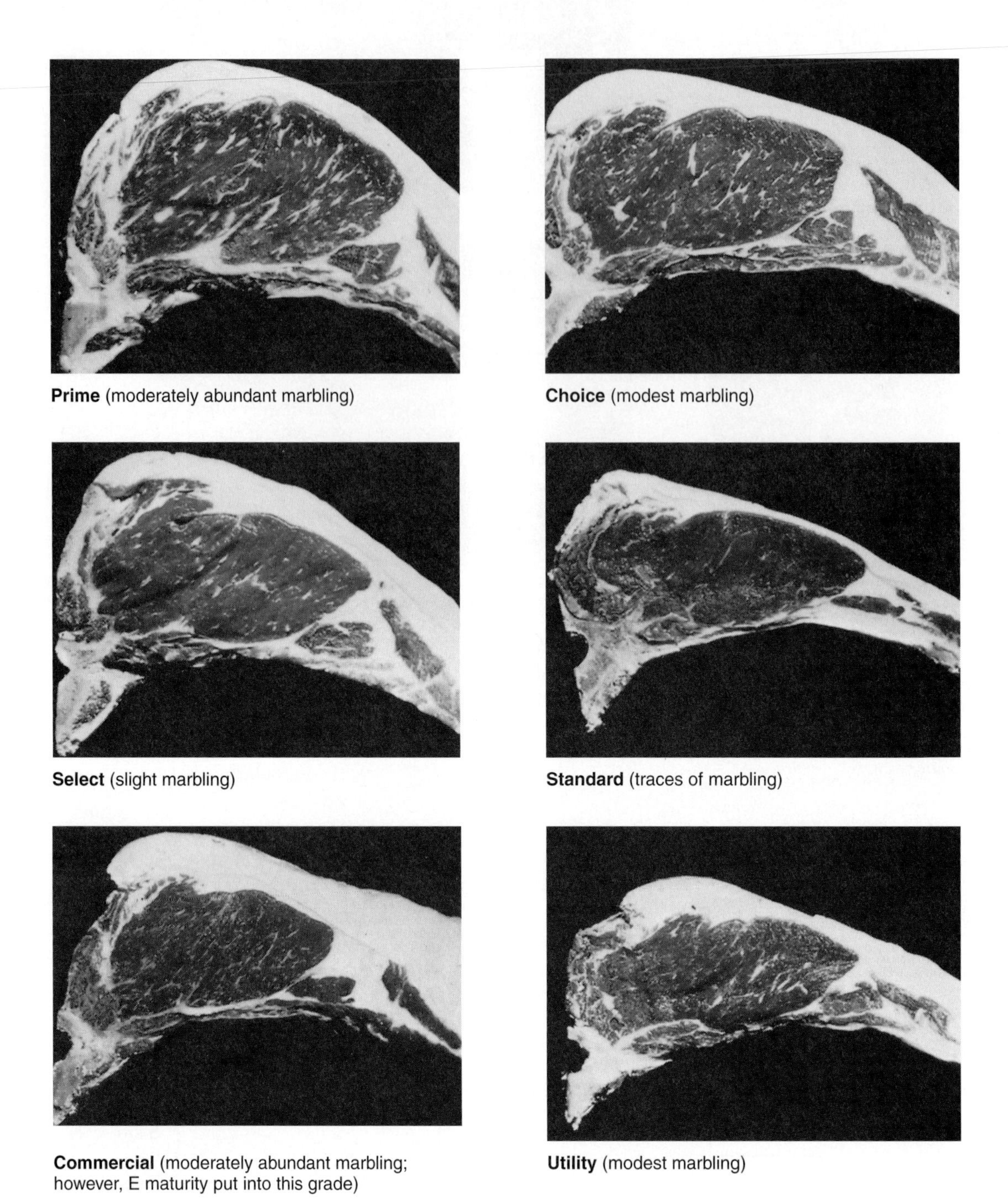

FIGURE 9.5 Exposed ribeye muscles (between twelfth and thirteenth ribs) showing several carcass quality grades resulting from combinations of marbling and maturity.
Source: American Meat Science Association, copyright 1997.

that C maturity starts at 42 months of age. Most cows are harvested after 4 years of age and graded Commercial, Utility, Cutter, or Canner regardless of the amount of marbling.

Color of the lean is evaluated along with bone maturity to determine the final maturity classification. Color of the lean becomes darker with advancing age of the animal. An A maturity carcass is usually a bright, cherry red whereas an older (C, D, or E) maturity carcass is usually a darker, coarser-textured lean. Occasionally, an A or B maturity has dark lean ("dark cutter"), typically resulting from the animal being stressed shortly before slaughter. Dark cutters, with A or B maturity, are often given a quality grade lower than what the marbling score would indicate.

Yield Grades

Yield grades (also called cutability) refer to pounds of boneless, closely trimmed retail cuts (BCTRC) from the round, loin, rib, and chuck. Marketing communications generally use yield grades designated numerically from 1 through 5. Yield grades, however, are often reported in tenths in research reports and carcass contests. A 700-lb carcass with a 3.0 yield grade would have 350 lbs of BCTRC (700 × 0.50), whereas a 700-lb carcass with a 4.0 yield grade would have 334 lbs of BCTRC (700 × 0.477). Table 9.2 shows the yield grades and their respective percentages of BCTRC.

Yield grades are determined from these four carcass characteristics:

1. Amount of fat, measured in tenths of inches, over the ribeye muscle or longissimus dorsi (Fig. 9.6).
2. Kidney, pelvic, and heart (KPH) fat, which is usually estimated as a percentage of carcass weight.
3. Area of ribeye muscle (REA), which is measured in square inches (Fig. 9.7).
4. Hot carcass weight. Carcass weight reflects amount of intermuscular fat. Generally, as the carcass increases in weight, the amount of intermuscular fat increases as well.

Fatness is the primary factor in determining yield grades. The fat measurement, over the ribeye muscle, measures most differences in fatness in the carcass (Fig. 9.8).

TABLE 9.2 Carcass Yield Grades and the Yield of BCTRC[a]

Yield Grade	% BCTRC	Yield Grade	% BCTRC
1.0	54.6	3.6	48.7
1.2	54.2	3.8	48.2
1.4	53.7	4.0	47.7
1.6	53.3	4.2	47.3
1.8	52.8	4.4	46.8
2.0	52.3	4.6	46.4
2.2	51.9	4.8	45.9
2.4	51.4	5.0	45.4
2.6	51.0	5.2	45.0
2.8	50.5	5.4	44.5
3.0	50.0	5.6	44.1
3.2	49.6	5.8	43.6
3.4	49.1		

[a]BCTRC = boneless, closely trimmed retail cuts from the round, loin, rib, and chuck.

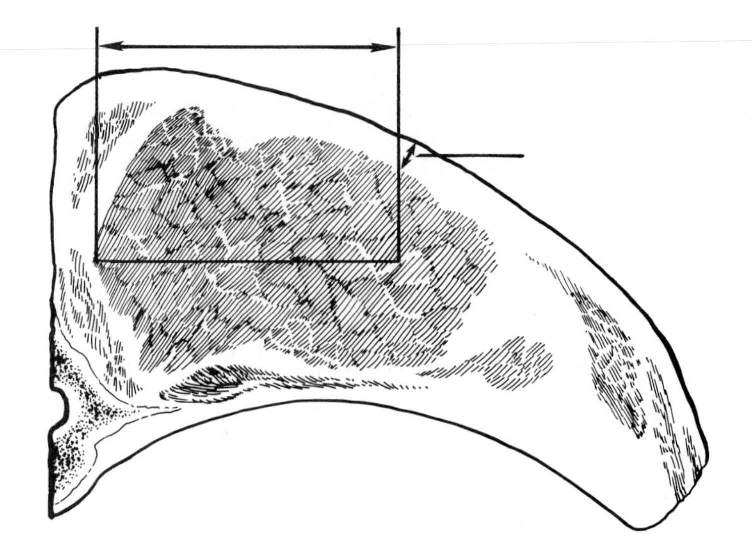

FIGURE 9.6 Location of fat measurement over the ribeye (longissimus dorsi muscle).
Source: Colorado State University.

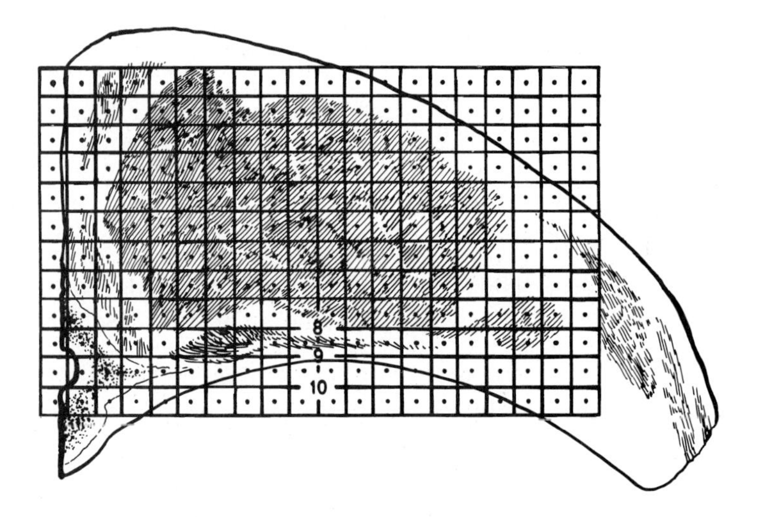

FIGURE 9.7 Plastic grid is placed over the ribeye muscle to measure area. Each square represents 0.1 in^2.
Source: Colorado State University.

Yield grades have been developed by processing carcasses into boneless, closely trimmed retail cuts (% BCTRC). Formulas have been determined for predicting yield grades from carcass measurements. The following formulas are used to compute numerical yield grades, or %BCTRC:

% BCTRC = 51.54 − 5.784 (inches of fat at 12th–13th rib)
− 0.462 (% kidney, heart, and pelvic fat)
− 0.0093 (lb hot carcass weight)
+ 0.740 (square inches of ribeye muscle)

Yield grade = 2.50 + 2.50 (inches of fat at 12th–13th rib)
+ 0.20 (% kidney, heart, and pelvic fat)
+ 0.0038 (lb hot carcass weight)
− 0.32 (square inches of ribeye muscle)

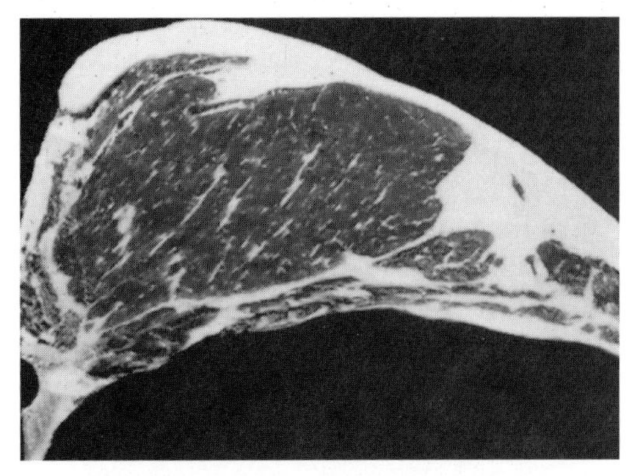

(a) **Yield Grade 1** (Fat 0.2 inch, ribeye area 13.9 sq. inch).

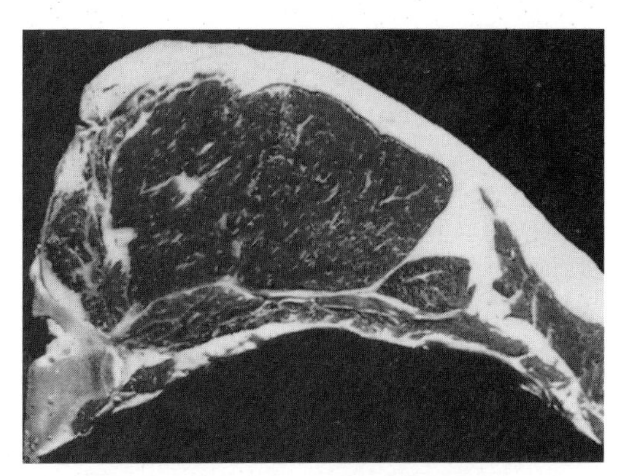

(b) **Yield Grade 2** (Fat 0.4 inch, ribeye area 12.3 sq. inch).

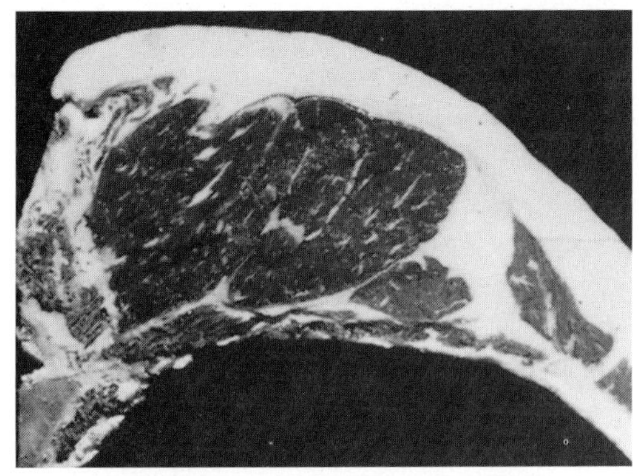

(c) **Yield Grade 3** (Fat 0.6 inch, ribeye area 11.8 sq. inch).

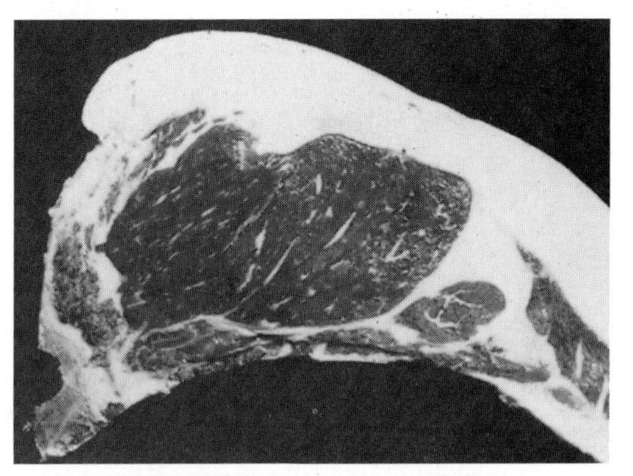

(d) **Yield Grade 4** (Fat 0.9 inch, ribeye area 10.5 sq. inch).

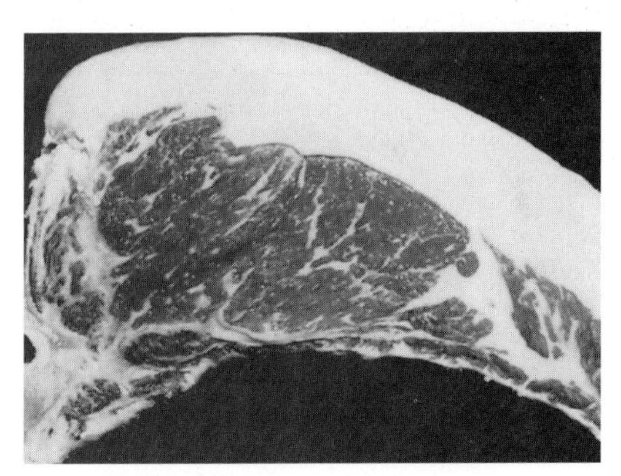

(e) **Yield Grade 5** (Fat 1.1 inch, ribeye area 10.9 sq. inch).

FIGURE 9.8 The five yield grades of beef shown at twelfth and thirteenth ribs.
Source: American Meat Science Association, copyright 1997. (a) **Yield Grade 1** (Fat 0.2 in., ribeye area 13.9 in.^{2}). (b) **Yield Grade 2** (Fat 0.4 in., ribeye area 12.3 in.^{2}). (c) **Yield Grade 3** (Fat 0.6 in., ribeye area 11.8 in.^{2}). (d) **Yield Grade 4** (Fat 0.9 in., ribeye area 10.5 in.^{2}). (e) **Yield Grade 5** (Fat 1.1 in., ribeye area 10.9 in.^{2}).

An alternative, more simplified method may also be used to determine the yield grades of carcasses or the potential yield grades of live fed cattle. This method uses the increments for the carcass characteristics shown in Table 9.3. For example, consider a carcass with 0.50 in. of fat, REA of 11.30 sq. in., KPH of 3.50%, and a carcass weight of 625 lbs. The starting base is a preliminary yield grade of 2.00, with 0.50 in. of fat ($5 \times 0.25 + 2.00 = 3.25$). A REA of 11.3 ($11.0 - 11.3 = 0.30$) is one-third of 1.0, so the increment would be -0.11 from $3.25 = 3.14$. KPH fat is the same as the base, so KPH has no influence on yield grade in this example. The carcass weight is 625 lbs ($625 - 600 = 25$, which is one-quarter of 100 lbs), so the increment is $4 \times 1/4 = 10.1$, added to 3.14 equals a final yield grade of 3.24 or a yield grade of 3.2.

Distribution of Beef Grades

The grading of beef carcasses by USDA meat graders is a voluntary program available to meat packers. Each packer may choose to have all, none, or some of their beef carcasses graded and marked with grade stamps (Fig. 9.9). When carcasses are officially graded, the grade is rolled or stamped onto the carcass (Fig. 9.10).

Carcass quality and yield grade stamps should not be confused with the inspection stamp also shown in Figure 9.9. The inspection stamp verifies the wholesomeness of the meat from a food safety standpoint. Most packers have at least some of their beef marked with grade stamps, usually the carcasses that are most desirable in yield and quality grades.

TABLE 9.3 Shortcut Method to Determine Yield Grade[a]

Carcass Trait	Base	Increment
Fat	0.00 in.	0.10 in. = ±0.25
REA	11.0 sq.in.	1.0 sq.in. = ±0.33
KPH fat	3.5%	1.0% = ±0.20
Carcass weight	600 lbs	100 lbs = ±0.40

[a]Assumes a starting yield grade base of 2.00.

FIGURE 9.9 Inspection marks and grade marks that are placed on the beef carcass. USDA choice represents one of several designated quality grades. Yield grade 2 represents one of the five yield grades.
Source: USDA.

INSPECTION MARK

FIGURE 9.10 Carcasses that are quality graded have the quality grade stamp "rolled" over the entire length of the carcass.
Source: American Simmental Association.

TABLE 9.4 Quality and Yield Grading of Beef Carcasses, 1980–2000

	Quality Grades (%)[a, c]			Yield Grades (%)[a]				
Year	Prime	Choice	Select[b]	1	2	3	4	5
1980	5.9	89	4	2	29	58	10	1.4
1985	3.6	93	3	4	41	50	5	0.6
1990	2.2	82	15	7	46	44	3	0.2
1993	2.0	68	29	10	48	40	2	0.2
1995	2.4	63	34	11	47	40	2	0.2
1997	2.1	52	35	14	50	35	1	0.1
2000	3.0	52	36	10	41	36	2	0.2

[a]Percent of cattle graded (~1/3 ungraded up to 1992, ~15% ungraded 1993–97).
[b]Name of grade changed from Good to Select in 1987.
[c]Less than 1% of graded steers and heifers are identified as Standard.
Source: USDA.

Table 9.4 shows the distribution of grades for beef actually graded and stamped during recent years. In 2000, for example, 82% of steer and heifer carcasses were graded and stamped for both quality grade and yield grade, 4% were stamped for yield grade only, 7% were stamped for quality grade only, and 8% were sold ungraded. During the last few years, the increases in yield grades 1 and 2 were more likely due to an increase in Select carcasses than to cow-calf producers and feeders producing leaner carcasses.

Value of the Current Carcass Grades

Discussion and assessment of the USDA beef carcass grades continues in the industry, with an emphasis on the quality grades. Some producers and industry leaders argue that the quality grades should be eliminated or revised for several reasons. The primary concern is that too much emphasis is placed on marbling given the findings of a substantial number of research studies—that only 10% to 20% of beef palatability differences are accounted for by differences in marbling. Yet marbling accounts for 90% of the variation in carcass grades of young cattle. Palatability is a complex trait that is heavily influenced by flavor and tenderness. While marbling is a reasonable indicator of flavor, the current quality grade system is not an effective measurement of tenderness.

Complicating the situation is the fact that nearly 75% of graded beef carcasses from fed steers and heifers have marbling scores of slight or small. The current grading system is not particularly effective at sorting palatability differences in this relatively narrow range.

Marbling is difficult to identify in the live animal; therefore, the feeder may extend feeding time in an attempt to produce a higher percentage of cattle grading Choice. This may give a higher economic return to the cattle feeder because of the price differential of Choice over Select. However, the practice is questionable because it dictates a longer feeding program and higher gain costs, particularly during the last 30 days of the feeding program. Feeding for higher marbling encourages overfattening of cattle, which, in turn, produces a less desirable consumer product. The excess fat produced (industry target of no more than 16.5% of carcass weight as fat) was estimated in 2000 to cost more than $42 per slaughter steer or heifer. Furthermore, a multitude of studies show that feeding periods beyond 140–160 days simply do not yield sufficient changes in marbling to offset the costs associated with increased days on feed.

Future of Instrument Grading

The current USDA grading system depends on subjective evaluations by trained graders. The use of more precise technologies to either augment or replace human evaluators has received considerable attention. Several technologies have been developed and evaluated—the Tenderness Classification System (MARC), the Colorimeter System (Colorado State, Ohio State, and South Dakota State Universities), and Beef Cam™ (Colorado State University and Hunter Laboratory).

The Tenderness Classification System uses a shear force measure that can be collected at commercial packing plant chain speeds (400 head per hour). This system involves removal of a 1-in.-thick rib steak from each carcass to be subsequently cooked for 7 minutes and then tested for shear force. The entire process is approximately 10 minutes in duration and explains between 46–56% of the variation in cooked beef tenderness. This system has not been implemented by industry due to the product loss and extra costs associated with implementation.

The Colorimeter System uses marbling, hump height, and colorimeter readings to help predict eating quality. Hump heights are used to sort out carcasses from higher percentage *Bos indicus* cattle. Colorimeter measurements require a three-second reading of the ribeye muscle where darker-colored muscle is considered to yield tougher product. Wulf et al. (1999) found that this three-factor system accounts for 39% of variation in eating quality (marbling—12%, hump height—8%, color reading—24%).

The Hunter Lab Beef Cam™ System uses video-image analysis to discern differences in lean and fat color as a means to predict palatability. Muscle and fat color are used to assess amount of marbling, physiological maturity of lean tissue, muscle pH, structural status of sarcomeres and connective tissue, feeding management, and calpastatin activity. Using images of up to 250,000 data points per measurement, Beef Cam™ sorts carcasses into certified palatable versus not certified as palatable. Initially, Beef Cam™ was very accurate in assessment of certified tender carcasses (~ 95%) but less accurate in those that were rejected (50–60%). However, improvements in the prediction equation have reduced the inaccuracies to a level of approximately 25% for rejected carcasses. The Nolan Ryan Beef Company has adopted this technology.

Additionally, the use of video-image technology has been utilized to determine variations in carcass composition. The combination of the capabilities of the Canadian Computer Vision System (CVS) and the Australian Dual Component Video Image Analysis Scan yielded an improved video-image analysis program. At the packing plant level, these systems can provide highly accurate measures of ribeye area and the assignment of USDA yield grades to carcasses to the tenth of a yield grade unit. Video-image analysis was approved by USDA to augment official Yield Grade scoring. These systems have been adopted by Excel Corporation and are being considered by several other large packing companies. The most noticeable change for producers is that these systems will allow yield grade scores to be reported in one-tenth increments. Instead of reporting a Yield Grade 3 (old system), the report would now indicate a 3.7, for example.

The use of objective assessment of carcasses for both eating quality and yield offers a powerful tool in the delivery of beef products more clearly matched to the demand of various market targets and allows for the transfer of highly accurate and meaningful data to producers and feeders.

Feeder Cattle Grades

The USDA feeder grades for cattle are intended to predict feedlot weight gain and the slaughter weight end point of cattle fed to a desirable fat-to-lean composition. The three criteria used to determine feeder grade are frame size, thickness, and thriftiness.

Frame sizes are used to predict compositional variation and are differentiated into large, medium, and small. Cattle in these categories would be expected to reach U.S. Choice at about 0.50 in. of backfat at different weights. For example, large-, medium-, and small-framed steers would reach this market target at weights of greater than 1,250, 1,100–1,250, and less than 1,100 pounds, respectively (Table 9.5).

Thickness scores are used to help distinguish between cattle of differing levels of muscularity and are categorized into four classifications (No. 1, No. 2, No. 3, and No. 4). No. 1 feeders are typically from beef-type breeds, are moderately thick, and show particular thickness through the forearm, stifle, and gaskin. No. 2 feeders are considered slightly thick, while No. 3 feeders are thinly muscled. No. 4 feeder cattle are still thrifty but less muscular than No. 3s. Dairy cattle typically fall into categories 3 and 4. Any feeders that are unthrifty or double-muscled are classified as USDA Inferior. These are cattle not expected to perform normally and may be of any combination of thickness and frame size. Frame and muscle classifications are combined to create 12 grades (in addition to Inferior). These grades are reported as Large No. 1, Large No. 2, and so forth. Images of these feeder cattle grades are provided in Figures 17.21 and 17.22.

Although frame size and ability to gain weight in the feedlot are apparently related in the sense that large-framed cattle usually gain more rapidly than the other frame-sized cattle, frame size appears to more accurately predict carcass composition or yield grade at different slaughter weights than that of gaining ability.

TABLE 9.5 Slaughter Weights of Large-, Medium-, and Small-Framed Slaughter Cattle at 0.50 Inches of Fat

	Slaughter Weight	
Frame Size	Steers (lbs)	Heifers (lbs)
Large	>1,250	>1,150
Medium	1,100–1,250	1,000–1,150
Small	<1,000	<1,000

Marketing Cows and Bulls

A frequently overlooked source of cash flow is market cows and bulls that have been deemed to be unacceptable as breeding animals. The profitable marketing of these cattle can make the difference between profit and loss for a cow-calf enterprise. In fact, almost 16% of the gross revenue to the average cow-calf enterprise originates from the sale of market cows and bulls.

Once a producer understands the potential value of market cows and bulls, it is important to implement a proactive management plan to ensure that this value is maximized. By utilizing appropriate husbandry, handling practices, and management protocols, beef producers can prevent quality defects. One of the core strategies in this process is to closely monitor herd health and to market in a timely fashion.

The 1999 market cow and bull quality audit found that defects were creating a loss of $69 per head. These losses originate mostly from a failure to market these animals in a timely fashion, from hide defects, and from undesirable composition.

A marketing alliance designed to specifically capture the value of culled breeding animals was established by a group of ranchers and the Nature Conservancy in Routt County, Colorado. The group formed *Yampa Valley Beef* with a specific goal of marketing to the food service establishments serving the tourist and ski trade. The group also has a direct sale approach to marketing gift baskets during the Christmas season.

Yampa Valley Beef puts a portion of its profit into a local land trust and uses its marketing to tie into a proactive conservation and open space theme.

There are multiple ongoing efforts to tie beef sales to a conservation and stewardship theme as a means to increase profitability so that family ranchers can stay on the land. Examples include Conservation Beef® in Helena, Montana, Lasater Beef in Matheson, Colorado, and the Kamuela Pride label, as part of the Hawaii Natural Meats program.

Markets for cows and bulls typically get stronger in the spring months following the large run of cows that are sold in October through December. As such, producers need to consider a short retained ownership period. If cheap feed sources are available, a 2- to 3-month retained ownership may well be advisable. Cattle-Fax data show that this strategy was effective in every year from 1980 to 1998.

MARKET CHANNELS

Market channels are the pathways through which cattle move from farm or ranch to feedlots, then to packing plants, and eventually to consumers. Figure 9.11 shows the marketing channels currently being used for cattle. The market channels available to cattle producers have evolved through a series of significant transitions, from the cattle drives of yesteryear to the utilization of informational technologies available today. Thus, the history of cattle and beef marketing has been characterized by change.

The auction market is most important in the marketing of feeder cattle, slaughter calves, slaughter cows, and slaughter bulls. Direct selling (also referred to as country) is most important for slaughter steers and heifers. In this situation, slaughter cattle are being marketed directly from the feedlot to the packer. Although use of the terminal market is declining compared with previous years, it is still an important market for some midwestern cattle producers. Figure 9.11 does not show the market channels for breeding cattle that are marketed primarily through auctions (private or consignment) and that are sold directly to commercial producers through private treaty. Producers should understand the various market channels and how they differ in terms, availability, marketing costs, and buyer competition. Table 9.6 illustrates the market channels utilized by various-sized cow-calf enterprises. Small and midsized herds rely more heavily on auction markets than do large herds that are more likely to utilize direct, forward contracting, and carcass basis sales.

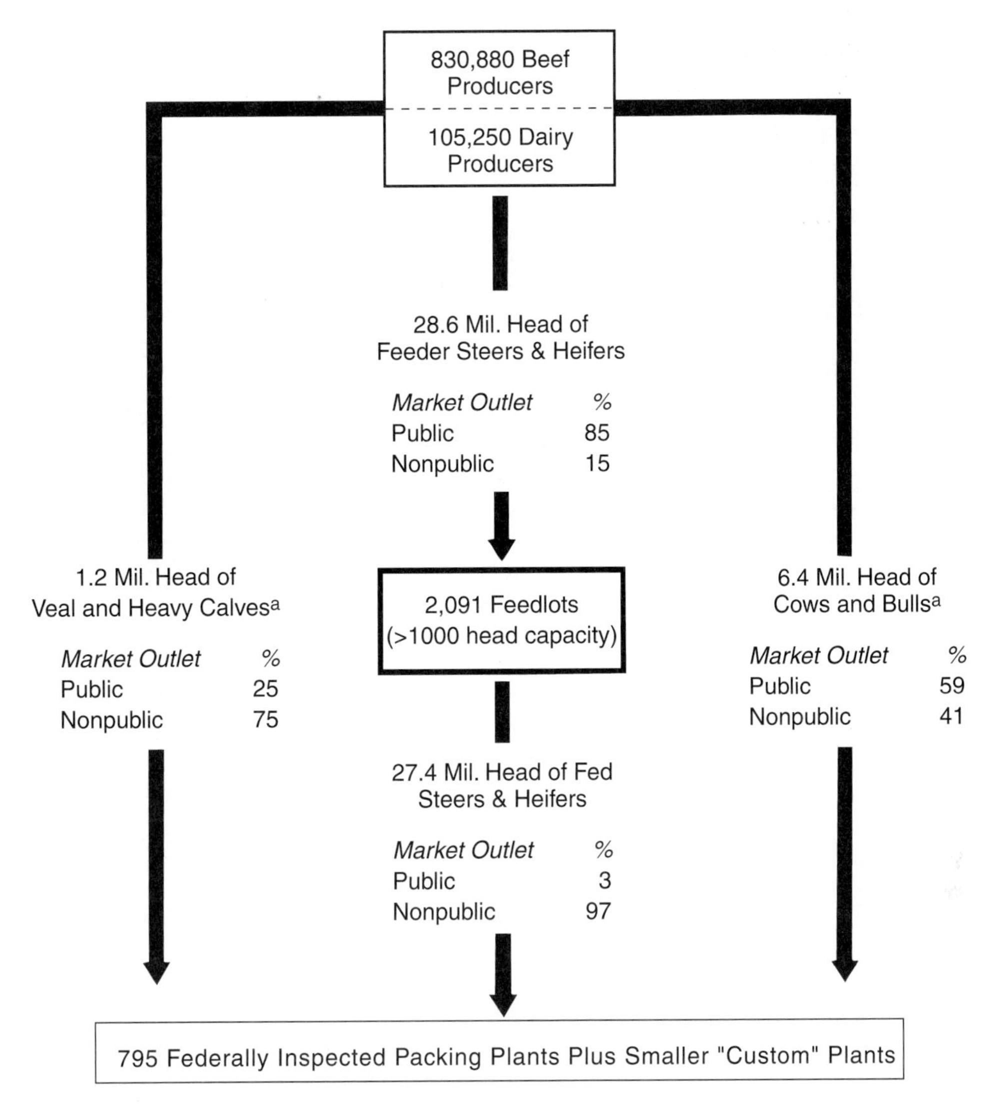

FIGURE 9.11 Marketing channels for cattle in the United States.
Source: Packers and Stockyards Statistical Report, USDA.

TABLE 9.6 Percent of Operations Marketing Cattle by Various Methods

	Herd Size			
Method	**<50**	**50–99**	**100–299**	**>300**
Auction	86.9	84.8	78.9	49.9
Direct-video	0.4	0.4	1.4	7.0
Direct-private treaty	9.5	10.1	13.8	28.7
Consignment	1.2	1.3	1.1	0.5
Forward contract	0.2	0.3	2.6	5.6
Carcass basis	1.0	1.9	1.3	6.8
Other	0.8	1.2	0.9	1.5

Source: NAHMS (1997).

Auction Market (Public)

The *auction market* is made up of approximately 2,000 livestock auctions, sometimes called *sale barns,* located throughout the United States. Cattle auctions are more numerous in areas with the greatest cow-calf numbers because primarily feeder cattle, cull cows, and bulls are sold through them. Large auctions are owned by corporations, whereas most small auctions are under individual proprietorship. Holding pens, scales, and sale area are under the same ownership.

The major cattle auction markets in 2000, each of which had cattle receipts over 190,000 head, are more frequently located in the Great Plains states, but they are also scattered throughout the United States (Table 9.7 and Figs. 9.12 and 9.13).

TABLE 9.7 Top 10 Public Markets for Cattle Receipts, 1996–2000

	Cattle Receipts—1996 (no. of head)	Cattle Receipts—2000 (no. of head)
LaJuanta, CO	403,891	347,827
Torrington, WY	277,880	304,599
Lexington, KY	205,206	286,831
Pratt, KS	310,835	268,073
West Plains, MO	254,662	241,883
Billings, MT	292,468	234,497
San Angelo, TX	245,058	170,463
Dodge City, KS	228,956	163,146
Springfield, MO	199,587	123,873
Dalhart, TX	NA	118,848

Source: USDA

FIGURE 9.12 Traditionally cow-calf producers have utilized local auction market as a means to sell feeder cattle and breeding stock that has been culled from the herd.

FIGURE 9.13 Auction day at Livestock Exchange, Inc. in Brush, Colorado. This auction market has pen space for more than 12,000 head of cattle. Pens immediately behind the Sales Pavilion (center of photo) contain hay and water for the cattle prior to the sale. Pens to the right are dry holding pens for the cattle after they sell. *Source:* Livestock Exchange, Inc., Brush, Colorado.

Cattle owners consign their livestock to the auction, where the cattle are sorted into uniform sale groups when a considerable amount of variation in type and grade exists. At sale time, cattle enter a sale ring and an auctioneer accepts incremental bids from the audience until no one advances the bid. Most cattle are sold by the pound, and their weights are typically displayed electronically during the auction.

Auction sales costs are usually by the head, with a percentage commission charge on the gross revenue ranging from 1.5% to 4% depending on the market and the class of livestock sold. Additional representative costs per head are feed (cost plus 25%), brand and health inspection ($0.64), insurance ($0.10), and a beef checkoff program fee ($1).

Terminal (Public) Market

Terminal markets are structured similarly to local auction barns with the primary exception that multiple agents are available at the terminal market. In 2000, there were five major *terminal markets* in operation, a marked reduction from the 80 terminal markets that existed in the 1930s. The importance of the terminal market declined rapidly when many packing plants moved away from terminal markets and closer to large commercial feedlots. In fact, the importance of the terminal markets has continued to decline throughout the 1990s (Table 9.8). Some terminal markets sell only feeder cattle, while others market both feeder and slaughter cattle.

TABLE 9.8 Terminal Markets—Cattle Receipts (1991–2000)

	1991	1996	2000
Oklahoma City, OK	499,000	554,165	484,392
Sioux Falls, SD	418,000	239,619	272,093
South St. Paul, MN	236,000	172,792	144,548
South St. Joseph, MO	NA	118,262	101,449
West Fargo, ND	NA	108,765	65,758

Source: USDA

Most terminal markets are similar in terms of their structure and activities. The stockyards company, usually a corporation, owns the physical facilities (land, buildings, pens, and scales). It charges a yardage fee to producers consigning their cattle for using the yards and other services, such as for feed and cattle handling. Cattle are consigned to a commission firm designated by the producer. The commission firm represents the producer as the seller and charges a fee for its services. Although there is a wide range of buyers on the terminal market, the primary buyers are feeder cattle buyers, packer buyers (salaried by different packing firms), order buyers (purchasing for someone else for a fee), and yard dealers (sometimes known as traders, speculators, or scalpers whose objective is to resell cattle at higher prices). The selling price of cattle on the terminal market is determined by private treaty; that is, the commission representative and the buyer (one at a time in the pen of cattle) bargain until a firm price is agreed upon. Some terminal markets have incorporated the auction as another marketing alternative; however, it is primarily feeder cattle that are sold through the auction.

Both terminal and auction markets, which handle livestock involved in interstate movement, are considered public markets and as such are subject to regulations specified by government personnel of Packers and Stockyards (P&S), an agency of the Agricultural Marketing Service of the USDA. P&S ensures that markets are properly bonded, prescribes rules for fair trade and competition, and periodically checks the scales.

Cooperative Marketing

Small producers often find themselves in the most difficult position in regards to marketing because an individual producer cannot attract on-farm buyers due to the small number of cattle that can be offered for sale at any one time. Therefore, cooperative efforts become essential to broadening marketing opportunities.

The formation of marketing associations allows groups of small producers to join together to function as a "big" producer. An example is the Buckingham Cattlemen's Association program in Virginia. These producers agree to a standardized health, genetic, and management practice as a means to create larger pools of feeder cattle. As a result, they have been able to attract not only more attention from buyers but price premiums as well. Compared with conventional marketing of small groups of calves via local auctions, the association has averaged premiums of $8.29, $6.12, and $5.08/cwt for 500-, 600-, and 700-lb steers, respectively.

Country Markets (Nonpublic)

Country markets include order buyers, commission representatives, and direct marketing. Order buyers and commission representatives are intermediaries between the seller and the ultimate buyer, whereas direct marketing does not involve the services of a marketing middle person. Examples of direct marketing are commercial producers who sell their feeder cattle directly to cattle feeders and packer buyers who purchase slaughter cattle directly from the feedlot. This latter marketing method is the most significant country market since a high percentage of slaughter steers, heifers, and calves are purchased in this manner.

Direct, or spot, marketing of fed steers and heifers is accomplished primarily in three ways: cattle are sold on a liveweight basis per 100 lbs, where buyers base their live price on (1) estimated carcass weight, quality grade, and yield grade; (2) dressed beef (flat price on carcass weight regardless of quality grade or yield grade); and (3) grade and yield (priced on quality grade and carcass weight). Examples of these three methods of selling fed cattle are shown in Table 9.9. In liveweight and dressed weight pricing, sellers are paid for the average performance of their cattle. As a result, the highest performing cattle receive a subpar price while the poorest performing animals receive a premium.

TABLE 9.9 Methods of Selling Fed Cattle[a]

Method	Example		
Live Cash Sale			
Cash bid of $74			
Liveweight	1,150 lbs		
	0.04% shrink		
	46 lbs		
	1,150 lbs		
	−46 lbs		
Payweight	1,104 lbs		
	× $74 per cwt		
	$ 816.96		
Freight	(generally, when cattle are sold live, packer pays the freight)		
	$ 816.96 net		
Dressed Beef Sale			
Beef bid of $116/cwt of carcass			
Liveweight	1,100 lbs	1,100 lbs	
	× 64% actual dressing percentage	× 61% actual dressing percentage	
	704 lb carcass wt	671 lb carcass wt	
	×$ 116 per cwt	×$116 per cwt	
Freight	$ 816.64	$778.36	
	−4.50 (transport–120 mi)	−4.50 (transport–120 mi)	
	$ 812.14 net received	$773.86	
	$38.28 advantage for higher dressing percentage.		
Grade and Yield	**Choice**	**Select**	**Standard**
Prices/cwt: Choice ($116), Select ($101), Standard ($94)			
Liveweight	1,200	1,200	1,200
	× 0.62	× 0.62	× 0.62
Dressing percentage	744 lbs	744 lbs	744 lbs
	× $116	× $101	× $ 94
Carcass weight	$ 863.04	$751.44	$699.36
	−4.50	−4.50	−4.50
Freight			
Net received	$ 858.54	$746.94	$694.86
% of total head in each quality grade	× 60	× 35	× 5
Net received by quality grade	$ 515.12	$261.43	$ 34.74
Average net received (per head for the 100 head)	= $ 811.29		

[a]Net dollars are not comparable for the three methods because liveweights are different.
Source: Cattle-Fax and USDA.

Direct, or spot, marketing involves a process whereby the feedlot develops a show list of cattle to be presented to order buyers at the beginning of each week. Buyers evaluate the cattle, estimate their value, and then either accept or reject the sellers' asking price. Date, time, and cost of delivery are typically a part of the negotiation. It is becoming more common for a majority of the transactions in any given week to be completed within a 1- to 3-day range.

Approximately 45% of fed cattle are currently sold on a carcass basis, an increase from the approximately 20% in 1971. Nebraska, Iowa, and Colorado account for about two-thirds of all carcass grade and weight marketings of steers and heifers. Packers in Texas, Iowa, Nebraska, Minnesota, and Wisconsin, account for approximately three-fifths of all cows and bulls purchased on a carcass basis.

Some producers prefer carcass grade and weight (yield) marketing, indicating that it is a more accurate method than dressing percentage and carcass grade in live animals. Other producers are fearful of this method of marketing, feeling they lose control of the marketing process when weighing, grading, and occasional condemnation of carcasses occurs behind "closed doors."

Nondirect Markets

Nondirect, or nonspot, transactions occur when the cattle are priced based on future performance and committed to a packer price to the time that cattle are ready for harvest or to be placed on the "show list." The primary forms of this approach are marketing agreements, forward contracting, and packer feeding.

Marketing agreements involve a longer-term relationship for the ongoing delivery of cattle where the number of animals, the date and conditions for delivery, performance specifications, and pricing method are predetermined. These arrangements are typical of some alliances.

Forward contracting may be part of a marketing agreement or a one-time transaction. In this system, price is either fixed or based off some publicly reported future price. The base price is often determined from the futures market. In this case, a buyer and seller agree to a differential, or basis, from the futures market for a specified contract month. Premiums and discounts are then applied. Packer feeding is the situation where a packer owns outright or in a partnership, part of the cattle scheduled for delivery to the plant. Packer feeding accounts for less than 5% of all fed cattle marketings.

Grid-Pricing

Another pricing mechanism that is becoming increasingly available is referred to as grid-pricing. Grid-pricing is a more complex approach that applies a series of premiums and discounts based on Quality Grade, Yield Grade, carcass weight, and level of carcass defects to a preestablished base price. While similar to Grade and Yield pricing, grid-pricing typically determines base price from some average price in the week prior to delivery. The Choice-Select mix, and other premiums and discounts are calculated from the plant average.

In grid-pricing, the individual performance of each animal in a particular lot is evaluated and rewarded accordingly. The goal of most grids is to reward cattle that grade Choice or better, have USDA Yield Grades of 1 or 2, have carcass weights between 600 and 900 lbs, and are free from any other defects such as dark cutter. Producers must be careful to balance the pursuit of premiums with other factors that influence profitability. Producers who enjoy success in marketing cattle via a grid system are able to avoid the discounts (< 3%), optimize pounds sold, and produce cattle that are capable of being 70% Choice and 70% Yield Grades 1 and 2.

In most grid-pricing systems, premiums are paid for cattle that have USDA Quality Grades in excess of a plant or regional average, and that have a higher percentage of USDA Yield Grades 1 and 2. The specific premiums paid depend on the emphasis in the product line—superior palatability, superior cutability, or a combination of the two. Table 9.10 illustrates the relationship of Quality and Yield grade premiums/discounts in a majority of the available grids.

A series of discounts are also applied to carcasses that fall outside of a predetermined carcass weight range, that have excessively large or small ribeye area, that fail to make A maturity, or that have undesirable lean color or other defects. As pointed out in Table 9.11, the discounts are typically larger on a per weight basis as compared with the premiums. As a

TABLE 9.10 Relationship of Marbling and Leanness in a Grid-Pricing System

Yield (muscling: leanness)	Quality (marbling)					
	Prime	Certified Program	Choice	Select	Standard	Other
YG1	$$$$$	$$$$	$$$	$$	– $$$$$	– $$$$$$
YG2	$$$$	$$$	$$	$	– $$$$$	– $$$$$$
YG3	$	$	– $	– $$	– $$$$$	– $$$$$$
YG4	– $$$$$	– $$$$$	– $$$$$	– $$$$$	– $$$$$$	– $$$$$$
YG5	– $$$$$$	– $$$$$$	– $$$$$$	– $$$$$$	– $$$$$$$	–$$$$$$$$

Source: Adapted from American Gelbvieh Association.

TABLE 9.11 National Carcass Premiums and Discounts for Slaughtered Steer and Heifers for the Week of June 18, 2001

	Value Adjustments ($/cwt)			
Quality	**Range**			**Simple Average**
Prime	2.00	–	13.00	6.57
Choice	0.00	–	0.00	0.00
Select	−18.00	–	−9.00	−15.13
Standard	−28.00	–	−12.00	−21.99
Certified programs				
Avg. Choice/Higher	0.00	–	4.38	1.18
—				
Bullock/Stag	−33.00	–	−2.00	−23.53
Hardbone	−40.00	–	−11.37	−25.94
Dark cutter	−40.00	–	−19.00	−28.99
Cutability[a]				
Yield Grade/Fat/inches				
1.0–2.0, <.1″	1.00	–	8.00	4.00
2.0–2.5, <.2″	1.50	–	3.00	2.08
2.5–3.0, <.4″	0.00	–	2.00	1.17
3.0–3.5, <.6″	−1.00	–	0.00	−0.11
3.5–4.0, <.8″	−1.00	–	0.00	−0.11
4.0–5.0, <1.2″	−20.00	–	−1.00	−13.38
5.0/up, >1.2″	−25.00	–	−15.00	−21.00
Weight				
400–500 lbs	−40.00	–	−15.00	−29.25
500–550 lbs	−30.00	–	0.00	−16.22
550–600 lbs	−10.00	–	0.00	−2.44
600–900 lbs	0.00	–	0.00	0.00
900–950 lbs	−2.50	–	0.00	−0.29
950–1,000 lbs	−15.00	–	0.00	−8.22
Over 1,000 lbs	−30.00	–	−10.00	−19.88

Based on individual packer's quality, cutability, and weight buying programs. Values reflect adjustments to base prices, dollars per cwt, on a carcass basis.

[a]If yield grades are not available, yield differentials may be based on fat at the 12th rib using a constant of average ribeye area and muscling for carcass weight and KPH. Superior or inferior muscling may adjust lean yield.

Source: USDA Livestock and Grain Market News.

result, cattle producers are advised to focus on avoiding the discounts as their first priority, and to then focus on the fine-tuning of their programs to better capture premiums.

In a grid-pricing formula, the Yield Grade price spreads tend to stay relatively stable over time. The greatest pitfall comes with Yield Grades 4 and 5 cattle that may incur a \$10–\$20/cwt carcass discount. However, the Choice-Select price spread may move \$15 to \$20 per cwt over the course of a single year. This spread is typically highest in the late fall and lowest in the early spring, although market conditions may shift these trends in some years (Fig. 9.14). Understanding the seasonal pattern of premiums and discounts is important in determining market strategy.

Monitoring the Choice-Select price differential is a key to making timely marketing decisions based on the genetic marbling potential of cattle. If the variation in this price spread ranges from \$0–\$20 per cwt over a period of time, then the difference between a Choice 700-lb carcass versus one that is Select ranges from \$0 to \$140 per head.

Because premiums and discounts are applied to a base price, the negotiation to determine base price is of importance. Understanding how the grade base is calculated is also important because it determines how the Choice-Select spread will be applied to price calculation. Packing plants in the northern Great Plains tend to run a higher percent Choice than those in the southern Great Plains. The Select discount is calculated by multiplying the grade-base percentage by the Choice-Select spread.

For example, if the plant average is 60% Choice and the Choice-Select spread is \$5.00/cwt, the Select discount is $.60 \times \$5.00 = \$3$/cwt. The difference between the spread and the Select discount equals the Choice premium ($\$5.00 - \$3.00 = \$2.00$). As the grade base increases, and/or as the Choice-Select spread widens, the discounts associated with Select grading carcasses become more severe. Figure 9.15 illustrates this relationship.

Table 9.12 shows the performance of two pens of cattle under a standard grid-pricing structure. Pen 1 receives a \$53.99 advantage because they avoided discounts and had more desirable Quality and Yield grades.

Figure 9.16 documents the trend in fed cattle marketing strategies shifting toward formula, contract, or alliance structure. Packer-owned cattle are not included but, if they were, the percentage of cattle marketed via nonspot market alternatives would increase to nearly 50%.

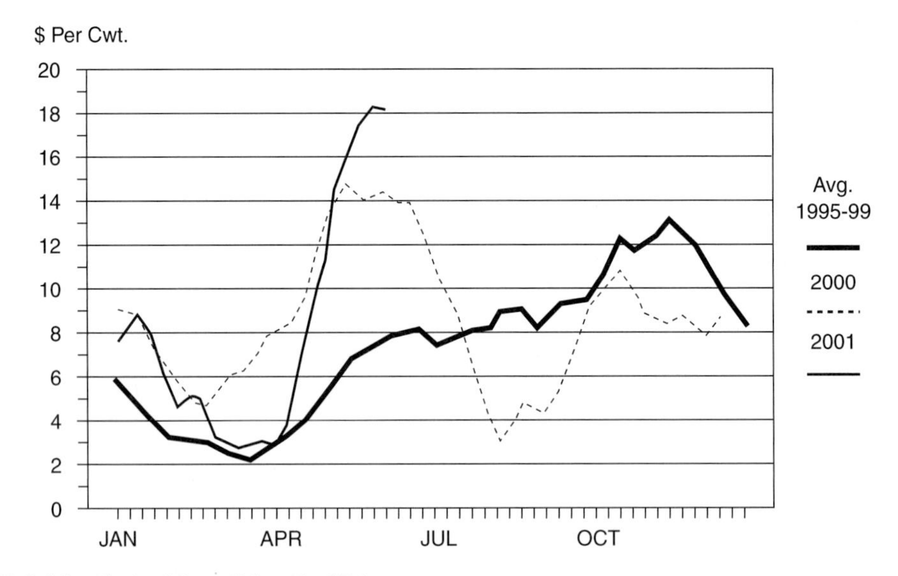

FIGURE 9.14 Choice Minus Select Beef Prices.
Source: Livestock Marketing Information Center.

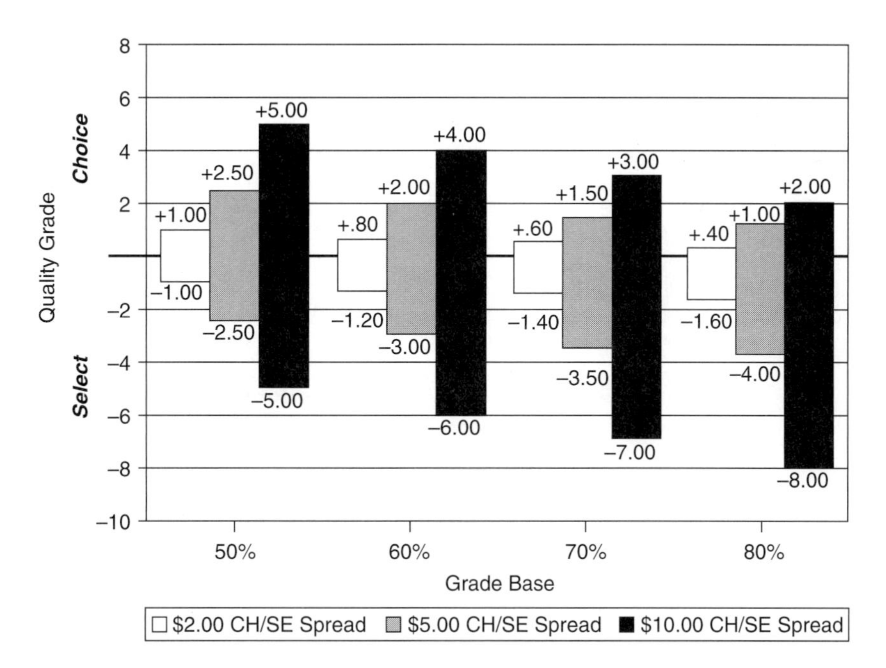

FIGURE 9.15 Effect of grade base and choice-select spread on the premiums and discounts associated with quality grade.
Source: Field, Tatum, and Kimsey (1998).

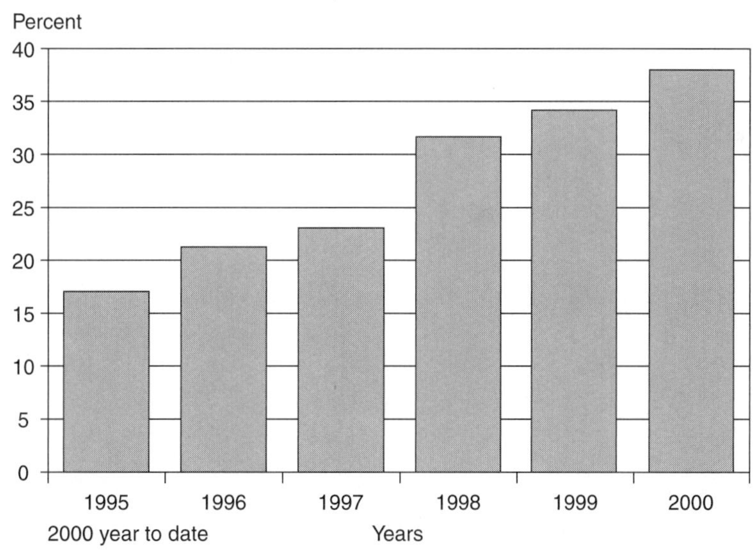

FIGURE 9.16 Percentage of fed cattle movement from formula, contract, and alliance pricing systems. (No packer-owned cattle are included; if they were added, the 2000 amount would increase by 10%.)
Source: Cattle-Fax.

TABLE 9.12 Performance of Two Pens of Cattle under a Grid-Pricing System

Example Program	Marketing Specifications	Premiums and Discounts: Quality Grade	Premiums and Discounts: Yield Grade
Live price	$67/cwt	Prime + $4.40 over Choice	YG1 +$3.00
Base carcass price	$105.51/cwt	Certified program + $2.50 over Choice	YG2 +$2.00
Base dressing %	$63.5	Choice + $2.80	YG3 −$1.00
Out discount[a]	$20.00/cwt	Select − $4.20	YG4 −$15.00
Choice/Select or Prime	$7.00/cwt	Standard − $15.00	YG5 −$20.00
Grade base	60% Choice		

Carcass Performance of Two Pens of Cattle	Pen 1	Pen 2
% Prime	9	0
% Choice	76	50
% Select	6	40
% No roll	0	10
% of Choice for certified program	9	0
% YG 1	51	3
% YG 2	30	53
% YG 3	19	44
% YG 4	0	0
% YG 5	0	0
% Outs[b]	0	17
Dressing percentage	64.8	62.0

Computation of Price Under a Grid System	Pen 1	Pen 2
Base price	$105.51	$105.51
Out allowance	.52	.52
	106.03	106.03
Prime	+0.49	0.00
Choice	+2.13	+1.40
Select	−0.25	−1.68
No roll	0.00	−1.50
Certified	+0.22	0.00
Quality Adjustment	+2.59	−1.78
YG 1	+1.53	+0.09
YG 2	+0.60	+1.06
YG 3	−0.19	−0.44
YG 4	0.00	0.00
YG 5	0.00	0.00
Yield Adjustment	+1.94	+0.71
Outs Adjustment	0.00	−3.40
Adjusted price ($/cwt)	110.56	101.56
Ave. carcass weight (lbs)/liveweight(lbs)	803/1,238	821/1,329
Gross return ($/hd)	887.80	
Difference ($/hd)	+53.99	833.81

[a]Dark cutters, carcass weights outside the range of 600 to 950 lbs.
[b]Weight nonconformance.

TABLE 9.13 Gross Revenues for Various Carcass Weights at Differing Prices

	Carcass Price ($/cwt)					
Carcass Weight	106	104	102	100	98	96
650	689	676	663	650	637	624
700	742	728	714	700	686	672
750	795	780	765	750	735	720
800	848	832	816	800	784	768
850	901	884	867	850	833	816

Note: Heavy carcasses can receive a lower price and still generate more gross revenue than a higher-priced, but lighter, carcass.
Source: Field, Tatum, and Kimsey (1998).

Cattle-Fax estimates that these marketing channels, plus packer-owned cattle, will likely increase to 45–50% by 2005.

As cow-calf producers and feeders evaluate marketing alternatives, such as grid-pricing, several factors must be considered. An entire set of cattle should not be subjected to a new pricing system without the knowledge of how they might perform. Therefore, data should be collected on a sample of cattle to determine the risks that may be incurred. Producers should also determine if there are any up-front fees or membership charges that are associated with participation in the grid.

Producers are also advised not to become so focused on capturing premiums that they forget about the importance of feedlot performance and weight to achieve profitability. As Table 9.13 illustrates, heavier carcasses that receive lower prices on a per pound basis may yield more gross revenue than lighter weight carcasses that receive significant premiums.

Electronic Marketing

Electronic marketing and market information transfer have significantly changed the process of cattle marketing. By utilizing the various forms of electronic transmission of information, buyers and sellers can make decisions based on a host of relevant data. Unlike the comparatively long and slow marketing process of the past, today large numbers of cattle can be sold within only a few hours. Approximately two million cattle were sold in the United States via video satellite or Internet auction in 2000.

Most electronic marketing systems share the following characteristics: (1) pricing is determined at a single location or over a single communication system, (2) the cattle are not moved from the seller's farm until they are sold, and (3) buyers can participate without being at the location where pricing occurs.

To date, the most successful form of marketing cattle by electronic means has been the video auction. Video auctions are conducted by showing buyers a videotape (2–4 minutes) taken at the farm or ranch of the cattle being offered for sale. The audio portion of the tape describes weight, location, and background information on the cattle. The seller also states the weighing location and various options of delivery dates. A satellite hookup allows the videotape to be shown to buyers gathered in several locations, then an auctioneer sells the cattle to the highest bidder. Some individuals or companies can receive the video auction by tuning their own satellite dish to the correct channel. After the sale, buyer and seller arrange for the shipping of the cattle.

Web-based livestock marketing is a relatively recent innovation. While some sites are simply listings of cattle for sale, others conduct real-time Internet auctions where several hundred buyers might be logged on to a site to bid for the available cattle.

Other approaches have been to combine information services that range from market reports to whole-herd recordkeeping with marketing programs such as on-line trading, auctions, and alliance support systems. Some services even offer on-line purchases of farm inputs such as tags, animal health care products, and feed supplements.

The literally frantic development of Web-based companies that began in the late 1990s will certainly create both success and failure. However, those companies that emerge from the transformation will hold a significant place of power as brokers of information and have access to supply/marketing chains. Several local or regional livestock auction companies have seen this trend as an opportunity to enlarge their services into the electronic arena and better meet the needs of their customers.

MAJOR FACTORS AFFECTING CATTLE PRICES

The major goal of the beef industry—efficient production of a highly palatable product with a profitable return to those who produce it—implies that two primary factors affect profitable beef production: (1) efficiency of production (or cost of production) and (2) the price received for the product. Although there are limits to improving efficiency of production, the current beef industry has the opportunity to make tremendous strides in cost-effective productivity, especially the cow-calf segment. Even at optimum levels of production efficiency (when numbers and weights are at the least cost), if market price is not high enough to cover costs, profitability will not be realized. Producers need to understand the advantages and disadvantages of different markets as well as the factors affecting the supply and demand for beef. These factors determine the price structure for beef, allowing producers to identify ways of predicting and possibly changing beef prices.

Cattle prices are the most widely discussed topic in the beef industry. The factors affecting cattle prices are complex, influenced by supply, demand, psychology, and several other factors. Because of this complexity, many misunderstandings are generated among various beef industry segments, some of which have resulted in serious accusations, lawsuits, price freezes, and consumer boycotts.

Cattle prices result from a free market system in which supply and demand factors determine the price of cattle. Producers find themselves as "price-takers" rather than "price-makers," yet they purchase most of their goods and services in a reverse pricing structure. Even within the beef industry, the pricing structure changes from the producer who says, "What will you pay me?" to the retailer who establishes a price for beef.

Supply and Demand

The various factors affecting the supply and demand of beef and ultimately its price are shown in Figure 9.17. Numbers of animals and pounds per animal eventually reflect the tonnage of beef produced. Poultry and pork are the meats that compete the most with beef. It is obvious that most of these factors have an effect on pounds of beef produced and the retail price of beef.

Psychology (Market Perception)

The supply and demand of beef have the most significant influence on cattle prices; however, how buyers and sellers perceive the market can also influence prices. The perceptions of a few

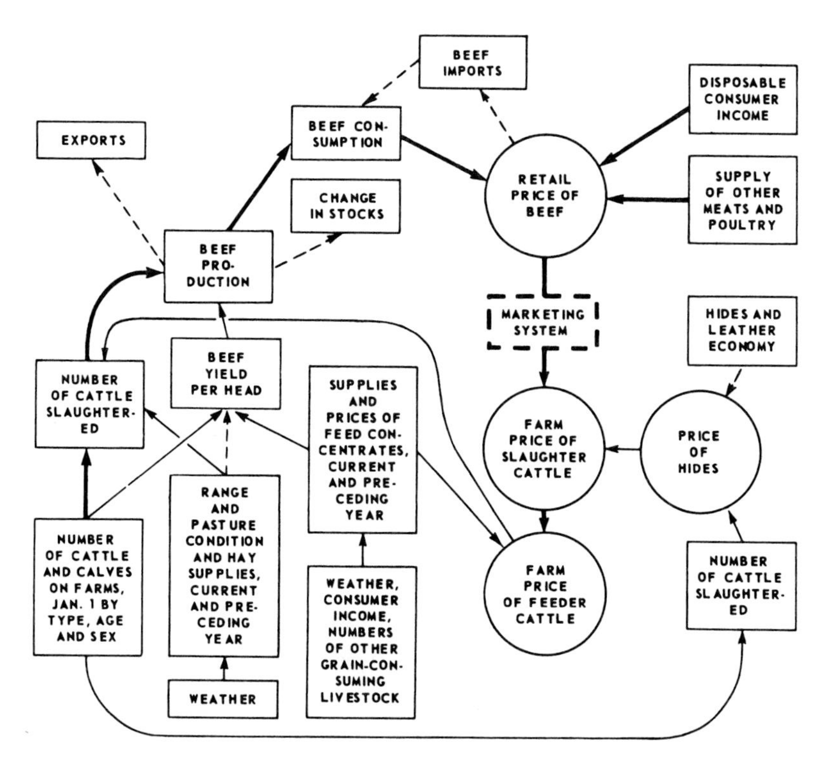

FIGURE 9.17 The demand and supply structure for beef. Arrows show direction of influence. Heavy arrows indicate major paths of influence. Light, solid arrows indicate less important actors of influence, while dashed arrows are pathways of minor influence.
Source: USDA.

individuals can be communicated rapidly to large numbers of people who impact the market. Significant and unexpected political, social, or international trade issues can alter the marketplace strictly due to the level of uncertainty. The markets should be viewed as fluid processes where trends and unpredictability are hallmarks.

Beef Cattle Cycles and Prices

During the past 100 years, there have been several beef cattle cycles with characteristic peaks and valleys in cattle numbers and prices. These periodic changes in cattle numbers have occurred primarily in response to changes in profitability and weather (drought). During the peaks in cattle numbers, the supply of beef was excessive and beef prices dropped; whereas during the low points in cattle numbers, the supply of beef was limited and the price of beef increased. These cycles for the past years are shown in Figure 9.18. The cyclic nature of prices is shown in Figure 9.19.

Typically, price cycles lag behind inventory number cycles. This means that cattle numbers cannot be changed rapidly even though prices are high. The most apparent problem of the beef cattle cycles has been wide price fluctuations due to wide swings in oversupplying and undersupplying the beef market. There is a biological time lag to increase cattle numbers (Fig. 9.20) in that it takes about three years from saving more heifer calves until they can produce slaughter progeny. The biological time frame is much shorter for poultry (Fig. 9.21) than for beef cattle.

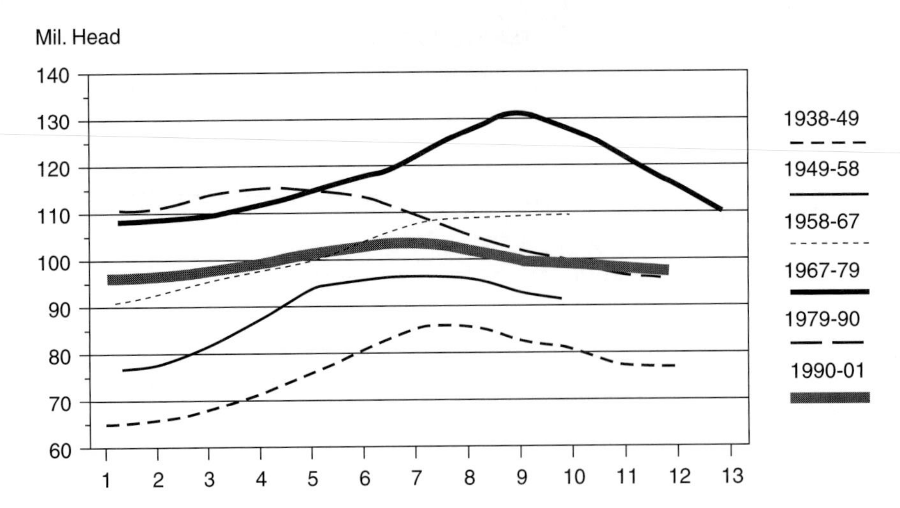

FIGURE 9.18 Total U.S. cattle inventory (Jan. 1) by various cycles.
Source: Livestock Marketing Information Center.

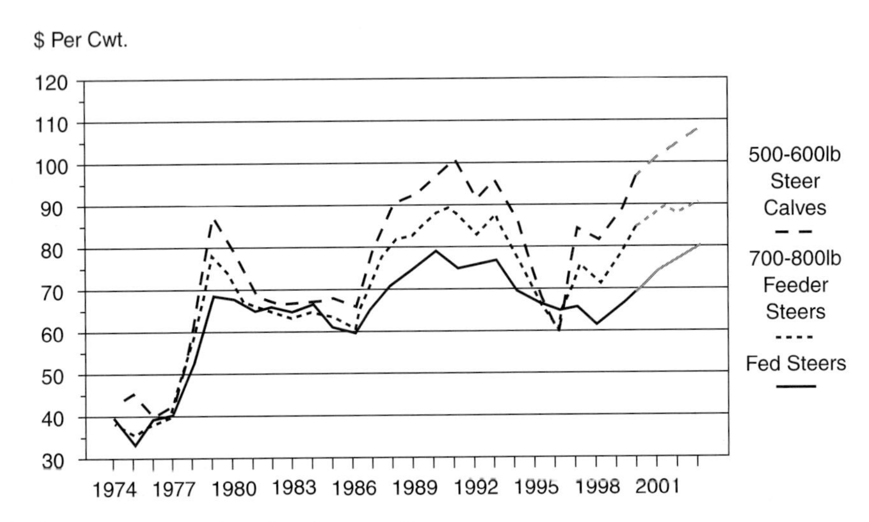

FIGURE 9.19 Average annual cattle prices over time.
Source: Livestock Marketing Information Center.

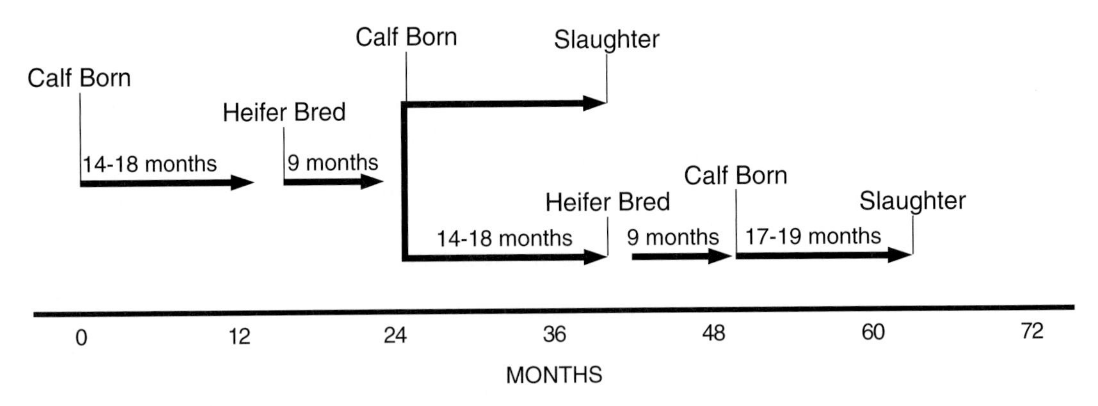

FIGURE 9.20 Biological time line for cattle supply.
Source: USDA.

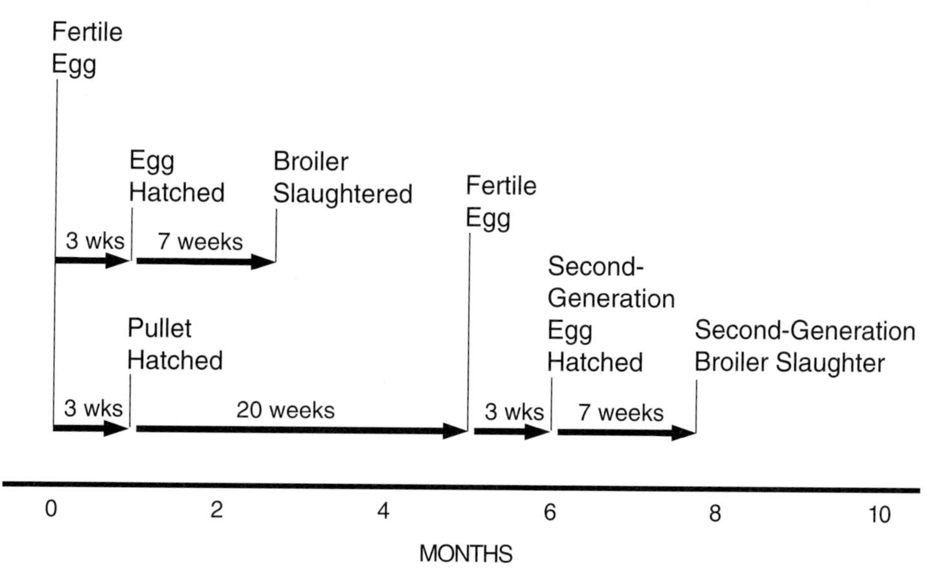

FIGURE 9.21 Biological time line for broilers.
Source: USDA.

Some market analysts argue that the cattle cycles of the past will not repeat themselves in the same way. They say that rapid increases in cattle numbers have occurred in the past because of the availability of cheap cattle, cheap land, cheap feed, and/or cheap money. From 1980 to 1992, none of these factors existed. Yet high cattle prices increased cattle numbers and the resulting large beef supply depressed cattle prices in 1993–1996 that resulted in a reduction in herd inventories. This implies that even in a mature beef industry there would be changes in cattle numbers that would result in years of unprofitable beef prices as well as years in which beef prices would be profitable. Note that the cattle inventory in 1990–2001 was less changed than in previous cycles, but that numbers did decline following the price trough of 1996–1997 (Figs. 9.18 and 9.19).

Relatively high prices for fed steers and heifers usually result in an increased number of cows and heifers retained for breeding and thus fewer of them being slaughtered. Conversely, as prices decline fewer females are kept as replacements. This is demonstrated in Figure 9.22.

Historical changes on total cow numbers are provided in Figure 9.23. Cow inventories are not expected to approach the levels of 1982 in the near future. While the swings in cow numbers will not be as dramatic as in past cycles, there is evidence that cycles are shortening and the cyclic nature of the industry will continue.

The relationship of profitability and cattle inventory is illustrated in Figure 9.24. Price, and to some extent profit, responds inversely to supply. For example, the declining inventory from 1985 to 1991 was accompanied by relatively high cow-calf profits. Rising inventories in 1980–1983 and 1993–1996 resulted in declining returns.

Factors such as drought, lower calf crop percentages, competitive meat prices, grain prices, and the global and national economies also significantly affect cattle prices.

Carcass Weight

Tonnage of beef produced is a function of the number of head slaughtered and average carcass weight. Carcass weights have been increasing over the past several years resulting, in part, in much higher production per cow (Fig. 9.25). Note that despite the dramatic increases

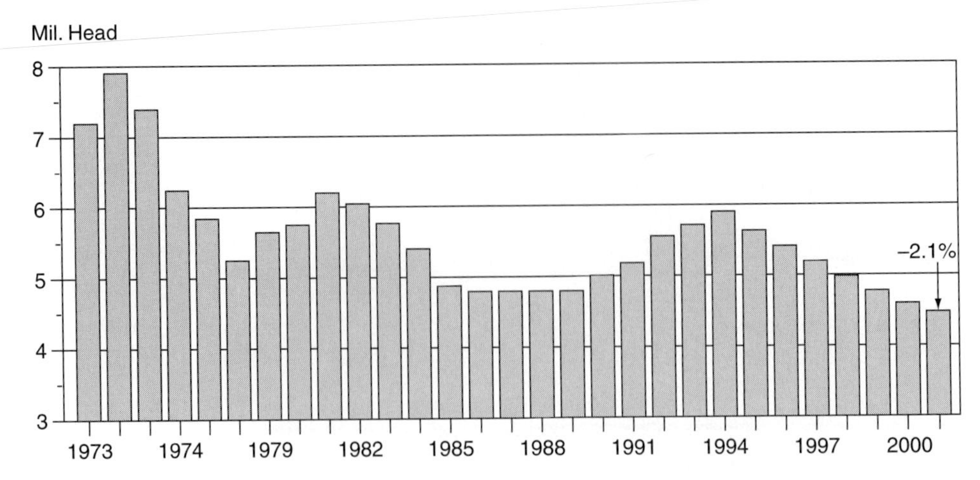

FIGURE 9.22 Variation in number of heifers held as replacement females is heavily influenced by cattle prices.
Source: Livestock Marketing Information Center.

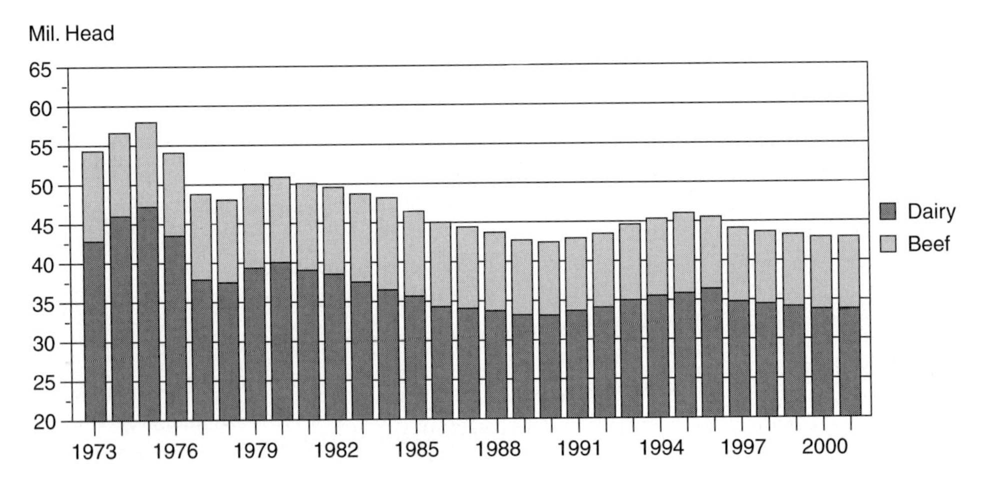

FIGURE 9.23 Historical changes in beef and dairy cow numbers.
Source: Livestock Marketing Information Center.

in individual animal productivity, profits were variable over the same time period. The increase in average carcass weights has allowed the beef industry to maintain or even increase beef production while the total cattle inventory has declined (Fig. 9.26).

There is a relationship between carcass weight and cattle prices. Heavy carcass weights do not necessarily directly cause lower beef prices, but they increase the total beef supply, which, in turn, tends to lower beef prices. When market prices begin to drop below producers' breakeven price, feeders typically feed their cattle longer anticipating a price increase. This usually has an antagonistic effect on prices by increasing the tonnage of beef and the number of Yield Grade 4–5 slaughter cattle. As cattle prices increase and fed cattle become profitable, feeders usually market their cattle at lighter weights, desiring to market as many cattle as possible before a price break occurs.

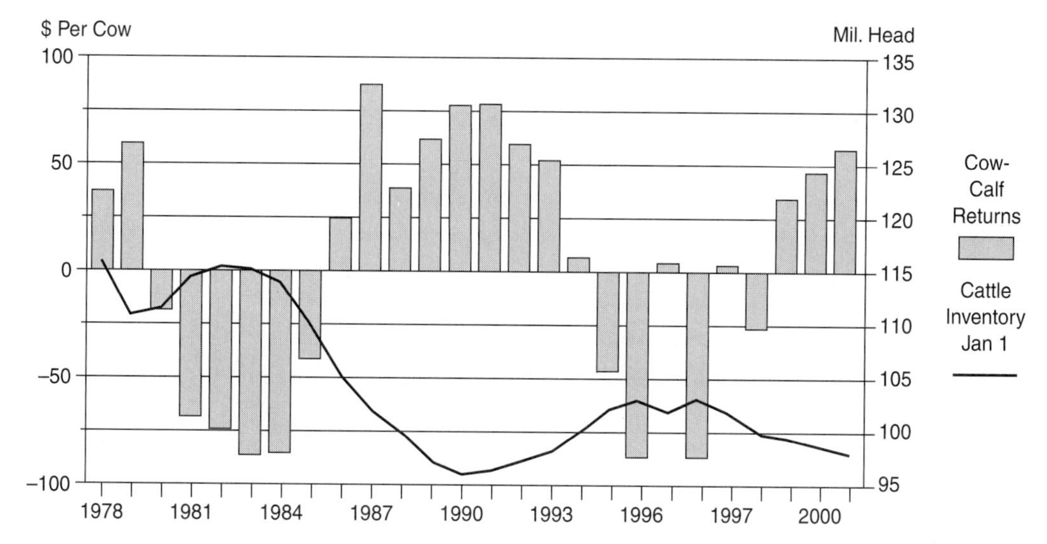

FIGURE 9.24 The relationship between cow-calf enterprise profitability and cattle inventory.
Source: Livestock Marketing Information Center.

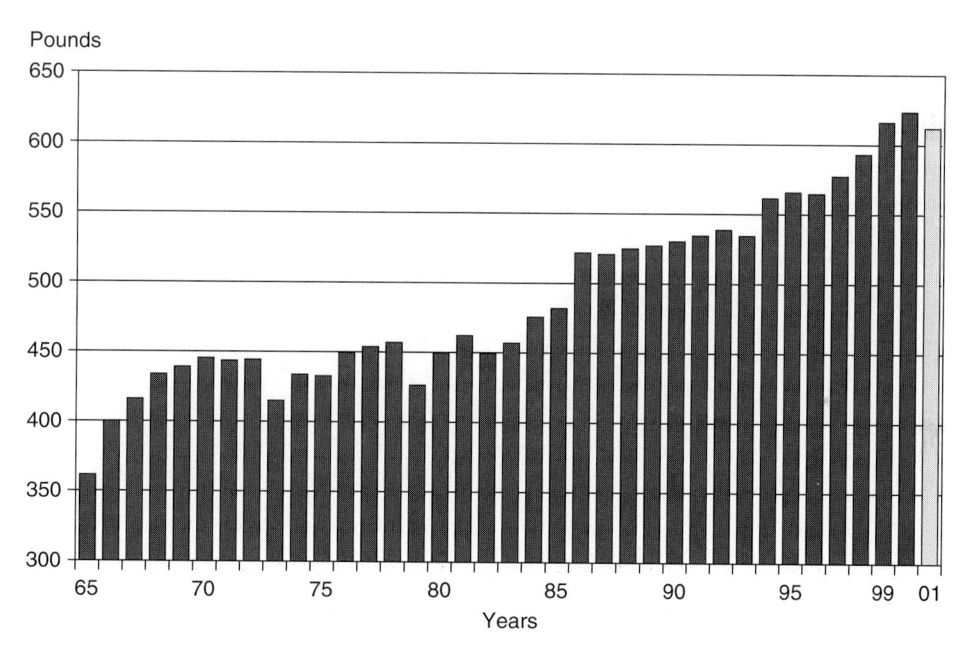

FIGURE 9.25 U.S. beef production (carcass weight) per beef cow.
Source: USDA.

By-Product Value

In addition to the value of the carcass, by-product value (e.g., hide and offal) contributes to the total value of slaughter cattle. The value of by-products has ranged from $45 to $100 per head from 1980 to 2000. The hide accounts for approximately two-thirds of the total by-product value. Hide values can fluctuate dramatically over a short time period and thus affect live cattle prices.

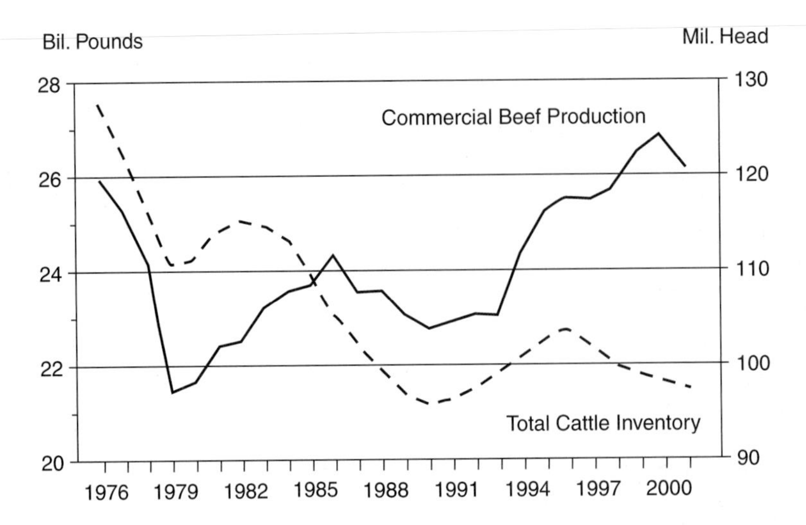

FIGURE 9.26 Changes in commercial beef production and total cattle inventory.
Source: Livestock Marketing Information Center.

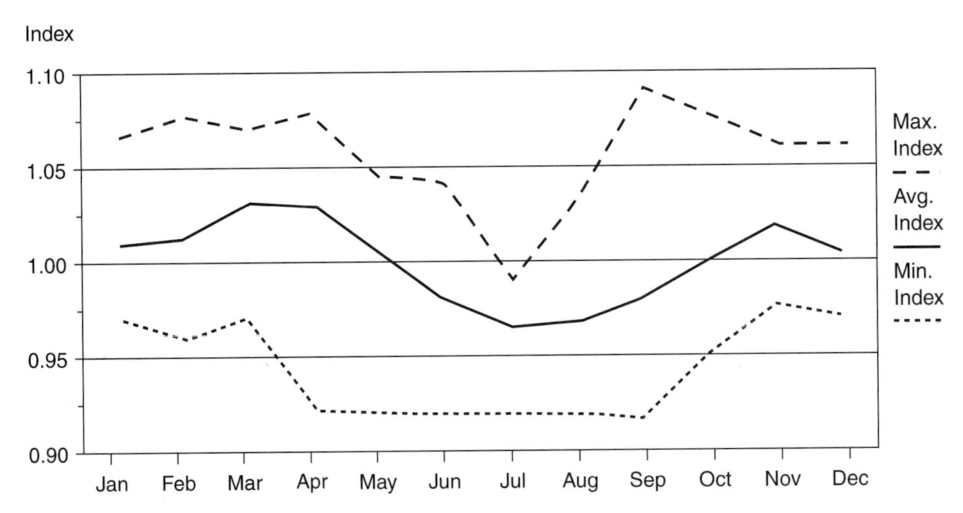

FIGURE 9.27 Seasonal variation in fed steer prices in the Southern Plains as measured by price index.
Source: Livestock Marketing Information Center.

Seasonal Prices

Between 1990 and 1999, there were seasonal differences in prices of fed steers and heifers, with the highest prices coming in March and April (Fig. 9.27). Some cattle feeders take advantage of this seasonal price difference by marketing 12–15-month-old calves that reach desirable slaughter weights prior to the price declines that typically occur beginning in late May and June. However, cattle producers should not depend on these seasonal price cycles for market steers to hold true for all years. Cow prices are consistently higher in the spring months than in the fall months (Fig. 9.28). This is true for most years because the heaviest culling time for cows is in the fall after calves are weaned from the cows.

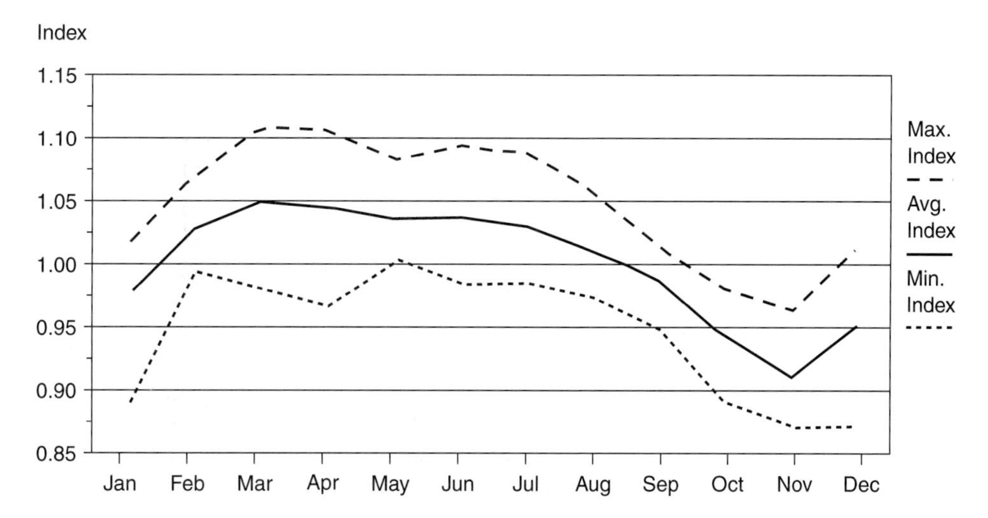

FIGURE 9.28 Seasonal variation in utility cow prices in the southern Plains as measured by price index. *Source:* Livestock Marketing Information Center.

Shrink

Shrink is a loss of weight affecting live cattle, carcasses, or retail cuts. It is used most frequently in marketing feeder cattle and slaughter cattle. Shrink is usually expressed as the percentage of liveweight loss. Major factors affecting shrink are (1) time in transit, (2) distance, (3) weather, (4) handling methods, and (5) type of feed fed. It is not uncommon for feeder cattle to shrink from 2% to 8% and slaughter steers to shrink from 2% to 6%. This affects dressing percentage (yield), which is the ratio of carcass weight to liveweight.

Buyers and sellers must know how to account for shrink differences in order to determine fair market prices (Tables 9.14 and 9.15). Buyers of feeder cattle should also know the costs involved in the cattle regaining lost weight.

An example from Table 9.14: If 500-lb calves sold for $80/cwt, then the total value is $400 per head. If the shrink is 4%, then the delivered weight is 480 lbs; thus the value of the weight delivered is $83.33/cwt ($400 ÷ 480 lbs).

TABLE 9.14 Buying Cattle—Changing Prices to Compensate for Shrink

Asking Price	% Shrink				
	2%	3%	4%	6%	8%
$100.00	$102.04	$103.09	$104.16	$106.38	$108.69
96.00	97.95	98.96	100.00	102.12	104.34
92.00	93.87	94.84	95.83	97.87	100.00
88.00	89.80	90.72	91.67	93.62	95.69
84.00	85.71	86.60	87.50	89.36	91.30
80.00	81.63	82.47	83.33	85.11	86.96
76.00	77.55	78.35	79.17	80.85	82.61
72.00	73.47	74.23	75.00	76.60	78.26
68.00	69.39	70.10	70.83	72.34	73.91
64.00	65.31	65.98	66.67	68.09	69.57
60.00	61.22	61.86	62.50	63.83	65.22

TABLE 9.15 Selling Cattle—Net Price after Shrink Allowance

Asking Price	% Shrink				
	2%	3%	4%	6%	8%
$100.00	$ 98.00	$ 97.00	$ 96.00	$ 94.00	$92.00
96.00	94.08	93.12	92.16	90.24	88.32
92.00	90.16	89.24	88.32	86.48	84.64
88.00	86.24	85.36	84.48	82.72	80.98
84.00	82.32	81.48	80.64	78.96	77.28
80.00	78.40	77.60	76.80	75.20	73.60
76.00	74.48	73.72	72.96	71.44	69.92
72.00	70.56	69.84	69.12	67.68	66.24
68.00	66.64	65.96	65.28	63.92	62.56
64.00	62.72	62.08	61.44	60.16	58.88
60.00	58.80	58.20	57.60	56.40	55.20

An example from Table 9.15: A cattle feeder hears a price quote of $72.00/cwt for 1,000-lb market heifers. The first impression is a total value of $720/head. However, in this price a 4% shrink is usually implied. Thus, the total dollars received would be $691.20 (960 lbs × $72/cwt or 1,000 lbs × $69.12/cwt). The $69.12/cwt comes from $72/cwt being decreased 4%.

Managing the factors that contribute to shrink can bolster profits. A few guidelines are as follows:

1. Cattle on green pasture shrink more than those consuming a drier ration. Also, feed changes near time of transport can increase shrink loss.
2. Weaning is a stressful time for calves and adding transport to the stress can increase shrink losses and morbidity/mortality rates. Preconditioned calves typically fare better than do those calves that are weaned and marketed back to back.
3. Gathering cattle gently and minimizing their stress is a key to minimizing shrink.
4. Cattle should not be overcrowded into trucks and trailers, and the footing should be good.

ASSESSING MARKETING COSTS

A good marketing plan includes knowing the costs of production and the breakeven prices. Part of the cost assessment involves marketing costs and the costs of different marketing alternatives. Direct markets and public markets as marketing alternatives were discussed earlier in the chapter. Costs for these various markets can be determined.

Direct marketing costs are primarily the costs associated with selling at the point of production. Examples include the sale of feeder cattle at the farm or ranch and the sale of fed cattle at the feedlot to the packer. A direct marketing cost worksheet is shown in Table 9.16.

Public marketing costs are those associated with auctions and terminal markets. Table 9.17 shows a cost worksheet for public marketing.

KEEPING CURRENT ON BEEF PRICES

Table 9.18 lists some comparative cattle and product prices and their sources. It is not uncommon for prices to fluctuate on a daily or even an hourly basis. Thus, it is important to know current prices to make valid marketing decisions and ensure maximum dollar returns.

TABLE 9.16 Direct Marketing Cost Worksheet

Marketing Component		1	2	3
Row	Item	Unit	Rate or Cost (per unit)	Total Cost (per cwt)
1.	Average weight per head	lb	______	
2.	Number of head		______	
3.	Asking price	$/cwt		______
4.	Terms of sale			
5.	Overnight stand: Estimated shrink	%	______	
6.	Shrinkage cost (line 3 × line 5)	$/cwt		______
7.	Weigh-up location: Estimated shrink	%	______	
8.	Shrinkage cost (line 3 × line 7)	$/cwt		______
9.	Pencil shrinkage	%	______	
10.	Shrinkage cost (line 3 × line 9)	$/cwt		______
11.	Cutbacks	%	______	
12.	Cutback price offer	$/cwt	______	
13.	Cutback cost (line 3 − line 12) × line 11			______
14.	Other costs:	______	______	______
		______	______	______
15.	Total direct sales costs:			
16.	Shrinkage costs only (sum lines 6, 8, and 10)			______
17.	Shrinkage cost, cutbacks, and			______
	other (sum lines 13, 14, and 16)		______	
				______

Source: Selected livestock enterprise budgets for Colorado, Colorado State University.

Market information on cattle numbers and prices is available on a daily or weekly basis from many local newspapers and radio stations. These sources generally give market reports on local auctions, direct sales, and terminal markets in certain areas. Several major weekly publications (e.g., *Drovers Journal, Record Stockman,* and *Western Livestock Journal*) give detailed market numbers and price coverage. There are additional sources of market information that provide more current prices if they are needed.

Cattle-Fax

Cattle-Fax is a member-owned market information, analysis, and research organization that serves cattle operators in all segments of the U.S. beef industry. It maintains the largest private database on the U.S. cattle industry. The mission of Cattle-Fax is to collect, analyze, and disseminate information so that its members and clients can make more profitable market and management decisions (see address in Appendix).

Cattle-Fax members receive up-to-the-minute cattle marketing information from regional analysts via toll-free WATS lines or information networks located throughout the country. The organization also provides written report updates; these are mailed to members weekly and are current, intermediate, and long-term market evaluations (Fig. 9.29).

FORECASTING BEEF PRICES AND MANAGING PRICE RISKS

One of the most significant variables affecting profitable beef production is the ability to project future market prices. For example, the cattle feeder currently purchasing a pen of feeder cattle will be selling them at an unknown price some 100–150 days in the future.

TABLE 9.17 Public Marketing Cost Worksheet

Marketing Component		1	2	3
Row	Item	Unit	Rate or Cost (per unit)	Total Cost (per cwt)
1.	Average weight per head	lb		
2.	Expected public market price	$/cwt	______	
3.	Number of head		______	
4.	Market charges:		______	
5.	Selling commission	$/hd		
6.	Yardage	$/hd	______	
7.	Feed	$/hd	______	
8.	Insurance	$/hd	______	
9.	Brand inspection	$/hd	______	
10.	Health inspection	$/hd	______	
11.	Veterinary services	$/hd	______	
12.	Meals and lodging	$/hd	______	
13.	Other	$/hd	______	
14.	Total market charges		______	______
	[(sum line 5 thru 13)/(line 1/100)]			
15.	Transportation:			
16.	Miles to market	miles		
17.	Rate per mile	$/mile	______	
18.	Total transportation costs		______	
	(line 16 × line 17)/line 3			
	[line 18/(line 1/100)]			
19.	Transit shrinkage:			______
20.	Without fillback:			
21.	Estimated shrinkage	%		
22.	Shrinkage cost (line 2 × line 21)		______	
23.	With fillback:			______
24.	Estimated shrinkage	%		
25.	Estimated fillback	%	______	
26.	Shrinkage cost		______	
	[(line 2 × line 24)−(line 2 × line 24 × line 25)]			______
27.	Total public marketing costs:			
28.	Without fillback (sum lines 14, 18, and 22)			______
29.	With fillback (sum lines 14, 18, and 26)			______

Source: Selected livestock enterprise budgets for Colorado, Colorado State University.

Price forecasting is formulating an outlook (or opinion) on where the market will be at some point in the future. The most important factors analyzed by price forecasting are supply, demand, psychology (especially in the short term), and current data (how fed cattle are moving out of the feedlots).

No person or group is right all the time in forecasting prices. The objective is to be accurate most of the time by recognizing changes as they occur and adjusting forecasts as necessary.

Table 9.19 identifies how some price changes can be managed.

THE FUTURES MARKET

The futures market is a highly standardized, regulated method of exchanging forward contracts. In the cash cattle market, it is not uncommon to forward-contract a prescribed number of cattle at a mutually agreed-upon price for delivery at a predetermined date in the future.

TABLE 9.18 Comparative Market Prices for Cattle and Cattle Products at a Given Point in Time

Beef Industry Segment	Product	Price[a] (June, 2001)	Reference Source[b]
Purebred (seedstock)	Bulls (14–30 months)		
	—to purebred breeders	$1,000–$25,000/head	2, 3
	—to commercial producers	$900–$3,000/head	2, 3
	Semen		
	—registered herds	Up to $100/unit	3, 6
	—commercial herds	$5–$15/unit	3, 6
Commercial Cow-Calf	Calves		
	—under 1 month	$100–$200/head	2
	—400–500 lb (6–10-month-old steers)	$105–$125/cwt	1, 2, 5
	Slaughter cows (900–1,200 lbs); good condition	$40–$48/cwt	1, 2, 5
	Slaughter bulls (1,200–2,000 lbs); good condition	$50–$60/cwt	1, 2
	Cow-calf pairs	$650–$950/pair	2
Yearling	Feeder steers (600–700 lbs)	$93–$98/cwt	1, 2, 5
	Feeder heifers (500–600 lbs)	$92–$102/cwt	1, 2, 5
Feeder	Fed steers (Choice, YG3, 1,050–1,250 lbs)	$76–$79/cwt	1, 2, 5
	Fed heifers (Choice, YG3, 900–1,100 lbs)	$76–$79/cwt	1, 2, 5
Packer	700–850 lb Choice steer carcasses (YG 3)	$121–$125/cwt	1, 2, 5
	550–700 lb Choice heifer carcasses (YG 3)	$121–$125/cwt	1, 2, 5
	Hide and offal	$9.33/cwt (liveweight)	1, 2, 5
Retailer	Average price of all retail cuts (Choice)	$3.45/lb	4
Consumer	Boneless top sirloin	$3.49/lb	4
	T-Bone steak (bone in)	$7.99/lb	4
	Ground beef	$1.99/lb	4

[a]Price shown gives a range that would include most of the cattle or product.
[b]Sources of market prices: (1) *Livestock Market News* (USDA); (2) Weekly livestock publications (e.g., *Western Livestock Journal, Drovers Journal*); (3) Breed Journals (e.g., *Angus Journal, Hereford World, Limousine Journal*); (4) *Livestock and Poultry Situation and Outlook* (USDA); (5) Cattle-Fax and Beef Business Bulletin; (6) AI price lists.

The language of a futures and options market contract specifies the number and kind of cattle and the month of delivery. Price is the only variable that is determined when a contract is bought or sold.

Trading of futures and options contracts is regulated by the exchange trading the contract and by the Commodity Futures Trading Commission—the government agency overseeing all futures activities. The Chicago Mercantile Exchange (CME) involves both fed cattle and feeder cattle contracts. The CME notes the following advantages of the futures and options markets in helping cattle producers manage market risks: the ability to (1) lock in profits, (2) enhance business planning, and (3) facilitate financing. Self-study guides, videos, and computer programs are available from the CME for those interested in a more detailed study of the futures and options markets (see address in the Appendix).

Oct. 19, 2001
Issue 42
Vol. XXXIII

Panhandle supplies peak

Smaller November and December inventories to provide relief

Fed cattle supplies in the Panhandle region will peak during October. This region has had the most difficult time moving cattle consistently and has the largest front-end supply problem in our data. The beginning inventory for the Panhandle increased contra-seasonally from September to October and reached its largest total for the year. Normally the Panhandle monthly beginning inventory peaks in March and July. But large feeder cattle placements during May and June compounded with lighter weight placements earlier in the year created the increase in available fed cattle numbers that are expected during the fourth quarter.

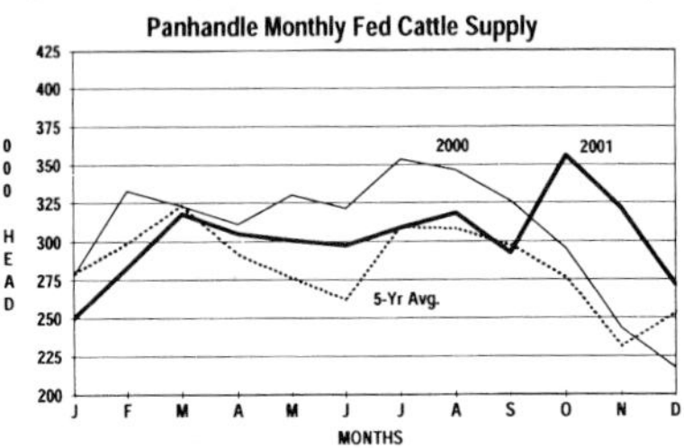

Smaller placed against totals for November and December will provide some supply relief in Texas. The placed against total for November in Cattle-Fax member feedlots is 82,000 head smaller than the placed against total for October. The placed against total for December is another 16,000 head smaller. Net placed against supplies from October to the beginning of December decline nearly 100,000 head, which will reduce the December monthly beginning inventory by more than 20% compared to October.

Panhandle shipments will likely increase during early November as feedlots work to cleanup front-end supplies. Several more weeks of large trade volumes and adequate slaughter totals will be needed to get this done and prevent the carryover from covering up the potential supply relief that the smaller November and December placed against totals can provide. — *Dave Weaber*

Currentness

Very current

Current

Caution

Uncurrent

Steer carcass weights were 826 pounds — up 13 pounds from a year ago.

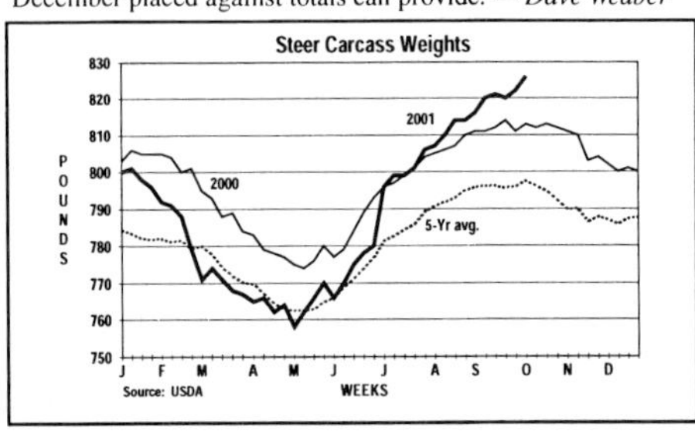

This week in brief

Market highlights

Fed cattle prices slipped $1-2/cwt. this week with a live trade of mostly $66/cwt. and dressed values primarily at $106/cwt. Volume in the South was good while Northern trade was moderate at best. Cutout levels declined, although not as much as expected. Through Thursday, Choice product fell by $.73 and $.76 to $118.27/cwt. and $115.65/cwt. As for Select, the Light product was down $1.16 to $106.29/cwt. and the Heavy appreciated by $.20 to close at $106.81/cwt. Feeder cattle prices were steady to $1/cwt. lower with calf values dropping $2-4/cwt. Commercial cows and bulls were steady to $1/cwt. lower.

Basis to continue to narrow

The nervousness in the market and continued unsettling news will effectively eliminate some of the large premiums in the futures markets during the next several weeks and months. The weak basis (futures well above cash) has contributed to the poor marketing rate and record heavy carcass weights that have burdened the market for the last 60-90 days. Periods of uncertainty have lead to a stronger basis where futures markets don't carry premiums as large as may have been established earlier.

Record hay production forecast

The USDA has projected total hay production in 2001 to be 162.303 million tons, up seven percent from a year ago and at record levels. Production increased in all four regions. The production of alfalfa and alfalfa mixtures is forecast at 81.6 million tons, up two percent from both the August estimate and last year. Other hay production is forecast at a record high 80.7 million tons 12 percent above last year's production.

Australian prices fall on BSE news

Australian cattle prices fell to a 13-week low as Japanese consumers who bought $1.5 billion of Australian beef last year turned away from beef products regardless of origin because of recent BSE concerns in Japan. The benchmark Eastern States Young Cattle Price fell 6.4 percent to $0.77 per pound on Friday. This was 11 percent below a record-high $0.88 per pound on September 20, two days before Japan's first mad-cow case was confirmed.

FIGURE 9.29 Cattle-Fax weekly update.
Source: Cattle-Fax.

TABLE 9.19 Methods of Managing Price Risks

Method	Comments
Good management practices	Obtain a combination of low costs and high productivity, thus lower breakeven prices. Producers with lower breakeven prices can be profitable even when market prices are relatively low.
Continuous buying and selling during the year	Producer obtains close to average price for the year. Avoids selling in just the low markets.
Buying and selling on the same market	Producer has a better knowledge of marketing costs and returns. There are some unknowns in using several markets not previously utilized.
Distributing cattle geographically	Market prices vary in different regions of the United States. The high and low markets average out. This also reduces weather risks.
Partnerships	Spreads the risk over two or more production segments of the beef industry, yet allows producers to share potential profits.
Retained ownership	Producer can take advantage of genetic superiority for added productivity beyond usual point of sale.
Contractual arrangements	Provides a market price at a future date so that net return can be projected prior to delivery of the cattle
Combination of several of the above	

Source: Adapted from Cattle-Fax.

Those people who use the CME fall into two general groups—hedgers and speculators. Hedgers are typically beef producers who produce, feed, and market cattle and who seek to transfer the price risk. Speculators may or may not own or produce cattle, but they buy or sell contracts in the hope of making a profit on price changes. They plan to use their knowledge to make profits. Commodity prices can be very volatile, so the potential for profit and loss is greater in commodity trading than in many other kinds of investments. Contracts traded on the CME are shown in Table 9.20. The specifications for such contracts change periodically, or a broker should be consulted.

What Is a Futures Contract?

A futures contract is a standardized agreement to buy or sell a commodity at a predetermined date in the future. The contract specifies the following information:

Commodity—live cattle or feeder cattle.
Quantity of the commodity (pounds of livestock as well as the range or weight for individual animals).
Quality of the commodity (specific U.S. grades).
Delivery point (location at which to deliver the commodity, or a *cash settlement* in the case of feeder cattle).
Delivery date (within a month of the contract termination date).

A futures contract does not specify the price at which the commodity will be bought or sold. Rather, the price is determined on the floor of the exchange as floor brokers execute buy and sell orders from all over the country. The prices they bid reflect the supply and demand for the commodity as well as expectations of whether the price will increase or decrease.

TABLE 9.20 Specifications for Live and Feeder Cattle Contracts

	Contract Type		
Specification	**Live (Fed) Cattle**	**Feeder Cattle**	**Stocker Cattle**
Size	40,000 lbs of 55% Choice, 45% Select, USDA grade live steers	50,000 lbs of 700–899 lbs medium and large frame #1	25,000 lbs of 500–599 lbs medium and large frame #1 feeder steers
Contract months	Feb., Apr., Jun., Aug., Oct., Dec.	Jan., Mar., Apr., May, Aug., Sep., Oct., Nov.	Jan., Feb., Mar., Oct., Nov., and Dec.
Delivery points or settlement	Stockyard delivery: Sioux City (IA), Omaha, Norfolk, Ogallala, North Platte (NE); Dodge City, Pratt (KS); Amarillo (TX); Guymon (OK); and Clovis (NM). Packing plant delivery: Any plant on the approved list or to an approved plant within 150 miles of the originating feedlot	Cash settled to the CME with a composite weighted average price for feeder steers	Cash settled

What Is an Option?

An *option* is a choice—that is, the right but *not* the obligation to buy or sell a futures contract at a specific price on or before a certain expiration date. There are two different types of options: *puts* and *calls.* Each type offers opposite pricing alternatives and the opportunity to take advantage of futures price moves without actually having a futures position.

Hedging

Hedging is a risk-management tool that permits producers to establish a buying or selling price for their livestock months before they are ready for sale or purchase. In hedging, a producer is taking a position opposite the cash position or what the producer owns. The cattle feeder who owns cattle, for example, sells a futures contract at a favorable price to cover the animals on feed. Packers wishing to cover future purchases buy futures contracts, as do cattle feeders planning to buy yearlings. The contract runs to some specified maturity date, usually months in advance. At that point, the producer can either (1) buy back the futures contracts to offset the contracts originally sold or (2) deliver the cattle to one of the specified delivery points (Table 9.20).

Hedging guarantees the selling price but it does not necessarily guarantee a profit. Profitability depends on placing the hedge at a price that exceeds all costs. It is extremely important for a person to know costs before becoming involved in hedging. The two major types of costs are (1) production costs and (2) basis. The latter is the difference between the futures market price and the producer's local cash price. The spread between these prices includes transportation, shrink, and other marketing costs as well as differences in supply and demand. Producers should know their basis and watch future quotations daily even if they do not have a hedged position, because this tells them what they could lock in through the use of a hedge.

There are times when futures are favorable for hedging and other times when they are not. The management rule is this: If producers cannot figure a satisfactory profit, they should not hedge but the producer should take a chance on the cash market instead.

The decision to hedge depends on the market situation and the degree of risk the producer is willing to assume. Each producer has a different degree of risk-carrying ability. Some cattle producers do not like hedging because they would rather gamble for higher profits (but they are also gambling for higher losses).

Choosing the day to place a hedge is an extremely difficult management decision. The use of a professional broker who can provide advice and account service is typically advisable. Before placing the hedge, a producer needs to consider all the market facts on supply and demand and how prices might be affected. After all the facts have been examined, the producer will come to one of two possible conclusions:

1. Higher prices are coming, so the producer will wait to place the hedge at a higher price.
2. Lower prices are likely, so the producer will go ahead and "lock in" a price on the futures market.

VALUE-BASED MARKETING

Value-based marketing has been debated in the beef industry for years. Value-based marketing has been defined as the process of adding value to a product. Many value differences among animals are bypassed in the beef industry that pays average prices for a commodity product. Thus, value-based marketing is unlikely to be used until the beef industry becomes more integrated or until beef becomes less of a commodity and more brand-name products. Although some value-based marketing does occur, it is not significant in terms of volume or value of the industry as a whole. Retained ownership of calves is a type of value-based marketing for commercial cow-calf producers. It is discussed in Chapter 5.

The 2000 Beef Quality Audit estimated that the beef industry lost approximately $100 for every market steer or heifer in the United States (Table 9.21). The study determined that the goals to be achieved by 2005 to improve overall quality of fed cattle were (1) eliminate USDA standards; (2) eliminate Yield Grades 4 and 5; (3) eliminate injection-site blemishes from whole-muscle cuts including the chuck; (4) eliminate side-branded hides; (5) reduce horns to less than 5% of the fed cattle supply; (6) develop and implement a standardized electronic animal identification system; (7) develop an information system that allows each producer to conduct a quality audit for his/her own herd; (8) assure that 100% of seedstock animals are accompanied by meaningful genetic data; (9) assure that 100% of cattlemen complete BQA training; (10) eliminate major and critical bruises that result in devaluation of subprimals; (11) improve the transportation of cattle; and (12) improve continually the eating quality of beef.

TABLE 9.21 Benchmark Value—Losses for Quality Challenges Identified in the NBQA—2000

	$ Loss/Fed Steer and Heifer
Excess external and seam fat	42.80
Inappropriate muscling	8.16
Palatability—marbling	23.35
Hardbones, bullocks, carcass pathology, offal condemnations, blood splash, calloused ribeyes, yellow fat	7.86
Injection-site lesions	3.59
Bruises	0.75
Inappropriate weight carcasses	6.46
Hide damage—branding	1.70
Total	100.00

Source: NCBA (2001).

A strategy workshop was held to determine the best strategies and tactics to achieve these goals. These strategies were (1) assist producers with use of selection and management techniques to produce cattle that fit customer expectations for marbling, red meat yield, weight, and other value-determining attributes; (2) assist producers with the process of collecting and analyzing data and sharing and utilizing information; (3) enhance an already commendable record in regards to the production of safe, nutritious, and wholesome beef; (4) assure delivery of predictable and uniform lots of cattle by more correctly managing implants, nutrition, horns, castration, sorting, and health programs while refining selection strategies to meet specific market windows; (5) assure that the needs of case-ready product marketing efforts can be met by improving the yield, consistency, and palatability of beef; (6) implement new production technologies only after carefully considering the consumer demand-perception, economic, environment, and animal welfare consequences; (7) encourage continued use of cattle-marketing systems that identify, categorize, and assign price-to-product attributes that affect consumer satisfaction by appropriately rewarding and discounting performance; (8) identify breeding, management, and sorting systems that optimize production, palatability, cutability, and profitability; and (9) encourage postharvest product enhancement technologies to assure the delivery of suitably tender and flavorful products to consumers while simultaneously managing the preharvest production process to achieve the same objectives.

The tactics were (1) develop and implement a voluntary, industry-driven, standardized electronic individual animal identification system that is tied to a seamless system of transmitting information up and down the production, processing, and distribution chain; (2) merchandize and purchase only those seedstock that are accompanied by objective performance information relative to economically important traits (production and end-product); (3) eliminate side brands; (4) eliminate horns via selection or early dehorning; (5) castrate early; (6) match implant strategies to cattle types to optimize product quality with economic returns; (7) develop management/production practices to reduce variation in weight and cut sizes within a lot; (8) utilize health management and nutrition protocols that contribute to improved quality attributes; (9) match a vast majority of the fed cattle to carcass weight targets of 650–850 lbs; (10) handle and transport cattle in a safe and humane manner; (11) train 100% of beef and dairy producers, veterinarians, transport providers, and others with an impact on cattle in beef quality assurance principles and procedures as well as humane handling practices; (12) move all injections to the neck region and eliminate intramuscular injections; (13) reduce immediately those genetic and management practices that contribute to production of USDA standards, Yield Grade 4s and 5s, dark cutters and nonconforming carcass weights and cut sizes; (14) change the Quality Grade and Yield Grade mix to 6% Prime, 27% Upper Two-Thirds Choice, 32% Low Choice, and 35% Select, and to 15% Yield Grade 1, 26% Yield Grade 2A, 27% Yield Grade 2B, 24% Yield Grade 3A, and 8% Yield Grade 3B; (15) participate in partnerships and coordinated market chains to foster communications and the delivery of products that meet consumer demands; and (16) continue to support and encourage development of branded beef product concepts and value-added, further processed beef items.

Historically, one of the major bottlenecks to value-based marketing was that boxed beef subprimals could have up to 1 in. of external fat and still meet boxed beef specifications. Today, approximately 75% of boxed beef is sold as closely trimmed (≤ 1/8 in.). The challenge now is to find a means to communicate these value differences between high-yielding versus average or subpar yielding cattle all the way through the production system.

Research at Oklahoma State University evaluated value differences between close-trim and commodity-trim boxed beef. They found that a 750-lb, Choice, Yield Grade 2 carcass fabricated into closely trimmed, boxed beef is worth $46.47 more than the same carcass fabricated into boxed beef with commodity-trim (Table 9.22).

TABLE 9.22 Fed Cattle Prices Based on Commodity Versus Closely Trimmed Boxed Beef Yields

	Commodity Trim		Close Trim	
Yield Grade	USDA Choice	USDA Select	USDA Choice	USDA Select
1	$77.22	$74.25	$82.70	$79.35
2	74.04	71.21	77.99	74.86
3	71.38	68.39	73.93	70.98
4	69.70	66.76	70.95	68.15

Source: Dolezal (1997).

The most clean-cut approaches to passing information, risk, and profits throughout the chain have come via the form of alliance structures. (Students are advised to review Chapter 8.) One example is Ranchers' Renaissance, a Colorado-based, consumer-focused, vertically coordinated beef cattle production system that involves seedstock, cow-calf, stocker, feeder, packer, and retail partners.

The group of people who comprise Ranchers' Renaissance have joined together to follow specific protocols to produce and process cattle that hit specifications, meet consumer wants, and capture efficiencies. Furthermore, they share information, costs, and benefits. The final product is a branded entity sold under the label of Cattlemen's Collection™.

Branded Beef Products

Beef has been marketed for decades as a commodity while other competing products have been sold under brand names and marketed by separate companies. When consumers purchase chicken, for example, they select from such brand names as Holly Farms, Frank Perdue, Tyson, and others. An established brand name creates consumer recognition and loyalty for that specific product name.

During the 1980s, branded beef products made their appearance primarily because consumer demand for beef appeared to be slipping and the beef industry was emphasizing the need for improvements in marketing. Several branded products were marketed to consumers who were looking for (1) "lite beef" with less fat and cholesterol (e.g., Laura's Lean), (2) organic or natural beef from cattle not fed antibiotics or growth stimulants (e.g., Coleman Natural Beef, Maverick Ranch Natural Lite), and (3) high palatability beef (e.g., Certified Angus Beef, Certified Hereford Beef). In 1988, when the major packers and retailers began trimming fat, many of the "lite beef" markets were lost. Thus, most name brands featuring less fat were abandoned or became less significant as their distinctiveness was lost compared with commodity beef. However, products such as Laura's Lean have carved out a sustainable market niche. Natural-product companies such as Coleman, Natural Beef, Maverick Ranch, and Bradley-3R Meats have also found sustained success. USDA has certified and verified the claims of 54 branded beef products as to a variety of live and carcass attributes (Table 9.23 and Fig. 9.30).

The branded product leader is Certified Angus Beef (CAB), which sold 590 million pounds in 2001, with 54% sold at retail and 29% via food service. The Certified Angus Beef program is based on a set of live animal and carcass specifications designed to assure palatability. Trends in acceptance percentage and total sales volume are outlined in Table 9.24.

TABLE 9.23 USDA Certified and Verified Beef Programs (N = 54)

Factors Used in Requirements	Programs Specifying Requirement (%)
Live requirements	
>50% black-hided	64
≥50% Red Angus	18
Belgian Blue	2
Piedmontese	2
Quality factors	
≥Slight abundant marbling	9
Moderate or higher marbling	33
Small or higher marbling	25
Slight or higher marbling	25
Yield factors	
Y.G. 4.9 or lower	13
Y.G. 3.9 or lower	27
Y.G. 2.9 or lower	4
Moderately thick or higher muscling	76
≥ 11.0 ribeye area	9
Carcass weight > 600 lbs	4
Carcass weight 600–900 lbs	5
Carcass weight 600–950 lbs	5
Carcass weight < 850 lbs	2
Other	
No dark cutters or blood splash	93
Hump height ≤ 2 in.	84
"Angus" in the brand name	65

Source: USDA (2001) (www.ams.usda.gov).

FIGURE 9.30 Branded beef products are increasingly popular and they have helped transform the beef industry into a customer focused business.

TABLE 9.24 Certified Angus Beef Percent Acceptance: Total Volume Certified and Pounds Sold

Year	Acceptance[a]	Sales (mil lbs)
1987	23.8	43.3
1989	19.5	71.0
1991	15.3	79.8
1993	17.9	123.4
1995	17.4	226.5
1997	17.2	331.8
1999	20.1	493.0
2001	18.5	593.0

[a]Percent of cattle that meet the live specifications also meeting the carcass specifications.
Source: Certified Angus Beef.

It is interesting to note that acceptance rates have been relatively stable over the past eight years, while the tonnage of sales has increased dramatically. The ability to effectively market the product while building a supply chain that consists of 65 licensed feedlots has given the CAB program tremendous momentum. Furthermore, CAB has implemented a genetic research and education effort with a goal to increase adoption of relevant technologies, to substantially enhance carcass data delivery and interpretation, and to provide industrywide communication and education about the value of Certified Angus Beef®. In the end, their goal is to achieve 80% Angus-based beef cattle population by 2005 and CAB acceptance rates of 30% by 2007.

Through these efforts, prices for Angus and Angus-cross feeder calves have risen relative to the total supply. Premiums are reported to range from $1.42/cwt to $5.58/cwt for Angus calves. Black-hided beef cattle are also likely to receive at least some premium regardless of breed.

The major obstacles to increasing market share for CAB are the increase in black Continental breed cattle, the implementation of more aggressive growth implant programs, and the lack of direct economic incentive to cow-calf producers.

The Value of Information

As the industry moves to a stronger consumer orientation, the need to know more about cattle in terms of their feedlot, health, and carcass performance increases significantly. The knowledge level of cattle feeders about the cattle they feed is surprisingly low (Table 9.25). There is also a tremendous amount of variation in the level of information known with a range of 0 to 100 and, thus, the high standard deviations. Only the vaccination schedule and whether cattle came from a single versus a commingled source were known for greater than 50% of the cattle being fed by the survey respondents.

However, feedlot managers were interested in knowing more about the cattle they feed as well as expressing a willingness to pay more for cattle with known information in some categories (Table 9.26). Feeders wanted more information ($\geq$ 90%) about yield grade, nutritional management, quality grade, feedlot gain, breed composition, vaccination schedule, implant history, morbidity and mortality history, and nonconformance history of the herd of origin. Two-thirds or more of the respondents expressed a willingness to pay higher prices for cattle with a known history in the following traits—vaccination schedule, quality grade, sire and associated performance data, feedlot gain, breed composition, yield grade, and implant history.

TABLE 9.25 Percent of Cattle Currently Being Fed Where Information Is Known Relative to Each Trait[a]

Trait	Information Is Known (%)	SD
Genetic origin		
Breed composition	49.3	34.9
Sire and associated performance data	31.1	34.7
Seedstock supplier	31.5	34.8
Health/management		
Vaccination schedule	55.5	35.7
Health products used (brands)	45.3	38.2
Implant history	48.7	35.5
Age of castration	31.8	41.3
Single versus multiple herd	50.8	39.4
Weaning age	37.7	34.6
Nutritional management	46.0	32.5
Herd history—feedlot and carcass		
Feedlot gain	36.6	33.4
Morbidity/mortality	25.1	31.2
Cost of gain	29.9	31.8
Quality grade	29.5	31.1
Yield grade	30.1	31.5
Dressing percent	29.7	33.7
Outs	28.8	34.3

[a]Survey of 31 licensed Certified Angus Beef™ feedlot managers.
Source: Behrends, Field, and Conway (2001).

TABLE 9.26 Percent of Feed Yard Managers Who Would Like to Have Information About Various Traits Relative to the Cattle They Feed and Their Willingness to Pay for the Information

	Desire Information			Willing to Pay		
Trait	Yes (%)	No (%)	Maybe (%)	Yes (%)	No (%)	Maybe (%)
Genetic						
Breed composition	93.5	6.5	0.0	72.4	20.7	6.9
Sire and associated performance data	87.1	12.9	0.0	79.3	17.2	3.5
Seedstock supplier	83.9	16.1	0.0	34.5	62.0	3.5
Health/management						
Vaccination schedule	93.5	6.5	0.0	83.3	13.4	3.3
Health products used (brands)	90.3	9.7	0.0	40.0	56.7	3.3
Implant history	93.5	6.5	0.0	53.4	43.3	3.3
Age of castration[1]	66.7	30.0	3.3	17.8	79.3	6.9
Single versus multiple herd	80.7	19.3	0.0	36.7	56.7	6.6
Weaning age	74.2	22.6	3.2	23.3	70.0	6.7
Nutritional management	96.8	3.2	0.0	53.3	46.7	0.0
Herd history—feedlot and carcass						
Feedlot gain	96.8	3.2	0.0	76.7	20.0	3.3
Morbidity/mortality	90.3	9.7	0.0	66.7	30.0	3.3
Cost of gain	74.2	25.8	0.0	40.0	53.3	6.7
Quality grade	96.8	3.2	0.0	80.0	13.3	6.7
Yield grade	100.0	0.0	0.0	70.0	23.3	6.7
Dressing percent[1]	83.3	16.7	0.0	44.8	51.7	3.5
Outs[1]	90.0	10.0	0.0	44.4	51.9	3.7

[1]Only 30 of 31 participants responded to these questions.
Source: Behrends, Field, and Conway (2001).

As marketed specifications become increasingly detailed, the value of information is highly likely to increase. The ability to capture, synthesize, and share information about genetic, health, and management history should be a priority for seedstock and cow-calf producers. Those with information are going to have more power in the marketplace than those without.

ADVERTISING AND PROMOTING RETAIL BEEF

Many types of consumer products are advertised and promoted to remain competitive with other consumer products. Prior to 1980, however, beef received increased consumer acceptance with minimal advertising or promotion. During the 1980s and 1990s, though, a different consumer environment emerged. Beef producers were challenged by an environmentalist movement, a consumer-oriented government, and the animal welfare movement, which advocated vegetarianism and associated health problems with beef consumption. The result was a poor consumer image for beef, and the beef industry identified ways to improve this image.

In 1986, the beef industry became the last major agricultural commodity group to initiate a checkoff program to fund a marketing program. The Cattlemen's Beef Promotion and Research Board (Beef Board) was created to manage the checkoff dollars originating from the $1 per head collected each time an animal was sold.

Checkoff dollars are used for promotion in the form of consumer advertising via national television, print, radio, and billboard campaigns, retail point-of-purchase materials, recipe labels and related materials, food service campaigns, and sponsorship of the national Beef Cook-Off. Checkoff monies also fund educational programs for children, teachers, nutritional and medical professionals, and consumer editors.

Furthermore, the checkoff provides funds for a significant research program, new product development and marketing, and efforts targeted at international markets. The 2000 budget for beef checkoff activities is outlined in Table 9.27.

Beef producers are highly dependent on consumers, who consume beef at relatively high levels as long as disposable income is high and the price is competitive with other meats. Producers must realize that their financial support of education, research, promotion, and

TABLE 9.27 2000 Cattlemen's Beef Board Budget

Activity	$
Revenues	
Assessments	48,068,845
Interest	1,607,962
Other	24,465
Total	49,701,272
Expenses	
Promotion	29,028,696
Research	4,284,065
Consumer information	6,809,609
Industry information	1,948,653
Foreign marketing	4,801,628
Producer communication	1,854,930
Program evaluation	136,411
Program development	84,744
Total	48,948,736

advertising programs is essential if beef demand is to be sustained or increased. This support appears to be crucial in maintaining a desirable consumer image of beef. Consumer attitudes drive the beef business. The beef industry must advertise and promote beef if it wants consumers to have a positive image about the beef industry and its products.

MANAGEMENT SYSTEMS HIGHLIGHTS

1. Understanding the specifications of a particular market target enables producers to better meet the needs of customers and thus sustain or increase market share.
2. Selecting the best marketing channel requires an in-depth understanding of production costs, costs associated with various marketing alternatives, and the degree of risk associated with each option.
3. Retained ownership beyond the cow-calf enterprise is most profitable when producers have knowledge about the feedlot and carcass performance of their cattle.
4. Producers should study price and inventory cycles, the effects of seasonability, and other factors affecting price to assure the proper timing of purchases and sales.
5. Participating in a supply chain structure designed to meet specific market targets is a potential method to capture profitability.

REFERENCES

Publications

Assessment of the Cattle and Hog Industries: Calendar Year 2000. 2001. Grain Inspection, Packers and Stockyard Administration. Washington, DC: USDA.

Barnes, K., Smith, S., and Lalman, D. 2000. *Managing Shrink and Weighing Conditions in Beef Cattle.* Fact sheet F-327. Stillwater, OK: Oklahoma State University Extension Service.

Behrends, L., Field, T. G., and Conway, K. 2001. *The Value of Information as Perceived by Feedlot Managers.* Animal Sciences Research Report. Fort Collins, CO: Colorado State University.

Cattle Industry Reference Guide. 2001. Englewood, CO: Cattle-Fax.

Dolezal, H. G. 1997. *Grid Pricing—The Known and Unknown.* Oklahoma State University. www.ansci.okstate.edu.meats/grid.

Field, T. G., Tatum, J. D., and Kimsey, K. M. 1998. *Grid-Pricing: A Workbook.* Fort Collins, CO: Colorado State University; Englewood, CO: NCBA.

Hedrick, H. B., Aberle, E. D., Forrest, J. C., and Judge, M. D. 1994. *Principles of Meat Science.* San Francisco: W. H. Freeman.

Improving the Consistency and Competitiveness of Market Cow and Bull Beef and Increasing the Value of Market Cows and Bulls. 1999. Fort Collins, CO: Colorado State University; Englewood, CO: NCBA.

Improving the Quality, Consistency, Competitiveness, and Market Share of Beef. 2000. NCBA, Colorado State University, Texas A&M University, Oklahoma State University, and West Texas State University.

Livestock Market News. Washington, DC: USDA.

Marousek, G. E., Stodick, L. D., Carlson, P., and Gibson, C. C. 1994. Economics of Value-Adding Rangeland Beef Cattle Enterprises. *Rangelands* 16: 1.

Murphey, C. E., Hallett, D. K., Tyler, W. E., and Pierce, J. C. 1960. Estimating Yields of Retail Cuts from Beef Carcasses. *J. Anim. Sci.* 19: 1240 (Abstr.).

Myers, J. 1999. *The Buckingham Cattlemen's Association Program.* Proceedings of the BIF Annual Research Symposium, Roanoke, VA.

National Animal Health Monitoring System. 1997. *Part 1: Reference of 1997 Beef Cow-Calf Management Practices.* USDA–Veterinary Services.

Outlook and Strategies: 2001 and Beyond. Englewood, CO: Cattle-Fax.

Packers and Stockyards Statistical Report. 2000. Washington, DC: USDA.

Petritz, D. C., Erickson, S. P., and Armstrong, J. H. 1982. *The Cattle and Beef Industry in the United States: Buying, Selling, Pricing.* Lafayette, IN: Purdue University Cooperative Extension Service CES Paper 93.

Retained Ownership. 1999. Englewood, CO: Cattle-Fax.

Roeber, D. L., Mies, P. D., Smith, C. D., Belk, K. E., Field, T. G., Tatum, J. D., Scanga, J. A., and Smith, G. C. 2001. National Market Cow and Bull Quality Audit—1999: A Survey of Producer-Related Defects in Market Cows and Bulls. *J. Anim. Sci.* 79: 658–665.

Schieflebein, D. 1997. *The Basics of Grid Pricing.* American Gelbvieh Association.

Self-Study Guide to Forward Pricing with Livestock Options. 1990. Chicago: Chicago Mercantile Exchange.

Self-Study Guide to Hedging with Livestock Futures. 1990. Chicago: Chicago Mercantile Exchange.

Steiner, R., Wyle, A. M., Vote, D. J., Cannell, R. C., Belk, K. E., Scanga, J. A., Wise, J. W., O'Connor, M. E., Tatum, J. D., and Smith, G. C. 2001. *Video Image Analysis Determination of Percentage Subprimal Yield of Beef Carcasses.* Animal Science Research Report. Fort Collins, CO: Colorado State University.

Wulf, D. M. 1999. *Techniques to Identify Palatable Beef Carcasses: MARC Tenderness Classification, SDSU Colorimeter and Near-Infrared Spectrophotometry Systems.* Proceedings of the Range Beef Cow Symposium XVI, Greeley, CO.

Visuals

"Beef Grading: Quality" (27 min.) and "Beef Grading: Yield" (31 min.). CEV, P.O. Box 65265, Lubbock, TX 79464-5265. www.CEV-INC.com.

"Gambling on the Grid" (1998; 20 min.). Colorado State University, Ft. Collins, CO and NCBA, 5420 South Quebec Street, Greenwood Village, CO 80111.

"Improve Cattle Handling for Greater Profits" (slide/cassette version, new edition). Livestock Conservation Institute, 1100 Jourie Blvd., Suite 143, Oak Brook, IL 60521.

"Livestock Grading and Slaughter Cattle" (43 min.); "Livestock Grading and Feeder Cattle (25 min.), and "Slaughter Cow Evaluation" (26 min.). CEV, P.O. Box 65265, Lubbock, TX 79464-5265. www.CEV-INC.com.

"Options on Live Cattle" (22 min.) and "Livestock Producers Talk Futures and Options" (20 min.). Chicago Mercantile Exchange, 30 S. Wacker Dr., Chicago, IL 60606.

Computer Programs

Computer-Assisted Retailer Decision Support (CARDS) Systems (1991). Texas A&M University, 432 Kleberg Center, College Station, TX 77843.

Options and Alternatives (PC-compatible). Chicago Mercantile Exchange, 30 S. Wacker Dr., Chicago, IL 60606.

Retained Ownership (1999). Software for Cattlemen. Cattle-Fax, P.O. Box 3947, Englewood, CO 80155.

Websites

Agricultural Marketing Service: www.ams.usda.gov
California's Western Video Market, Cottonwood, CA: www.wvmcattle.com
Cattle-Fax: www.cattle-fax.com
CattleinfoNet, Sebastian, FL: www.cattleinfonet.com
CattleSale.com, Boise, ID: www.cattlesale.com
Chicago Mercantile Exchange: www.CME.com
Drovers Journal: www.agcenter.com
Livestock Marketing Information Center: www.Lmic1.co.nres.usda.gov
Producers' Video Auction, Fort Worth, TX: www.producersvideoauction.com
Superior Livestock Auction, Brush, CO and Fort Worth, TX: www.superiorlivestock.com

The Global Beef Industry

Cattle, including domestic water buffalo, contribute food, fiber, fuel, and power to many of the nearly 6 billion people throughout the world. Beef producers in the United States are significantly influenced by such global events as international trade, drought, hunger, population changes, political pressures, business opportunities, new sources of breeding stock, and disease problems. It is important for U.S. beef producers to understand the global beef industry, not only because they are affected by it but also because more of them will be involved in future international beef activities.

NUMBERS, PRODUCTION, CONSUMPTION, AND PRICES

World Cattle Numbers

There are approximately 1.35 billion cattle in the world. The rate of increase in world cattle during recent years has been approximately 1–2% per year. As indicated in Table 10.1, the leading countries in cattle numbers are India (219 mil), Brazil (167 mil), China (105 mil), the United States (98 mil), and Argentina (55 mil).

World Meat Production

Although cattle numbers are abundant in India, the available beef meat (3 lbs) per capita is very low because cattle are sacred to Hindus and, as such, they do not consume beef (Table 10.1). Cattle and buffalo are used heavily for draft purposes and meat from these animals is considered a by-product. Buffalo supply 75–90% of the agricultural power in several developing countries. This power indirectly provides food—such as rice and other cereal grains—for these heavily populated areas of the world.

TABLE 10.1 World Cattle Numbers, Meat Production, and Beef Production—2000

			Meat Production (carcass weight)			
Country[a]	Human Population (mil)	Cattle[a] (mil hd)	Total (mil lbs)	Beef/Veal (mil lbs)	Beef as % of Total	Beef/Veal Consumption (lb Carcass Weight per Capita)
Argentina	36.6	55.0	9,233	6,392	69	152
Australia	18.7	26.7	8,045	4,381	54	79
Bangladesh	126.9	23.6	934	374	40	3
Brazil	167.9	167.5	31,704	14,237	45	77
Canada	30.8	12.8	8,888	2,777	31	71
China	1,274.0	104.6	141,776	11,071	8	9
Colombia	41.6	26.0	3,198	1,662	52	35
Ethiopia	61.1	35.0	1,399	628	45	10
France	58.9	20.5	14,017	3,504	25	56
Germany	82.2	14.6	13,702	3,004	22	31
India	998.1	218.8	10,639	3,178	30	3
Indonesia	209.2	12.1	4,262	780	18	4
Kenya	29.5	13.8	899	562	62	18
Madagascar	15.5	10.4	595	326	55	20
Mexico	97.4	30.3	9,920	3,118	31	51
Myanmar	45.1	10.9	983	225	23	5
Nigeria	108.9	19.8	1,968	657	33	5
Pakistan	152.3	22.0	3,861	787	20	5
Russian Fed.	147.2	27.5	9,475	4,686	49	10
South Africa	39.9	13.7	2,938	1,300	44	31
Sudan	28.9	37.1	1,479	652	44	13
Tanzania	32.8	14.4	718	494	69	14
Turkey	65.5	11.1	3,055	771	25	21
Ukraine	50.6	10.6	3,790	1,770	47	27
United Kingdom	58.9	11.1	7,729	1,560	20	46
United States	276.2	98.1	82,950	27,133	33	100
Uruguay	3.3	10.8	1,300	998	77	135
Venezuela	23.7	15.8	2,199	793	36	33
World Total	5,978.4	1,350.0	513,077	126,002	24	

[a] Countries listed have at least 10 million head of cattle.
Source: FAO, FAS-USDA.

World Beef Consumption

Per capita beef supply and consumption are highest in Argentina, Uruguay, the United States, Australia, Brazil, and Canada (Table 10.1). Figure 10.1 identifies leading countries in per capita beef consumption and other countries whose consumption is highly dependent on imports. Cattle numbers in Australia and Argentina exceed their human population, so these countries are large exporters of beef even though their per capita beef consumption is high. Meat and beef production are relatively high in China, however, per capita consumption is low because of the extremely large population in that country.

World Beef Prices

Table 10.2 shows that supply, demand, and general economic conditions greatly influence beef prices in different parts of the world. While the data in Table 10.2 are from 1994, the relationships are valid. Of the cities shown, Brazilia has the lowest beef prices, while Tokyo has

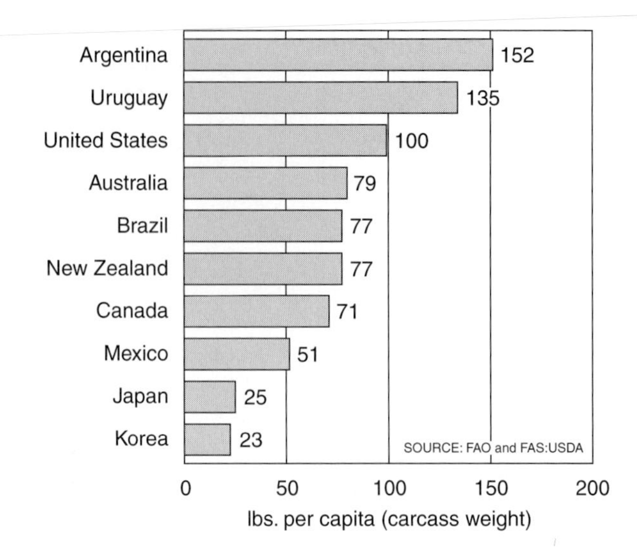

FIGURE 10.1 Per capita beef consumption in selected countries. *Source:* FAO and FAS-USDA.

TABLE 10.2 Time (Hour:Min) and U.S. Dollars Required to Purchase One Pound of Specific Food Items

	Boneless Sirloin Steak		Boneless Pork Roast		Whole Broiler		Total Food Basket[a]	
City	Time	$/lb	Time	$/lb	Time	$/lb	Time	$/15 items
Bern	1:00	15.01	0:21	5.23	0:08	1.19	4:03	45.94
Bonn	0:27	5.39	0:14	2.73	0:05	1.05	2:42	27.31
Brazilia	1:04	1.21	1:16	1.43	0:29	0.55	2:37	9.22
Buenos Aires	0:50	2.27	1:23	3.75	0:24	1.10	11:14	18.78
Canberra	0:22	3.03	0:10	1.40	0:07	1.03	3:07	15.86
London	0:37	4.91	0:25	3.23	0:08	1.08	3:52	20.57
Madrid	0:52	5.63	0:36	3.94	0:13	1.40	5:26	25.65
Mexico City	0:53	2.40	0:55	2.64	0:17	0.84	7:19	14.65
Ottawa	0:19	3.19	0:15	2.49	0:05	0.89	2:31	18.71
Paris	0:51	6.37	0:30	3.79	0:23	2.86	5:20	27.46
Pretoria	0:43	2.22	0:26	1.33	0:17	0.87	7:53	15.20
Rome	0:42	5.30	0:24	3.06	0:14	1.76	4:42	24.50
Seoul	2:11	9.90	0:25	1.87	0:13	1.00	8:56	29.74
Stockholm	0:46	8.44	0:48	8.75	0:18	3.29	4:46	38.39
Tokyo	1:14	21.14	0:21	6.00	0:09	2.57	4:57	59.21
Washington DC[b]	0:21	3.39	0:24	3.84	0:04	0.63	2:35	17.64

[a]Total of 15 specific food items.
[b]About $1.00 higher than other major U.S. cities.
Source: USDA *Food Review,* (1994).

the highest. Consumers in Seoul would spend the highest percent of their weekly labor to purchase l lb of beef. Notice that in all cities, consumers spend considerably less per pound for chicken than they do for beef. Only in isolated cases is beef cheaper than pork. By comparing dollars spent versus the labor time required to purchase 1 lb of meat products or the total food basket, the wage differences can be observed. For example, even though consumers in Bern and Tokyo spend the most for the total food basket, these consumers require less labor to make that purchase than do residents of Buenos Aires, Madrid, Mexico City, Paris, Pretoria, and Seoul.

TABLE 10.3 U.S. Exports and Imports of Major Animal Products—2000

Commodity	Export Value (mil $)	Import Value (mil $)
Live animals (excluding poultry)	608	1,555
Cattle and calves	237	1,000
Meat and meat products (excluding poultry)	5,451	3,275
Beef and veal	3,108	2,135
Pork	1,268	773
Edible offal, variety meat	745	87
Other	330	380
Poultry meat	1,744	40
Other products		
Tallow, grease, fat, lard	421	64
Sausage casings	66	65
Hides and skins	1,479	146
Wool and mohair	21	72
Dairy products	996	1,540
Eggs and egg products	172	20
Feathers and down	1,007	131
Bull semen	59	19
Total for animals and products	10,601	7,262
Total for agricultural products	48,440	37,714

Source: USDA-FAS.

TABLE 10.4 U.S. Beef and Veal Imports/Exports and Their Relationship to Domestic Production and Disappearance

	Beef and Veal Imports and Exports as a Percent of U.S. Beef Production				
Year	Beef/Veal Export (mil lbs)	Beef/Veal Imports (mil lbs)	Net Beef/Veal Imports	Exports as a % of Production	Imports as a % of Production
1965	54	942	888	0.27	4.56
1970	40	1,816	1,776	0.18	7.55
1975	53	1,782	1,729	0.21	6.69
1980	177	2,085	1,908	0.80	8.70
1985	332	2,091	1,759	1.37	8.02
1990	1,006	2,356	1,350	4.36	9.67
1994	1,611	2,369	758	6.53	9.32
1996	1,877	2,073	196	7.53	7.90
1998	2,171	2,643	472	8.34	9.95
2000	2,516	3,032	516	9.32	11.02

Source: Livestock Marketing Information Center (2001).

INTERNATIONAL TRADE

Table 10.3 indicates the importance of beef cattle and beef products in international trade. Beef and hides have the highest export value and beef has the highest import value when compared with other animal products. Table 10.4 describes the relationship of imports and exports to U.S. domestic beef production.

The United States has been a significant importer of beef, primarily for use as ground product, for the past half century. The volume of imported beef peaked at its highest level in 2000. However, because of steadily growing beef exports, the net beef import levels were relatively low. The dramatic growth of beef exports, primarily in the form of high-value cuts during the 1990s, has shifted the United States from being a significant net importer toward sustaining a position as close to being a net exporter.

Imports

The 3 billion lbs of beef and veal that are imported represent approximately 9% of total U.S. production (Fig. 10.2). Figure 10.3 illustrates the value of cattle, beef, and beef by-products imported into the United States. The United States imports beef to meet the deficit for ground and processed beef as well as to maintain trade relations with other countries. The major suppliers of beef to the United States are listed in Table 10.5. Australia is the highest volume beef supplier to the U.S. market. Canada ranks a close second as a U.S. supplier. This is of interest in terms of what part of the market is being targeted by each. Canada exports more whole-muscle retail cuts while Australia supplies mostly lower-valued manufacturing beef.

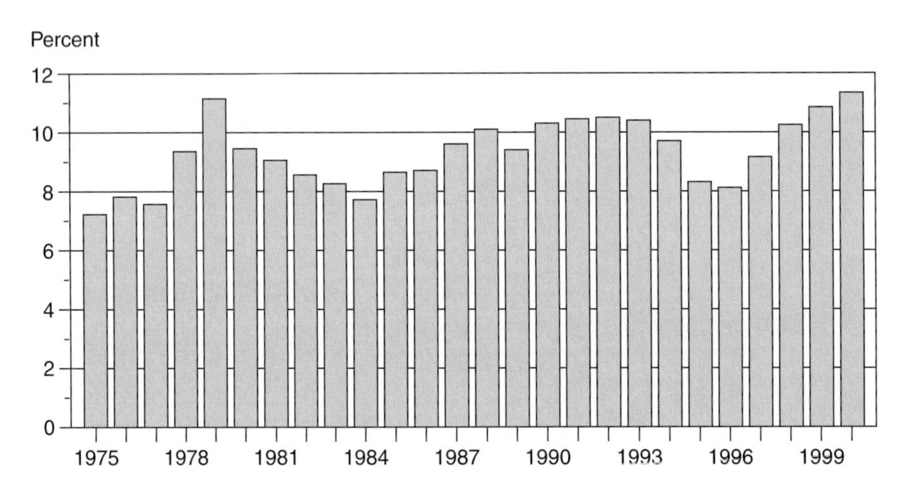

FIGURE 10.2 U.S. beef and veal imports as a percentage of production (carcass weight).
Source: Livestock Marketing Information Center.

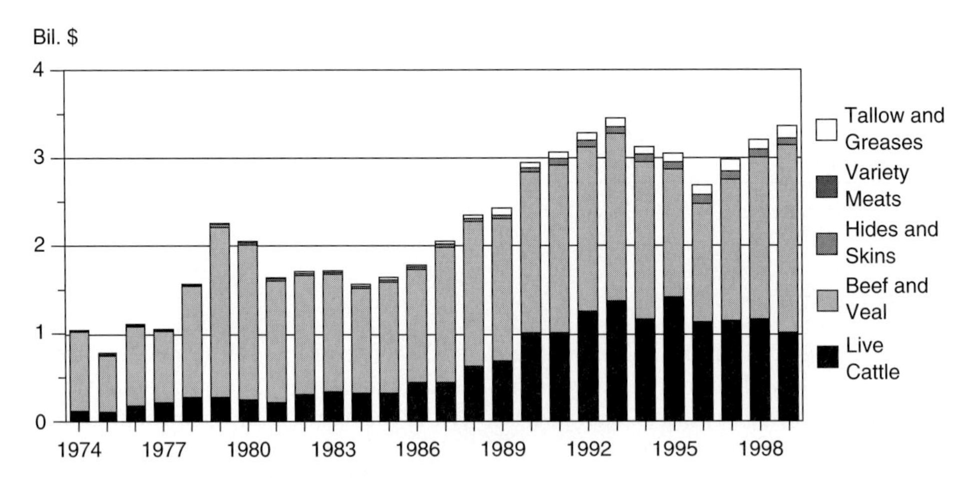

FIGURE 10.3 Values of U.S. imports of beef, hides, and by-products.
Source: Livestock Marketing Information Center.

TABLE 10.5 U.S. Beef Imports Leading Suppliers—2000

Country	Carcass Wt (1 mil lbs)	Live Cattle (1,000 head)	Exports to U.S. as % of Total Production	Exports to U.S. as % of Total Exports
Australia	754		17.2	28.2
Canada	739	965	26.6	70.3
New Zealand	470		34.2	48.2
Brazil	95		0.06	8.0
Argentina	81		1.2	10.1
Mexico	1	1,223	<.001	5.8
Total	2,248	2,188		

Source: USDA, Livestock Marketing Information Center.

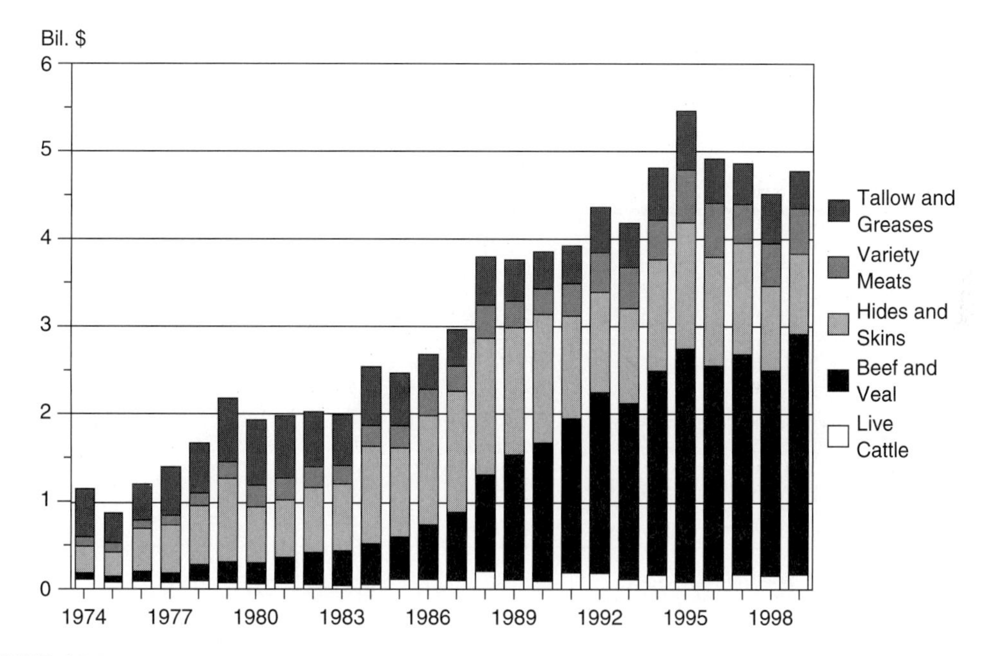

FIGURE 10.4 Values of U.S. exports of beef, cattle, hides, and by-products.
Source: Livestock Marketing Information Center.

Exports

The beef export market is important to the United States beef industry for several reasons. First, hides, variety meats, and fat have higher values in foreign markets than in the U. S. market. Second, there is great demand for U.S.-fed beef in countries where the standard of living is increasing. Third, U.S. exports have an influence on U.S. cattle prices.

Figure 10.4 shows the value of beef products in the export market. Hides and beef comprise the biggest percentage of export revenues. Beef exports have expanded rapidly since 1975 (Fig. 10.5). The primary reason for a marked increase in exports has been the promotional activities of the U.S. Meat Export Federation in several countries. Japan is the largest purchaser of U.S. beef (50%), although Canada, Mexico, and South Korea have been expanding as U.S. beef markets in recent years (Table 10.6).

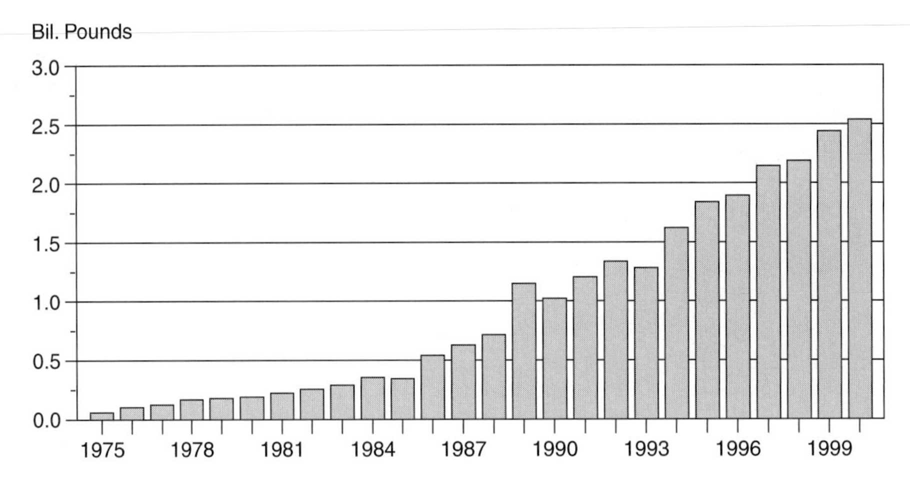

FIGURE 10.5 U.S. beef and veal exports (carcass weight).
Source: Livestock Marketing Information Center.

TABLE 10.6 Value of U.S. Beef Exports to Primary Customers

	Million $						
Year	Total Value	Japan	Canada	Mexico	Republic of Korea	Russian Federation	Hong Kong
1988	1,484	841	104	85	27	NA	13
1990	1,878	960	317	112	117	0.8	15
1992	2,512	1,487	365	260	217	0.5	24
1994	2,759	1,624	376	281	240	24	30
1996	3,046	1,927	331	198	254	60	64
1998	2,804	1,573	294	452	148	48	49
2000	3,567	1,802	313	586	507	15	52

Source: U.S. Meat Export Federation, USDA.

The export market is increasingly important to the U.S. cattle industry. The equivalent of 3.3 million head of fed steers and heifers were exported in 2000 and accounted for the vast majority of sales of beef variety meats, hides, and by-products.

Reaping economic resources from the export market is not without risk. The BSE and FMD outbreaks in the British and European cattle herds and the *E. coli* 0157:H7 problem in Japan in 1996 are examples of situations that can have significant impacts on the worldwide market. From October 2000 to March 2001, beef consumption in the 15 nations of the European Union had fallen 25% due to the combined effects of BSE and the foot and mouth disease outbreak. Furthermore, changing economic and political conditions in the world community can exert dramatic influence on export markets.

To counter these potential risks, United States exporters must continue to strive for perfection in assuring their clients that their beef products are safe and wholesome. Furthermore, customer service and product value must be exceptional to maintain a competitive market position.

TABLE 10.7 Beef Semen Export Sales[a]

Year	Units (1,000)	Value ($1,000)
1980	299	1,141
1985	149	652
1990	276	1,174
1995	495	1,808
1998	847	2,292
2000	740	2,187

[a] The vast majority of sales are to South America, Canada, Australia, and New Zealand.
Source: National Association of Animal Breeders.

For example, the United States banned the importation of British beef and cattle, increased monitoring of the domestic beef herd for BSE, and banned feeding of ruminant-derived bonemeal and protein to other ruminants as a result of the BSE problem in Britain. The United States has developed better hazard analysis critical control point protocols and food safety intervention techniques to reduce microbiological contamination of beef. Sustained efforts in these and similar areas should yield beneficial results in the world market. The International Beef Quality Audit results point to the need for the U.S. industry to develop a true export mentality, to work more aggressively on quality issues to maintain competitiveness, and to respond proactively to the innovations of competitors.

The sale of genetics from U.S. sires to worldwide customers was valued at nearly $2.2 million in 2000. South American sales accounted for more than half of the total (Table 10.7).

A discussion of major beef producing and consuming nations follows. Nations are categorized as beef suppliers (Argentina, Australia, Brazil, European Union, and New Zealand), NAFTA partners (Canada, Mexico, and the United States), and beef buyers (China, Japan, Korea, and the Russian Federation).

BEEF SUPPLIERS

Argentina

Population (2000)—36.6 mil
Number cattle per capita—1.5
Number cattle per sq. mi.—51
Carcass beef produced—6.39 bil lbs
Carcass beef exported—794 mil lbs

Percent population in agriculture—11%
Land area—1.07 mil sq. mi.
Cattle population—55.0 mil
Per capita beef consumption—152 lbs (carcass weight)
Exports as a percent of production—12.5%

Argentina's landscape can be divided into four major regions: (1) northern Argentina, which has a hot and humid climate; (2) the Pampa, a fertile plain that spreads west beyond Buenos Aires; (3) the Andine, which is the mountainous western region; and (4) the Patogonia, a dry windswept plateau in southern Argentina.

The Pampa and northern Argentina produce most of the beef and wheat that are the country's leading farm products. The vast majority of Argentine beef is grass fed. As such, poor per head production efficiencies coupled with a marginal rural marketing and processing infrastructure combine to restrict Argentina's competitiveness. Furthermore, tight credit supplies make it difficult to improve the highway, rail, and port facilities. Argentine beef exports are largely directed at Chile, Germany, and the United States.

Historically, foot and mouth disease (FMD) has limited potential export markets to neighboring South American trade partners or to canned-cooked products in FMD-free regions of the world. In 1997, the United States opened trade to fresh/frozen beef from FMD-free zones within Argentina. However, FMD outbreaks in 2000–2001 have closed the door to fresh/frozen beef from Argentina and Uruguay. The disease outbreak will cost Argentina and Uruguay about one-third of their export volume valued at between $50 to $100 million.

Argentina, if it achieves FMD-free status, has the potential to become a major player in world trade of manufacturing and ground beef. The current strategy is to use a national vaccination program to manage FMD. At a cost of $0.70 per head, the program will require an expenditure of $35 million. Argentina will be particularly competitive in lower-priced beef items with regards to trade in South Korea, Taiwan, Mexico, and Japan. The threat is greatest to market share losses for Australia and New Zealand.

Argentina joined with Brazil, Paraguay, and Uruguay in 1995 to form the South American trade union MERCOSUR. Brazil is Argentina's largest trading partner but MERCOSUR is yet in its infancy and potentially susceptible to political and economical instability.

To achieve its beef trade potential, Argentina must attain a number of economic and policy reforms. Monetary policy will be critical in determining how great a role Argentina will play in the export market.

Some Argentine beef producers have begun to invest in feedlots as a means to add value to their cattle, especially for the Japanese and U.S. markets. Others continue to focus on carving out market share by extolling the virtues of natural, grass-fed beef. Such diversification makes Argentina a potentially significant competitor. Already they have targeted organic markets in the European Union and premium programs in the United States.

However, Argentina must address the following to capture profits in international trade:

1. Overcome poor overall herd productivity, despite implementation of contemporary management practices in some herds, particularly those of large scale.
2. Continue to improve genetic potential of the herd for growth rate and carcass characteristics.
3. Overcome poor profitability at the cow-calf level.
4. Increase cash flow into the processing, packing, and transportation sectors.
5. Develop an export mentality along with the necessary beef promotional and marketing efforts required to open new markets.
6. Ensure that a high percentage of beef carcasses originates from animals young enough to classify as A maturity.

Australia

Population (2000)—18.7 mil	Percent population in agriculture—4.6%
Number cattle per capita—1.43	Land area—2.97 mil sq. mi.
Number cattle per sq. mi.—9.0	Cattle population—26.7 mil
Carcass beef produced—4.38 bil lbs	Per capita beef consumption—79 lbs (carcass weight)
Carcass beef exported—2.84 bil lbs	Exports as a percent of production—65.0%

Australia is approximately the same size as the United States, excluding Alaska. However, the U.S. population is nearly 15 times that of Australia. Because of the dry conditions that prevail over the interior, Australians tend to be most concentrated in cities along the southeastern coastal region. In fact, more than 80% of the population resides in the states of Victoria, New South Wales, and Tasmania, plus the cities of Adelaide, South Australia, and

Brisbane, Queensland. The remainder of the continent is sparsely inhabited. A vast majority of the cattle, sheep, wheat, cotton, fruit, and vegetable production occurs in the eastern half of the country.

Australia is an export dependent economy. For example, Australia exports more than 60% of its total beef production. About 40% of beef exports are sold as frozen with the remainder as chilled product. About 60% of beef exports are from grass-fed cattle. The remainder (grain-fed) is directed at the Japanese market. The primary markets for Australian beef are Japan (36%), the United States (34%), Korea (9%), and the Pacific Rim (9%). Almost all of the beef imported by the United States is utilized as hamburger. Australian exports of beef to the United States are dramatically influenced by the cattle cycle. When U.S. production peaks occur, the demand for imported product declines. For example, imports of Australian beef in 1995 were the lowest in three decades due to the rising levels of U.S. domestic production.

Australia also has a strong live cattle export market, moving nearly 850,000 head in 1999. The three primary destinations for these cattle are Philippines (32%), Egypt (29%), and Indonesia (19%). About two-thirds of the live export trade originates from western Australia and the Northern Territory.

Traditionally, the Australian cattle production system has produced grass-fed beef. However, due to increasing demand for higher-quality products, its feedlot industry has been developed over the past decade (Fig. 10.6). The total feedlot capacity is 850,000 head with about 680 accredited feedlots. As a means of comparison, the largest cattle feeding company in the United States has a 375,000-head capacity. However, the high incidence of drought, the high cost of feed grains, and the volatility of the cattle market limit the potential for feedlot expansion in Australia. Most of the feedlots are located in New South Wales and southern Queensland.

The Australian beef industry faced significant economic duress during the 1990s. The United States captured market share from the Australians in both Japan and Korea. In 2001, Australian beef recaptured about 2% market share in Japan.

The Australian industry is characterized as being highly innovative in many areas. For example, the Australians have embraced the concepts of quality assurance and national identification systems. In fact, as of July 2001, cattle sold for harvest must be accompanied by a National Vendor Declaration form. This document communicates the status of sale in regards to the use of hormonal growth promotant, accreditation of the quality assurance plan, source of cattle, withholding periods, and feeding protocols.

However, only recently has there been an effort to implement a grading system. The Meat Standards Australian program is a cut-based system that differentiates product for tenderness and classifies product into appropriate cooking-style categories. The Australian industry will continue to push its beef products as "lean, clean, and green." Their strategy to utilize the pastoral image as a marketing tool is a sound approach. There are major efforts underway to improve product quality, production systems, and processing techniques in Australia with the intent to compete with the United States for share of high-value markets.

Brazil

Population (2000)—167.9 mil	Percent population in agriculture—16.7%
Number cattle per capita—1.0	Land area—3.29 mil sq. mi.
Number cattle per sq. mi.—51	Cattle population—167.5 mil
Carcass beef produced—14.2 bil lbs	Per capita beef consumption—77 lbs (carcass weight)
Carcass beef exported—1.36 bil lbs	Exports as a percent of production—9.6%

(a)

(b)

FIGURE 10.6 Traditional Australian grass-finishing programs (a) are being replaced by grain-finishing systems (b).

Brazil has a larger cattle inventory than the United States but produces less than one-half of the tonnage of beef. It takes Brazil about three times as long to move cattle through the production and processing chain as it does in the United States. These challenges are several confronting the Brazilian beef industry's ability to compete on a global scale. However, there are several distinct advantages—the significant grasslands and the availability plus low cost of labor.

Brazil is the largest country in South America and has 5.8 million farms. The cattle distribution in Brazil can be described regionally. In the west central region are 30% of the cattle with the remainder located in the southeast (23%), northeast (17%), north (15%), and south (15%). Breed types tend to be of European and British origin in the southern states, while central Brazil is populated by cattle of *Bos indicus* influence.

While the predominate production system is grassland-based, approximately 1 million head of cattle are finished in confinement or semiconfinement. Brazilian beef exporters have set a goal of exporting more than 1 million metric tons annually. If this goal is attained, they will overtake Australia as the world's largest exporter. However, they are currently exporting only about two-thirds of that number with the European Union as the major buyer.

FMD also presents a barrier to the world market for Brazilian beef. The Brazilian government has worked to achieve FMD-free status for the states of Rio Grande do Sul and Santa Catarina as the result of an intensive vaccination program. If successful, this would affect the trade status of 120 million head of cattle.

Infrastructure issues and disease outbreaks are limiting the growth of Brazil's exports outside of South America. However, both Brazil and Argentina should be viewed as sleeping giants.

European Union

Population (2000)—375.0 mil	Percent population in agriculture—4.5%
Number cattle per capita—0.2	Land area—1.1 mil sq. mi.
Number cattle per sq. mi.—75	Cattle population—82.3 mil
Carcass beef produced—15.9 bil lbs	Per capita beef consumption—36 lbs (carcass weight)
Carcass beef exported—1.3 bil lbs	Exports as a percent of production—8.2%

The European Union (EU) is comprised of 15 member nations: Austria, Belgium, Luxembourg, Denmark, Finland, France, Germany, Greece, Ireland, Italy, Netherlands, Portugal, Spain, Sweden, and the United Kingdom. The countries of the EU share common agricultural policies including high support price programs, protection against competition from imports, and export restrictions (subsidies).

Since 1989 U.S. beef has been banned from the EU market because of the use of growth promotants. Trade representatives from the United States argue that the ban is not scientifically based and is a trade barrier disguised as a food safety concern. In July 1997, the World Trade Organization ruled in favor of the United States and ordered the EU to lift its ban or pay damages to the United States. The EU appealed the decision but the decision in favor of the United States was upheld. The EU refuses to submit to WTO authority and moved to permanently ban trade in beef grown with the use of hormonal growth implants.

In response, the United States has initiated a system of "carousel" retaliation against a variety of EU products via the imposition of stiff tariff levels.

The beef industry in the EU has been defined by two events—BSE and FMD. While both of these events originated in Great Britain, the diseases spread to multiple sites in Continental Europe.

The effects of the BSE situation have been devastating for farmers, beef demand, and consumer confidence. The EU has nearly lost its export market, prices have fallen by 15–30%, per capita consumption of beef is down by 10% (the equivalent of U.K. annual beef production), and live cattle trade is restricted. The loss of the export market is demonstrated by the drop in tonnage exported by France of 141,000 metric tons from 1996–2000. In an effort to offset the economic consequences, government purchases of certified BSE-free beef have been instituted to the point that more than 1 million tons of beef are expected to be in EU storage.

In an effort to control BSE, the EU is banning the feeding of meat and bonemeal to all farm animals, conducting testing of all animals over 30 months of age, adding the entire intestines of beef animals to the list of risk materials to be removed from the food chain, and removal of all animals over 30 months of age from the food chain unless they have been tested for BSE.

In early 2001, an outbreak of FMD originated in England that eventually spread to Ireland, France, and Holland. While the vast majority of the cases were in England, the outbreak coupled with lagging consumer confidence as a result of the BSE outbreak caused further erosions in beef demand. More than 200,000 head of livestock were slaughtered as a containment measure. The loss in export market has been estimated at $12 million per week.

While much of beef production occurs on relatively small, family-owned farms in the EU, the government has had a significant influence on decision-making for an extended period of time. European consumers will pressure policymakers for upgraded food safety measures. Coupled with pressure over the effects of animal agricultural practices on the environment and animal welfare, it is likely that full trace-back systems will be required. This will only accelerate the movement to coordinated food systems, not only in the EU, but worldwide.

New Zealand

Population (2000)—3.8 mil	Percent population in agriculture—7.9%
Number cattle per capita—2.5	Land area—59 thousand sq. mi.
Number cattle per sq. mi.—159	Cattle population—9.4 mil
Carcass beef produced—1.3 bil lbs	Per capita beef consumption—77 lbs (carcass weight)
Carcass beef exported—1.01 bil lbs	Exports as a percent of production—78%

New Zealand has been described as the "best farm on earth." Agriculture is very productive but is limited by landmass and proximity to major markets. To an even greater extent than Australia, New Zealand's beef industry is export driven. The climate on both the north and south islands allows production of vast quantities of forage. The feedlot industry is very small with only one feedlot of any consequence (capacity of 10–15,000 head). New Zealand exports are targeted to the United States, Japan, and Korea. The four largest markets for New Zealand beef are the United States (64%), Canada (10%), Japan (6%), and Taiwan (5%). Marketing efforts are focused on creating an image of environmental compatibility with an emphasis on grasslands. This "green and clean" message has worked well with modern consumers.

The typical New Zealand livestock enterprise consists of 2,830 sheep and 230 beef cattle (Fig. 10.7). However, there is tremendous range in enterprise conditions from improved pasture under irrigation to dryland range conditions. Herd productivity would be comparable, if not superior, to the average U.S. cow-calf enterprise.

The dairy industry is a major supplier of beef with milking cows outnumbering beef females by more than double. Of the total export value, dairy (48%) and red meat (42%) products are the leaders. With only 1.47 million beef cows, more than 50% of exported beef originated from the dairy sector. The mix of cattle contributing to the export trade is steers (28%),

FIGURE 10.7 Upland grazing on the South Island of New Zealand.

heifers (8%), cows (42%), and bulls (22%). As such, a high percentage of beef imported from New Zealand is used in hamburger trade. Without much opportunity to expand production, New Zealand is likely to experience a steady, although relatively small, position in the world market.

NAFTA PARTNERS

The North American Free Trade Agreement was implemented on January 1, 1994, to eliminate tariffs and other trade barriers between the United States, Canada, and Mexico. The interrelationship of the beef industries of these nations will have significant impact on world markets.

Canada

Population (2000)—30.8 mil	Percent population in agriculture—2.6%
Number cattle per capita—0.4	Land area—3.85 mil sq. mi.
Number cattle per sq. mi.—3.3	Cattle population—12.8 mil
Carcass beef produced—2.77 bil lbs	Per capita beef consumption—71 lbs (carcass weight)
Carcass beef exported—1.2 bil lbs	Exports as a percent of production—43%

As a North American Free Trade Agreement partner, Canada's beef industry is of significant interest. The fourth largest nation on earth, Canada has a relatively small human population. Nearly 80% of Canadians live within 200 miles of the U.S.–Canadian border. Similar to the United States, the average cowherd in Canada is small at approximately 45 head.

The Canadian beef industry is concentrated in the western provinces while a majority of the population lives in the east. More than 70% of the Canadian beef cowherd is located in the three most western provinces—Saskatchewan (25%), Alberta (42%), and British Columbia (6%). The growth of the feedlot industry has centered in southern Alberta, which produces two-thirds of Canadian fed cattle. The cost of producing fed cattle in this region is considered among the lowest in North America. Both Excel and IBP have established packing operations in Alberta. Almost 60% of Canadian beef slaughter occurs in Alberta.

Beef produced in western Canada requires two transport days to reach the population centers of eastern Ontario. The evolving market will likely yield a scenario where eastern Canadian beef buyers will be supplied by the United States, while cattle and beef produced in the western provinces will be marketed into the United States. Beef shipped from Omaha can arrive in Toronto or Montreal in half the time required from a Calgary origination.

Canada is a net exporter of both live cattle and beef. However, live cattle exports peaked in 1996 and have declined since (1.5 billion to 96.5 billion in 2000). This equates to about 3.75% of U.S. cattle marketings. Beef exports to the United States have grown rapidly with Canada challenging Australia as the leading beef supplier. Of Canada's total beef exports, 76% is destined for the U.S. market. Table 10.8 shows the trends in cattle and beef exports to the United States.

While most of Canada's export attention has been focused on the United States, Canadian exporters are looking to China as a future market target. It is estimated that by 2010, the United States will only take about 60% of Canada's beef exports, down from a high of 90% in 1997. The primary suppliers and customers relative to Canada's beef imports and exports are described in Table 10.9.

TABLE 10.8 Trade in Cattle and Beef Between the United States and Canada, 1970–2000

	Live Cattle[1]		Beef and Veal[2]	
Year	Import to Canada	Export to U.S.	Import to Canada	Export to U.S.
1970	NA	NA	11.6	80.6
1975	NA	NA	7.9	21.4
1980	NA	NA	10.9	94.4
1985	NA	NA	19.7	194.7
1990	30.15	904.68	152.0	178.9
1992	56.91	1,273.22	188.4	279.8
1994	92.41	1,010.30	212.1	392.6
1996	40.72	1,509.14	213.0	515.5
1998	117.23	1,313.48	192.7	674.6
2000	349.54	964.70	192.9	739.4

[1] 1,000 head.
[2] Million lbs, carcass weight equivalent.
Source: Livestock Marketing Information Center, USDA.

TABLE 10.9 Canadian Beef Exports and Imports

Export Customer	Percent	Import Supplier	Percent
United States	77	United States	47
Japan	5	Australia	18
Korea	2	Argentina	13
Mexico	2	Uruguay	11
Others	14	Others	11

Source: Canfax (2001).

TABLE 10.10 Trade in Cattle and Beef between the United States and Mexico

	Live Cattle[1]		Beef[2]	
Year	Import to Mexico	Export to U.S.	Import to Mexico	Export to U.S.
1985	NA	NA	3.5	NA
1990	210.3	1,261.2	62.8	.66
1995	14.6[3]	1,653.4	64.4	4.63
1997	236.4	669.4	235.0	8.64
2000	126.7	1,222.6	394.1	11.47

[1] 1,000 head.
[2] Million lbs (carcass weight).
[3] Devaluation of the peso.
Source: Livestock Marketing Information Center.

Mexico

Population (2000)—97.4 mil
Number cattle per capita—0.31
Number cattle per sq. mi.—40
Carcass beef produced—4.18 bil lbs
Carcass beef exported—2.2 mil lbs
Percent population in agriculture—24.1%
Land area—756.1 thousand sq. mi.
Cattle population—30.3 mil
Per capita beef consumption—51 lbs (carcass weight)
Exports as a percent of production—< 1.0%

Mexico can be described as having three geographic regions—the northern arid and semiarid lands, the central high plateau, and the tropical regions. The northern Mexican states account for a majority of cattle exports to the United States. Seventy percent of the USDA-approved packing plants are also located in the northern states. The northern states of Mexico are similar to the southern border states of the United States and account for about 30% of the Mexican herd. Breeding programs in this region are characterized by the use of Hereford, Angus, Charolais, and Simmental cattle as well as Zebu crosses.

The central region accounts for 26% of the national herd but the genetic base is largely Criollo or Zebu. The tropical region has 44% of the total cattle inventory where Zebu and Continental-Zebu crosses are most typical.

Mexico is a net importer of beef and an exporter of feeder cattle (Table 10.10). Mexico has emerged as the second largest market for U.S. beef and variety meat; accounting for 18% of beef exports. A strengthening economy, as well as the fact that Mexico's domestic beef industry is not able to meet domestic demand are the reasons for a strong U.S. export market. As the middle class continues to grow in Mexico, the demand for modern food retailing and family-style foodservice establishments has also grown.

The growth of Mexico's beach resort industry, as well as the HRI trade in the major urban centers, also offers export opportunities focused on the higher-valued middle meats.

Three challenges have faced the Mexican beef industry—drought, an unstable valuation of the peso, and economic uncertainty resulting from political instability. A drought coupled with a devalued peso resulted in historic high import of Mexican feeder cattle in 1995. Rainfall in 1996 and 1997 allowed the rebuilding of herd numbers. But, since that time, dry weather has created rising numbers of cattle being exported to the United States.

The beef exported from Mexico to the United States is miniscule. As such, Mexico experiences a beef trade deficit with the United States of approximately $200 million annually. If the Mexican economy can find a position of positive growth, its importance as a trading partner will increase. The rise in oil prices could be a substantial benefit to the Mexican economy. If investment into the packing, processing, and value-added capability of the Mexican industry fails to occur, then Mexico could be left in the unenviable position of providing raw good inputs (feeder cattle) while importing higher-priced, value-added beef.

United States

Population (2000)—276.2 mil	Percent population in agriculture—2.1%
Number cattle per capita—0.35	Land area—3.6 mil sq. mi.
Number cattle per sq. mi.—27	Cattle population—98.1 mil
Carcass beef produced—27.1 bil lbs	Per capita consumption—100 lbs (carcass weight)
Carcass beef exported—2.51 bil lbs	Exports as a percent of production—9.3%

The United States exports nearly 10% of its annual beef production resulting in net imports accounting for approximately 2% of its beef measured as a percentage of production (Figs. 10.8 and 10.9). Contrary to the rhetoric of those opposed to participating in the international market, exports have increased steadily over the past 25 years (Fig. 10.8). In 1980, the United States exported beef from 273,000 head of cattle. In 2000, the export market had grown to the equivalent of 3.3 million head. International sales accounted for nearly 12% of IBP's net sales.

Net imports reached a peak of more than 10% of annual production in 1979 and have fallen to a steady level of between 1% and 2% (Fig. 10.9).

The international market is critical to the United States industry. The improvements in beef trade have added $2.53/cwt to fed steer price and $3.87/cwt to feeder cattle prices (Marsh, 2000). The total value of the export of cattle, beef, and beef by-products was more than six times greater in 2000 than it was in 1975. Exports were valued at $5.5 billion in 2000 (Table 10.11). The percentage of the total accounted for by each commodity was beef and veal—55%, hides—24%, variety meat—10%, tallow and fat—6%, and live cattle—5%.

The 1994 International Beef Quality Audit identified the strengths and weaknesses of U.S. beef by studying 19 countries from five regions where significant growth potential for U.S. beef sales existed. These regions were Asia, Europe, ASEAN, the Middle East, and North America.

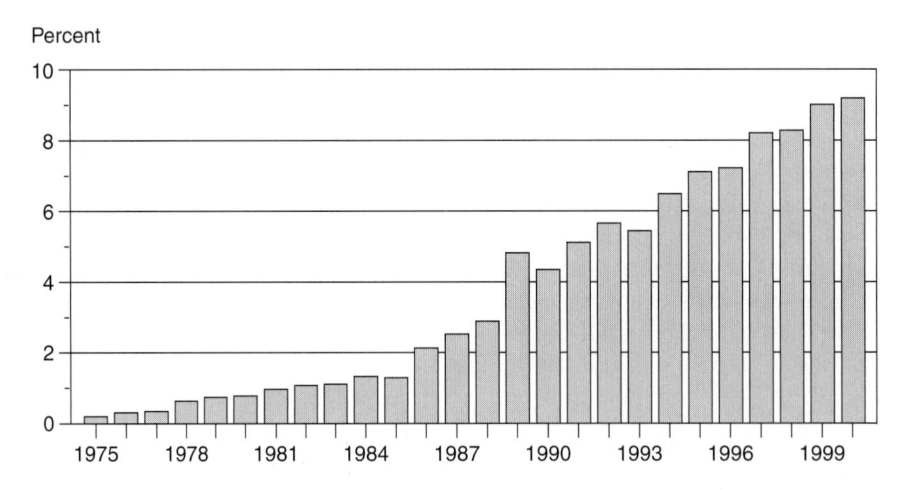

FIGURE 10.8 U.S. beef and veal exports as a percentage of production (carcass weight basis).
Source: Livestock Marketing Information Center.

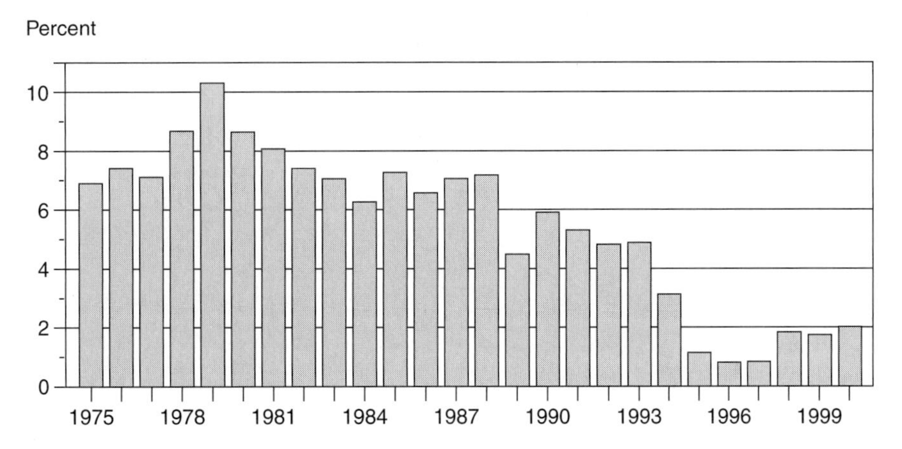

FIGURE 10.9 U.S. beef and veal net imports as a percentage of production (carcass weight basis).
Source: Livestock Marketing Information Center.

TABLE 10.11 Value of Beef, Cattle, and By-Product Exports—1965–2000 (mil $)

Year	Cattle/Calves	Beef/Veal	Hides/Skins	Variety Meat	Tallow, Greases, Lard	Total
1965	17.1	22.6	100.2	56.0	225.6	421.5
1970	29.3	24.6	134.3	69.5	244.1	501.8
1975	77.2	70.1	269.9	109.9	331.8	858.9
1980	54.6	249.3	650.9	243.8	727.2	1,925.8
1985	122.3	467.2	1,035.6	248.6	582.3	2,456.0
1990	88.5	1,580.2	1,458.8	297.5	407.9	3,832.9
1995	86.0	2,646.4	1,444.0	613.3	661.2	5,450.9
2000	271.3	3,049.2	1,310.4	555.4	337.1	5,523.5

Source: Livestock Marketing Information Center.

The primary advantages of U.S. beef in the world market included (1) ability to purchase individual items in volume, (2) tenderness and flavor, (3) high value for the price, (4) overall product quality, (5) confidence in product safety, and (6) positive image of the United States and its grading system.

Problems identified by foreign buyers, which lowered their desire to purchase U.S. beef, included (1) external fat exceeds purchase criteria, (2) inconsistent and/or heavy box or whole-muscle weight, (3) inadequate customer service, (4) excessive seam fat, (5) inadequate shelf life, (6) excessive packaging purge, (7) poor box condition, (8) leaks in vacuum packaging, (9) poor overall workmanship, and (10) inadequate labels both in terms of language and information.

Australia and New Zealand, in particular, have a stronger level of service orientation in the export market than does the United States. As competition increases for gaining share of the world's beef market, attention to customer service, improved quality, and better value delivery will be required. It will become increasingly important to invest in international marketing if the U.S. beef industry wants to ensure profitability.

BEEF BUYERS

China

Population (2000)—1.27 bil	Percent population in agriculture—71.1%
Number cattle per capita—0.1	Land area—33.69 mil sq. mi.
Number cattle per sq. mi.—3	Cattle population—104.6 mil
Carcass beef produced—11.8 bil lbs	Per capita beef consumption—9 lbs (carcass weight)
Carcass beef exported—66 mil lbs	Exports as a percent of production—< 1%

With the world's largest human population, China is a tempting target for the world's beef export market. The Chinese population and farmland are both concentrated in the eastern one-third of the country. Only 13% of China's land is arable. However, due to the very long growing season in the major farming region, China has remained self-sufficient in food production. Traditionally, beef was a by-product from the use of cattle as draft animals.

Throughout the 1990s, China maintained policies supportive to increasing beef production. Cattle and buffalo numbers grew by 21% from 1994 to 1996. From 1996, cattle numbers only increased by 1%. Agricultural policy favors the use of forages and pasture to support the growing national cattle herd. As cattle husbandry skills and management systems are developed, beef production is expected to rise in the future. However, beef is still a minor industry as it accounts for less than 1% of the total meat consumed in China. An underdeveloped transportation and marketing infrastructure will hinder extended growth of beef production. Still, the Chinese are developing a grading system, investing in genetic improvement efforts, and securing capital to improve processing and fabrication facilities.

Many analysts expect that Chinese beef imports will increase as demand for scarce resources limits growth of domestic beef production and as disposable income rises due to economic growth. Hong Kong, which has now been transferred from the British Commonwealth to Chinese control, is a significant market for U.S. beef. Despite the declines in beef consumption that occurred in 1996 due to concerns over food safety fueled by the BSE problem in England and the *E. coli* outbreak in Japan, the United States was able to increase exports to Hong Kong. Frozen beef sales have been particularly strong due to the added convenience. U.S. beef holds the dominant position in Hong Kong's large HRI trade (84%).

The realization of the huge market potential that exists in China will be dictated by political events. As China defines her role in the world community, the struggle between those desiring modernization and those of a more conservative philosophy will determine the degree of market opportunity. Hong Kong will serve as the initial test. Promised democratized rule, Hong Kong's future rests on the autocratic whims of the Chinese mainland.

The United States took a major step by establishing normal trade relations with the Chinese in 2000. As part of the agreement, beef tariffs will be reduced from 45% to 12% over a five-year period. China also agreed to accept the USDA inspection system. China is a significant market not only for beef but also for beef offal items (Fig. 10.10).

China imports 51% of its beef and 32% of beef variety meats from the United States. Australia is the primary competitor with about 30% of the total beef and variety meat import sales to China.

Economists are forecasting significant growth in the Chinese economy with real growth of nearly 8% annually from 2000–2005. Coupled with a declining unemployment level, the Chinese market becomes even more attractive to the world's beef exporters. In fact, the U.S. Meat Export Federation forecasts that beef imports into China will triple by 2007 as demand outdistances domestic production. Much of the increased demand is expected to focus on higher-quality beef. Given the low levels of productivity and lack of grain feeding in China, the United States is poised to capture a large share of this increasing market.

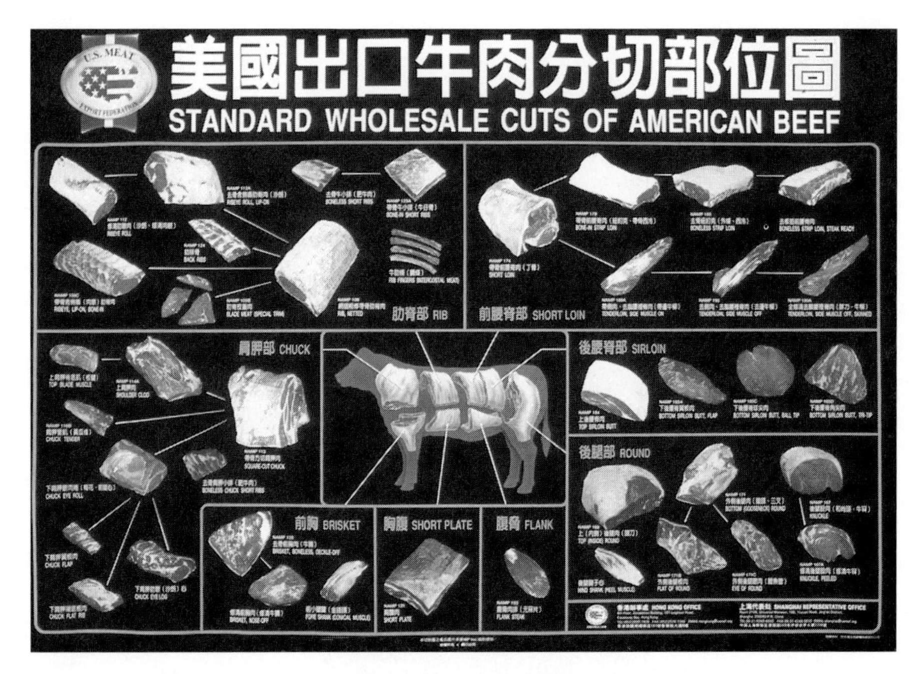

FIGURE 10.10 Beef cuts promotional materials in the Chinese market.
Source: U.S. Meat Export Federation.

Japan

Population (2000)—125.7 mil	Percent population in agriculture—5.3%
Number cattle per capita—0.04	Land area—1,459 sq. mi.
Number cattle per sq. mi.—34	Cattle population—4.6 mil
Carcass beef produced—1.17 bil lbs	Per capita beef consumption—25 lbs (carcass weight)
Carcass beef exported—4.2 mil lbs	Exports as a percent of beef produced—< 1%

The opening of the Japanese beef market in April of 1991 was a landmark in the development of the U.S. beef export market to the Pacific Rim. The combination of Japan's economic strength and growing demand amongst Japanese consumers for beef products has made Japan the focus of many beef exporting nations. About one-half of the value of U.S. beef exports is purchased by Japanese customers.

However, the complexity of the distribution system and high tariffs on imported beef have made market development a difficult task. Nonetheless, the market is sufficiently lucrative that Japanese beef distribution companies have integrated backwards by investing in cattle feeding and processing facilities in Australia primarily but also in several other countries as well. The Japanese controlled about 50% of feedlot cattle in Australia in the late 1990s.

Japanese consumers are heavily concentrated into urban areas where living space is at a premium. Approximately one-half of Japan's population lives in three urban centers—Tokyo, Osaka, and Nagoya. Because of space limitations, the degree of home perishable food storage is minimal. Consumers traditionally purchase food in one-day increments to assure freshness. Shelf life, color, flavor, and presentation continue to be important influences on the buying decisions of Japanese customers. Because of the important role that food plays in Japanese culture, products with highly perceived quality are required to capture market share.

FIGURE 10.11 Japanese consumers typically prefer domestically produced beef over imported products. Importers must understand and meet Japanese consumer expectations if they are to increase market share.

U.S. exporters also had to learn to effectively do business with Japanese customers. Japanese people are very conscious of preserving harmony in both professional and social encounters. Business negotiations require a great deal of patience, respect for experience, awareness of the focus on group rather than individual needs, and the requirement of establishment of personal trust prior to formal business arrangements. Perhaps one of the most significant barriers to expanded U.S. beef trade is the process of adapting Americans to the culture of international clients.

About two-thirds of U.S. beef is purchased at retail with most of the remainder marketed via foodservice. The foodservice industry in Japan has continued to grow and accounts for 51% of total beef sales. Chilled and frozen beef each account for about one-half of Japan's beef imports.

The most significant competition in the Japanese beef market is Australia. The United States and Australia account for 95% of Japan's beef imports with the United States holding a 48.5% share compared with Australia's 44.5%. Prior to 1996, Australia had maintained the dominant position. However, since 1996, the United States has held the lead for beef sales. U.S. beef sells at a premium to Australian products, but not to Japanese beef (Table 10.12).

Japanese consumers continue to prefer domestic beef as compared with imports (Fig. 10.11). In USMEF-sponsored surveys, consumers rated Japanese beef higher than U.S., Australian, and Canadian products for quality, value, freshness, and price. In the same survey, Australian beef was ranked higher for both quality and value than U.S. beef.

Despite trade liberalization, Japan imposes a 38.5% tariff on chilled and frozen beef imports. This tariff places significant downside pressure on the price of imported beef. As a means of comparison, the Japanese government earns half as much from the tariff as Australian producers do from raising, feeding, and processing beef products for the Japanese

TABLE 10.12 Japanese Retail Beef Price Index

Product	Index
U.S. beef	100
Japanese wagyu	241
Japanese dairy	152
Australian beef	81

Source: USMEF.

market. The United States has a competitive advantage because of the significant volume of beef production that allows for export of specific beef cuts as opposed to whole carcasses.

Japanese consumers are hyperconscious of food safety. In fact, beef consumption drops in Japan every time a food safety problem occurs almost anywhere in the world. Marketing efforts that actually identify the producer, farm, and production practices of the beef purchased are taking hold in Japan.

As Japanese consumers become more value, price, and food safety focused, they will demand greater assurances from those who supply beef to their families. The USMEF refers to current Japanese consumer trends as the "3 K's"—Kirei (clean), Kenko (healthy), and Kodawari (good value).

Domestic Japanese beef production has been declining as the migration from rural to urban areas increases while the average age of Japanese farmers continues to climb. From 1995 to 2000, almost 25% of Japanese beef producers exited the business. Still the average beef herd is small with 24 head of cattle per farm. If Japan liberalizes dairy trade, dairy cow numbers are also expected to decline. As a result, in the future only about 30% of beef demand in Japan will be met by domestic production.

Japan is destined to rely on food producers outside its borders. With an average farm size of approximately 3 acres, Japan cannot produce enough food for a population of more than 125 million people. However, the Japanese market requires suppliers with an excellent food safety track record, the ability to adapt to Japanese culture, and the staying power to evolve into a trusted trade partner.

Republic of Korea

Population (2000)—46.5 mil
Number cattle per capita—0.05
Number cattle per sq. mi.—66
Carcass beef produced—613 mil lbs
Carcass beef exported—23 mil lbs
Percent population in agriculture—8.6%
Land area—38.2 thousand sq. mi.
Cattle population—2.5 mil hd
Per capita beef consumption—23 lbs (carcass weight)
Exports as a percent of production—3.7%

Korea has become a major market for U.S. beef, accounting for about 14% of both value and tonnage of exports. The Korean market is heavily focused on HRI trade as food service accounts for better than 60% of Korean beef consumption. The United States holds about 57% of the market share with Australia, New Zealand, and Canada as the major competition.

The recovery of the Korean economy, improved exchange rates, and a decline in domestic production, coupled with consistent promotion and relationship building, have led to optimistic future market opportunities for U.S. beef in the Korean market.

However, Korean consumers continue to be very sensitive to food safety issues. Coverage of the BSE and FMD crises in Europe only heightened the instability of consumer confidence. Then in 2000, buckshot was found in beef that had originated in the United States. The incident received major national media coverage in Korea and illustrates the need for implementation of superior quality assurance efforts to protect both domestic and international consumers.

Liberalization of trade in 2001 caused many Korean beef producers to dramatically decrease their herd inventories. Domestic cattle inventories declined 18.5%. Domestic cattle production in Korea originates almost entirely from small farms producing less than five head per farm. The predominant breed of domestic cattle is Hanwoo. In an effort to protect domestic cattle producers during the phase-in of trade liberalization, the Korean government established increased price supports, country-of-origin labeling of beef in restaurants, development of specialized marketing infrastructure to promote Hanwoo as a branded product, and incentives to improve the quality of Korean beef.

However, trade liberalization opens a number of opportunities to increase sales of U.S. beef in Korea. Most significantly, the separate retail distribution system for imported beef will be eliminated. Previous to the change, U.S. beef was only offered in one retail store for every 10 retail outlets that merchandised Korean beef. This change will open 30,000 retail shops for the sale of imported beef.

High-quality beef sales to Korea are expected to nearly double from 2000 to 2007. Even more dramatic increases in the short term are likely to occur in the variety meat trade. This trend will be heightened by the decline in domestic beef production, particularly in regards to outside skirts and oxtails, traditional Korean favorites.

Service, on-time delivery, and maintaining business relationships will be key to increasing sales in Korea.

Russian Federation

Population (2000)—147.2 mil
Number cattle per capita—0.2
Number cattle per sq. mi.—4
Carcass beef produced—4.05 bil lbs
Carcass beef exported—11 mil lbs

Percent population in agriculture—10.5%
Land area—6.59 mil sq. mi.
Cattle population—27.5 mil
Per capita beef consumption—10 lbs (carcass weight)
Exports as a percent of production—< 1%

The breakup of the former USSR created dramatic political and economic upheaval for the citizens of the Soviet states. Cattle numbers in the Russian Federation have declined by nearly 30 million head since 1991. Production of beef, pork, and poultry is down 58% over the same period. Simultaneously, personal incomes have declined which in turn has yielded reduced capacity to increase purchase of meat products.

There are signs of a slow recovery in the Russian economy. However, these trends will not translate to improved productivity and profitability in the beef sector. The production, marketing, and processing infrastructure continues to decline and serves as a significant barrier to Russia's ability to be self-sufficient in beef production. Beef production is shifting from the traditional, large collective farms to smaller, individually owned farms. But, cost of production is high and feed shortages, both in pasture and supplement, are barriers to increased growth in the beef sector.

Beef consumption has declined by about half since the early 1990s, and variety meats command a large portion of the total consumption. There is some increase in demand for high-quality beef to supply the urban restaurant trade. Despite the problems with FMD and BSE, the EU continues to be a leading supplier of beef to Russia. The willingness of the EU to subsidize exports to Russia has created a barrier for expansion of sales by other nations. Ukrainian beef accounts for about one-quarter of Russia's beef, although supplies are of poor quality and inconsistently delivered. South America and Mongolia are new competitors in the market.

The U.S. trade with Russia is likely to remain focused on variety meats. Without improvements in economic performance that result in higher consumer disposable incomes, improved agricultural, marketing, and processing infrastructure, and a more competitive currency, the Russian Federation will remain a market place of limited real potential. However, if an economic turnaround can be achieved, Russia is a market of strong potential.

ROLE OF BEEF IN WORLD DIET

The role of animal agriculture in feeding the increasing world population is an often hotly debated topic. Despite the recent declines in the growth of the human population, agriculture will have to increase its productivity to adequately provide nutrition for people. Population trends are illustrated in Figure 10.12. Regional population concentrations are compared over time in Table 10.13. These primary geographical areas are projected to increase their share of the world population—Sub-Saharan Africa, the Near East and northern Africa, and Asia (China excepted). Furthermore, nearly 85% of the world's population is expected to reside in less developed countries by 2020 as compared with the approximately 73% world population share held by less developed nations in 1970.

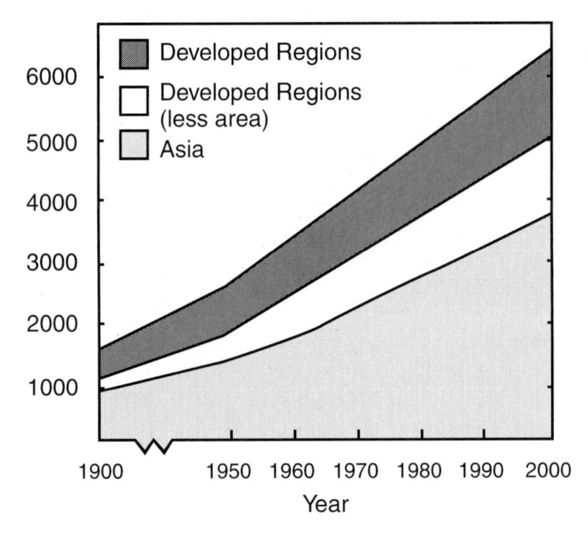

FIGURE 10.12 Past, present, and projected world population, 1900 to 2000.
Source: USDA.

TABLE 10.13 Share of World Population

Region	1970	1996	2020
Less developed countries	72.9%	79.7%	83.6%
More developed countries	27.1	20.3	16.4
Sub-Saharan Africa	7.8	10.3	13.5
Near East and North Africa	3.9	5.1	6.4
China	22.5	21.3	18.9
Other Asia	29.7	33.2	35.0
Latin America and Caribbean	7.7	8.5	8.5
Eastern Europe and newly independent states	9.5	7.2	5.8
Rest of the world	18.9	14.5	12.0

Source: World Population Profile, USAID, U.S. Dept. of Commerce.

TABLE 10.14 Worldwide Consumption and Cost of Dietary Calories

	Spending on Food as a Share of Private Consumption Expenditure[a] (%)	Cost per 100 Calories (U.S. Cents)	Average Daily Calorie Consumption (N)	Share of Total Calorie Consumption from:			
				Cereal and Roots (%)	Meat and Fish (%)	Vegetables and Oil (%)	Milk (%)
High Income							
Japan	20.8	39	2,887	42	13	9	4
Switzerland	18.2	29	3,381	24	18	10	12
France	15.5	16	3,632	26	19	11	11
Ireland	20.1	11	3,837	32	13	12	11
Canada	10.5	11	3,092	25	16	13	8
United States	8.7	10	3,732	25	26	13	10
Middle Income							
Peru	41.6	9	1,882	51	7	7	4
Mexico	33.7	8	3,177	48	8	10	5
Algeria	35.7	3	2,897	58	3	13	6
Ecuador	33.9	3	2,583	40	6	20	5
Low Income							
Tanzania	71.0	1	2,018	69	4	5	2
Nepal	67.3	1	1,957	78	1	4	4
Ethiopia	51.8	1	1,610	74	3	3	2

[a]Does not account for food consumed away from the home or food produced at home.
Source: USDA-ARS (1996).

Hunger, particularly chronic persistent hunger, continues to be a problem in regions of the world where economic and political turmoil predominate. Illiteracy rates are high, poverty levels are significant, and the marketing, transportation, processing, and financial infrastructure is poorly developed. In less developed countries, per capita total caloric intake, total protein intake, and intake of calories/protein from animal sources are lower than for consumers in more privileged countries (Table 10.14).

Animal products, including beef, are nearly always preferred when they are available and consumers have a relatively high level of disposable income. Availability of animal products is influenced by the degree of consumer affluence, the percent of the population working in agriculture, the cultural norms of a particular society, and the degree of agricultural production, processing, and marketing infrastructure development.

Cereal grains are the most important source of energy in worldwide diets. Energy derived from cereal grains, however, is twice as important in developing countries (some exceptions) as in developed countries. Beef and other animal products provide approximately 30% of the dietary calories for consumers in developed countries but only 9% for those residing in developing countries. Animal products supply nearly 60% of the protein for consumers in developed countries but only 22% for consumers in developing nations.

Beef and other meats have been adopted into the diet because their assortment of amino acids more closely matches the needs of the human body than does the assortment of amino acids in foods originating from plant sources. Furthermore, vitamin B_{12}, which is required in the human diet, may be obtained from foods of animal origin but not those of plant origin. Beef and other animal foods also efficiently provide a variety of micronutrients (iron, zinc, etc.), offer a high degree of nutrient density (a high proportion of the recommended daily allowance of several nutrients relative to the number of calories per serving), and the unique taste desired by consumers.

Animals are included into agricultural systems of worldwide cultures because much of the earth's landmass is unsuitable for growing cultivated crops. Approximately two-thirds of the world's 323 billion acres are permanent pasture, range, and meadows; of the total land area, only 3.4 billion acres are suitable for producing cultivated plants that can be directly consumed by humans. Most of the untillable acres can produce roughage in the form of grass, forbs, and browse that is digestible by grazing ruminants. Cattle and sheep are the most important domesticated species for this purpose. These "walking food factories," traveling without the input of fossil fuel, can harvest and convert forages that are not digestible by humans to high-quality protein food. Appropriately managed, the grazing ruminant provides a highly sustainable and low-input source of food in a variety of climatic conditions.

There are concerns about nutritional efficiency, water and land resources, and competition for nonrenewable resources as they relate to animal agriculture. Animal agriculture has been criticized for wasting resources that could otherwise be used to feed persons with inadequate diets. The practice of providing livestock and poultry with feed that could be eaten by humans and utilizing land resources to produce crops that could be consumed by humans seems to be most often criticized. Consideration must be given to economic systems and consumer preferences to understand why these practices are perpetuated.

Agriculturalists produce what consumers want to eat, as reflected by the prices consumers can and are willing to pay. In most countries, as per capita income rises, consumers increase their consumption of meat and animal products, which are generally more expensive on a per pound basis than products derived from cereal grains (Fig. 10.13). Food producers, processors, and services are responsive to consumer demands. The desire for lower fat content in the diet led to the provision of 1% and skim milk products, the reduction in fat from retail beef, and changes in the USDA grading standards. However, in developed countries taste is equally, if not more important, than price and nutrition in consumer food selections.

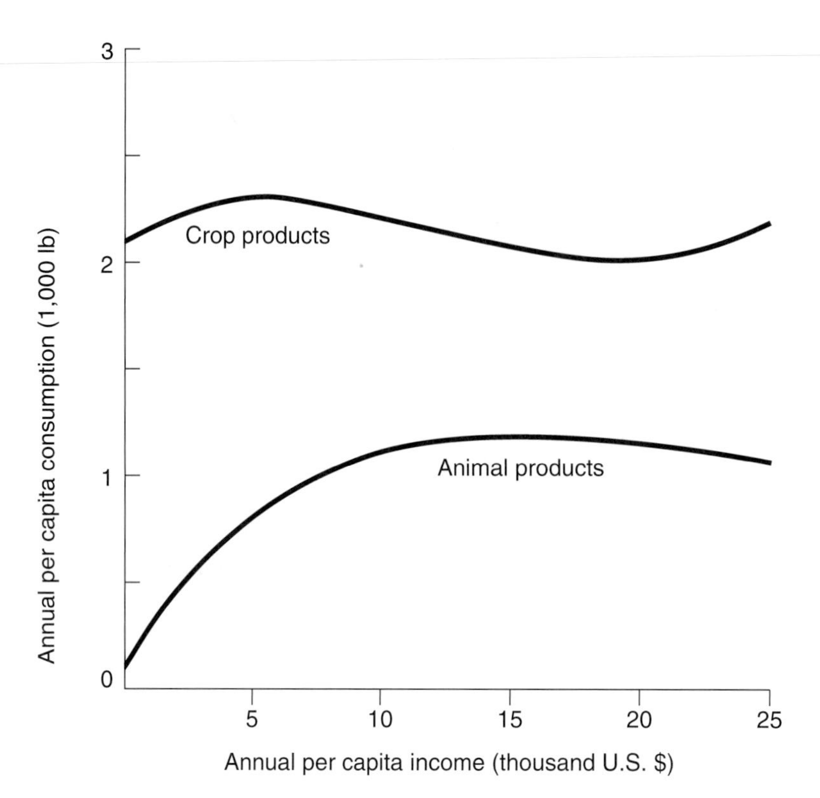

FIGURE 10.13 Animal product consumption as related to per capita income.
Source: Courtesy of FAO (Production Yearbook).

Management practices are also sensitive to consumer demand. Cattle can produce large quantities of meat without grain feeding. Consumers in Argentina, South America, Australia, and New Zealand often prefer beef from grass-fed animals. Consumers in the United States, Canada, and the Pacific Rim typically prefer grain-fed beef. Only 20% of the concentrated feedstuffs (grain and protein meals) fed to livestock are utilized by beef cattle. Less than 2 lbs of concentrate is used per pound of liveweight produced, which is lower than for broilers or hogs. The amount of grain feeding in the future will be dictated by cost and availability of feed grains, consumer disposable income and taste preferences, and the cost of non-feed inputs (e.g., fossil fuel, governmental regulation, etc.).

Because of the relatively limited supply of tillable land, utilizing grazing animals to harvest forage from pastures and rangelands is likely to continue to be an important agricultural activity. Only Brazil and India appear to be dramatically overstocked as measured by comparing the percent of the world's pasture and cattle inventory on a nation-by-nation basis (Table 10.15). China and Australia have the potential for increased production, although seasonal and environmental variability, proximity to markets, and infrastructure development will likely inhibit growth to some extent. The United States, Argentina, Mexico, and many of the remaining top 15 beef cattle producers appear to have optimized the cattle herd to the available pasture resources (Table 10.15).

It has been argued that people in the United States and elsewhere should consume less food from animals and more foods from plants. The moral side of the issue argues that people elsewhere in the world starve when those in the United States could meet their own food needs by eating foods from plants rather than feeding plants to animals. The United States could then send the balance of the plant-derived foods abroad. Grain could be provided at minimum cost to developing countries because of the U.S. grain surplus. Yet there are prob-

TABLE 10.15 Percent of World Human Population, Cattle Population, and Pasture (2000–2001)

Country[a]	Human Population (% of total)	% in Agriculture[b]	% of Country Classified Pasture	% of World's Pasture	% of World Cattle
India	16.6	55.8	3.83	0.32	16.2
Brazil	2.8	16.7	21.88	5.39	12.4
China	21.3	67.1	42.89	11.67	7.7
United States	4.6	2.2	26.11	6.98	7.3
Argentina	0.6	11.0	51.89	4.14	4.1
Sudan	0.4	62.4	46.29	13.21	2.7
Ethiopia	1.0	84.4	20.00	0.58	2.6
Mexico	1.6	24.2	39.03	2.33	2.2
Russian Federation	2.4	10.4	51.71	2.59	2.0
Australia	0.3	4.6	50.70	12.20	1.9
Colombia	0.6	21.4	39.09	1.19	1.9
Bangladesh	2.6	56.7	4.61	0.02	1.7
Pakistan	2.5	52.1	6.49	0.14	1.6
France	0.9	3.4	19.32	0.29	1.5
Nigeria	1.8	34.1	42.86	1.14	1.4

[a] Top 15 countries for beef cattle numbers.
[b] Percent of all economically active people who are employed in agriculture.
Source: FAO.

lems associated with providing low-cost food to other countries: economic pressures on the American farmer who cannot make an adequate living eventually lead to increased taxes; inadequate distribution of food in developing countries because of poor transportation and marketing distribution systems; and negative effects on the agricultural production of countries receiving the grain.

There is evidence that the development of animal agricultural systems in developing countries could initiate a self-sustaining, more productive agriculture requiring only small inputs of fossil-fuel energy. Efficient utilization of grazing animals in specific integrated farming systems could provide energy inputs from draft and transportation and increase food and other useful by-products from renewable sources of energy.

Increases in U.S. animal production have been remarkable during the past half-century. For example, pounds of liveweight beef marketed per breeding female increased from 220 lbs in 1925, to 435 lbs in 1975, to 485 lbs in 1985, and to 625 lbs in 2000. The increase in productivity occurred primarily because people had an incentive to progress under a free enterprise system. They learned how to improve their standard of living using the available resources.

Remarkable changes in agricultural efficiency in the United States have resulted from a stable political system, compulsory and free education, and industrial development. The percent of disposable income spent for food was 30% in 1950 and less than 11% in 2001. It has been argued that the same agricultural efficiency and relative food prices could be achieved by many developing countries by more effective use of their resources.

Our advanced technologies should be shared with other people throughout the world. The achievements have been built on knowledge developed through meaningful experience and research; extension of knowledge to producers; and development of an industry to provide inputs in transportation, processing, and marketing. Dwindling dollars presently being put into agricultural research and extension of knowledge may not provide the technology needed for future food demands in the United States.

Population growth and agricultural development will determine in part whether the people of the world will be adequately fed. These challenges will be most difficult in the developing countries. About 30% of the world human population and 32% of the ruminant animal population live in developed regions of the world. However, the ruminants of these same regions produce two-thirds of the world's meat and 80% of the world's milk. In developed regions, a higher percentage of animals are used as food producers, and these animals are higher in productivity on a per animal basis than in developing regions. This is the primary reason for the higher level of human nutrition in developed countries.

With assistance, many developing regions of the world may be able to achieve levels of plant, animal, and eventually human food productivity similar to those of developed regions. Except perhaps in India, there are abundant supplies of animal feed resources, not competitive with human food production, to support expansion of animal populations and production. It has been estimated that through changes in resource allocation, an additional 8 billion acres of arable land (twice what is presently being used) and 9.2 billion acres of permanent pasture and meadow (23% more than is presently being used) could be put into production in the world. This, plus the potential increase in productivity per acre and per animal presently reflected in the developed countries, demonstrates the magnitude of world food production potential. This potential, however, cannot be realized without proper planning and increased incentive to individual producers.

In the long run, each nation must assume responsibility for more of its own food supply by efficient production, barter, or purchase, and by keeping future food production technology ahead of population increases and demand. The extensive untapped resources that exist throughout the world can greatly enhance food production, including an ample supply of animal products. The greatest resource is each human being who has the opportunity to be productive and self-reliant.

MANAGEMENT SYSTEMS HIGHLIGHTS

1. The beef industry generated $5.5 billion in export value. Beef and veal (~ $3.1 billion) and hides ($1.3 billion) account for the majority of this value.
2. The leading customers for beef from the United States are Japan, Mexico, Republic of Korea, and Canada.
3. The United States exports approximately 10% of its total beef production, the equivalent of 3.3 million head of fed cattle.
4. The ability of the U.S. beef industry to generate wholesome products that meet the specifications of foreign buyers is important to the growth of the export market.
5. Demand for animal products increases as worldwide consumers experience increases in per capita disposable income.
6. The human resource is the most important component in worldwide management systems.

REFERENCES

Publications

Action, Expanding Foreign Markets for U.S. Meat Products (periodical). Denver, CO: U.S. Meat Export Federation.

Agricultural Trade Highlights. Washington, DC: USDA-FAS.

Argentina Economic Trends Report. 1996. U.S. Embassy, Buenos Aires, Argentina.

Argentina Livestock and Beef Market Situation. 1996. Secretariat for Agriculture, Buenos Aires, Argentina.

Australian Livestock and Meat Corporation Annual Report. 1998–99. Sydney, Australia.

Aylward, L. 2000. International Intrigue. *Meat and Poultry* (Oct.), pp. 34–36.

Baldwin, R. L. 1980. *Animals, Feed, Food, and People: An Analysis of the Role of Animals in Food Production.* Boulder, CO: Westview Press.
Band, S., Goddard, M., and Hygate, L. 1992. *Targeting the Japanese Beef Market.* Meat Research Corporation, Sydney, Australia.
Canada's Beef Industry. 2001. Canfax Research Services, Calgary, Alberta, Canada.
Cattle and Beef: Impact of the NAFTA and Uruguay Round Agreement on U.S. trade. 1997. Washington, DC: U.S. International Trade Commission.
Econamou, A. 1999. Can Aussie Beef Match U.S. Beef? *Beef Improvement News* (Nov./Dec.). Sydney, Australia, pp. 11–12.
FAO Statistical Databases. 2001. Rome, Italy: FAO.
FAPRI 2000 International Agricultural Outlook. 2000. Iowa State University, Ames, IA; University of Missouri, Columbia, MO.
Foreign Agricultural Trade of the United States. 2001. Washington, DC: USDA-ERS.
Garcia, J. J. *Argentina's Position Relative to the Global Beef Market.* 1996. Master's thesis. Colorado State University, Fort Collins, CO.
Goodsir, G. 2000. The Competitors. *Meat and Poultry* (Oct.), pp. 26–32.
International Agricultural Trade Reports. 1998–2001. Washington, DC: USDA-FAS.
Japanese Business and Etiquette and Protocol. 1990. EPISTAT International, Inc., Washington, DC.
Livestock, Dairy, and Poultry Situation and Outlook. 2001. Washington, DC: USDA-ERS.
Livestock Marketing Information Center. 2001. Denver, CO: USDA-ES.
Livestock Roundup. 1995–96. Livestock Marketing Information Center. Denver, CO: USDA-ES.
Long-Term World Agricultural Commodity Baseline Projection. 1995. Washington, DC: USDA-Commodity Economic Division.
Marsh, J. 2000. It's a Matter of Net Trade. *BEEF* (Dec.). Minneapolis, MN, p. 38.
Meat and Livestock—A Vital Industry. 1990. Australian Livestock and Meat Corporation, Sydney, Australia.
Morgan, J. B., Smith, G. C., Belk, K. E., and Neel, S. W. *International Beef Quality Audit.* 1994. Colorado State University and U.S. Meat Export Federation.
National Vendor Declaration for Cattle. 2001. Meat and Livestock Australia, Sydney, Australia.
New Zealand Meat Producers Board Annual Report. 1996. Wellington, New Zealand.
1996 Beef and Pork Exports: Impact on Domestic Industries. 1997. Denver, CO: CF Resources, Inc.
Peck, C. 2000. *Hanging Tough. BEEF* (Aug.). Minneapolis, MN, pp. 13–16.
Peck, C. 2001. *Struggling to Rise. BEEF* (Jan.). Minneapolis, MN, pp. 79–82.
Perkins, Doug. March 1997. Don't Cry for Argentina. *BEEF,* pp. 9–10.
Price, C. 1996. *The U.S. Foodservice Industry Looks Abroad.* Global Food Markets. Washington, DC: USDA-ERS.
Robinson, J. 2000. *China's Promise. National Cattlemen* (Aug.). Englewood, CO, pp. 24–26.
U.S. Meat Export Analysis and Trade. 1995–2001. USDA-AMS, Des Moines, IA; Iowa State University, Ames, IA; USMEF, Denver, CO.
Variety Meats from the USA—A Buyer's Guide. 1996. Denver, CO: U.S. Meat Export Federation.
World Agricultural Production. 2001. Washington, DC: USDA-FAS.
World Food Review. 1994. Washington, DC: USDA.
World Livestock Situation. 1997. FAS circular. Washington, DC: USDA.
World Population Profile. 1996. U.S. Department of Commerce. Washington, DC: USAID.

Websites

Canadian Cattlemen's Association: www.cattle.ca
FAO databases: www.apps.fao.org
Livestock Marketing Information Center: www.lmic1.co.nrcs.usda.gov
Meat and Livestock Australia: www.mla.com.au
New Zealand Meat Board: www.meatnz.com.nz
USDA Agricultural Marketing Service: www.ams.usda.gov
USDA Foreign Agriculture Service: www.fas.usda.gov
U.S. Meat Export Federation: www.usmef.org

Reproduction

A live calf born and weaned to each breeding female each year is the primary objective for successful cattle reproduction. However, cows are not managed as individuals but as a herd, so the economic evaluation of total herd reproductive performance is most critical. Reproductive efficiency in cattle, as measured by the number of calves born and weaned each year per 100 females in the breeding herd, is considered the most important economic factor in cattle production. Reproduction is at least twice as important as growth or carcass for cow-calf producers who sell their calves at weaning. It is essential for producers to understand the reproductive principles, for it is the focal point of overall animal productivity.

Producers can economically manage their herds for optimum reproductive rates. Management decisions that result in breeding animals having early puberty, high conception rates, minimum calving difficulty, and early rebreeding are most critical. Keeping the animals healthy, providing adequate nutrition, selecting genetically superior animals, and giving attention at parturition are some of the most important factors in minimizing reproductive losses.

STRUCTURE AND FUNCTION OF THE REPRODUCTIVE ORGANS

The Reproductive Organs of the Cow

Figure 11.1 shows the reproductive organs of the cow and their location in the body. The reproductive organs consist of a pair of ovaries (suspended by ligaments just behind the kidneys) and a pair of funnel-shaped tubes (infundibulum), which are part of the oviducts that lead directly into the uterine horns. The two uterine horns merge together to form the uterine body. Collectively, the two uterine horns and the uterine body comprise the uterus (womb). The uterus leads into the cervix, which has a folded surface surrounded by muscles. The cervix opens into the vagina, a relatively large canal or passageway that leads posteriorly to the external parts—the vulva and the clitoris. The urinary bladder empties into the vagina through the urethral opening.

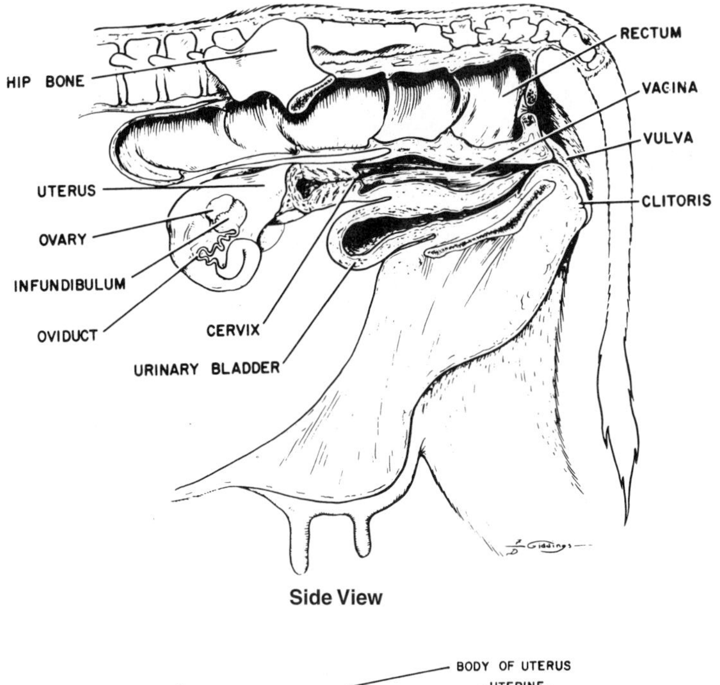

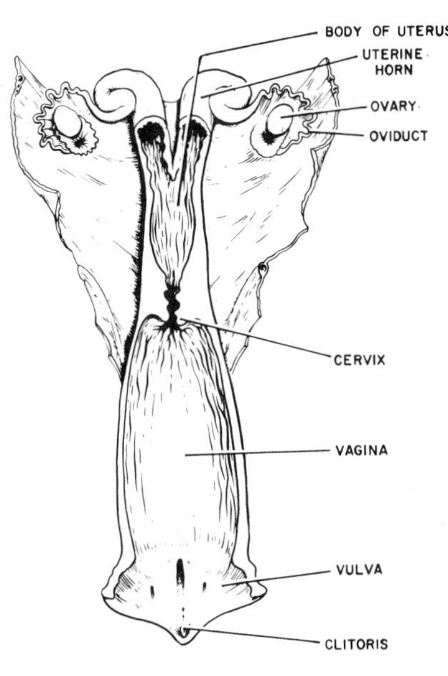

FIGURE 11.1 Reproductive organs of the cow. *Source:* Colorado State University.

The ovaries produce eggs (female sex cells, also called ova) and hormones (Fig. 11.2). Each egg or ovum is individually generated by follicles within the ovary (Fig. 11.3). All of the follicles that a female will ever produce are present at birth as immature follicles. The majority of these follicles (> 99%) grow to various stages, cease growth, deteriorate, and are absorbed (Fig. 11.4). Less than 1% of follicles will mature (Graafian follicles), reaching a maximum size of about 0.4 in. in diameter, and go on to ovulate. During ovulation the egg is released from the mature follicle and the follicle develops into a "yellow body," or corpus luteum, which becomes a vitally important structure if pregnancy occurs.

FIGURE 11.2 Bull sperm and cow egg, each magnified several hundred times. The sperm are located at the bottom and on the left-hand side of the egg. Note the comparative sizes. Each are single cells and each contains one-half the chromosome number typical of the other body cells.
Source: Animal Reproduction Laboratory, Colorado State University.

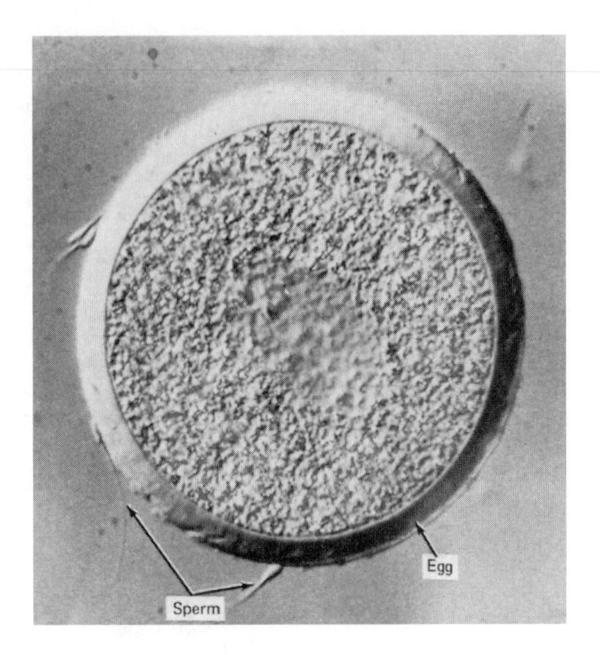

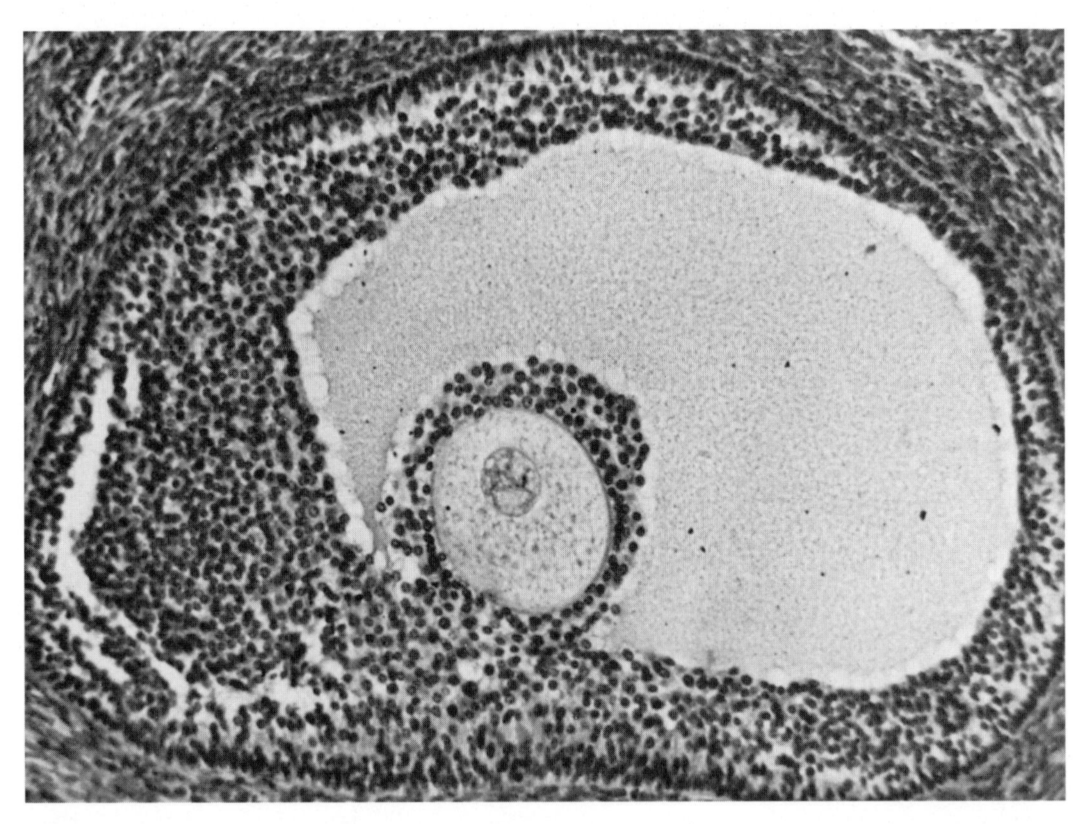

FIGURE 11.3 The large structure, outlined with a circle of dark cells, is a follicle located on a cow's ovary (magnified several hundred times). The smaller circle, near the center, is the egg. The large light gray area is the fluid which fills the follicle. When the follicle ruptures, the fluid will wash the egg into the oviduct (ovulation).
Source: Animal Reproduction Laboratory, Colorado State University.

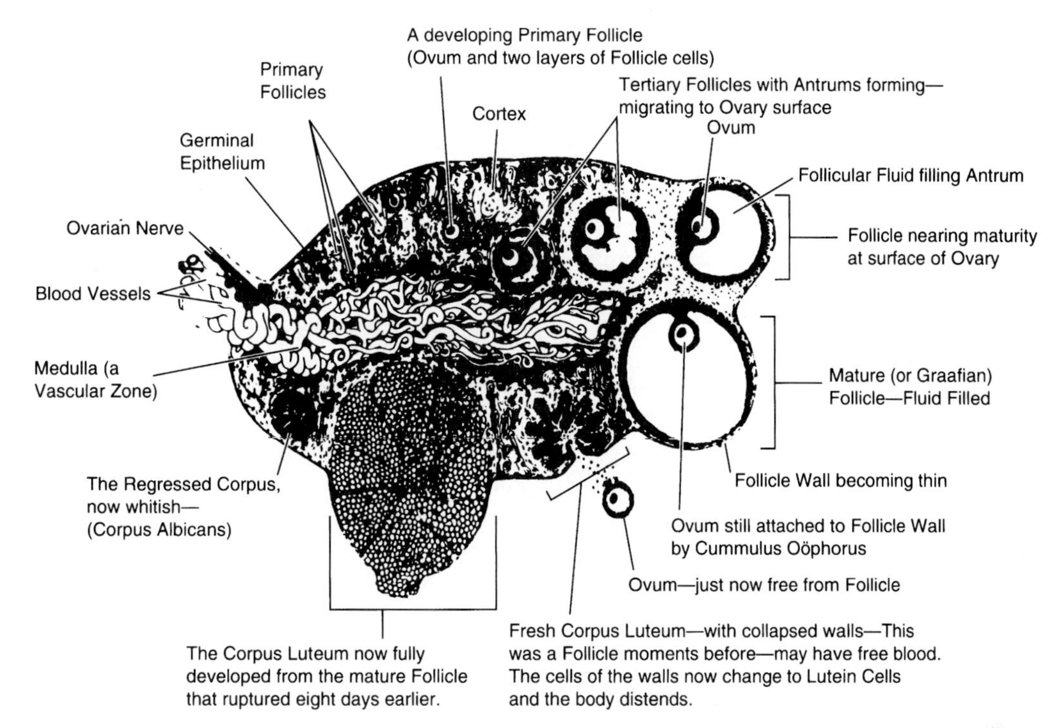

FIGURE 11.4 A cross section of the bovine ovary showing how a follicle develops to full size and then ruptures, thus allowing the egg to escape (ovulation). The follicle then becomes a "yellow body" (corpus luteum) which is actually orange-colored in cattle. The corpus luteum degenerates in time and disappears. Many follicles cease development, stop growing, and disappear without ever reaching the mature stage. From Bone, J. F. *Animal Anatomy and Physiology.* 4th ed. Corvallis: Oregon State University Book Stores. Copyright © 1975.

The oviducts receive the ova immediately after the ova leave the ovaries through the open end of the oviduct (called the infundibulum). Ova are tiny, 200 μm or less in diameter (200 μm = 1/5 mm), which is approximately the size of a dot made by a sharp pencil.

The uterus has a relatively short body and longer uterine horns. Spermatozoa pass through the uterus to the upper ends of the oviducts to intercept and fertilize the ovum (see Fig. 11.2). The fertilized ovum (embryo) travels to the uterus, develops into a fetus (see Fig. 17.2), implants into the uterine wall and placenta, and remains in the uterus until parturition (birth).

The cervix is a narrow opening from the vagina to the uterus. The cervical passageway changes from one that is so tightly closed it is nearly sealed during pregnancy, or nearly closed when the animal is not in estrus (heat), to a relatively open, very moist canal at the height of estrus.

The vagina serves as the organ of copulation at mating and as the birth canal at parturition. Its mucosal surface changes during the estrous cycle from very moist when the animal is ready for mating to almost dry, even sticky, between periods of heat. The urethra from the urinary bladder joins the posterior ventral vagina; from this juncture to the exterior vulva, the vagina serves the double role of a passage for the reproductive and urinary systems.

A highly sensitive organ, the clitoris is located at the lower tip of the vagina. The clitoris is the homologue of the penis in the male (e.g., it comes from the same embryonic source as the penis). There is some research indicating that clitoral stimulation or massage following artificial insemination will increase conception, but it has not been well verified.

Ovarian hormones (progesterone and estrogen) fluctuate during the estrous cycle in the mature heifer or cow. Estrogen is produced by the follicle in response to follicle-stimulating hormone (FSH) production from the anterior pituitary. More specifically, FSH stimulates follicular development on the ovary and thus production of estrogen. The increased levels of estrogen in the bloodstream stimulate the female to display behavioral estrus. Luteinizing hormone (LH) is produced by the same cells of the anterior pituitary that produce FSH. For the most part, luteinizing hormone and FSH levels parallel one another in the bloodstream throughout the estrous cycle. Luteinizing hormone has three primary functions. One is to stimulate ovulation of a properly developed follicle during estrus; the second is to control the formation of the corpus luteum; and the third is to stimulate progesterone production from the luteal cells of the corpus luteum. The ovarian hormones estrogen and progesterone feed back to control LH and FSH secretion by controlling the secretion of gonadotropin-releasing hormone (GnRH) from the hypothalamus.

When estrogen levels increase, just prior to estrus and during estrus, it stimulates a surge of GnRH and thus a significant release of FSH and LH that results in follicle maturation and ovulation. Following ovulation, the cells that once produced estrogen have been destroyed and are luteinizing to form the luteal cells of the corpus luteum (CL). Thus, estrogen production diminishes rapidly. Formation of a mature CL occurs during the next 4 to 5 days of the estrous cycle. As the corpus luteum matures, progesterone production increases to a peak on about day 10 to 12 of the estrous cycle and remains elevated until approximately day 17 of the estrous cycle. This elevated level of progesterone feeds back to the hypothalamus to diminish GnRH secretion. Thus, LH and FSH secretions also decrease. If the uterus of the cow has not recognized a pregnancy by day 17, the uterus begins to produce the hormone prostaglandin $F_{2\alpha}$. Prostaglandin ($PGF_{2\alpha}$) acts upon the corpus luteum to cause CL regression. As a result of CL regression, progesterone levels decrease abruptly removing the inhibition on GnRH, LH, and FSH secretion. As a result, GnRH secretion increases thus, LH and FSH secretions increase and the increased level of FSH in the bloodstream stimulates new follicular development and increased estrogen production resulting once again in estrus (Fig. 11.5).

Ovulation occurs in the cow approximately 24 to 30 hours after the beginning of estrus. During ovulation, the follicle ruptures due to the interaction of hormones rather than bursting as a result of built-up pressure inside. When ovulation occurs, the follicular fluid surrounding the ovum carries the ovum into the infundibulum of the oviduct. After the ovum is released from the follicle, the remaining cells of the follicle luteinize during the next 4 to 5 days to form the corpus luteum. The corpus luteum is about twice the size and about the same shape as the mature follicle but no longer contains the egg. The importance of the corpus luteum is to produce high levels of progesterone that maintain the optimum uterine environment for embryo development to allow the embryo to signal its presence. If an embryo does not signal its presence by approximately day 17 of the estrous cycle, the uterus begins to secrete prostaglandin $F_{2\alpha}$ to stimulate regression of the corpus luteum and initiation of a new estrous cycle.

If pregnancy occurs, the corpus luteum remains present on the ovary and continues to produce progesterone at high levels for the duration of pregnancy. These high levels of progesterone suppress GnRH with LH and FSH production resulting in low levels of estrogen production from the ovary. Thus, cows do not display heat during the pregnancy.

The process of follicle development, ovulation, and corpus luteum development and regression is illustrated in Figure 11.4. The cow will exhibit estrus approximately 12 hours (a range of 10–27 hours); the average length of the estrous cycle is 21 days (a range of 19–23 days). The major changes in ovarian structure and function, in relation to the estrous cycle, are shown in Figure 11.5.

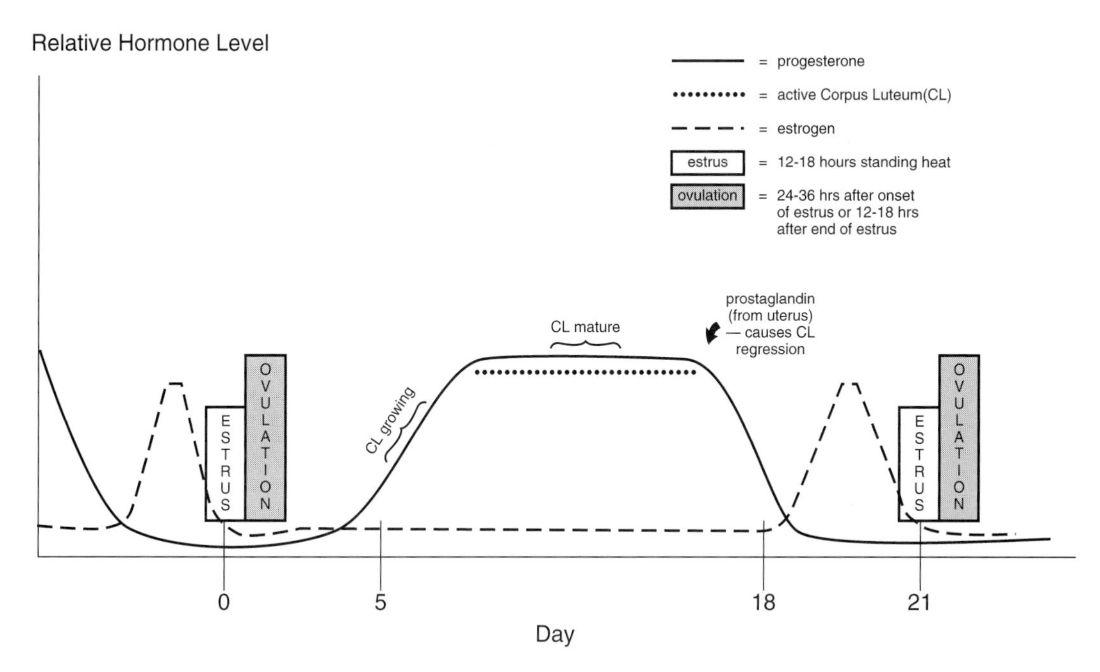

FIGURE 11.5 The major hormone and ovarian changes in the cow's estrous cycle.
Source: Colorado State University.

The changing length of day also influences the estrous cycle, the onset of pregnancy, and the seasonal fluctuations in male fertility. Changing day length acts both directly and indirectly on the animal—it acts directly on the hypothalamus by influencing the secretion of hormones and indirectly by affecting plant growth, thus altering the amount and quality of available nutrition. Increasing length of day is associated with increased reproductive activity in both males and females.

The Reproductive Organs of the Bull

Figure 11.6 shows the reproductive organs of the bull. They consist, in part, of two testes (testicles) that are bean-shaped organs held in the scrotum. Male sex cells (called sperm or spermatozoa) are formed in tiny seminiferous tubules of the testis. Sperm from each testicle then pass through very small tubes into an epididymis, which is a tube that is attached to the exterior of the testicle. Each epididymis leads to a larger tube, the vas deferens (also called the deferent duct or ductus deferens). The two vas deferens converge from the left and right sides of the body to connect with the urethral canal at its upper end, very near to where the urinary bladder opens into the urethra. The urethra is a large canal that leads through the penis to the outside of the body.

The seminal vesicles and prostate gland are accessory sex glands that are found at the base of the urinary bladder. The left and right parts of the seminal vesicles, which lie against the urinary bladder, consist of glandular tissue that supplies secretion that moves through the exit tube of each seminal vesicle into the urethra. The prostate gland is composed of a group of some 12 or more glandular tubes, each of which empties into the urethra. Another accessory sex gland, the bulbourethral (Cowper's) gland, which also empties its secretion into the urethral canal, is posterior to the prostate.

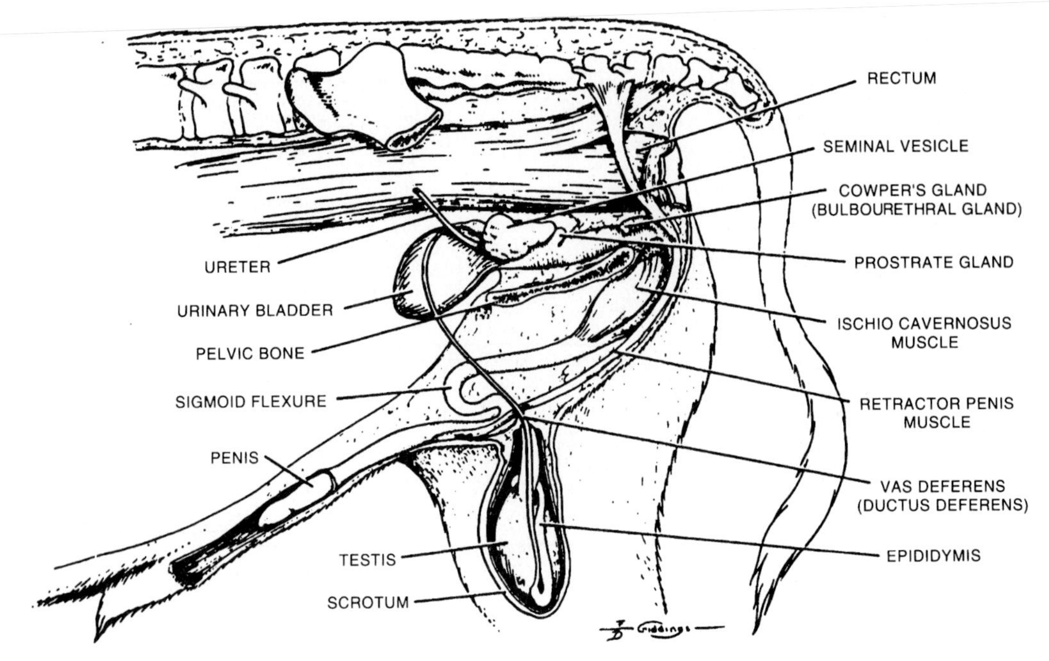

FIGURE 11.6 Reproductive organs of the bull.
Source: Colorado State University, Dennis Giddings.

The testicles (testes) produce (1) sperm cells that fertilize the ova and (2) the hormone testosterone that gives the bull its masculine characteristics. Thus, if both testicles are removed (castration), the male loses his sperm-producing capacity and his masculine appearance. The castrated bull calf (steer) is not only sterile, but the masculine characteristics (heavy neck and shoulders; deep bellow) will not develop. If the calf is castrated while he is immature, the vas deferens, seminal vesicles, prostate, and bulbourethral glands all but cease further development. If castration is done in the mature animal, the remaining genital organs tend to shrink in size and function.

The scrotum is a two-lobed sac that contains and protects the two testicles. The scrotum also regulates temperature of the testicles, maintaining them at a temperature lower than body temperature (3°F–7°F lower). When environmental temperature is low, the tunica dartos muscle of the scrotum contracts, drawing the testicles toward the warmer temperature of the body; when the environmental temperature is high, this muscle relaxes, permitting the testicles to drop away from the body and its warmth. This heat-regulating mechanism of the scrotum begins at about the time of puberty. When the environment is so hot that the testicles cannot cool sufficiently, the formation of sperm is impeded and a temporary condition of lowered fertility is produced.

Occasionally during fetal development, one or both of the testicles fail to descend from the body cavity into the scrotum. The animal, known as a cryptorchid, is sterile if both testicles are retained but usually fertile if one testicle descends. Masculine characteristics develop because testosterone is produced.

Within each testicle, sperm cells are generated in the seminiferous tubules and testosterone is produced in the cells between the tubules, which are called interstitial cells or cells of Leydig (Fig. 11.7).

Sperm cells from the testicle undergo maturation in the epididymis. In passing through this long tube, the sperm acquire more capacity to fertilize ova. Sperm taken from the part of

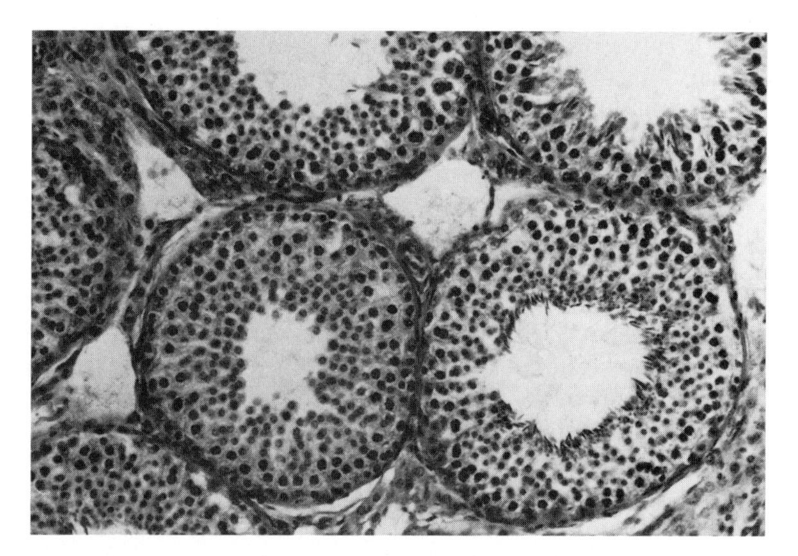

FIGURE 11.7 A cross section through the seminiferous tubules of the testis of the bull. The tubule in the lower right-hand corner demonstrates the more advanced stages of spermatogenesis as the spermatids are formed near the lumen (opening) of the tubule.
Source: Animal Reproduction Laboratory, Colorado State University.

the epididymis nearest the testicle are not likely to be able to fertilize ova, whereas those taken from areas farther along this long, winding tube increasingly show the capacity to fertilize.

In the sexually mature bull, sperm reside in the epididymis in vast numbers. Sperm confined to the epididymis for long periods of time degenerate and are absorbed in the part of the epididymis farthest from the testicle.

The vas deferens are transportation tubes that carry sperm-containing fluid from each epididymis to the urethra. The vas deferens join the urethra near its origin as the urethra leaves the urinary bladder. In the mature bull, the vas deferens is about 0.3 in. in diameter, except in its upper part where it widens to form the ampulla, about 4–6 in. long and a half-inch wide. The ampulla of the vas deferens is profusely supplied with nerves from the pelvic plexus of the sympathetic nervous system.

Under the excitement of anticipated mating, the secretion loaded with spermatozoa from each epididymis is propelled into each vas deferens and accumulates in the ampulla of the deferent duct. This brief accumulation of semen in the ampulla is an essential part of sexual arousal. The sperm reside briefly in the ampulla until the moment of ejaculation when the contents of each ampulla are pressed out into the urethra, through the penis, enroute to their deposition in the female tract.

The urethra is a large, muscular canal extending from the urinary bladder and runs posteriorly through the pelvic girdle and curves downward and forward through the full length of the penis. Very near the junction of the bladder and urethra, the tubes from the seminal vesicles and the tubes from the prostate gland join this large canal. The bulbourethral glands join the urethra at the posterior floor of the pelvis.

The bulbourethral glands are located on either side of the pelvic urethra, just posterior to the urethra-penis where the urethra-penis dips downward in its curve. The bulbourethral glands are covered by fibrous tissue. In the bull, they are about ½–1 in. in size. The secretion from the bulbourethral glands is thick and viscous, very slippery and lubricating, and whitish in color.

The seminal vesicles, prostate, and bulbourethral glands are known as the accessory sex glands. Their primary functions are to add seminal volume and nutrition to the sperm-rich fluid coming from the epididymis. The bull has a semen volume of 3–10 ml per ejaculate with 4–18 billion sperm in each ejaculate.

The penis serves as (1) a passage for urine, (2) a passage for the products originating in other reproductive organs, and (3) an organ of copulation. It is a muscular organ characterized especially by its spongy, erectile tissue that fills with blood under considerable pressure during periods of sexual arousal, making the penis rigid and erect.

The penis of the bull is about 4–5 ft in length and 1.5 in. in diameter, tapering to the free end, and it is s-shaped when relaxed. This s-curve, or Sigmoid flexure, is eliminated when the penis is erect. The s-curve is restored after copulation when the relaxing penis is drawn back into its sheath by a pair of retractor muscles. The free end of the penis is termed the glans penis. Only a small portion of the penis of the bull extends beyond its sheath during erection. The full extension awaits the thrust after entry into the vagina has been made.

The testicles produce hormones under stimuli coming from the anterior pituitary (AP) situated at the base of the brain. The AP elaborates two hormones important to male performance—luteinizing hormone (LH) and follicle-stimulating hormone (FSH). These two hormones are known as gonadotropic hormones because they stimulate the gonads (ovary and testicle). LH stimulates testosterone production by the interstitial cells. FSH stimulates cells in the seminiferous tubules to develop into functional spermatozoa. There is some evidence that FSH also influences testosterone levels. It secretes releasing factors through portal vesicles that affect the anterior pituitary and its production of FSH and LH.

Puberty in the Heifer

Age at puberty is a major determinant of lifetime reproductive efficiency of cows. *Puberty* in the heifer is attained when she will stand to be mounted and she ovulates for the first time. This may occur in some heifers when they are 6–8 months of age. Occasionally, these heifers become pregnant prior to weaning if they are not separated from late-weaned bull calves. Usually, the heifers that show early estrous cycles discontinue these cycles until they approach 12–13 months of age. Research studies have shown that this first estrus is not accompanied by an ovulation in nearly 25% of the heifers.

Puberty is influenced primarily by age and weight of the heifers in addition to length of daylight (photoperiod), breed, biological type (mature cow weight), and other factors. Weight may be a more important factor than age, so heifers should reach an adequate weight by 13–14 months of age if they are to calve at 2 years of age. The target breeding weight of the heifers should be 60–65% of the average mature cow weight in the herd. For example, if the cow weight was 1,150 lbs (BCS = 5), the desired heifer weight would be approximately 700 to 750 lbs.

Puberty in the Bull

Puberty in the bull occurs when viable sperm are first produced. It occurs at approximately 12 months of age, although it can vary in individual bulls several months before or after this age depending on biological type (primarily frame size and potential mature weight), nutrition, and health status. A semen collection can verify puberty; however, scrotal circumference is a good predictor of puberty. Most bulls have reached puberty when scrotal circumference measures approximately 26 cm. However, this does not mean the bulls are satisfactory breeders at this scrotal circumference as sperm evaluation (number, activity, and abnormalities) and other reproductive traits must be evaluated.

TABLE 11.1 Breed Comparisons: Testicle Size of Yearling Bulls and Its Relationship to Age of Puberty in Heifers

				Scotal Circumference of Yearling Bulls	
Breed	No. Heifers	Heifer Age at Puberty (days)	No. Bulls	Average (cm)	Range (cm)
Gelbvieh	81	341	22	34.8	30.2–42.2
Brown Swiss	126	347	19	34.3	31.0–39.6
Red Poll	95	352	20	33.5	29.7–37.1
Angus	24	372	79	32.8	26.2–38.4
Simmental	157	372	28	32.8	26.2–39.1
Hereford	27	390	55	30.7	26.2–36.1
Charolais	132	398	31	30.4	25.4–37.6
Limousin	161	398	20	30.2	24.4–34.3
Average		368		32.3	

Source: MARC Beef Research Program, Progress Report No.1.

There are scrotal circumference differences both between and within breeds as noted in Table 11.1. This table also shows that as breed average scrotal circumference increases, heifers reach puberty earlier. The correlation between heifer age at puberty and yearling bull scrotal circumference is 0.98 among breed averages.

BREEDING

Natural Service

The term *natural service* implies the bull is responsible for breeding cows rather than the producer using artificial insemination. Approximately 95% of beef cows are bred by natural service.

To help eliminate reproductive losses due to poor fertility, bulls should be evaluated for breeding soundness 30–60 days prior to the breeding season. A breeding soundness evaluation (BSE) consists of a (1) physical examination of the bull with emphasis on the reproductive system, (2) measurement of scrotal circumference, and (3) evaluation of at least one semen sample to evaluate sperm morphology and motility. The BSE should be done by an experienced veterinarian or a qualified reproductive physiologist.

Surveys show that approximately 30% of cow-calf producers in the United States semen test and take scrotal circumference measurements as part of a prebreeding evaluation of their bulls (NAHMS, 1997, 1998). Reproductive efficiency could improve if more producers regularly evaluated the breeding soundness of their bulls. Approximately 30% of all tested bulls are classified as questionable or unsatisfactory breeders.

Physical Examination

A *physical examination* should include observation of all conditions that might interfere with the bull's ability to locate cows in heat and breed them. A review of previous disease problems and any recent stressful conditions is useful. For example, high temperatures resulting from infection and extremely low temperatures causing frostbite of the testicles can cause structural defects in sperm cells.

A visual appraisal for unthriftiness, body condition, and structural soundness is also made. Bulls should be in good body condition, neither too thin nor too fat. Feet, legs, and eyes are among the most important structural traits that should be sound.

The reproductive system is examined by rectal palpation. Size, shape, and consistency of the prostate, seminal vesicles, and ampullae are noted, and the internal inguinal rings are examined for size. The external genitalia (scrotum, penis, and prepuce) are examined for any structural abnormalities or adhesions when the semen sample is collected. The pelvic area can also be measured.

Scrotal Circumference

Scrotal circumference is primarily related to sperm production and semen volume. Scrotal circumference has been shown to be related to age of puberty in heifers (Table 11.1). This relationship also indicates that as scrotal circumference in bulls increases, their daughters reach puberty at earlier ages. Scrotal circumference is measured in centimeters (cm) by using the following simple procedure. The testicles are palpated gently but firmly down into the bottom of the scrotum. A scrotal tape is placed up over the testicles-scrotum and tightened loosely around the scrotal neck close to the body wall. The tape is then slid slowly and carefully downward. The sliding loop will enlarge as the size of the testicles-scrotum increases until the largest circumference is attained (Fig. 11.8). No additional tension should be placed on the free end of the tape. Once the tape has reached the largest circumference, it will drop off and the size can be read directly from the tape.

Table 11.2 provides the guide for determining acceptable scrotal circumference for bulls of different ages as recommended by the Society for Theriogenology. Figure 11.9 shows the relationship of increased scrotal size to improved semen characteristics. Note a threshold effect where there appears to be little advantage to having scrotal circumferences above 36–38 cm in yearling bulls. Figure 11.10 illustrates the distribution of scrotal circumference measures in the Angus breed. Note that a large majority of the bulls fall in the 32 to 40 cm range.

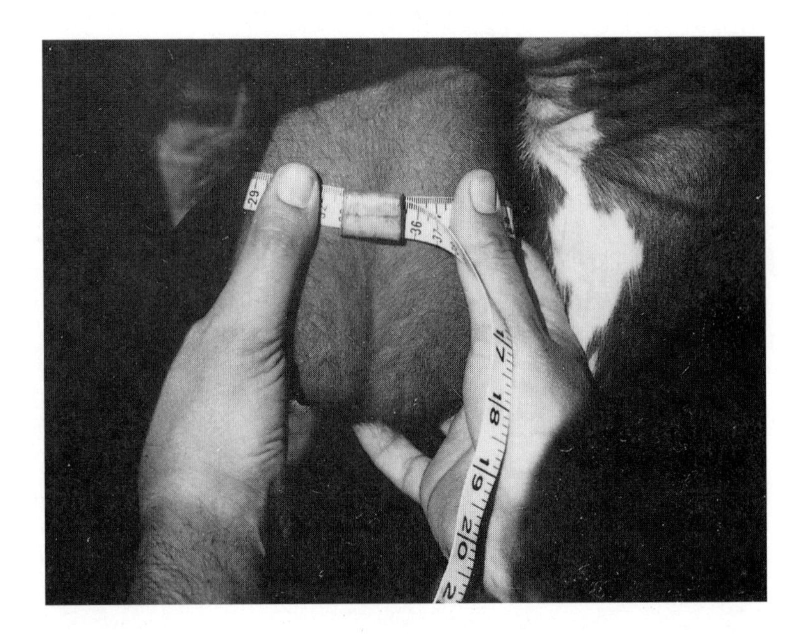

FIGURE 11.8 Measuring scrotal circumference.

TABLE 11.2 Evaluation Criteria for Scrotal Circumference, Semen Morphology, and Motility for Breeding Soundness Evaluation in Bulls

Category	Threshold
Scrotal circumference	30 cm at ≤15 months of age 31 cm at >15 ≤18 months 32 cm at >18 ≤21 months 33 cm at >21 ≤24 months 34 cm at >24 months
Sperm morphology	≥70% normal sperm
Sperm motility	≥30% individual motility and/or "Fair" gross motility

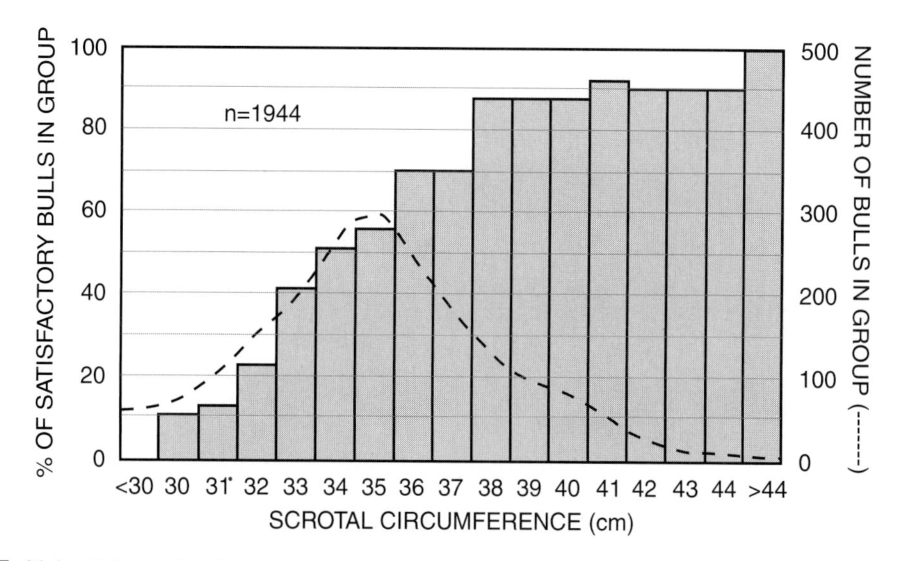

FIGURE 11.9 Relationship between scrotal circumference and semen characteristics in yearling bulls.
Source: Coulter.

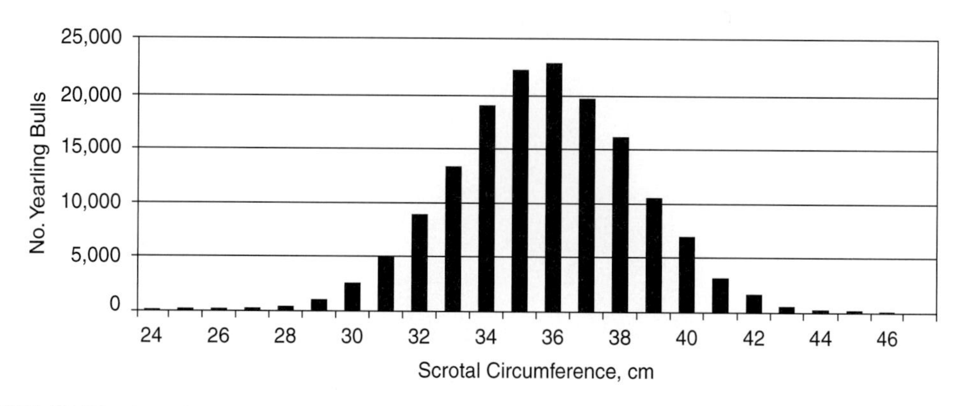

FIGURE 11.10 Distribution of scrotal circumference in yearling Angus bulls.
Source: American Angus Association.

Semen Evaluation

A *semen sample* is collected through the use of an electroejaculator. Temperature control of the semen sample is necessary to properly evaluate motility. Also, the sample must be kept clean and free of water, urine, or chemical disinfectants. The semen sample is evaluated for motility and morphology. Motility can be estimated by observing the mass movement of sperm under the microscope. Sperm cell motility is more critically evaluated by observing the movement of individual sperm cells.

Morphology or sperm cell structure is evaluated by placing a small amount of semen and stain on a slide, gently mixing them, and spreading a thin film on a glass slide. A special microscope is used to evaluate the sperm cells and at least 100 cells are counted and classified as normal or as one of several abnormal types. Sperm cell abnormalities fall into two classifications: primary abnormalities, which are generally the more severe morphologic defects, and secondary abnormalities. Secondary abnormalities, such as droplets, are usually temporary and are a reflection of sexual immaturity. Some of these abnormalities are shown in Figure 11.11.

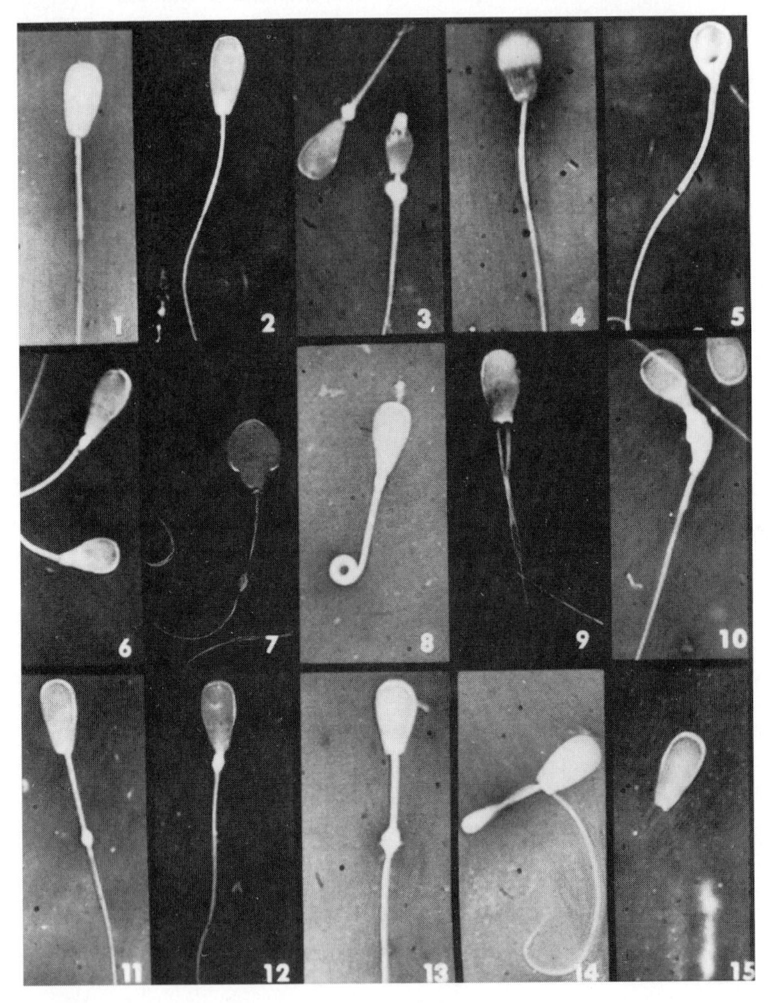

FIGURE 11.11 Some of the more common primary and secondary abnormalities of bull sperm. 1 & 2—Normal bovine sperm; 3—head defects; 4—large head; 5—round head; 6—pear-shaped head; 7—head defects and abnormal midpiece; 8—coiled tail; 9—double midpiece; 10—swollen midpiece; 11 & 13—distal droplet (droplet at midtail); 12—proximal droplet (droplet near head); 14—bent tail; and 15—separated head. *Source: Successful Farming Magazine.*

Although many other factors are related to high conception rates, the number of abnormal sperm cells can be an additive factor in a bull's infertility if they become too numerous. There is evidence that sperm cell abnormalities may begin to affect fertility when 25% or more of the cells are abnormal.

As previously cited, Table 11.2 gives the evaluation criteria for scrotal circumference, semen morphology, and semen motility as recommended by the Society for Theriogenology (American Veterinary Society for the Study of Breeding Soundness). A breeding soundness examination form can be completed that classifies the bull into one of three categories: (1) satisfactory potential breeder, (2) unsatisfactory potential breeder, or (3) classification deferred. Bulls that are classified as "satisfactory potential breeders" must pass the physical examination and equal or exceed the minimum thresholds in each of the categories identified in Table 11.2.

When evaluating scrotal circumference and semen characteristics of young bulls near one year of age, it should be kept in mind that bulls of this age will generally improve in scrotal size and semen quality with further maturity. Scrotal circumference at 12 months of age is approximately three-quarters of the potential size at maturity (Fig. 11.12). Development of the epididymis and other sperm-carrying ducts improves in most bulls up to 15 months of age. Due to immaturity, some young bulls receive "classification deferred" on the BSE at 12–14 months of age even though many of these bulls will pass a BSE at a later time (usually a few weeks). This is why a retest is common for a certain percentage of yearling bulls.

Bulls passing a breeding soundness examination may later experience fertility problems. Significant changes in scrotal temperature, caused by disease and extremely high or low ambient temperatures, can cause reduced fertility, temporary sterility, or permanent sterility. Scrotal frostbite, scrotal sunburn, severe insect bites on the scrotum, and certain diseases or infections can have a detrimental effect on semen quality. For example, foot rot in the bull can cause two breeding problems: (1) a physical difficulty in breeding females because of lameness and (2) an elevated body temperature that may reduce semen quality and fertility. Some of the environmental insults to testicular function may take from 2 to 12 months to restore normal fertility.

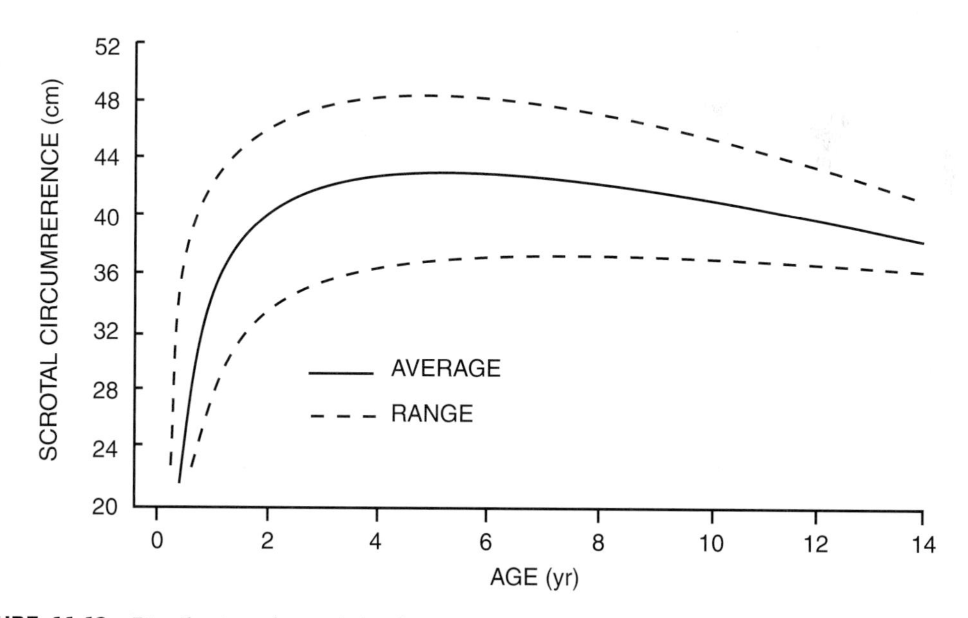

FIGURE 11.12 Distribution of scrotal size, by age, in 1,275 Angus bulls. Note the range in size above and below the average.
Source: Coulter.

Nutrition in young bulls has a significant effect on testicular function. Diets adequate in the basic nutrients are needed to help initiate puberty. However, there are several studies showing that feeding high-energy diets can lower semen production and semen quality (Coulter and Kastelic, 1997). Feeding programs for growing yearling bulls should utilize moderate-energy diets that achieve a target liveweight at the beginning of the breeding season (e.g., 1,000–1,200 lbs at 14–16 months of age). Bull performance tests, or developing show bulls where weaned calves are fed to gain more than 3 lbs per day or where body fat is increased beyond minimum levels, increase the risks of impaired fertility in the bulls.

Libido (sex drive), mating ability, and mating capacity are important factors affecting calf crop percentage; however, there is no simple, successful measurement of these characteristics in individual bulls. It is recommended that virgin, yearling bulls be mated to one or two females prior to the beginning of the breeding season. Observation of their mating behavior may allow producers to avoid serious problems during the subsequent breeding season. The mating experience of the yearling bull may help him be a more successful breeder early in the breeding season.

Cow-to-Bull Ratios

Producer surveys show that cow-to-bull ratios of 18 to 1 for yearling bulls and 26 to 1 for mature bulls are common (NAHMS, 1997). Many cow-calf producers could reduce production costs by increasing the cow-to-bull ratios in their herd. Individual bulls can successfully impregnate 50 to 60 cows in a 60-day breeding season; however, the typical *cow-to-bull ratio* is quoted as 30:1. This ratio is used in single-sire herds as a base unless the mating capacity of individual bulls is known. Environmental conditions may direct an upward or downward adjustment of the 30:1 ratio. Dominance of individual bulls in multiple-sire breeding herds exists, and individual bulls will rank differently over several years as to the proportion of calves they sire.

Yearling bulls (14–18 months of age) can be used without reducing pregnancy rate if yearling bulls are properly managed. Average cow-to-bull ratio should be somewhat less (25:1) and the yearling bull should be well fed and not allowed to get too thin.

Bulls can be used naturally in a synchronization program and calf crop percentage can be kept high. In this situation, one bull should be used on one group of 15 to 30 cows kept in a relatively small pen or drylot. It has been well demonstrated that a bull can settle a relatively large number of cows in a 5-day period; however, there is considerable variation among bulls. Giving the bull a 10-day rest between cycles is recommended.

ARTIFICIAL INSEMINATION

Artificial insemination (AI) is a process whereby semen is deposited in the female reproductive tract by artificial techniques rather than by natural mating. Successful AI was first accomplished in cattle in the early 1900s.

The primary advantage of AI is the extensive use of outstanding bulls to optimize genetic improvement. For example, a bull used in natural service typically sires 10–50 calves per year over a productive lifetime of 3–8 years. In an artificial insemination program, a bull can produce 200–400 units of semen per ejaculate, with four ejaculates typically collected per week. If the semen is frozen and stored for later use, hundreds of calves can be produced by a single sire (one calf per 1.5 units of semen), and many of these offspring can be produced long after the sire is dead. AI also can be used to control reproductive diseases, and sires can be used that have been injured or are dangerous when used naturally.

Artificial insemination and length of breeding season can affect the number of resulting pregnant cows. A long breeding season (120 days or more, under natural service) extends available bull power, whereas a short breeding season (45 days) places heavy demand on available bulls. Cow-to-bull ratios would vary under these two extremes. Length of breeding season can affect AI technicians. Heat checking and technician efficiency can become lax with extended breeding seasons. AI technicians can tire and affect conception rates when large numbers of cows are bred in a few days with or without estrous synchronization.

AI programs can lower calf crop percentage unless high levels of management are given to heat checking, semen quality and handling, and insemination technique. A producer considering implementing an AI program may want to shorten a long breeding season by gradually reducing the number of days of natural service over a period of 2–3 years. For some operations, a 25-day AI plus a 25–50-day natural service breeding season seems to work well. Other operators may use a 4-day AI program plus a 50- or 60-day natural service breeding.

Costs and returns should be carefully assessed before implementing an AI program. There are very few research studies that compare the economics of AI with well-planned natural service programs. The number of cows that are artificially inseminated in the United States is not likely to increase rapidly above the current level of 5% unless AI becomes more cost effective or more convenient to implement. The most widespread use of AI involves yearling heifers that have been developed in feedlots. Heat detection and insemination are easier to implement under these conditions.

Semen Collection and Processing

There are several different methods of collecting semen. The most common method is the artificial vagina (Fig. 11.13), which is constructed similar to the natural vagina. The semen is collected when a bull mounts an estrous female or by training him to mount another animal or object (Fig. 11.14). When the bull mounts, his penis is directed into the artificial vagina by the person collecting the semen and the semen accumulates in the collection tube (Fig. 11.15).

Semen can also be collected by using an electroejaculator, in which a probe is inserted into the rectum and an electrical stimulation causes ejaculation. This is used most commonly in bulls that are not easily trained to use the artificial vagina or where semen is collected infrequently.

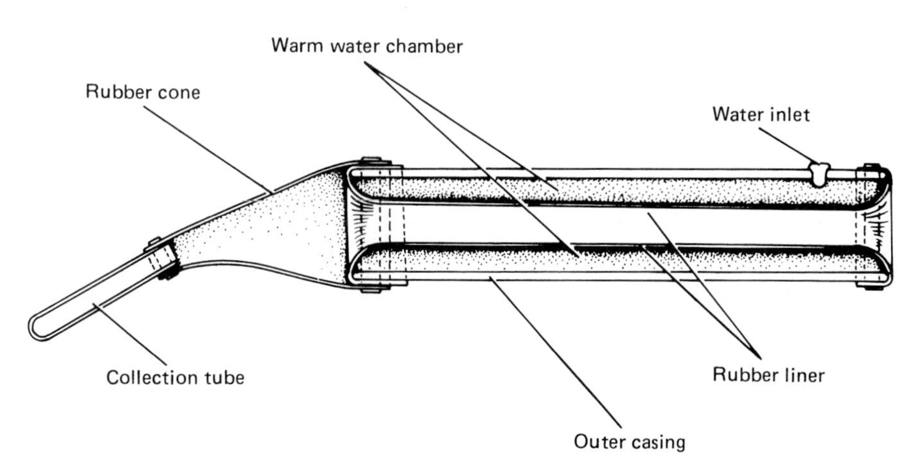

FIGURE 11.13 An artificial vagina used in collecting semen from the bull.
Source: Colorado State University.

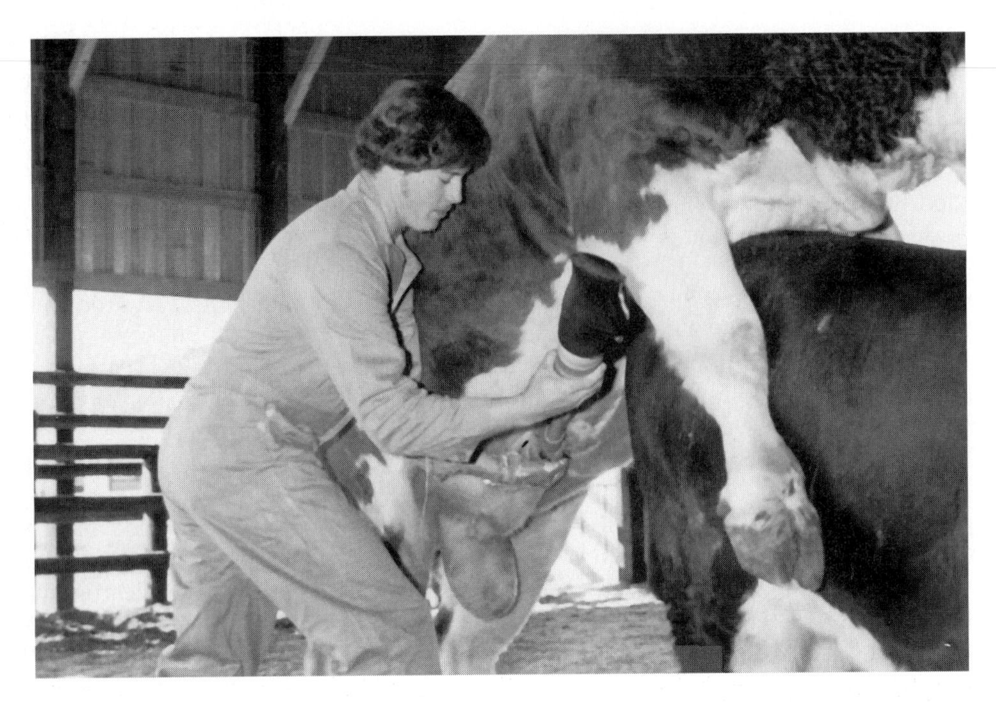

FIGURE 11.14 Collecting semen from a bull using the artificial vagina.
Source: Animal Reproduction Laboratory, Colorado State University.

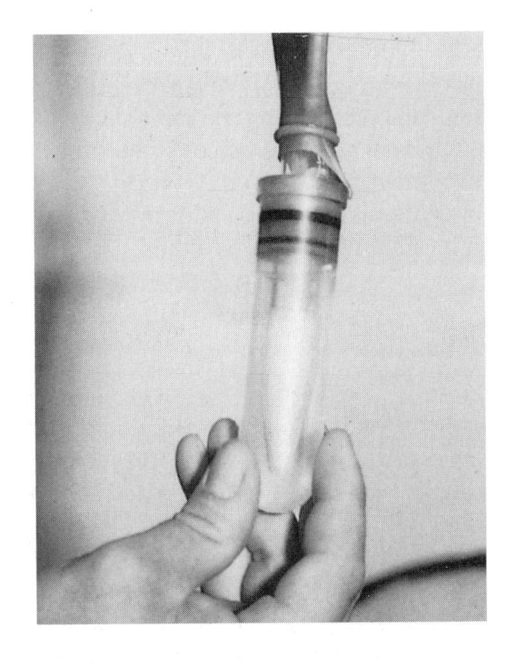

FIGURE 11.15 Approximately 4 ml of bull semen in the collection tube attached to the artificial vagina. Note that the collection tube is immersed in water to control the temperature. The livability of the sperm is decreased when they are subjected to sudden temperature changes.
Source: Animal Reproduction Laboratory, Colorado State University.

If semen is collected too frequently, the number of sperm per ejaculate decreases. Semen from the bull is typically collected twice a day for 2 days a week. After the semen is collected, it is evaluated for volume, sperm concentration, motility of the sperm, and sperm abnormalities (Fig. 11.16). The semen is usually mixed with an extender that dilutes the ejaculate to a greater volume. This greater volume allows a single ejaculate to be processed into several

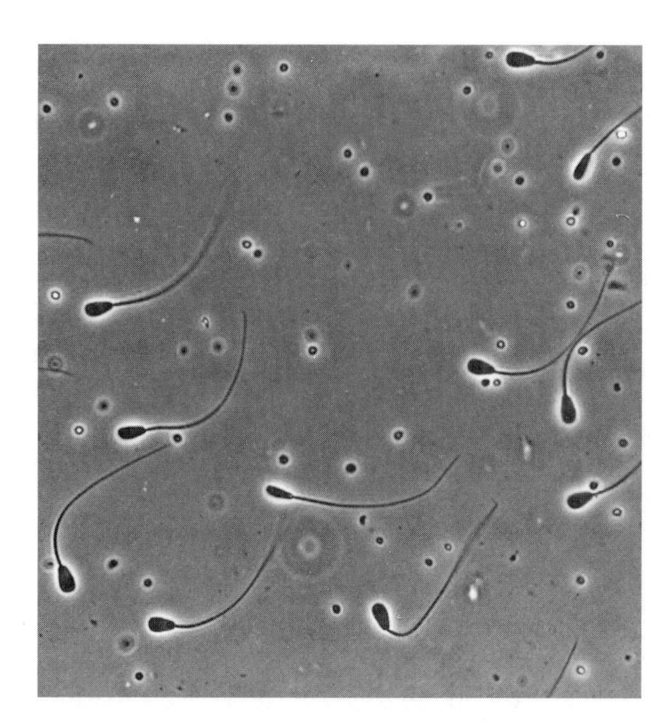

FIGURE 11.16 Normal bull semen.
Source: Animal Reproduction Laboratory, Colorado State University.

units of semen, where one unit of semen is used each time a female is inseminated. The extender is usually composed of nutrients such as milk and egg yolk, a citrate buffer, antibiotics, and glycerol. The amount of extender used is based on the projected number of viable sperm available in each unit of extended semen. For example, each unit of semen for insemination in cattle should contain at least 10 million motile, normal spermatozoa.

Some semen is used fresh. However, because semen can only be stored fresh for 2 to 3 days, most of it is frozen in liquid nitrogen and stored in plastic straws. Straws are the most common method of storing and using frozen semen. The semen can be stored in this manner for an indefinite period of time and still retain its fertilization capacity. Most cattle are inseminated with frozen semen rather than fresh semen.

Sexing Semen

Research has identified a process where approximately 90% X-chromosome bearing sperm and 90% Y-chromosome bearing sperm can be sorted from one another (Seidel, 1997). The process utilizes flow cytometry that sorts at a speed of 10 million cells per hour (about a normal AI dose). While the technique is promising, there are two concerns: (1) sperm viability is slightly compromised and (2) the number of sperm sorted is lower than the number needed for routine artificial insemination. As a result of these factors, pregnancy rates from the use of sexed semen are about 80% of what is normally expected from an AI breeding program.

However, in 2000, sexed-semen technology was introduced commercially to the British dairy industry as a result of a joint venture between XY, Inc., a Fort Collins research firm, and a major British breeder cooperative. Because of the technology cost—about $50 per sorted dose, plus semen cost—adoption will be restricted to the seedstock industry. The technology is expected to become less expensive in the future and thus could have a significant impact on both the beef and dairy industries.

The benefits of sexed semen include making it possible to select which breeding females have replacement heifer offspring, producing all male progeny from terminal sires, and breeding first-calf heifers such that they give birth to female offspring and thus lowering dystocia rates.

Insemination of the Female

Prior to insemination, the frozen semen is thawed. Semen that is thawed should not be refrozen and used later because conception rates are reduced.

High conception rates using AI depend on the female cycling and ovulating; accurately detecting estrus; using semen that has been properly collected, extended, and frozen; thawing and handling the semen satisfactorily at the time of insemination; insemination techniques; and avoiding stress and excitement to the animal being inseminated.

Semen handling technique is a critical component of a successful AI program. The temperature in a liquid nitrogen tank warms as semen is moved toward the top of the tank. For example, at approximately 3 to 4 in. from the top of the tank, the temperature warms above the critical control point and thawing may be initiated.

If semen is transferred between tanks, the exchange should occur quickly and away from direct sunlight. When removing individual straws from the tank at breeding time, minimize the amount of time the semen rack is elevated in the neck of the tank. Furthermore, it is important to frequently monitor the level of nitrogen in the tank and to keep the tank on a pallet to avoid corrosion.

Detecting Estrus

Estrus must be detected accurately because it signals the time of ovulation and determines the proper timing of insemination. The best indication of estrus is the condition called *standing heat,* in which the female stands still when mounted by a male or another female.

Cows are typically checked for estrus twice daily, in the morning and evening. They are usually observed for at least 30 minutes to detect standing heat. Other observable signs are restlessness, roughed up tailhead, being followed by a group of bull calves, attempting to mount other cows, pink swollen vulva, and a clear mucous discharge from the vagina. Some producers use sterilized bulls or hormone-treated cows as heat checkers in the herd. These animals are sometimes equipped with a head harness that greases or paint-marks the cow's back when she is mounted. Other estrous detection aids include chalking or painting tailheads, use of color indicative patches (Kamar or Bovine Beacons) or electronic transmitting patches (HeatWatch).

Research suggests that the distribution of when cows express estrus is skewed to the early morning hours. The percent of cows expressing estrus from midnight to 6:00 A.M.; 6:00 A.M. to noon; noon to 6:00 P.M.; and 6:00 P.M. to midnight was 43%, 22%, 10%, and 25%, respectively.

Proper Timing of Insemination

Females should have a body condition score (BCS) of 5 or 6 to assure a high pregnancy rate. The length of estrus and time to ovulation is quite variable in cows and heifers. This variability poses difficulty in determining the best time for insemination. An additional challenge is that the egg and sperm are short-lived when put into the female reproductive tract. It is estimated that the ova is viable for 6–10 hours following ovulation, and sperm are viable for 24–30 hours in the female reproductive tract. Also, estrus is sometimes expressed without ovulation occurring and occasionally the reverse of this occurs.

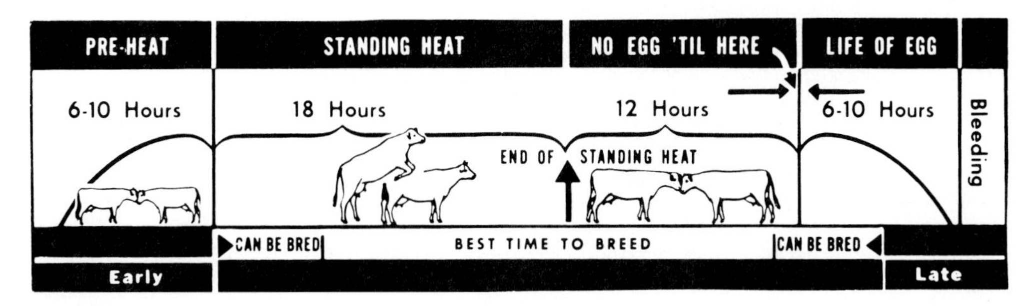

FIGURE 11.17 Identifying the preferred time for AI breeding.
Source: American Breeders Service.

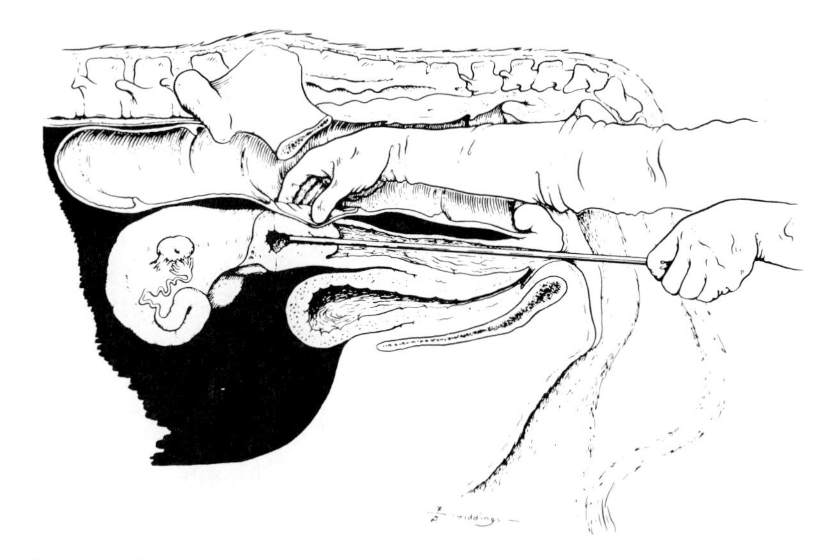

FIGURE 11.18 Artificial insemination of the cow. Note that the insemination tube has been manipulated through the cervix. The inseminator's forefinger is used to determine when the insemination rod has entered into the uterus.
Source: Colorado State University.

Time of insemination should occur 6–8 hours prior to ovulation because sperm require 2–6 hours in the female tract before they are fully capable of fertilization. Cows found in estrus in the morning are usually inseminated that evening, and cows in heat in the evening are inseminated the following morning. Because ovulation occurs 24–30 hours after the onset of heat, insemination should occur near the end of estrus (Fig. 11.17).

Cows are penned and inseminated in a chute that restrains the animal. The most common insemination technique in cattle involves the inseminator having one arm in the rectum to manipulate the insemination tube through the cervix (Fig. 11.18). The insemination tube is passed just through the cervix and the semen is deposited into the body of the uterus.

The number of cows inseminated each year is not well documented. In the United States, it is estimated that approximately 3–5% of beef cows (or about 1.5 million) are inseminated each year. In contrast, approximately 75% of U.S. dairy cows (approximately 7 million) are inseminated each year.

TABLE 11.3 Beef Semen Sold Domestically, Exported, and Custom Frozen

	Units of Semen (thou)								
	Domestic Sales			Export Sales[a]			Custom Frozen Sales		
Breed	2000	1996	1980	2000	1996	1980	2000	1996	1980
Angus	702	557	418	195	104	155	810	600	252
Red Angus	77	52	29	357	116	12	166	78	16
Simmental	72	102	192	44	64	29	68	79	264
Polled Hereford	39	68	189	54	45	52	73	47	159
Charolais	18	18	28	14	20	10	65	45	28
Brahman	15	27	22	26	46	63	64	29	18
Limousin	14	20	30	13	12	7	109	107	70
Gelbvieh	11	15	NA	7	5	NA	52	51	NA
Maine Anjou	4	7	22	—	—	—	108	98	NA
Other	43	9	64	24	11	—	790	123	170
Total	995	903	1,034	734	438	299	2,305	1,762	1,472

[a] Dollar value for export sales in 2000 was $2.2 mil.
Source: National Association of Animal Breeders.

The amount of semen that is sold domestically, exported, and custom frozen for the various breeds is shown in Table 11.3. It is not known how much beef semen is used to breed dairy cows. Semen is available commercially from several artificial insemination companies (see the beef AI organizations listed in the Appendix).

SYNCHRONIZATION OF ESTRUS

Estrous synchronization involves controlling or manipulating the estrous cycle so that females in a herd express estrus at approximately the same time. It is a useful part of an AI program because checking heat and breeding animals, particularly in large pasture areas, is time consuming and expensive. Estrous synchronization is a tool used in successful embryo transfer programs and it can be used with natural service where bulls are used intensively in breeding cows for a few days. Synchronization of estrus is used in approximately 12% of the U.S. cowherds (NAHMS, 1998).

Success of a synchronization program depends on many factors, including facilities, available labor, the body condition, postcalving interval, and fertility level of the cows and heifers. Also important are herd health, high-quality semen, qualified inseminators, females that are cycling, and accurate detection of estrus. But one of the most serious considerations affecting a herd's success with estrous synchronization is selecting the proper method. Several of the more common synchronization methods are presented based on a review by Geary (1997).

Prostaglandin

In 1979, prostaglandin was cleared for use in cattle. *Prostaglandins* are naturally occurring fatty acids that have important functions in several of the body systems. The prostaglandin that has a marked effect on the reproductive system is prostaglandin $F_{2\alpha}$. Lutalyse®, Estrumate®, and Prostamate® are commonly used prostaglandins.

Earlier in the chapter, it is pointed out that the corpus luteum (CL) controls the estrous cycle in the cow by secreting the hormone progesterone. Progesterone prevents the expression of heat and ovulation. Prostaglandin destroys the CL, thus eliminating the source of progesterone. About three days after the injection of prostaglandin, the cow will be in heat. For prostaglandin to be effective, the cow must have a functional CL. It is ineffective in heifers that have not reached puberty or in noncycling mature cows. In addition, prostaglandin is ineffective if the CL is immature or has already started to regress. Prostaglandin is, then, only effective in heifers and cows that are in days 5–18 of their estrous cycle. Because of this relationship, prostaglandin is given in either a one- or two-injection system separately or in combination with other products that influence the estrous cycle.

One-Injection System with 10 Days of Breeding

Days 1–5: The first 5 days are a conventional AI program of heat detection and insemination.
Day 6: At this point, the herd owner must decide whether to proceed with the prostaglandin injection. If the percent cycling is satisfactory and drug costs are considered reasonable, the breeder injects the females not already detected in heat with prostaglandin.
Days 6–11: The synchronized AI breeding season continues with five more days of heat detection and insemination (Fig. 11.19).
Days 27–33: Cattle that synchronize but do not settle on the first service will return to heat. Repeat services can be performed at this time to produce additional AI calves.

This is the most popular protocol that uses only prostaglandin to synchronize estrus and can result in more than 90% of cyclic cows being bred the first 10 days of the breeding season. This system can be shortened by not inseminating during days 1–5. However, this will lower heat detection to 65–75%. In any event, if less than 20% of the females express heat in the first five days, success will be limited. Timed inseminating should not be used with this system.

One Injection of Prostaglandin with Breeding after Injection

Inject all cows and check heat and breed females 12 hours after standing heat (Fig. 11.19). Approximately 75% of the cycling cows would be expected to display heat during the next 4–5 days following injection (Fig. 11.19). For commercial herds, bulls can be turned in with the herd on day six.

Two-Injection System

By using two injections of prostaglandin, it is possible to get all cycling cows and heifers in heat within a five-day period. The following schedule describes how this is accomplished:

Day 1: All animals are injected with prostaglandin.
Days 11–14: All animals are again injected with prostaglandin. The second injection must be given on one of these days. While day 11 was traditionally recommended, more recent research suggests a better response if the second injection is administered on day 14.
Within 5 days following the second injection: All cycling cows and heifers should come into heat and can be heat detected and inseminated (Fig. 11.19). Most repeat services will occur approximately 20–25 days after the first heat is detected.
Days 32–38: Most repeat services occur between these days.

There are two different two-injection programs for synchronization with prostaglandin that allows breeding after each injection or only after the second injection. Timed insemination is not recommended with the two-shot system.

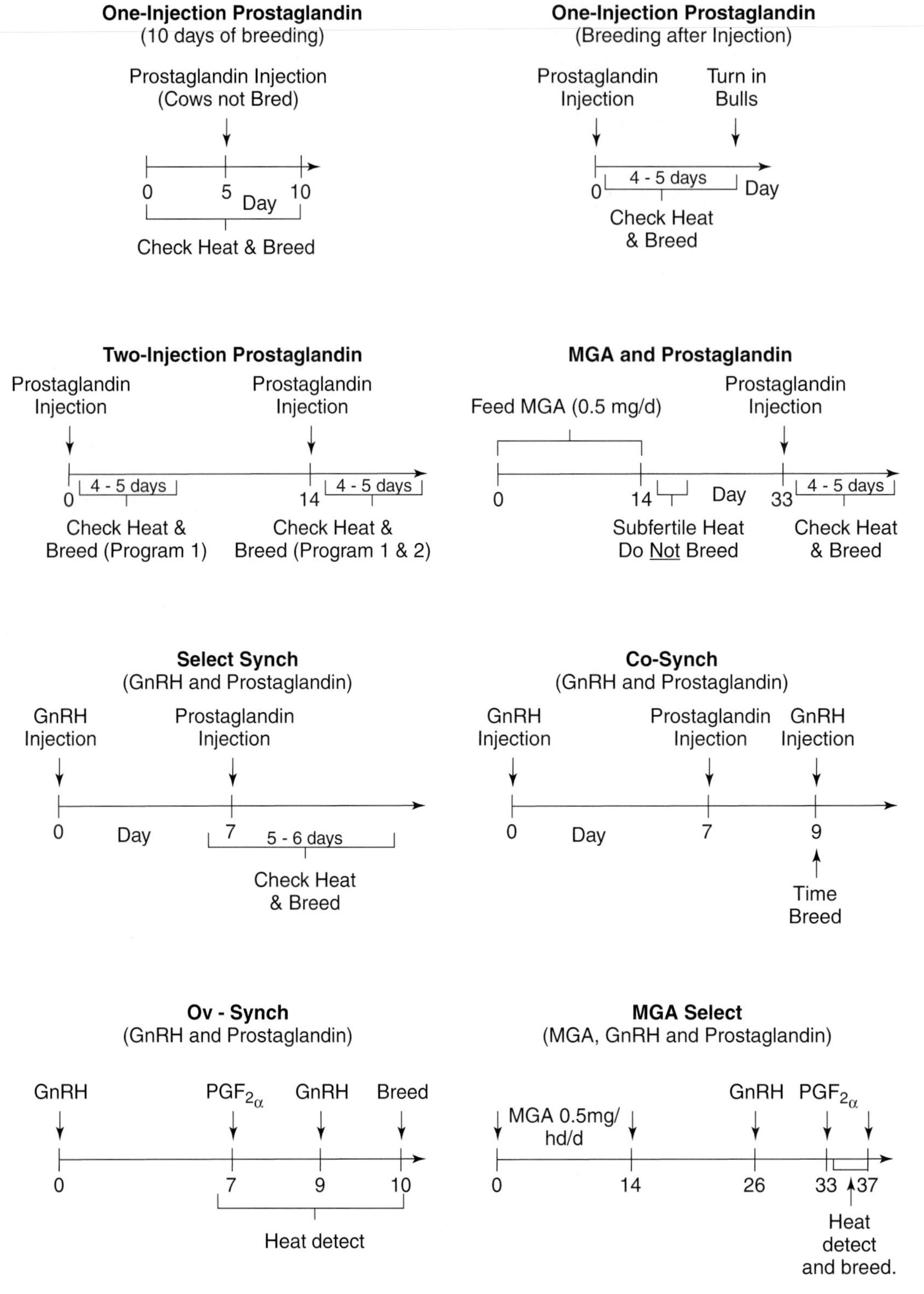

Figure 11.19 Protocols from several estrous synchronization programs.
Source: Geary (1997).

MGA and Prostaglandin

The use of *MGA (melengestrol acetate)* in conjunction with prostaglandin results in an inexpensive, easily administered synchronization system. This system best fits the breeding of heifers being raised in semiconfinement. The MGA-prostaglandin system showing the most promise is the following 31–36-day system:

Days 1–14: Feed 0.5 mg MGA per head per day. The MGA prevents heifers from coming into heat during this period. MGA can be fed as a topdressing or can be mixed in with the feed. It is important to provide enough bunk space for all heifers to have an opportunity to eat all the required amount of MGA.

Day 15: Stop feeding MGA. Heifers should come into heat within the next seven days. The fertility at this heat is very low, so do not inseminate.

Day 31: Inject prostaglandin. Heifers and cows respond to the prostaglandin by coming into heat within the next five days.

Days 32–38: Inseminate according to estrus. While the synchronized estrus will be relatively tight, some females may not show heat until approximately four days after injection. Some heifers may not initiate a cycle, so semen is wasted if timed insemination is used (Fig. 11.19). For natural service, the prostaglandin shot can be waived with bull turnout occurring on that day.

Work at Virginia Tech suggests that a tighter estrus synchronization is achieved if the prostaglandin is given on day 33, 19 days following the end of MGA feeding.

Prostaglandin is not a wonder drug; it will work only in well-managed herds where a high percentage of the females are cycling. Biologically, it has been well demonstrated that prostaglandin can synchronize estrus. Producers, however, must weigh the cost against the economic benefit. Caution should be exercised in administrating prostaglandin to pregnant cows as it may cause abortion. Estrous synchronization in cattle may not be advisable in areas where inadequate protection can be given to young calves during severe blizzards. Also, herd health programs must be excellent to prevent high losses from calf scours and other diseases that become more serious where large numbers of newborn calves are grouped together.

Select Synch (GnRH and Prostaglandin)

This system involves an injection of a gonadotropin-releasing hormone (GnRH) to each female followed seven days later with an injection of prostaglandin (Fig. 11.19). Estrous detection begins 24 hours prior to the prostaglandin injection and continues for the next five to six days. Most females will display standing estrus on day 2 through day 4 following the prostaglandin injection. Some producers administer a second GnRH injection to females that have not been detected in heat 72 hours past the prostaglandin injection. These females are then mass inseminated at the time of the second injection.

Response to this estrus synchronization system, like the other systems, will vary from herd to herd. Data from seven different herds showed a 70% estrous response and a 60–65% average conception rate.

Co-Synch (GnRH and Prostaglandin)

This is a slight variation to the Select Synch system. It involves administering the GnRH injection on day 0, prostaglandin on day 7, and a second GnRH injection 48 hours after the prostaglandin injection coupled with mass or timed mating (Fig. 11.19). The GnRH injection initiates a fertile ovulation in cows that have not yet exhibited estrus. Calf removal for 48 hours from the time of the prostaglandin injection until the timed insemination increases pregnancy rates by approximately 10%.

Ov-Synch (GnRH and Prostaglandin)

This is a variation of the Co-Synch system that involves an injection of GnRH on day 0, prostaglandin administration on day 7, and a second GnRH injection given 48 hours later. Sixteen to 18 hours after the second GnRH females are mass mated. Pregnancy rates of about 50% can be expected. By using 48-hour calf removal, this rate can be increased to 60%. This system is labor-intensive in that cows may go through the chute four times.

MGA Select

This system combines the MGA-PGF system with the Select Synch program as a means to tighten synchrony, initiate reproductive activity in anestrous cows, and improve pregnancy rates. MGA is fed at the rate of 0.5 mg/herd/day for 14 days. Twelve days later GnRH is given. Seven days following GnRH, prostaglandin is administered, then heat detection and breeding (five days) is initiated.

Synchronization with Natural Service

Some producers cannot economically use AI but would like the benefits of estrous synchronization. Research has shown that natural service can be used with estrous synchronization if managed properly. One bull per 15 to 20 females in a small pasture (or drylot), and rotated every 24 hours with a rested bull, is recommended during the synchronization period of 4–5 days. Observations have shown that 5 to 20 females may be serviced by one bull in a 24-hour period. Pregnancy rates during the synchronized period have ranged from 60–80% and 75–95% during a 30-day breeding season.

Bulls can be used in all synchronization programs. The most popular programs are to produce MGA for 14 days, wait 17 days, and then place the bulls with the females. The advantages of this program are low drug costs, no heat detection, and less demand on the bulls in a short time period.

Economic Assessment

Cost comparisons for different estrous synchronization programs and AI are shown in Table 11.4. Natural service costs can be in the $15–$75 range depending primarily on the purchase price of bulls and the cow-to-bull ratio (Chapter 5). Caution should be exercised in comparing these costs because they are calculated at a fixed point in time. While the costs can be used as guidelines, current costs should be calculated for each operation at the time estrous synchronization and AI are being evaluated.

EMBRYO TRANSFER

The primary function of *embryo transfer* is to increase the reproductive rate of valuable females by tenfold or more in a given year and fivefold or more in a cow's lifetime. Even greater increases in reproductive rates can be expected as new technologies are improved. Other uses of embryo transfer are to obtain offspring from infertile cows and to export or import breeding stock to reduce disease risk.

TABLE 11.4 Comparison of Estrous Synchronization Programs Using AI

Program	Estimated Synchronized Pregnancy Rate (%)[a]	Estimated Cost/Pregnant Female ($)
Prostaglandin:		
One injection of prostaglandin (10-day breeding)	50	34
One injection of prostaglandin with heat detection	37	36
Two injections with breeding after each injection	50	37
Two injections with breeding after the 2nd injection	50	39
MGA/Prostaglandin (heifers):		
With heat detection	55	35
Without heat detection (mass AI)	55	36
Select Synch	55	37
Co-Synch (mass AI):	50	53
With calf removal	60[b]	44

[a] Pregnancy rate is defined as the number of pregnant females divided by the number of females that received the synchronization treatment. These costs can be compared to $32 for natural service without synchronization.
[b] Preliminary data in 2 herds revealed that 48-hour calf removal (from prostaglandin injection until the 2nd GnRH injection) can increase pregnancy rates 10%.
Source: Geary (1997).

Embryo transfer is sometimes referred to as *ova transplant* or *embryo transplant.* In this procedure, an embryo in its early stage of development is removed from its own mother's (the donor's) reproductive tract and transferred to another female's (the recipient's) reproductive tract. The first successful embryo transfers were accomplished in rabbits in 1890 and in cattle in 1951. Commercial embryo transfer companies have been established in the United States and several foreign countries. More than 400,000 embryos are transferred annually throughout the world with about one-half of the embryos utilized in the United States. Approximately one-half of the embryos are fresh with the remaining half frozen.

Superovulation is the production of a greater-than-normal number of eggs. Females that are donors for embryo transfer are injected with fertility drugs, which usually cause several follicles to mature and ovulate. The two most common methods of superovulation are using pregnant mare serum gonadotropin (PMSG) or follicle-stimulating hormone (FSH), with the latter usually producing more usable embryos. FSH injections are given over 3–4 days with prostaglandin $F_{2\alpha}$ (e.g., Lutalyse) administered usually on the third day.

A nonsurgical embryo collection procedure occurs when a flexible rubber tube (Foley catheter) with three passageways is passed through the cervix and into the uterus (Fig. 11.20). A rubber balloon, which is built into the anterior end of the tube, is then inflated to about half the size of a golf ball so that it expands to fill the uterine lumen and to prevent fluid from escaping around the edges. There are two holes in the tube anterior to the balloon that lead into separate passageways, one for fluid entering the uterus and the other for fluid draining from the uterus. A balanced salt solution (to which antibiotics and heat-treated serum are added) is placed in a container and held about 3 ft above the cow. The container is connected to the Foley catheter by an inflow tube. A second tube is connected to the other passageway of the catheter to drain off the medium that has washed (flushed) the ova and embryos out of the uterus. The solution is then collected in tall cylinders holding about two pints of fluid. The embryos are isolated by filtering the collected fluid through a cup-sized container with a fine stainless-steel filter in the bottom and then searching for them with a microscope.

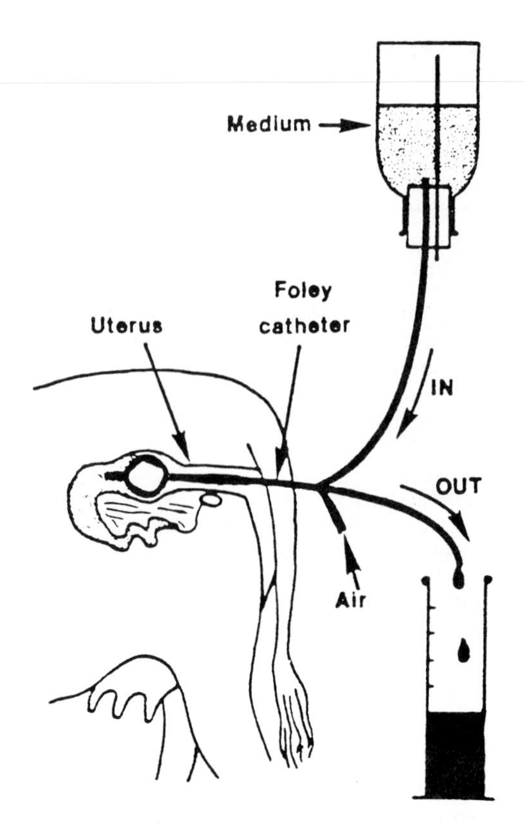

FIGURE 11.20 Nonsurgical collection of embryos.
Source: Colorado State University.

FIGURE 11.21 A Limousin donor cow with her 11 embryo transfer calves, the result of one superovulation.
Source: Dr. Peter Elsden.

At approximately 12 hours and 24 hours after first standing estrus, the donor is bred artificially. Usually frozen-thawed semen is used, and 2–4 straws of high-quality semen is recommended for each insemination. The recipients are usually synchronized with a prostaglandin. The recipient cow must be in estrus within 1.5 days of the donor's estrus for best results. Recipients are selected for calving ease, high milk production, and excellent health status.

An ideal response is 5 to 12 usable embryos per donor per superovulation treatment; however, a range of 0 to more than 20 embryos can be expected (Fig. 11.21). On average, two to four calves will result per superovulated donor if fertile donors are utilized in a well-managed embryo transfer program. A potential donor cow usually produces more embryos if she is 3–10 years of age, has calved regularly each year of her productive life, usually conceives in two services or less, has exhibited regular estrous cycles, comes from fertile blood lines, did not retain her placenta, and has no history of calving difficulties. The bull, to be bred to the donor cow, should have a history of excellent semen production and a high conception rate determined by his semen having been used in an AI program.

The key to the justification of embryo transfer is identifying genetically superior cows and bulls. Procedures for identifying genetically superior cattle are given in detail in Chapters 4 and 12. Much of the guesswork can be eliminated by making repeat matings where genetically superior calves have previously been produced. Embryo transfer is usually confined to seedstock herds, where genetically superior females can be more easily identified and high costs can be justified. Embryo transfer calves have an estimated average breakeven price of $1,500–$2,500 at one year of age (costs for embryo transfer are $500–$1,000 above the usual production costs). Seedstock producers must evaluate the marketability of the embryo transfer calves. Some genetically superior embryo transfer calves may not sell for a sufficient amount to cover costs. Some embryo transfer calves that are not genetically superior may be merchandised for high prices because of a demand that has been previously created.

Donor selection that combines expected high levels of financial return, genetic superiority, and high reproductive potential indicates that the number of donors worthy of an embryo transfer program are relatively few. This "relatively few" number could be justified into several thousand head, which is less than 1% of the total United States cowherd.

Although embryo transfer was done surgically in the past, today new nonsurgical techniques are used. The embryos are transferred by way of an artificial insemination gun shortly after being collected. The recipient females need to be in the same stage (within 36 hours) of the estrous cycle for highly successful transfers to occur. Large numbers of females must be kept for this purpose, or estrous synchronization of a smaller number of females will be necessary. The embryos can be frozen in liquid nitrogen and remain dormant for years or decades.

Although the conception rate is lower for frozen embryos and higher for fresh embryos, ongoing research is narrowing the difference in conception rate. In cattle, approximately 85% of frozen embryos are normal after thawing; however, 40–55% of those normal embryos result in confirmed pregnancies at 60–90 days. In contrast, fresh embryos transferred the same day of collection have a pregnancy rate of 55–65%.

While there is no decline in the embryo recovery rate the first three times a donor is superovulated, additional superovulation treatments result in a reduced number of embryos for some donors. Most donors can be superovulated three or four times a year. Season, breed, and lactating versus dry donors appear to have little effect on the success of embryo transfer. Surveys have shown that 2% of the recipients abort between 3 and 9 months of gestation, 4% of the embryo transfer calves die at birth, and another 4% die between birth and weaning—representing a total loss of 10% between 3 months gestation and weaning. These losses may be similar to those experienced in a typical natural service program.

Recent advances in reproduction technology include the birth of the first test-tube calf (fertilization occurred outside the cow's body) in 1981 and the first identical twin calves resulting from embryo splitting. Splitting one embryo to produce two identical offspring is one of the initial steps in the cloning of cattle and other intriguing aspects of genetic manipulation (Fig. 11.22).

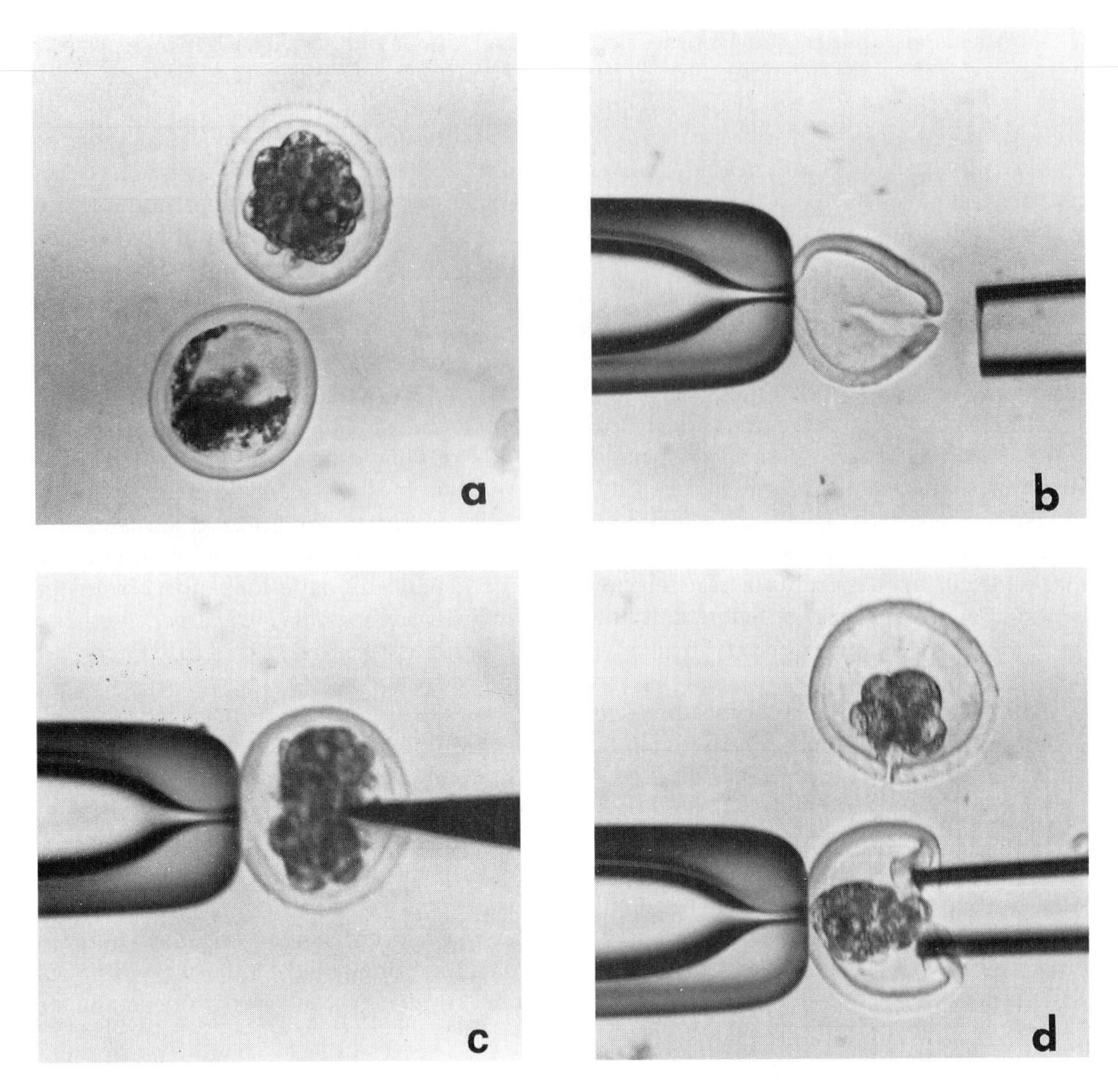

FIGURE 11.22 The process of embryo splitting. (a) A morula (40 to 70 cells of the fertilized egg) and an unfertilized egg; (b) the unfertilized egg with the cellular contents sucked out of it; (c) the morula is divided into two groups of cells with the microsurgical blade; (d) one-half of the cells are left in the morula while the other one-half of the cells are placed inside the other unfertilized egg. The result is two genetically identical embryos ready for transfer.
Source: Williams et al., 1983 NAAB Conference on Beef AI and Embryo Transfer.

CLONING

Cloning is a technique of molecular biology that produces a duplicate of the DNA or genes of an individual animal. Dolly is the noted sheep produced in 1997 in Scotland, being the first mammal cloned from an adult cell. Using a nuclear transfer procedure, the nucleus from a mammary gland cell was inserted into an empty egg cell. Cell division was initiated and the embryo was transplanted into another recipient ewe. Dolly was born genetically identical to the donor of the adult cell. This genetic and reproductive manipulation opens the door to the numerous possibilities of research in improving the productivity of cattle.

ABS Global, Inc., successfully cloned a Holstein bull calf in 1997 and reportedly has since repeated the result in beef cattle. A team of Australian geneticists cloned 470 cattle embryos from a single embryo. In 1998, a Holstein bull calf, "Mr. Jefferson," was cloned in Virginia. This clone was produced by nuclear transfusion from a fetal cell rather than an adult cell link, as was the case with Dolly.

However, cloning continues to be a controversial technology and only through an effective consumer education program will it ever become commonplace in the arena of food production.

PREGNANCY

When the sperm and egg unite (fertilization), conception occurs and pregnancy (gestation) is initiated. The fertilized egg initiates a series of cell divisions (see Fig. 11.23; also refer to Fig. 17.1). The new organism migrates through the oviduct to the uterus within 2–5 days, at which time it has developed to the 16- or 32-cell stage. The chorionic and amniotic membranes develop around the new embryo and attach it to the uterus. The embryo (later called the fetus) obtains nutrition and discharges wastes through these membranes. This period of attachment (from the 30th to the 35th day of pregnancy) is critical; if the uterine environment is not favorable, embryonic mortality will occur. It is vital that management practices protect the female early in pregnancy by providing feed of sufficient quality and by minimizing stress. The female should enter the breeding season in a thrifty, gaining condition and maintain that condition throughout the first weeks of pregnancy.

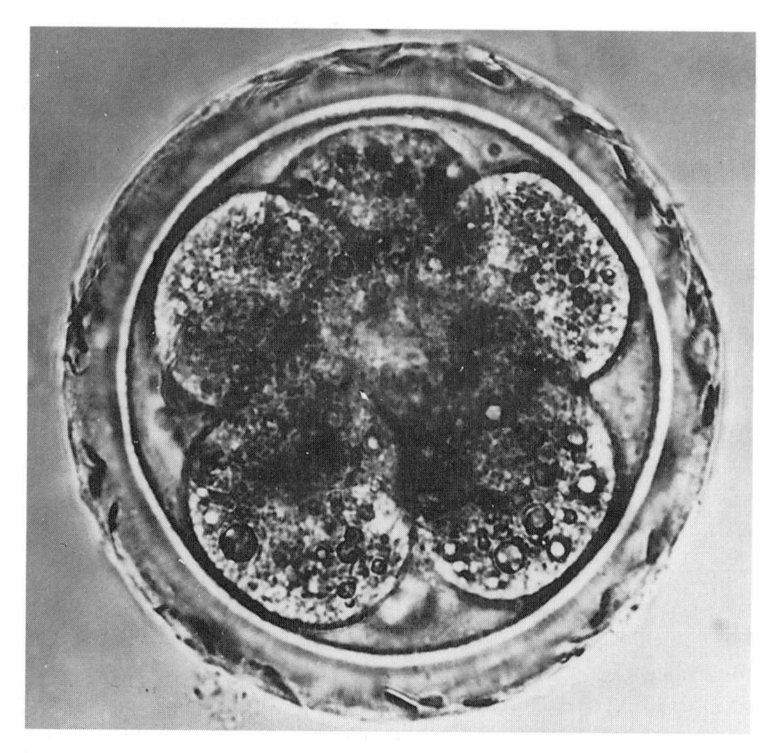

FIGURE 11.23 A bovine embryo in the 6-cell stage of development (magnified about 500 times). *Source:* Animal Reproduction Laboratory, Colorado State University.

The *embryonic stage* in the life of an individual embryo is defined as that period in which the body parts differentiate to the extent that the essential organs are formed. At approximately 45 days, the embryo becomes designated as a fetus. The *fetal period,* which lasts until birth, is mainly a time of growth. Gestation length in cattle is approximately 285 days with a range of 275–295 days. The gestation table that appears on the inside front cover of this book makes it easy to calculate projected calving dates when breeding dates are known.

Determination of Pregnancy

In some cow-calf operations, it is economically feasible to determine pregnancy in heifers and cows prior to their anticipated time of calving. This allows nonpregnant females to be evaluated for culling because they will not produce a calf to offset production costs. Approximately 35% of the U.S. cow herds are pregnancy tested each year (NAHMS, 1998).

Pregnancy in cows is most commonly determined by physical examination. The examiner's arm is inserted into the cow's rectum and the reproductive tract is palpated for pregnancy indications. The palpator wears a protective covering (e.g., a rubber or plastic sleeve) over the arm and hand. A lubricant such as liquid soap is used on the sleeve for ease of entry into the cow's rectum.

Palpation takes only a few seconds when the cow is properly restrained and the palpator is experienced. The restraining chute should have a front wall or gate and a bar just above the hocks. The latter prevents the animal from kicking the palpator. A gate located at the rear of the chute allows the examiner to enter and exit with ease. The gate, which swings across the chute, also provides a front entrance for cows coming into the chute behind the palpator. Facilities should accommodate the safety of the palpator and the proper handling of the animals.

Palpators must have a thorough knowledge of the female reproductive system and the changes that occur during pregnancy. They should be experienced in palpation and should be palpating cows on a regular basis. Technique and practice are important in the accurate diagnosis of pregnancy. Individuals palpating only a few cows once a year are likely to encounter a large number of errors. The use of ultrasound technology via transrectal screening can greatly enhance the accuracy of pregnancy detection once the technician is very well trained.

A skilled and practiced palpator can determine pregnancy as early as 30 days after breeding. Palpation at this early stage, however, should be accompanied by good breeding records that tell the palpator the approximate breeding date of the animal. Most palpators prefer breeding to end approximately 45 days prior to palpation.

Table 11.5 gives the fetal size and some other identifying characteristics used in pregnancy diagnosis.

Gestation Length and Losses

Historically, gestation length has been reported as 283 days for the European breeds. However, the introduction of the Continental breeds extended gestation length several days, so the average gestation length in the total population of cattle today is 285–286 days (see Fig. 11.24). The Brahman breed has a longer gestation length, averaging 290 days.

Losses during gestation are usually low—averaging between 2–3% reduction in calf crop percentage. The embryo may fail to implant for various reasons or abortion can occur during any stage of pregnancy. Abortion may result from physical trauma or disease problems. Seri-

TABLE 11.5 Fetal Size and Characteristics Used in Determining Pregnancy

Days of Gestation	Fetal Size		Identifying Characteristics
	Weight	Length (in.)	
30	1/100 oz	2/5	One uterine horn slightly enlarged and thin; embryonic vesicle size of large marble. Uterus in approximate position of nonpregnant uterus. Fetal membranes of a 30–90 day pregnancy may be slipped between fingers
45	1/8–1/4 oz	1–1 1/4	Uterine horn somewhat enlarged, thinner walled and prominent. Embryonic vesicle size of hen's egg
60	1/4–1/2 oz	2 1/2	Uterine horn 2 1/2 to 3 1/2 in. in diameter; fluid filled and pulled over pelvic brim into body cavity. Fetus size of a mouse
90	3–6 oz	5–6	Both uterine horns swollen (4–5 in. in diameter) and pulled deeply into body cavity (difficult to palpate). Fetus is the size of a rat. Uterine artery 1/8–3/16 in. in diameter Cotyledons 3/4–1 in. across
120	1–2 lbs	10–12	Similar to 90-day but fetus more easily palpated. Fetus is size of small cat with the head the size of a lemon. Uterine artery 1/4 in. in diameter. Cotyledons more noticeable and 1 1/2 in. in length. Horns are 5 to 7 in. in diameter
150	4–6 lbs	12–16	Difficult to palpate fetus. Uterine horns are deep in body cavity with fetus size of large cat—horns 6–8 in. in diameter. Uterine artery 1/4–3/8 in. in diameter. Cotyledons 2–2 1/2 in. in diameter
180	10–16 lbs	20–24	Horns with fetus still out of reach. Fetus size of a small dog. Uterine artery 3/8–1/2 in. in diameter. Cotyledons more enlarged. From sixth month until calving, a movement of fetus may be elicited by grasping the feet, legs, or nose
210	30–40 lbs	24–32	From 7 months until parturition, fetus may be felt. Age is largely determined by increase in fetal size
240	40–60 lbs	28–36	Uterine artery continues to increase in size—210 days, 1/2 in. in diameter; 240 days, 1/2–5/8 in. in diameter; 270 days, 1/2–3/4 in. in diameter
270	60–100+ lbs	28–38	

Source: A. M. Sorenson, Jr. and J. R. Beverly, *Determining Pregnancy in Cattle* (College Station, TX: Agricultural Extension Service, 1979).

ous diseases, such as brucellosis, leptospirosis, IBR (Infectious Bovine Rhinotracheitis), and others, can reduce the calf crop percentage by as much as 50%. Herd health programs (discussed in Chapter 16) should be planned on the basis of herd needs and surrounding conditions. If herd health is not given serious attention, gestation losses will be much higher than the 2–3% normally expected.

There are fewer death losses in crossbred calves than in purebred calves, demonstrating the effect of heterosis.

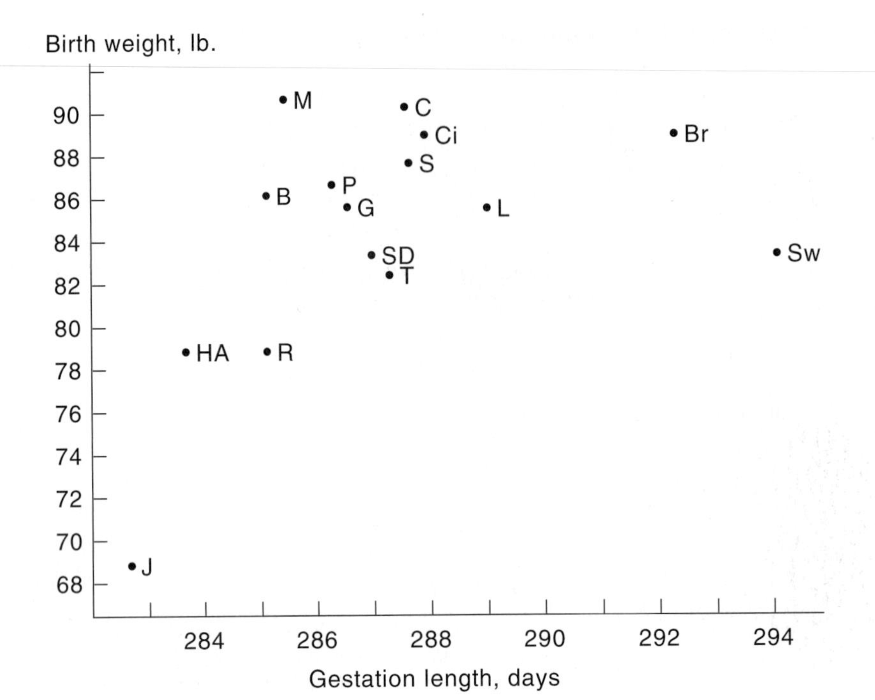

FIGURE 11.24 Relationship of gestation length to birth weight for various sire breeds. Breeds are (B) Brown Swiss, (Br) Brahman, (C) Charolais, (Ci) Chianina, (G) Gelbvieh, (HA) Hereford-Angus, (J) Jersey, (L) Limousin, (M) Maine-Anjou, (P) Pinzgauer, (R) Red Poll, (S) Simmental, (SD) South Devon, (Sw) Sahiwal, and (T) Tarentaise.
Source: MARC.

TABLE 11.6 Calf Losses by Days Following Calving

Days After Calving	% All Losses	Major Category	% Losses from Major Category
1–3	68%	Dystocia	61%
4–6	6	Accidents	31
7–9	5	Scours	38
10–42	12	Pneumonia	32
43–weaning	9	Missing	33

Source: U.S. Livestock and Range Research Station, Miles City, Montana.

CALVING

Factors Affecting Calf Losses at or Shortly after Birth

Parturition (birth) marks the termination of pregnancy. Calf losses at birth are typically the second most important reason for low-percent calf crops. Calving difficulty (dystocia) accounts for most calf deaths within the first 24 hours of calving (Table 11.6), and most calving difficulty occurs in 2-year-old heifers. Approximately 17% of heifers in U.S. cowherds are given assistance at calving, while only 3% of the cows require assistance (NAHMS, 1997).

Angus

Barzona

Beefmaster[1]

Brahman[2]

Brangus[3]

Plate A

Charolais[4]

Chianina

Gelbvieh[5]

Hereford

Limousin[6]

Plate B

Longhorn[7]

Rangemaker[5]

Piedmontese[5]

Red Angus[5]

Red Brangus[8]

Plate C

Santa Gertrudis[9]

Saler[10]

Shorthorn[11]

Simmental[5]

South Devon[5]

Plate D

Plates A–D Sources: National Beefmaster Association[1], J.D. Hudgins[2], International Brangus Breeders Association[3], American International Charolais Breeders Association[4], Leachman Cattle Company[5], North American Limousin Foundation[6], Texas Longhorn Breeders Association of American[7], Eagles Nest Ranch[8], Santa Gertrudis Breeders International[9], Wilson Livestock[10], American Shorthorn Association[11].

Plate E

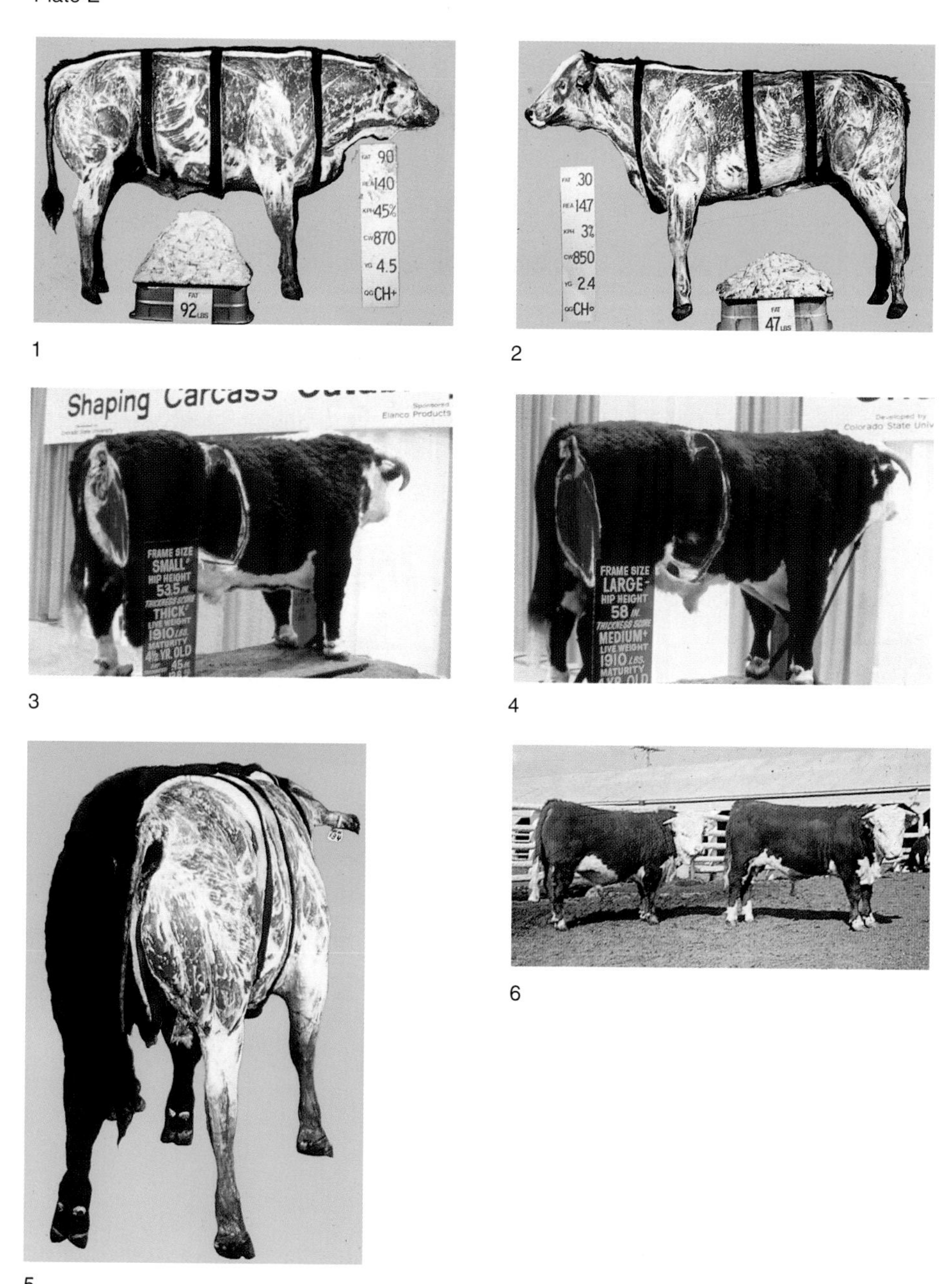

Plate E compares body composition of different yield grades and muscling in both live animals and their carcasses. Note the differences in trimmed fat from a Yield Grade 4 (1) versus a Yield Grade 2 (2). The Black stripes are hide and fat that remains at key locations on the body. Muscling differences exist independent of frame size. Small framed cattle can be heavily muscled (3) while large framed cattle can be average to below average in muscularity (4). A rear view of a market steer illustrates how the fat increases in thickness fom the middle of the back to the edge of the loin (5). Two yearling Hereford bulls of approximately the same weight are different in fat and lean compositin (6). The bull on the left would sire market steers having Yield Grade 4 or 5 carcasses, while the bull on the right would sire Yield Grade 1 or 2 steers. This assumes that the bulls would be bred to similar frame-size cows, and the steers slaughtered at approximately 1,150 pounds. Compare to plates F and G.

Plate F

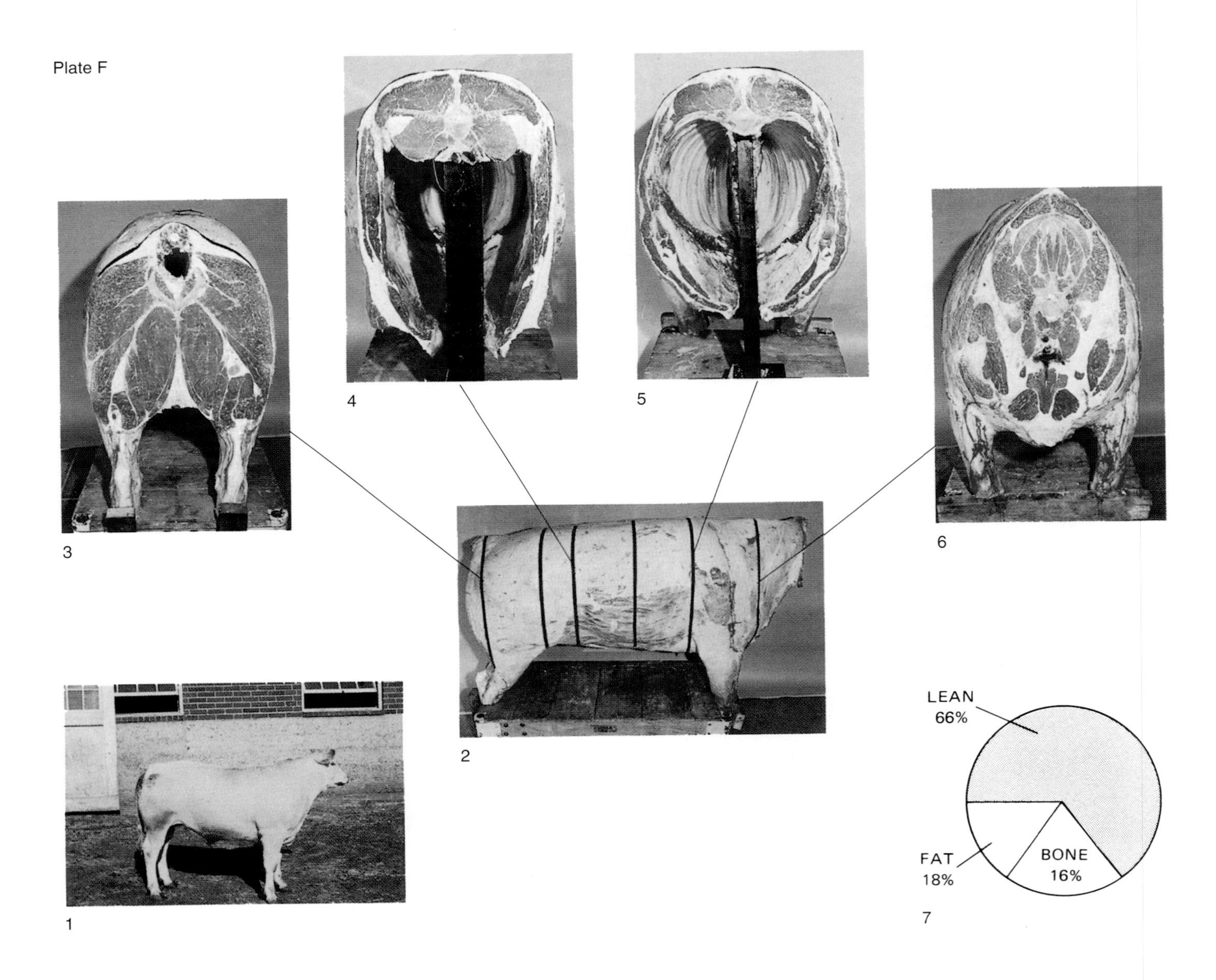

Plate G

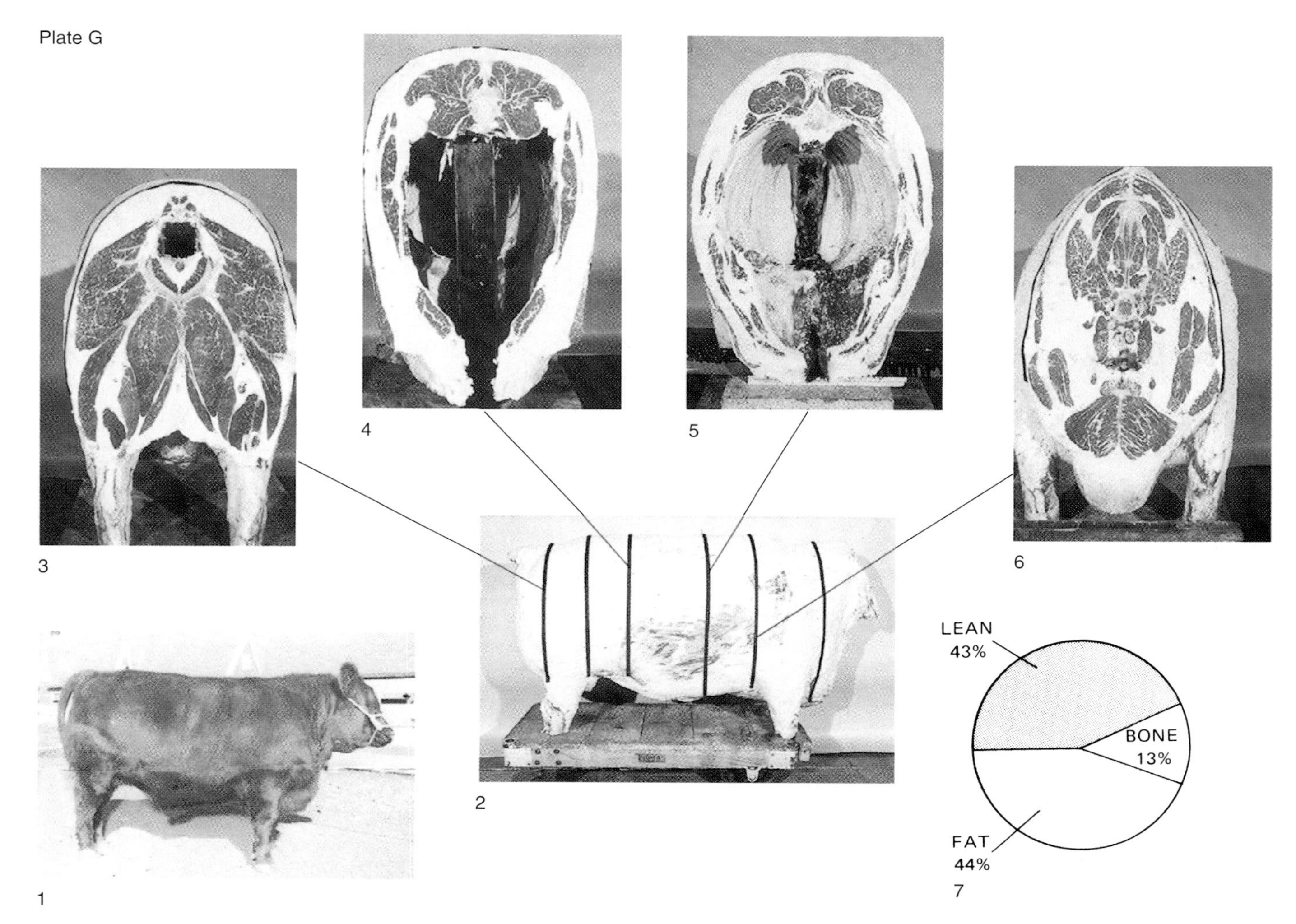

Plates F & G compare fat to lean composition on a Yield Grade 2 steer (plate F) and a Yield Grade 5 steer (plate G): (1) live animal—side view; (2) numbered ribbons indicate where following cross sections were made; (3,4,5, & 6) cross sections from rump, hip, mid-carcass, and shoulder, (7) the percentages of fat, lean, and bone found in this carcass.

Plate H

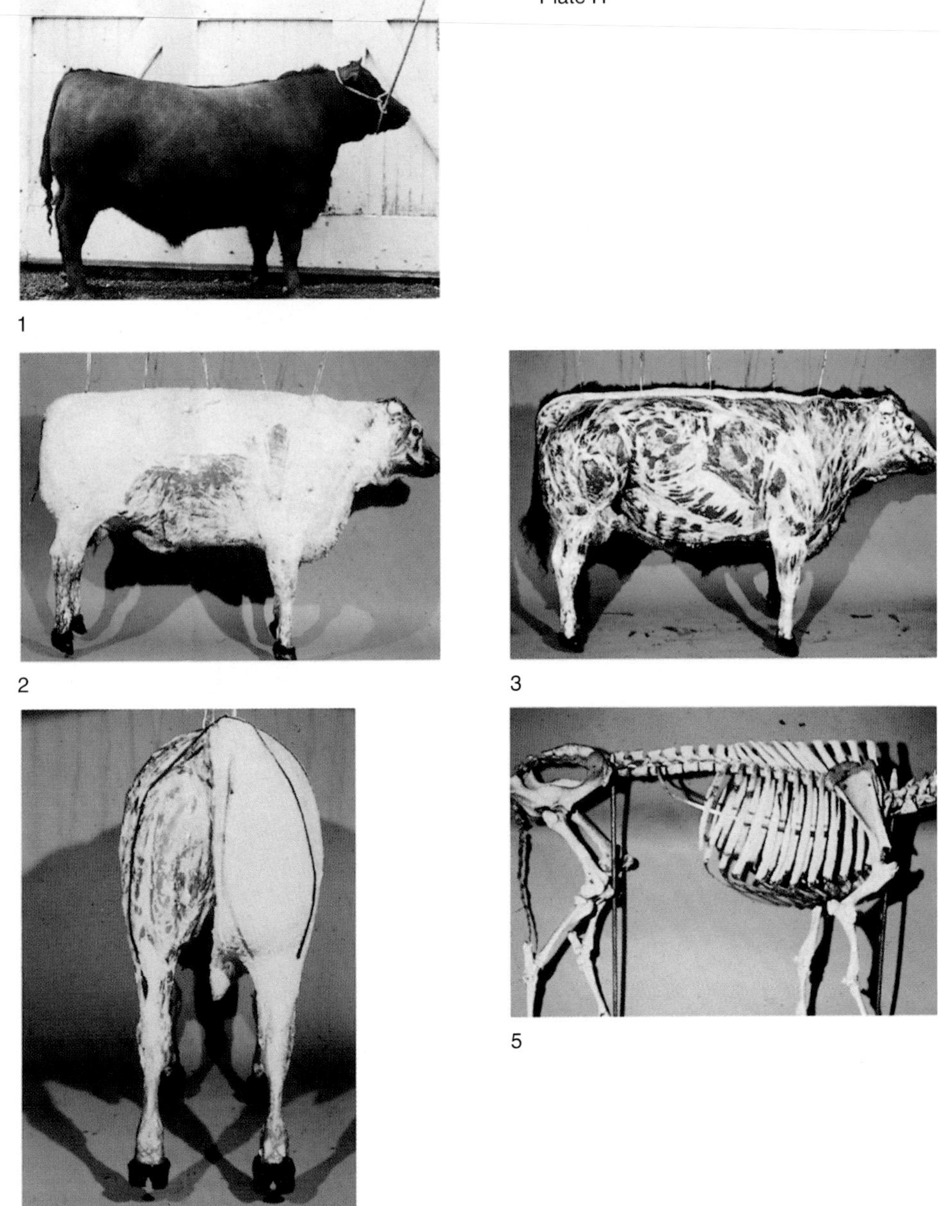

Plate H shows the composition and appearance of a steer carcass at various stages of removal of hide, fat, and muscle: (1) live steer with hair clipped from one side; (2) frozen steer with hide removed and fat exposed; (3) fat removed from one-half of steer's body; (4) rear view with fat removed from the left side. Note cod and twist fat on the right side; (5) skeleton of the beef animal after all the muscle has been removed.

Handling Calving Difficulties

The challenges posed by calving difficulty are twofold. The cow-calf producer must know how to handle calving difficulties when they occur as well as how to prevent or minimize them. In most cases, the problem is simply that the calf is too large at birth, or the birth canal (pelvic opening) of the cow is too small, or both. Research conducted by the Meat Animal Research Center (MARC) has shown a fourfold increase in calf death loss when calving difficulty is experienced. Even though it is considered the most important single factor contributing to calving difficulty, calf birth weight has accounted for less than 10% of the variation observed in calving difficulty in cows 4 years old and older. The relationship is much higher in younger cows.

Most calving difficulty losses can be prevented with timely and correct obstetrical assistance. The majority of calves that die at birth or shortly thereafter are normal calves whose lungs have functioned for varying lengths of time.

Cows about to calve generally exhibit certain behavior patterns and physical characteristics. For example, they may have a red, swollen vulva, show expulsion of the water sac (water bag), become separated from the herd, turn their heads toward their flank, and have a full, tight udder. Manual assistance should be given if the cow has not calved in an hour after she first starts the abdominal presses.

The membranes that form around the embryo in early pregnancy and attach to the uterus are collectively called the *placenta.* This organ of pregnancy produces hormones—estrogens and progesterones. A proper balance of estrogens and progesterones in the initiation of parturition is attained when the former predominate in quantity. The uterine muscles become sensitive to the hormone oxytocin, which is produced in the hypothalamus and released by the posterior pituitary. Under the stimulus of oxytocin, the weak, rhythmic contractions of the uterus that prevail through most of pregnancy become pronounced and cause labor pains, and the parturition process is initiated. Parturition is a synchronized process. The cervix, until now tightly closed, relaxes. The relaxation of the cervix, along with the pressure generated by the uterine muscles on the contents of the uterus, permits the passage of the mature fetus into and through the vagina. Another hormone, relaxin, also aids in parturition. Relaxin, which originates in the corpus luteum or placenta, helps to relax the cartilage and ligaments in the pelvic region and thereby to increase the size of the pelvic opening.

At the beginning of parturition, the calf typically assumes a position that will offer the least resistance as it passes into the pelvic area and through the birth canal. The normal presentation of the fetus is where the front feet are extended with the head between them (see Fig. 11.25). Occasionally, a calf may present itself in one of several abnormal positions (see Figs. 11.25 and 11.26); in this case, assistance must be given at parturition or the offspring and/or the dam may die. An abnormally small pelvic opening or an unusually large fetus can cause mild to severe parturition problems. Some examples of providing assistance to the cow or heifer are given in the following section. The procedure used in simple cases of calving difficulty is shown in Fig. 11.27; the procedures used in more difficult cases are shown in Figs. 11.25 and 11.26.

Assisting the Delivery

The following ten steps are followed to assist relatively easy dystocia problems. The term *frontward* indicates how to proceed if the calf is positioned in the birth canal with its front feet and head first. The term *backward* is used to indicate the procedure for the calf presenting with its rear legs first. The steps that do not indicate calf position apply to both presentations. (The equipment used to restrain the female during calving assistance is discussed in Chapter 18.)

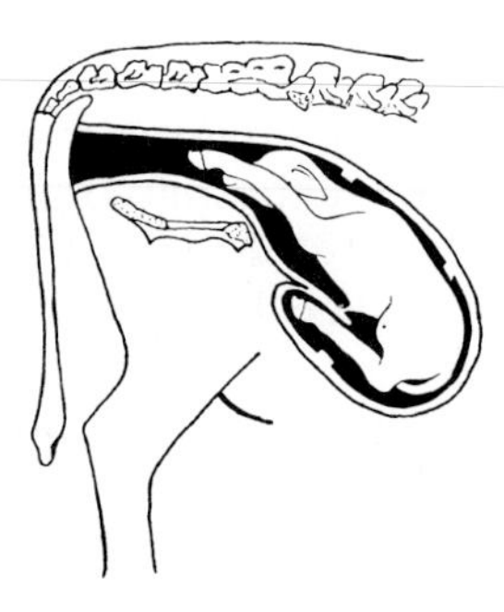

Normal Position

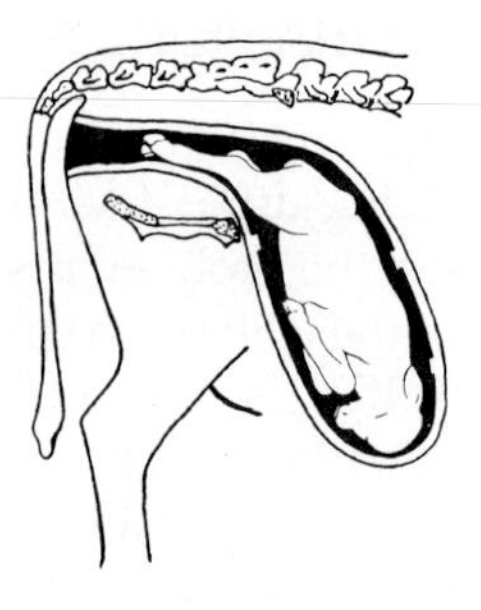

Posterior presentation *of the calf. Delivery may often proceed without complications. Assistance may be important if labor is prolonged. Death of the calf can occur due to rupture of the navel cord and subsequent suffocation.*

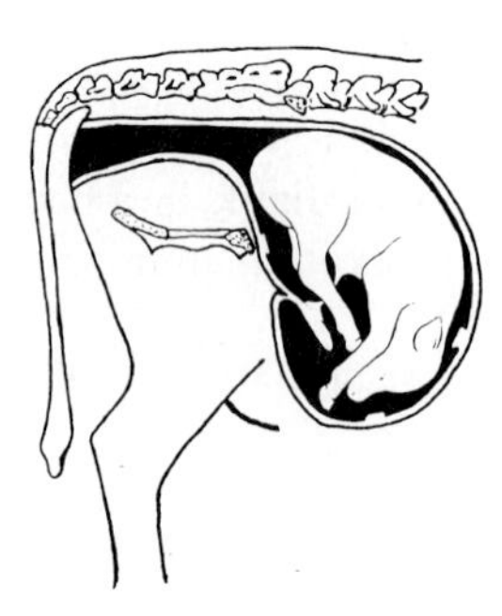

Posterior presentation *with rear legs extended under the calf's body (breech presentation). May be corrected by pushing the calf forward and grasping the legs one by one. As each leg is drawn into the birth canal, keep the hock pointed toward the cow's flank and the hoof to the midline.*

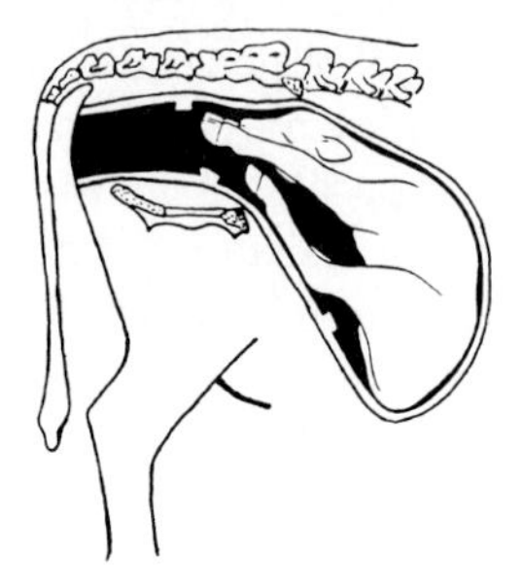

Anterior presentation *with the rear legs extended beneath the body (dogsitting* posture*). A very serious type of malpresentation. If allowed to progress into advanced labor, fetal death may result. Early professional attention may be required.*

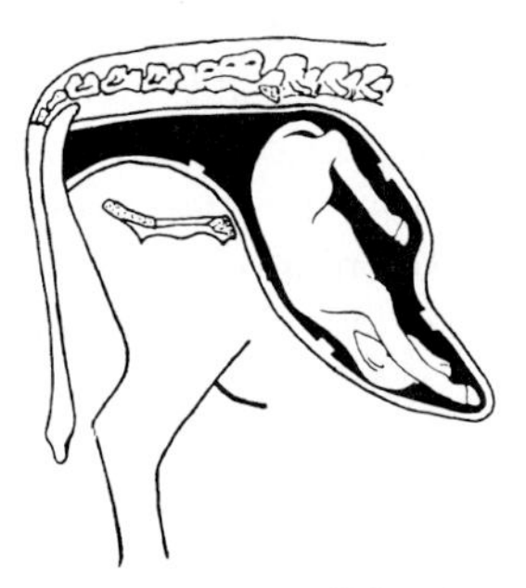

Posterior presentation *with fetus in an upside down position. This situation can be caused by twisting of the uterus or rotation of the calf. Delivery must never be attempted in this position, and professional assistance often is required.*

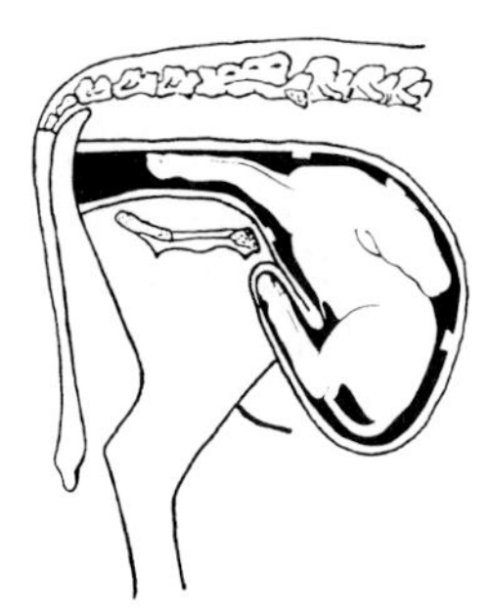

Anterior presentation *with the head and neck turned back over the body. Secure the legs with chains. Push the calf back into the body. This often brings the head into normal position.*

FIGURE 11.25 Normal and abnormal presentations of the calf at parturition, with guidelines in assisting the difficult deliveries.
Source: Texas A & M University.

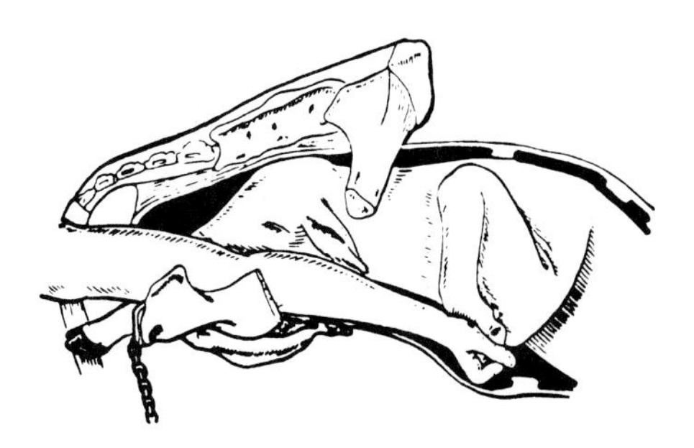

Correction of a simple leg flexion in an anterior presentation. The calf is first pushed forward and the retained foot is grasped in the cupped hand. The foot is carried outwards and then forward in an arc over the pelvic rim. More difficult cases may require that a snare be attached to the retained fetlock to help extend the leg.

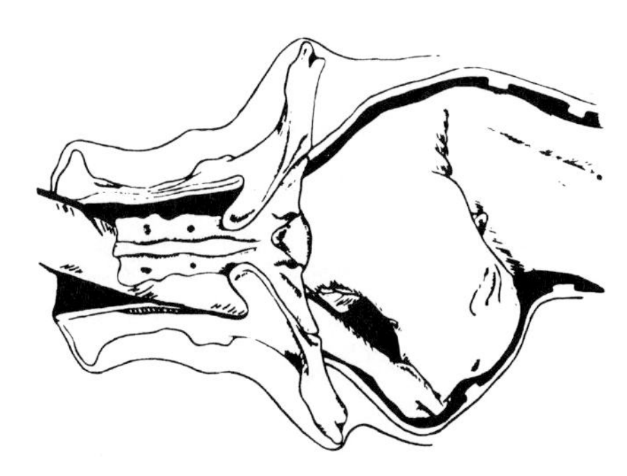

Correction of head and neck deviation in anterior presentation. Correcting difficult cases may require pulling the head and neck around with the hand. The calf is first pushed forward and the hand quickly moved to grasp the calf's muzzle. The head is pulled around and in line with the birth canal. In some cases it may be necessary to apply a snare to the lower jaw for additional traction. Excessive pressure should be avoided, as the jaw is not strong and will fracture easily. A safer and more common practice is looping the obsterical chain around the poll, under the ears and through the mouth in a "war bridle" manner. The attachment allows greater traction but will permit the calf's mouth to gape. Care should be taken to avoid cutting the birth canal with the calf's sharp incisor teeth.

Correction of the hock flex in posterior presentations. The calf is first pushed forward. The hand grasps and cups the calf's foot, then draws it back as the hock is flexed. The foot in the cupped hand is lifted over the pelvic rim and into the vagina. An alternate method in more difficult cases is to place a snare around the pastern, attached at the front of the leg. The snare is then pulled between the digits of the foot so that when traction is applied the fetlock and pastern are flexed. The calf is pushed forward and the foot is guided over the pelvic rim as an assistant pulls the snare.

FIGURE 11.26 Abnormal presentations of the calf at parturition, with guidelines in assisting the difficult deliveries.
Source: Texas A & M University.

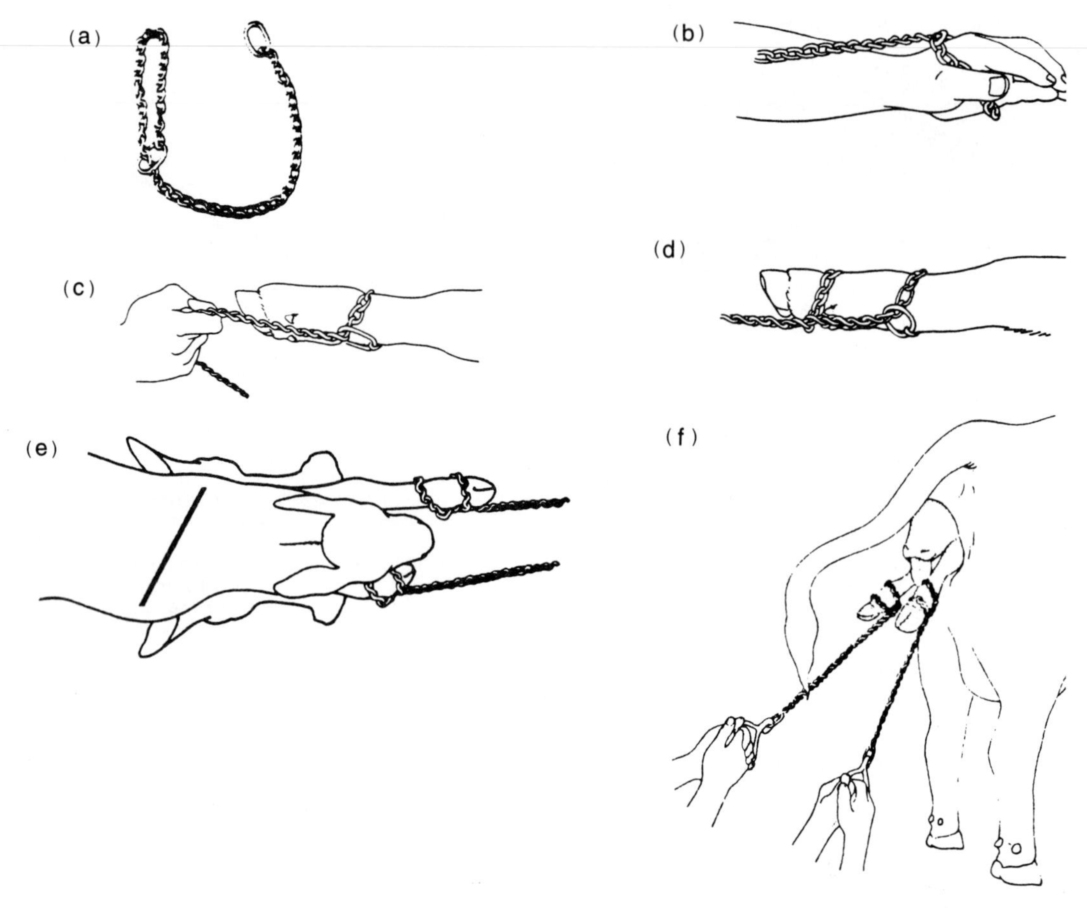

FIGURE 11.27 Diagrams of providing simple assistance to a beef female experiencing difficulty calving. Content of this figure was adapted from Battaglia, R. A. 2001. *Handbook of Livestock Management Techniques*. Upper Saddle River, NJ: Prentice-Hall.

All backward presentations should be considered emergencies. In this situation, the umbilical cord is cramped between the fetus and pelvis early in delivery, slowing the flow of blood and nutrients to the fetus. If delivery is not rapid once the fetus has entered the birth canal, brain damage or death can result.

1. Make a loop in the obstetrical chain by passing the chain through the oblong ring at the end of the calving chain (Fig. 11.27a).
2. Slip the loop over the gloved and well-lubricated hand in order to allow for easy application and maneuverability in the birth canal or uterus (Fig. 11.27b).
3. Attach the loop of the chain to one leg of the calf and slide it up on the cannon bone 2–3 in. above the dewclaws. It may be necessary to maintain a slight tension on the chain so that it does not slip off the leg (Figure 11.27c).
4. Half-hitch the chain between the dewclaw and hoofhead. The half-hitch can be made on your hand outside the cow and then applied to the leg. The hitch in a hard pull helps to distribute the stress imposed on the bones over the two locations instead of just one, reducing the possibility of bone fracture (Fig. 11.27d). Do not apply a single loop of the calving chain around both legs at the same time. The danger of breaking one or both legs is very high, especially if the birth is difficult and a hard pull is used.

5. Repeat steps 1–4 for the second leg.
6. Before applying the handles to the calving chains, make sure that the chains will pull from the bottom of the leg (dewclaw side). This will ensure that the legs will be pulled straight and not at an angle.
7. Attach two handles to the calving chains and pull gently, making sure that the loop and half-hitch of the chain have not slipped from the desired position on the calf's leg.
8. Some calves can be delivered by pulling both legs evenly; however, it is best to pull alternately on one leg and then the other a few inches at a time. When the legs are "walked out" in this manner, the shoulders or hips are allowed to pass through the pelvic girdle one at a time (Fig. 11.27e).

Calf forward: If the shoulders become lodged in the pelvis, apply traction to the calf's head to reduce the compaction of the head against the sacrum (top) of the birth canal and to reduce the dimensions of the shoulder and chest region. The chain can be applied by making a loop as described in step 1 and applying it to the nose and muzzle of the calf (between the muzzle and the eyes). Care should be taken when pulling on the nose to prevent the nose or jaw from being broken.

Calf backward: In a breech delivery, the extraction of the fetus is against the normal direction of hair growth. The birth canal should be liberally lubricated with an obstetrical or mild household soap dissolved in warm water. It may be necessary to rotate the fetus about one-eighth of a turn to take full advantage of the greatest diameter of the cow's pelvis. If delivery proves extremely difficult, a Caesarean section is probably necessary and should not be delayed.

9. Once the calf's legs are exposed, the calf should be pulled downward (toward the cow's hocks) at a 45-degree angle (Fig. 11.27f). As the calf's head and shoulders come through the birth canal, the chance of uterine or cervical lacerations is greatest. This damage may lead to infection and future reproductive problems. Because pressure dilates the cervix and birth canal, traction that is applied gradually can usually prevent such damage. If assistance is given to a cow too early, the slow application of traction would not interfere with normal cervical dilation and would minimize the potential of damaging the birth canal.
10. One or two individuals using chains with manual strength should be able to pull a calf out. If the birth is extremely difficult, however, it may be necessary to use a mechanical calf puller. When this situation arises, it is best to seek experienced help. A calf puller that is used incorrectly can cause permanent damage to both cow and calf. It has been demonstrated that the leverage exerted by calf pullers can pull over a ton of dead weight.

Preventing Calving Losses

Dystocia should be prevented as cows experiencing dystocia have calf losses 4–5 times greater than calves born without difficulty. Calf survival is less when colder temperatures are combined with precipitation, so calving later in the spring may be part of a good preventative program.

Personal attention to calving (particularly with heifers) should not become lax toward the end of a long calving season, when calving competes with the planting of crops and nighttime hours. When management inputs are lowered, greater calf losses occur. The Konefal method of daytime calving claims that feeding cows at 11:00 A.M. to noon and again at 9:30–10:00 P.M. keeps the cows busy feeding and results in most calves arriving in the daytime hours. Research data verify this relationship, though not all cows respond the same. High birth weight, particularly in 2-year-old heifers, is the primary cause of dystocia. Birth weights have been increasing significantly in the U.S. cattle population over the past several years, primarily because producers have been using more bulls from larger breeds and bulls from breeds where more selection pressure for growth and frame size has occurred for several years. Birth weight appears to be more highly influenced by genetics than by environment within the same year. High levels of feed (15–18 lbs TDN) during gestation only increase birth weights 3–5 lbs over cows fed low levels of feed (7–8 lbs TDN). There is also little difference in calving difficulty in cows and heifers on different levels of feed during gestation. There may be different birth-weight levels from cattle of similar genetic backgrounds raised in contrasting regions of the United States. These differences have been reported to be as large as 7–20 lbs. The reason for the differences is not known, though it has been suggested that nutrition and temperature may affect birth weight. Large yearly differences in birth weights within the same herd, where breeding is essentially the same, suggest that environment plays a significant role in these differences.

There is evidence that the level of crude protein in the gestation of the pregnant heifer can cause differences in the birth weight of calves and in the degree of calving difficulty. Bellows (1982) reports birth weights of 73 lbs and 84 lbs and calving difficulties of 42% and 58% in heifers fed 86% and 145% of the NRC crude protein, respectively. Additional research on protein levels does not show this large a difference in calving difficulty. Recommendations are to feed heifers a balanced ration and to avoid feeding lower protein levels to circumvent calving problems. Producers should be aware that they can overfeed protein by not only feeding excessive amounts of protein supplement but also by feeding large amounts of certain forages such as alfalfa hay.

It has been shown that selection for yearling weight in 30-lb increments has increased birth weight by almost 2.5 lbs. This is because of the high genetic correlation (R = 0.60) between birth weight and growth rate (yearling weight). There are both breed of sire (Fig. 11.28) and breed of dam (Fig. 11.29) effects on birth weight and thus on calving difficulty. Part of the breed effect on birth weight is due to differences in gestation length (Fig. 11.24).

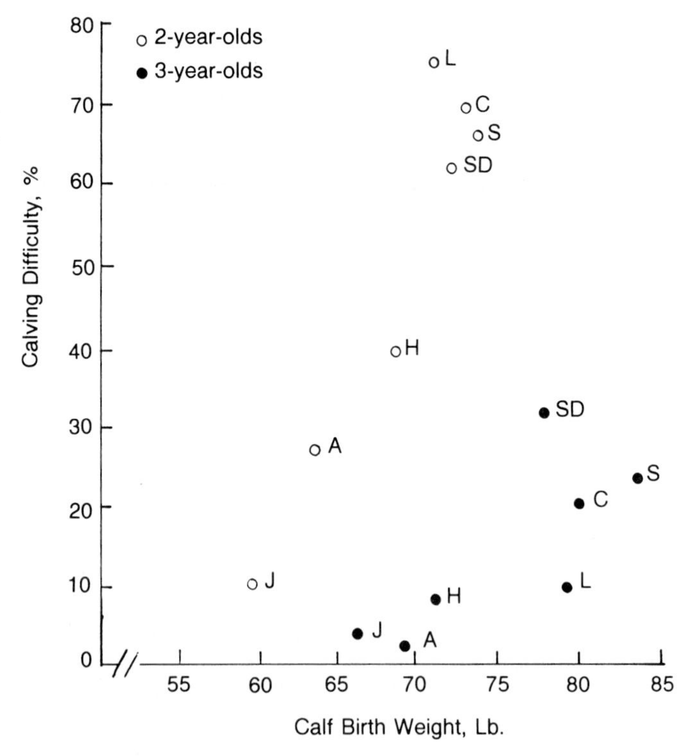

FIGURE 11.28 Relationship between calf birth weight and calving difficulty. (Different breeds of bulls, (A) Angus, (C) Charolais, (H) Hereford, (J) Jersey, (L) Limousin, (S) Simmental, and (SD) South Devon used on Angus and Hereford cows.)
Source: MARC.

Some producers feel that the shape of the calf affects calving difficulty—that longer, slimmer calves have less difficulty than shorter, thicker calves. Although most research data do not support this claim, some observations do support shape of the calf as being important in affecting calving difficulty at heavier birth weights. Figure 11.30 verifies some of these observations where longer, slimmer Chianina- and Brahman-sired calves demonstrated less calving difficulty than thicker-shaped (hips and shoulders) Charolais- and Maine-Anjou-sired calves. The birth weights of all four sire breeds were very similar.

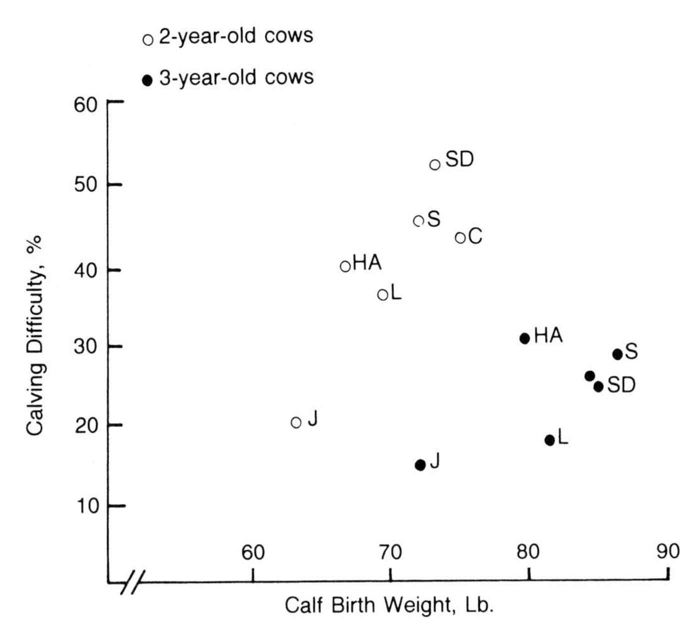

FIGURE 11.29 Relationship between calf birth weight and calving difficulty in (C) Charolais, (HA) Hereford-Angus, (J) Jersey, (L) Limousin, (S) Simmental, (SD) South Devon, and crossbred cows.
Source: MARC.

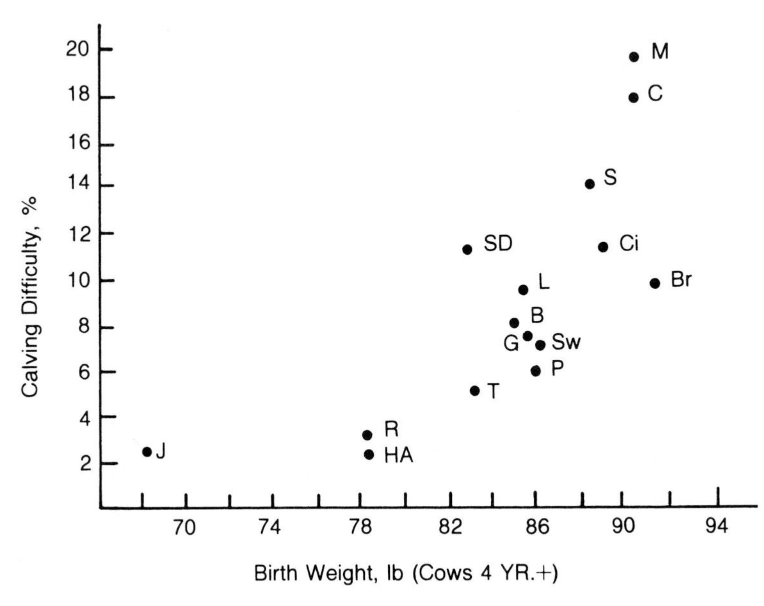

FIGURE 11.30 Relationship of birth weight to calving difficulty, posing question about importance of shape of the calf. Designated sire breeds are the same as those previously identified in Figures 11.28 and 11.29.
Source: MARC.

Birth weight and birth-weight EPD should be included in a selection program, primarily through bull selection (see Chapter 4). The level of birth weight acceptable to various producers will differ. Some cowherds can easily handle 85–100-lb birth weights, whereas other herds experience marked calving difficulty with 80-lb calves. Although cows in a particular herd might easily have 90–100-lb calves, 2-year-old heifers in that same herd would likely have difficulty with calves over 80 lbs. Therefore, more care must be used in selecting bulls for heifers than for cows. Caution should be exercised in saving replacement heifers that have large birth weights even though their dams did not experience dystocia. These replacement heifers will likely have calves with large birth weight, and experience dystocia when they calve as 2-year-olds.

Natural service bulls should be selected with moderate birth weights and AI bulls should have progeny test data for birth weights and calving ease. Weaning weight and yearling weight ratios can be relatively high without encountering serious birth-weight problems. However, most bulls with very high growth potential usually sire heavier birth-weight calves. The following index is effective in selecting for growth and against birth weight:

$$\text{Index} = \text{yearling weight} - 3.2\,(\text{birth weight})$$

The index should be used for selection on a within-herd basis.

There is little justification to selecting bulls from breeds known for extremely light birth weights because serious reductions in weaning weight and feedlot gain will result. Furthermore, calves with very low birth weights experience higher rates of morbidity and mortality. Pounds of calf weaned per cow exposed will typically be lowered. Rather, the actual birth weight of a bull and his birth-weight EPD, weaning weight, and yearling weight EPDs should be evaluated. Some breeds known for calving ease in the past are now producing bulls with birth weights in excess of 100 lbs. These bulls may work satisfactorily on cows but not on heifers in most cow herds.

Another factor affecting birth weight is age of cow. Two-year-old heifers have calves weighing an average of 8 lbs less than calves from mature cows. This relationship needs to be taken into account when evaluating birth-weight records. Bull calves weigh about 5 lbs more than heifer calves, which is why more bull calves require assistance at birth.

In general, the size of the pelvic opening (birth canal) ranks as an important factor affecting calving difficulty in the dam. It is generally true that dams with larger pelvic openings have less trouble at calving. However, within all increments in size of pelvic area, calving difficulty increases with increased birth weight.

It is a questionable practice to include pelvic measurements in a female selection program. The end results do not apparently justify the cost. It has also been determined that cows with larger pelvic areas generally give birth to heavier calves. Pelvic area is largely a function of the size of the female, and larger females have heavier calves. It is not known for certain whether selection for increased pelvic size can offset the increased birth weight.

There is evidence that selection for skeletal size (e.g., hip height) will account for most differences in pelvic area. Thus, the larger the skeletal size, the larger the pelvic opening. However, the larger the skeletal size, the greater the mature weight, which results in larger calves at birth and the larger mature weight may not be profitable to maintain. The conflict between pelvic area and birth weight exists, whether pelvic area is measured directly or indirectly.

Many beef producers believe that adequate exercise in the beef female is needed to prevent calving difficulty. The reasoning is that muscle tone and strength are needed for the female to expel the calf properly. Research at Miles City does not support this concept. Cows kept in feedlots during the last three months of gestation had essentially no difference in calving difficulty when compared with cows that exercised daily (see Table 11.7). Cows that were forced to exercise by walking to water required approximately 30% more feed to maintain the same body weight compared with the cows in the feedlot. Exercise might benefit beef females

TABLE 11.7 Effect of Exercise on Calving Difficulty

Activity	No.	Gestation Length (days)	Birth Wt (lbs)	Assisted (%)	Calving Difficulty (score)
Forced[a]	30	280	74	24%	1.3
Restricted[b]	31	280	71	26	1.5

[a] Walked 2 miles daily during the last 90 days of gestation.
[b] Held in feedlots during the last 90 days of gestation.
Source: U.S. Livestock and Range Research Station, Miles City, Montana.

kept in confinement facilities for extended periods of time, however, the confinement raising of beef cows is not common.

Calf Losses After Birth Until Weaning

Calf losses from 24 hours after birth until weaning are usually in the range of 3–6%. Higher losses may occur when severe weather or disease problems exist. In the Great Plains and in some western regions of the United States, for example, spring blizzards have caused calf death losses from 10–50%.

However, disease appears to cause the largest number of losses in most herds, particularly infectious calf scours and pneumonia. The prevention and treatment of these and other diseases are discussed in detail in Chapter 16.

Management Programs to Reduce Calf Losses

Management programs have been implemented in several herds of cattle to reduce calf death losses by 16–22%. In one case involving several herds, the calf death loss was significantly reduced by implementing the following management practices and calving techniques:
(1) improved calving facilities, (2) improved sanitation, (3) treatment of sick calves with nutrient and electrolyte drench, (4) closer observation during calving, (5) a herd vaccination program, (6) bull selection for lighter birth weights, and (7) improved nutrition of the cow herd. Thus, the producer who gives careful attention to calving losses can reduce them significantly.

REBREEDING

Heifers that calve early in their first and second calving seasons continue to calve early throughout their lifetime. Heifers calving late in the calving season are more likely to be open as 3-year-olds, especially in a short breeding season of 60 days or less. Heifers can be selected for early pregnancy, at pregnancy test time, when more heifers have been saved earlier as potential replacements.

Heifers typically have a longer postpartum interval than cows. This appears to be an age-of-cow effect; however, there is evidence that nutrition plays a role as well (Table 11.8). Heifers, fed together with cows, typically do not get an equal share of feed. Thus, energy restriction delays the onset of estrus. Feeding of 2- and 3-year-old cows separately from mature cows from calving through breeding will usually shorten the postpartum interval in the young cows. Current management practices of feeding young cows and mature cows separately occurs in approximately one-third of the U.S. cow herds (NAHMS, 1997).

A longer postpartum interval in heifers emphasizes the need to have heifers calve at the beginning of the calving season. Breeding heifers 2–3 weeks prior to the cow herd and

TABLE 11.8 Effects of Gestation Feed Level on Reproduction in Heifers and Cows

Gestation Feed[a]	Dam	Postpartum Interval (days)	In Heat by Beginning Breeding Season (%)	October Pregnancy[b] (%)
Low	Heifer	100	17%	50%
	Cow	58	93	83
High	Heifer	77	47	78
	Cow	60	88	81

[a] Low = 8.0-lb TDN; high = 15.0-lb TDN fed during last 90 days of gestation.
[b] After a 45-day AI period.
Source: U.S. Livestock and Range Research Station, Miles City, Montana.

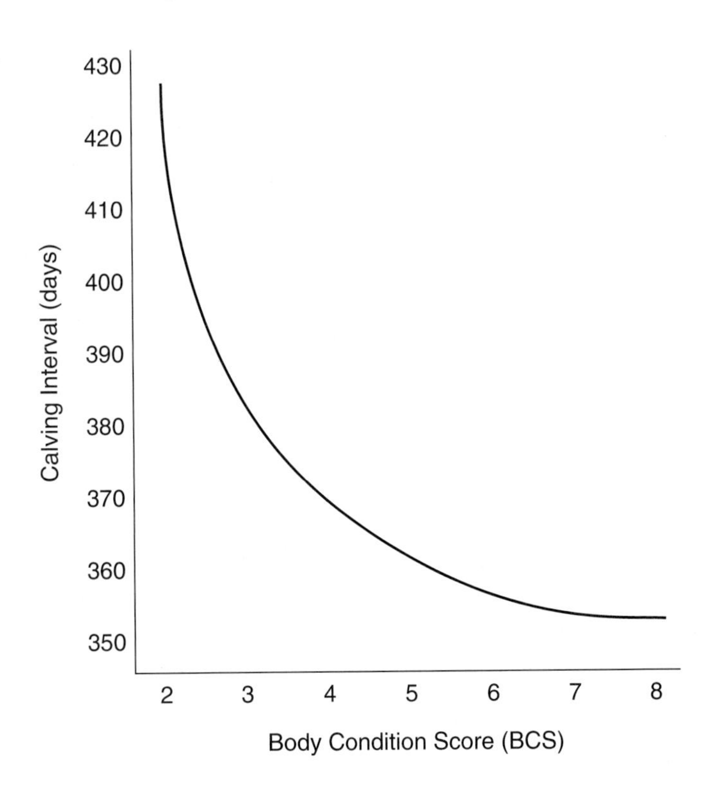

FIGURE 11.31 Relationship of body condition score to calving interval.
Source: University of Missouri Extension Service.

selection for early pregnancy can accomplish this objective. Breeding heifers earlier than the cow herd lengthens the total breeding and calving seasons. This should be analyzed carefully so that the benefits are greater than the possible additional costs.

The postpartum interval is highly related to the body condition score (BCS) of the cows. Cows that have longer postpartum intervals will also have longer calving intervals (see Fig. 11.31). Research has shown that about half the number of cows in "thin" condition will be cycling 60 days after calving as compared with cows in "good" body condition. However, cows in too high or fat body condition reflect an uneconomical feed level that may also negatively affect reproductive efficiency.

The body condition scoring system is described in Table 11.9 where the numerically higher scores identify cows that have more fat on the back, over the ribs, around the tailhead, and in the brisket. Figure 11.32 gives a visual perception of cows representing several body condition scores. Only about 3% of cow-calf producers condition score their cows (NAHMS, 1997),

TABLE 11.9 System of Body Condition Scoring (BCS) for Beef Cattle

BCS	Description
1	*Emaciated*—Cow is extremely emaciated with no palpable fat detectable over spinous processes, transverse processes, hipbones, or ribs. Tailhead and ribs project quite prominently.
2	*Poor*—Cow still appears somewhat emaciated, but tailhead and ribs are less prominent. Individual spinous processes are still rather sharp to the touch but some tissue cover over dorsal portion of ribs.
3	*Thin*—Ribs are still individually identifiable but not quite as sharp to the touch. There is obvious palpable fat along spine and over tailhead with some tissue cover over dorsal portion of ribs.
4	*Borderline*—Individual ribs are no longer visually obvious. The spinous processes can be identified individually on palpation but feel rounded rather than sharp. Some fat cover over ribs, transverse processes, and hipbones.
5	*Moderate*—Cow has generally good overall appearance. On palpation, fat cover over ribs feels spongy and areas on either side of tailhead now have palpable fat cover.
6	*High moderate*—Firm pressure now needs to be applied to feel spinous processes. A high degree of fat is palpable over ribs and around tailhead.
7	*Good*—Cow appears fleshy and obviously carries considerable fat. Very spongy fat cover over ribs and around tailhead. In fact, "rounds" or "pones" beginning to be obvious. Some fat around vulva and in crotch.
8	*Fat*—Cow very fleshy and overconditioned. Spinous processes almost impossible to palpate. Cow has large fat deposits over ribs and around tailhead, and below vulva. "Rounds" or "pones" are obvious.
9	*Extremely fat*—Cow obviously extremely wasty and patchy and looks blocky. Tailhead and hips buried in fatty tissue and "rounds" or "pones" of fat are protruding. Bone structure no longer visible and barely palpable. Animal's mobility might even be impaired by large fatty deposits.

Source: Richards et al., 1986. *J. Anim. Sci.* 62: 300.

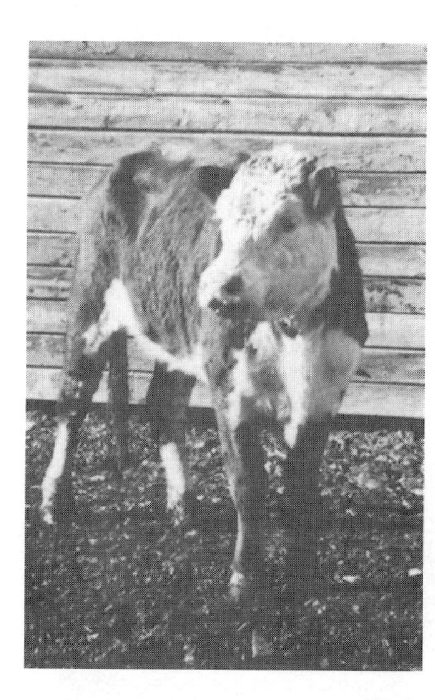

Body condition score 2 (thin condition)

FIGURE 11.32 Examples of body condition scores in Hereford cows.
Source: Colorado State University.

(continued)

Body condition score 4 (borderline condition)

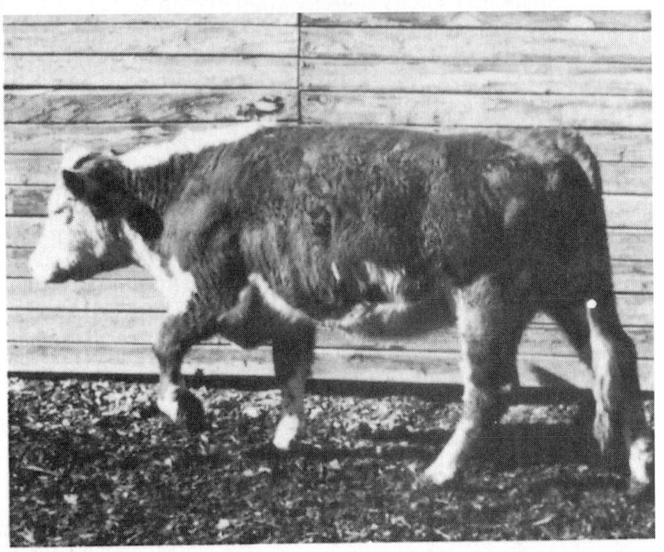

Body condition score 6 (optimum moderate condition that also includes BCS 5)

Body condition score 8 (overly fat condition)

FIGURE 11.32 (Continued)

although it appears more producers are observing the body condition of their cows and making needed changes in their genetic selection and breeding programs.

Extreme calving difficulty and delayed obstetrical assistance extend the postpartum interval. Results from Miles City have shown that a 10-minute increase in duration of labor can lengthen the interval from calving to first estrus by 2 days, reduce the percentage of females exhibiting estrus the first 21 days of the breeding season by 7%, and decrease the percentage that become pregnant during a 45-day AI period by 6%. Pulling the calf is recommended if the cow has not had her calf one hour after the second stage of labor begins. The second stage starts at the first abdominal press and ends with the birth of the calf.

MANAGEMENT SYSTEMS HIGHLIGHTS

1. Reproduction is the most important component of the biological system, the cow-calf management system, and in most situations the total beef industry management system. A live calf is absolutely the first step in attaining profitability.
2. Reproduction technologies need to be evaluated economically before implementing them. The statement that "economics trumps technology every time" should be a principle of caution.
3. Relatively high levels of reproductive performance are usually desired, however, optimum levels of reproduction performance are more cost effective and profitable than having maximum production as a goal.
4. Nutritional management and health management are critical components of assuring profitable levels of reproductive performance.

REFERENCES

Publications

A.I. Management Manual. 3rd ed. 1991. DeForest, WI: American Breeders Service.

Assam, S. M., et al. 1993. Environmental Effects on Neonatal Mortality of Beef Calves. *J. Anim. Sci.* 71: 282.

Beal, W. E. 2001. Synchronization of Estrus in Beef Heifers. *American Red Angus Magazine* (April). Denton, TX.

Bellows, R. A., Short, R. E., and Richardson, G. V. 1982. Effects of sire, age of dam, and gestation feed levels on dystocia and postpartum reproduction. *J. Anim. Sci.* 58: 18.

Beverly, J. R. 1979. *Recognizing and Handling Calving Problems.* College Station, TX: MP-1203, Texas A&M University.

Chenoweth, P. J., Spitzer, J. C., and Hopkins, F. M. 1992. *A New Breeding Soundness Evaluation Form.* Proceedings of the Society for Theriogenology, San Antonio, TX. pp. 63–70.

Coulter, G. H. 1997. *Bull Fertility: BSE, Abnormalities, etc.* Proceedings of the Range Beef Cow Symposium, Rapid City, SD: CO, NE, SD, and WY Extension Services.

Coulter, G. H., and Kastelic, J. P. 1997. *The Testicular Thermoregulation—Management Interaction in the Beef Bull.* Proceedings of the BIF Conference, Dickinson, ND.

Cupps, P. T. 1991. *Reproduction in Domestic Animals.* New York: Academic Press.

Curtis, J. L. 1990. *Cattle Embryo Transfer Procedure.* Manhattan, KS: John L. Curtis.

Geary, T. 1997. *Synchronization Programs Update.* Proceedings of the Range Beef Cow Symposium XV, Rapid City, SD: CO, NE, SD, and WY Extension Services.

Geary, T. W., Downing, E. R., Bruemmer, J. E., and Whittier, J. C. 2000. Ovarian and Estrous Response of Suckled Beef Cows to the Select Synch Estrous Synchronization Protocol. *Prof. Anim. Sci.* 16: 1.

Hafez, E. S. E. (ed.). 1993. *Reproduction in Farm Animals.* 6th ed. Philadelphia: Lea & Febiger.

Healy, V. M., et al. 1993. Investigating Optimal Bull:Heifer Ratios Required for Estrus-Synchronized Heifers. *J. Anim. Sci.* 71: 291.

Holmes, P. R. 1989. *Reproductive Efficiency in the Beef Herd.* MSDAGVET, Granville, NSW, Australia.

Investing in the Future. Replacement Heifer Management Resource Manual. 1991. Indianapolis, IN: Elanco Health Products, Eli Lilly and Co., Lilly Corporate Center.
Martin, L. C., et al. 1992. Genetic Effects on Beef Heifer Puberty and Subsequent Reproduction. *J. Anim. Sci.* 70: 4006.
National Animal Health Monitoring System (NAHMS). *Part 1: Reference of 1997 Beef Cow-Calf Management Practices. Part III: Reference of 1997 Beef Cow-Calf Production Management and Disease Control* (Jan. 1998). USDA, APHS, VS.
Odde, K. G. 1990. A Review of Synchronization of Estrus in Postpartum Cattle. *J. Anim. Sci.* 68: 817.
Patterson, D. J., et al. 1992. Management Considerations in Heifer Development and Puberty. *J. Anim. Sci.* 70: 4018.
Ritchie, H. D., and Anderson, P. T. 1992. *Calving Difficulty in Beef Cattle. Part I. Factors Affecting Dystocia. Part II. Management Considerations.* Stillwater, OK: Beef Improvement Association. BIF-FS6a and BIF-FS6b.
Schillo, K. K., Hall, J. B., and Hillman, S. M. 1992. Effects of Nutrition and Season on the Onset of Puberty in the Beef Heifer. *J. Anim. Sci.* 70: 3994.
Seidel, G. E., Jr. 1997. *Current Status of Sexing Bovine Semen.* Proceedings of the BIF Research Symposium, Dickinson, ND.
Seidel, G. E., Jr., and Seidel, S. M. 1991. *Training Manual for Embryo Transfer in Cattle.* Rome, Italy: Food and Agriculture Organization of the United Nations.
Selk, G. 2001. Heat Detection and Beef AI Programs. *American Red Angus Magazine* (April). Denton, TX.
Senger, P. L. 1997. *Pathways to Pregnancy and Parturition.* Pullman, WA: Current Conceptions, Inc.
Sprott, C. R., Harris, M. D., Richardson, J. W., Gray, A. W., and Forrest, D. W. 1998. Pregnancy to Artificial Insemination in Beef Cows as Affected by Body Condition and Number of Services. *Prof. Anim. Sci.* 14: 231.
Staigmiller, R. B., Short, R. E., and Bellows, R. A. 1990. *Developing Replacement Beef Heifers to Enhance Lifetime Productivity.* Research for Rangeland Based Production. Montana Agricultural Experiment Station and USDA-ARS (Fort Keogh Livestock and Range Research Lab).
Stevenson, J. S. 1999. *Estrus Synchronization and Induction Protocols in Beef Cattle.* Proceedings of the BIF Annual Research Symposium, Roanoke, VA.
Stevenson, J. S., Thompson, K. E., Forbes, W. L., Lamb, G. C., Greiger, D. M., and Corah, L. R. 2000. Synchronizing Estrus and (or) Ovulation in Beef Cows after Combinations of GnRH, Norgestomet, and Prostaglandin $F_{2\alpha}$ with or without Timed Insemination. *J. Anim. Sci.* 78: 1747.
Thibier, M. 1996. The 1995 Statistics on the World Transfer Industry. *Embryo Transfer Newsletter* 14: 275.
Whittier, J. 1990. *Body Condition Scoring of Beef and Dairy Cows.* Extension Guide 2230. Columbia, MO: University of Missouri.

Visuals

"Beef Reproduction I," (1995; 23 min.). CEV, P.O. Box 65265, Lubbock, TX 79464.
"Beef Reproduction II," (1995; 43 min.). Semen processing, estrous cycle, AI, pregnancy diagnosis. CEV, P.O. Box 65265, Lubbock, TX 79464.
"Beef Reproduction III," (1995; 21 min.). Reproductive tract at various stages of pregnancy; calving. CEV, P.O. Box 65265, Lubbock, TX 79464.
"Cash in on More Calves. Handling Calving Difficulties" (1986; 25 min.). Montana State University, Bozeman, MT.
"Cow Condition Scoring" (1989; 24 min.). Elanco Health Products, Eli Lilly and Co., Lilly Corporate Center, Indianapolis, IN 46285.
"Embryo Transfer" (1995; 40 min.). CEV, P.O. Box 65265, Lubbock, TX 79464.
"Investing in the Future. Replacement Heifer Management" (1991; 19 min.). Elanco Health Products, Eli Lilly and Co., Lilly Corporate Center, Indianapolis, IN 46285.

Genetics and Breeding

Beef cattle traits are determined by heredity (genetic makeup) and by the environment to which cattle are exposed. A knowledge of breeding principles and genetic resources allows commercial cow-calf operators to produce cattle that will profitably meet industry production and market specifications.

GENETIC PRINCIPLES

Cells and Chromosomes

Bodies of cattle are composed of cells, tissues, organs, and systems. Cells have numerous different functions depending on the type of tissue involved. One part of the cell, the *nucleus*, is especially important because it contains the genetic material that directs the function of the cell. Within the nucleus are rod-shaped, threadlike bodies called *chromosomes*. Chromosomes occur in pairs, with each species having a consistent number, and cattle having thirty pairs (Fig. 12.1). Each chromosome has the intrinsic ability to duplicate itself (mitosis). When a cell divides for growth or repair, the thirty pairs (sixty chromosomes) duplicate themselves. Therefore, in the division process, each of the resulting two cells has the same number of chromosomes.

The testicle of the bull and the ovary of the cow produce the sperm and egg through a cell division process called meiosis. In this division, chromosomes pair up in the center of the cell, with one chromosome of each pair randomly going into each of the two new cells. Thus, the sperm and the egg each contain thirty chromosomes, or one-half the number in other body cells. Fertilization—the union of the sperm and egg—restores the typical chromosome number. This random segregation and recombining of chromosome pairs is the major source of new genetic combinations available for selection.

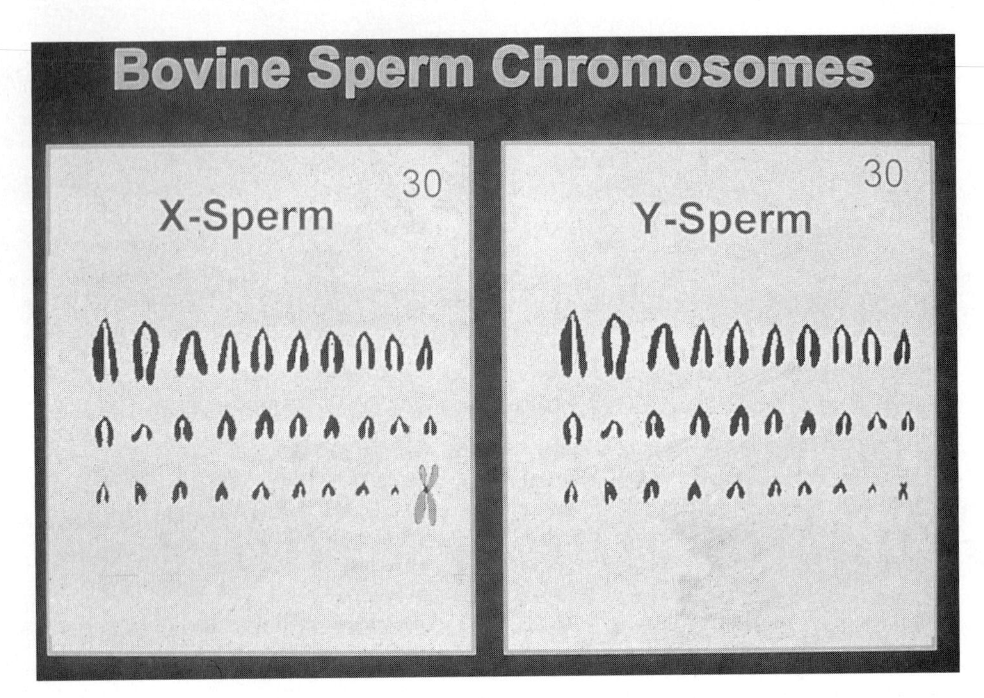

FIGURE 12.1 The thirty chromosomes of the bull and cow, including the X and Y chromosomes.
Source: Duane Garner and the Animal Reproduction Biology Laboratory, Colorado State University.

Sex Determination

The genetic sex of the calf is determined at the time the sperm and egg unite at fertilization. There is one pair of the thirty chromosomes known as *sex chromosomes*. They are designated as the X and Y chromosomes, where the cow has two X chromosomes and the bull has an X and a Y (Fig. 12.1). The sex of the calf depends on whether the sperm fertilizing the egg has an X or a Y chromosome. Since the cow transmits only an X chromosome to each of her offspring through the ovum, an egg fertilized by an X-bearing sperm will produce an XX calf, which is a heifer. An egg fertilized by the Y-bearing sperm produces an XY calf, which is a bull. The bull, then, determines the sex of the calf.

There is evidence that for the first few weeks after conception, all mammals start forming the basic female sex organs. If there is a Y chromosome present, it contains a gene that removes the female parts of the original embryo and the male anatomy begins to develop.

Since the bull can transmit either an X or a Y chromosome to each calf, the probability of him transmitting either one is 50%. The sex ratio of all the calves would be expected to be 50% bulls and 50% heifers. There is evidence, however, that shows the ratio of 105 bull calves to every 100 heifer calves at birth and even a higher ratio of bulls at conception. The reason for these differences is not known.

The bull possesses a smaller amount of genetic material than the cow because the Y chromosome is only about 40% as long as the X chromosome. About 6% of the total genetic material is carried on the sex chromosomes in the cow, whereas in the bull it is 4%. Traits carried on the sex chromosomes are called *sex-linked traits*. Little is known about sex-linked traits in cattle, although several of these traits are known in humans (hemophilia, muscular dystrophy, and red-green color blindness).

Genes

Genes are specific units of inheritance that are located on or as a part of the chromosome. This gene location is called a *locus,* Latin for "a place." Since chromosomes occur in pairs, so do genes. Pairs of genes occupying the same loci (position) on the chromosomes are called *alleles.* These genes control metabolic function and body development, which the cattle producer measures in a variety of ways such as reproductive performance, body weight gain, and carcass composition. Most beef cattle traits (body weight gain) are controlled by many pairs of genes located on several chromosome pairs. The exact number of pairs of genes existing in cattle is not known; however, there appears to be thousands of pairs.

Inheritance with One Pair of Genes

An example of a trait controlled by a single pair of genes is black and red color as found in Angus and Red Angus cattle. It is a simple example to show how basic genetic principles work. If B = gene for black color and b = gene for red color, then the following three combinations of genes can exist on the chromosomes:

B B   B b   b b

Animal BB is black in color. Genetically, it is homozygous black, which means that identical genes occupy the same locus on each chromosome. Bb is black in color. Genetically, it is heterozygous black, which means that different genes occupy the same locus on each chromosome. In this case, the black gene (B) is dominant to the recessive red gene (b). Dominance means that this gene will express itself while it masks the expression of the recessive gene. Animal bb is red in color. Genetically, it is homozygous red.

Various matings of the preceding three types of animals for black and red color will produce different phenotypes and genotypes (see Table 12.1). The *phenotype* is what a producer sees or measures in the animal, while the *genotype* is the actual gene composition of the animal. As indicated in the table, cattle with the same phenotype can differ in genotype, thus breeding performance can differ as well. For example, Bb × Bb (all parents are black) gives a 3:1 phenotypic ratio (3 black, 1 red) and three different genotypes in the calves (BB—homozygous black, Bb—heterozygous black, and bb—homozygous red).

TABLE 12.1 Different Phenotypes and Genotypes with Black and Red Gene Inheritance Resulting from Mating the Following Bulls and Cows

Bull		×	Cow(s)		=	Calves	
Genotype	Phenotype		Genotype	Phenotype		Genotypes	Phenotypes
BB	Black	×	BB	Black	=	all BB	black
BB	Black	×	Bb	Black	=	½ BB	black
						½ Bb	black
BB	Black	×	bb	Red	=	all Bb	black
Bb	Black	×	Bb	Black	=	¼ BB	black
						½ Bb	black
						¼ bb	red
Bb	Black	×	bb	Red	=	½ Bb	black
						½ bb	red
bb	Red	×	bb	Red	=	all bb	red

TABLE 12.2 Traits Controlled or Largely Influenced by One Pair of Genes

Trait	Type of Gene Action
Black, red color	Black (B) dominant to red (b)
Color in Shorthorns	Red (R) has no dominance over white (r)
Color dilution[a]	Dilution (D) dominant to nondilution (d)
Pigmentation, albino	Normal pigmentation (A) dominant to albino (a)
Polled, horned condition	Polled (P) dominant to horned (p) in British breeds
Shorter dwarf, normal size	Normal size (D) dominant to dwarf (d)
Hypotrichosis (short hair or hairlessness), normal	Normal (H) dominant to hypotrichosis (h)
Hydrocephalus, normal	Normal (H) dominant to hydrocephalus (h)
Osteopetrosis (marble bone disease), normal	Normal (O) dominant to osteopetrosis (o)
Syndactyly (mule-foot), normal	Normal (S) dominant to mule-foot (s)
Arthrogryposis (palate-pastern syndrome), normal	Normal (A) dominant to palate-pastern (a)
Double muscling,[b] normal	Normal (D) dominant to double muscling (d)

[a]Black color is diluted gray when DD or Dd exists with BB or Bb; red color is diluted to yellow when DD or Dd exists with RR or Rr.
[b]Recessive inheritance shown in this table is typical of the British breeds. In other breeds (e.g., Piedmontese), the double-muscling gene appears to be dominant. Other pairs of genes also modify the expression of double muscling.

Not all genes have dominant or recessive effects. An excellent example is color in Shorthorns, which is controlled primarily by a single pair of genes:

R = red gene RR = red
r = white gene Rr = roan
rr = white

Red is not dominant to white in this case. The white gene will express itself. Other genes are involved that determine the amount of red and white in roan animals (Rr). There are numerous coloration and spotting patterns in cattle that involve several pairs of genes (Olson, 1994). Color is not generally considered an economically important trait, although some solid-colored cattle appear to be preferred in the marketplace.

Table 12.2 identifies several common cattle characteristics that appear to be controlled by a single pair of genes. The genetic defects or abnormalities identified in this table are discussed in more detail later in this chapter.

Inheritance with Two Pairs of Genes

The following example using two pairs of genes permits a producer to sense more clearly what is occurring:

Example

B = black gene P = polled gene
b = red gene p = horned gene
Assume bull is B b P p and cows are B b P p

TABLE 12.3 Genotypes and Phenotypes in Calves Resulting from Mating Animals with Two Heterozygous Gene Pairs (BbPp × BbPp)

		Eggs			
		BP	Bp	bP	bp
Sperm	BP	BBPP[a] black, polled[b]	BBPp black, polled	BbPP black, polled	BbPp black, polled
	Bp	BBPp black, polled	BBpp black, horned	BbPp black, polled	Bbpp black, horned
	bP	BbPP black, polled	BbPp black, polled	bbPP red, polled	bbPp red, polled
	bp	BbPp black, polled	Bbpp black, horned	bbPp red, polled	bbpp red, horned

[a]Genotype. (Number of different genotypes) 1 (BBPP), 2 (BBPp), 2 (BbPP), 4 (BbPp), 1 (BBpp), 1 (bbPP), 2 (Bbpp), 2 (bbPp), 1 (bbpp).
[b]Phenotype. (Number of different phenotypes) 9 black, polled; 3 black, horned; 3 red, polled; 1 red, horned.

Since one chromosome of each pair goes into the formation of sperm and eggs, there are four genetically different kinds of sperm and four genetically different kinds of eggs that can be formed. When fertilization occurs, each of the genetically different kinds of sperm has an equal chance of uniting with each of the genetically different kinds of eggs. With these various matings, different genotypes and phenotypes can result in the offspring (Table 12.3).

A different type of gene action will give the same genotypic ratio, but the phenotypic ratio will be different. An example is color in Shorthorns with polled, horned condition:

RrPp (roan, polled bull) × RrPp (roan, polled cows)

Phenotypic ratio of the calves:

3 red, polled	1 red, horned
6 roan, polled	2 roan, horned
3 white, polled	1 white, horned

Genotypic ratio of the calves:

Same as in previous example with black, red color and polled, horned condition (1:2:2:4:1:1:2:2:1).

Where both pairs of genes show no dominance, the phenotypic and genotypic ratios would be the same (1:2:2:4:1:1:2:2:1).

The inheritance of the horned or polled condition is actually more complex than previously presented for British breeds of cattle. Scurs (hornlike growths) are inherited differently and the African horn gene is involved in the inheritance of horns for Zebu-type cattle such as Brahman. The primary genes involved in these three situations are:

P = polled gene	S = scur gene	A = African horn gene
p = horned gene	s = absence of scur gene	a = absence of African horn gene

Table 12.4 shows the expected phenotypes resulting from the different genetic combinations. Note that the phenotypic expression is different in cows and bulls when the Ss or Aa gene combination occurs. The inheritance of scur may involve more genes than presented here; however, the one pair of genes explains most of the inheritance involved. Horned cattle may carry the gene for scur but it is visually hidden by the horn gene and horn growth.

Continuous Variation and Many Pairs of Genes

Most economically important traits in beef cattle, such as growth rate and carcass composition, are controlled by hundreds of pairs of genes; therefore, it is necessary to expand one's thinking beyond inheritance involving one pair and two pairs of genes. Consider even a simplified example of twenty pairs of heterozygous genes (one gene pair on each pair of twenty chromosomes) affecting yearling weight. The estimated number of genetically different gametes (sperm or eggs) and the genetic combinations are shown in Table 12.5. Remember that for one pair of heterozygous genes, there are three different genetic combinations; for two pairs of heterozygous genes, there are nine different genetic combinations.

Most beef cattle would likely have some gene pairs heterozygous and some homozygous, depending on the mating system being utilized. Table 12.6 shows the number of gametes

TABLE 12.4 Inheritance for Scurs[a] and African Horn Gene

Genotype	Phenotype: Cows	Phenotype: Bulls
PP (orPp) SS	Scurred polled	Scurred polled
PP (or Pp) Ss	Smooth polled	Scurred polled
PP (orPp) ss	Smooth polled	Smooth polled
pp SS (or Ss or ss)	Horned	Horned
PP (or Pp) AA	Horned	Horned
PP (or Pp) Aa	Horned	Polled
PP (or Pp) aa	Polled	Polled
pp AA (or Aa or aa)	Horned	Horned

[a]Scurs are incompletely developed horns that are usually not attached to the skull. They vary in size from scablike growths to nearly the size of horns.

TABLE 12.5 Number of Gametes and Genetic Combinations with Varying Numbers of Heterozygous Gene Pairs

No. Pairs of Heterozygous Genes	No. Genetically Different Sperm or Eggs	No. Different Genetic Combinations (Genotypes)
1	2	3
2	4	9
n	2^n	3^n
20	$2^n = 2^{20}$ = approx. 1 mil	$3^n = 3^{20}$ = approx. 3.5 bil

TABLE 12.6 Number of Gametes and Genetic Combinations with Eight Pairs of Genes with Varying Amounts of Heterozygosity and Homozygosity

Bull Cow	Aa aa	Bb Bb	Cc CC	Dd Dd	Ee Ee	FF FF	GG gg	Hh Hh	Total
No. different sperm for bull[a]	2	2	2	2	2	1	1	2	64
No. different eggs for cow[b]	1	2	1	2	2	1	1	2	16
No. different genetic combinations possible in calves[c]	2	3	2	3	3	1	1	3	324

[a]Total = $2 \times 2 \times 2 \times 2 \times 2 \times 1 \times 1 \times 2 \ldots = 64$.
[b]Total = $1 \times 2 \times 1 \times 2 \times 2 \times 1 \times 1 \times 2 \ldots = 16$.
[c]Total = $2 \times 3 \times 2 \times 3 \times 3 \times 1 \times 1 \times 3 \ldots = 324$.

and genotypes, where eight pairs of genes are either heterozygous or homozygous and each gene pair is located on a different pair of chromosomes.

GENE MAPPING

Each chromosome is composed of deoxyribonucleic acid (DNA) and proteins. The structure of DNA has been identified as a highly wound spiral double helix of nucleoprotein. The individual subunits of the DNA molecule consists of a five carbon sugar (deoxyribose), a phosphoric acid, and a nucleotide base of either adenine, guanine, cytosine, or thymine (Fig. 12.2). A gene is an ordered sequence of nucleotide bases in a segment of DNA. Genetic linkage occurs when the pieces of DNA (genes) are carried on the same chromosome in relatively close proximity. These linked genes do not assort independently from one another when the sperm and egg are formed. This genetic link forms the basis for mapping the genes.

There are several techniques that are used to identify genes or markers on the various chromosomes. Mapping analyses have identified markers that appear to be associated with marbling, tenderness, ribeye area, double-muscling, disease resistance, growth traits, ovulation rate, and twinning. Most economically important traits are influenced by a multitude of genes, which complicates the gene identification process.

There are approximately 3 billion pieces of information in the bovine genome and these are influenced by both the environment and thousands of intracellular proteins. The sheer volume of information complicates the mapping of a species genome. Furthermore, once the genome is reasonably well understood, it is quite possible that genes and markers will have different effects in different breeds of cattle and in the multitude of environments in which they exist.

Agricultural Research Service (USDA) scientists predict that to fully map the bovine genome will require an additional investment of more than $20 million. They also suggest that these funds are not likely to be generated until 2003 at the earliest. Even private investment has been difficult to attract and maintain. For example, in 2001, the most well-known DNA/biotechnology firm serving the U.S. beef industry, Genomics FX, closed its doors. Investors withdrew their financial support due to uncertain markets for the technologies being developed and unsatisfactory results from the research division.

Green (1996) identified several values of using the bovine gene map—(1) it provides a basis to search for difficult to improve quantitative traits, for example, disease resistance, carcass palatability, and reproductive efficiency; (2) dominance and epistatic effects can be better understood; (3) more effectively planned mating systems; and (4) DNA "bar-coding" of cattle for complete industry traceback and data collection purposes.

Gene mapping is a slow and laborious process. Therefore, the beef industry should not expect this to happen in a short period of time. However, an exciting new era of cattle breeding is emerging. A more detailed discussion of the use of marker-assisted selection (MAS) can be found later in this chapter.

MATING STRATEGIES

Mating strategies can preserve genetic superiority and create hybrid vigor. Genetic improvement is realized in many herds by utilizing a combination of selection and mating strategies. Strategies of mating are identified primarily by animal performance or the genetic relationship of the animals being mated.

Mating strategies can be determined on an individual animal basis where specific sires are chosen to mate with specific dams with the goal of realizing a desired outcome in the progeny. The available approaches to this strategy include *assortative mating* (positive or negative) and *random mating*.

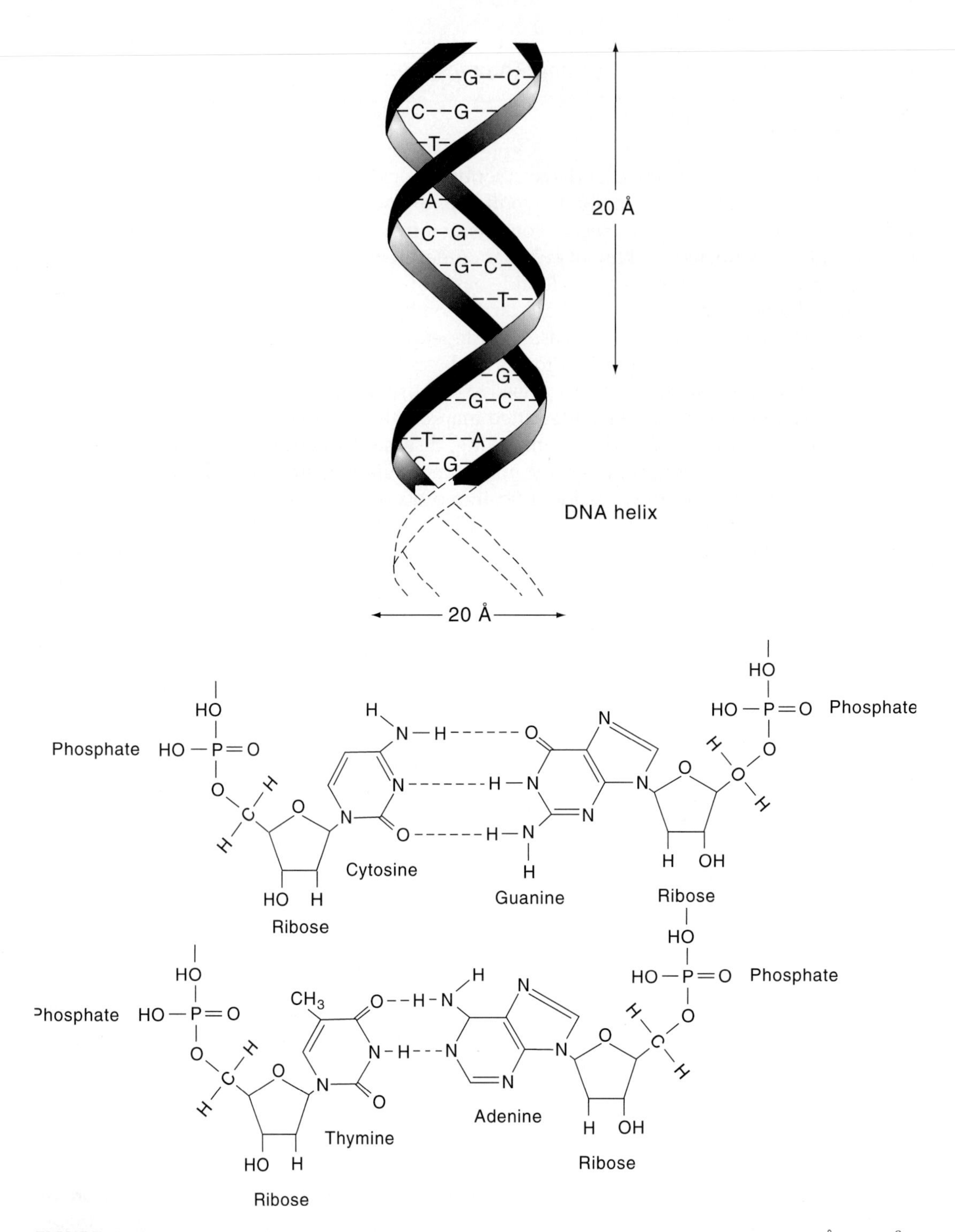

FIGURE 12.2 Deoxyribonucleic acid (DNA) helix and structure of nucleotides. One angstrom (Å) is 10^{-8} centimeter.

Assortative mating takes the approach of pairing similar phenotypes/genotypes (positive) or dissimilar types (negative). This approach requires breeders to make reasoned decisions based on objective performance records. An example might be to mate sires known for high marbling to females that also have high breeding values for intramuscular fat deposition. This approach is typically taken when rapid genetic progress is desired, so attempts are made to produce progeny with extreme performance in one or several traits. It should be noted that positive assortative mating also increases variability.

When breeders mate the highest marbling sires to the lowest marbling females they would be using negative assortative mating. This approach results in the production of progeny with intermediate levels of performance compared with the parents. This strategy slows the speed of genetic change but does reduce variation in the resulting progeny.

Random mating is another option that can be very effective in seedstock herds, particularly when a herd has been highly selected over time. This approach is the most typical strategy in many commercial herds where detailed cow records are unavailable and selection pressure is mostly focused on picking sires. This strategy recognizes and accepts the random nature of genetic inheritance.

The two major systems of mating, based on relationship, are inbreeding and outbreeding. *Inbreeding* is the mating of animals more closely related than the average of the breed or population. The two different forms of inbreeding are as follows:

1. *Intensive inbreeding*—the mating of closely related animals whose ancestors have been inbred for several generations.
2. *Linebreeding*—a mild form of inbreeding where inbreeding is kept relatively low while a high genetic relationship to an ancestor or a line of ancestors is maintained.

Outbreeding involves the mating of animals less closely related than the average of the breed or population. The five different forms of outbreeding are as follows:

1. *Outcrossing*—the mating of unrelated animals within the same breed.
2. *Grading up*—the mating of purebred sires to nondescript or grade females and their female offspring generation after generation.
3. *Linecrossing*—the crossing of rather distinct lines (that may or may not be inbred) of the same breed.
4. *Crossbreeding*—the mating of animals of different established breeds. (A detailed discussion is provided in Chapter 13.)
5. *Species cross*—the crossing of animals of different species (e.g., cattle to bison).

Since mating systems are based on the relationship of animals being mated, it is important to understand more detail about genetic relationship. Proper pedigree evaluation also involves understanding relationships. *Relationship* is best described by whether the genes of an animal (or animals) exist primarily in a heterozygous or homozygous condition. Figure 12.3 shows the mating systems and their relationship to a homozygous or heterozygous condition. For an excellent discussion on computation of relationship, see *Understanding Animal Breeding* by Bourdon (2000).

Inbreeding is the mating of relatives with the goal of increasing homozygosity. Inbreeding is utilized for several reasons. First of all, inbreeding can be utilized to increase uniformity. This is particularly effective for traits controlled by a limited number of genes such as coat color or the absence of horns. For traits affected by many gene pairs (polygenic), uniformity is enhanced only at very high levels of inbreeding. The development of inbreds with the intent of crossing highly inbred lines to maximize hybrid vigor has been effectively utilized by the crop industry. Inbreeding is also a useful strategy in the identification of which animals are carriers of genetic defects.

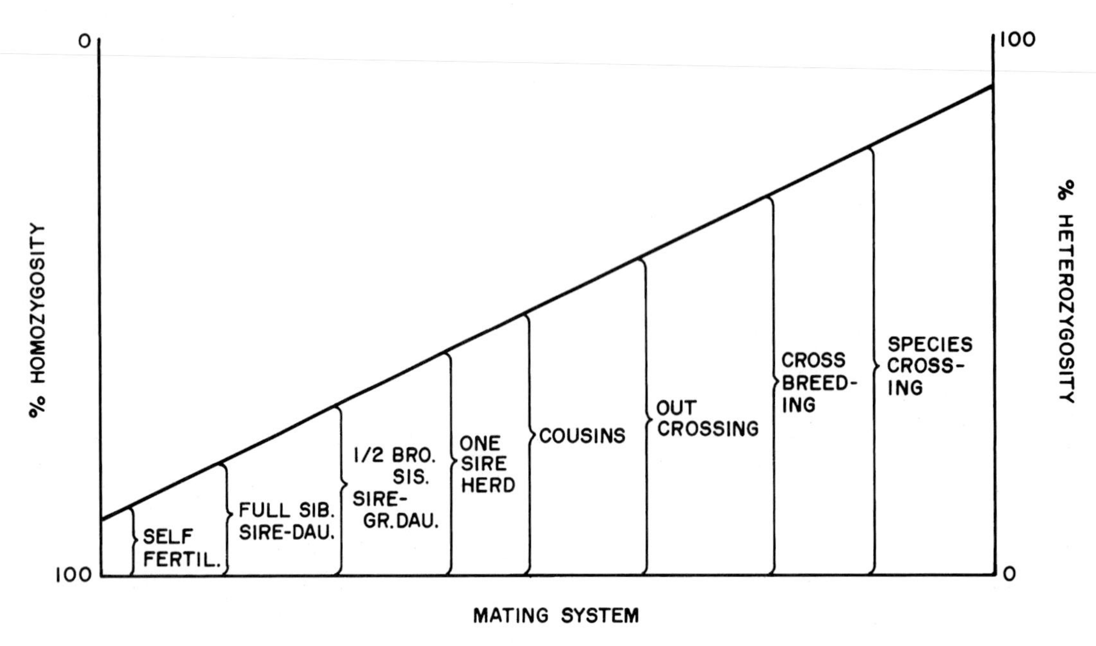

FIGURE 12.3 Relationship of mating systems to the amount of heterozygosity or homozygosity. Self-fertilization is currently not an available mating system in cattle.
Source: Colorado State University.

The drawbacks to intensive inbreeding include the production of progeny with deleterious recessive alleles (increasing homozygosity not only pairs desirable genes, but also those that are undesirable) and reduced general performance as a result of inbreeding depression.

Development of Inbred Lines

There are numerous, genetically different inbred lines that can be produced in a given population such as a breed. The number of different inbred lines is 2^n where n is the number of heterozygous gene pairs. For example, with two pairs of heterozygous gene pairs, there are 2^2 or 4 different inbred lines that are possible. For example, with AaBb genes, the different, completely homozygous lines that can result are AABB, AAbb, aaBB, and aabb.

There have been research projects involving cattle in which highly inbred lines were produced. It was expected that the crossing of these inbred lines would produce results similar to hybrid corn (Fig. 12.4). The cattle-breeding project at the San Juan Basin Research Center (SJBRC) at Hesperus, Colorado, developed some of the most highly inbred cattle in the world. Figure 12.5 shows the arrow pedigree of one inbred bull produced in the Royal line of Hereford cattle.

There have been several cattle produced at the SJBRC in the Royal, Brae Arden, Prospector, and Tarrington lines of inbred Herefords, whose inbreeding coefficients have ranged from 40–60%. These and other inbreeding studies indicate the following:

1. Increased inbreeding is usually detrimental to reproductive performance and preweaning and postweaning growth. Inbred cattle are more susceptible to environmental stresses. While 60–70% of the inbred lines show the detrimental effects of increased inbreeding, 30–40% of the lines show no detrimental effects with some lines demonstrating improved productivity.

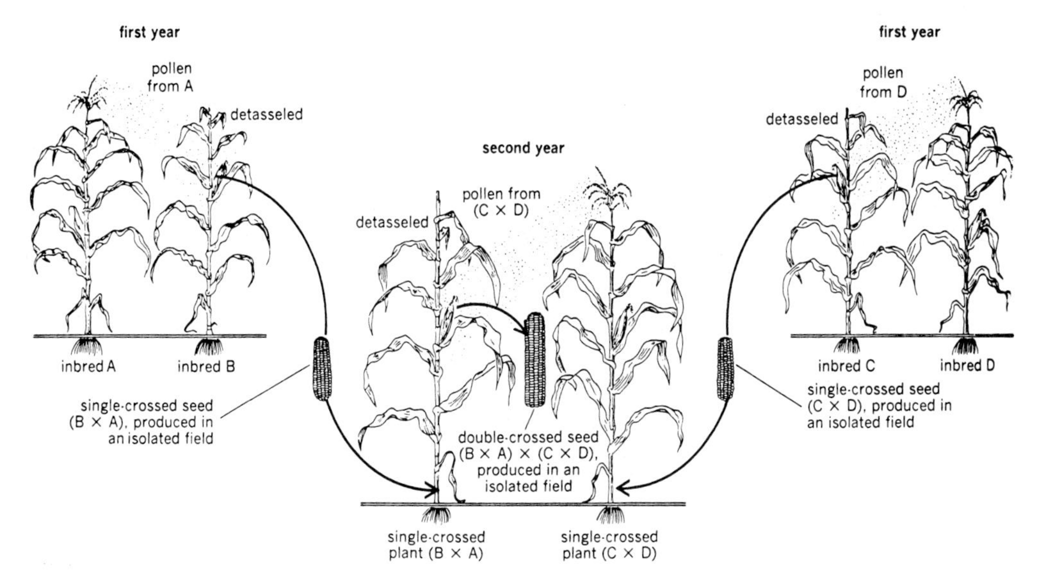

FIGURE 12.4 Crossing of inbred lines of corn to eventually produce the double-crossing hybrid corn seed utilized widely today. Note the relative size of the corn ears from the inbred lines compared with the resulting hybrid cross.
Source: USDA.

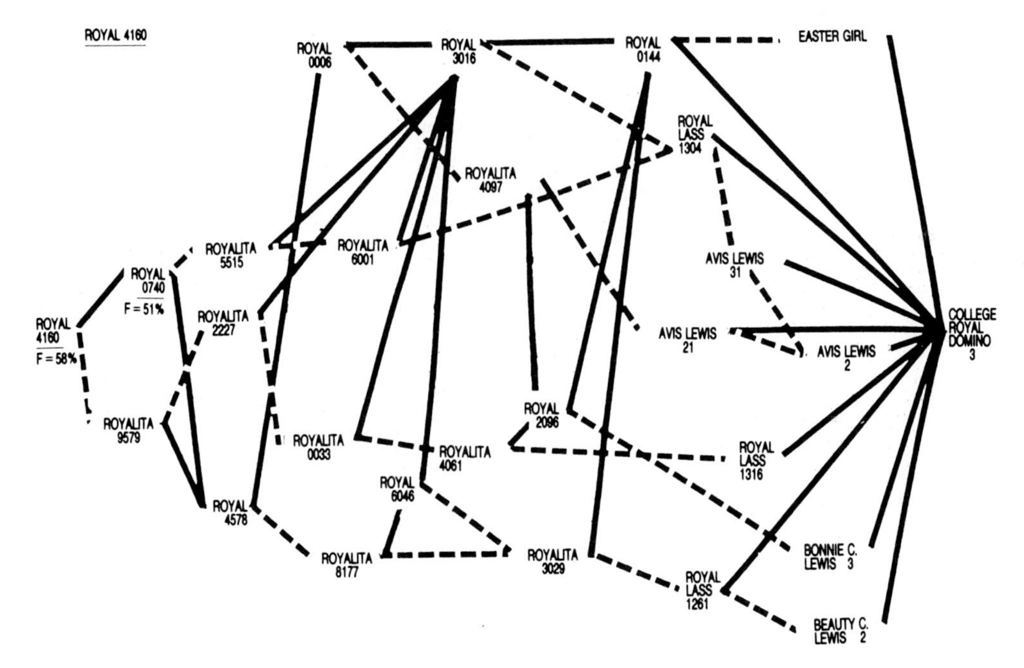

FIGURE 12.5 Arrow pedigree of Royal 4160 which has an inbreeding coefficient of 58%. Solid lines represent the genetic contribution of the bulls, while the cows' contribution is represented by a broken line.
Source: CSU Expt. Sta. (San Juan Basin Research Center) General Series 982.

2. At the SJBRC, the inbred lines showed a yearly genetic increase of 2.6 lbs in weaning weight over a 26-year period, while the line crosses made a 4.6-lb increase over the same time period. Heterosis is demonstrated in the line crosses, and the 4.6-lb increase is typical of what breeders might expect from using intense selection in an outbred herd.
3. Inbreeding quickly identifies some desirable genes as well as undesirable genes, particularly the hidden, serious recessive genes.
4. Inbred bulls with superior performance are most likely to have superior breeding values that will result in more uniform progeny with high levels of genetically influenced productivity.
5. Crossing of inbred lines results in heterosis.

It is not recommended that breeders develop their own lines of highly inbred cattle. Inbreeding depression will usually negatively affect the economics of a cattle operation.

Linebreeding

Some seedstock producers use inbreeding in well-planned linebreeding programs. Consideration should be given to implementing linebreeding when breeders have difficulty introducing sires from other herds that are genetically superior to those they are raising. The primary objective of a linebreeding program is to maintain a high genetic relationship of current progeny to an outstanding ancestor (usually a sire). Inbreeding is kept at a relatively low level. Figure 12.6 shows an example of linebreeding in which the bull Rito 2RT2 is only 9% inbred, yet has a relationship of 31% to his ancestor, Band 234 of Ideal 3163, which contributes genes to him by three different pathways. If 3163 appeared only once in Rito 2RT2's pedigree, the relationship between the two bulls would be only 12% if 3163 was three generations removed from 2RT2.

Outbreeding

Species crossing is occasionally used to introduce new sources of genetic material in developing new breeds or to make cattle more adaptable to certain environmental conditions. Bison crossed with cattle is an example of a species cross. *Bos indicus* × *Bos taurus* are common crosses in the Gulf Coast states, though some question this as a true species cross since they both have the same number of chromosomes and interbreed freely. Table 12.7 shows how cattle species fit into the zoological classification.

Outcrossing involves the mating of unrelated animals and is the most widely used mating system by both commercial and seedstock producers. Outcrossing results in a higher level of heterozygosity. With respect to deleterious recessives, outcrossing does not eliminate these undesirable genes, but rather masks their effects and maintains them in the population disguised in the heterozygote. So long as these recessives occur in low frequencies, they typically have limited impact. The usefulness of outcrossing is primarily dependent on the effectiveness of selection (selection differential × heritability).

Crossbreeding

There are two primary reasons for using crossbreeding in cattle: (1) breed complementarity and (2) heterosis. Breed complementarity implies using breeds in a crossbreeding program so that their strengths and weaknesses complement one another. No one breed is superior in all desired production characteristics; therefore, planned crossbreeding programs, using breed complementarity, can significantly increase herd productivity.

BRACKET PEDIGREE

BAND 234 OF IDEAL 3163
Q A S TRAVELER 23-4
SIRE
Q A S BLACKBIRD EVE 601 1
R R TRAVELER 5204
SHOSHONE VANTAGE JB23
ERISKAY OF ROLLIN ROCK 3302
ERISKAY OF ROLLIN ROCK 7003
RITO 2RT2
BAND 234 OF IDEAL 3163
TEHAMA BANDO 155
DAM
TEHAMA BLACKCAP G373
RITA 0B5 OF 8E23 BANDO
Q A S TRAVELER 23-4
RITA 8E23 OF 5H28 TRAVELER
RITA 5H28 OF 3A12 RITO 9J9

ARROW PEDIGREE

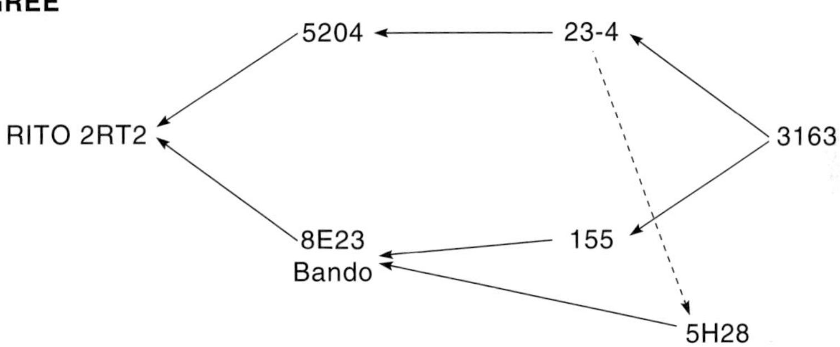

Inbreeding of RITO 2RT2

ancestor	n	n^1	1	(1+Fa)	
23-4	1	2	1	—	$(1/2)^4 = 6.2$
3163	2	2	1		$(1/2)^5 = 3.1$
					9.3%

Relationship of RITO 2RT2 to 3163

ancestor	n	n^1	#	(1+Fa)	
3163	3	0		—	$(1/2)^3 = 12.5$
3163	3	0		—	$(1/2)^3 = 12.5$
3163	4	0		—	$(1/3)^4 = 6.1$
					31.1%

FIGURE 12.6 Pedigree showing linebreeding.
Source: American Angus Association.

TABLE 12.7 Zoological Classification of Cattle

Kingdom *Animalia:* Includes all animals in the animal kingdom
Phylum *Chordata:* Animals that have a backbone (vertebrae) or the rudiment of a backbone (chorda)
Class *Mammalia:* Warm-blooded, hairy animals that suckle their young from a milk-producing mammary gland
Order *Artiodactyla:* Even-toed, hoofed mammals
Family *Bovidae:* Ruminants having hollow, up-branched horns and a placenta type of numerous cotyledons. This family also includes wild cattle, bison, and buffaloes[a]
Genus *Bos:* Ruminant quadrupeds, including domestic and wild cattle having a stout body and curved, hollow horns that stand out laterally from the skull
Species *taurus:* Includes most cattle found in the United States, including their European ancestors
Indicus: These are the Zebu (humped) cattle, including the Brahman breed in the United States

[a]The domesticated species are *Bibos banteng* (the Bali cattle found primarily in Indonesia), *Bibos frontalis* (the Mithan cattle in southeastern Asia), *Poephagus grunniens* (the Yak of Tibet), and *Bubalus bubalus* (all breeds of river and swamp buffalo that are concentrated primarily in Asia). The genus and species of the American Bison is *Bison bison.*

TABLE 12.8 Computation of Heterosis for Weaning Weight

Breed or Category	Low Breed Productivity	High Breed Productivity
Angus	410	510
Hereford	390	490
Average of the two breeds	400	500
Average of crossbreeds	420	525
% heterosis	20/400 = 5%	25/500 = 5%

Crossbreeding, when properly managed, allows for the effective use of heterosis (hybrid vigor), which is used extensively in the poultry, swine, and corn industries. Heterosis is the change in productivity in the crossbred progeny compared with the average of breeds used in the cross. The calculation of heterosis for weaning weight is shown in Table 12.8, which also indicates the importance of using highly productive cattle from breeds being utilized in the crossbreeding program. Highly productive crossbred cattle result from crossing highly productive purebred cattle, usually through purebred bulls. The use of crossbreeding in commercial breeding programs is discussed in Chapter 13.

TRAITS AND THEIR MEASUREMENT

One of the challenges for cattle breeders is to identify those traits that are most important to profitability, and to focus selection pressure on them. More than 70 different traits have been identified that either have or would have associated genetic prediction estimates. This much information is overwhelming. Dr. Bruce Golden at Colorado State University has identified a concept known as *economically relevant traits* (Table 12.9). These are traits that have a direct impact on profitability via influence on either cost of production or gross revenues. Many traits are associated with economically important traits, but they are indirect measures. These indirect predictor traits are termed *indicator traits* (Table 12.9).

Most economically important traits of beef cattle can be classified under (1) reproductive performance, (2) weaning weight, (3) yearling weight, (4) feed efficiency, (5) carcass merit, (6) longevity, (7) conformation, (8) freedom from genetic defects, (9) disposition, and (10) environmental adaptability.

TABLE 12.9 Economically Relevant Traits and Associated Indicator Traits

Economically Relevant Traits	Indicators
Weight at sale:	
Weaning wt.	205-day wt., 365-day wt.,
Weaning maternal	carcass wt., birth wt., fat thickness, ribeye area
Yearling wt.	
Carcass wt.	
Lb of retail yield	
Likelihood of calving ease	Calving ease score, birth wt., gestation length
Feed requirements for maintenance	Mature cow wt., body condition score, milk production, internal organ wt.
Productive life or stayability	Calving records, days to calving, milk production, calving interval
Likelihood of heifer pregnancy	Pregnancy diagnosis, scrotal measures
Tenderness	Shear force, marbling, color analysis
Feed efficiency	Feed consumption
Docility	Docility or chute scores

Source: Golden et al. (2000).

Reproductive Performance

Reproductive performance, measured as percent calf crop, has the highest economic importance when compared with other beef cattle traits (such as growth or carcass). Most cow-calf producers have a goal for percent calf crop weaned (e.g., the number of calves weaned compared with the number of cows in the breeding herd) of 85% or higher. Beef producers also desire each cow to calve every 365 days or less and to have a calving season for the entire herd of less than 90 days. All of these are good objective measures of reproductive performance. Some reproductive traits (e.g., calving interval) have low heritabilities, whereas other reproductive traits (e.g., scrotal circumference and birth weight) have high heritabilities. Reproductive traits with low heritabilities can be most effectively improved by changing the environment (e.g., providing adequate nutrition and maintaining good herd health practices). However, the development of genetic predictors of heifer pregnancy rate and lifetime female productivity will aid in the improvement of reproductive performance.

Reproductive performance can be improved through breeding methods by crossbreeding to obtain heterosis for percent calf crop weaned, using bulls with relatively light birth weights (heritability of birth weight is 40%) which decreases calving difficulty, and by selecting bulls that have a relatively large scrotal circumference. Scrotal circumference has a high heritability (50%), and bulls with a larger scrotal size (over 32 cm for yearling bulls) produce a larger volume of semen and have half-sister heifers that reach puberty at earlier ages than heifers related to bulls with a smaller scrotal size. Scrotal circumference is a threshold trait as there is no apparent advantage in increasing the size beyond 38–40 cm.

Work by Doyle et al. (2000) and Evans et al. (1999) suggests that heifer pregnancy is approximately 12–14% heritable. Stayability, or the likelihood of cows producing at least five calves, is also sufficiently heritable to warrant selection pressure. Rebreeding rate as a 2-year-old was found to be quite lowly heritable and, as such, is best improved via management.

Reproductive tract scores (1–5) can be taken by rectal palpation of the reproduction tract when heifers are 13–15 months old. Scores of 4 and 5 indicate the heifer is cycling or close to cycling. Heifers with immature ovaries and reproductive tracts receive a score of 1–2. Heritability of reproductive tract score is 30% and will respond to selection.

Weaning Weight

Weaning weight, as measured objectively by using scales, reflects the milking and mothering abilities of the cow and the preweaning growth rate of the calf. Weaning weight is commonly expressed as adjusted 205-day weight, where weaning weight is adjusted for age of calf and age of dam. This adjustment puts all weaning weight records on a comparable basis, since older calves will weigh more than younger calves and mature cows (5–9 years of age) will milk heavier than young cows (2–4 years of age).

The weaning weight of a calf is usually computed by dividing the calf's adjusted weight by the average weight of other calves in the herd and then expressed as a ratio. For example, a calf with a weaning weight of 550 lbs in a herd averaging 500 lbs has a ratio of 110. This calf's weaning weight ratio is 10% above the herd average. Ratios can be used primarily for selecting cattle within the same herd and contemporary group. Comparing ratios between herds is misleading from a genetic standpoint because most of the differences are caused by differences in environment. Bulls raised in different herds can be compared for genetic differences in weaning weights by comparing their weaning weight EPDs. Weaning weight will respond to selection because it has a heritability of 30 percent.

Yearling Weight

Yearling weight measures weaning weight and postweaning gain to a weight that approaches slaughter weight. Postweaning growth may take place on a pasture or in a feedlot. Usually animals with relatively high postweaning gains make efficient gains at a relatively low cost to the producer.

Postweaning gain in cattle is usually measured in pounds gained per day after a calf has been on a feed test for 100–120 days. Weaning weight and postweaning gain are usually combined into yearling weight, that is, the adjusted 365-day weight. It is computed as follows:

$$\text{Adjusted 365-day weight} = (160 \times \text{average daily gain}) + \text{adjusted 205-day weight}$$

Average daily gain for 140 days and adjusted 365-day weight both have high heritabilities (40%), so genetic improvement can be quite rapid when selection is based on postweaning growth or yearling weight.

Feed Efficiency

Feed efficiency is usually measured by pounds of feed required per pound of liveweight gain. Specific records for feed efficiency can only be obtained by keeping records on the amount of feed consumed by individual animals. With the possible exception of some bull-testing programs, determining feed efficiency on an individual animal basis is usually not economically feasible.

The interpretation of feed efficiency records can be rather confusing depending on the endpoint to which animals are fed. The feeding endpoint can be a certain number of days on feed (e.g., 140 days), to a specified slaughter weight (e.g., 1,200 lbs), or to a carcass compositional endpoint (e.g., low Choice quality grade or 0.40 in. of fat over the rib eye). Most differences shown by individual animals in feed efficiency are related to pounds of body weight maintained through feeding periods and daily rate of gain or feed intake of each animal. Cattle fed from a similar initial feedlot weight (e.g., 600 lbs) to a similar slaughter weight (e.g., 1,200 lbs) will demonstrate a high correlation between rate of gain and efficiency of gain. In this situation, cattle that gain faster will require fewer pounds of feed per pound of gain. Thus, a breeder can select for rate of gain and thereby make genetic improvement in feed efficiency.

However, when cattle are fed to the same compositional endpoint (approximately the same carcass fat), the difference in amount of feed required per pound of gain is relatively small.

The heritability of feed efficiency is high (45%), so selection for more efficient cattle can be effective. It is logical to use the genetic correlation between gain and efficiency when possible because of the expense of obtaining individual feed efficiency records.

Carcass Merit

Carcass merit is presently measured primarily by carcass weight, tenderness, quality grade, and yield grade (the latter two are described in detail in Chapter 17). Many cattle breeding programs seek to produce cattle that will quality grade Choice and have yield grades from 1.5 to 3.4. Visual or objective measurements of backfat thickness on the live animal and measurements of hip height can assist in predicting the yield grade at certain slaughter weights. Visual estimates can be relatively accurate in identifying actual yield grades if cattle differ by as much as one yield grade (see Color Plates E, F, G, and H in Chapter 13).

The most accurate measure of quality grade results from carcass evaluation. Traditionally, steer and heifer progeny of different bulls are fed and harvested to best identify genetic superiority or inferiority of bulls for both quality grade and yield grade. Heritabilities of most beef carcass traits are high (over 40%), so selection can result in marked genetic improvement for these traits.

A recent development in the assessment of carcass traits is the use of ultrasound technology (Fig. 12.7). The correlations between ultrasonic measures of carcass traits on an individual animal and that animal's own carcass performance measured in the cooler are reasonably high. Correlations between ultrasonic measures of marbling, fat thickness, and ribeye on yearling Angus bulls and cooler measures on steer progeny for the same traits are approximately 0.70 (Wilson et al., 1999). Given these relationships, the use of ultrasound to provide data from which genetic predictors can be estimated is a useful tool.

FIGURE 12.7 The use of chute-side ultrasonic measurement of body composition is a means to enhance genetic selection for carcass traits.

Tenderness can be measured by the Warner-Bratzler Shear test, which measures the pounds of force needed to cut through cores of meat, or measured by a trained panel of people who evaluate the palatability characteristics (juiciness, flavor, and tenderness) of meat. However, these methods of tenderness evaluation are slow and expensive. Nonetheless, the American Simmental Association released shear force EPDs on a limited number of sires in 2000.

More objective measurements for carcass tenderness and muscling are needed. Tenderness is not highly correlated to marbling nor is muscling highly correlated to actual ribeye area or ribeye area per cwt. of carcass. Visually appraising muscling by evaluating stifle thickness is more highly correlated to trimmed box beef yield than is ribeye area.

Longevity/Stayability

Longevity, which measures the length of productive life, is an especially important trait for cows. Bulls are usually kept in a herd for only a few years, or inbreeding may occur. Some highly productive cows remain in the herd until 15 years of age or older, while other highly productive cows have been culled from the herd prior to reaching three or four years of age. Little direct selection for longevity in cows exists because few cows remain highly productive past the age of 10 years. Some cows are culled because of problems such as skeletal unsoundness, poor udders, or unhealthy eyes or teeth. Most cows that leave the herd early have poor reproductive performance (open or late pregnancy).

Some producers need to improve their average herd weaning weights as rapidly as possible rather than improve longevity. In this situation, a relatively rapid turnover of cows is needed.

Some selection for longevity occurs because highly productive cows that stay in the herd for a long period of time leave more numbers of potential replacement heifers. Some beef producers attempt to identify bulls that have highly productive, relatively old dams. Certain conformation traits, such as skeletal soundness and udder soundness, may be evaluated to extend the longevity of production. Crossbreeding will increase longevity as crossbred cows usually retain their teeth longer and in better condition than straightbred cows.

A few breeds have included a stayability EPD in their sire summary. This EPD predicts genetic differences in the likelihood or probability that daughters of bulls will remain in production until they are at least six years of age. Cows culled from the herd prior to six years of age are costly because in most herds it takes two to four calves produced per cow to cover the cost of the replacement heifer minus the salvage value of the cow. Longevity/stayability has a heritability of approximately 20% so genetic change can be made in this trait by including it in a selection program.

Conformation

Conformation is the form, shape, and visual appearance of an animal. How much emphasis to put on conformation in a beef cattle selection program has been and continues to be controversial. Some producers feel that putting a productive animal into an attractive package contributes to additional economic returns. It is more logical, however, to place more selection emphasis on traits that will produce additional numbers of calves and pounds of lean growth for a given number of cows. Placing some emphasis on conformation traits such as skeletal (particularly feet and legs), udder, and eye and teeth soundness is justified (Figs. 12.8 and 12.9). Conformation differences, such as in fat accumulation or predisposition to fat, can be used effectively to make meaningful genetic improvement in carcass composition.

Most conformation traits are medium to high (30–60%) in heritability, so selection for them will result in genetic improvement.

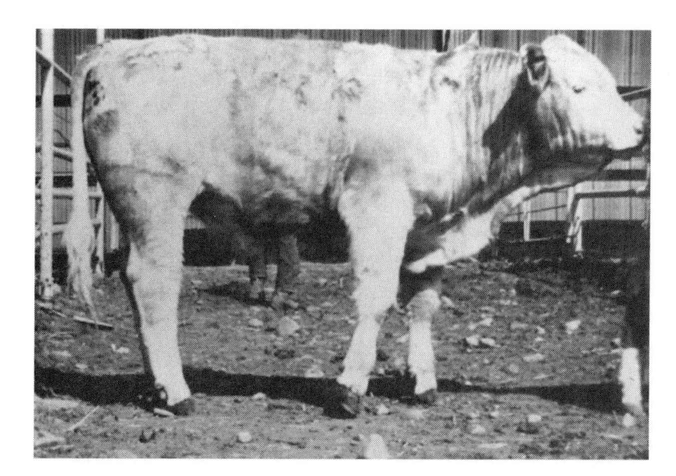

FIGURE 12.8 Bulls that are postlegged (hind legs too straight) are likely to become stifled. This affects their longevity as breeding animals since they would have to be culled from the herd.
Source: Colorado State University.

FIGURE 12.9 Soundness of feet and legs, eyes, and testicles can be evaluated visually. Reproductive soundness, including scrotal circumference, is best measured more objectively.
Source: Richard Sell, Booker, Texas.

Genetic Defects

Genetic defects, including those identified previously in the sections on longevity and conformation, need to be considered in breeding productive beef cattle. Cattle have numerous known hereditary defects as more than 200 different defects have been identified. Most of them, however, occur infrequently and are of minor concern. Some defects increase in their frequency, so selection needs to be directed against them. Most of these defects are determined by a single pair of genes that is usually recessive (Table 12.2). When one of these hereditary defects occurs, it is a logical practice to cull both the cow and the bull.

Some genetic defects that are observed in cattle today include the following:

Achondroplasia (bulldog dwarfism)—The homozygote may be aborted at six to eight months' gestation and has a compressed skull, nose divided by furrows and a shortened upper jaw giving the bulldog facial appearance. Inherited as an incomplete dominant. The most common type of *dwarfism* is *snorter dwarfism,* in which the skeleton is quite small and the forehead has a slight bulge. Some snorter dwarfs exhibit a heavy, labored breathing sound. This defect was common in cattle in the 1950s but has decreased significantly since that time. Also occurring is the long head dwarf (simple recessive) and the compress (comprest) dwarf inherited as incomplete dominance.

Alopecia—A lethal abnormality very similar to hypotrichosis. It takes a laboratory analysis to distinguish between the two. It has only been observed in Polled Herefords since 1988. Calves have kinky, curly hair that is soon lost in patches around the head, neck, and shoulders. Skin changes and anemia occur in all cases with death occurring before seven months of age due to anemia. Believed to be a simple recessive.

Ankylosis results from an abnormal union of any of the joints in the calf's body. Cleft palate frequently occurs in this recessively inherited defect.

Arthrogryposis (palate-pastern syndrome) is a defect in which the pastern tendons are contracted and the upper part of the mouth is not properly fused together (cleft palate).

Brachynathia inferior (parrot mouth)—Cattle have a short lower jaw (simple recessive inheritance).

Cryptorchidism is the retention of one or both testicles in the body cavity.

Dermoid (feather eyes)—Skinlike masses of tissue occur on the eye or eyelid. Animals may become partially or completely blind. Mode of inheritance is polygenic.

Double muscling is evidenced by an enlargement of the muscles, with large grooves between the muscle systems especially noticeable in the hind leg. Double-muscled cattle usually grow slowly, and their fat deposition in and on the carcass is much less than that of normal beef animals.

Hydrocephalus—A bulging forehead due to fluid that has accumulated in the brain area is the typical symptom of *hydrocephalus.* Calves with arthrogryposis, hydrocephalus, or osteopetrosis usually die shortly after birth.

Hypotrichosis (hairlessness)—Partial to almost complete lack of hair. Inherited as a simple recessive.

Hypotrichosis ("rat-tail")—A form of congenital hypotrichosis characterized by the colored hair anywhere on the body being short, curly, malformed, and sometimes sparse. An abnormal tail switch occurs. Calves have slower rates of gain between weaning and yearling ages compared with normal calves from similar mating. Defect has occurred from mating of Simmental with Angus. This abnormality is controlled by interaction between two loci where at least one gene is for black color and must be heterozygous at the other locus involved.

Neuraxial Edema—Calves are normal size at birth but may not be able to get up or lift their head. Muscle spasms of neck and legs may last for one to two minutes (simple recessive).

Osteopetrosis (marble bone disease) is characterized by the marrow cavity of the long bones being filled with bone tissue. All calves with osteopetrosis have short lower jaws, a protruding tongue, and impacted molar teeth.

Progressive Bovine Myeloencephaly (weaver calf)—Calves start developing a weaving gait at six to eight months of age and get progressively worse until death at 12 to 20 months (simple recessive).

Protoporphyria (photosensitivity)—Cattle are sensitive to sunlight and develop scabs and open sores when exposed to sunlight. Inherited as a simple recessive.

Syndactyly (mule-foot) is a condition in which one or more of the hooves are solid in structure rather than cloven. Mortality rate is high in calves with syndactyly.

When an abnormal calf is born, the producer should first try to determine whether the abnormality is genetic. Some abnormalities are environmentally caused, such as *crooked calf disease.* The crooked-legged condition makes one suspicious of a genetic abnormality, however, it is caused by the cow eating the lupine plant during a certain stage of pregnancy (Chapter 15).

A producer should not panic if a genetically abnormal calf is born in the herd. A commercial producer might consider selling the cow and replacing the bull. A seedstock producer may need to take a more serious look at the problem and take additional steps to eliminate or reduce the incidence of the genetic problem.

The genetic abnormalities that have caused the greatest concern in recent years were identified earlier in the chapter (Table 12.2). Several of these abnormalities reached unexpected levels of occurrence due to a selection preference for the carrier animal or a noted bull being used extensively through AI. In the latter situation, a carrier bull can sire several thousand calves before the abnormality expresses itself. Obviously, one-half of the bull's calves would carry the undesirable recessive. However, this would not be known until one of his carrier daughters was mated to another animal carrying a similar undesirable gene. Even then, the chance of an abnormal calf occurring is one chance in four for each carrier × carrier mating.

Occasionally, it is deemed desirable and necessary to progeny test a bull from unknown or known carrier parentage. It is generally accepted that the level of testing should be at the 1% level of probability; that is, there is only one chance in a hundred that a bull could be tested this extensively and still be carrying the undesirable recessive gene.

Using the inheritance of the mule-foot abnormality (Table 12.2), a bull (S?) could be mated to eight females showing the mule-foot condition (ss) or sixteen known carriers (Ss) or thirty-two of his own daughters (½ ss and ½ Ss). If no mule-foot calves were identified, the bull would be tested to the 1% level of probability (Table 12.10).

TABLE 12.10 Probability of Detecting a Bull as a Carrier of an Undesirable Recessive Gene, Using Three Kinds of Test Matings

	Mating Bull to . . .		
No. Matings	Double Recessive	Known Heterozygote	Own Daughter
1	0.5	0.25	0.125
2	0.75	0.44	0.235
3	0.87	0.58	0.330
4	0.94	0.68	0.414
5	0.97	0.76	0.487
6	0.98	0.82	0.551
7	0.99	0.87	0.607
8	0.996	0.90	0.656
9	0.998	0.925	0.699
10	0.999	0.944	0.737
15	—	0.987	0.865
20	—	0.997	0.931
25	—	0.9993	0.965
30	—	—	0.982
35	—	—	0.991
40	—	—	0.995
50	—	—	0.999

Disposition

Disposition or *docility* refers to the level of calmness or excitability of cattle. This trait is best evaluated when producers are near cows during calving or when working cattle through handling and restraining facilities. Obvious differences in the behavior of cattle exist that affect their disposition (Chapter 18).

There is evidence that both the genetic makeup of cattle and the environmental conditions to which they are exposed determine disposition. There are breed differences, line of breeding differences within a breed, and individual differences in disposition. Cattle with very poor dispositions are sometimes culled to prevent human injury or to decrease repair costs of fences and working facilities. How producers handle their cattle also has a significant impact on their disposition (Fig. 12.10).

Adaptability

The Brahman breed and Brahman crossbreds are used widely throughout the Southeast and Gulf Coast areas because of the breed's adaptability to high environmental temperatures and high humidity. Most of the *Bos taurus* breeds are not adapted to these subtropical conditions, thus, their performance is lower than the Zebu cattle.

Bos taurus cattle can function much better in colder environments than the *Bos indicus* breeds. Newborn Brahman calves are unable to maintain thermostability in cold (5°C) environmental temperatures.

Adaptability of different biological types will vary depending on the environment where they are expected to perform. For example, cows of large mature weight and high milk production will likely have below-average reproductive performance in a low-cost grazed forage

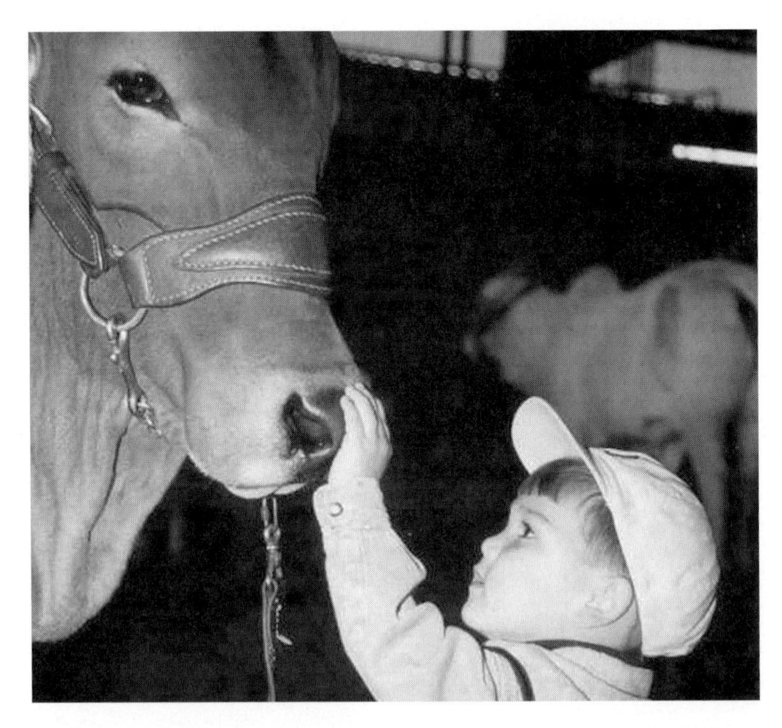

FIGURE 12.10 Disposition in cattle has a genetic basis, however, the major influence on disposition is how the cattle are handled. Gentle dispositions are practicularly important when children are involved. *Source:* J. D. Hudgins, TX.

environment. In this situation, there would be high breakeven prices of weaned calves. The cows are adapted to an environment where large amounts of harvested and (or) supplemental feed must be provided to achieve satisfactory levels of reproductive performance.

IMPROVING BEEF CATTLE THROUGH BREEDING METHODS

Genetic improvement in beef cattle occurs by selection and the choice of mating system. Significant improvement by selection results when the selected animals are superior to the herd average and when the heritabilities of traits are medium to high (above 30%). Factors affecting the rate of genetic improvement from selection include genetic variation, heritability, selection differential, and generation interval.

Genetic Variation

Most economically important traits in beef cattle are influenced by many pairs of genes. This *genetic variation,* along with different environmental effects, causes considerable variation in cattle traits. Reports of yearling weights of bulls in excess of 1,700 lbs, ribeye areas larger than 20 sq. in., and 205-day weaning weights over 800 lbs demonstrate that high levels of productivity are possible with exceptional genetic and environmental conditions. These extremely high levels of productivity are usually not economically feasible because maximum profitability is reached before maximum levels of productivity (Chapter 3).

Differences in productivity are noticeable even under similar environmental conditions (Fig. 12.11). It is not uncommon for a commercial cow-calf producer to find weaning weight differences of over 100 lbs in calves within the same herd.

As noted earlier, many economically important traits in beef cattle show continuous variation primarily because they are controlled by many pairs of genes. As these genes express themselves and as the environment influences these traits, producers usually observe and measure large differences in the performance of animals for any one trait. For example, if many calves are weighed at weaning (approximately 205 days of age) in a single herd, there would be considerable variation in the calves' weights. Distribution of weaning weights of the calves would be similar to the example shown in Figure 12.12. The bell-shaped curve distribution demonstrates that most of the calves are near the herd average (440 lbs), with relatively few calves having extremely high or low weaning weights when compared at the same age.

FIGURE 12.11 These two cows had the same productive opportunity being in the same herd for their entire lifetimes. The cow on the left produced 11 calves with an average weaning weight of 533 lbs, while the cow on the right had an average weaning weight of 384 lbs on 11 calves. With calf prices at 75 cents per pound, the cow on the left had a higher gross return of $1,229 compared with the other cow. These differences currently exist in most herds today.
Source: Dwight F. Stephens.

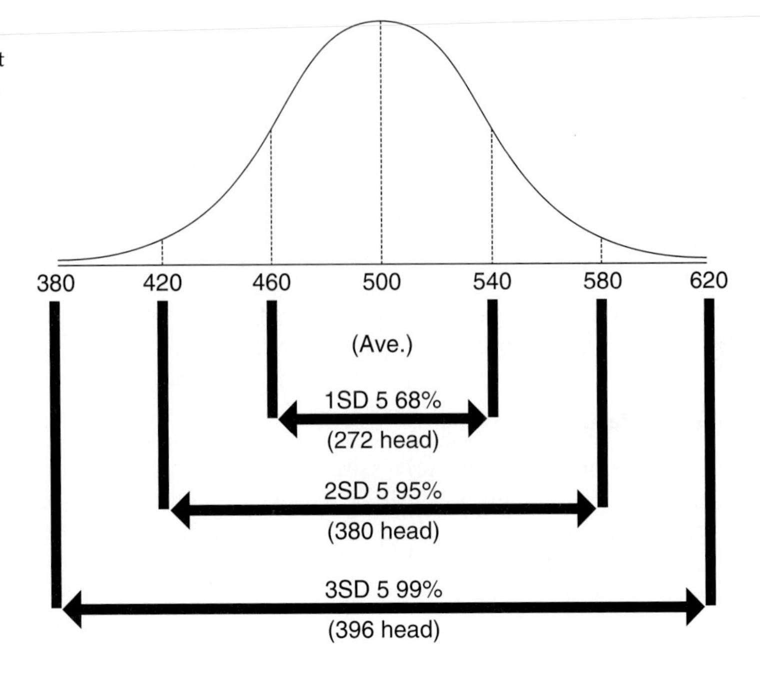

FIGURE 12.12 A normal bell-shaped curve for weaning weight showing the number of calves in the area under the curve. (400 calves in the herd.) One standard deviation (SD) equals 40 lbs. *Source:* Colorado State University.

Figure 12.12 shows the use of the statistical measurement of *standard deviation (SD),* which describes the variation or differences in a herd (in this case, the average weaning weight is 500 lbs and the calculated standard deviation is 40 lbs). Using herd average and standard deviation, the variation in weaning weight shown in Figure 12.12 can be described as follows:

500 lbs ± 1 SD(40 lbs) = 460–540 lbs
(68% of the calves are in this range)

500 lbs ± 2 SD(80 lbs) = 420–580 lbs
(95% of the calves are in this range)

500 lbs ± 3 SD(120 lbs) = 380–620 lbs
(99% of the calves are in this range)

One percent of the calves (4 calves in a herd of 400) would be on either side of the 380–620-lb range. Most likely, two calves would be below 380 lbs and two calves would weigh more than 620 lbs.

Weaning weight is a phenotype because its expression is determined by the genotype (the genes received from the sire and the dam) and the environment to which the calf is exposed. Weaning weight has a complex inheritance: There are many pairs of genes involved and, at the present time, the individual pairs of genes cannot be identified like certain traits controlled by one or two pair of genes. Consider the following formula:

Phenotype = genotype + environment

The phenotype will more closely predict the genotype if producers expose their animals to a similar environment; however, the latter must be within economic reason. This similarity between phenotype and genotype, where many pairs of genes are involved, is predicted with the estimate of heritability. The primary method of making genetic improvement in traits controlled by many pairs of genes involves both heritability and selection of animals with superior phenotypes. The application of this method is discussed in detail in the following section.

Phenotypic standard deviations for beef cattle traits are shown in Table 12.11.

TABLE 12.11 Averages and Phenotypic Standard Deviations for Selected Beef Cattle Traits

Trait	Average	Standard Deviation
Birth weight	90 lbs	10 lbs
Body condition score (BCS)[a]	5 units	0.7 units[a]
Scrotal circumference	33 cm	2.5 cm
Weaning weight	525 lbs	45 lbs
Daily gain (feedlot)	3.0 lbs/day	0.3 lb
Yearling weight	985 lbs	70 lbs
Yearling hip height	49 in	1.4 in
Mature cow weight	1250 lbs	125 lbs
Slaughter weight	1210 lbs	95 lbs
Dressing percentage	60.6%	1.9%
Carcass weight	735 lbs	63 lbs
Marbling score[b]	5.0 units	0.59 units
Retail product (0 fat trim)	65.8%	3.2%
Fat thickness (12th rib)	0.26 in	0.12 in
Ribeye area	12.5 sq in	1.3 sq in
Tenderness (WBS)	11.2 lbs	2.4 lbs

[a]Based on a visual score (range of 1–9).
[b]4.0–4.9 = slight; 5.0–5.9 = small.
Source: Adapted from MARC (data from 12 breed groups and three composite populations).

Heritability

A *heritability estimate* is a percentage figure that indicates what proportion of variation in a trait is due to heredity. Heritability identifies that portion of variation that is passed on from parent to offspring. Table 12.12 lists heritability estimates for some economically important traits for beef cattle.

Selection Differential

Selection differential is sometimes referred to as *reach* because it shows how superior (or inferior) the average of selected individuals is to the average of the group from which they were selected.

Predicting Genetic Change

An example of a selection differential is shown in the following calculation of genetic change:

$$\text{Genetic change per year} = \frac{\text{heritability} \times \text{selection differential}}{\text{generation interval}}$$

Bulls

Selected bulls average	545 lbs
Of bulls in herd	400 lbs
Selection differential =	145 lbs
Heritability	0.30
Total genetic superiority =	43.5 lbs
Only half passed on (43.5 ÷ 2) =	21.7 lbs

TABLE 12.12 Heritability Estimates for Important Beef Cattle Traits

Trait(s)	Heritability[a] (%)
Reproductive	
Age at puberty	40%
Weight at puberty	50
Scrotal circumference	50
Breeding soundness examination (BSE)	10
Primary sperm abnormalities	30
Secondary sperm abnormalities	2
Reproductive tract score	30
First service conception rate	25
Stayability	20
Calving ease (direct and maternal)	15
Weight at puberty	40
Gestation length	40
Birth weight	40
Pelvic area	50
Body condition score	40
Calving interval	10
Multiple births	5
Heifer pregnancy	14
Growth	
Weaning weight	30
Milk production	20
Postweaning ADG (feedlot)	45
Postweaning ADG (pasture)	30
Efficiency of feedlot gain	45
Maintenance (MEm)	50
Yearling weight	40
Yearling hip height	40
Mature weight	50
Carcass	
Carcass weight (at similar age)	40
Marbling	35
Fat thickness	45
Percent retail product	25
Ribeye area	25
Tenderness	25
Shear force (WBS)	40
Sensory panel	10
Other Traits (functional, convenience, longevity)	
Cancer eye susceptibility	30
Horn fly resistance	60
Pink eye susceptibility	25
Pulmonary arterial pressure	40
Disposition (docility)	40
Longevity/stayability	20
Sheath area (Zebu)	45
Udder attachment	20
Teat size	50

[a]Heritabilities below 20% are considered low; those 20–39% are considered medium; and those 40% and higher reflect highly heritable traits.

Heifers

Selected heifers	460 lbs
Average of heifers in herd	400 lbs
Selection differential =	60 lbs
Heritability	0.30
Total genetic superiority =	18.0 lbs
Only half passed on (18 ÷ 2) =	9.0 lbs

Mating bulls to heifers

Fertilization combines the genetic superiority of both parents
21.7 lbs + 9.0 lbs = approximately 30 lbs

Generation Interval

Generation interval is the average age of the parents when the calves are born. The selected heifers in the previous example represent only about 15 percent of the total cowherd, so the 30 lb estimate is for one generation. This much selection would have to be practiced over approximately 6 years to replace the entire cowherd. Thus, 30 lbs ÷ 6 years = 5 lbs per year. This example demonstrates how generation interval affects the yearly genetic change.

Multiple Trait Selection

The example with weaning weight represents genetic change if selection is for only one trait. If selection is practiced for more than one trait, genetic change is $1\sqrt{n}$, where n is the number of traits in the selection program. If four traits are in the selection program, the genetic change per trait would be $1\sqrt{4} = \frac{1}{2}$. This means that only one-half of the progress would be made for any one trait, compared to giving all the selection to one trait. This reduction in genetic change per trait should not discourage producers from multiple trait selection. Maximizing genetic progress in a single trait selection usually is associated with lowering productivity in some other economically important traits.

Genetic and Phenotypic Correlations

Traits are genetically and phenotypically correlated (Table 12.13). Phenotypic correlations measure how two traits change in the same or opposite directions because of genetic and environmental influences on the traits. A genetic correlation between two or more traits means that some of the same genes affect each trait. Genetic correlations varying from +1.0 to −1.0 may help or hinder a genetic improvement program. Correlations less than +0.20 and less than −0.20 are considered low and relatively unimportant, whereas correlations greater than +0.50 and greater than −0.50 are considered important.

The genetic correlation between rate of gain and feed per pound of gain (from similar beginning weights to similar end weights) is negative. This is desirable from a genetic improvement standpoint because animals that gain faster require less feed (primarily for maintenance). This relationship is also desirable because rate of gain is easily measured whereas feed efficiency is an expensive trait to measure. Yearling weight and mature weight are positively correlated with birth weight; that is, as yearling weight increases, birth weight and mature weight also increase. This may pose a potential problem in that birth weight to a large extent reflects calving difficulty.

TABLE 12.13 Genetic Correlations, Phenotypic Correlations, and Heritabilities for Several Beef Traits[a]

Trait[b]	AP	WP	CE	BW	WW	MM	PWG	F/G	YW	HH	SC	MCW	BCS	CWT	BF	REA	MRB	CUT	WBS
AP	**0.40**	−0.41	—	−0.16	−0.14	—	0.16	—	−0.05	−0.11	−0.70	—	−0.02	0.17	—	—	—	0.30	—
WP	0.43	**0.50**	—	–	—	—	0.07	—	—	—	−0.39	—	–	0.07	—	—	—	0.08	—
CE	—	—	**0.15**	−0.74	−0.21	−0.15	−0.54	—	−0.29	—	0.01	−0.23	–	−0.31	—	—	—	−0.02	—
BW	−0.08	0.29	−0.28	**0.40**	0.46	−0.14	0.32	−9.56	0.55	0.27	0.04	0.67	−0.02	0.60	−0.27	0.31	0.31	0.05	−0.01
WW	−0.12	—	0.04	0.50	**0.30**	−0.16	0.44	−0.50	0.81	0.53	0.19	0.57	0.14	0.71	0.24	0.49	−0.09	0.57	−0.83
MM	—	—	—	0.34	—	**0.20**	0.07	—	−0.25	−0.29	0.19	—	–	—	—	—	—	—	—
PWG	—	—	−0.09	0.18	0.44	—	**0.45**	−0.64	0.81	0.41	0.28	0.10	–	0.87	0.19	0.32	0.11	0.18	0.30
F/G	—	—	—	−0.12	0.00	—	0.67	**0.45**	−0.60	0.22	−0.04	—	—	0.50	−0.24	—	—	—	—
YW	−0.11	—	0.01	0.38	0.71	—	0.81	−0.46	**0.40**	0.40	0.36	0.72	0.25	0.91	0.32	0.51	−0.33	0.87	0.72
HH	−0.60	—	—	0.34	0.60	0.64	0.39	−0.04	0.63	**0.40**	0.14	—	—	—	0.36	—	—	—	—
SC	0.11	—	—	0.13	0.34	—	0.36	0.12	0.39	0.29	**0.50**	—	—	0.30	0.78	—	—	0.25	—
MCW	—	—	−0.03	0.34	0.57	—	0.10	−0.15	0.72	—	—	**0.50**	−0.18	0.21	—	—	—	—	—
BCS	−0.90	—	—	0.01	0.19	—	—	—	0.30	0.30	0.10	0.35	**0.40**	—	—	—	—	—	0.29
CWT	—	—	—	0.41	0.56	—	0.68	—	0.85	—	—	0.21	—	**0.40**	0.37	0.48	0.25	0.00	0.10
BF	—	—	—	−0.07	0.16	—	0.18	0.14	0.30	—	0.27	—	—	0.29	**0.45**	0.01	0.35	−0.56	0.26
REA	—	—	—	0.17	0.23	—	0.28	—	0.35	—	—	—	—	0.45	0.10	**0.30**	0.21	0.45	−0.28
MRB	—	—	—	−0.02	−0.02	—	0.13	—	0.14	—	—	—	—	0.15	0.2	0.06	**0.35**	−0.25	−0.31
CUT	—	—	—	0.05	0.43	—	0.16	—	0.85	—	—	0.25	—	0.17	0.51	0.45	−0.25	**0.35**	−0.16
WBS	—	—	—	0.05	0.06	—	0.02	—	—	—	—	—	—	−0.02	−0.01	−0.02	−0.20	0.03	**0.40**

[a]Heritabilities are on the diagonal, genetic correlations to the right; phenotypic correlations are to the left of the diagonal. Correlations are positive (+), unless shown as negative (−).

[b]AP (age at puberty); WP (weight at puberty); CE (calving ease); BW (birth weight); WW (weaning weight); MM (maternal milk); PWG (postweaning gain); F/G (lb feed per lb gain); YW (yearling weight); HH (yearling hip height); SC (scrotal circumference); MCW (mature cow weight); BCS (body condition score in cows); CWT (carcass weight); BF (carcass backfat); REA (ribeye area); MRB (marbling); CUT (cutability or % retail cuts); WBS (Warner Bratzler Shear).

Sources: Koots et al. (1993); Gregory et al. (1995).

Correlations that pose challenges to producers are:

1. Marbling is negatively correlated to fat thickness in the carcass.
2. Yearling weight is positively correlated to birth weight.
3. Fat thickness in steer carcasses is negatively correlated to reproductive performance of half-sister females.

Research studies and observations in progressive cow herds show that improvement can be made in selecting for these economically important traits that have challenging correlations. For example, it is possible to combine moderate birth weights (65–80 lbs) with rapid feedlot gains (≥ 3.0 lbs/day). Another example shows that it is feasible to produce carcasses that have Choice quality grades and also have Yield Grade 2 carcasses. Optimum combinations of these traits where extremes are avoided should be the selection goal.

Artificial Insemination

Figure 12.13 shows the rate of genetic improvement for yearling weight under different selection schemes. The most significant points include the following:

1. A marked difference between herd A (no AI, males and females selected for yearling weight) and herd D (bulls obtained from a herd in which no selection is practiced for yearling weight).
2. The small difference between herd B (bulls selected from herd A for yearling weight and females selected for yearling weight within herd) and herd C (bulls selected from herd A) but selection not practiced for yearling weight on females. Effective sire selection will account for 80–90 percent of the genetic change in a herd.
3. The greatest change in herd F, where the top 1 percent of the sons of the top 1% of the bulls were used to build generation upon generation of AI selection. The rate of improvement in herd F is 1.5 times that of herd A, where natural service is used. The primary reason for the greater improvement in herd F is a larger selection differential.

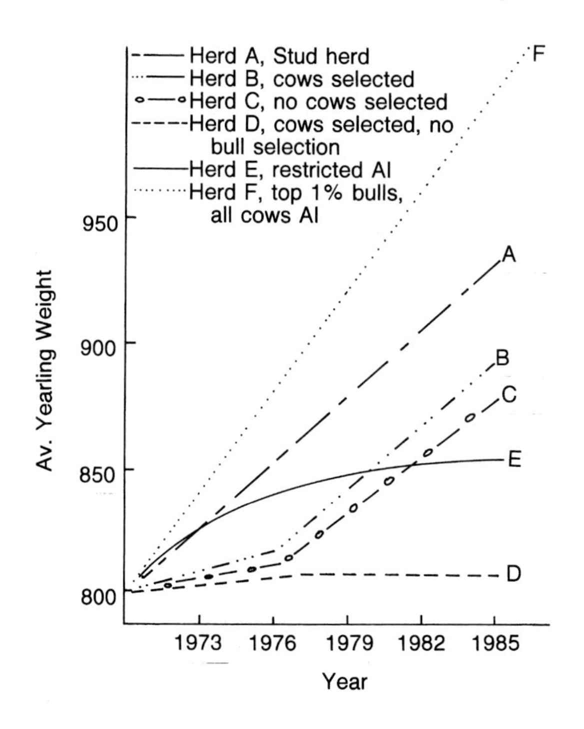

FIGURE 12.13 Expected genetic change in different types of beef herds. Adapted from Magee, Mich. State Univ. Qtr. Bul. 48:4.

4. Rapid genetic change may only be desirable for a limited period of time—until an optimum level of performance is reached that maximizes profit.

SELECTION PROGRAMS

Types of Selection

There are three types of selection: (1) tandem, (2) independent culling level, and (3) selection index.

Tandem is selection for one trait at a time. When the desired level is achieved in that one trait, then selection is practiced for a second trait. Tandem is the least effective of the three selection methods. Because of negative genetic correlations, the prolonged selection for one trait may yield deficiencies in another. While progress in the selected trait may be significant in the short run, it is very difficult to hold a trait at an optimal level over time.

Independent culling level establishes minimum culling levels for each trait in the selection program. Even though it is not as effective as the selection index, independent culling level is the most common type of selection in use today. Its major disadvantage is that an animal slightly below the minimum culling level in one trait but highly superior in other traits would be culled (Table 12.14). For example, in this case, bull B is culled for being 5 lbs under the minimum weaning weight despite being superior in the other three traits.

The *selection index* is the most effective selection type when selecting for two or more traits. As noted earlier, caution should be exercised about genetic change when many traits are included in the selection program. A selection index is not easily constructed; it requires highly involved statistical methods in order to put the heritabilities, economic values, genetic correlations, and variabilities of several traits into a single formula. An example of a selection index is—Index = yearling weight − 3.2 (birth weight) as mentioned in Chapter 11. This index is effective in selecting for an optimum combination of higher yearling weights and lower birth weights.

Perhaps the longest application of a selection index to a cattle population is that utilized by Landcorp Farming, Ltd., in New Zealand. The breeding objective, defined in 1976, and applied continuously to the herd was:

$$H = 0.53 \times L \times D_p (4.8F - 1) + 0.06 \times M \times D_m$$

Where:

H = net lifetime income per cow
.53, .06 = the net income (1976 NZ$/kg carcass) from the slaughter of young stock and cull cows, respectively
L = market weight (kg) of progeny at 30-months of age
D_p, D_m = dressing percentage of young stock and cows, respectively
F = net fertility
M = sale weight (kg) of cows

TABLE 12.14 Independent Culling Level Selection in Yearling Bulls

		Bull				
Trait	Culling Level	A	B	C	D	E
Birth weight (lb)	85 (max)	105	82	85	83	80
Weaning weight (lb)	500 (min)	559	495	505	570	600
Postweaning gain (lb/day)	3.0 (min)	3.7	3.6	3.1	3.3	3.5
Scrotal circumference (cm)	30 (min)	34	37	31	29	35

In the calculation, 4.8F−1 is the total number of saleable calves per the lifetime of a female minus her own replacement. To account for nutrient intake, the gross income from young stock and cows was adjusted 11% and 32%, respectively. Following the development of the breeding objective, selection indexes were formulated to predict overall breeding values for each animal. Selection of replacement bulls and heifers took place at one year of age and were based on the following:

$$I = 40.4F_D + .0398\ MWW + (-.2274WW_I) + 6191\ YW_I$$

Where:

I = income
F_D = fertility of the dam
WW_I = weaning weight of the individual
YW_I = yearling weight of the individual

The results of this breeding approach were genetic trend increases of direct and maternal weaning weight of .72 and .33 lb per year. Yearling weight breeding values rose 1.67 lbs annually, and fertility—defined as the number of calves weaned per cow exposed to mating—increased the equivalent of one additional calf per 100 head of females bred. The overall index breeding value increased $4.32 (1976 NZ) per year. Thus, as a result of selecting for the defined breeding objective, each new heifer entering the herd was worth $4.32 more than the heifers entering the herd in the preceding year (Enns and Nicoll, 1997).

Selection Methods

The three primary methods of selection—(1) pedigree, (2) individual appearance or performance, and (3) progeny testing—are shown in Figure 12.14.

Pedigree information is most useful before animals have expressed their own individual performance or before their progeny's performance is known. Pedigree is also useful in assessing genetic abnormalities and traits expressed much later in life (longevity) and in selecting for traits expressed only in one sex (e.g., bull selection for milk production). In evaluating pedigree information, only the information from the animal's closest relatives should be used because of their higher genetic relationship to the animal.

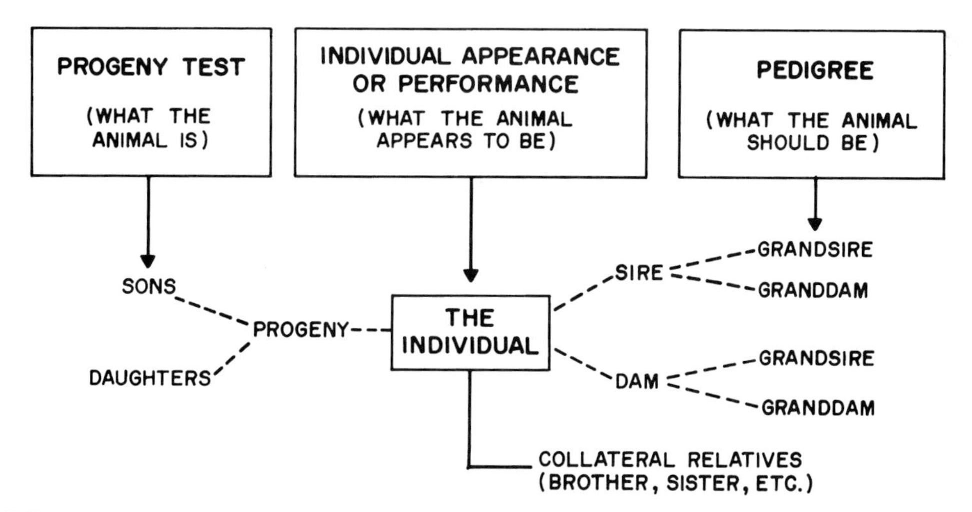

FIGURE 12.14 The selection methods (how they tie together and a brief appraisal of each). *Source:* Colorado State University.

Selection on *individual appearance and performance* should be practiced on traits that are economically important and have high heritabilities. One primary advantage of selection on individual performance is that it permits rapid generation turnover and thus shortens the general interval.

Progeny testing is more accurate and useful than the other two selection methods. Such tests are particularly useful in selecting for carcass traits (where good carcass indicators are not available in the live animal), sex-limited traits (such as mothering ability), and traits with low heritabilities. The disadvantages of progeny testing are (1) only a limited number of animals can be progeny tested, (2) a longer generation interval is required to obtain progeny information, and (3) there is decreased accuracy in poorly conducted tests. Progeny testing is usually limited to bulls because a sufficient number of calves cannot be obtained on a cow to give an adequate progeny test.

In a properly conducted progeny test, bulls are assigned cows at random after the cows have been classified by age, line of breeding, and known performance. A bull bred to a selected group of cows can give biased information that does not accurately reflect his breeding ability. An adequate progeny test involves a minimum number of progeny, typically 20 head per sire through weaning and 10 head through postweaning and carcass evaluation. The greater the numbers of progeny, the greater the validity of the test. In addition, progeny test data will be biased and the accuracy greatly reduced if cows and calves are not fed and managed uniformly until all data are collected.

All three selection methods can be used effectively in a cattle breeding program. Outstanding young bulls can best be identified by using individual performance and pedigree information on the sire, dam, and half-sibs. These outstanding young sires then need to prove themselves genetically based on a valid progeny testing program. Results of the progeny tests will determine which sires to use extensively in later years.

NATIONAL SIRE EVALUATION

At one time, genetic comparisons were limited to within-herd comparisons, central bull test stations, and highly structured progeny tests. These efforts, while useful, did not yield the desired accuracy and breadth of prediction across populations. With the advent of improved statistical programs, such as Best Linear Unbiased Prediction (BLUP), genetic evaluation systems on a national level could be conducted. These contemporary genetic prediction efforts have been embraced by nearly all of the beef breed associations (Fig. 12.15).

FIGURE 12.15 The production and distribution of national sire summaries has revolutionized the ability of beef cattle breeders to make appropriate selection and mating decisions.

A more detailed discussion of using EPDs in beef cattle selection is presented in Chapter 4 and should be reviewed at this time.

National sire evaluation programs (NSEP) are based on a set of guidelines established by the Beef Improvement Federation. The key to NSEP development is the widespread use of AI and the use of common sires across herds to allow for the creation of linkages between a large number of herds (Fig. 12.16). Furthermore, it is absolutely critical that breeders collect and report accurate data on as many animals as possible for the traits in question. Finally, it is critical that breeders report accurate contemporary groupings to assist in the calculation of the most reliable genetic predictions possible.

To calculate an EPD for a particular trait, performance information can be utilized from the individual, its siblings, its ancestors, and best of all, from its progeny. Because selection of replacement animals occurs prior to their actually becoming parents, the use of several different EPD formats is made. These include *parent EPDs* for those animals with progeny data and *non-parent EPDs* for those animals without corresponding progeny data. Non-parent EPDs can be calculated on those animals that have individual data on a trait but also for those animals where the data is not yet available.

For example, a yearling weight EPD can be computed for a weaning calf by utilizing only pedigree data. These are referred to as *pedigree estimates.*

An emerging trend in sire evaluation is to begin to create and report economic indexes. The dairy industry has successfully utilized this approach for a number of years and its application in the beef industry can help to aggregate a number of EPDs into one value reported as differences in economic performance.

One example is the grid merit EPD reported by the American Gelbvieh Association in 2001. Grid merit combines the EPDs for carcass weight, ribeye area, marbling, and backfat thickness into a dollar value that can then be utilized to make comparisons between potential parents. For example, assume that bull A has a grid merit EPD of +16 and bull B has an estimate of +7. Bull A's progeny would be expected to yield \$9 (16 − 7) more in per head revenue than those of sire B if progeny are marketed on the Gelbvieh alliance grid. By combining information from a variety of traits into one value, selection is focused on the economically

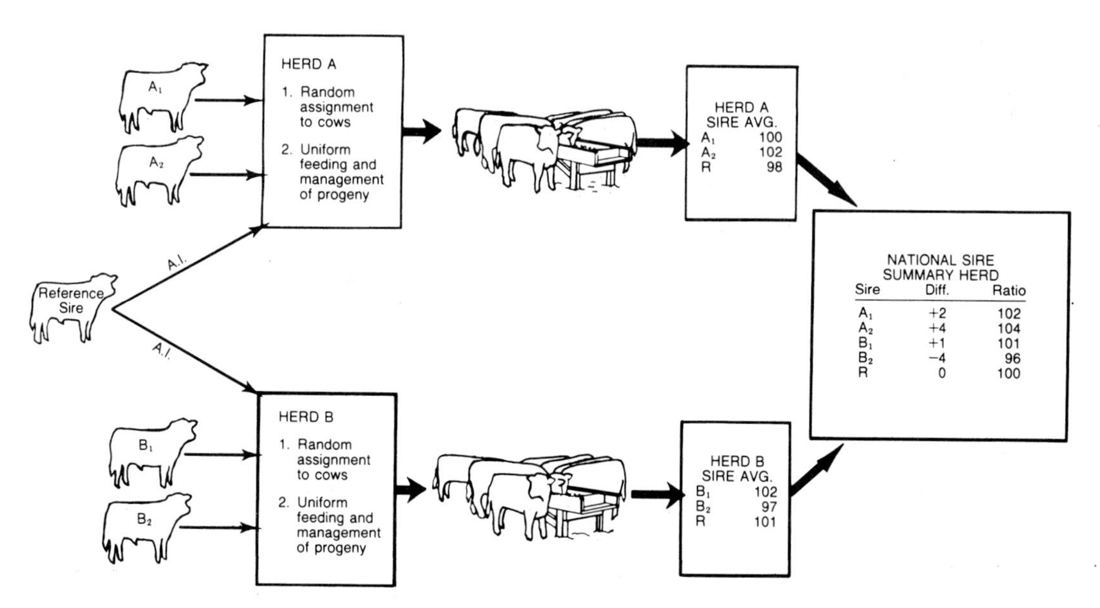

FIGURE 12.16 Basic breeding and data analysis plan for National Sire Evaluation.
Source: USDA.

TABLE 12.17 Possible Change Values for EPDs of Various Traits and Associated Accuracies

BIF Accuracy	Birth Weight	Weaning Weight	Yearling Weight	Maternal Milk	Fat Thickness	Ribeye Area	Marbling Score
0	2.97	11.74	15.07	8.82	0.03	0.31	0.24
.2	2.38	9.39	12.06	7.05	0.02	0.25	0.19
.4	1.78	7.04	9.04	5.29	0.02	0.19	0.14
.6	1.19	4.69	6.03	3.53	0.01	0.12	0.10
.8	0.59	2.35	3.01	1.76	0.01	0.06	0.05
.9	0.30	1.17	1.51	0.88	0.00	0.03	0.02
1.0	0.00	0.00	0.00	0.00	0.00	0.00	0.00

Source: Red Angus Sire Summary (2001).

important trait—in this case, grid revenue—as opposed to being focused on a multitude of indicator traits.

Breeders are cautioned to avoid intensive single-trait selection even on an index basis. Keeping focus on the whole system continues to be an important component of a profitable breeding plan.

Associated with each EPD is a measure of accuracy. The range of values for accuracy is 0 to 1, with values closer to 1 indicative of estimates in which a high level of confidence can be given. In essence, accuracy increases as more data are used in making the estimate. High accuracies indicate that the estimate is more reliable and that it is less likely to change as more information becomes available.

Another way to report accuracy is with possible change values (Table 12.17). Assume that an animal has an EPD for yearling weight of +60 and an associated accuracy of .80. Given the possible change value of 3.01 from Table 12.16, we can be 68 percent confident that the true genetic value of the animal for yearling weight lies in the range of +60 ± 3.01.

Understanding and utilizing accuracy information allows breeders to make appropriate selection decisions based on their own tolerance for risk.

EPDs have traditionally been restricted to within-breed comparisons. However, this is a limiting factor for those producers who utilize multiple breeds in planned crossbreeding systems. Based on research conducted at the Meat Animal Research Center, a set of adjustments have been developed to allow for rough comparisons of EPDs on animals from different breeds (Table 12.18). The adjustments are made from an Angus base.

Assume that a breeder is attempting to compare Angus, Pinzgauer, and Hereford bulls, all of which have a birth weight EPD of +3.0. It would be incorrect to assume that they are similar genetically for the trait. The adjustment factors of +4.6 and +9.0 need to be added to the EPDs of the Hereford and Pinzgauer, respectively. Because Angus is the base, nothing would be added. This would result in estimates of +7.6, +12.0, and +3.0 for the Hereford, Pinzgauer, and Angus bulls. At this point, a more accurate comparison can be made.

MARKER-ASSISTED SELECTION

DNA information in the beef cattle industry can be applied to parentage testing, marker-assisted selection, and gene transfer. The current application of this technology is limited.

However, *parentage identification* via DNA testing is rapidly becoming the industry standard. The use of DNA testing allows for retrospective selection. For example, if a herd is

TABLE 12.18 Across Breed EPD Adjustments, 2000

Breed	Birth Weight	Weaning Weight	Yearling Weight	Milk
Angus	0	0	0	0
Brahman	14.5	38.8	4.5	28.0
Charolais	12.0	44.8	73.4	8.8
Gelbvieh	7.8	16.6	−2.7	13.6
Hereford	4.6	3.6	2.2	−10.0
Limousin	7.2	29.7	34.6	3.7
Maine Anjou	9.1	32.9	29.4	12.2
Pinzgauer	9.0	31.2	34.8	9.9
Red Angus	3.7	6.7	13.0	NA
Saler	6.7	33.0	48.4	12.6
Shorthorn	8.7	31.5	49.9	14.5
Simmental	7.8	25.4	40.2	19.7
South Devon	8.2	25.4	57.5	13.8
Tarentaise	5.1	33.2	25.6	20.6

Source: MARC (2000).

using multiple sire pastures and retaining ownership of progeny to the packing plant, this technology allows a producer to sort out above- and below-average performers in feedlot and carcass traits and to then use parentage identification via DNA analysis to determine which sires to keep and which to cull.

Marker-assisted selection (MAS) is based on the identification of specific regions of chromosomes where genes affecting economically important traits are located, and then identifying MAS individuals with favorable allelic combinations. Unfortunately, most traits of interest are affected by a multitude of genes. Furthermore, many times we are limited to the use of markers associated with a gene of interest as opposed to selecting directly for the gene.

MAS will be most effective in large populations of animals, when selection is desired before the trait is measured, during early generations of selection, when heritability is relatively low, and when quantitative trait loci have large effects (Davis and DeNise, 1998). Furthermore, it is very unlikely that MAS will replace quantitative estimates such as breeding values and expected progeny differences. Instead, MAS will be used to augment existing genetic prediction tools.

Finally, gene transfer, or the use of laboratory techniques to move a desired gene from one individual to another, will also have usefulness. However, its impact in beef cattle will likely be limited to pharmaceutical production or possibly using transgenics to enhance disease resistance. Despite its strong appeal in the popular press, DNA technologies are yet a long way from widespread commercial application in the beef industry.

The following areas of knowledge must be developed before the potential of DNA technology is unleashed:

1. Creation of extensive maps of the genomes of beef cattle and microbial species of importance.
2. Development and sequencing of extensive expressed sequence tag libraries to allow for the sharing of knowledge about gene function in specific tissues or in response to disease agent and other environmental factors.
3. Creation of bioinformatic databases that allow for the integration and use of data from multiple sources.
4. Ethics and public policy issues must be considered and discussed (Wyse, 2001).

MANAGEMENT SYSTEMS HIGHLIGHTS

1. Effectively managing the biological system of cattle requires understanding and applying genetic principles.
2. Optimum levels of genetic selection and heterosis can be combined that contribute to low production costs, high levels of profitability, and the production of highly palatable consumer beef products.
3. Economically important traits must be objectively measured if the management system is highly successful.
4. Genetic design of consumer beef products is needed for Total Quality Management (TQM) so consumers can have consistent and positive eating experiences. However, single-trait selection should be avoided.
5. The use of genetic prediction and emerging DNA technologies will heighten the ability to more accurately select and propagate desired genotypes.

REFERENCES

Publications

American Gelbvieh Association. 2001. *Gelbvieh Unveils Grid Merit EPD.* News release (May 4).

American Simmental Association. 2001. *Sire Summary Report.*

Beef Improvement Federation. 1996. *Guidelines for Uniform Beef Improvement Programs.* Northwest Research Extension Center, 105 Experiment Farm Road, Colby, KS 67701.

Bourdon, R. M. 2000. *Understanding Animal Breeding.* Upper Saddle River, NJ: Prentice Hall.

Brethour, J. R. 2000. Using Serial Ultrasound Measures to Generate Models of Marbling and Backfat Thickness Changes in Feedlot Cattle. *J. Anim. Sci.* 78: 2055.

Brinks, J. S., and Knapp, B. W. 1975. *Effects of Inbreeding on Performance Traits of Beef Cattle in the Western Region.* Fort Collins, CO: Colorado Agricultural Experiment Station Tech. Bull. 123.

Davis, G. P., and DeNise, S. K. 1998. The Impact of Genetic Markers on Selection. *J. Anim. Sci.* 76: 2331.

Doyle, S. P., Golden, B. L., Green, R. D., and Brinks, J. S. 2000. Additive Genetic Parameter Estimates for Heifer Pregnancy and Subsequent Reproduction in Angus Females. *J. Anim. Sci.* 78: 2091.

Enns, R. M., and Nichoil, G. B. 1997. *Index Selection in Practice—a New Zealand Case Study.* Proceedings of the BIF Research Symposium, Dickinson, ND.

Evans, J. L., Golden, B. L., Bourdon, R. M., and Long, K. L. 1999. Additive Genetic Relationships Between Heifer Pregnancy and Scrotal Circumference in Hereford Cattle. *J. Anim. Sci.* 77: 2621.

Golden, B. L., Garrick, D. J., Newman, S., and Enns, R. M. 2000. *Economically Relevant Traits: A Framework for the Next Generation of EPDs.* Proceedings of the BIF Research Symposium, Wichita, KS.

Green, R. D. 1996. *How Will DNA Technology Impact Beef Cattle Selection?* Proceedings of the 2nd Congress on Genetics and Reproduction of the Brazilian Zebu Breeders Assoc., Uberaba, Brazil (Oct. 27–29).

Gregory, K. E., and Cundiff, L. V. 1980. Crossbreeding in Beef Cattle: Evaluation of Systems. *J. Anim. Sci.* 51: 1224.

Gregory, K. E., Cundiff, L. V., and Koch, R. M. 1995. *Composite Breeds to Use Heterosis and Breed Differences to Improve Efficiency of Beef Production.* Clay Center, NE: U.S. MARC.

Griffin, D. B., Savell, J. W., Recco, H. A., Garrett, R. P., and Cross, H. R. 1999. Predicting Carcass Composition of Beef Cattle Using Ultrasound Technology. *J. Anim. Sci.* 77: 889.

Hickman, C. G. (ed.). 1991. *Cattle Genetic Resources. World Animal Science B7.* New York: Elsevier Science Publishers B.V.

Koots, K. R., Gibson, J. P., Smith, C., and Wilton, J. W. 1993. 1. Analyses of Published Genetic Parameter Estimates for Beef Production Traits. 2. Phenotypic and Genetic Correlations. Personal communication.

Legates, J. E. 1990. *Breeding and Improvement of Farm Animals.* New York: McGraw-Hill.

Long, C. R. 1980. Crossbreeding for Beef Production: Experimental Results. *J. Anim. Sci.* 51: 1197.
Olson, T. 1994. The genetics of coat color inheritance in cattle. *American Hereford Journal,* Oct., p. 32.
Reverter, A., Johnston, D. J., Graser, H. U., Wolcott, M. L., and Upton, W.H. 2000. Genetic Analysis of Live-Animal Ultrasound and Abattoir Carcass Traits in Australian Angus and Hereford Cattle. *J. Anim. Sci.* 78: 1786.
Schalles, R. R. 1986. *The Inheritance of Color and Polledness in Cattle.* Bozeman, MT: American Simmental Association.
Tess, M. W., and Thrift, F. A. 1992. *Genetic Aspects of Beef Production in the Southern Region.* Auburn, AL: Auburn University.
Willham, R. L. 1982. Genetic Improvement of Beef Cattle in the United States: Cattle, People, and Their Interaction. *J. Anim. Sci.* 45: 659.
Wilson, D., Rouse, G., Hays, C., and Hassen, A. 1999. *Genetic Evaluation of Angus Ultrasound Measure.* Proceedings of the BIF Research Symposium, Roanoke, VA.
Wyse, R. 2001. *A Livestock Genomics Institute.* Proceedings of the Intl. Livestock Congress, Houston, TX.

Visuals

"Basic Animal Microgenetics" (1996; 43 min.). CEV, P.O. Box 65265, Lubbock, TX 79464.
"Cattle Abnormalities" (no. 8861; 16-mm color film; 26 min.). Bureau of Audio-Visual Instruction, University of Wisconsin Extension, Box 2093, Madison, WI 53701.
"Like Begets Like" (1994; 8.5 min.). American Angus Association, St. Joseph, MO.

Websites

Commercial genetic services: www2.perkin-elmer.com/ab/aggen/
Listing of mapped cattle genes: http://locus.jouy.inra.fr
Meat Animal Research Center (Clay Center): http://sol.marc.usda.gov
Blood typing and DNA services: www.immgen.com
Genetic testing services: www.absglobal.com/genmark.htm
Cattle gene mapping information: http://bos.cvm.tamu.edu/bovarkdb.html

Cattle Breeds

A *cattle breed* is defined as a race or variety related by descent and similar in certain distinguishable characteristics. More than 250 breeds of cattle are recognized throughout the world, and several hundred other varieties and types have not been given breed names.

Some of the oldest breeds introduced in the United States were officially recognized as breeds during the mid- to late 1800s. Most of these breeds originated from crossing and combining existing strains of cattle. When a breeder or group of breeders decided to establish a breed, identifying that breed from other breeds was of paramount importance. Major emphasis was placed on readily distinguishable visual characteristics, such as color, color pattern, polled or horned condition, and rather extreme differences in size, form, and shape.

Breeds in the United States are categorized as either *Bos taurus* or *Bos indicus.* Even though *taurus* and *indicus* reflect different species, this classification is questioned by several authorities. Species crosses are usually infertile or highly infertile, which is not the case in *Bos taurus* or *Bos indicus* crosses. Typical *Bos taurus* breeds are Bristish breeds (e.g., Angus, Hereford, Shorthorn) and Continental breeds (e.g., Simmental, Limousin, Charolais). The most common *Bos indicus* breeds in the United States are the Gray and Red Brahman with the Nellore, Gyr, and Indu-Brazil being less numerous as they are recent imports from Brazil in the early 1980s.

New cattle breeds, such as Brangus, Santa Gertrudis, and Beefmaster, were formed in the United States during the 1950s by crossing Brahman with one or more of the British breeds. Other breeds and composites of breeds are currently being developed. These breeds are developed by combining the desirable characteristics of several existing breeds. In most cases, however, the same identifying characteristics already mentioned have been emphasized in breeding programs to give new breeds visual identity.

Soon after some of the first breeds were developed, the word *purebred* was attached to them. Herd books and registry associations were established to assure the "purity" of each breed and to promote and improve each breed. Purebred refers to purity of ancestry as established by the pedigree, which shows that only animals recorded in that particular breed have been mated to produce the animal in question. Purebreds, therefore, are cattle from various breeds that have individual pedigrees recorded in their respective breed registry association.

When viewing a herd of purebred Angus, Charolais, or another breed, the uniformity (particularly that of color or color pattern) is noted. Because of this uniformity of one or two characteristics, the word *purebred* has come to imply genetic uniformity (homozygosity) of all characteristics. A breed may be homozygous for a few qualitative traits (e.g., color and horns or polled), but they are highly heterozygous for quantitative traits (e.g., birth weight, weaning weight, and others). High levels of homozygosity occur only after several generations of close inbreeding (e.g., father-daughter, brother-sister). This close inbreeding has not occurred in the cattle breeds but only in a few research herds. If breeds were highly homozygous, they could not be improved or changed genetically even if changes were desired.

WHAT IS A BREED?

The genetic basis of cattle breeds is not well understood by many beef producers. Often the statement is made, "There is more variation within a breed than there is between breeds." The validity of this statement needs to be carefully examined. Considerable variation does exist within a breed for most of the economically important traits. Figure 13.1 depicts the variation in 205-day weight for calves in a particular breed. The bell-shaped curve is formed by connecting the high points of each weight classification. The breed average is represented by the solid line that separates the bell curve into equal halves. Most calves are near the breed average; however, at the outer edge of the bell curve are high- and low-weaning weight calves. Note that there are fewer calves at these extremes. A specific way to describe this variation using the standard deviation is discussed in Chapter 12.

Figure 13.2 compares three breeds of cattle for weaning weight. The breed averages are different; however, the variation within each breed is comparable among all three. Although both Figures 13.1 and 13.2 are hypothetical examples, they are based on realistic samples of data obtained from the various breeds of cattle. Figure 13.3 shows breed comparisons for retail product produced from cattle slaughtered at the same age. Note in this comparison that there is more variation between the extremes in breed averages (Jersey and Charolais) compared with the variation within any one breed.

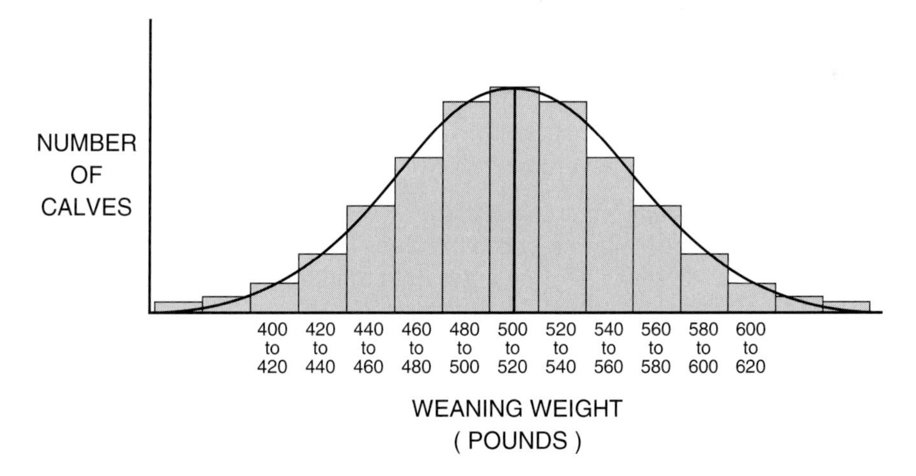

FIGURE 13.1 Variation or differences in weaning weight in beef cattle. The variation shown by the bell-shaped curve could be representative of a breed or a large herd. The dark vertical line in the center is the average or the mean and, in this example, is 510 lbs.
Source: Colorado State University.

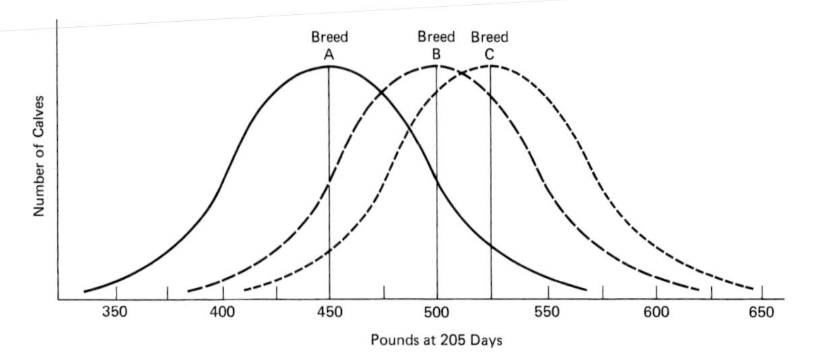

FIGURE 13.2 Comparison of breed averages and the variation within each breed for three breeds of beef cattle. The vertical lines are the breed averages. Note that some individual animals in Breed B and Breed C can be lower in weaning weight than the average of Breed A.
Source: Colorado State University.

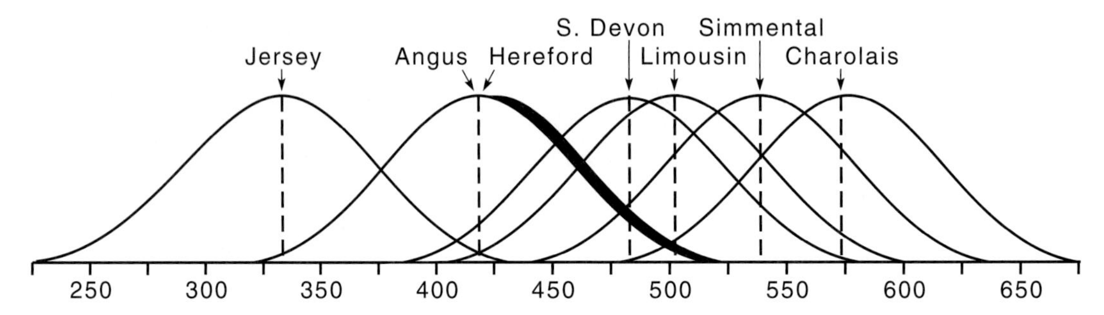

FIGURE 13.3 Retail product (pounds) at 457 days of age for cattle slaughtered from several breeds. Retail product is boneless, closely trimmed (to 0.3 in. fat) cuts from the carcass.
Source: Cundiff (MARC).

Major U.S. Beef Breeds

Shorthorn, Hereford, and Angus were the major beef breeds in the United States during the early 1900s. During the 1960s and 1970s, the number of cattle breeds in the United States remained relatively stable at 15 to 20. Today, more than 80 breeds of cattle are available to United States beef producers. However, only 10 to 15 make a significant contribution to the total number of cattle in the United States. Why the large importation of the different breeds from several countries? Following are several possible reasons:

1. Feeding larger amounts of grain to more cattle, a practice started in the 1940s, resulted in many overfat cattle—cattle that had been previously selected to fatten on grazed forage. Therefore, a need was established for grain-fed cattle that could produce a higher percentage of lean to fat at the desired slaughter weight.
2. Economic pressures to produce more weight in a shorter period of time demonstrated a need for cattle with more milk and more growth.
3. An opportunity was available for some promoters to capitalize on merchandising a certain breed as being the ultimate in all production traits. This opportunity could be merchandised easily because there was little comparative information on breeds.

Color Plates A, B, C, and D identify the most numerous breeds of beef cattle in the United States. Table 13.1 gives some distinguishing characteristics and brief background information for each of these breeds. Other cattle breeds that contribute to the beef supply are described in Table 13.2.

TABLE 13.1 Background and Distinguishing Characteristics of the Most Numerous Beef Breeds in the United States—Registrations of at least 10,000 in 2000

Breed[a]	Distinguishing Characteristics	Background
Angus	Black color, polled	Originated in Aberdeenshire and Angushire of Scotland. Imported into the United States in 1873
Beefmaster	Various colors; horned	Developed in the United States from Brahman, Hereford, and Shorthorn breeds. Selected for its ability to reproduce, produce milk, and grow under range conditions
Brahman	Various colors, with gray predominant; horned. They are one of the Zebu breeds that have the hump over the top of the shoulder. Most Zebu breeds also have large, drooping ears and loose skin in the throat and dewlap	The Brahman is a Zebu breed developed in the United States from cattle imported from India and Brazil in the early 1900s. These cattle are heat tolerant and well adapted to the harsh conditions of the Gulf Coast region
Brangus	Black; polled	U.S. breed developed around 1912—3/8 Brahman and 5/8 Angus
Charolais	White color; heavy muscled; horned or polled	One of the oldest breeds in France. Brought into the United States soon after World War I, but its most rapid expansion occurred in the 1960s
Gelbvieh	Golden colored or black; horned or polled	Originated in Austria and West Germany. Developed as a dual-purpose breed (draft, milk, and meat)
Hereford	Red body with white face; horned	Introduced into the United States in 1817 by Henry Clay. Followed the Longhorn in becoming the traditionally known range cattle
Limousin	Golden red color or black; heavy muscled; polled or horned	Introduced into the United States in 1969, primarily from France
Maine-Anjou	Red and white spotted; horned	A large breed developed in France and introduced into Canada in late 1968
Polled Hereford	Red body with white face; polled	Bred in 1901 in Iowa by Warren Gammon who accumulated several naturally polled cattle from horned Hereford herds
Red Angus	Red color; polled	Founded as a performance breed in 1954 by selecting the genetic recessives from black Angus herds
Salers	Uniform mahogany red; medium to long hair; horned	Raised in the mountainous area of France, where they were selected for milk, meat, and draft power
Santa Gertrudis	Red color; horned	First U.S. breed of cattle; developed on the King Ranch in Texas—5/8 Shorthorn and 3/8 Brahman

(continued)

TABLE 13.1 (Continued)

Breed[a]	Distinguishing Characteristics	Background
Shorthorn (includes polled and milking)	Red, white, or roan in color; horned or polled	Introduced into the United States in 1783 under the name "Durham." Most prominent in the United States around 1920. Illawarra is an Australian breed of dairy cattle similar to the Milking Shorthorn. Illawarra and Milking Shorthorn have been crossed with the U.S. Shorthorn in recent years
Simmental	Yellow to red and white color pattern; polled or horned	A prominent breed in Switzerland and France. First bull arrived in Canada in 1967. Originally selected as a dual-purpose breed for milk and meat

[a]Names and addresses of the major breed associations are listed in the Appendix.

TABLE 13.2 Beef Breeds, Dairy Breeds, Composites (Synthetic or Hybrid Breeds), Buffalo (Bison), and Other Cattle Types That Contribute to the U.S. Beef Supply

Breed or Type[a]	Distinguishing Characteristics	Brief Background
American Breed NA	Various colors	Brahman (1/2), Charolais (1/4), Bison (1/8), Hereford (1/16), and Shorthorn (1/16)
Amerifax 500	Solid red or black in color; polled	Originated in the United States in 1977. Purebreds are 5/8 Angus and 3/8 Beef Friesian
Ankina NA	Black; polled	Originated from crossing Angus and Chianina breeds
Ankole Watusi 120	Claimed as the world's longest horned cattle	Originated in Africa. Approximately 120 purebreds in the United States in 1990. Registry formed in 1983
Barzona 600	Dark reddish-brown; polled or horned	Developed by the Bard Family in 1942 in the intermountain desert area of Arizona. Selected for hardiness traits under rigorous range conditions. Foundation stock came from Angus, Africander, Hereford, and Santa Gertrudis breeds
Beefalo (also referred to as Cattalo) 786	Various colors; polled or horned	Originated in California where the bison was crossed with domestic cattle. Term *Cattalo* or *Catalo* originated in Canada, where similar crosses were made earlier
Beef Friesian NA	Similar black and white markings as Holstein, although more black color	Imported to the United States from Ireland in 1971. Selected for more muscling than United States Holstein

TABLE 13.2 (Continued)

Breed or Type[a]	Distinguishing Characteristics	Brief Background
Belgian Blue 875	White, blue, or black and white; horned and polled; heavily muscled	Imported from Belgium through Canada in 1986
Belted Galloway 1,100	Black with white belt; polled	Imported from Scotland in 1950
Blonde d'Aquitaine 625	Fawn-colored, sometimes with a reddish tinge; heavily muscled; horned	Originated in France, and live cattle imported to the United States in 1973. Semen imported prior to that time
Bonsmara NA	Dark red; horned and polled	South African breed developed by Jan Bonsma and imported to the United States in 1996. Composite of 5/8 Africander, 3/16 Hereford, and 3/16 Shorthorn; tropically adapted
Braford 2,964	Usually red color with white markings. Brockled face and pigmented eyes are common. Brahman-type ears are apparent; polled or horned	United States composite; composition of Brahman, Hereford, Polled Hereford, and Zebu breeds varies
Brah-Maine NA	Dark red to black and white; horned	United States composite; 5/8 Maine-Anjou and 3/8 Brahman
Brahmental NA	Various colors	Originated in United States by crossing Brahman and Simmental. Must have a minimum of 1/4 Brahman and a minimum of 3/8 Simmental
Brahmousin 190	Red with black pigmentation; horned	United States composite; 5/8 Limousin and 3/8 Brahman
Bralers 250	Red color; horned	United States composite; 5/8 Salers and 3/8 Brahman
Braunvieh 4,500	Various shades of brown, lighter colored around muzzle and inside legs; horned	Imported from Switzerland in 1983. Similar to Brown Swiss, however, they have been selected to be more dual purpose (meat and milk)
British White 400	White with black markings (ears, eyes, nose, and feet); polled	Imported from England; claimed to have the longest documented history of any specific breed of cattle
Buffalo (Bison) 1,000	Not typically considered a breed, however, registry associations have been formed	Bison are wild oxen belonging to the cattle family. They are native to North America and numbered more than 20 million head in the 1850s
CASH NA	Various colors	Composite breed combining Charolais, Angus, Swiss, and Hereford
Charbray NA	Various colors; horned	Cross between Charolais and Brahman
Chargrey NA	Various colors	Developed in Australia by crossing Charolais and Murray Grey breeds
Charswiss NA	Various colors	Cross between Brahman and Brown Swiss

(continued)

TABLE 13.2 (Continued)

Breed or Type[a]	Distinguishing Characteristics	Brief Background
ChiAngus NA	Black; polled	United States composite; maximum of 3/4 and 1/4 of Chianina and Angus, respectively
Chianina 6,280	White or black color with black eyes and nose; horned or polled; extremely tall cattle	An old breed originating in Italy. Acknowledged as the largest breed, mature bulls weighing more than 3,000 lbs
Chiford NA	Marked like Hereford with pale fawn, cream to dark red color; polled or horned	Out of registered Chianina bull or cow and Hereford or Hereford cross cow or bull. No more than 75% Chianina
Chimaine NA	Various colors; horned or polled	United States composite; any percentage up to and including 3/4 Chianina
Composites (synthetics or hybrids) NA	Various colors; polled or horned	There are several different composites or synthetics currently being developed (e.g., MARC I, MARC II, MARC III, CASH, Leachman Hybreds, RX_3, and others). They are not considered breeds because a breed association is not currently associated with them where pedigrees are recorded. Several of the breeds listed (e.g., Brangus, Beefmaster, Santa Gertrudis) could be considered composites or synthetics—in a broad sense, several other breeds could be classified as composites
Criollo/Corriente 2,868	Multiple color patterns; horned	Descendants of cattle brought to the New World by the Spanish explorers, likely related to Florida Cracker cattle. Mostly used for roping and rodeo, demand for approximately 40,000 per year
Cracker Cattle	Various colors	Not designated as a breed. However, a breeder's association was formed to preserve descendants of the Spanish cattle brought to Florida. Also called *Florida Scrub* or *Piney Woods* cattle
Dairy Breeds	Holstein (black and white), Jersey (fawn with black switch), Guernsey (light red and white), Brown Swiss (solid blackish or brown), Ayrshire (mahogany and white spotted), and Red and White (red and white)	These breeds have been selected for milk production with meat production considered a by-product of the dairy industry. Of the the approximate 9 million dairy cows, the Holstein is the most numerous
Devon 100	Dark red (North Devon) to light red or brown (South Devon); horned	North Devon is an old breed originating in England. South Devon is more of a dual-purpose breed. The Devon first came to America with the Pilgrims in 1623

TABLE 13.2 (Continued)

Breed or Type[a]	Distinguishing Characteristics	Brief Background
Dexter NA	Claimed as smallest breed in the United States (cows, 650 lbs; bulls, 850 lbs)	Originated in Ireland; introduced into the United States in 1912
Galloway 801	Polled, with the majority solid black in color, some dun-colored, and others white with black noses, ears, and feet. The Belted Galloway is black with a distinctive white belt; polled	An old breed from the Scottish province of Galloway. Long, burly hair has made it adaptable to the harsh climates of the North. The Belted Galloway has a separate breed association. It has the same origin as the Galloway but had an infusion of the belted cattle in the seventeenth or eighteenth century
Gelbray 300	Various colors; polled or horned	Originated in the United States by crossing the Gelbvieh and Brahman breeds. Gelbvieh (maximum 3/4; minimum 5/8) and Brahman (maximum 3/8; minimum 1/4). Registered in American Gelbvieh Association
Hays Converter NA	Various colors	Developed in the 1950s by Harry Hays, former Canadian Minister of Agriculture, by crossing Hereford, Holstein, and Brown Swiss breeds
Herens NA	Rusty brown to black in color with some white on the udder; horned	Mountain cattle from Switzerland
Irish Blacks NA	Black; polled	Originated in Ireland
Leachman Hybreds NA	Red color; polled	Combines Red Angus, Gelbvieh, Simmental, Salers, and South Devon breeds into several different composites
Longhorn	Multicolored; characterisically long horns	Came to West Indies with Columbus. Brought to the United States through Mexico by the Spanish explorers. Longhorns were the noted trail-drive cattle from Texas into the Plains states
Mandalong Special NA	Primarily a solid, reddish-tan color; polled large but refined in bone	Developed in Australia by combining Brahman, Shorthorn, Charolais, Chianina, and British White breeds
Marchigiana 81	Light gray to almost white in color. Tail switch is dark colored; horned	Originated in the Marchi and surrounding areas of Italy. Resembles Chianina in color and conformation, which implies possible intermixing with them
MARC I NA		A composite developed at the Meat Animal Research Center. Combines Swiss, Limousin, Charolais, Hereford, and Angus breeds

(continued)

TABLE 13.2 (Continued)

Breed or Type[a]	Distinguishing Characteristics	Brief Background
MARC II NA		A composite developed at the Meat Animal Research Center. Combines Hereford, Angus, Simmental, and Gelbvieh breeds
MARC III NA		A composite developed at the Meat Animal Research Center. Combines Red Poll, Hereford, Angus, and Pinzgauer breeds
Murray Grey 700	Silver gray; polled	Developed in Australia from Shorthorns crossed with Angus
Normande 250	Coat colors may be yellowish or blond, very dark brown, or almost black, and white. Colors occur in spots and speckles	Originated in France. First 7/8 Normande calf was born in the United States in 1976
Piedmontese 574	Extremely heavily muscled. Color varies; horned	Imported from Italy
Pinzgauer 926	Reddish chestnut color with white markings on rump, back, and belly; horned	A hardy breed developed in the Pinza Valley of Austria and in areas of Germany and Italy. Introduced into North America in 1972
Ranger NA	Various colors	Developed in the 1940s–1960s in several western United States cattle herds. Crosses of Hereford, Red Angus, Shorthorn, Scotch Highland, and Brahman are the primary breeds represented
Red Brangus NA	Red; polled	No specific percentage of Brahman and Angus, though generally considered about 50% of each
Red Poll 579	Red color; polled	Introduced into the United States in the late 1800s. Originally a dual-purpose breed but now considered a beef breed
Romagnola 650	Gray with black muzzle; horned; heavily muscled	Originated in Italy; imported to the United States in 1974
Romosinuano NA	Red (light and dark); polled	*Bos taurus* breed that is adapted to tropical conditions; developed in Costa Rica and Columbia; introduced into the United States as embryos in 1992 from Costa Rica and Venezuela
RX_3 1,000	Red color; polled	A composite breed combining 1/4 Hereford, 1/2 Red Angus, and 1/4 Red and White Holstein
Salorn 300	Various colors	United States composite; 5/8 Salers and 3/8 Longhorn

TABLE 13.2 (Continued)

Breed or Type[a]	Distinguishing Characteristics	Brief Background
Santa Cruz 8,875		United States composite developed by the King Ranch in Texas, involves crosses of Santa Gertrudis, Red Angus, and Gelbvieh
Scotch Highland 1,400	Golden color with long, shaggy hair; horned	Bred in the highlands of Scotland. Imported to the United States in 1894
Senepol 1,442	Mostly polled; color varies from light tan to dark red	Developed in the Virgin Islands in the 1900s by crossing the Red Poll and N'Dama cattle
Simbrah NA	Various colors; polled or horned	Purebreads are 5/8 Simmental and Brahman. Registered in the American Simmental Association
Tarentaise 2,000	Solid wheat-colored hair ranging from light cherry to dark blond; horned	Mountain cattle derived from an ancient Alpine strain in France. Originally a dairy breed where maternal traits were emphasized
Tuli NA	Color ranges from silver through golden brown to rich red (red predominates); polled or horned	An early maturing, medium-sized pure African Sango (*Bos taurus*) breed. Claimed to have heat and tick resistance similar to *Bos indicus* breeds
Wagyu 800	Solid, dull black to solid light brown with reddish tinge; horned or polled	The native cattle of Japan are called Wagyu where three breeds are recognized: Japanese Black (predominate), Japanese Brown, and Japanese Polled. Some United States producers are using Wagyu to access the Japanese market where highly marbled beef is economically rewarded
Welsh Black 200	Black; horned	Imported from Wales in 1963
White Park 300	White color with characteristic black ears and muzzle; polled or horned	Imported from England in 1940. United States breed association formed in 1975
Zebu NA	Various colors; horned; characteristic hump over shoulders	Originated in India. Imported from Mexico to the United States in the 1800s and in 1926. Later importations came from South America. There are approximately fifty strains of Zebu (*Bos indicus*). Six strains (Gyr, Gray Zebu, Nelore, Guzerat, Indu-Brazil, and Red Zebu) are most numerous in the United States
Others[b]		

[a]Names and addresses of major breed associations are listed in the Appendix. Number of 2000 registrations listed under breed name.

[b]There are other breeds or types, such as Aubrac, Buelingo, Charmaine, Corriente, Dutch Belted, El Monterey, Gascone, Geltex, Kerry, Lincoln Red, Lineback, Loala, Luing, MRI, Norwegian Red, Parthenais, Simbrangerford, Texon, Water Buffalo, and other composites with specific names that are not identified in the table, but little is currently known about them.

TABLE 13.3 Major U.S. Beef Breeds Ranked by Annual Registration Numbers (in thousands)

Breed[a]	Year						Date Association Formed
	2000	1997	1995	1990	1985	1980	
	thousand head						
Angus	260.9	239.5	224.8	159.0	175.5	257.6	1883
Hereford/Polled Hereford[b]	84.9	105.6	112.9	170.5	180.0	353.2	1881
Limousin	48.8	61.5	79.3	71.6	44.5	13.8	1968
Charolais	42.7	49.2	55.0	44.8	23.2	23.0	1957
Simmental	43.1	47.9	71.0	79.3	73.0	66.1	1969
Red Angus	39.6	35.7	30.0	15.4	12.0	12.5	1954
Beefmaster	32.3	56.6	47.1	38.4	32.1	30.0	1961
Gelbvieh	26.3	30.1	33.8	22.8	16.1	NA	1971
Brangus	26.9	27.7	31.0	32.1	30.3	24.5	1949
Shorthorn	18.6	15.5	20.0	18.0	16.7	19.4	1872
Brahman	18.0	15.1	15.4	13.0	29.9	36.4	1924

[a]Only breeds with more than 15,000 annual registrations for 2000 are listed.
[b]Hereford and Polled Hereford are combined. Prior to 1996 they were separate breed associations and some animals were registered in both associations.
Source: Breed associations.

The relative importance of the various breeds' contributions to the total beef industry is best estimated by the registration numbers of the breeds (Table 13.3). These twelve breeds would be considered the major breeds of beef cattle in the United States, since they contribute most of the genetics to the commercial cow-calf segment.

Although registration numbers are for purebred animals, they reflect commercial cow-calf producers' demand for different breeds. Breed rankings based on registration numbers that reflect industry demand were verified with a Cattle-Fax survey in 1995 showing that the top five breeds used in commercial cow-calf operations were Angus, Hereford/Polled Hereford, Simmental, Charolais, and Limousin. Numbers of cattle belonging to various breeds in this country have changed over the past years. It is reasonable to expect that in future years some breeds will become more numerous while others will decrease significantly in number. These changes will be influenced by economic conditions as well as by genetic changes currently being made in the economically important traits to meet industry demand.

BREED EVALUATION FOR COMMERCIAL PRODUCERS

Breeds exist primarily to be used as genetic resources by commercial producers. Most commercial beef producers use crossbreeding because they can take advantage of the heterosis in addition to genetic improvement from selecting within two or more breeds. A crossbreeding system should be determined by availability of breeds (and biological types within breeds) and how well adapted they are to the commercial producers' feed supply, market demands, and other environmental conditions (Fig. 13.4). A good example of adaptability is the Brahman breed, which is more heat and humidity resistant than most other cattle breeds. Because of its higher resistance, the level of productivity in the southern and Gulf regions of the United States is higher for the Brahman, Brahman crosses, and other breeds that include Zebu breeding.

FIGURE 13.4 Cattle are expected to produce under some harsh environments—in this situation a limited supply of forage. Some cattle are more adaptable than others.

Most commercial producers travel less than 150 miles to purchase their bulls for natural service. Therefore, producers should critically assess the breeders, breeds, and biological types within a 150 mile radius of their individual operations. This assessment, in most cases, should have a greater priority than determining only the breeds and crossbreeding systems to use.

Breed Differences

Breeds should be chosen for a crossbreeding system based on how well the breeds and their respective biological types complement each other along with achieving cost-effective levels of heterosis. Breed differences have been evaluated most extensively in the Germ Plasm Evaluation (GPE) program at the U.S. Meat Animal Research Center (MARC) located at Clay Center, NE. Topcross performance of more than 25 different sire breeds has been evaluated in five cycles of the GPE program (Table 13.4).

Hereford-Angus reciprocal F_1 crosses (Hereford sires mated to Angus dams and vice versa) were produced using semen from the same sires. Data were pooled over cycles by adding the average differences between Hereford-Angus reciprocal crosses and other breed groups (two-way and three-way F_1 crosses) within each cycle to the average of Hereford-Angus reciprocal crosses over the three cycles. Females produced by these matings were all retained to evaluate age and weight at puberty and reproduction and maternal performance through 7 or 8 years of age. Data are presented for breed crosses (two-way and three-way) grouped into several biological types based on relative differences (X = lowest; XXXXXX = highest) in growth rate and mature size, lean-to-fat ratio, age at puberty, and milk production (Table 13.5).

TABLE 13.4 Sire Breeds in the MARC Evaluation Program[a]

Breed Cross Groups	Cycle I (1970–1972)	Cycle II (1973–1974)	Cycle III (1975–1976)	Cycle IV (1986–1990)	Cycle V (1992–1996)
F_1 crosses from Hereford or Angus dams[b]	Hereford Angus Jersey South Devon Limousin Simmental Charolais	Hereford Angus Red Poll Brown Swiss Gelbvieh Maine-Anjou Chianina	Hereford Angus Brahman Sahiwal Pinzgauer Tarentaise	Hereford[c] Angus[c] Longhorn Salers Galloway Nellore Shorthorn Piedmontese Charolais Gelbvieh Pinzgauer	Hereford Angus Tuli Boran Brahman Belgian Blue Piedmontese
Three-way crosses out of F_1 dams[d]	Hereford Angus Brahman Devon Holstein	Hereford Angus Brangus Santa Gertrudis			

[a]The Germ Plasm Evaluation program at the U.S. Meat Animal Research Center, Clay Center, NE.
[b]For example, Jersey sire × Hereford dam.
[c]Hereford and Angus sires, originally sampled in 1969, 1970, and 1971, have been used throughout the program. In Cycle IV, a new sample of Hereford and Angus sires produced after 1982 were used and compared with the original Hereford and Angus sires.
[d]For example, Brahman sire × (Jersey × Hereford) dam.
Source: USDA (MARC).

Although the information in Table 13.5 is useful, it should not be considered the final answer for decisions on breeds to use. First, a producer needs to recognize that this information reflects breed averages; therefore, there are individual animals and herds of the same breed that are much higher or lower than the average ranking given. This demonstrates that there are several different biological types within the same breed. Also, the level of nutrition was high so that the breeds with high milk production and large mature weights would also have high reproductive rates.

Figures 13.5–13.17 show considerable genetic variation between and within breeds for several economically important traits. In these figures, the averages for the various breed crosses are shown on the lower horizontal axis. The spacing on the vertical axis is arbitrary but the ranking of the biological types (separate bars) from the top to bottom reflects increasing increments. Breed rankings within each biological type are noted within each bar.

These frequency curves are shown for selected breeds to compare to the average of Hereford and Angus (the base comparison). The frequency curves reflect the expected distribution of breeding values for the breeds assuming a normal distribution where ±1, ±2, and ±3 genetic standard deviations represent 68%, 95%, or 99.6% of the observations around the mean, respectively. For example, in Figure 13.8, the Jersey with the heaviest birth weight would not be as heavy as the Charolais or Maine-Anjou with the lightest birth weight. Also, some Hereford or Angus calves would have birth weights lighter than the heaviest Jersey calves or heavier than the lightest Charolais or Maine-Anjou birth weights. The Jersey average for birth weights would be 3.2 standard deviations in breeding value (approximately 20 lbs) different from the Hereford-Angus, while the breed difference in average birth weight for

TABLE 13.5 Breed Crosses Grouped into Six Biological Types on the Basis of Four Major Criteria

Breed Crosses Grouped into Biological Types	Traits Used to Identify Biological Types[a]			
	Growth Rate and Mature Size	Lean-to-Fat Ratio[b]	Age at Puberty	Milk Production
Jersey (J)	X	X	X	XXXXX
Longhorn (Lh)	X	XX	XXX	XX
Hereford-Angus (HA)	XXX	XX	XXX	XX
Red Poll (RP)	XX	XX	XX	XXX
Devon (D)	XX	XX	XXX	XX
Shorthorn (Sh)	XXX	XX	XXX	XXX
Galloway (Gw)	XX	XXX	XXX	XX
South Devon (Sd)	XXX	XXX	XX	XXX
Tarentaise (T)	XXX	XXX	XX	XXX
Pinzgauer (P)	XXX	XXX	XX	XXX
Brangus (Bn)	XXX	XX	XXXX	XX
Santa Gertrudis (Sg)	XXX	XX	XXXX	XX
Beefmaster (Bf)	XXX	XX	XXXX	XX
Sahiwal (Sw)	XX	XXX	XXXXX	XXX
Brahman (Bm)	XXXX	XXX	XXXXX	XXX
Nellore (N)	XXXX	XXX	XXXXX	XXX
Braunvieh (B)	XXXX	XXXX	XX	XXXX
Gelbvieh (G)	XXXX	XXXX	XX	XXXX
Holstein (Ho)	XXXX	XXXX	XX	XXXXXX
Simmental (S)	XXXXX	XXXX	XXX	XXXX
Maine-Anjou (M)	XXXXX	XXXX	XXX	XXX
Salers (Sa)	XXXXX	XXXX	XXX	XXX
Piedmontese (Pm)	XXX	XXXXXX	XX	XX
Limousin (L)	XXX	XXXXX	XXXX	X
Charolais (C)	XXXXX	XXXXX	XXXX	X
Chianina (Ci)	XXXXX	XXXXX	XXXX	X

[a]Breed crosses grouped into several biological types based on relative differences (X = lowest; XXXXXX = highest).
[b]Steers were slaughtered at 15 months of age.
Source: Adapted from MARC Research Progress Reports.

Hereford-Angus is 3.7 standard deviations (approximately 24 lbs) different from the Charolais or Maine-Anjou breed average.

The statement, "There is more variation within a breed, than between breeds," is not necessarily true. If one says "between breed averages," then the magnitude of variation is about the same within a breed as compared with between breed averages. This gives evidence that there is tremendous selection potential both within and between breeds.

A comparison of Figures 13.5–13.17 shows that there are genetic antagonisms between traits when comparing breed averages. Breeds that excel in retail product from birth to market age (Fig. 13.14) sire progeny with heavier birth weights (Fig. 13.8), greater calving difficulty, reduced calf survival, and reduced rebreeding in dams; produce carcasses with lower marbling (Fig. 13.16) but very acceptable meat tenderness (Fig. 13.17); tend to reach puberty at an older age (Fig. 13.5); and generally have heavier mature weights (Fig. 13.11). Heavier mature weight and high milk production increase output per cow (e.g., calf weight and

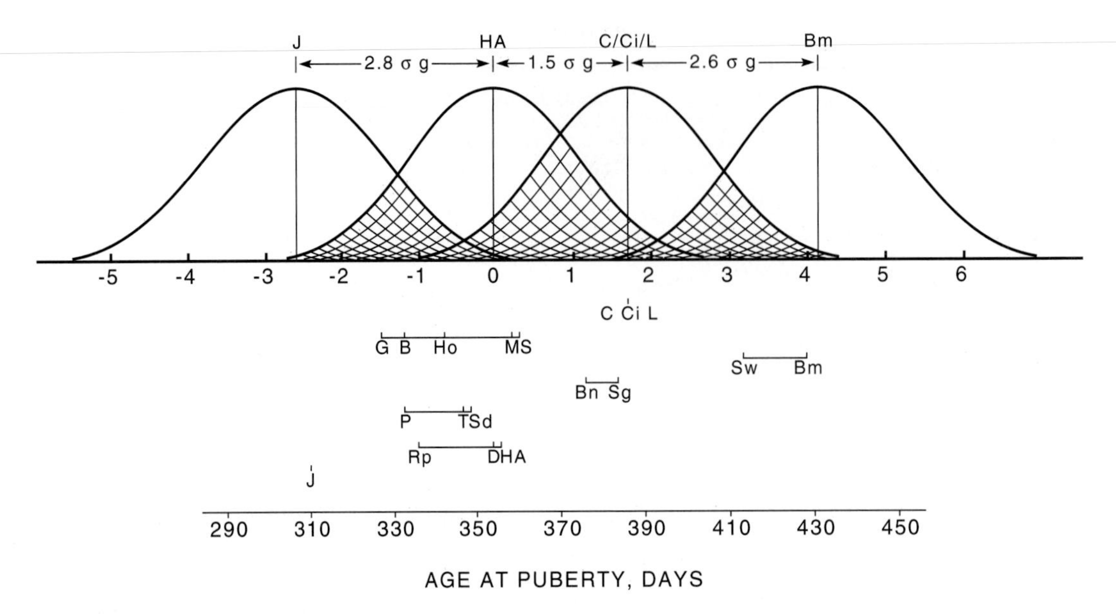

FIGURE 13.5 Variation between and within breeds for age at puberty. See Table 13.5 for breed abbreviations.
Source: MARC.

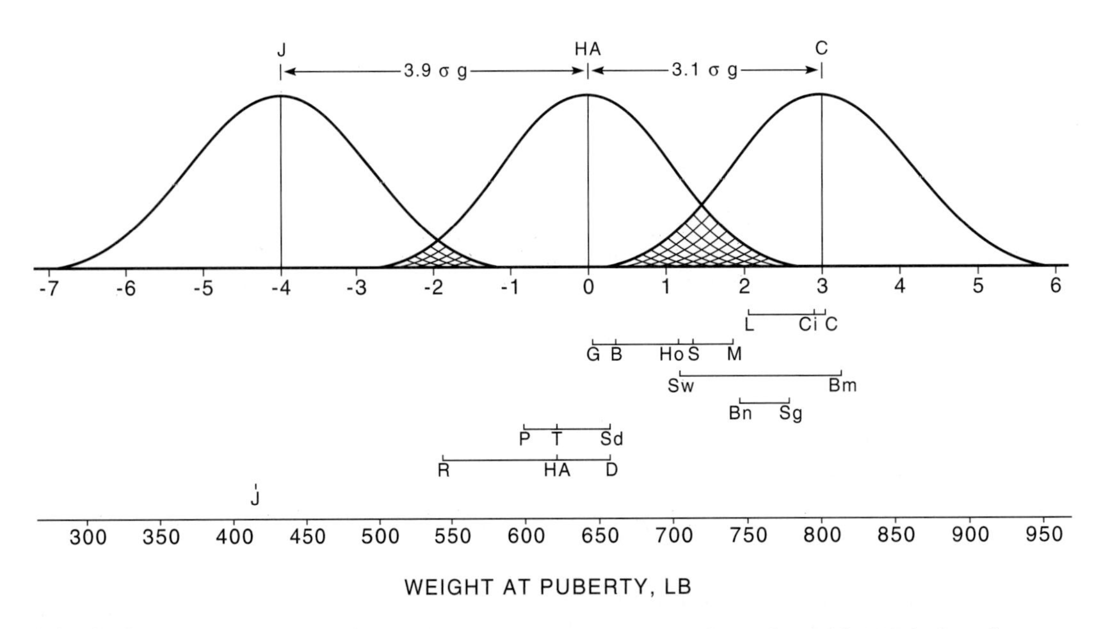

FIGURE 13.6 Variation between and within breeds for weight at puberty. See Table 13.5 for breed abbreviations.
Source: MARC.

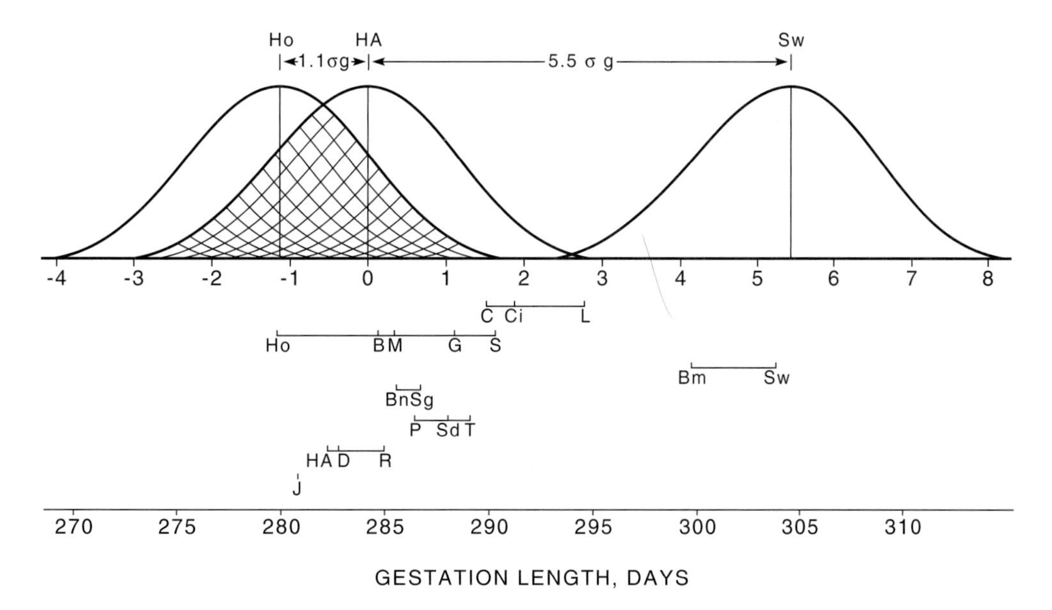

FIGURE 13.7 Variation between and within breeds for gestation length. See Table 13.5 for breed abbreviations.
Source: MARC.

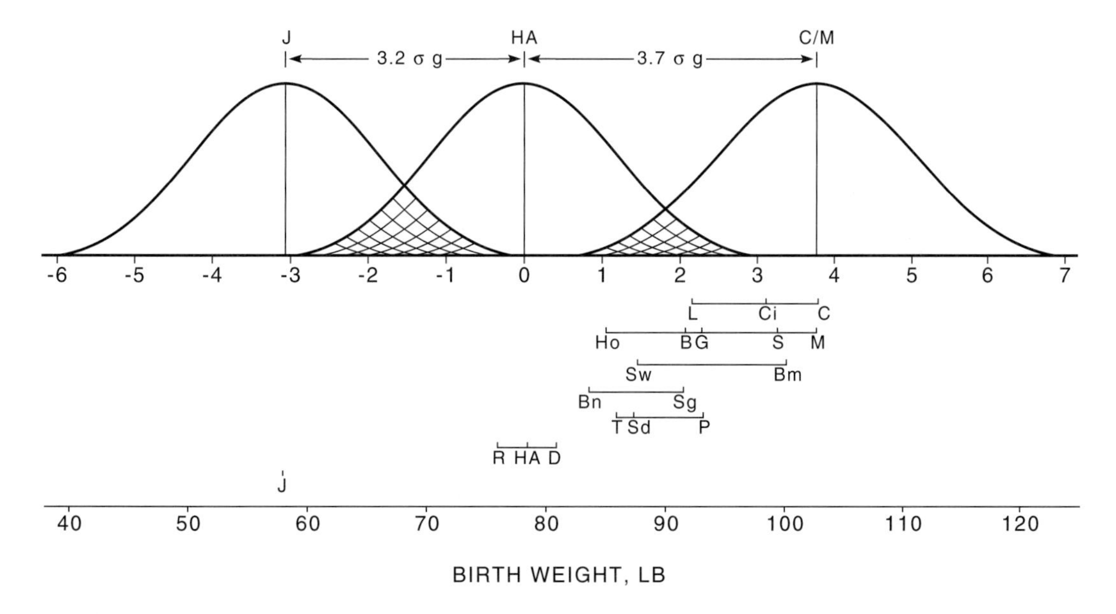

FIGURE 13.8 Variation between and within breeds for birth weight. See Table 13.5 for breed abbreviations.
Source: MARC.

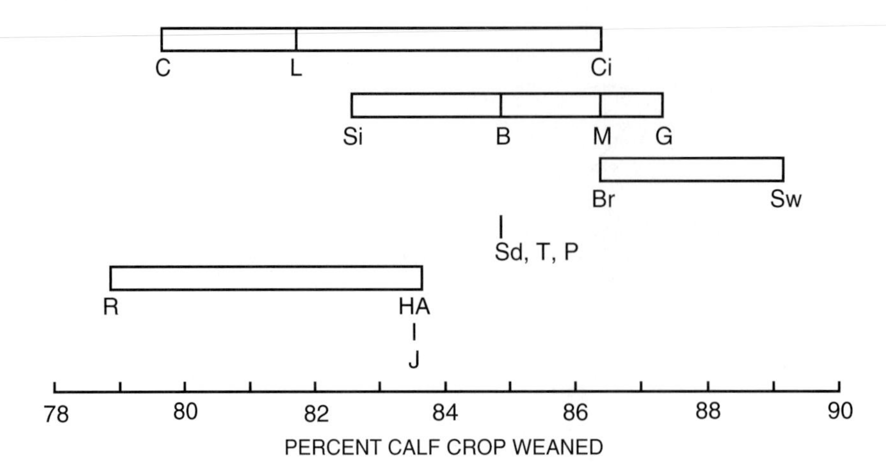

FIGURE 13.9 Breed of sire of dam means for percentage calf crop weaned per cow exposed to breed. See Table 13.5 for breed abbreviations.
Source: Cundiff (1991).

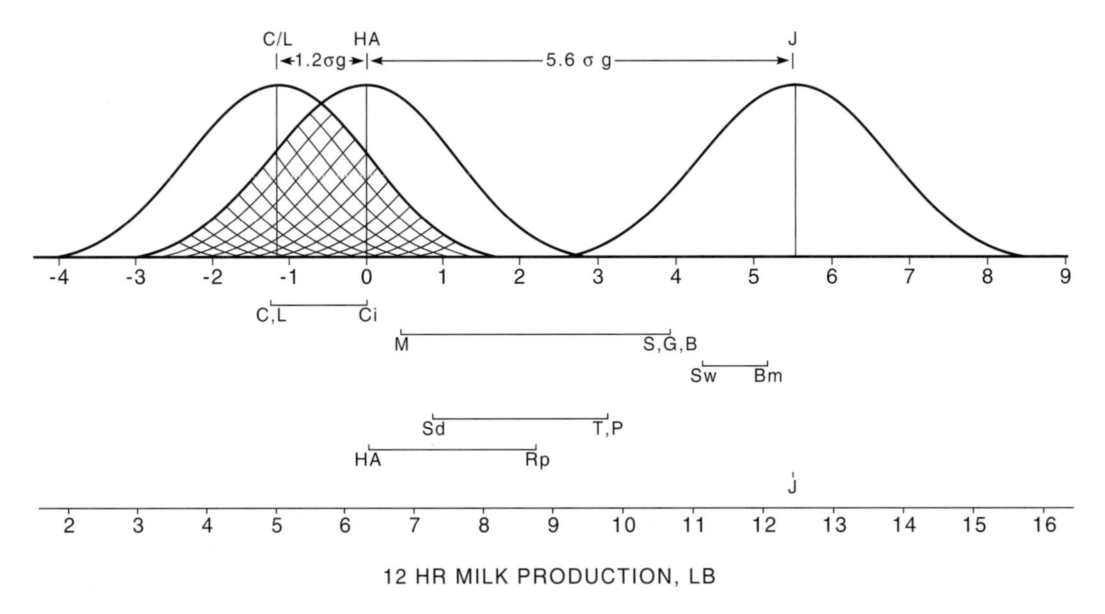

FIGURE 13.10 Variation between and within breeds for milk production. See Table 13.5 for breed abbreviations.
Source: MARC.

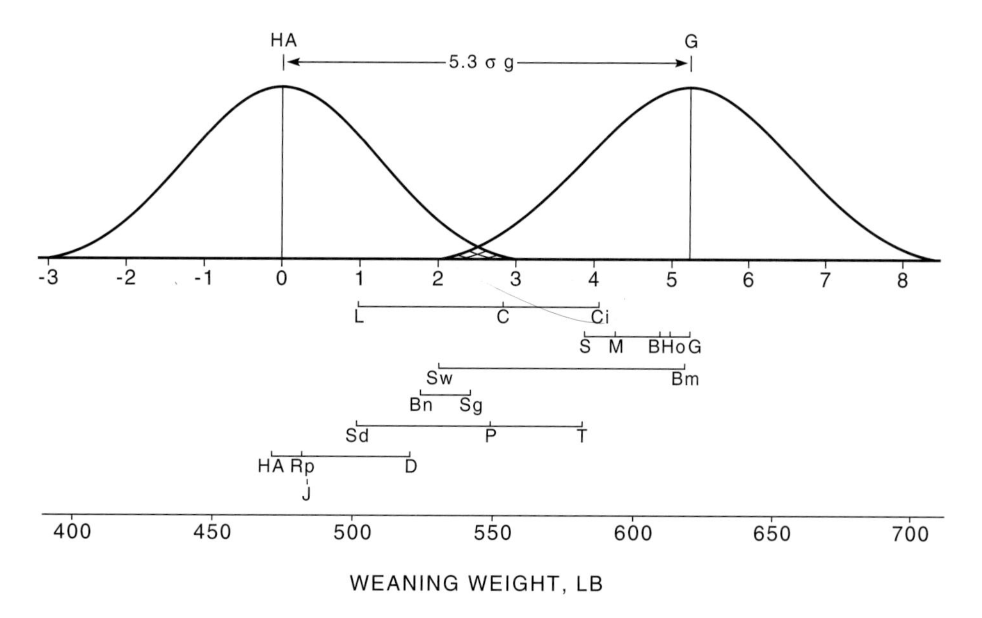

FIGURE 13.11 Variation between and within breeds for weaning weight. See Table 13.5 for breed abbreviations.
Source: MARC.

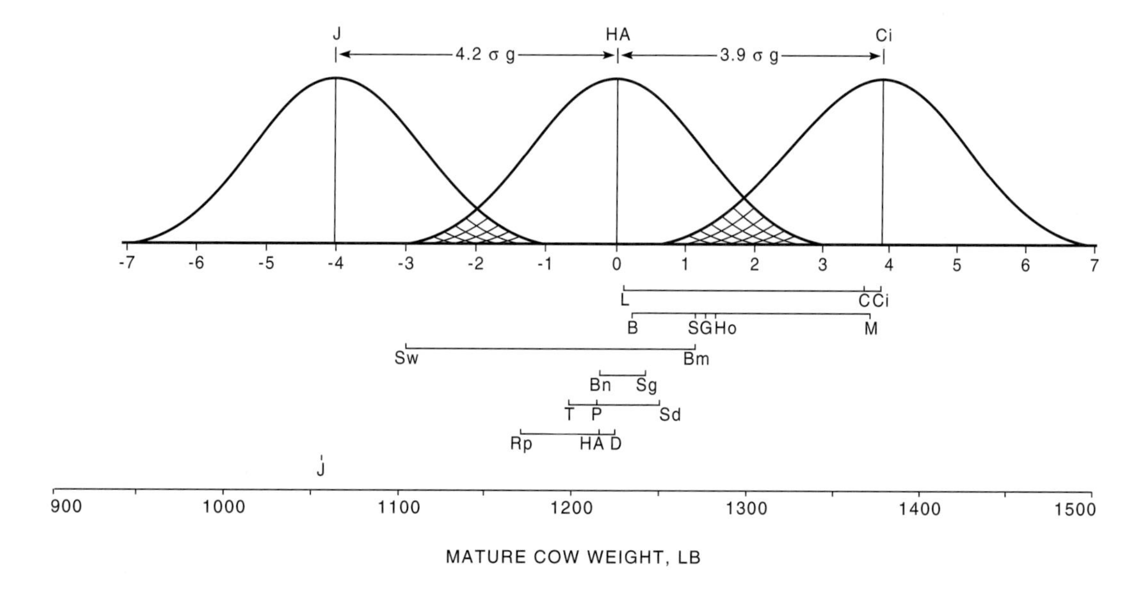

FIGURE 13.12 Variation between and within breeds for mature cow weight. See Table 13.5 for breed abbreviations.
Source: MARC.

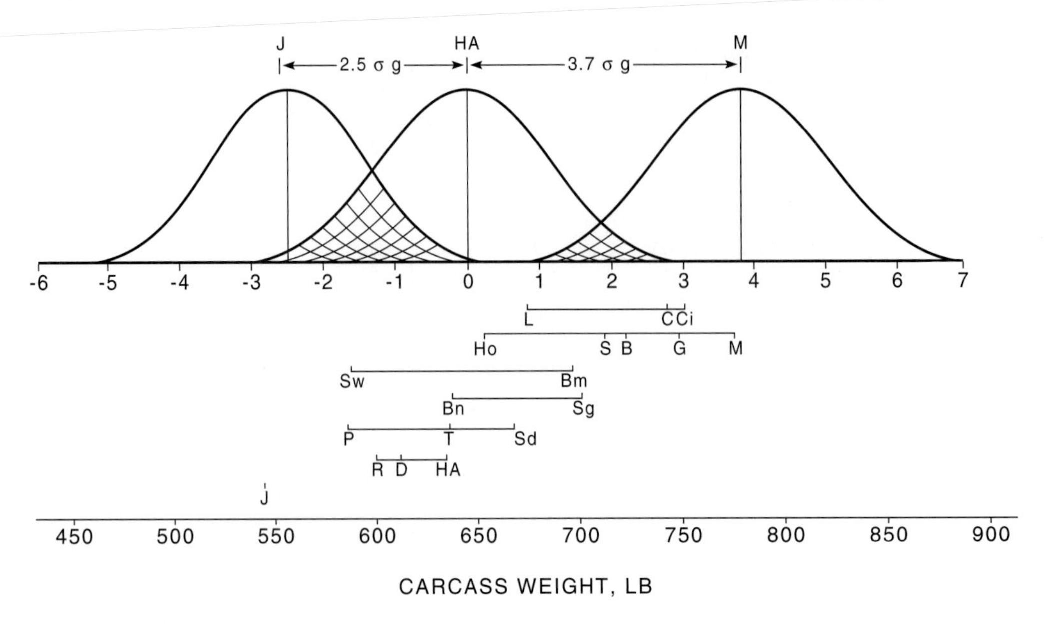

FIGURE 13.13 Variation between and within breeds for carcass weight at 15 months of age. See Table 13.5 for breed abbreviations.
Source: MARC.

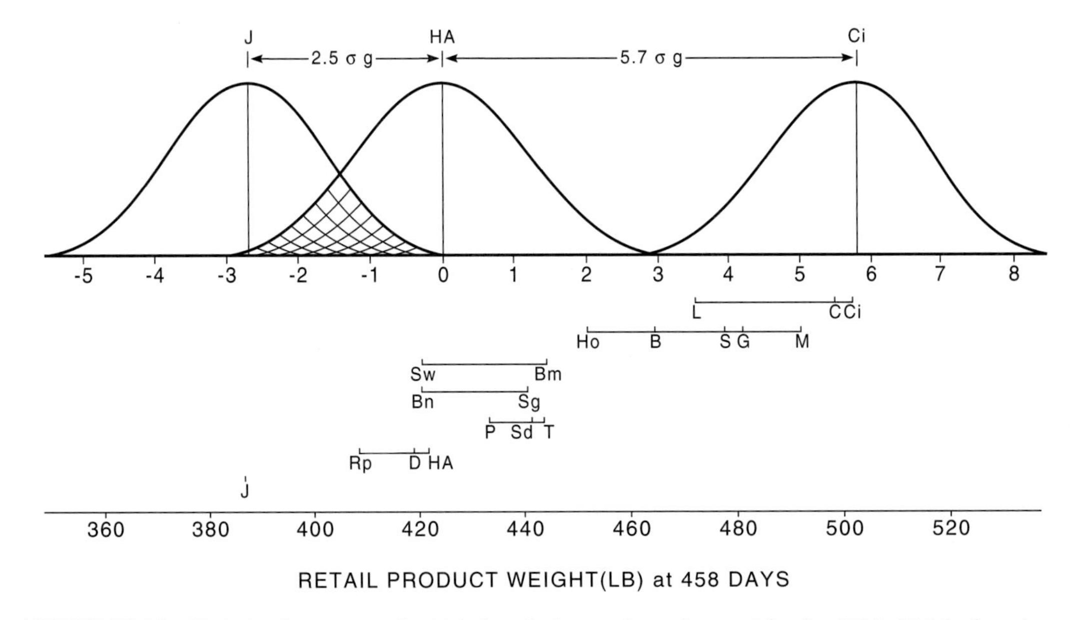

FIGURE 13.14 Variation between and within breeds for retail product weight. See Table 13.5 for breed abbreviations.
Source: MARC.

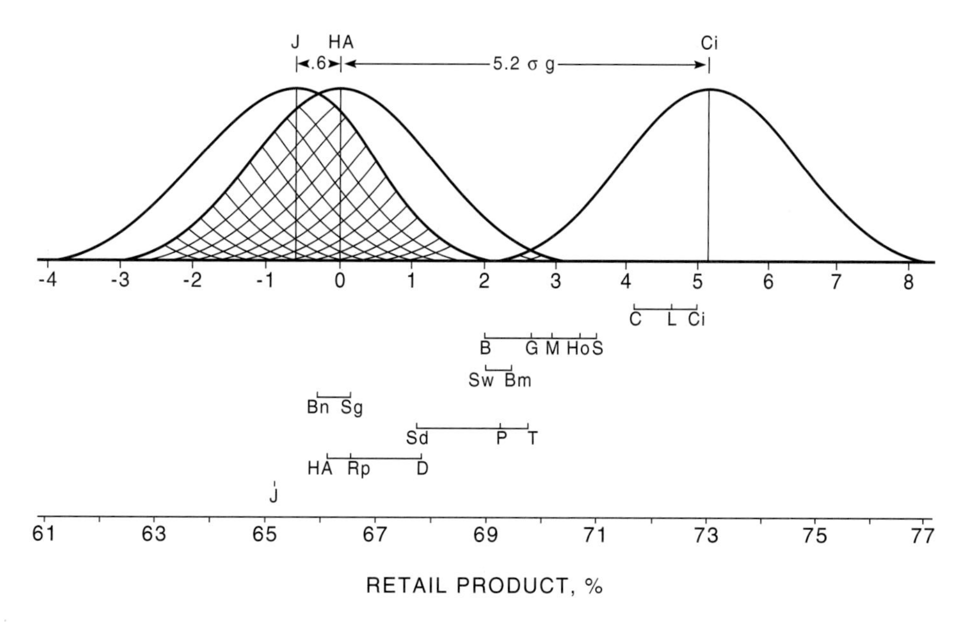

FIGURE 13.15 Variation between and within breeds for percent retail product. See Table 13.5 for breed abbreviations.
Source: MARC.

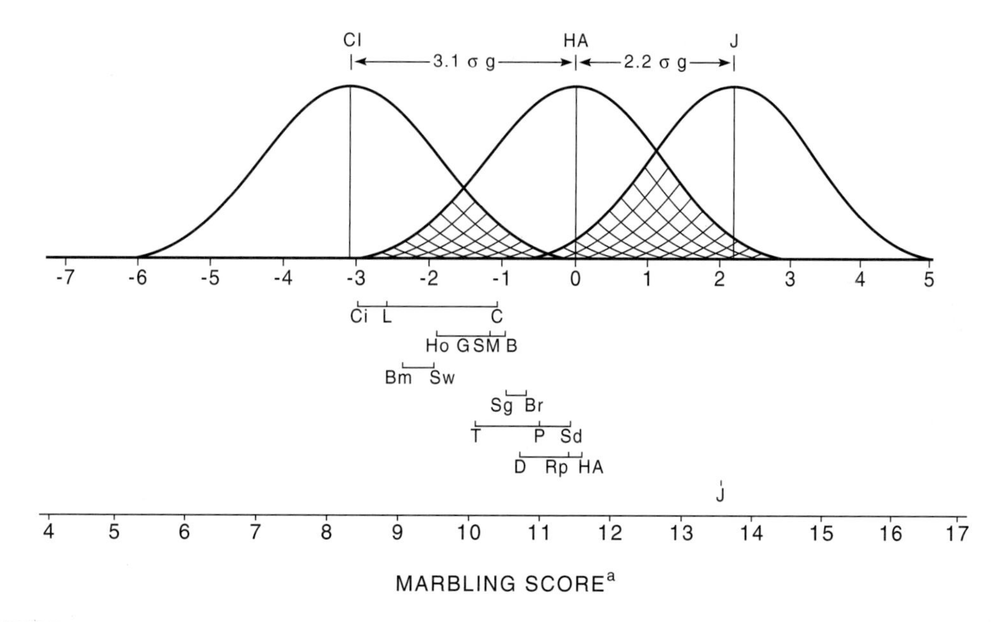

FIGURE 13.16 Variation between and within breeds for marbling. See Table 13.5 for breed abbreviations.
Source: MARC.
[a]Minimum scores: Select = 8; Choice = 11.

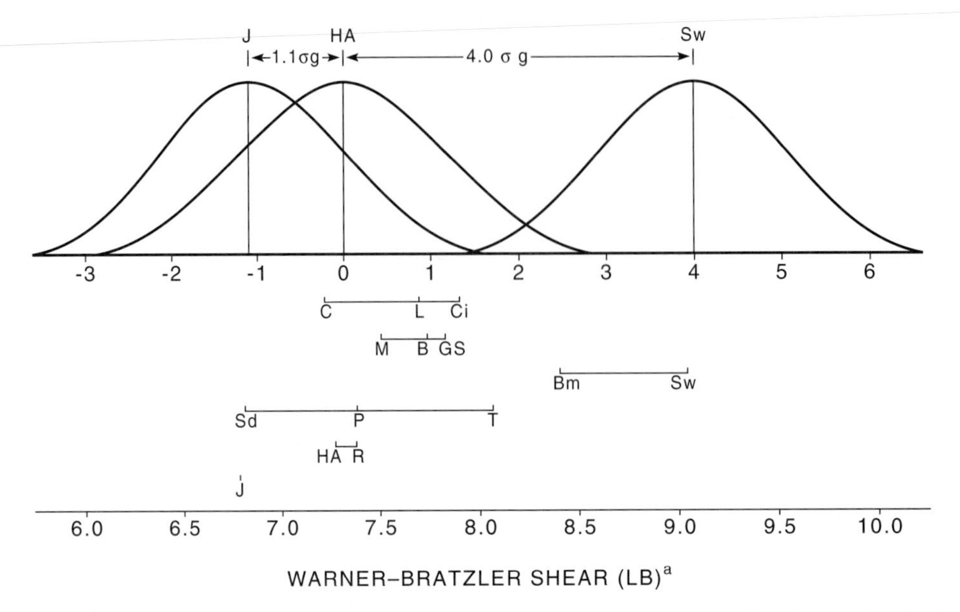

FIGURE 13.17 Variation between and within breeds for tenderness (the higher the shear value the less tender). See Table 13.5 for breed abbreviations.

[a]Shear values less than 8.0 lbs are considered satisfactory in tenderness.

Source: MARC.

slaughter weight of cull cows), but they also increase nutrient requirements for maintenance and lactation so that differences in life cycle biological efficiency are generally small. An economic evaluation in most cow-calf environments usually favors an optimum or balanced combination of these traits where extremes are avoided.

A careful analysis of the information in Table 13.5 and Figures 13.5–13.17 shows that no one breed is superior in all these economically important traits. This gives an advantage to commercial producers using a crossbreeding or a composite breeding program if they select breeds whose superior traits complement each other. An excellent example of breed complementation is shown by the Angus and Charolais breeds; when crossed, they complement each other in terms of both quality grade and yield grade. The superior marbling (quality grade) of the Angus is combined with the numerically low yield grade of the Charolais.

Some caution should be exercised in the breed comparisons noted in Figures 13.5–13.17 as these breeds were given an excellent forage and nutrition program. The breeds could rank differently in an environment characterized by a less plentiful, but a lower-cost supply of forage and feed.

Producers need to use some of the previously described methods to identify superior animals within the breed (Chapter 4). It should also be recognized that these average breed rankings can change with time, depending on the improvement programs used by leading breeders within the same breed. Those traits that have high heritabilities would be expected to change most rapidly, assuming the same selection pressure for each trait.

Breeds are genetic resources for commercial producers to obtain the most profitable biological types and to utilize heterosis by crossing the breeds. Producers should understand how breed average compares for the economically important traits and realize that each breed can contain several different biological types. Table 13.6 shows 27 biological types, where three biological types exist for each of the three traits.

TABLE 13.6 Number of Biological Types in a Breed When Considering Three Traits[a]

Biological Type Category	Trait		
	Mature Cow Weight	Milk Production	Muscling
Heavy (H)	H	H	H
Medium (M)	M	M	M
Light (L)	L	L	L
Number of biological types in each trait	3	3	3

[a]Total number of biological types is 27 (3 × 3 × 3).
Source: Colorado State University.

Using Breeds in Crossbreeding Systems

Before a producer implements a crossbreeding system, it is important to address the following questions:

1. What is the most appropriate grazing system for the enterprise? What are the issues associated with duration, frequency, and timing of the grazing system that may limit or be in conflict with the mating system alternatives?
2. What proportion of total income is derived from the cow-calf enterprise, and what times of the year do other enterprises compete for a producer's time?
3. What is the cost and availability of labor?
4. How much complexity can be effectively handled by management?
5. How important is simplicity and convenience to implementing the breeding program?
6. What is the cost and availability of bulls from desired breeds or composites?
7. What is the feasibility and cost effectiveness of utilizing artificial insemination?
8. What is the marketing plan for the enterprise?

Producers have a multitude of breeding system options from which to choose. The most simple approach is straightbreeding throughout the herd. The primary advantage is convenience. However, when this choice is made, a producer gives up two very powerful genetic tools—heterosis and breed complementarity. Effective crossbreeding systems take advantage of both.

Heterosis is the performance advantage of crossbred progeny compared with the average of the parental breeds involved in the cross for a particular trait (Fig. 13.18). Heterosis is expressed in individuals (crossbred progeny), a maternal influence (crossbred or composite cows), and in a smaller number of traits via paternal effect (crossbred or composite sires). Heterosis is highest for the traits related to reproductive rate, cow lifetime productivity, and calf survival rate. Moderate positive effects are also obtained in the growth traits (Table 13.7).

The concept of breed complementarity is founded on the premise that no one breed excels in all economically important traits. As such, the systematic matching of breed strengths and weaknesses can yield progeny with optimal and highly desirable trait combinations.

However, implementation of an effective and profitable crossbreeding system requires thorough planning. It should also be recognized that a crossbreeding system usually increases the level of management and must be implemented under the resource and forage constraints of a particular farm or ranch.

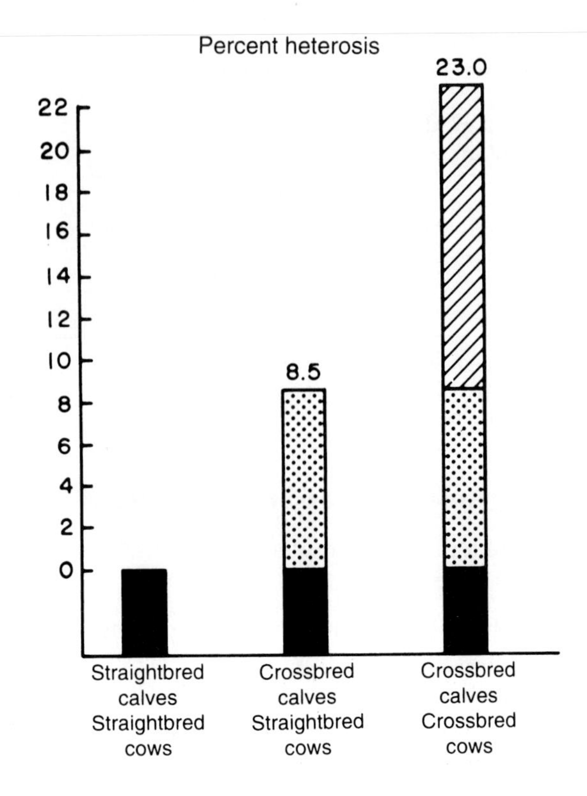

FIGURE 13.18 Heterosis, resulting from crossbreeding, for pounds of weaned calf per cow exposed to breeding.
Source: USDA.

TABLE 13.7 Average Levels of Heterosis for Traits of Beef Cattle

Trait	Individual Heterosis (%)	Maternal Heterosis (%)	Total Heterosis (%)
Calving	0	6	6
Weaning rate	0	8	8
% reaching puberty by 15 mo.	15	—	15
Survival to weaning	3	1	4
Calving difficulty	2	0	2
Birth weight	4	2	6
Weaning weight	5	6	11
Yearling weight	4	—	4
Feed conversion	−2	—	−2
Carcass weight	3	—	3
USDA quality grade	2	—	2
USDA yield grade	5	—	5
Trimmed retail cuts	3	—	3
Milk production	—	—	0
Calf wean. wt./cow exposed	—	—	18
Cow forage intake	—	—	2
Cow longevity	—	—	38
Cow lifetime productivity	—	—	23

Source: Adapted from Kress et al. (1999).

TABLE 13.8 Biological Traits of Importance in Cattle Production

	Heterosis	Heritability
Traits at the ranch		
Percent calf crop weaned	high	low
Cow longevity	high	low
Sale weight	moderate	moderate
Cow lifetime productivity	high	low
Traits at the feedyard		
Growth rate	moderate	moderate
Disease resistance	moderate/high	low
Traits at the product level		
Palatability/tenderness	low[a]	moderate/high
Cutability	low[a]	moderate/high

[a]Breed complementarity is very useful in balancing these two traits.
Source: Field and Cundiff (2000).

The goal is to assure that the cow herd is well matched to the environment with particular attention to the forage resource. Crossbreeding is not a "silver bullet" for cattle breeders. In fact, crossbreeding may yield unfavorable outcomes. These failures occur because:

1. Overuse of beef cattle breeds that have too much growth, milk, birth weight, and/or mature size.
2. The mating system was too complex or was not implemented in a systematic manner.
3. Seedstock producers failed to develop the expertise and service orientation to assist their clients in development of effective crossbreeding systems.
4. The use of poor-quality bulls in a crossing scheme will not yield desirable results. Selection of bulls must be based on an objective set of criteria that allows for the identification of superior sires. The use of selection in conjunction with heterosis is most desirable (Table 13.8).

For the cow-calf producer, many of the benefits of heterosis are measured at weaning time as pounds of calf weaned per cow exposed. This is due to the combined effects of crossbreeding on fertility, calf survival rate, and growth. These cumulative benefits from several crossing systems are listed in column two of Table 13.9. Note that the greatest advantage is derived from crossbred calves weaned from crossbred cows.

When choosing a crossbreeding system, producers are advised to focus on two general targets—producing profitable cows well suited to the forage production of a particular ranch, and production of profitable progeny that fit the demands of a particular consumer market.

As a result, replacement heifer production considerations and the simplicity of the crossing system need to be evaluated (Table 13.9). Furthermore, one of the major challenges facing commercial cow-calf producers is that of overcoming the genetic antagonisms between maternal, growth, and carcass traits. The utilization of breed complementarity is very important in resolving these antagonisms (Table 13.9).

Figure 13.19 illustrates the process of matching cow biological type to a variety of environments. Ranges in annual precipitation are used to delineate acceptable levels of milk and mature size. When rainfall is limited (< 12 in.), the options in cow size and level of lactation are limited to cows with mature weights of 1,100 lbs and milk production less than 15 lbs/day. At the upper end of the precipitation scale, larger cows capable of higher milk production may be acceptable. In essence, as forage availability becomes more limited, the risk of reproductive failure due to mismatched biological type increases.

TABLE 13.9 Comparison of Various Crossbreeding Systems for Heterosis, Breed Complementarity, Replacement Production, and Simplicity

System	Heterosis	Breed Complementarity	Production of Replacements	Simplicity
Rotation				
Two-breed	16	—	+	+
Three-breed	20	—	+	—
Two-breed w/F_1 sires	22	+	+	+
Static terminal sire				
Three-breed	20	+	+	—
Buy purebred females	24	++	—	++
Buy crossbred F_1 females	28	++	—	++
Rotate sire breed				
Three-breed	16	—	+	+
Rotational terminal sire				
Two-breed	21	+	+	+
Composite				
Two-breed	14	+	+	++
Three-breed	17	+	+	++
Four-breed	20	+	+	++
Multiple-sire breed				
Two-breed with crossbred females	10	—	+	+

Source: Adapted from Bourdon (2000).

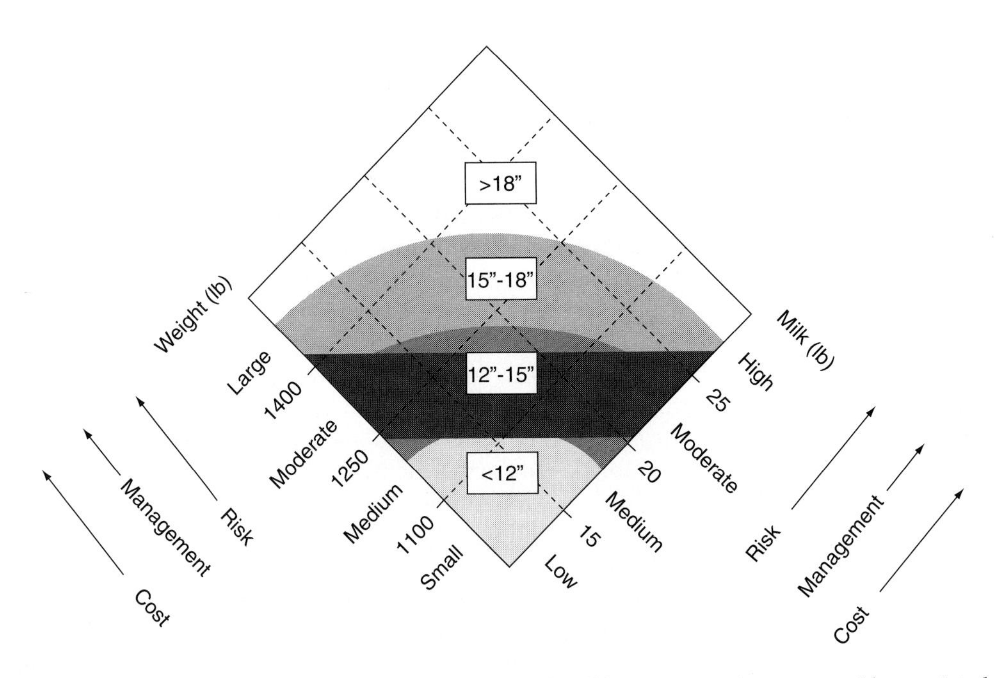

FIGURE 13.19 Matching cow biological type (weight and milk) to range environment, with associated risk, management, and cost. Ranges in inches (12"–15") are annual precipitation and/or represent availability of winter feed resource.
Source: Kress and MacNeill (1999).

Consistent high levels of heterosis can be maintained generation after generation of crossbreeding systems, just as those shown in Figures 13.20, 13.21, 13.22, and 13.23 are utilized. The attributes of the breeding systems diagrammed in Figures 13.20–13.23 are listed in Table 13.10.

In many environments, the use of heat-adopted breeds is required. In these cases, it would be possible that a British breed bull would be used as a terminal sire. The same situation might occur where a high percentage of Continental-breed herd was attempting to hit a quality/marbling based target. In this case, a British purebred or British-cross F_1 might be deemed appropriate.

There are numerous herds in the United States having 20–40 cows per herd involving only one breeding pasture. Table 13.11 shows several simplified crossbreeding systems that can be used. One system involves the rotation of two breeds, another system rotates three breeds, while the third system is simplified by using a composite breed. In each system, a new bull is introduced after every two years to avoid mating heifers back to their sire.

The single-sire rotation is expected with yield 59% of maximum individual heterosis and 47% of maximum maternal heterosis. This compares with 72% of maximum individual heterosis and 56% of maximum maternal heterosis that can be obtained from a two-breed rotation in a large herd or through artificial insemination. The single-sire rotation in a three-breed rotation should produce 27% of maximum individual and 60% of maternal heterosis.

2-Breed Rotation

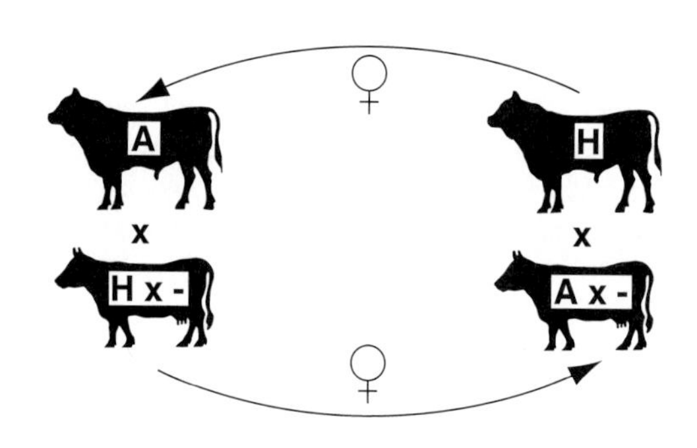

1. Requires two breeding pastures or AI.
2. Utilizes individual and maternal heterosis (67% of maximum).
3. Allows limited use of complementarity.
4. Replacement females produced within the system and need to be identified by breed of sire.
5. Genetic improvement determined primarily by genetic potential of Angus and Hereford sires.
6. Breeds should be similar for size and milk production.
7. Expected to increase calf production per cow by 16%.

3-Breed Rotation

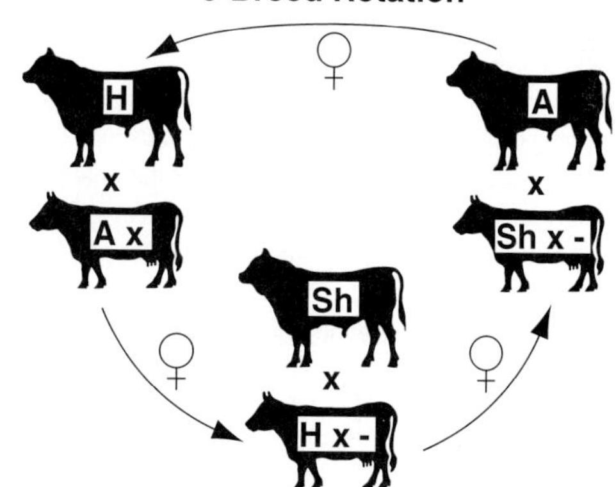

1. Requires three breeding pastures or AI.
2. Utilizes individual and maternal heterosis (86% of maximum).
3. Allows limited use of complementarity.
4. Replacement females produced within the system and need to be identified by breed of sire.
5. Genetic improvement determined primarily by genetic potential of Hereford, Angus, and Shorthorn sires.
6. Breeds should be similar for size and milk production.
7. Each crossbred cow should be mated to the breed of sire to which she is most distantly related.
8. Expected to increase calf production per cow by 20%.

FIGURE 13.20 Two-breed and three-breed rotational crossbreeding systems.
Source: Adapted from Kress and MacNeal (1999).

Rotational Terminal Sire

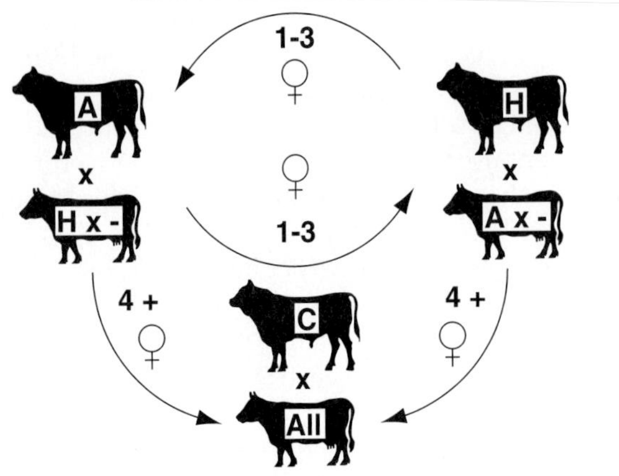

1. Requires three breeding pastures or AI.
2. Approximately 45% of females in rotation and 55% of females in terminal portion of system.
3. Utilizes individual and maternal heterosis.
4. Maximizes complementarity in 55% of herd.
5. Roughly 70% of progeny marketed are from terminal sire breed.
6. AI and sexed semen would make this system more efficient.
7. Genetic improvement determined primarily by genetic potential of Angus, Hereford, and Charolais sires.
8. 1- to 3-year-old females are bred in the rotational part of the system and 4-year-old and older cows are bred in the terminal part of the system.
9. Expected to increase calf production per cow exposed by 21% for 2-breed rotation and 24% for 3-breed rotation.

Static Terminal Sire

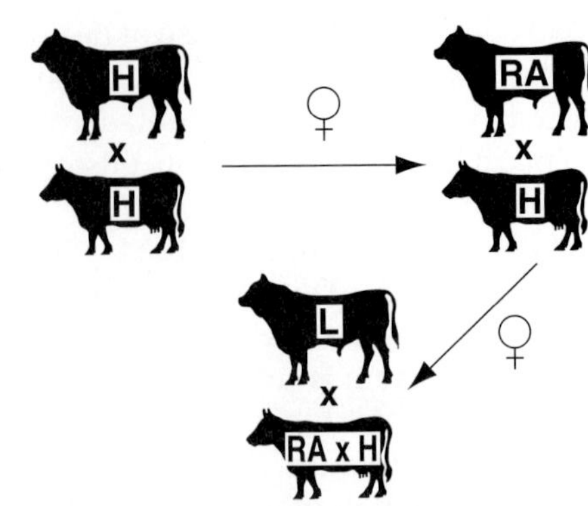

1. Requires three breeding pastures or AI.
2. Approximately 25% of females in H x H matings, 30% of females in RA x H matings, and 45% of females in terminal sire matings.
3. Individual heterosis in 75% of system and maternal heterosis in 45% of system.
4. Maximizes complementarity in 45% of herd.
5. Roughly 60% of progeny marketed are from terminal sire breed.
6. AI and sexed semen would make this system more efficient.
7. Genetic improvement determined primarily by genetic potential of Hereford, Red Angus, and Limousin sires
8. Expected to increase calf production per cow exposed by 20%.

FIGURE 13.21 Examples of terminal crossbreeding systems.
Source: Adapted from Kress and MacNeal (1999).

Single-sire rotations can increase productivity and profitability of small beef herds. Choice of biological type is important in addition to obtaining the heterosis by crossing the breeds. Biological type and performance should be similar for all the breeds that are used. This is important as bulls will be bred to both heifers and cows for calving ease and the biological type must match the most economical feed and production environment.

While commercial producers can use crossbreeding, purebred breeders cannot do so without affecting breed purity. Therefore, commercial producers have an advantage over purebred breeders in being able to utilize more genetic resources to make genetic improvements. Selection and crossbreeding must be used effectively to optimize productivity, otherwise market targets and high levels of profitability will not be achieved (Fig. 13.25).

Although traits with a low heritability respond very little to genetic selection, they show a marked improvement in a sound crossbreeding program. Carcass traits associated with growth rate (carcass weight per day of age, fat thickness, and retail product weight at the same age) will show some heterosis. Commercial producers need to (1) select sires carefully

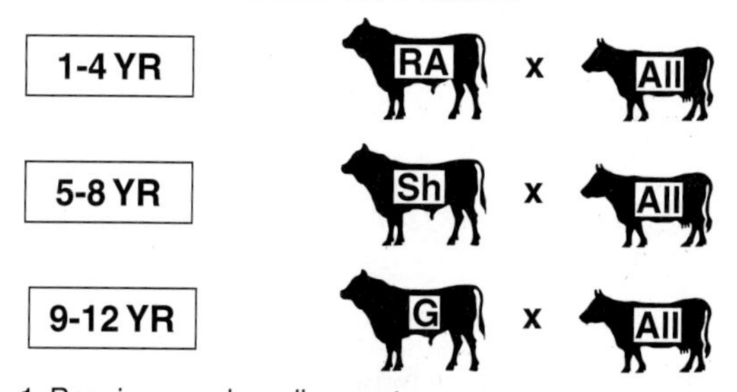

1. Requires one breeding pasture.
2. Start with available and adapted female breed (Hereford for example).
3. Utilizes individual and maternal heterosis.
4. Allows limited use of complementarity.
5. Replacement females from within the system and do not need to identify cows by breed of sire.
6. Genetic improvement determined primarily by genetic potential of Red Angus, Shorthorn and Gelbvieh sires.
7. Each sire breed could be used for two to four years.
8. This system can be considered an approximation to a 3-breed rotation.
9. Expected to increase calf production per cow exposed by 16%.

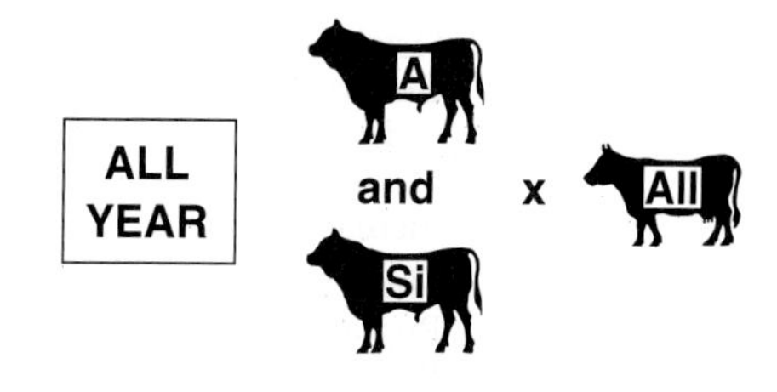

1. Requires one breeding pasture.
2. Utilizes individual and maternal heterosis.
3. Allows limited use of complementarity.
4. Replacement females are crossbreds from within the system and do not need to be identified by breed of sire.
5. Genetic improvement determined primarily by genetic potential of Angus and Simmental sires.
6. This system can be considered an approximation to a composite.
7. Expected to increase calf production per cow exposed by 10%.

FIGURE 13.22 Examples of less management-intensive crossbreeding systems.
Source: Adapted from Kress and MacNeal (1999).

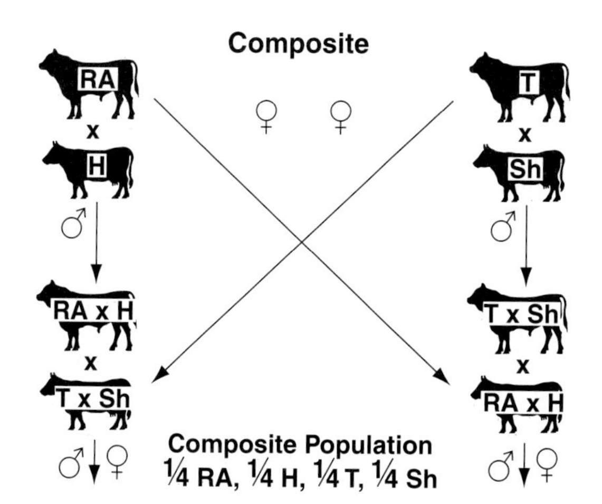

1. Requires one breeding pasture after composite development.
2. Utilizes individual and maternal heterosis.
3. Allows limited use of complementarity.
4. Replacement females and bulls from within the system and do not need to be identified by breed of sire.
5. Genetic improvement determined primarily by genetic potential of selected sires in composite.
6. Expected to increase calf production per cow exposed by 20%.

FIGURE 13.23 Example of a composite breeding system.
Source: Adapted from Kress and MacNeal (1999).

FIGURE 13.26 The Brahman-Hereford crossbred cow (pictured) is a popular cow in several southern states. These cows can be bred to several breeds of bulls. Pictured is an Angus sired calf.
Source: Courtesy of L. I. Smart, Louisiana State Univ.

in order to improve traits with a high heritability, and (2) use a well-planned crossbreeding program in order to use heterosis effectively while matching biological types to their most cost-effective environments (Fig. 13.26).

BREED EVALUATIONS FOR LOW COST PRODUCTION

Comparing breeds is a sensitive issue among cattle breeders and cow-calf producers. These comparisons are usually avoided unless the discussion is with people having similar viewpoints. However, comparing breed averages is needed to initially assess the genetic input into total beef management systems. Equally important, if not more important, is selecting the biological types within the breeds that match low-cost production, contribute to consistently higher net profit, and also produce highly palatable consumer products.

There are beef industry leaders who propose reducing the number of breeds to four or five to improve the uniformity and consistency of costs, profits, and beef products. Also a reduction in the number of biological types within breeds would be needed to improve the uniformity and consistency.

A review of the research literature on evaluating low-cost commercial cow-calf operations that also have excellent feedlot and cattle performance of their cattle, strongly suggests that (1) Angus, Red Angus, and Hereford should be numbered among the breeds to be utilized in an integrated management system, and (2) that the identification of moderate or optimum biological types within the breeds is equally as important as the breeds selected. The following information gives evidence to support these statements.

Angus × Hereford was the leading breed cross in U.S. cow herds among more than 200 different breed combinations. Red Angus showed the largest increase in cow-calf herds when compared with nine other leading beef breeds. Angus and Hereford breeds ranked as the No. 1 or No. 2 most popular breeds in all five regions in the United States (1994 and 1995 Cattle-Fax Cow-Calf Producer Surveys).

Calving Difficulty (Dystocia)

Angus had the least (32%) and Herefords the third least (49%) calving difficulty in 2-year-old heifers of nine different breeds (Gregory et al., 1995).

Feed Efficiency

Hereford and Angus breeds ranked first and second, respectively, when comparing nine breeds for gain efficiency (liveweight gain/Mcal of ME, g) at endpoints of "small" marbling score or 4% fat in the longissimus dorsi muscle (Gregory et al., 1995).

Carcass Traits

Angus were highest (77%) and Hereford third highest (60%) in percent USDA Choice carcasses; Angus were highest and Herefords second highest in sensory panel tenderness when compared with steers from nine pure breeds slaughtered at 438 days of age (Gregory et al., 1995).

Several hundred head of straightbred Hereford steers demonstrated that this breed sample was excellent in tenderness, juiciness, and flavor (1992, 1993, and 1995 Colo. State Univ. Beef Program Reports).

There are numerous observations from several cow-calf operations with low production costs on their calves and yearlings retained through the feedlot and sold on the grid that support the concept of optimum production. Many of these cows are Angus × Hereford crosses, have a moderate biological type (mature weight of 1,000–1,150 lbs), milk production that produces 500–550-lbs weaning weights, and a weaned calf breakeven price of < $0.65/lb. These same feeder cattle gain rapidly in the feedlot and when harvested at 0.4 in. of fat, they produce optimum carcass weights of 700–800 lbs.

Longevity

Angus-Hereford cross females (compared with several breed crosses) excelled in lifetime productivity based on pounds of calf weaned per cow exposed (MacNeil, 1993, Fort Keogh Research Report).

However, the British breeds are less competitive with other breeds in terms of percent retail yield, muscularity, and growth rate.

BREED EVALUATIONS TO IMPROVE CONSUMER MARKET SHARE

The primary components of beef palatability are tenderness, flavor, and juiciness. While most consumers rate tenderness as the most important attribute, the combination of all three attributes is necessary to improve or stabilize consumer market share.

Marbling has a positive relationship to juiciness, flavor, and tenderness; therefore, increased levels of marbling will improve the palatability of beef. Differences in marbling account for approximately 10 to 15% of the differences in tenderness. While this relationship between marbling and tenderness is not very high, it is sufficient in magnitude to continue its inclusion in the quality grades. Additional measures of tenderness are needed. Currently the Warner-Bratzler Shear (WBS) measures tenderness even though it is a rather slow, expensive process. There are large breed differences in WBS values (Fig. 13.17). Even more important are the sire differences for WBS within a breed. It has been shown that more uniform and consistently tender beef can be produced by using bulls whose progeny had low (highly tender) WBS values.

Breeds vary in muscle fiber color, which is related to the ability to deposit marbling. Heavily muscled breeds, such as Belgian Blue and Piedmontese, have primarily white muscle fibers. The muscle metabolism of these breeds use glucose and they deposit very little intramuscular fat. Breeds such as Wagyu and Angus have primarily red muscle fibers. Their muscle metabolism involves most fatty acids, which includes the ability to deposit relatively large amounts of marbling.

A literature review of the breed differences for genetic merit relative to carcass traits (Marshall, 1994) allows a ranking of breeds for marbling score and pounds of retail product yield (Table 13.12). If the goal is to meet market specifications by optimizing marbling and retail yield, it becomes obvious that the use of multiple breeds is important. As Table 13.12 illustrates, those breeds excelling in marbling are less proficient at retail yield and vice versa.

A comparison of various breed crosses and their ability to hit USDA Yield and Quality Grade targets is provided in Table 13.13. These comparisons illustrate the need for a 25% to 50% Continental influence in combination with 75% to 50% British input to minimize non-conformance (MARC II and MARC III). This same performance could be achieved in 75% of the progeny produced by terminal sires bred to rotational-cross or composite females.

TABLE 13.12 Top 10 Breeds for Marbling Score and Retail Product

Rank	Breed	Marbling Score[a]	Retail Product (lbs)	Breed	Retail Product (lbs)	Marbling Score[a]
1	Jersey	614	363	Charolais[d]	482	519
2	Red Angus	574	389	Charolaise[e]	460	524
3	Angus	564	389	Piedmontese	460	506
4	Shorthorn	562	431	Salers	453	511
5	South Devon	550	416	Chianina	453	444
6	Hereford × Angus[b]	547	396	Holstein	453	521
7	Hereford × Angus[c]	539	436	Maine-Anjou	451	496
8	Santa Gertrudis	534	418	Simmental	444	506
9	Pinzgauer	530	411	Nellore	440	501
10	Galloway	525	400	Gelbvieh	438	503

[a]Slight = 400–499; small = 500–599, etc.
[b]Original H × A in MARC germ plasm evaluation.
[c]Current H × A in MARC germ plasm evaluation.
[d]Current Charolais in MARC germ plasm evaluation.
[e]Original Charolais in MARC germ plasm evaluation.
Source: Marshall (1994).

TABLE 13.13 Conformance of Various Breed Crosses and Composites to Yield and Quality Grade Targets in Steers Produced at the U.S. Meat Animal Research Center

	MARC I[a]	MARC II[b]	British	Continental	MARC III[c]
≥ 70% Yield Grade (YG) 1 & 2	83.1	56.1	37.6	89.3	52.5
≥ 70% Quality Grade (QG) Ch & up	43.1	54.7	69.6	30.4	66.0
% Non-conform YG	16.9	33.9	62.4	10.7	47.5
% Non-conform QG	56.9	45.3	30.4	69.6	34.0
Deviation from acceptable non-conform (30%)					
YG	OK	3.9	32.4	OK	17.5
QG	26.9	15.3	0.4	39.6	4.9
Total	26.9	19.2	32.8	39.6	22.4

[a]MARC I =1/4 Charolais, 1/4 Limousin, 1/4 Braunvieh, 1/8 Angus, 1/8 Hereford.
[b]MARC II = 1/4 Gelbvieh, 1/4 Simmental, 1/4 Hereford, 1/4 Angus.
[c]MARC III = 1/4 Pinzgauer, 1/4 Red Poll, 1/4 Hereford, 1/4 Angus.
Source: Field and Cundiff (2000) (Adapted from MARC data).

FIGURE 13.27 Several breed associations have begun to promote the use of first generation cross breeding stock. For example, both the American Gelbvieh Association and the American Simmental Association have programs that combine the respective Continental breeds with Angus or Red Angus in a 50:50 cross.

In an effort to capture the market potential of a 50% Continental, 50% British breed combination, both the American Simmental Association and the American Gelbvieh Association have developed programs that promote the production of F_1 bulls (Fig. 13.27).

INTEGRATED PRODUCTION SYSTEMS

When grading economic efficiency in integrated production systems (cow-calf through carcass) involving five breeds (Angus, Hereford, Limousin, Simmental, and Charolais) and their two- and three-breed rotational crosses, Angus and Herefords were the most economically efficient purebreds when valued by carcass weight. When steers were slaughtered at a constant fat finish (low choice), purebred Angus and Herefords and their crosses spent the fewest days on feed and had the lowest costs. Angus and Herefords graded Choice at lighter weights (Tess, 1993, Montana Ag Research).

Angus × Hereford cows gave the highest economic performance of five biological types of cows under either (1) no resource restraints or (2) fixed forage resource base. In the latter situation, net profit varied by $22,000 under 2,700 AUM of range forage (Davis, 1994. *J. Anim. Sci.* 72: 2591).

Researchers at MARC have indicated that if cows of large size and high milk production are not given high levels of feed, postpartum interval will increase and conception rates will decline. The F_1 cows of each breed cross in each cycle of the MARC program were given the same feeding regime. Breeds having large mature weight and high milk production were provided a nutritional regime for high reproductive rates in addition to meeting requirements for growth, maintenance, and lactation (Cundiff et al., 1991).

There are real challenges in combining breeds and biological types to obtain an optimum combination of maternal (primarily reproduction), growth, and carcass traits. Cundiff (MARC) noted the following:

> "Unfortunately, breeds (and sires within breeds) that excel in growth and retail product also: (1) Sire progeny with higher birth weights, increased calving difficulty, reduced calf survival, and reduced breeding of dams, (2) tend to be older at puberty, and (3) have heavier mature weight that increases maintenance feed requirements."

These antagonisms between maternal traits and growth and (or) carcass traits can be resolved by either (1) selecting for an optimum combination of these traits while avoiding the extremes, or (2) using terminal-cross bulls on cows that are superior for maternal traits and above average for feedlot and carcass traits. There are advantages and disadvantages for each of these two systems that can be evaluated by cow-calf producers by determining how they affect long-term profitability.

While breed average evaluations are important criteria to evaluate, these are just the beginning of evaluating and using the genetic resources. Selecting the biological types within the various breeds that best match genetics to low-cost production environments while producing retail products that are high in palatability.

Optimum levels of selection and heterosis are challenging to determine because there are few research studies that focus on the evaluation of genetic optimums. For example, how does a producer determine if a two-breed composite cow with approximately 12% heterosis is more profitable than a three- to four-breed crossbred cow that achieves 18–20% heterosis? There is evidence that a two-breed should utilize similar biological types, however it is not clear if three- to four-breed crossbred cows should be composed of similar widely divergent biological types.

Melton and Colette (1993) raise questions about what economic evaluations best rank and evaluate breeds. They state that simple measures of efficiency (e.g., output:input ratios) or annual net returns may not measure relative values and breed rankings. They suggest that the computation of multigenerational net present values, under alternative price scenarios and prevailing production conditions, be used to rank breeds.

Commercial cow-calf producers can best evaluate breeds of cattle and biological types within breeds by measuring their contribution to low breakeven prices of calves, yearlings, and carcasses. The combination of breakeven price evaluations with high continuing net profits gives assurance in making correct management decisions.

MANAGEMENT SYSTEMS HIGHLIGHTS

1. Understanding between-breed variation assists managers in their quest to effectively match cattle to a particular set of resources. Selection within a breed is most critical to fine-tuning the system.
2. Effective crossbreeding programs require that producers understand breed strengths and weaknesses.
3. Design of a composite mating system is appropriate in some cases. Breeders are advised to be wary of maximizing heterosis if it is not cost-effective.
4. The benefits of planned crossbreeding systems are numerous and are best realized in combinations of approximately 50% British breed and 50% Continental breed influence for most situations.
5. Daughters of terminal sire bulls should not be maintained as replacement females.

REFERENCES

Publications

American Gelbvieh Association. 2001. *The SmartCross™*. Handbook. Westminster, CO.

Beef Research Progress Reports. No. 1 (Apr. 1982); No. 2 (Dec. 1985); No. 3 (June 1988); No. 12 (May 1990); and No. 4 (May 1993). Clay Center, NE: U.S. Meat Animal Research Center (MARC).

Bourdon, R. M. 2000. *Understanding Animal Breeding.* Upper Saddle River, NJ: Prentice-Hall.

Briggs, H. M., and Briggs, D. M. 1980. *Modern Breeds of Livestock.* New York: Macmillan.

Cundiff, L. V., Gregory, K. E., Koch, R. M., and Dickerson, G. E. 1986. *Genetic Diversity Among Cattle Breeds and Its Use to Increase Beef Production Efficiency in a Temperate Environment.* Proceedings of the Third World Congress on Genetic Application to Livestock Production, Lincoln, NE. pp. 271.

Cundiff, L. V., Gregory, K. E., and Koch, R. M. 1991. Reproduction and maternal characteristics of diverse breeds of cattle used for beef production. *The American Beef Cattlemen,* Jan., p. 4.

Daley, D., Field, T., and Taylor, B. 1995. *New Cattle Breeds and Lines: Composites, Synthetics, and Hybrids.* Fort Collins, CO: Colorado State University.

Davis, K. C., Tess, M. W., Kress, D. D., Doornbos, D. E., and Anderson, D. C. 1994. Lifecycle Evaluation of Five Biological Types of Beef Cattle in a Cow-Calf Range Production System. II. Biological and economic performance. *J. Anim. Sci.* 72: 2591.

Field, T. G., and Cundiff, L. V. 2000. *Designing Breeding Systems That Work.* BEEF: Cow-calf 2000. Minneapolis, MN.

Franke, D. E. 1997. Postweaning Performance and Carcass Merit of F_1 Steers Sired by Brahman and Alternative Subtropically Adapted Breeds. *J. Anim. Sci.* 75: 2604.

Green, R. D., Field, T. G., Hammett, N. S., Ripley, B. M., and Doyle, S. P. 1999. *Can Cow Adaptability and Carcass Acceptability Both Be Achieved?* Proceedings of the Western Section of American Society of Animal Science, Provo, UT.

Gregory, K. E., Cundiff, L. V., and Koch, R. M. 1995. *Composite Breeds to Use Heterosis and Breed Differences to Improve Efficiency of Beef Production.* Clay Center, NE: MARC, USDA-ARS.

Hickman, C. G. 1991. *Cattle Genetic Resources: World Animal Science, B7.* New York: Elsevier Science Publishers B.V.

Kress, P. D., and MacNeill, M. D. 1999. *Crossbreeding Beef Cattle for Western Range Environments.* 2nd ed. Ardmore, OK: Samuel Roberts Noble Foundation.

Lamberson, B. 1990. *Crossbreeding Systems for Small Herds of Beef Cattle.* Extension Guide G2040. Columbia, MO: University of Missouri.

MacNeil, M. 1993. *Comparison of Breed-Type for Lifetime Productivity.* Fort Keogh Research Report.

Marshall, D. M. 1994. Breed Differences and Genetic Parameters for Body Composition in Beef Cattle. *J. Anim. Sci.* 72: 2745.

Mason, I. L. 1988. *A World Dictionary of Livestock Breeds, Types, and Varieties.* 3rd ed. Wellingford, Oxon, UK: CAB International.

Melton, B. E., and Colette, W. A. 1993. Potential Shortcomings of Output:Input Ratios as Indications of Economic Efficiency in Commercial Beef Breed Evaluations. *J. Anim. Sci.* 71: 579.

O'Connor, S. F., et al. 1997. Genetic Effects on Beef Tenderness in *Bos indicus* Composite and *Bos taurus* Cattle. *J. Anim. Sci.* 75: 1822.

Payne, W. J. A., and Hodges, J. 1997. *Tropical Cattle: Origins, Breeds and Breeding Policies.* Ames, IA: Iowa State University Press.

Randel, R. D. 1994. Factors Affecting Calf Crop. Chap. 2 in *Unique Characteristics of Brahman and Brahman Based Cows.* Boca Raton, FL: CRC Press.

Rouse, J. E. 1970. *World Cattle I: Cattle of Europe, South America, Australia, and New Zealand.* Norman: University of Oklahoma Press.

Rouse, J. E. 1970. *World Cattle II: Cattle of Europe and Asia.* Norman: University of Oklahoma Press.

Rouse, J. E. 1973. *World Cattle III: Cattle of North America.* Norman: University of Oklahoma Press.

Smith, G. C. 1993. *Assuring the Consistency and Competitiveness of Correct Biological Types of Cattle.* Proceedings of the 42nd Annual Florida Beef Cattle Shortcourse, Univ. of Florida, Gainesville, FL.

Tess, M., Lamb, M., and Robison, O. 1993. Comparison of Breeds and Mating Systems for Economic Efficiency in Cow-Calf Production. Montana Ag Research. 10:22.

Van Vleck, L. D., Splan, R. K., and Cundiff, L. V. 1999. *Genetic Correlations Between Carcass Traits and Heifer Productivity Traits.* Proceedings of the BIF Annual Research Symposium, Roanoke, VA.

Walker, Hayes, III. 1989. *Blue Book of Beef Breeds.* Allen, KS: PAW Publishing.

Visuals

"Cattle Breed Identification: British, Continental, Zebu, Composites" (1991; several videos, each 20–30 min.). CEV, P.O. Box 65265, Lubbock, TX 79464-5265.

Websites

Oklahoma State University (excellent overview of beef cattle breeds): www.ansi.okstate.edu/BREEDS/index.htm

See various breed association websites in Appendix.

Nutrition

Beef cattle nutrition involves the chemical and physiological processes by which cattle utilize available feeds to maintain body functions, reproduce, lactate, grow, and produce a desirable end product. Cattle have inherited certain genetic potentials; however, the development of these genetic potentials into economically important traits is highly dependent on the environment to which cattle are exposed. Nutrition is a significant part of this environment.

NUTRIENTS

A *nutrient* is any feed constituent that functions in the support of life. Beef cattle need nutrients to live and the feeds they consume provide the nutrients. Nutrients are composed of at least 20 of the more than 100 known chemical elements. These 20 elements and their chemical symbols are calcium (Ca), carbon (C), chlorine (Cl), cobalt (Co), copper (Cu), fluorine (F), hydrogen (H), iodine (I), iron (Fe), magnesium (Mg), manganese (Mn), molybdenum (Mo), nitrogen (N), oxygen (O), phosphorus (P), potassium (K), selenium (Se), sodium (Na), sulphur (S), and zinc (Zn).

The six basic classes of nutrients—water, carbohydrates, fats, proteins, minerals, and vitamins—are found in varying amounts in animal feeds.

Water

Water contains hydrogen and oxygen. The terms *water* and *moisture* are used interchangeably. Typically, *water* refers to drinking water, while *moisture* is used in reference to the amount of water in a given feed or ration. The remainder of the feed, after accounting for moisture, is referred to as *dry matter.* Moisture is found in all feeds ranging from 10% in air-dry feeds to over 80% in fresh green forage. Cattle consume several times more water than dry matter each day and will die from lack of water more quickly than from lack of any other nutrient. Water in feed is no more valuable than water from any other source. This is important to understand to properly assess feeds that vary in their moisture content.

Water has important body functions because it enters into most metabolic reactions, assists in transporting other nutrients, helps maintain normal body temperature, and gives the body its physical shape (it is the major component of cells).

Carbohydrates

Carbohydrates contain carbon, hydrogen, and oxygen in either simple or complex forms. The more simple carbohydrates, such as starch, supply the major energy source for cattle rations, particularly feedlot rations. The more complex carbohydrates, such as cellulose, are the major components of the cell walls of plants. These complex carbohydrates are not as easily digested as simple carbohydrates.

Fats

Fats and oils, also referred to as *lipids,* contain carbon, hydrogen, and oxygen, though there is more carbon and hydrogen in proportion to oxygen than with carbohydrates. Fats are solid and oils are liquid at room temperature. Fats contain 2.25 times as much energy per pound as do carbohydrates.

Fats are comprised of fatty acids and glycerol:

Example: $3C_{17}H_{35}COOH + C_3H_5(OH_3) = C_{57}H_{110}O_6 + 3H_2O$
Stearic acid (a fatty acid) + glycerol = stearin (a fat) + water

There are saturated and unsaturated fats, depending on their particular chemical composition. Saturated fatty acids have single bonds tying the carbon atoms together (e.g., –C–C–C–C–), whereas unsaturated fatty acids have one or more double bonds (e.g., –C=C=C–C–). The term *polyunsaturated fatty acids* is applied to those having more than one double bond. While more than 100 fatty acids have been identified, only three or four have been determined as dietary essentials. The two apparent functions of the essential fatty acids are as (1) precursors of prostaglandins and (2) structural components of cells.

Proteins

Proteins always contain carbon, hydrogen, oxygen, and nitrogen and sometimes iron, phosphorus, and/or sulphur. Protein is the only nutrient class that contains nitrogen. Proteins in feeds, on the average, contain 16% nitrogen, which is why feeds are analyzed for the percent of nitrogen in the feed (the percentage is multiplied by 6.25 to convert nitrogen to percent protein).

Proteins are composed of various combinations of some 25 amino acids. Amino acids are called the "building blocks" of the animal's body because they make up the muscle mass plus other parts. Amino acids carry the amino group (NH_2) in each of their chemical structures. There are many different combinations of amino acids that can be structured together.

Minerals

Chemical elements other than carbon, hydrogen, oxygen, and nitrogen are called *minerals.* They are inorganic because they contain no carbon, whereas organic nutrients do contain carbon. Some minerals are referred to as *macro* (required in larger amounts) and others are called *micro* or *trace* (required in smaller amounts).

Minerals have many body functions, including bone structure, body water balance, oxygen transport and transfer, and metabolic reactions.

Vitamins

Vitamins are organic substances containing carbon, hydrogen, and oxygen. There are sixteen known vitamins that are important in animal nutrition. Vitamins are required by beef cattle in very small amounts. They function in regulating certain body processes toward normal health, growth, and reproduction. Vitamin A is typically the only vitamin of concern in most beef cattle rations.

PROXIMATE ANALYSIS OF FEEDS

The nutrient composition of a feed cannot be determined accurately by visual inspection. The value of a feed can be measured by a proximate analysis, which separates feed components into groups according to their feeding value. The analysis is based on an analysis of a feed sample, so it is no more accurate than how representative the sample is of the entire feed source.

The groups measured in a proximate analysis are water, crude protein, crude fat (sometimes referred to as *ether extract*), crude fiber, nitrogen-free extract, and ash (minerals). Figure 14.1 shows these components resulting from a feed that had a laboratory analysis of 88% dry matter, 13% protein, 4% fat, 10% crude fiber, and 56% nitrogen-free extract (NFE) on a *natural* or *air-dry basis.* The analysis might be reported on a dry-matter basis of 12% moisture, 14.8% protein, 4.5% fat, 11.4% crude fiber, and 63.6% nitrogen-free extract. Therefore, caution needs to be exercised in interpreting the proximate analysis results because laboratories report their analytical values in different ways.

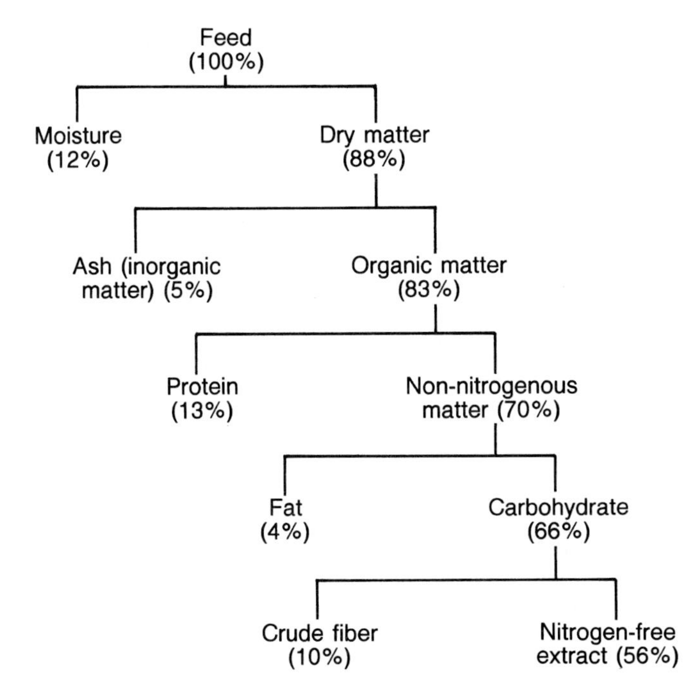

FIGURE 14.1 Proximate analysis showing the inorganic and organic components of a feed on a natural or air-dry basis.
Source: Colorado State University.

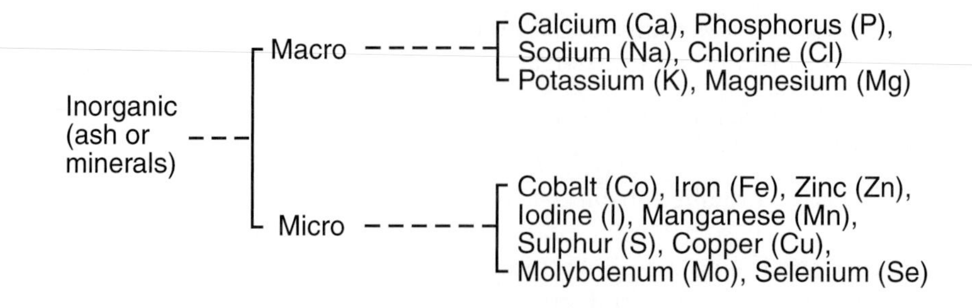

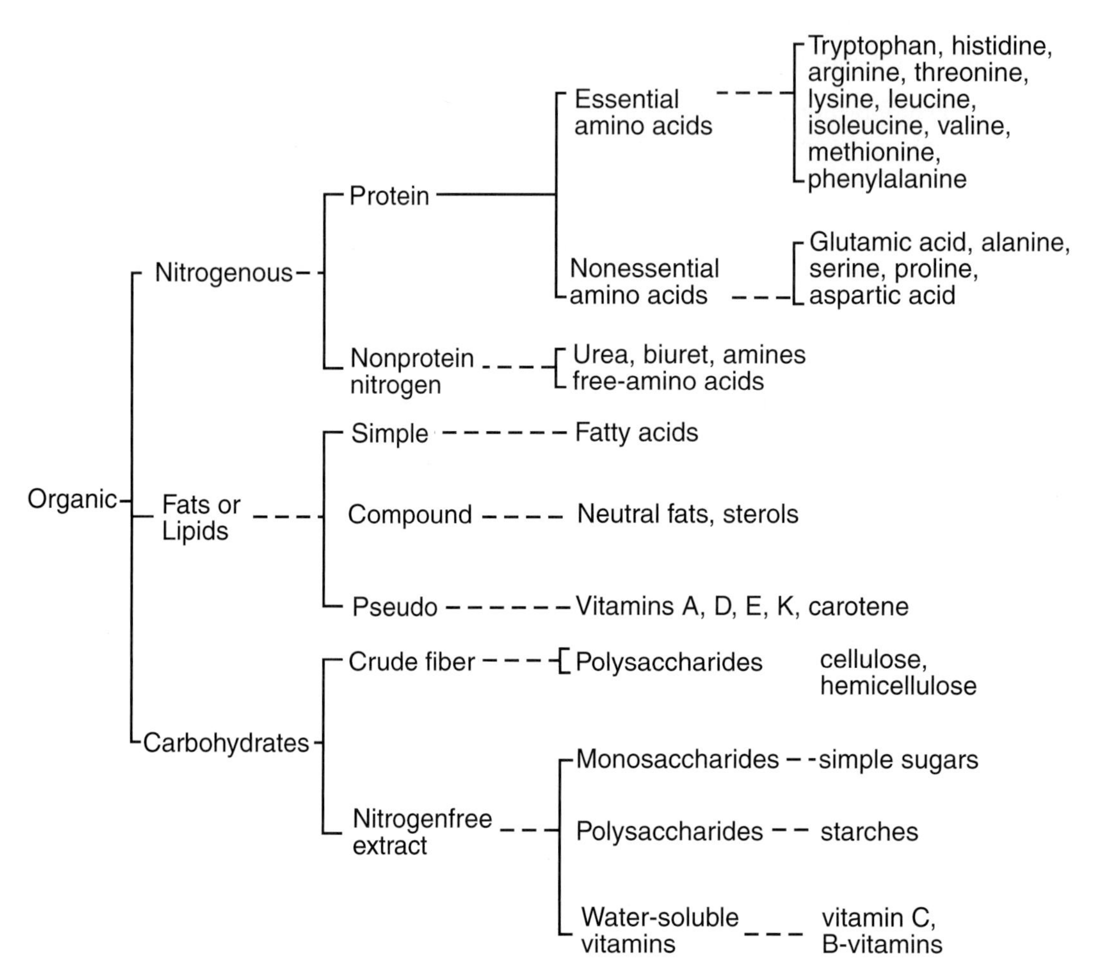

FIGURE 14.2 Chemical analysis scheme of inorganic and organic nutrients.
Source: Colorado State University.

The proximate analysis for the six basic nutrients does not distinguish the various components of a nutrient. For example, the ash content of a feed does not tell the amount of calcium, phosphorus, or other specific minerals. Figure 14.2 gives the chemical analysis for organic and inorganic nutrients. There are specific chemical analyses for each of these nutrients in a feed if such an analysis is needed.

THE RUMINANT DIGESTIVE SYSTEM

Beef cattle are *ruminants,* which means they ruminate or chew a bolus of feed called a *cud.* Cattle eat the feed and form it into boluses that pass to the stomach. After eating for a while, cattle usually lie down, regurgitate the boluses and proceed to chew the feed, breaking it into smaller pieces. A beef cow normally ruminates approximately 8 hours each day. As the animal ruminates, large amounts of saliva (35–40 gallons per day) are secreted from the salivary glands in the mouth. This amount varies with the diet being fed. For example, forage diets stimulate increased saliva production. Saliva is highly alkaline and it neutralizes (buffers) the large amount of acid produced in the ruminant stomach during digestion.

The *digestive tract,* sometimes referred to as the *alimentary canal* or *tract,* includes those organs from the mouth to the anus through which feed passes (see Fig. 14.3). Table 14.1 identifies the primary parts of the digestive tract and some of their capacities.

The process of digestion is twofold: (1) to physically break the feed into smaller pieces and (2) to chemically break more complex nutrients into simpler forms that can be absorbed by the

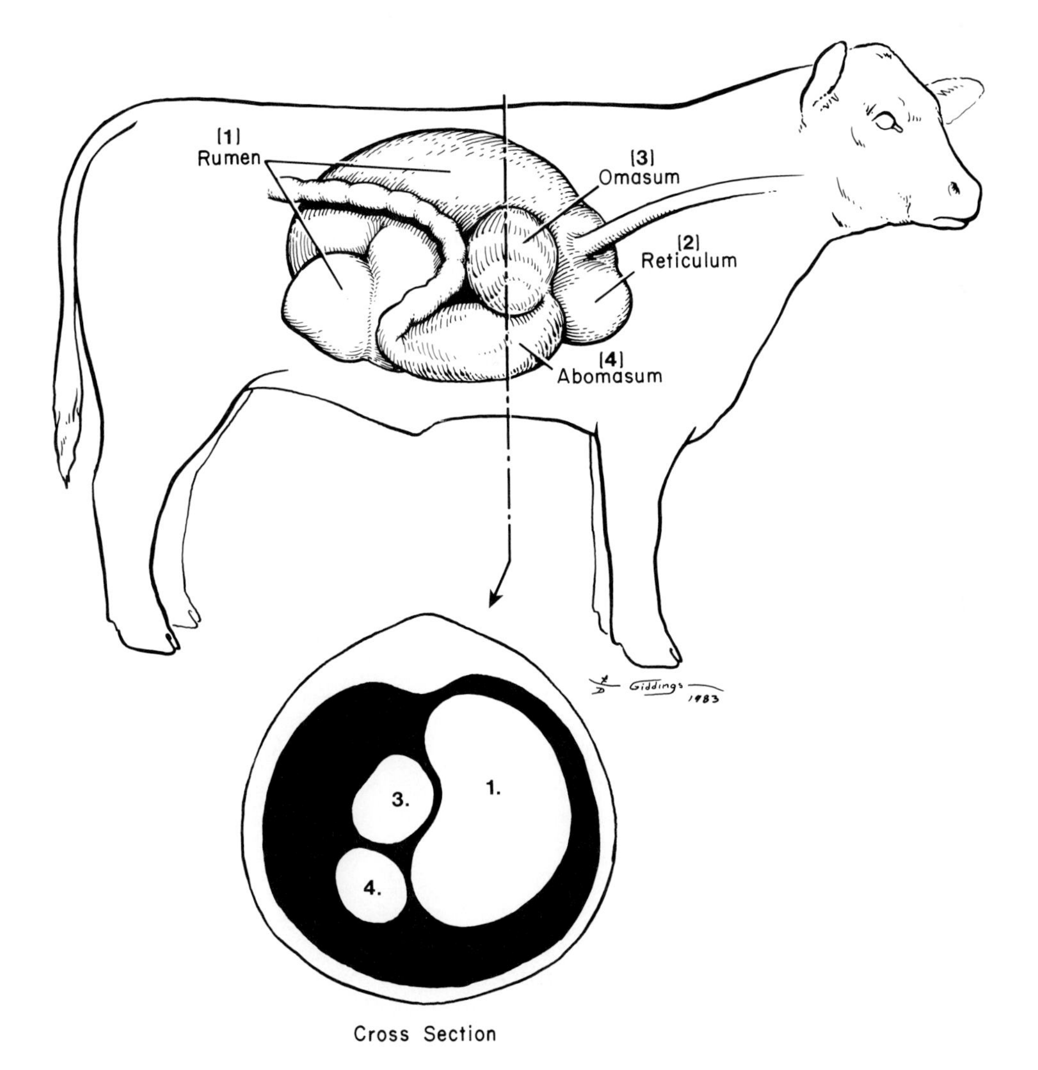

FIGURE 14.3 Beef cattle digestive tract.
Source: Colorado State University.

TABLE 14.1 Parts and Capacity of the Digestive Tract of a Cow

Part	Capacity (gal)
Mouth	—
Esophagus	—
Stomach (four compartments)	50
Rumen (pauch)	(40)
Reticulum (honeycomb)	(2)
Omasum (manyplies)	(4)
Abomasum (true stomach)	(4)
Small intestine	15
Large intestine	10
Anus	—

digestive tract and utilized on the cellular level by the animal. Simply stated, proteins are converted to amino acids; starch and other polysaccharides to simple sugars; fiber to organic acids and salts of organic acids; fats to fatty acids and glycerol; and vitamins and minerals to many soluble forms. Much of the gross physical breakdown of feed occurs at the time of chewing and particularly during rumination. The omasum is lined with many folds or plies (manyplies) of tissue that are believed to exert a squeezing and grinding action on the feed. There are several digestive juices and enzymes throughout the digestive tract that chemically break feed down into more readily usable nutrients.

It is most important to understand how the ruminant stomach functions. The rumen (Table 14.1) is a large storage area that functions as a fermentation vat. Living in the rumen fluid are billions of microorganisms, such as protozoa, bacteria, and yeasts. The kinds and numbers of these microorganisms change depending on the type of feed the animal is eating. This is why producers need to monitor how animals adjust to changes in feeds (e.g., changing an animal from a high hay or grass ration to a high grain ration).

Microorganisms manufacture many essential amino acids when there are adequate supplies of nitrogen and simple carbohydrates in the ration. The beef animal then digests the microorganisms, which become a protein source to the animal. Microorganisms also synthesize several B vitamins, which therefore do not have to be critically assessed in beef cattle ration formulation. Microorganisms also assist in breaking down complex carbohydrates (crude fiber), which permit the ruminant to consume large amounts of grass and hay.

In the rumen fermentation process, microorganisms convert some simple carbohydrates (starch) to volatile fatty acids such as acetic, propionic, and butyric. In this process, methane gas is produced, which the animal releases primarily through belching. Occasionally, the gas-releasing mechanism does not function properly and gas accumulates in the rumen causing a condition called bloat. Death will result if gas pressure builds to a high level and interferes with adequate respiration.

DIGESTIBILITY OF FEEDS

Digestibility refers to the percentage of a nutrient in a feed that is absorbed from the digestive tract. Different feeds and nutrients vary greatly in their digestibility. Many feeds have been subjected to digestion trials, where feeds of known nutrient composition have been fed to cattle. Feces have been collected and the nutrients in the feces analyzed. The difference between nutrients fed and nutrients excreted in the feces is considered the digestibility of the feed.

ENERGY EVALUATIONS OF FEEDS

Energy needs of cattle account for the largest portion of feed consumed. Several systems have been devised to evaluate feedstuffs for their energy content. The most common energy system is *total digestible nutrients (TDN),* typically expressed in pounds, kilograms, or percentages. TDN is calculated after obtaining the proximate analysis and digestibility figures for a feed. The formula for calculating TDN is as follows:

$$\text{TDN} = (\text{digestible crude protein}) + (\text{digestible crude fiber}) + (\text{digestible nitrogen-free extract}) + (\text{digestible crude fat} \times 2.25)$$

TDN is roughly comparable to *digestible energy (DE)* but it is expressed in different units. TDN and DE both tend to overvalue roughages compared to the net energy system.

Even though there are some apparent shortcomings in using TDN as an energy measurement of feeds, it works well in balancing rations for cows. There is more precision in the energy measurement of feeds in using the net energy (NE) system, so it should be used for growing cattle and in feedlot rations. The TDN system overestimates the energy value of roughages for weight gain compared to NE. NE measures energy values in megacalories per pound or kilograms of feed. The calorie basis, which measures the heat content of feed, is as follows:

Calorie (Cal)—amount of energy or heat required to raise 1 g of water 1°C.
Kilocalorie (Kcal)—amount of energy or heat required to raise 1 kg of water 1°C.
Megacalorie (Mcal)—equal to 1,000 kilocalories or 1,000,000 calories.

Figure 14.4 shows various ways that energy in feeds is utilized by cattle. Also shown are various energy measurements of feeds that reflect the types of energy utilization by the beef animal. Gross energy (GE) is the quantity of heat (calories) released from the complete burning of the feed sample in an apparatus called a bomb calorimeter. GE has little practical value in evaluating feeds for cattle because the animal does not metabolize feeds in the same manner as a bomb calorimeter. For example, oat straw has the same GE value as corn grain. Digestible energy (DE) is GE of feed minus fecal energy. Metabolizable energy (ME) is GE of feed minus energy in the feces, urine, and gaseous products of digestion.

NEm (net energy for maintenance) and *NEg* (net energy for gain) are more commonly used for formulating rations for feedlot cattle than any other energy system. NEm in feedlot cattle rations is the amount of energy needed to maintain a constant body weight. Animals of known weight, fed for zero energy gain, have a constant level of heat production. The NEg measures the increased energy content of the carcass after feeding a known quantity of feed energy. All feed fed above maintenance is not utilized at a constant level of efficiency. Higher rates of gain require more feed per unit of gain since composition of gain varies with rate of gain.

The TDN and NE systems are compared in Table 14.2. Information in the table is based on feeding a simple ration of ground ear corn (90%) and supplement (10%) to a yearling steer (772 lbs) for different rates of gain.

FEEDS: CLASSIFICATION AND COMPOSITION

Classification

Feeds are naturally occurring ingredients in the rations of cattle that are used to sustain life. The terms *feeds* and *feedstuffs* are generally used interchangeably, though the latter term is more inclusive. Feedstuffs can include certain nonnutritive products, such as additives (to

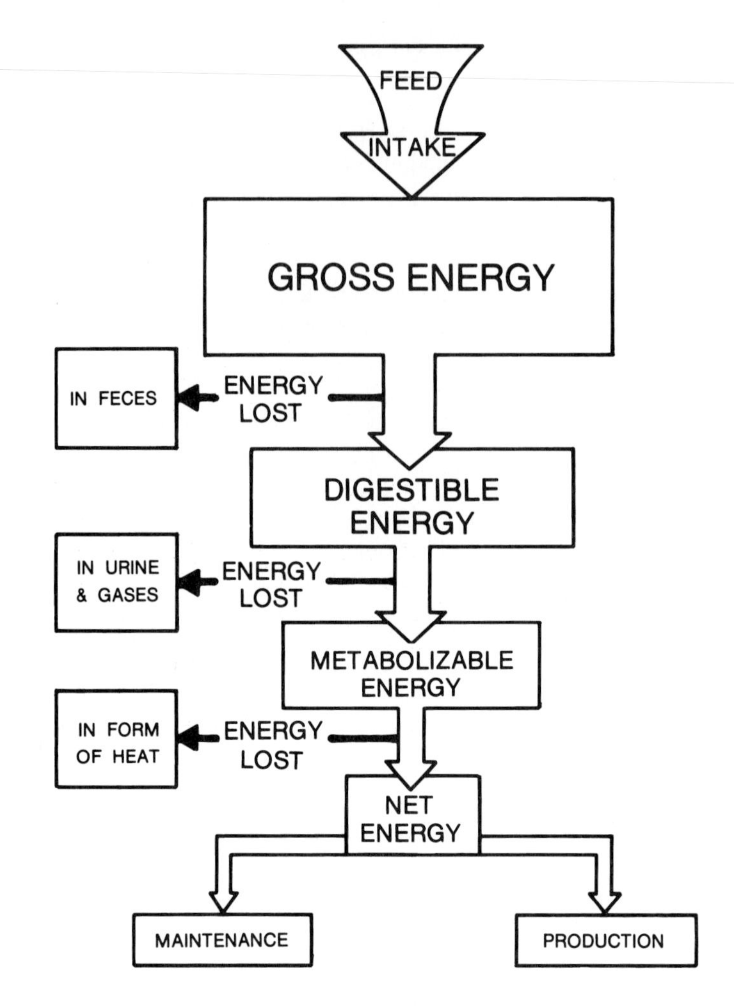

FIGURE 14.4 Measures of energy and energy utilization by cattle. *Source:* Colorado State University.

TABLE 14.2 Comparison of the TDN and NE Energy Systems for a Yearling Steer with Varying Rates of Gain

	Rate of Gain (lbs)		
Energy System	0	2	2.9
TDN (lbs)	6.4	12.8	15.0
NEm (Mcal)	6.2	6.2	6.2
NEg (Mcal)	—	4.3	6.5
Total lbs feed			
TDN basis	8.2	16.5	19.3
NE basis	7.2	15.0	19.0
Feed per lb gain (NE)	—	7.5	6.6
Feed per lb gain (TDN)	—	8.2	6.6

promote growth and reduce stress, for flavor and palatability, to add bulk, or to preserve other feeds in the ration).

The National Research Council (NRC) classification of feedstuffs is as follows:

1. Dry roughages and forages: hay (legume and nonlegume), straw, fodder, stover, and other feeds with greater than 18% fiber (hulls and shells)
2. Range, pasture plants, and green forages
3. Silages (corn, legume, and grass)
4. Energy feeds (cereal grains, mill by-products, fruits, nuts, and roots)
5. Protein supplements (animal, marine, avian, and plant)
6. Mineral supplements
7. Vitamin supplements
8. Nonnutritive additives (antibiotics, coloring materials, flavors, hormones, and medicants)

Roughages and forages are used interchangeably, although roughages usually imply a bulkier, coarser feed. In the dry state, roughages have more than 18% crude fiber. The crude fiber is primarily a component of cell walls, which is not highly digestible. Most roughages are also relatively low in TDN; there are exceptions, though, such as corn silage, which has over 18% crude fiber and approximately 70% TDN.

Feedstuffs that contain 20% or more protein are classified as *protein supplements* (e.g., soybean meal and cottonseed meal). Feedstuffs having less than 18% crude fiber and less than 20% protein are classified as *energy feeds.* Concentrates, such as the cereal grains, are typical energy feeds, which is reflected by their high TDN values.

Nutrient Composition

Feeds are analyzed for their *nutrient composition.* The ultimate goal of nutrient analysis of feeds is to predict the productive response of cattle when they are fed rations of a given nutrient composition.

The nutrient compositions of some of the more common feeds utilized for cattle are shown in Table 14.3. The information in the table represents averages of numerous feed samples. Feeds are not constant in composition so analyses should be obtained whenever economically feasible. Since an analysis is not always feasible or possible because of lack of available laboratories and insufficient time, feed analysis tables are the next best source of reliable information on nutrient composition of feeds. It is not uncommon to expect the following deviations of actual feed analysis from the table values for several feed constituents: crude protein (± 15%), energy values (± 10%), and minerals (± 30%).

Digestible protein is included in many feed composition tables, but because of the large contribution of body protein to the apparent protein in the feces, digestible protein is more misleading than crude protein. For this reason, crude protein is more commonly used in formulating rations.

Digestible protein (DP) can be calculated from crude protein (CP) content by using the following equation (% DP and % CP are on a dry-matter basis):

$$\% \text{ DP} = 0.9 (\% \text{ CP}) - 3$$

Four measures of energy value (TDN, NEm, NEg, and NEl) are shown in Table 14.3. TDN has been the industry standard in the past and many cow-calf producers are familiar with it as a measure of energy value. However, the most recent NRC standards utilize the net energy system as a more precise tool for ration formulation.

TABLE 14.3 Nutrient Composition of Selected Feeds Commonly Used in Beef Cattle Rations

Feed	Dry Matter (%)	TDN (%)	NEm[a] (Mcal/cwt)	NEg[b] (Mcal/cwt)	NEl[c] (Mcal/cwt)	Crude Protein (%)	Bypass (%)	Ca (%)	P (%)
Alfalfa hay (early bloom)	90	59	59	28	59	19	20	1.41	0.26
Alfalfa (dehydrated)	92	61	62	31	61	19	60	1.42	0.25
Alfalfa (fresh)	24	61	62	31	61	19	18	1.35	0.27
Barley grain	89	84	92	61	87	12	28	0.06	0.38
Beet tops (ensiled)	25	52	52	16	51	12	—	1.38	0.22
Beet pulp	91	75	79	50	77	11	44	0.65	0.08
Bermudagrass (hay, coastal)	89	56	56	23	56	10	20	0.47	0.21
Birdsfoot trefoil (grazed)	22	66	68	38	67	21	20	1.78	0.25
Bluegrass (grazed, Kentucky)	36	69	71	43	70	15	20	0.37	0.30
Bromegrass (grazed, early vegetation)	30	64	65	36	65	15	22	0.45	0.34
Bromegrass (hay)	89	55	55	21	55	10	33	0.40	0.23
Clover, red (hay)	88	55	55	21	55	15	28	1.50	0.25
Corn (whole grain)	88	87	96	64	90	9	58	0.02	0.29
Corn (grain rolled)	88	89	99	67	93	9	52	0.02	0.30
Corn silage (mature)	34	72	75	47	74	8	28	0.28	0.23
Corn stover (no ears or husks)	80	59	59	28	59	5	30	0.35	0.19
Cottonseed meal	92	80	86	56	83	46	50	0.21	1.19
Dicalcium phosphate	96	—	—	—	—	—	—	22.00	18.65
Fescue KY31 (fresh)	29	64	65	36	65	15	20	0.48	0.37
Fescue KY31 hay (mature)	88	52	52	16	51	11	30	0.45	0.26
Lespedeza (grazed early)	25	60	60	30	60	16	50	1.20	0.24

TABLE 14.3 (Continued)

Feed	Dry Matter (%)	TDN (%)	NEm[a] (Mcal/cwt)	NEg[b] (Mcal/cwt)	NEl[c] (Mcal/cwt)	Crude Protein (%)	By-pass (%)	Ca (%)	P (%)
Lespedeza (hay)	92	54	54	20	54	14	60	1.10	0.22
Limestone (ground)	98	—	—	—	—	—	—	34.00	0.02
Meadow hay (native, intermount)	90	50	50	12	49	7	23	0.61	0.18
Molasses (cane)	76	75	79	50	77	5	0	1.00	0.10
Oats (grain)	89	76	81	52	78	13	18	0.05	0.41
Oats (silage)	35	60	60	30	60	12	21	0.45	0.31
Prairie hay (midbloom)	91	50	50	12	49	7	37	0.40	0.15
Sagebrush (grazed)	50	50	50	12	49	13	—	1.00	0.25
Sorghum stover (no heads)	87	55	55	21	55	5	—	0.49	0.12
Sorghum silage	32	59	59	28	59	9	30	0.48	0.21
Soybean meal (solvent)	91	84	92	61	87	49	35	0.38	0.71
Soybean (straw)	88	42	43	0	40	5	—	1.59	0.06
Sudangrass (hay)	88	57	57	25	57	9	30	0.50	0.22
Timothy (hay, full-bloom)	88	57	57	25	57	8	30	0.43	0.20
Wheat (hard grain)	89	88	98	65	91	14	28	0.05	0.43
Wheat (grazed early)	21	71	74	46	73	20	16	0.35	0.36
Wheat (straw)	91	42	43	0	40	3	60	0.16	0.05
Wheatgrass, crested (early)	37	60	60	30	60	11	25	0.46	0.32
Wheatgrass, crested (full-bloom)	50	55	55	21	55	10	33	0.39	0.28
Wheatgrass, crested (hay)	92	54	54	20	54	10	33	0.33	0.20

[a] Net energy maintenance.
[b] Net energy growth.
[c] Net energy lactation.
Source: Adapted from 2001 Feed Composition Guide, *BEEF.*

MANAGEMENT FOR HIGH NUTRIENT CONTENT OF FEEDS

How feeds are grown, harvested, and preserved can have a significant effect on total nutrients produced per unit of land, nutrient content of feeds, and safety of the feeds at the time they are being fed to cattle. Some major management practices utilized to enhance nutrient content and utilization of feeds are presented here. Hays and crop residues are covered in Chapter 15.

Silages

Silages are produced from green forage crops that are compressed and stored under anaerobic (oxygen-free) conditions in upright or horizontal silos. The feed is preserved through an acid fermentation process that prevents mold and other spoilage from occurring.

The most common crops used for silage are corn, sorghums, grasses, legumes, and some small grains. These must contain enough fermentable carbohydrate to allow sufficient lactic acid production to register a pH of 4. With some forages, grain may be added to achieve this effect. Haylage is a forage that is intermediate between hay and silage in terms of moisture content (40–60%), whereas the moisture content of silage is usually 60–75%.

Silage comprises a large part of the feeding program of many beef cattle operations, primarily because TDN and beef produced per acre is greatest with silage. Caution should be exercised, however, because silage is one of the most variable feedstuffs in nutrient content and storage loss. The nine factors that interact to affect silage quality are type of silage crop, weather, stage of maturity, moisture, additives, fineness of chopping, packing and filling the silo, silo structure, and feed-out of the silage.

Moisture content of silage is the single most important factor affecting silage quality. Ensiling above 72–74% moisture can produce seepage, undesirable butyric acid, high fermentation losses, and reduced intake of silage by cattle. Ensiling below 50–55% moisture creates problems in eliminating air (poor compaction) and achieving sufficient fermentation and low pH.

An optimum stage of maturity should be selected that will make the best compromise between increased dry-matter yield and decreased digestibility. Table 14.4 gives the recommended maturity stages for several silage crops. A fine chop—one-fourth to one-half inch—is recommended for most crops. Hollow-strawed cereals need to be chopped finer than solid-stemmed corn and sorghum. Crops that are chopped too coarse will have more air trapped when ensiled and more silage will be refused by cattle when it is fed.

Type of silo can affect the quality of silage as losses can be 30% or higher (Table 14.5). Silo construction cost and silage losses need to be compared.

TABLE 14.4 Stage of Maturity Recommendations for Several Silages

Crop	Recommended Stage of Maturity
Alfalfa	Late bud to one-tenth bloom
Cereal grains (barley, oats, and wheat)	Soft-dough kernels
Corn	Kernels fully dented
Grasses (summer and perennial)	Prior to head emergence
Sorghum	Soft-dough kernels

TABLE 14.5 Silo Type and Silage Losses

Type of Silo	Dry-Matter Loss (%)
Oxygen-limiting	3–8%
Concrete upright	5–15
Trench or bunker	12–25
Open stack	20–40

Packing and filling the silo should be done to eliminate air pockets while also exposing a small surface area. After filling a trench or bunker silo, the surface should be sealed to keep out air, rain, and snow and thereby reduce spoilage loss. Producers should grow those silage crops that are best suited agronomically for their area, selecting crops that can economically produce the greatest amount of TDN per acre. Feed-out should allow the silage to be fresh and not exposed to air for extended periods of time. This is especially critical when the weather is warm, when the growth of mold and other spoilage organisms is stimulated.

There are four general categories of silage additives: (1) acids for direct acidification, (2) preservatives (sterilants and fatty acids), (3) feedstuffs (molasses, NPN, grain), and (4) fermentation aids (enzymes, inoculants, and antioxidants). Before using an additive, however, the costs and returns should be carefully evaluated. Claims made for additives should be based on well-documented research. Additives should not be used as a substitute for poor management.

The amounts of grain and moisture in silage have major influences on its feed value, and both should be used to determine the dollar value of silage. The value of corn silage as a standing crop can be determined in the following three ways: (1) leave a few rows for later grain harvest to determine grain yield per acre, (2) harvest by hand a small plot (1/100 acre) and determine grain yield, or (3) use relationship of 1 ton of 30% dry-matter silage for each harvested foot of plant, not counting the tassel. The last option assumes an average stand, so adjustments would have to be made for light or heavy stands.

Corn grain yield estimates in silage can be used to estimate the value of the silage. For example, if it is estimated that there are 6 bushels of number 2 corn in a ton of wet silage (30–35% dry matter), then the cash corn price times the 6 bushels plus harvesting costs would give a reasonable estimate on the value of the silage (e.g., corn selling for \$3 per bu × 6 bu corn per ton = \$18 + \$4 harvest cost = \$22 for 35% dry-matter silage).

If the amount of grain in the silage is not known, a rough estimate of silage value is 8 times the market value of the corn grain. If the silage is already in storage, it is worth about 10 times the market price of corn.

Sorghum silage should be priced similar to corn of the same moisture content. Forage sorghums with fairly high grain yield usually have 80–90% the value of corn silage per unit of dry matter. Sudan and Sudan-sorghum crosses or varieties, with low grain yields, may have only 65–80% the value of corn silage per unit of dry matter.

Grains and Concentrates

Concentrates are high-energy feeds (mostly feed grains and their by-products) that contain less than 20% protein and less than 18% fiber. Corn and sorghum (milo) are the most common feed *grains* in cattle rations, with barley, oats, wheat, and other concentrates being of lesser importance.

Table 14.6 shows the relative feeding values of several grains. The relative value, using the information in the table, is calculated in the following manner:

$$\text{Relative value of grain} \times \text{corn cost per lb} = \text{relative cost of grain per lb}$$

For example, to determine the price to pay for barley relative to corn if corn is \$3.10 per bu or 5.5¢ per lb (price of corn), then 5.5¢/lb × 90% = 4.95¢/lb or 48 lb per bu × 4.95¢ per lb = \$2.38 per bu for barley. Barley would be purchased if the price is less than \$2.38 per bu compared with corn at \$3.10 per bu.

The maximum level recommendations for the grains listed in Table 14.6 are for high-concentrate rations. They would be higher for oats, rye, and wheat if grain was being limit-fed in the ration.

TABLE 14.6 Relative Feeding Values of Various Grains for Cattle

Grain	Relative Value (% lb-for-lb basis)	Maximum Level (%)	Bushel Weight (lbs)
Barley	88–90%	100	48
Corn	100	100	56
Corn and cob meal	85–95	100	70
Milo	85–95	100	56
Oats	70–90	25	32
Rye	80–85	20	56
Wheat	100–105	40	60

Feedlot cattle consume most of the feed grains in the beef production system, although concentrates can be fed to cows during 1–3 months of the calving-breeding seasons, when energy requirements are highest and grains are a cost-effective source of energy.

Feed grains are harvested and stored in a relatively dry condition (less than 20% moisture); otherwise, the grain will spoil and lose its nutrients and feeding acceptability to cattle. Another factor associated with spoilage of grains is the production of *mycotoxin* from molds growing in feedstuffs. The major mycotoxin of concern in regards to corn is *aflatoxin*. These pathogens can be produced on growing plants or in storage conditions. Aflatoxin may cause reduced growth rate, feed consumption, and liver damage in cattle consuming contaminated feeds. The maximum allowable level of aflatoxin is 20 ppb for immature cattle, 100 ppb for breeding cattle, and 300 ppb for finishing cattle. Utilization of an on-farm quality assurance program that establishes a feed sampling and testing protocol is highly recommended. Most grains are processed in one of the following ways: (1) dry process (ground or rolled), (2) steam flaking, (3) oxygen-limited reconstitution (adding water, then storing in an airtight area), (4) early harvested (ground and ensiled in trench or airtight silos), (5) early harvested with acid treatment of grain, and (6) pelleted. Corn is the only grain that can be fed satisfactorily in whole form to feedlot cattle. Whole corn is usually fed to feedlot cattle on high (80–90%) concentrate rations. It is essential that milo be processed to improve its digestibility. Other advantages in processing grains are improved gain and efficiency of gain and increased feed intake by cattle. Using high-moisture grains reduces field losses during the time of harvesting.

Increased animal performance must be tempered with energy costs in processing. Currently, steam flaking is the most intensive grain-processing alternative; however, it is feasible only in large-sized feedlots (20,000 head and above).

Caution should be exercised if wheat is fed by keeping it below 40% of the total ration. It can cause compaction if ground too fine and acidosis if cattle are not accustomed to it. However, much higher levels of steam-flaked wheat have been successfully fed by increasing the roughage and ionophore levels in the finishing ration.

Protein Supplements

Protein supplements are dry or liquid feedstuffs that contain 20% or more protein. They may be of plant origin (soybean meal, cottonseed meal, linseed meal, corn gluten meal, etc.) or animal origin (blood meal, dried skim milk, etc.). Because of concerns about BSE, the feeding of ruminant by-products to cattle is prohibited.

Molasses is commonly used as the base for liquid supplements. Soybean meal is the most widely used protein supplement in the United States. Cottonseed meal, however, is commonly used in the South, since most U.S. cotton is grown in that area.

Urea is not a protein supplement but it is a source of nitrogen (42–45%) for protein synthesis by rumen bacteria. Urea has a protein equivalent of 262–281%, since 1 lb of urea contains as much nitrogen as 2.62–2.81 lbs of protein. Urea works well with other plant proteins to lower the cost of protein in the ration. It works best in feedlot rations that are relatively high in energy because rumen organisms need a readily available energy source to convert the urea nitrogen into microbial protein. Urea should not be utilized as a supplement for low-quality forages, since assuring a consistent intake of urea is difficult to achieve.

General guidelines for using urea or other sources of NPN (nonprotein nitrogen) are as follows: (1) a maximum of one-third of total nitrogen in the ration should come from urea; (2) urea should not represent more than 10–15% of a typical protein supplement; (3) it should not be more than 1% of a diet or 3% of a concentrate mixture; and (4) it should not be more than 5% of a supplement when used with low-grade roughages. Urea poisoning can occur when its level in the ration is too high or when it is not mixed properly.

Vitamin Supplements

The only *vitamin supplement* of general practical importance to beef cattle is vitamin A (precursor is carotene). Vitamin A or carotene content in feeds depends largely on maturity, harvest conditions, and length and conditions of storage. Where roughages being fed contain a bright green color or are being fed as immature fresh forage (e.g., pasture), there will probably be sufficient vitamin A value in these roughages to meet the animal's requirement. If there is a question about vitamin A availability a feed analysis may be needed, or vitamin A supplementation is usually economical. Vitamin A is generally included in all commercial protein supplements for growing–finishing cattle.

Mineral Supplements

Mineral supplements can be simple or complex, the latter occurring when several trace minerals are included. In most areas of the United States, mineral supplementation of salt, phosphorus, and calcium is adequate. Trace minerals needed in feeds are largely determined by their level in the soil on which feeds are grown or by other environmental factors. However, there are areas in the United States where mineral deficiencies or toxicities do exist (see Fig. 14.5). In these areas, additional mineral supplementation should be included and methods to reduce toxicity effects should be employed. Trace minerals, if necessary, are usually supplied in a trace mineralized salt.

Determining the appropriate mineral supplementation program requires a critical analysis of both the nutritional needs of cattle and the cost of delivering appropriate mineral supplements. Water and forage samples should be analyzed on at least an annual basis to provide trend lines and provide a starting place to identify potential problems. Late fall is a good time to collect the samples for spring calving herds to allow sufficient time to make adjustments, if needed.

Soil tests are also useful in helping to monitor mineral issues. Minerals may be antagonistic to one another. For example, high levels of molybdenum may interfere with copper absorption.

The mineral requirements for cattle vary with their stage of production. The two months leading up to calving as well as the same time period following parturition is a critical time

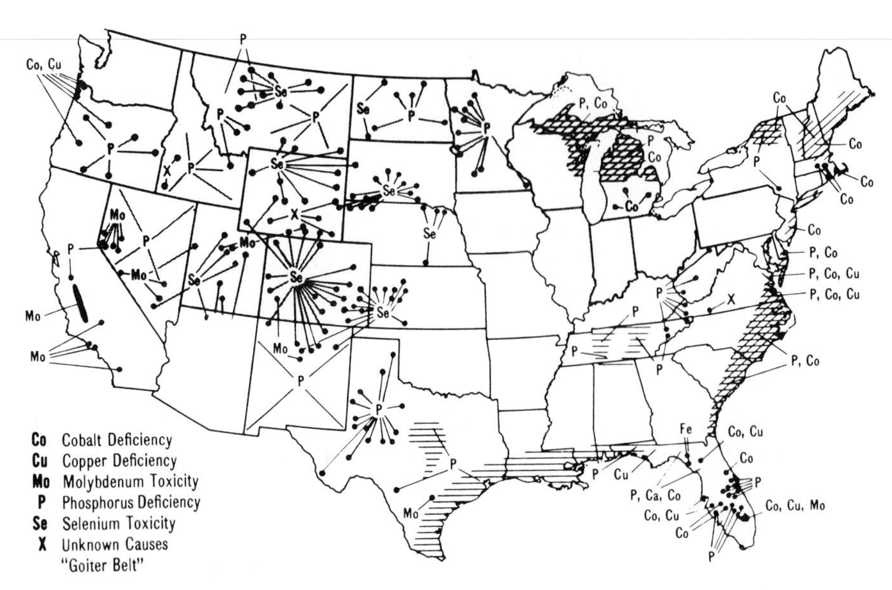

FIGURE 14.5 Areas of mineral deficiency or toxicities.
Source: USDA.

TABLE 14.7 Disorders Associated With Minerals in Beef Cattle

Mineral	Disorder
Calcium	Milk fever (hypocalcemia) —paralysis, circulatory collapse, coma and death; most cases within 24 hr. following calving
Magnesium	Grass tetany (hypomagnesemia) —staggering, incoordination, excitability, profuse salivation; typically occurs in lactating cattle grazing lush pasture, especially small grains (oats, barley)
Manganese	Deficiencies associated with suppressed estrus expression, reduced fertility, and poor skeletal growth
Copper	Deficiencies associated with increased retention of placentas, retarded growth, poor hair coat, diminished fertility
Zinc	Deficiencies associated with reduced immune response, poor fertility, reduced feed efficiency and growth; increased incidence of foot rot
Selenium	Toxicity —alkali disease, blind staggers, death
Cobalt	Deficiency prohibits formation of vitamin B_{12} and leads to emaciation and anemia

Source: Adapted from multiple sources.

for mineral supplementation. The method of mineral feeding is also critical to assure that each animal in the herd has access to and consumes appropriate supplement levels.

Bioavailability or ease of absorption of mineral sources is also an important consideration. For example, the bioavailability of calcium from dicalcium phosphate is high but it is low from hay. The supplementation of copper as cupric chloride or cupric sulfate is highly available while cupric oxide is quite low in bioavailability.

The failure to provide appropriate (neither too high nor too low) mineral levels can lead to a number of disorders (Table 14.7).

NUTRIENT REQUIREMENTS OF BEEF CATTLE

In formulating rations, producers must know the nutrient requirements of beef cattle, in addition to knowing the nutrient composition of feedstuffs. The National Research Council (NRC) publications are the generally accepted standards that identify the nutrient requirements of beef cattle. Tables 14.8 and 14.9 show the nutrient requirements of several sexes of cattle within a few selected weights and levels of productivity. More extensive tables of nutrient requirements and feed compositions can be found in the National Research Council's *Nutrient Requirements of Beef Cattle.*

TABLE 14.8 Daily Nutrient Requirements for Breeding Cattle (Heifers, Cows, and Bulls)

Weight (lbs)	Daily Gain (lbs)	Dry-Matter Consumption (lbs)	Total Crude Protein (lbs)	TDN (lbs)	ME (Mcal)	Ca (g)	P (g)	Vitamin A (1,000 IU)
		Growing Heifer Calves and Yearlings (see Table 14.10)						
		Two-Year-Old Heifers—Last Third of Pregnancy						
1,000[a]	0.73	20.7	1.8	11.7	20.9	29	22	31
1,100[a]	0.80	22.3	1.9	12.6	22.5	31	22	33
1,200[a]	0.88	23.7	2.0	13.3	24.2	33	24	34
1,300[a]	0.95	25.2	2.1	14.1	25.6	34	25	35
1,400[a]	1.02	26.6	2.2	14.8	26.9	37	27	37
		Cows Nursing Calves—Average Milking Ability[b]—First 3–4 Months Postpartum						
1,000	0	23.0	1.9	12.5	20.9	24	17	36
1,200	0	26.0	2.1	13.9	23.4	27	19	41
1,400	0	29.0	2.3	15.4	25.7	30	21	45
		Cows Nursing Calves—Superior Milking Ability[c]—First 3–4 Months Postpartum						
1,000	0	25.4	2.6	14.9	24.9	35	24	37
1,200	0	28.4	2.8	16.4	27.3	37	25	42
1,400	0	31.3	3.0	17.8	29.7	40	27	47
		Dry Pregnant Mature Cows—Middle Third of Pregnancy						
1,000	0	16.7	1.2	8.2	13.4	14	14	23
1,200	0	19.5	1.4	9.5	15.6	17	17	26
1,400	0	23.3	1.6	11.4	18.7	21	21	30
		Dry Pregnant Cows—Last Third of Pregnancy						
1,000	0.9	21.0	1.6	10.9	18.3	23	14	25
1,200	0.9	24.1	1.8	12.6	20.9	27	17	28
1,400	0.9	27.0	2.1	14.2	23.9	32	21	32
		Bulls—Growth and Maintenance (Moderate Activity)						
1,400	1.0	26.8	2.0	15.0	24.6	26	23	48
1,600	1.0	29.7	2.2	16.6	27.2	29	26	53
1,800	0	28.9	2.0	14.0	23.0	27	27	51
2,000	0	31.3	2.1	15.2	24.5	30	30	55
2,200	0	33.6	2.3	16.3	26.7	33	33	60

[a]Mature weight potential.
[b]10 lbs of milk per day (equivalent of approximately 450 lbs of calf at weaning if there is adequate forage).
[c]20 lbs of milk per day (equivalent of approximately 650 lbs of calf at weaning if there is adequate forage).
Source: Nutrient Requirements of Beef Cattle. 1996. Washington, DC: National Research Council.

TABLE 14.9 Nutrient Requirements for Growing and Finishing Cattle

Live Weight (lbs)	TDN (% DM)	Daily Gain (lbs)	Dry-Matter Intake (lb/day)	Protein Intake (lbs)	Crude Protein (%)	NEm (Mcal/lb)	NEg (Mcal/lb)	Ca (%)	P (%)
1,100 lbs at finishing (28% body fat—feedlot steers and heifers) or maturity (replacement females)									
605	50	0.68	16.3	1.2	7.2	0.45	0.20	0.22	0.13
	70	2.86	16.9	2.1	12.7	0.76	0.48	0.49	0.24
	90	4.00	14.7	2.6	17.8	1.04	0.72	0.72	0.34
660	50	0.68	17.5	1.2	7.1	0.45	0.20	0.21	0.13
	70	2.86	18.0	2.2	12.3	0.76	0.48	0.45	0.23
	90	4.00	15.7	2.7	17.1	1.04	0.72	0.66	0.32
715	50	0.68	18.5	1.3	6.9	0.45	0.20	0.20	0.13
	70	2.86	19.1	2.2	11.5	0.76	0.48	0.42	0.21
	90	4.00	16.7	2.6	15.9	1.04	0.72	0.61	0.30
770	50	0.68	19.6	1.3	6.8	0.45	0.20	0.19	0.12
	70	2.86	20.2	2.2	10.9	0.76	0.48	0.39	0.20
	90	4.00	17.6	2.6	14.8	1.04	0.72	0.56	0.28
825	50	0.68	20.6	1.3	6.6	0.45	0.20	0.19	0.12
	70	2.86	21.3	2.2	10.3	0.76	0.48	0.37	0.19
	90	4.00	18.6	2.6	13.9	1.04	0.72	0.52	0.26
880	50	0.68	21.7	1.4	6.5	0.45	0.20	0.19	0.12
	70	2.86	22.4	2.2	9.8	0.76	0.48	0.34	0.18
	90	4.00	19.5	2.5	13.1	1.04	0.72	0.48	0.25
1,300 lbs at finishing (28% body fat—feedlot steers and heifers) or maturity (replacement heifers)									
715	50	0.76	18.5	1.3	7.3	0.45	0.20	0.22	0.13
	70	3.21	19.1	2.5	13.0	0.76	0.48	0.49	0.24
	90	4.48	16.7	3.0	18.3	1.04	0.72	0.72	0.34
780	50	0.76	19.8	1.4	7.1	0.45	0.20	0.21	0.13
	70	3.21	20.4	2.5	12.1	0.76	0.48	0.45	0.23
	90	4.48	17.8	3.0	16.9	1.04	0.72	0.66	0.32
845	50	0.76	21.0	1.4	6.9	0.45	0.20	0.21	0.13
	70	3.21	21.7	2.5	11.4	0.76	0.48	0.42	0.22
	90	4.48	18.9	3.0	15.7	1.04	0.72	0.60	0.30
910	50	0.76	22.2	1.5	6.7	0.45	0.20	0.20	0.13
	70	3.21	22.9	2.4	10.7	0.76	0.48	0.39	0.20
	90	4.48	20.0	2.9	14.6	1.04	0.72	0.56	0.28
975	50	0.76	23.4	1.5	6.6	0.45	0.20	0.20	0.13
	70	3.21	24.1	2.5	10.2	0.76	0.48	0.37	0.19
	90	4.48	21.0	2.9	13.7	1.04	0.72	0.52	0.26
1040	50	0.76	24.5	1.6	6.5	0.45	0.20	0.19	0.13
	70	3.21	25.3	2.4	9.6	0.76	0.48	0.34	0.19
	90	4.48	22.1	2.8	12.9	1.04	0.72	0.48	0.25

Source: Nutrient Requirements of Beef Cattle. 1996. Washington, DC: National Research Council.

RATION FORMULATION

Rations should be formulated or balanced to meet the nutrient requirements of animals in different stages of production. Another important consideration is the cost of the ration with an assurance that it is palatable and will not cause any serious digestive disturbances or toxic effects.

Rations should be balanced for energy (using TDN, ME or NEm, and NEg), crude protein, vitamin A, calcium, and phosphorus. Salt is usually provided free choice or mixed into the ration. An assessment should be made to determine whether trace minerals are needed.

Balancing Cow Rations Using TDN

TDN is a satisfactory energy basis for balancing rations for cows. Table 14.10 shows an example of balancing a cow's ration where silage is the primary source of feed. The calculations used in the table are as follows:

Line a: TDN requirement from Table 14.8.
Line b: Calculate lbs of corn silage dry matter (DM) needed by dividing lbs of TDN required by % TDN in corn silage ($13.9 \div 0.72 = 19.3$).
Line c: Calculate lbs of "wet" corn silage needed by dividing lbs of corn silage DM by % DM in corn silage ($19.3 \div 0.34 = 56.8$ lbs).
Line d: Crude protein (CP) requirement from Table 14.8.
Line e: Calculate lbs of CP supplied by corn silage by multiplying lbs of corn silage DM fed (line b) by % CP in corn silage ($19.3 \times 0.08 = 1.54$).
Line f: Calculate CP deficiency by subtracting lbs of CP supplied on *line e* from the CP requirement on *line d* ($2.10 - 1.54 = 0.56$).
Line g: Calculate lbs of CP supplement needed by dividing lbs of CP deficiency by % CP in the supplement ($0.56 \div 0.35 = 1.60$).

This ration is now balanced for TDN and CP; however, it does not meet the minimum dry-matter requirement of 26 lbs (Table 14.8). To meet the dry-matter requirement, one would increase the silage fed to approximately 77 lbs. Fulfilling the dry-matter requirement is not a major concern in feeding cows as long as energy, protein, mineral, and vitamin requirements are met. Salt, phosphorus, and possibly vitamin A would be supplied on a free-choice basis to meet the cow's nutrient requirements.

It is not uncommon to have dry matter vary in the same silage crop. Amount of wet silage fed should be adjusted as moisture content changes. If moisture changes are not considered, then critical nutrients will be oversupplied or undersupplied.

Using the same cow requirements as shown in Table 14.10, the CP deficiency may be filled using a legume such as alfalfa hay. The Pearson Square method can be used to determine the amount of alfalfa hay needed, along with corn silage, to meet the dry-matter and crude protein requirements. An example using Pearson's Square follows:

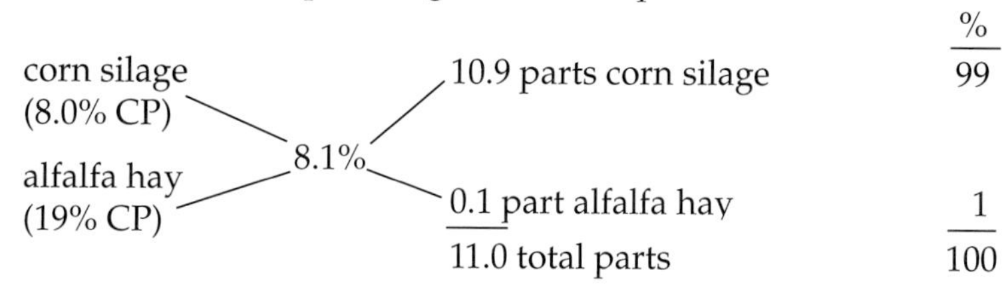

TABLE 14.10 Beef Cow Ration with Corn Silage and Protein Supplement

1,200-lb Cow (average milk, early lactation)	Amount
a. TDN required (lbs/day)	13.9
Corn silage needed:[a]	
b. lbs of silage (dry matter [DM]/day)	19.3
c. lbs of wet silage/day	56.8
Crude protein (CP) supplement needed:[b]	
d. CP required (lbs/day)	2.10
e. CP supplied by corn silage (lbs/day)	1.54
f. CP deficiency (lbs/day)	0.56
g. lbs of protein supplement/day	1.60

[a]Composition of corn silage taken from Table 14.3: dry matter, 34%; TDN, 72%; and crude protein, 8%.
[b]Assume a protein supplement of 35% CP.

TABLE 14.11 Checking the Pearson Square Method Calculations

Feed	Part of Ration %	Part of Ration lbs	Protein %	Protein lbs
Corn silage	99	25.7	8. 0	2.05
Alfalfa hay	1	0.3	19.0	.05
Total or average	100	26.0	8.1	2.10

TABLE 14.12 Comparative Energy Costs Using TDN and NEg

Feedstuff	Cost/Ton ($)	% TDN	Cost/lb TDN ($)	NEg (Mcal/lb)	Cost/Mcal NEg ($)
Alfalfa hay[a]	$110	59	$0.093	0.28	$0.196
Barley (48 lbs/bu)	90	84	0.054	0.61	0.074
Corn (56 lbs/bu)	90	87	0.055	0.64	0.070
Corn silage (28% dry matter)	25	72	0.017	0.47	0.026
Cottonseed meal	165	78	0.106	0.54	0.153
Dehydrated alfalfa	140	61	0.115	0.31	0.226
Soybean meal	205	84	0.122	0.61	0.168
Wheat (60 lbs/bu)	117	88	0.066	0.65	0.090

[a]Example of calculation: Alfalfa hay at $\frac{\$110/\text{ton}}{2{,}000} = \$0.055/\text{lb}$; TDN $= \frac{0.055}{0.59} = \$0.093/\text{lb}$; NEg $= \frac{0.055}{0.28} = \$0.14/\text{Mcal}$.

On the left-hand side of the square are the crude protein percentages of corn silage and alfalfa hay taken from Table 14.3. The 8.1% in the middle of the square is the crude protein requirement (expressed in %) of the 1,200-lb cow. This is calculated as follows:

$$\frac{\text{2.10-lb CP requirement}}{\text{26-lb dry-matter requirement}} = 8.1\%$$

On the right-hand side of the square are parts of corn silage and parts of alfalfa hay of the total ration. These parts are obtained by subtracting, diagonally, the smallest % from the largest %. Table 14.11 permits a check of the calculations.

Least-cost rations can be calculated using a computer program that can analyze many feed sources and their nutrient composition and cost. Some rough approximations of least-cost rations can be hand calculated by determining energy and protein costs on several readily available feeds. Since energy and protein costs make up the largest part of the total nutrient costs, using the information in Tables 14.12 and 14.13 can be helpful. The information in these tables is for example purposes only. Individual producers should substitute feeds common to their area, the energy and protein compositions of the feeds, and the current feed cost.

Formulating Feedlot Rations

The net energy system is typically used when formulating feedlot rations. In example 1, a ration is evaluated in terms of adequacy in net energy and crude protein for a 700-lb medium-framed steer expected to gain approximately 3 lbs per day during a 140-day finishing period.

TABLE 14.13 Calculation of Value of Protein Supplements on the Basis of Crude Protein

Supplement	% Crude Protein	Cost/Ton ($)	Cost/lb Crude Protein ($)
Alfalfa hay	19	$110	$0.289
Cottonseed meal	45	165	0.183
Dehydrated alfalfa	19	140	0.368
Soybean meal[a]	50	205	0.205

[a]Example of calculation: Soybean meal at $\frac{\$205/\text{ton}}{2{,}000} = \$0.1025/\text{lb}$; CP cost $= \frac{0.1025}{0.50} = \$0.205/\text{lb}$.

TABLE 14.14 Feed Fed to a 700-lb Steer

Feed	Lbs Fed	Dry Matter (%)	Dry Matter (lbs)
Corn silage (mature)	9.0	34.0	3.06
Corn	10.0	88.0	8.80
Chopped alfalfa hay	2.0	90.0	1.80
Beet pulp	1.5	91.0	1.36
32% protein supplement	1.0	90.0	0.90
Total			15.92

TABLE 14.15 Calculating the NEm and NEg in the Ration

Feed	Dry Matter (lbs)	NEm (Mcal)	NEg (Mcal)	Protein (lbs)
Corn silage	3.06	2.29	1.44	0.24
Corn	8.80	8.45	5.63	0.79
Chopped alfalfa hay	1.80	1.06	0.56	0.34
Beet pulp	1.36	1.07	0.68	0.12
32% protein supplement[a]	0.90	0.68	0.45	0.29
Total	15.92	13.55	8.76	1.78

[a]32% protein supplement contains 0.75 Mcal/lb (NEm) and 0.5 Mcal/lb (NEg).

The requirements for this steer (Tables 14.9 and 14.16) are 19.1 lbs dry matter, 2.2 lbs crude protein, 5.85 Mcal, NEm, and 6.35 Mcal, NEg. Table 14.14 shows pounds of feeds being fed and their conversion from an "as-fed" to a "dry-matter" basis. The conversion is calculated by multiplying the pounds "as fed" by its dry-matter content (Table 14.3).

Table 14.15 shows calculations for determining the megacalories for maintenance and gain and amount of protein in the ration (e.g., for corn silage, NEm $= 3.06 \times 0.75 = 2.29$; NEg $= 3.06 \times 0.47 = 1.13$; and CP $= 3.06 \times 0.08 = 0.24$). Values are obtained from Table 14.3.

The questions posed in example 1 are answered in the following steps:

Step 1: Determine NEm and NEg per pound of feed:

$$\text{From Table 14.15, the NEm per lb ration} = \frac{13.55}{15.92} = 0.85$$

$$\text{From Table 14.15, the NEg per lb ration} = \frac{8.76}{15.92} = 0.55$$

Step 2: Pounds of feed needed to meet the maintenance requirement of the 700-lb steer:

$$5.85 \text{ Mcal} \div 0.85 = 6.88 \text{ lbs}$$

Step 3: Pounds of feed needed to meet the requirement for 3.0 lbs daily gain:

$$6.35 \text{ Mcal} \div 0.55 = 11.54 \text{ lbs}$$

Step 4: Total lbs of feed the steer must eat to gain 3.0 lbs per day:

$$6.88 \text{ lbs} + 11.54 \text{ lbs} = 18.4 \text{ lbs}$$

Conclusion: If the steer has a capacity to eat 15.9 lbs of dry matter of the ration fed, then a 3.0 lbs per day gain will not be realized. The ration could be made more dense in energy by substituting corn grain for corn silage on a lb-for-lb basis until the desired gain is realized. The 1.73 lbs of protein supplied in the 15.9 lbs of dry matter fails to supply the 2.2 lbs protein requirement.

In example 2, we ask how the net energy system can be used to predict the amount of daily gain if the feed consumption is known. Using the content of example 1, what would be the expected gain if the 700-lb steer consumed 15.9 lbs of the ration in Table 14.14?

Step 1: Pounds of feed needed to meet daily maintenance requirement = 6.88 lbs (Step 2 in example 1).

Step 2: Pounds of feed remaining for gain:

$$15.9 \text{ lbs} - 6.88 = 9.02 \text{ lbs}$$

Step 3: Mcal of NEg supplied from the remaining amount of feed:

$$9.02 \text{ lbs} \times 0.55 \text{ (NEg per lb in ration)} = 4.96 \text{ Mcal}$$

Step 4: Daily gain expected from 4.96 Mcal of NEg for 700-lb steer (see Table 14.16):

4.96 Mcal produces approximately 2.4 lbs of gain

In example 3, let us suppose that the genetic ability of the 700-lb steer only allowed the daily consumption of 12.5 lbs of the ration. What would be the expected gain? The same steps as in example 2 would be followed.

Mcal of NEg for gain = 3.09 Mcal. Expected gain from 3.42 Mcal (Table 14.16)
= approximately 1.55 lbs per day.

Most feedlot rations are balanced for vitamin A, calcium, and phosphorus. Potassium becomes important as the level of concentrate increases or when nonprotein nitrogen is substituted for intact protein. Sulfur also becomes important when the level of nonprotein nitrogen increases in the ration. Most purchased protein supplements have trace minerals added. It is important to balance trace minerals for optimum performance.

Adjusting Rations for Weather Changes

The published nutrient requirements and predicted gains of cattle are the results of research studies in which the animals were protected from environmental extremes. However, environmental factors, especially temperature, can alter both performance and nutrient requirements. Thus cattle producers should be aware of critical temperatures that affect cattle performance, then consider changing their feeding program if economics so dictate.

The *thermoneutral zone (TNZ)* is the range in effective temperature where the rate and efficiency of performance is maximized. *Critical temperature* is the lower limit of the TNZ and is typified by the ambient temperature below which the performance of cattle begins to decline as temperatures become colder. Figure 14.6 shows that the maintenance energy requirement increases more rapidly during cold weather than does rate of feed intake. This results in a reduction of gain and more feed required per pound of gain, which typically causes cost per pound of gain to be higher.

Effective ambient temperatures below the lower critical temperature (below the TNZ) constitute cold stress, and those above the TNZ constitute heat stress. The term *effective ambient temperature* is an index of the heating or cooling power of the environment in terms of dry bulb temperature. It includes any environmental factor—such as radiation, wind, humidity, or precipitation—that alters environmental heat demand. Not all effective ambient temperatures have been calculated for cattle, although some combined environmental factors are available. The windchill index, for example, is shown in Table 14.17. An example in

TABLE 14.16 Net Energy Requirements for Growing-Finishing Cattle

Body Wt, lbs NEm Required	300 3.10 NEg		400 3.85 NEg		500 4.55 NEg		600 5.21 NEg	
Daily Gain (lbs)	**Steers**	**Heifers**	**Steers**	**Heifers**	**Steers**	**Heifers**	**Steers**	**Heifers**
				(Mcal/day)				
0.5	0.47	0.52	0.59	0.64	0.72	0.78	0.83	0.90
0.6	0.59	0.64	0.73	0.80	0.87	0.95	1.00	1.09
0.7	0.68	0.74	0.84	0.92	1.02	1.12	1.17	1.28
0.8	0.79	0.87	0.98	1.08	1.17	1.29	1.34	1.48
0.9	0.88	0.97	1.10	1.21	1.33	1.46	1.52	1.68
1.0	1.00	1.11	1.24	1.38	1.48	1.64	1.70	1.88
1.1	1.12	1.24	1.39	1.54	1.64	1.82	1.88	2.09
1.2	1.21	1.35	1.51	1.68	1.80	2.00	2.06	2.30
1.3	1.33	1.49	1.66	1.86	1.96	2.19	2.24	2.51
1.4	1.43	1.61	1.78	2.00	2.12	2.38	2.43	2.73
1.5	1.55	1.75	1.93	2.18	2.28	2.57	2.62	2.95
1.6	1.65	1.87	2.06	2.32	2.45	2.77	2.81	3.17
1.7	1.78	2.02	2.21	2.51	2.61	2.97	3.00	3.40
1.8	1.88	2.14	2.34	2.66	2.78	3.17	3.19	3.63
1.9	2.01	2.29	2.50	2.85	2.95	3.37	3.39	3.87
2.0	2.14	2.42	2.63	3.01	3.12	3.58	3.58	4.11
2.1	2.24	2.57	2.79	3.20	3.30	3.79	3.78	4.35
2.2	2.37	2.74	2.95	3.40	3.47	4.01	3.98	4.59
2.3	2.48	2.87	3.08	3.57	3.65	4.22	4.18	4.84
2.4	2.61	3.03	3.25	3.77	3.83	4.44	4.39	5.09
2.5	2.72	3.17	3.38	3.93	4.01	4.66	4.59	5.35
2.6	2.86	3.33	3.55	4.15	4.19	4.89	4.80	5.61
2.7	2.97	3.47	3.69	4.32	4.37	5.12	5.01	5.87
2.8	3.11	3.65	3.86	4.54	4.56	5.35	5.22	6.14
2.9	3.22	3.79	4.00	4.71	4.74	5.59	5.44	6.40
3.0	3.36	3.97	4.18	4.94	4.93	5.82	5.65	6.68
3.1	3.47	4.11	4.32	5.12	5.12	6.06	5.87	6.95
3.2	3.62	4.30	4.50	5.34	5.31	6.31	6.09	7.23
3.3	3.76	4.48	4.68	5.51	5.51	6.56	6.31	7.57
3.4	3.88	4.63	4.83	5.76	5.70	6.81	6.54	7.80

(continued)

TABLE 14.16 (Continued)

Body Wt, lbs NEm Required	700 5.85 NEg		800 6.47 NEg		900 7.06 NEg		1,000 7.65 NEg	
Daily Gain (lbs)	**Steers**	**Heifers**	**Steers**	**Heifers**	**Steers**	**Heifers**	**Steers**	**Heifers**
				(Mcal/day)				
0.5	0.93	1.01	1.02	1.11	1.12	1.22	1.21	1.32
0.6	1.12	1.22	1.24	1.35	1.35	1.47	1.46	1.59
0.7	1.31	1.44	1.45	1.59	1.58	1.73	1.71	1.88
0.8	1.51	1.66	1.67	1.83	1.82	2.00	1.97	2.17
0.9	1.71	1.88	1.89	2.08	2.06	2.27	2.23	2.46
1.0	1.91	2.11	2.11	2.33	2.30	2.55	2.49	2.76
1.1	2.11	2.34	2.33	2.59	2.55	2.83	2.76	3.06
1.2	2.31	2.58	2.56	2.85	2.79	3.11	3.02	3.37
1.3	2.52	2.82	2.78	3.12	3.04	3.40	3.29	3.68
1.4	2.73	3.06	3.01	3.39	3.29	3.70	3.56	4.00
1.5	2.94	3.31	3.25	3.66	3.55	4.00	3.84	4.33
1.6	3.15	3.56	3.48	3.94	3.80	4.30	4.12	4.66
1.7	3.37	3.82	3.72	4.22	4.06	4.61	4.40	4.99
1.8	3.58	4.08	3.96	4.51	4.33	4.92	4.68	5.33
1.9	3.80	4.34	4.20	4.80	4.59	5.24	4.97	5.67
2.0	4.02	4.61	4.45	5.09	4.86	5.57	5.26	6.02
2.1	4.25	4.88	4.69	5.39	5.13	5.89	5.55	6.38
2.2	4.47	5.16	4.94	5.70	5.40	6.23	5.84	6.74
2.3	4.70	5.44	5.19	6.01	5.67	6.56	6.14	7.10
2.4	4.93	5.72	5.45	6.32	5.59	6.90	6.44	7.47
2.5	5.16	6.01	5.70	6.64	6.23	7.25	6.74	7.85
2.6	5.39	6.30	5.96	6.96	6.51	7.60	7.05	8.23
2.7	5.63	6.59	6.22	7.28	6.80	7.96	7.35	8.61
2.8	5.87	6.89	6.48	7.61	7.08	8.32	7.67	9.00
2.9	6.11	7.19	6.75	7.95	7.37	8.68	7.98	9.40
3.0	6.35	7.50	7.02	8.29	7.67	9.05	8.30	9.80
3.1	6.59	7.81	7.29	8.63	7.96	9.43	8.61	10.20
3.2	6.84	8.12	7.56	8.98	8.26	9.81	8.94	10.61
3.3	7.09	8.44	7.83	9.33	8.56	10.19	9.26	11.03
3.4	7.34	8.76	8.11	9.68	8.86	10.58	9.59	11.45

Source: Nutrient Requirements of Beef Cattle. 1984. Washington, DC: National Research Council.

the table shows that with a temperature of 20°F and a wind speed of 30 mph, the effective ambient temperature is −16°F.

The low critical temperature for cattle depends on how much insulation is provided by the hair coat, whether the animal is wet or dry, and how much feed the cow consumes. Table 14.18 shows some low critical temperatures for beef cattle. For example, a cow being fed a maintenance ration may have a low critical temperature of 32°F when dry but a low critical temperature of 60°F when wet.

The coldness of a specific environment is the value that must be considered when adjusting rations for cows. *Coldness* is simply the difference between effective temperature (windchill) and low critical temperature. Using this definition for coldness instead of using the temperature reading on an ordinary thermometer helps explain why the wet, windy days in March may be colder for a cow than the extremely cold but dry, calm days of January.

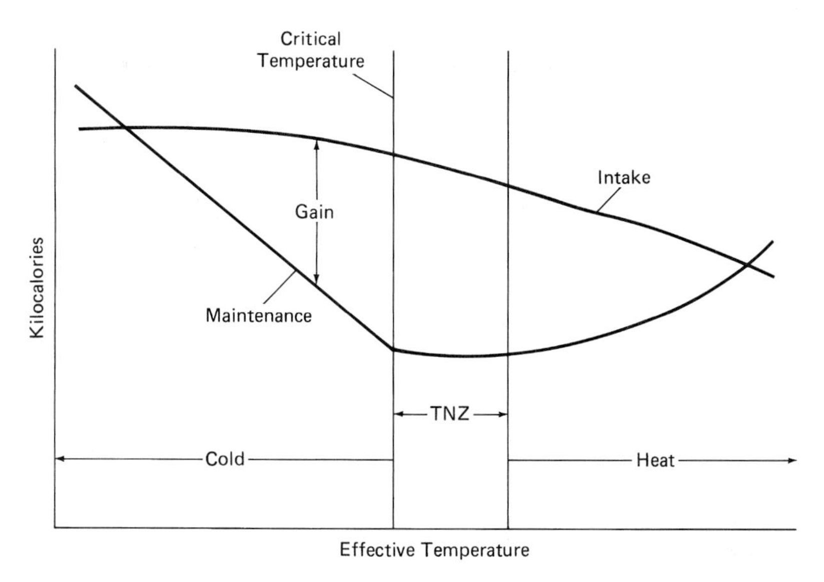

FIGURE 14.6 Effect of temperature on rate of feed intake, maintenance energy requirement, and gain. *Source:* Ames (1980).

TABLE 14.17 Windchill Factors for Cattle with a Winter Coat

Wind Speed (mph)	Temperature (°F) −10	−5	0	5	10	15	20	25	30	35	40	45	50
Calm	−10	−5	0	5	10	15	20	25	30	35	40	45	50
5	−16	−11	−6	−1	3	8	13	18	23	28	33	38	43
10	−21	−16	−11	−6	−1	3	8	13	18	23	28	33	38
15	−25	−20	−15	−10	−5	0	4	9	14	19	24	29	34
20	−30	−25	−20	−15	−10	−5	0	4	9	14	19	24	29
25	−37	−32	−27	−22	−17	−12	−7	−2	2	7	12	17	22
30	−46	−41	−36	−31	−26	−21	−16	−11	−6	−1	3	8	13
35	−60	−55	−50	−45	−40	−35	−30	−25	−20	−15	−10	−5	0
40	−78	−73	−68	−63	−58	−53	−48	−43	−38	−33	−28	−23	−18

TABLE 14.18 Estimated Low Critical Temperatures for Beef Cattle

Coat Description	Critical Temperature
Summer coat or wet	59°F
Fall coat	45°F
Winter coat	32°F
Heavy winter coat	18°F

The major effect of cold on the nutrient requirements of cows is increased need for energy, which usually means the total amount of daily feed must be increased. Feeding tables recommend that a 1,200-lb cow receive 16.5 lbs of good mixed hay to supply energy needs during the last one-third of pregnancy. How much feed should the cow receive if she is dry and has a winter hair coat but the temperature is 20°F with a 15 mph wind? The coldness is calculated

TABLE 14.19 Maintenance Requirement Multipliers for Various Climatic Conditions

	30°F		10°F		−10°F	
	Dry	Wet	Dry	Wet	Dry	Wet
	Beef cow wintering ration (0.60 Mcal NEm/lb DM)					
	Wind at 1.0 mph					
Dairy/*Bos indicus*	1.19	1.19	1.29	1.68	1.58	2.07
Continental	1.19	1.19	1.29	1.55	1.41	1.92
British	1.19	1.19	1.29	1.45	1.39	1.79
	Wind at 10.0 mph					
Dairy/*Bos indicus*	1.22	1.48	1.60	1.94	1.98	2.39
Continental	1.19	1.41	1.47	1.84	1.82	2.27
British	1.19	1.34	1.36	1.75	1.69	2.17
	Calf wintering ration (0.35 Mcal NEg/lb DM)					
	Wind at 1.0 mph					
Dairy/*Bos indicus*	1.19	1.47	1.50	1.93	1.87	2.39
Continental	1.19	1.37	1.36	1.80	1.69	2.23
British	1.19	1.28	1.29	1.69	1.55	2.09
	Wind at 10.0 mph					
Dairy/*Bos indicus*	1.41	1.69	1.85	2.20	2.29	2.72
Continental	1.30	1.61	1.71	2.10	2.12	2.59
British	1.21	1.54	1.60	2.01	1.98	2.48
	Finishing ration (0.62 Mcal NEg/lb DM)					
	Wind at 1.0 mph					
Dairy/*Bos indicus*	1.19	1.19	1.33	1.76	1.69	2.21
Continental	1.19	1.19	1.29	1.63	1.51	2.05
British	1.19	1.19	1.29	1.51	1.39	1.92
	Wind at 10.0 mph					
Dairy/*Bos indicus*	1.24	1.52	1.67	2.03	2.11	2.54
Continental	1.19	1.44	1.54	1.93	1.95	2.42
British	1.19	1.36	1.42	1.83	1.81	2.31

Source: NRC (1996).

by subtracting the windchill or effective temperature (4°F) from the cow's low critical temperature (32°F). Thus, the magnitude of coldness is 28°F. A rule of thumb (more detailed tables are available) is to increase the amount of feed 1% for each degree of coldness. A 28% increase of the 16.5-lbs (original) requirement would mean that 21.1 lbs of feed must be fed to compensate for the coldness. This example is typical of many feeding situations; however, if the cow was wet, the same increase in feed would be required at 31°F windchill (28°F of coldness). Similar relationships exist for growing–finishing cattle. Table 14.19 provides maintenance requirement multipliers for various types of cattle in wet/dry conditions under several temperature and wind scenarios. Thin-hided cattle have higher maintenance requirements across all conditions as compared with thicker-hided cattle. Maintenance requirements increase with wet conditions, increased wind, and falling temperatures.

COW-CALF NUTRITION

Following is a brief summary of the recommendations for feeding commercial cows. The basic goal is to produce the most pounds of calf per unit of land available at the least cost over a period of several years. Each operation is unique so recommendations should be adjusted to meet the needs of an operation.

Energy

Most pregnant cows gain 100–150 lbs during late summer through the winter months. This weight gain represents primarily the weight of calf, fluids, and membranes. More weight gain or adjusting the biological type of cows may be necessary if cows enter the winter period in poor body condition. During the time between weaning calves and the winter period, cows usually need only a salt-phosphorus supplement if they have sufficient feed to maintain their body weight. During the last two months of pregnancy, however, cows should gain 1–2 lbs per day unless they are in an above-average body condition. Most cows gaining less than 1 lb per day will not provide adequate nutrition for the developing calf. Some producers attempt to reduce the birth weight of the calf and subsequent calving problems by reducing feed intake during late pregnancy. This practice usually affects calf birth weight only by 2–4 lbs, but more importantly, these cows will produce less colostrum and will have a longer postpartum interval. Calves that receive an inadequate level of colostrum are more susceptible to disease. One of the most important nutritional management practices is to see that calves receive adequate quantities of colostrum within a few hours after birth. Cows in poor condition (< BCS 5) are less able to produce an adequate level of immunoglobulins in their colostrum and thus their calves are more susceptible to disease.

The first 90 days after calving is the most important nutritional period for the brood cow. Most calves are born during the late winter through the spring months, when most pastures are at their lowest point nutritionally. Supplemental feed is provided at this time so cows can produce an adequate supply of milk and still gain body weight. Cows that do not gain weight prior to breeding and during the breeding season are most likely to have poor reproductive performance. Producers may want to weigh a sample of their cows during this period to assess that weight gains are occurring and adjust the amount of feed being provided if necessary.

A logical management practice involves sorting cows into groups prior to calving through part of the breeding season and providing levels of feed consistent with cow age and body condition. First and second calf females and thin cows need to be separated from other cows to assure the feeding levels needed for high reproductive performance. Feed that has been evaluated to be of higher quality should be targeted to younger females and thin cows. Lower-quality feeds are then fed to nonlactating, midgestation older females in adequate to good condition.

Cow size should be adjusted to fit existing environmental conditions. Moderate-sized and moderate-milking cows are highly profitable in most cow-calf operations. Cow size and milk production may be less in arid and semiarid regions.

Summer forage, water, and salt will provide all of the cow's nutrient requirements if the nutrient sources are available in ample supply. Winter pasture will provide the cow's energy needs if it is present in adequate amounts and is not covered with snow. When adequate amounts of feed are not present, the cheapest source of supplemental energy should be provided. Some common sources include hay, grain, and silage. Supplementing a high-protein concentrate when energy is needed is not an economically sound practice. In this situation, protein is converted to energy.

Protein

Figure 14.7 illustrates that during spring and early summer, protein is adequate in most range and pasture grasses. The pregnant cow needs 6.5% protein in her ration; 9% is needed for a lactating cow and 11% for a growing calf or yearling. During the winter period, protein in mature grass frequently decreases to 4–5%.

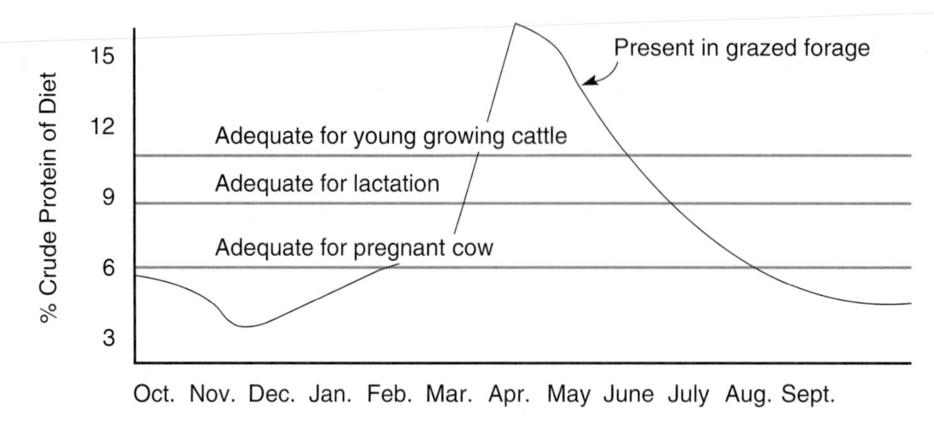

FIGURE 14.7 Crude protein content in a grazing cow's diet.
Source: Colorado State University.

TABLE 14.20 Recommended Levels of Crude Protein for Pregnant and Lactating Cows and Heifers

	Crude Protein in Supplement					
Item	50%	40%	30%	20%	15%	10%
Supplement fed to cows during last 2 months of pregnancy (lbs/cow/day)	0.6	1.0	1.5	2	3.5	7
Supplement fed during lactation and to pregnant heifers (lbs/cow/day)	1.3	1.8	2.3	4	11	20

Protein is usually more than adequate when feeding alfalfa hay. When grass hay is fed and questions arise about meeting the cow's requirements, a feed analysis should be made. Corn stalks, straw, and other aftermath feeds provide a cheap source of energy, but as a rule they need to be supplemented with protein, especially during the latter part of the grazing period. Underfeeding protein is common and results in decreased performance. Overfeeding protein will not harm animals, but the practice is extremely uneconomical. When a protein supplement is necessary, the cheapest source available should be fed. Often, cottonseed cake, alfalfa hay, or a feed grain is usually a more economical choice than purchasing a commercially prepared supplement. Supplements are available, however, in loose, block, liquid, and range cube forms. Table 14.20 gives a general idea of how much protein supplement to feed. Research at Kansas State University has demonstrated that supplementation to beef cows in adequate-to-good body condition can be delivered every 5 to 7 days without affecting weight and condition.

Many cow-calf producers and yearling operators make a mistake in feeding nonprotein nitrogen (NPN) as the supplemental source of protein when cattle are on an all-forage ration. NPN in the form of urea, nucleic acids, or biuret must be fed along with a readily available source of carbohydrates to be adequately utilized. This readily available carbohydrate source is not provided in adequate amounts with range forage alone. If adequate energy is not available, cattle show little benefit from feeding NPN and the money spent on NPN is not cost effective. Regardless of the source of NPN, it is of questionable benefit in meeting the beef cow's protein needs while grazing dry forage.

TABLE 14.21 Limit-Fed Corn-Based Diets to Meet the Nutrient Requirements of Mature Beef Cows

	Dry, Pregnant, Mature Cows, Middle 1/3 of Pregnancy			Dry, Pregnant, Mature Cows, Last 1/3 of Pregnancy			Mature Cows, Nursing Calves		
Cow Weight (lbs)	Hay	Corn	Soybean Meal	Hay	Corn	Soybean Meal	Hay	Corn	Soybean Meal
1,100	6.3	5.3	.4	6.3	7.6	.6	6.3	11.3	2.2
1,200	6.8	5.9	.2	6.8	7.9	.6	6.8	11.6	2.2
1,300	7.4	6.2	.2	7.4	8.2	.6	7.4	11.9	2.2
1,400	8.0	6.4	.2	8.0	8.5	.6	8.0	12.2	2.2

Source: Buskirk (1999).

Limit-Feeding Corn

When high levels of corn production hold prices at competitive levels compared with hay or other feedstuffs, feeding grain to cows may be a profitable option. Whole shelled corn can be limit-fed to mature cows and bred heifers conveniently and cheaply. Corn should never be totally substituted for forage as this will lead to a variety of digestive problems. As a rule of thumb, a minimum of 0.5% of body weight of hay should be fed. For example, a cow weighing 1,200 lbs would receive 6 lbs per day of hay on a dry-matter basis.

A feedstuffs analysis will determine how much protein and mineral supplementation will be required. Cows will need to be transitioned to the limit-fed concentrate over a period of 7–10 days by decreasing roughage and increasing concentrate incrementally over that time period.

To avoid problems with dominant cows preventing adequate consumption by other females, each cow will need 2 to 2.5 ft of bunk space or feeding area. Feeding on the ground is an alternative, but this method leads to increased wastage.

A suggested ration for mature cows at maintenance and free from environmental stress is provided in Table 14.21.

Minerals

Salt, in either a loose or block form, should be made available free choice to cattle at all times. The cheapest source should be utilized. Calcium, as shown in Figure 14.8, is present in adequate amounts throughout the year in range and pasture forages. Supplementation is unnecessary.

Figure 14.8 also illustrates that phosphorus is adequate during summer months but deficient for more than half the year. Inadequate phosphorus can cause reduced growth rates, decreased milk production, and delayed conception. Research shows varied responses to phosphorus supplementation; however, it is recommended as a risk management practice.

Phosphorus can be supplemented by mixing a phosphorus source with free-choice salt. The most common supplement is one or two parts salt to one part phosphorus source. Some sources of phosphorus include dicalcium phosphate and steamed bonemeal. Producers usually find that mixing their own salt-phosphorus supplement and providing it free choice is significantly cheaper than using a commercially prepared supplement.

On some pastures grass tetany is a problem (Table 14.7). This condition is most commonly observed early in the spring on fresh, green, rapidly growing grasses. Affected animals are

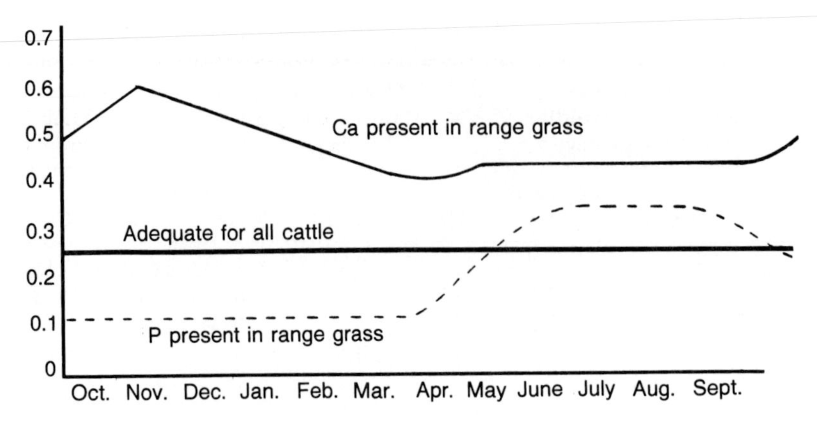

FIGURE 14.8 Calcium and phosphorous in a grazing cow's diet.
Source: Colorado State University.

often seen staggering. The best prevention is to mix magnesium oxide (MgO) in the salt supplement. This should be started several weeks ahead of the potential problem time and continued throughout the grazing period. If palatability is a problem, MgO can be mixed with soybean meal or grain. Usually no benefit is seen from supplementation of Mg when tetany is not a problem.

Except for a few isolated local deficiencies, all other essential minerals in the range cow diet are present in adequate amounts in the forage. Individual herd and area problems should be considered. As a general rule, cafeteria-style mineral feeding or the feeding of complex mineral mixtures is not necessary or beneficial.

Vitamins

Generally, if the forage being consumed contains green color, it may supply sufficient carotene to meet vitamin A requirements. In forages, beta carotene, which is the precursor of vitamin A, is present. Vitamin A is stored in the liver primarily during the spring and summer months, and this storage usually supplies the animals' needs during the winter. Vitamin A deficiency in cows occurs infrequently. Some pastures during periods of extended drought, however, may not supply adequate amounts of vitamin A. In these cases, 3–6 months hepatic supply is sufficient to meet the cow's requirements until the last few months before calving, at which time the cow may become deficient. She needs it primarily at this time to produce colostrum for the calf and for her own needs. Often a cow has ample vitamin A available, but the calf is born deficient. Colostrum must provide the vitamin A to make up the deficiency. Deficiencies make calves more susceptible to disease, and the deficient cow is more susceptible to retained placentas and infections. The calf's deficiency is supplied only if the colostrum contains adequate vitamin A and the calf receives an adequate amount of colostrum. Most veterinarians and nutritionists recommend an intramuscular injection of vitamin A for the calf at birth, considering this an inexpensive insurance policy. If producers are concerned that their brood cows are deficient, they may have them tested. Current recommendations suggest that blood from five to seven randomly selected cows be analyzed for plasma vitamin A levels. The cost of blood analysis should be evaluated as it may be more economical to inject all cows with vitamin A. In some cases, vitamin A can be fed in a supplement form.

In the cow, B vitamins are synthesized by rumen microorganisms. Other vitamins are usually found in more than adequate amounts in the cow's diet. There are several commer-

TABLE 14.22 Maximum Allowable Contents of Minerals and Solids

Substance	Maximum Level (parts per mil)
Aluminum	5
Arsenic	0.2
Cadmium	0.05
Chromium	1
Cobalt	1
Copper	0.5
Fluoride	2
Iron	5
Lead	.05
Mercury	.01
Nitrite	10
Selenium	.05–.10
Vanadium	0.1
Zinc	24
Nitrate + nitrite	100
Total dissolved solids	10,000
Magnesium + sodium sulfates	5,000
Alkalinity	2,000

Source: USDA.

cial products available for supplementation of these vitamins. Their use is questionable from both nutritional and economical standpoints.

Water

Clean, fresh water should be made available free choice to cattle at all times. If water is of questionable quality, it should be tested at a local laboratory. Suggested water content standards are listed in Table 14.22. Often when cattle are brought into drylot, they do not recognize water in automatic waterers. By running the water fast enough to make noise the first few days, cattle become more easily adapted to water consumption under drylot conditions.

The approximate daily water intake for beef cattle is shown in Table 14.23. Water intake of a given class of cattle, under a specific management program, is primarily a function of dry-matter intake and ambient temperature. Cattle prefer to drink water that is near the normal temperature of the rumen (100–102°F). Intake of water will nearly double as the air temperature moves from 50° to 90°. It is not necessary to cool water; in fact, doing so may reduce consumption. However, during cold weather, warming water will increase intake.

Creep Feeding Calves

Creep feeding allows calves to have supplemental feed (usually concentrates). Feed is provided in a creep feeder or some other physical barrier is provided to prevent cows from having access to the supplemental feed. Creep rations can be very simple and use of the cheapest concentrate is usually advisable. Addition of bulky feeds, such as oats, beet pulp, and dehydrated alfalfa pellets, will prevent digestive problems. A ration such as 50% corn and 50% oats or 50% barley and 50% oats is satisfactory; there is no advantage to feeding a complex creep ration. The creep feeder should be located near the area where cows and calves spend a considerable amount of time, such as near water or shade. The advantages and disadvantages of creep feeding are pointed out in Table 14.24.

TABLE 14.23 Approximate Total Daily Water Intake of Beef Cattle

Live Weight (lbs)	Daily Water Consumption at Various Temperatures[a] 40°F (gal)	50°F (gal)	60°F (gal)	70°F (gal)	80°F (gal)	90°F (gal)
	Growing Heifers, Steers, and Bulls					
400	4.0	4.3	5.0	5.8	6.7	9.5
600	5.3	5.8	6.6	7.8	8.9	12.7
	Finishing Cattle					
800	7.3	7.9	9.1	10.7	12.3	17.4
1,000	8.7	9.4	10.8	12.6	14.5	20.6
	Wintering Pregnant Cows					
1,100	6.0	6.5	7.4	8.7	—	—
	Lactating Cows					
900+	11.4	12.6	14.5	16.9	17.9	16.2
	Mature Bulls					
1,600+	8.7	9.4	10.8	12.6	14.5	20.6

[a]Water intake of a given class of cattle under a specific management program is primarily a function of dry-matter intake and ambient temperature. Water intake is quite constant up to 40°F.
Source: Nutrient Requirements of Beef Cattle. 1984. Washington, DC: National Research Council.

TABLE 14.24 Creep Feeding Considerations

Advantages	Disadvantages
1. Added weaning weight 2. When price of calves is higher relative to feed costs 3. Useful for calves with skeletal frame and growth potential, particularly if they will go into feedlot shortly after weaning 4. Purebred breeder can better measure growth potential (yearling weights) of bull calves. Added bloom and weight give an advantage to the purebred breeder's merchandising program 5. Less stress at weaning time because calves adjust to their postweaning ration sooner. Producer may want to creep feed only 2–3 weeks before weaning the calves 6. Most useful during periods of drought or low feed supply to maintain body condition of cows for a good reproductive rate the following year	1. Cost of added weight is usually higher than the return 2. Not recommended for replacement heifers—as it usually impairs future milk production. Therefore, producer needs to separate male and female calves if creep feeding is practiced 3. Creep-fed yearlings weigh only slightly more than noncreep-fed yearlings 4. Creep feed may be converted very inefficiently. Some studies show that the feed conversion may be as high as 15 lbs of creep feed per lb of gain. This poor conversion usually occurs when calves are on high milking cows and have access to ample supplies of high-quality forage 5. Market price of calves may be less if they are too "fleshy" or fat at sale time

Developing Replacement Females

Heifers of 500–700 lbs need a 4% higher level of protein in their diet than do mature animals. The concentrate fed should contain enough protein, phosphorus, additional energy, and vitamin A to meet the requirements of the heifer. Heifers should reach 85% of their mature weight by calving time. Again, it should be stressed that "starving out" calving difficulty is not effective. After calving, heifers should be separated from cows and fed to meet their additional requirements and weight gain. Implants and growth stimulants should not be used on replacement heifers.

The following calculations should be applied to the target weights and days that fit each producer's replacement heifer program:

Example:

Expected mature cow weight	1,000 lbs
	×0.65
Weight desired at breeding	650 lbs
Present weaning weight	−450 lbs
Gain needed	200 lbs
Days until breeding season	180
Gain needed	1.1 lbs per day

These 450-lb medium-framed heifers must gain 1.1 lbs per day for most of them to reach puberty prior to the breeding season. To accomplish this, they must consume about 11 lbs of dry matter a day that is composed of 80–90% roughage and about 7.7 lbs of TDN. A ration with this much TDN will require some concentrate to be supplemented in addition to the forage. Some protein, phosphorus, and vitamin A may be necessary as well, depending on the forage being consumed (refer to Table 14.9). Additional considerations related to replacement heifer development can be found in Chapter 5.

YEARLING-STOCKER CATTLE NUTRITION

Yearling-stocker cattle are typically grown on high-roughage rations. Most of these cattle are between 5 months and 25 months of age and weigh from 300 lbs to 1,000 lbs. Yearling-stocker cattle can be raised on grazed forage or in confinement (e.g., in a feedlot); the latter situation is referred to as *backgrounded cattle.*

Rations for yearling-stocker cattle are formulated using the net energy system. See Chapter 6 for a more detailed discussion of nutritional programs.

FEEDING FEEDLOT CATTLE

There are numerous feeding programs utilized by cattle feeders. Individual feeding programs depend on (1) the age, weight, and sex of the cattle; (2) whether feeds are raised or purchased; (3) the relative costs of available feedstuffs; (4) whether one or more groups of cattle will be fed per year using the same facilities; and (5) the length of feeding time.

Simple feeding programs involve feeding silage and grain with an added protein-mineral-vitamin supplement or self-feeding a ration where both rations can remain relatively constant for the entire feeding period. Complex feeding programs combine the altering of concentrate-to-roughage ratios several times (3 to 5) with formulating least-cost (both ingredient and gain) rations. In the more complex feeding programs, feed intake is maximized by feeding a ration several times a day while carefully monitoring feedbunks and the digestive patterns of cattle.

Calves are usually grown on higher roughage rations before being fed high-grain finishing rations; otherwise, cattle finish at lighter weights than the preferred slaughter weights. Other large-framed calves, weighing 500–700 lbs at weaning and possessing a high genetic potential for gain, are usually phased into high-energy rations as quickly as possible. A similar feeding program would be used for 700–900 lbs, thin to moderately fleshed yearling cattle.

Concentrate-to-roughage ratios can vary from 100% roughage to 100% concentrate to any combination of those two extremes. More conventional rations range from 40% roughage–60%

concentrate to 10% roughage–90% concentrate. The most common sources of roughage are silage (essentially 50% grain and 50% forage, on a dry-matter basis), hay, and corn cobs, while the most common concentrates or grains are corn (in the Midwest and northern Great Plains), milo (in the southern Great Plains and Southwest), and barley (in the northwestern United States). The supplement mixture of protein, minerals, vitamins, and possibly additives, usually supplied at 1–2 lbs per day, is also considered part of the concentrates.

Selecting the concentrate-to-roughage ratio to be fed is largely based on economics, assuming the feeder has the flexibility of making these choices. Table 14.25 gives an example of how to evaluate different concentrate-to-roughage choices at various feed prices.

Properly mixing rations and delivering correct amounts of feed are among the most important concerns in feeding cattle. Where high daily rates of gain are economically feasible, a high level of feed intake in a well-balanced, energy-dense ration is essential. Whether large amounts of feed are consumed on a daily basis by each animal in the feedlot is largely determined by the appetite of the animal. Appetite is influenced by genetic and environmental factors. Genetic selection for gain appears to be primarily a selection preference for cattle that have large appetites. Previous level of nutrition, which accounts for compensatory gain, health of cattle, weather, kinds of feed, and condition of feed in the feedbunk all influence the appetite of cattle.

Individuals monitoring feedbunks and observing recent fecal material assess the appetite of cattle and the condition of feed presented to cattle. This type of assessment is critical to proper management of the daily feeding program in the feedlot.

TABLE 14.25 Returns per Head for Yearlings in One-Lot/Year and Continuous-Feeding Systems

	Corn:Alfalfa							
	20:80		40:60		60:40		90:10	
Feed Prices	One Lot/Yr	Continuous Feeding	One Lot/Yr	Continuous Feeding	One Lot/Yr	Continuous Feeding	One Lot/Yr	Continuous Feeding
Corn at $2.50/bu								
Hay $30/ton	$ 48	$116	$ 53	$158	$ 53	$192	$ 55	$241
$60/ton	−17	5	11	74	30	138	50	227
$90/ton	−82	−105	−29	−11	8	84	46	216
Corn at $3.00/bu								
Hay $30/ton	39	100	36	126	33	144	30	171
$60/ton	−26	−10	−4	42	10	89	25	157
$90/ton	−91	−121	−45	−44	−12	36	21	146
Corn at $3.50/bu								
Hay $30/ton	29	83	20	92	13	96	5	101
$60/ton	−36	−27	−20	8	−10	41	1	90
$90/ton	−101	−138	−60	−76	−33	−14	−4	76
Corn at $4.00/bu								
Hay $30/ton	19	68	4	59	−8	46	−19	34
$60/ton	−46	−42	−36	−25	−30	−7	−24	20
$90/ton	−111	−155	−76	−109	−53	−62	−29	6
Corn at $4.50/bu								
Hay $30/ton	10	−51	−12	25	−28	−2	−44	−36
$60/ton	−55	−60	−52	−59	−51	−58	−49	−50
$90/ton	−120	−170	−92	−143	−73	−110	−53	−62
Corn at $5.00/bu								
Hay $30/ton	0	34	−28	−8	−48	−51	−69	−106
$60/ton	−65	−77	−68	−92	−71	−105	−73	−119
$90/ton	−130	−187	−108	−177	−93	−160	−78	−131

One of the greatest challenges for a feedlot manager is to develop an effective receiving regime for newly arrived feeder cattle. These animals have frequently experienced the stresses of transportation and weaning. As such, they are susceptible to disease and poor performance if not handled appropriately. A general set of guidelines for a receiving ration would be 14–16% crude protein, no more than 50% concentrates (more for yearlings), plus appropriate vitamin and mineral supplementation. Adopting feeder cattle to waterers and feedbunks is also important. In the first three days on feed, newly arrived calves eat for an average of 115 min (range of 15 to 225 min) spread over eating frequencies of 11.5 feedings per day (range of 5 to 21). Daily drinking time per day was 7.7 min (range of 1–31 min) spread over 6 daily trips to the water source (range of 2 to 18) (Buhman et al., 2000). The high degree of variability in consumption behaviors points to the need for close supervision of newly arrived feeders.

Hot and humid weather conditions also create challenges for feedlot operators. Heat stress not only affects feedlot and carcass performance but, under severe conditions, a high level of mortality may occur. The key to managing heat stress is to assure highly available sources of fresh water. Sprinklers may be used to wet cattle at a rate of a 1- to 2-min shower every half hour. Sprinklers should have a nozzle for every 8 to 10 head of cattle and a flow rate of 2.5 gal/min/nozzle. It is also useful to shift the feeding schedule to coincide with a period of 2 to 4 hours following daily peak temperature. Cattle should not be handled or mixed under these weather conditions. If cattle must be handled, the best time would be from midnight to approximately 8:00 A.M. Shade is also a factor, but the cost of these structures must be balanced with the risk of prolonged, extreme heat conditions.

MANAGEMENT SYSTEMS HIGHLIGHTS

1. The primary competitive advantage of beef cattle is their ability to utilize low-quality feedstuffs while providing high-quality protein.
2. Nutrient composition and animal nutrient requirements must be understood to assure the development of cost-effective rations that contribute to optimal levels of animal performance.
3. Timing of harvest, stage of maturity, feed storage conditions, weather patterns, and feed processing are several factors that can significantly affect feed values.
4. Adjusting rations for climatic conditions, such as wind and precipitation conditions (dry versus wet), is critical to assuming optimal annual performance under variable conditions.

REFERENCES

Publications

Albin, R. C., and Thompson, G. B. *Cattle Feeding: A Guide to Management.* Amarillo, TX: Trafton Printing.

Ames, D. R. 1980. Livestock Nutrition in Cold Weather. *Animal Nutrition and Health* (Oct.).

Ames, D. R. 1981. Feeding Beef Cows in Winter. *Angus Journal* (Jan.).

Anderson, D. C. 1978. Use of Cereal Residues in Beef Cattle Production Systems. *J. Anim. Sci.* 46: 849.

Bopp, S. G. 2001. Feeling the Heat. *Bovine Veterinarian* (May/June).

Buhman, M. J., Perino, L. J., Galyean, M. L., and Swingle, R. S. 2000. Eating and Drinking Behaviors of Newly Received Feedlot Calves. *Prof. Animal Scientist* 16: 241.

Buskirk, D. 1999. Limit Feeding Corn to Cows. *Cattle-Call.* Michigan State University Extension Newsletter 4: 3.

Cheeke, P. R. 1991. *Applied Animal Nutrition.* New York: Macmillan.

Church, D. C. 1991. *Livestock Feeds and Feeding.* 3rd ed. Englewood Cliffs, NJ: Prentice-Hall.

Cow-Calf Management Guide and Cattle Producers' Library. Cooperative Extension Service in WA, OR, ID, MT, WY, UT, NV, and AZ.
Ensminger, M. E., Oldfield, J. E., and Heinemann, W. W. 1990. *Feeds and Nutrition.* Clovis, CA: Ensminger Publishing Company.
Farmer, S. G., Cochran, R. C., Simms, D. D., Klevesahl, E. A., and Wickersham, T. A. 2000. *Effects of Frequency of Supplementation on Performance of Beef Cows Grazing Winter Pasture.* Proceedings of the Kansas State University Cattlemen's Day Conference, Manhattan, KS.
Great Plains Beef Cattle Feeding Handbook. Cooperative Extension Service in CO, KS, MT, NE, NM, ND, OK, SD, TX, and WY.
Great Plains Beef Cow-Calf Handbook. Cooperative Extension Service in CO, KS, MT, NE, NM, ND, OK, SD, TX, and WY.
Guyer, P., and Henderson, P. *Estimating Corn and Sorghum Silage Value.* Great Plains Beef Cattle Handbook. GPE-2401.
Jackson, J. J., Greer, W. J., and Baker, J. K. 2000. *Animal Health.* Danville, IL: Interstate Publishers.
Jurgens, M. H. 1993. *Animal Feeding and Nutrition.* 7th ed. Dubuque, IA: Kendall-Hunt.
Light Cattle Management Seminar. 1982. Fort Collins, CO: Colorado State University Veterinary Medicine and Animal Science Extension Service.
Meiske, J. 1983. Maximizing Ration Returns. *Feedlot Management* (Feb.).
National Research Council. 1981. *Effect of Environment on Nutrient Requirements of Domestic Animals.* Washington, DC: National Academy Press.
National Research Council. 1984. *Nutrient Requirements of Beef Cattle.* Washington, DC: National Academy Press.
National Research Council. 1996. *Nutrient Requirements of Beef Cattle.* 7th ed. Washington, DC: National Academy Press.
Preston, R. L. 2001. Typical Composition of Feeds for Cattle. *BEEF* (Jan.).
Ward, J. K. 1978. Utilization of Corn and Grain Sorghum Residues in Beef Cow Forage Systems. *J. Anim. Sci.* 46: 831.

Visuals

"Basic Livestock Nutrition" 2001 (video, 20 min.). CEV, P.O. Box 65265, Lubbock, TX 79464-5265.
"Decision Making: Energy Utilization for Feedlot Cattle" (video). CEV, P.O. Box 65265, Lubbock, TX 79464-5265.
"Ruminant Digestive Systems" 2001 (video, 41 min.). CEV, P.O. Box 65265, Lubbock, TX 79464-5265.
"Supplementing Beef Cattle" (27 min.). National Cottonseed Products Association, Inc., P.O. Box 12023, Memphis, TN 38182-0023.

Computer Programs

Taurus, A Ration Program for Beef Cattle. University of California, Davis, CA.

Managing Forage Resources

Forage from range, pasture, hay, and crop residues is the primary basis for the existence of beef cattle and their efficient production in the United States and elsewhere. More than 80% of the feed consumed by cattle is forage, and most of it is grazed. Forage can be converted by cattle from an unusable product for human consumption to a highly preferred consumer product with minimal energy expended. Beef producers are interested in having large amounts of highly palatable, nutritious forage available for grazing on a year-round basis. They are also interested in systems of grazing that optimize pounds of beef produced without sacrificing the future productivity of the range and pastures in the future.

Range and pasture areas in the United States are envisioned somewhat differently. Typically, the *range* area is conceived as the arid and semiarid lands of the seventeen western states (further broken down into the 11 western states and the major six states in the Great Plains area). *Pasture* areas can be considered as the irrigated pastures of the West in addition to the grazed areas of the eastern United States.

Rangeland has been defined by the Society for Range Management as land on which the native vegetation (climax or natural potential) is predominantly grasses, grass-like plants, forbs, or shrubs suitable for grazing or browsing and present in sufficient quantity to justify grazing or browsing use. Rangelands include natural grasslands, savannahs, shrublands, most deserts, tundra, alpine communities, coastal marshes, and wet meadows. While this definition is inclusive for all grazed forage, the discussion in this chapter uses the former definition that separates range and pasture areas.

GRAZED FORAGE RESOURCES

The total land area in the United States is approximately 2.2 billion acres. Approximately 1 billion acres are available for grazing by livestock, primarily cattle (see Table 15.1). An additional 63 million acres is used for hay production. The available range, pasture, and forage resources in the United States and their utilization are very diverse, primarily because of the different

TABLE 15.1 Agricultural and Nonagricultural Uses of Land in the United States (Lower 48 States)

Major Land Uses	Acres (mil)	% of Total
Agricultural		
Cropland	455	24.0
Cultivated cropland	349	18.4
Cropland used for pasture	68	3.6
Cropland idled	39	2.0
Pastureland and range	578	30.5
Forestland grazed	140	7.4
Total land grazed	786	41.5
Total agricultural land	1,180	62.0
Nonagricultural		
Forestland not grazed	412	21.7
Urban, transportation, and other built-up areas	99	4.7
Other land (includes 9 mil acres of small water areas)	117	6.2
Recreation and wildlife areas	96	5.1
Total nonagricultural land	724	38.0
Total land area	1,894	100.0

Source: USDA.

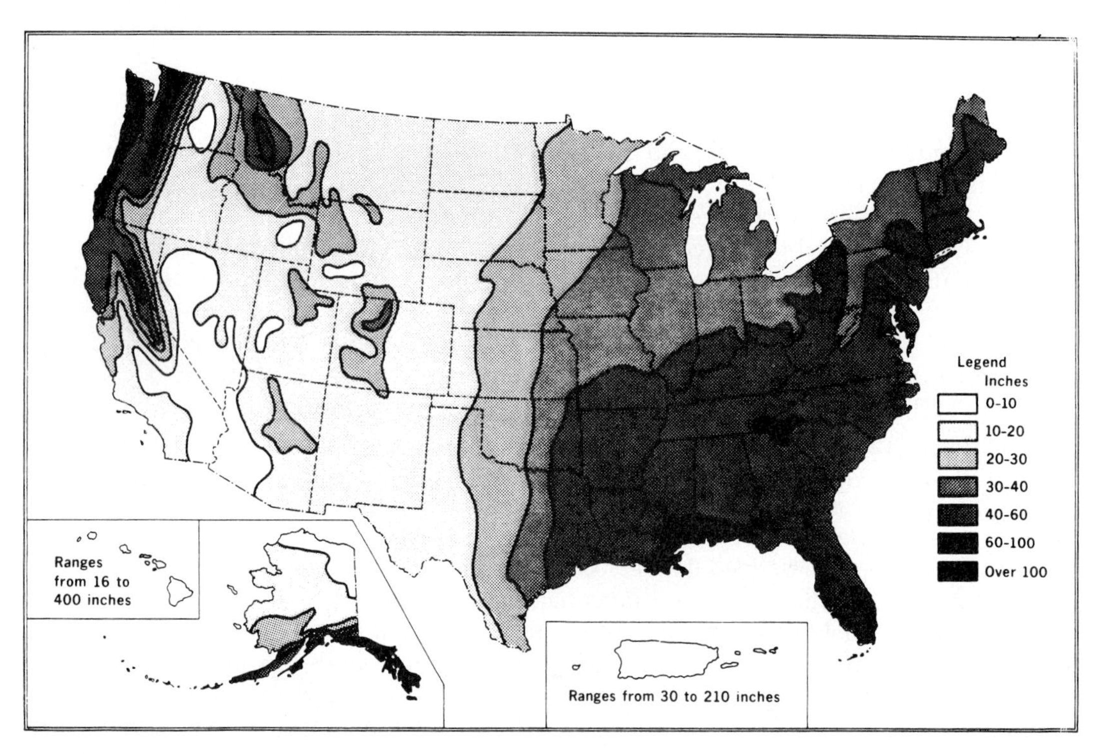

FIGURE 15.1 Average annual precipitation in the United States.
Source: U.S. Water Resources Council.

climatic and soil conditions. Beef cattle derive their nutrients from a ration composed of 83% forage and 17% concentrate. Stocker cattle and cows exist on a ration that is nearly 96% forage. During the relatively short feedlot phase (100–145 days), the ration is approximately 5–15% forage.

Annual precipitation, sometimes associated with other climatic influences, can have marked effects on forage supply (Fig. 15.1). In the West, high precipitation occurs in the

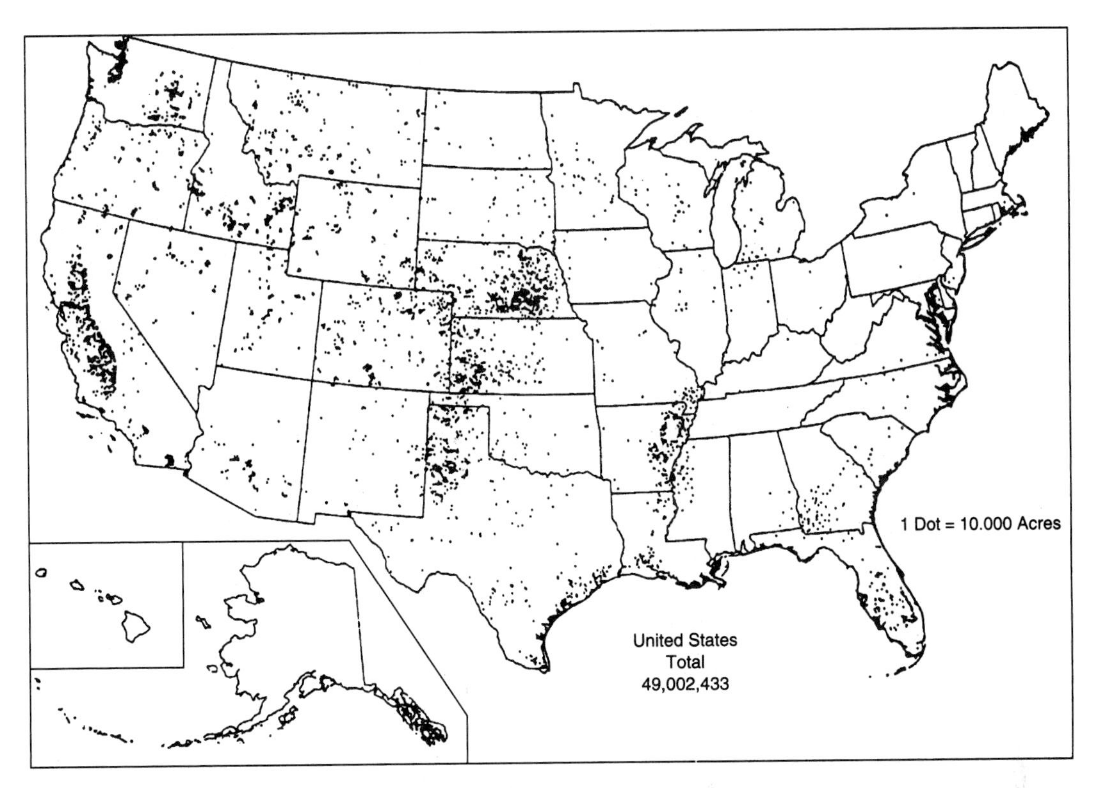

FIGURE 15.2 Irrigated land in farms. Each dot represents 10,000 acres.
Source: U.S. Department of Commerce.

mountains, where much of it accumulates as snow pack in the winter. The snow provides needed irrigation water during the spring and summer runoff. In addition to the snow pack runoff, irrigation water is pumped from deep wells. The irrigated land shown in Figure 15.2 provides pasture, hay, and crop residues for cattle.

Drought, in conjunction with overgrazing, can reduce the amount of desirable plant species, which in turn affects the future forage supply even in years of adequate precipitation. Extended periods of drought (more than a year) usually result in herd reductions, shipping in high-priced feed, or moving cattle to an area with a more abundant feed supply. Figure 15.3 illustrates the variation in hay production and utilization. Harvested forage supplies and prices are affected by variation in weather conditions. The major drawback of harvesting forage via mechanical means is cost. It has been estimated that between 50 and 70% of the annual cow cost for intermountain ranchers is tied to the haying enterprise. Another drawback of becoming overly reliant on harvested forages is that producers become more dependent on fossil fuels. The use of fossil fuel powered mechanization to harvest forage is not as cost-effective over time as the utilization of the grazing ability of ruminant animals.

PLANT TYPES AND THEIR DISTRIBUTION

The general classifications of plants used primarily for grazed forage are *grasses*, *forbs*, and *browse*. Some grazing areas contain all of these types of plants; more commonly, however, there are mixed species of one or two classes present. Since hundreds of different species for each plant class exist, only some of the more important species in the cow-calf and yearling production areas (Fig. 15.4) are discussed here.

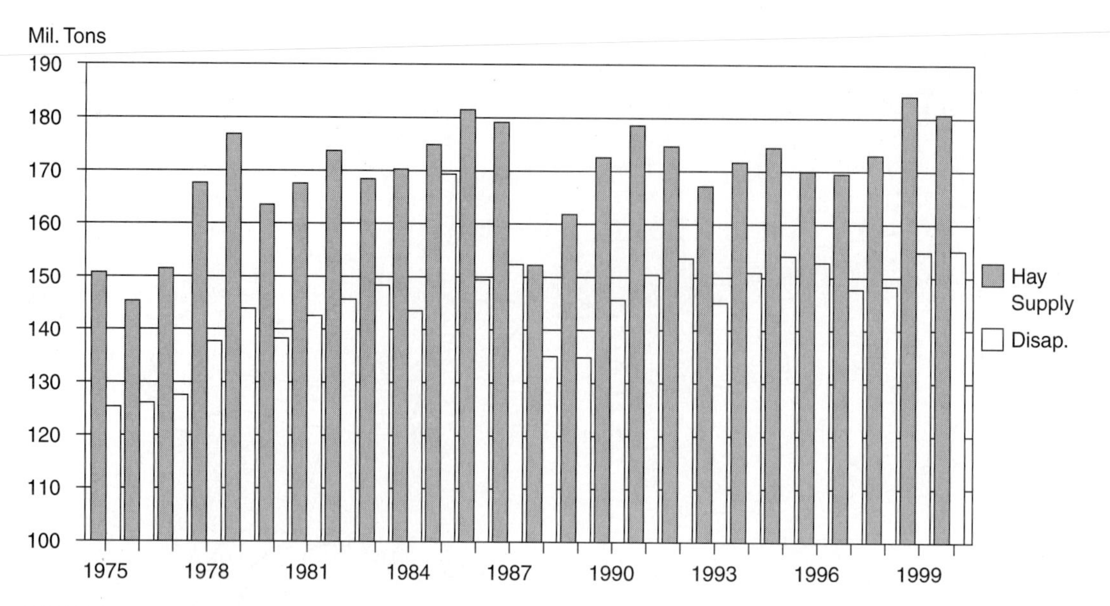

FIGURE 15.3 U.S. hay supply and disappearance.
Source: Livestock Marketing Information Center.

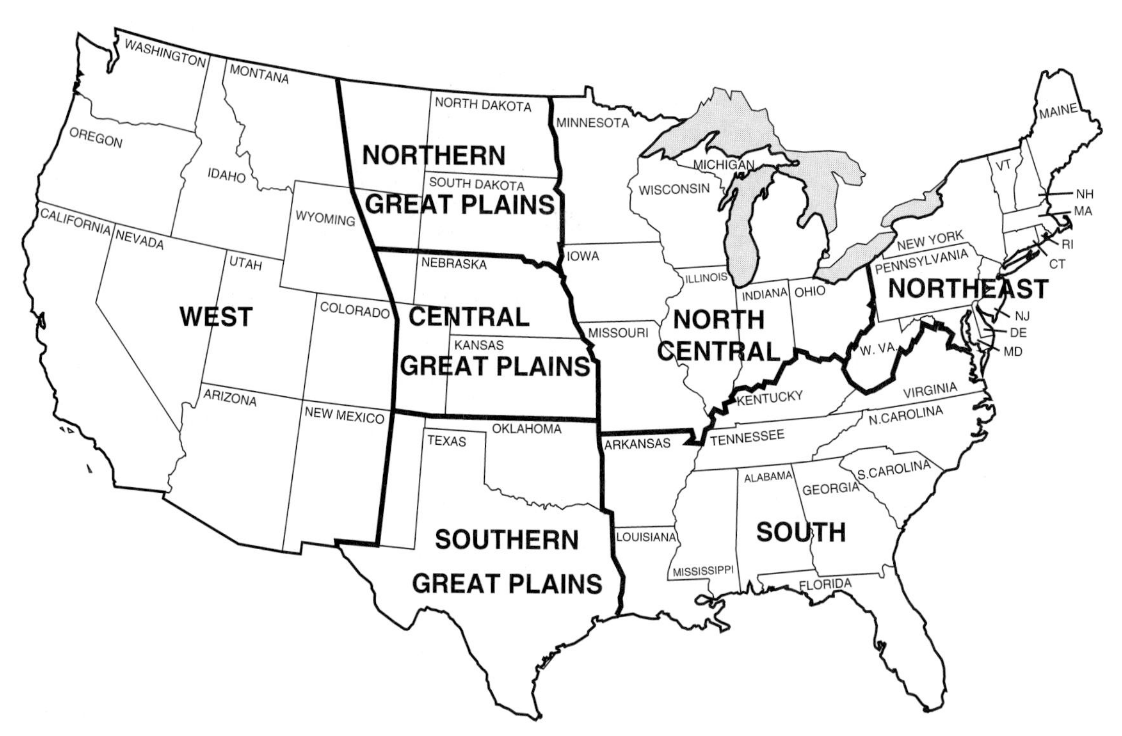

FIGURE 15.4 Cow-calf and yearling production areas.
Source: Colorado State University.

Forages are also classified as annuals or perennials and as warm-season or cool-season plants. Cool-season plants make most of their growth during late fall, winter, and early spring while warm-season plants have their major growth during late spring to early fall. Table 15.2 gives an overview of the significant grasses, forbs, and browse that contribute to beef production in the United States.

TABLE 15.2 A General Classification of the Major Forages in the United States

	Perennials		Annuals	
Category	Warm Season	Cool Season	Warm Season	Cool Season
Grasses	Bahiagrass Bermudagrass Big bluestem Blue grama grass Buffalo grass Carpetgrass Caucasian bluestem Dallis grass Eastern grama grass Indiangrass Johnsongrass Little bluestem Switchgrass	Crested wheatgrass Hardinggrass Intermediate wheatgrass Kentucky bluegrass Orchardgrass Reed canarygrass Russian wildrye Smooth bromegrass Tall fescue Tall wheatgrass Timothy Western wheatgrass	Browntop millet Corn Crabgrass Forage sorghum Foxtail millet Grain sorghum Pearl millet Sorghum-Sudan hybrids Sudangrass	Barley Oats Rescue grass Rye Ryegrass Triticale Wheat
Forbs, legumes	Kudzu Perennial peanut Sericea lespedeza	Alfalfa Alsike clover Birdsfoot trefoil Red clover White clover	Alyceclover Cowpea Korean lespedeza Soybean Striate lespedeza Velvetbean	Arrowleaf clover Bull clover Berseem clover Bigflower vetch Black medic Bur clover Button clover Caley pea Common vetch Crimson clover Hairy vetch Hop clover Lappa clover Persian clover Rose clover Subterranean clover Sweet clover Winter pea
Other forbs	Compass plant Halfshrub sundrop Pitchens sage Roundhead Slender lespedeza Weeds			
Browse		Four-wing saltbush Mountain mahogany Sagebrush Shadscale Winterfat Other brushes and shrubs		

Grasses

Figure 15.5 shows the distribution of some major grasses in the United States. Table 15.3 gives a more detailed description of major grasses. The most important native grasses are found in the western two-thirds of the United States, while tame pasture predominates from the eastern Great Plains toward the Atlantic seaboard.

The wheatgrasses, bluestems, grama grasses, buffalo grass, and switchgrasses are most common on western ranges. Orchardgrass, smooth bromegrass, reed canarygrass, the fescues, and bermudagrass (primarily coastal) account for most tame grass production in the United States. Tall fescue is the most widely grown cultivated pasture grass in the country, while smooth bromegrass is the most widely distributed cool-season grass in use.

Forbs

Forbs are herbaceous annual and perennial dicots and monocots other than grasses. Legumes are the most important forbs; numerous other forbs are called weeds by cattle producers. Many forbs are fair to good range forage plants. Some highly palatable range forbs are sensitive to overgrazing and so are used as indicators of range condition. Common perennial range forbs include halfshrub sundrop, pitchers sage, compass plant, and slender lespedeza.

Legumes are identified by their butterfly-like flowers. Legumes are usually higher in protein than most other grazed plants and are important to the nitrogen-fixation process that enhances soil fertility. The most important forage legumes are alfalfa, the trefoils, the lupines, sweet clover, kudzu, and true clovers. Alfalfa is widely distributed because it is adapted to environmental conditions in nearly all states.

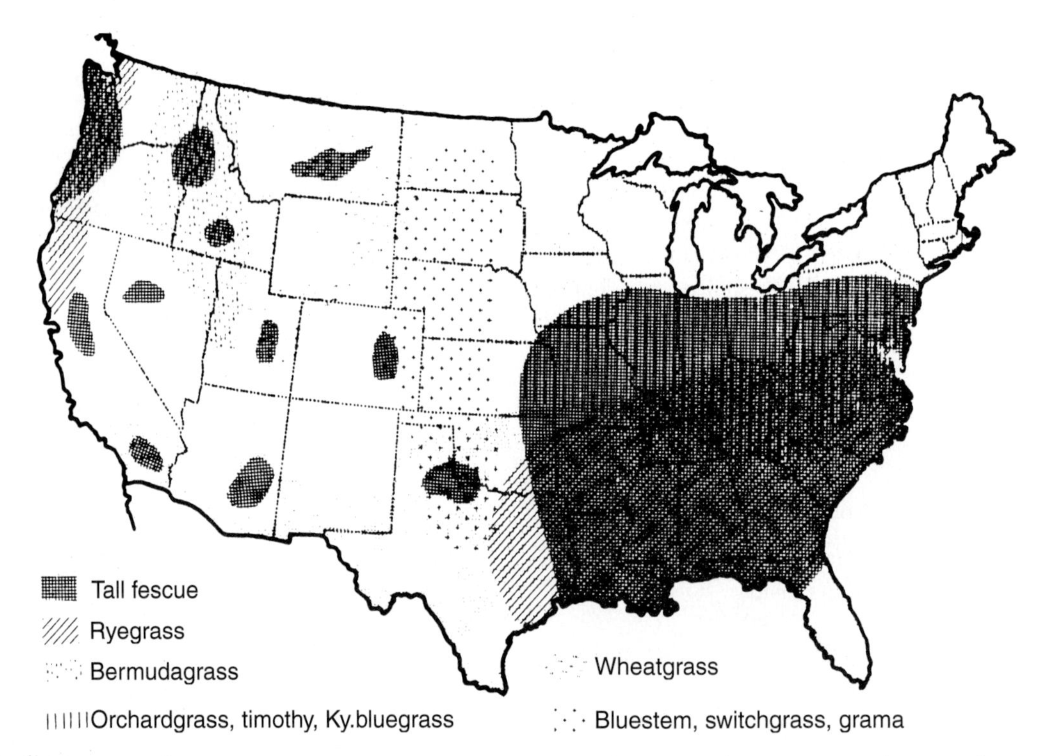

FIGURE 15.5 Distribution of some of the more important cool-season and warm-season grasses in the United States.
Source: Courtesy of the Amer. Soc. Of Anim. Sci., *J. Anim. Sci.* 57: 546, Suppl. 2.

TABLE 15.3 An Overview of the Production Requirements, Expected Yields, and Nutritional Value of Selected Grasses and Legumes

	Characteristic						Nutritional Values (%)			
Grass/ Legume	Winter Hardiness	Drought Tolerance	Soil	Annual Nitrogen	Harvest Maturity	Average Yield (tons/acre)	CP	ADF	NDF	Uses
Alfalfa	Very good	Excellent	Fertile, well-drained	0[1]	Late, bud-early bloom	4–8+	17–28	28–35	38–46	Grazing, hay, silage, greenchop
Red Clover	Very good	Good	Fertile, well-drained	0[1]	Early mid-bloom	3.5–5+	17–28	28–35	38–46	Grazing, hay
Birdsfoot Trefoil	Excellent	Very good	Not on droughty soils	0[1]	Early flower	4–6	17–24	28–35	38–46	Grazing, hay
Crown vetch	Very good	Very good	Fertile, well-drained	0[1]	Early to mid-bloom	2.5–3	17–24	28–35	38–46	Grazing, hay
Switchgrass	Very good	Excellent	Widely adapted	0[2]	Early boot	—	7–12	35–40	55–60	Grazing, hay
Big Bluestem	Very good	Excellent	Well-drained to somewhat poorly drained	0[2]	Early boot	—	7–12	35–40	55–60	Grazing, hay
Buffalograss	Very good	Excellent	Not sandy soils	—	Early boot	—	7–12	35–40	55–60	Grazing, hay
Coastal Bermudagrass	Poor	Very good	Heavy soils	300–400	10″ tall	15–20	11–14	35–40	55–60	Grazing, hay, greenchop
Bahiagrass	Poor	Excellent	Sandy	150–300	—	4–6	11–14	35–40	55–60	Grazing, hay
Tall Fescue	Good	Fair	Tolerates most soils	75–240	Early boot	3–5	11–14	35–40	55–60	Grazing, hay
Orchardgrass	Very good	Very good	Most soils, avoid poorly drained	75–240	Boot	3.5–5	11–14	35–40	55–60	Grazing, hay, silage
Smooth Bromegrass	Very good	Good	Well-drained, silt or clay loam	75–240	Late boot	3.5–5	11–14	35–40	55–60	Grazing, hay, silage
Timothy	Very good	Fair-good	Well-drained, productive silt or clay loam	75–240	Early head	2–3.5	11–24	35–40	55–60	Grazing, hay
Reed Canarygrass	Good	Excellent	Widely adapted, good in poorly drained soils	75–240	Early heading	3.5–6	11–14	35–40	55–60	Hay, grazing, silage
Kentucky Bluegrass	Very hardy	Poor	Heavy soil	150+	Boot	—	11–14	35–40	55–60	Grazing
Corn Silage	N/A	Good	Well-drained	100–200	½–¾ milk	25	6–10	24–33	43–52	Silage
Sorghum	N/A	Very good	Well-drained	40–60	Soft- to mid-dough	7–9	9–14	30–38	40–50	Silage
Sorghum-Sudan	N/A	Very good	Well-drained	50–100	24–30″ or boot	10–12	9–12	26–35	45–50	Silage, hay, grazing

[1]Nitrogen application not needed if stand is greater than 50% legume.
[2]None during stand establishment.
Source: Adapted from *Pioneer Forage Manual.*

Browse

Browse consists of perennial woody shrubs and vines used as forage by cattle. Some examples of range browse are sagebrush, shadscale, winterfat, mountain mahogany, and four-winged saltbush.

Cattle grazing on range that includes grass, forbs, and browse typically change their consumption pattern of each plant class as the grazing season progresses. Grass comprises a higher percentage of the diet in early summer. In late summer, there is an increase in the consumption of forbs and browse.

GRAZING MANAGEMENT

Grazing systems are extremely diverse—there are hundreds of combinations of climatic differences, plant species, terrain, types of cattle, and preferences of cattle producers. There is no best grazing system for all producers, so the most economical system that best fits each producer's needs and goals should be identified by the management team. However, management should always focus on the health and care of natural resources. The membership of the NCBA adopted a resolution of stewardship beliefs that includes the following 15 points of environmental stewardship:

1. Recognize the environment for its varying and distinct properties.
2. Manage for the whole resource, including climate, soil, topography, plant, and animal communities.
3. Realize that natural resources are ever changing and management must adapt.
4. Recognize and appreciate the interdependence of ecosystems.
5. Recognize that management practices should be site-and-situation specific, and must be locally designed and applied.
6. Recognize that successful management is an ongoing, long-term process and commit to stewardship, economic success, and business continuity.
7. Strive to develop a management framework that involves family, employees, and business associates so that the entire team is committed to common goals.
8. Monitor and document for effective practices.
9. Never knowingly cause or permit abuses that result in permanent damage on public or private land.
10. Develop ways to communicate and share the vast practical experience of other resource stewards.
11. Become involved in organizations that provide an effective way to educate and support individuals.
12. Solicit input from a variety of sources on a regular basis as a means to improve the art and science resource management.
13. Help develop public and private research projects to enhance the current body of knowledge.
14. Recognize that individual improvement is the basis for any change.
15. Communicate with diverse interests to resolve resource management issues.

Range and pasture grazing systems are primarily a function of three factors: (1) amount of forage produced, (2) forage quality, and (3) the efficiency by which the forage is harvested (Heitschmidt and Walker, 1983). All systems of grazing management are centered on one basic principle: controlling the frequency and severity of defoliation of individual plants. This principle needs to be carefully assessed over several grazing seasons as well as during a single grazing season to allow managers to build in as much freedom from risk as is possible.

Effective grazing systems are determined by stocking rates, intensity and frequency of grazing, grazing of single or mixed stands of forage, and other factors that can optimize efficient beef production. Grazing systems, to be effective over the long term, need to be economically and environmentally sound. Furthermore, they should be as simple to implement as is possible.

Some grazing intensities and frequencies are reflected in the following commonly used grazing terms:

Continuous—allowing grazing on a specific area throughout the entire grazing season (does not refer to level of use nor does it imply year-long grazing).
Cell system—the management of a larger acreage in a multi-pasture grazing system typically utilizing high-intensity, short-duration grazing. Such a system often incorporates a common water source and cattle handling facility central to all pastures.
Rotational—cattle are moved from pasture to pasture on a schedule (based on calendar, level of use, or stage of growth). These systems typically rest pastures after they have been grazed.
Deferred—not grazed until the primary forage plants have matured, usually the setting of seed.
Rested—no grazing for at least a 1-year period.
Rested-rotational—incorporates at least one rested pasture on a rotational basis.
Seasonal—grazed for the duration of a particular season of the year.
Short duration—heavy stocking rates for short periods; each unit may be grazed several times during a grazing season.

The objective of most grazing systems is to optimize animal performance and maintain or improve the production potential of the vegetation. The kinds of available animals and the type of vegetation, along with the seasons of grazing, help determine the stocking rate as part of the overall grazing plan. Matching dietary preferences and nutrient requirements of cattle to kinds and quality of vegetation is the first important step in determining the proper grazing management system.

The season of grazing is determined primarily by recognizing two important relationships. First, animal performance is greatest when the forage is grazed when it is growing most rapidly (high in quantity and high in nutrient quality) and before it matures. Second, to obtain range improvement, the desirable forage species must be allowed to reproduce. Rangelands generally provide flexibility in that they can be combined with tame pastures.

The optimum stocking rate (the animal demand per unit area over a period of time) is not a constant; it will vary from season to season and year to year. Grazing pressures that are too high for long periods of time will reduce the cattle's nutrient intake and thus reduce their performance as well as reducing the diversity of plant species on the site. The correct stocking rate for a particular pasture depends on the desired level of animal performance, the ability of the vegetation to withstand grazing, and the range or pasture improvement goals. Carrying capacity is the stocking rate applied to a site with a specific animal density for a specific time without damaging the resource. To determine carrying capacity for a site, producers need to know the following:

1. Annual pounds of forage produced per acre.
2. The percent of annual production consumed by cattle (continuous grazing—33% utilization, weekly controlled rotation—50% utilization, daily controlled rotation—70% utilization).
3. Daily animal consumption as determined by livestock performance goals.
4. Grazing season duration as determined by forage availability, rest period, etc. (Morrow, Univ. of Missouri).

TABLE 15.4 Animal Unit Values (AU) for Various Classes of Beef Cattle

Class of Cattle	Animal Unit
Cow and calf (1,000 lb, above-average milking ability, spring calving)	1.00
Calf (spring born, 3–4 months)	0.30
Replacement heifer (24–36 months)	1.00
Cow (1,000 lb, nonlactating)	0.90
Yearling cattle (12–17 months)	0.70
Weaned calves (<12 months)	0.50
Yearling bulls	1.20
Mature bulls	1.50

Source: Adapted from Blasi and Corah (1993).

Carrying capacity can be roughly calculated by the following formula:

$$\text{Carrying capacity} = \frac{\text{annual forage production} \times \text{seasonal utilization rate}}{\text{average daily intake} \times \text{length of grazing season}}$$

Many range managers estimate stocking rate using the animal unit month concept. This approach provides an inexact estimate but it is the standard in many regions of the world. An animal unit month (AUM) is a 1,000-lb mature cow of above-average milking ability, a calf at side less than 4 months of age with a 400-lb weaning weight, that consumes 680 lbs of dry matter in one months' time (Table 15.4).

If cows are heavier than 1,000 lbs, or wean calves heavier than 400 lbs, the standard AUM must be modified. A rule of thumb adjustment is to increase the AUM one-tenth for every 100 lbs of cow weight over 1,000 lbs and every 100 lbs of weaning weight beyond 400 lbs. For example, a 1,200-lb, spring calving cow weaning a 600-lb calf would require:

$$\text{Cow} = \frac{1{,}200}{1{,}000} = 1.2 \text{ AU}$$

$$\text{Calf} = .3 \text{ AU} + \frac{600 - 400}{1{,}000} = .5 \text{ AU}$$

$$\text{Pair} = 1.2 + .5 = 1.7 \text{ AU}$$

Estimates of carrying capacity for various plant communities of differing conditions on several soil types are illustrated in Table 15.5.

Grasses such as bermudagrass, bahiagrass, K. R. bluestem, buffalo grass, Kentucky bluegrass, and common curly mesquite are well suited for intensive grazing. Most taller grasses and bunch grasses do not withstand intensive grazing. Where mixed stands of these grasses exist, the species that eventually predominate will depend on the intensity and length of grazing.

Cattle are selective in their grazing for two reasons. First, they respond to topography, distance to water, and other factors. Salt distribution, water distribution, fencing, and forced movement of cattle are four practices utilized to prevent overgrazing by assuring more uniform animal distribution. Second, some species of plants in mixed stands are more palatable than others. Some grazing systems will prevent overgrazing of the more palatable species.

Currently, there is considerable interest in short duration grazing systems. Some of these systems are excellent in obtaining the best combination of animal and forage performance;

TABLE 15.5 Average Carrying Capacity for Various Range Sites and Conditions

Range Site	Range Condition			
	Excellent	Good	Fair	Poor
	AUM/Acre			
Tallgrass prairie				
Clay upland	1.30	.95	.65	.45
Limy upland	1.10	.90	.65	.45
Loamy lowland	2.30	1.70	1.15	.65
Loamy upland	1.50	1.10	.75	.50
Mixed-grass prairie				
Clay upland	.90	.70	.55	.25
Limy upland	.90	.70	.45	.35
Loamy lowland	1.35	1.05	.75	.50
Loamy upland	1.00	.70	.55	.35
Sand prairie				
Subirrigated	2.25	1.75	1.25	.85
Sandy	.90	.70	.50	.35
Choppy sands	.60	.50	.35	.25
Short-grass prairie				
Loamy upland	.60	.50	.40	.30
Loamy lowland	.80	.70	.60	.40

Source: Adapted from Blasi and Corah (1993).

FIGURE 15.6 A solar powered fence is used to separate a pasture into grazing cells. This photo was taken in early May and depicts the recovery of the pasture site on the right side of the fence line. The cows on the left side had been removed from the right-hand pasture site two weeks previously. Notice the re-growth that has occurred as a result of rest. The pasture area on the left would be expected to recover equally well following intensive but short-term grazing. A well-managed intensive grazing system can yield improved productivity and sustainability.

however, they may be expensive to implement because additional labor and fencing are required. Range improvement may occur under short duration grazing systems if moderate stocking rates, which will give acceptable levels of cattle performance, are utilized. Strong proponents of certain short duration systems have generated considerable controversy. Much of the attention has focused on the Savory grazing method (Fig. 15.6). Excellent discussions of the Savory grazing method and the basic principles of grazing systems can be found in Heitschmidt and Walker (1983) and Savory (1999).

FIGURE 15.7 Cattle would graze the foothill areas in the early spring and the wooded mountain pasture areas (middle of the photo) during the summer and early fall. In autumn and early winter, the cattle would be moved to lower elevations to graze forage aftermath from the hay meadows. This is a common grazing system in the intermountain area of the western United States.The area above timberline would not be grazed.

Range and pasture areas that permit year-round grazing with satisfactory animal performance are usually the most economical for cattle producers. This is true if input costs can be moderated. The primary objective is to have green forage (which is typically higher in nutrient content than dry, mature forage) available to cattle as long as possible during the year. In the intermountain area, such a grazing system might be as follows: early spring, crested wheatgrass; late spring to early summer, grazing cool-season grasses in the foothills area; late summer to fall, grazing warm-season grasses or cool-season grasses in the high mountain pastures (Fig. 15.7); and winter, grazing on the desert shrub areas at lower elevations.

Prior research (Hart et al., 1993) emphasized the importance of two requirements of sound forage management: (1) proper stocking rate, and (2) even livestock distribution. This research demonstrates that these two requirements can be achieved independent of grazing system. Also, these researchers noted that reduced pasture size and distance to water may be responsible for the claimed benefits of short duration and controlled rotation grazing systems. A careful economic evaluation must assess the short-term and long-term benefits of fencing for reduced pasture size, water development, and/or rotation grazing (intensive, less intensive basis, or not at all).

Effective grazing management requires the careful use of range management tools, timing, and knowledge about a particular geographic area. Some of the common misconceptions relative to grazing as noted by Reece (1991) are listed below:

1. Rangeland plants must be grazed to assure plant health or vigor.
2. Late-season grazing is more favorable to plant health than grazing early in the growing season.
3. Conventional deferred rotational grazing systems provide all pastures with equal recovery opportunity.
4. Stocking rates can always be increased once grazing systems are implemented.
5. Dormant vegetation cannot be damaged by overgrazing.
6. Increased stocking rates are required to increase total animal production per acre.
7. Continuous grazing is "bad."
8. There is one grazing strategy best suited to all range environments.
9. Production per animal is less critical than production per acre in generating maximum profits.

A review of the extensive literature on grazing management fails to yield a consensus opinion as to the most desirable grazing system. However, the following summary points from Valentine (1990) are useful:

1. No grazing system eliminates the need for appropriate stocking rates and the application of sound management principles.
2. Grazing systems must be developed on a site-specific basis.
3. A specialized grazing system is only as good as the ability of the range manager.
4. Because of changing environmental conditions, flexibility is key to the success of a grazing system.
5. If a rotational grazing system is utilized, managers should be aware of the following:
   a. The need and cost of fencing.
   b. The cost of assuring adequate water for livestock.
   c. Variation in grazing capacity of different pastures.
   d. The potential effects of drought on the system.
   e. Potential effects on wildlife.

In the end, excellent grazing management is driven by a set of clear objectives that may include improving profitability by lowering feed costs, by limiting reliance on expensive inputs such as fossil fuels, improving resource quality/health, and optimizing animal performance.

Range Monitoring

Range condition cannot be effectively accomplished without incorporation of the management objectives for a particular site. Range condition can be evaluated via two conceptual models. Ecological status is the present state of vegetation and soil as compared to the potential natural state of the site. The second concept requires the assignment of a resource value rating (RVR) to the forage relative to a specific use. RVR is most appropriate for improved range sites.

Unfortunately, the diverse and often combative views as to what the management goals for a particular range site should be have made it difficult to determine universally accepted procedures for monitoring and evaluating range condition. Nonetheless, it is critical for forage managers to develop a clear set of goals and objectives that help achieve the mission statement of the organization.

Effective range and pasture monitoring requires that baseline information is obtained via a pasture survey, the development of a landscape description and management objectives for each pasture, and monitoring of responses of plants to management changes. Four typical tools utilized to monitor a range site are present plant community assessment, photo points, relative degree of use, and grazing response index.

Present plant community assessment involves a description of the structure and appearance of the plant community present on a specific site, a listing of plant species and the percent occurrence of each, and determination of the length, initiation, and conclusion of the growing season for each of the major plant species. The sample should be taken near the time of peak growth. Future samples should be based on stage of growth rather than time of year to assure the best comparison.

Photo points are the collection of comparative photographs of the same site taken over time. These comparisons allow assessment of the visually apparent changes in forage and soil condition. Designated sites must be clearly delineated to allow accurate identification over time.

Relative degree of use evaluates the differences in grass, forb, and browse utilization between protected plots versus unprotected plots. Protected enclosures are obtained by building cages over a relatively small land area to eliminate animal access. Differences in plant condition and availability observed on caged versus uncaged plots are assumed to be due to animal utilization.

Grazing response index ties together three plant health indicators—frequency, intensity, and opportunity. Frequency, a time dependent event, is defined as the number of times a plant is grazed during the grazing period. Plants grazed once in a growing season are scored +1, those grazed twice in a growing season are assigned a neutral score (0), and plants grazed three times during a grazing season receive a negative mark (−1). When grazing exceeds 15 days on a pasture, the likelihood of plants being grazed more than once increases.

Intensity of use is stocking rate dependent. If less than 50% of the leaf area of the plants in a pasture is removed, then a score of +1 is given. If about one-half of the leaf area has been removed, then a neutral score of 0 is assigned. Finally, if more than 50% of the leaf area of a plant is removed, then a score of −1 is indicated.

Opportunity of the plant to grow or regrow depends on water availability, temperature, and other habitat factors. Maximum regrowth opportunity is assigned a score of +2, moderate opportunity for regrowth is viewed as neutral, and minimum regrowth opportunity is viewed negatively (−2). Monitoring the scores relative to frequency, intensity, and opportunity allows a manager a means to evaluate trends and take corrective action.

Range Improvements

Range improvements are based on the ecological principle of providing a competitive advantage for desirable plant species by increasing their access to water, nutrients, and sunlight. Simultaneously, undesirable plant types are controlled to limit their access to limited resources. Range improvement programs are typically implemented to increase quantity and (or) quality of forage, enhance livestock/wildlife production, enhance the ability to handle and care for livestock, improve water resources, and minimize erosion.

Range improvement techniques typically can be characterized as biological control (grazing, insects), mechanical control (cutting, bulldozing, chaining, mowing, disking), herbicidal control, prescribed burning, seeding, and fertilizing. The range improvement techniques best suited to a particular site are determined by a combination of geographical, ecological, economic, regulatory, and management expertise considerations. An example set of costs associated with prescribed burning on three Oklahoma range sites is provided in Table 15.6.

TABLE 15.6 Prescribed Burning Costs for Rangeland Improvement on Three Oklahoma Ranches

	Site		
Input	**Coffey[a,d]**	**Stuart-A[b,c]**	**Stuart-B[c,e]**
Fuel	$ 83.10	$ 225.50	$ 523.12
Fireguard protection	729.00	250.00	817.55
Labor for actual burn	936.00	694.00	740.00
Tool use cost—purchase basis	135.80		
Equipment cost—rental basis	596.69		
Grazing loss cost—lease risk	337.68		
Total cost	2,818.27	1,169.50	2,080.67
Total acres treated	608.00	5,000.00	5,976.00
Total cost/acre	4.64	0.23	0.35

[a]Timber.
[b]Grassland.
[c]Grassland/timber.
[d]Equipment purchased or rented.
[e]Equipment on hand and fully depreciated.
Source: Stevens, R., et al. 1997. *Rangelands,* 19: 2.

Ecosystem Descriptions

A single definition of the term *ecosystem* is not universally accepted. In the broadest context, an ecosystem would encompass the earth and all its resources. It is difficult to imagine the process of understanding all of the relationships that exist in such a large system. Yet, for managers of land, the process of understanding the relationships between humans, natural resources, economies, and agricultural resources is critical to long-term success. Typically, managers tend to view ecosystems on the context of a specific ranch and the adjoining lands.

The two primary ecosystem domains are the humid-temperate domain that comprises the eastern half of the continental United States plus the West Coast, and the dry domain that characterizes the western half of the United States. Domains are then subclassified into divisions based on climatic zone descriptions. Divisions are then broken into provinces based on vegetation descriptions. The polar and humid tropical domains will not be discussed as they have minimal carrying capacity of the U.S. beef cattle herd. The provinces of the humid-temperate domain and the dry domain and a brief outline of the land surface form, climate, vegetation, and soils of each province are detailed in Table 15.7 and Table 15.8, respectively.

TABLE 15.7 A Description of the Provinces of the Humid-Temperate Domain[a]

Land Surface Form	Climate	Vegetation	Soil
Map Code/Province: 212–Laurentian Mixed-Forest Province (147,300 mi²)			
Low-relief, rolling hills Elevation range of sea level to 2,400 ft.	Winter-mod. long to severe average annual temperature range of 35–50°F More than 120 days above 50°F Growing season 100–140 days Average precipitation of 24–45 in.	Forest-pine, birch, maple, beech, conifer, northern white pine, and eastern red cedar	Variable, including peat, mud, clay, silt, sand, and gravel
Map Code/Province: M212–Adirondack/New England Mixed Forest-Coniferous Forest-Alpine Meadow Province (43,600 mi²)			
Subdued glaciated mountains, plateaus Elevation range from 500–4,000 ft.	Warm summers, cold winters Average annual temperature range 37–52°F Growing season 100 days Evenly distributed precipitation; average of 35 in. 100 in. snowfall	Hardwood and mixed forest predominate	Spodosols—stony, cool, moist
Map Code/Province: 221–Eastern Broadleaf Forest (oceanic) (104,500 mi²)			
Diverse topography, including glaciated, plateau, hilly, and mountainous Elevation range of 1,000–3,000 ft.	Warm summers, cold winters Average annual temperature range 40–60°F Precipitation year-round; average 35–60 in.	Tall broadleaf trees, small trees, shrubs, buckeye, basswood, white oak, Northern red oak, and pine-oak forest	Alfisols, ultisols, and inceptisols

(continued)

TABLE 15.7 (Continued)

Land Surface Form	Climate	Vegetation	Soil
Map Code/Province: 222–Eastern Broadleaf Forest (continental) (270,000 mi²)			
Rolling to flat Elevation range of 80–1,650 ft.	Hot summers, most rainfall in growing season Average annual temperature range 40–65°F Average precipitation 20–50 in.	Oak-hickory forest; prairie	Alfisols, ultisols, mollisols
Map Code/Province: M221–Central Appalachian Broadleaf Forest-Coniferous Forest-Meadow Province (68,000 mi²)			
Subdued low mountains, valleys, and dissected plateaus with mountainous topography Elevation range from 300–6,000 ft.	Temperate climate Average annual temperature range 50–64°F Growing season range 100–220 days Average annual precipitation 35–80 in.	Mixed oak, pine forest, oak forest, Northeastern hardwood forest	Ultisols, inceptisols
Map Code/Province: 231–Southeastern Mixed Forest Province (193,000 mi²)			
Piedmont and irregular Gulf Coastal plains 50–80% of area with gentle slope Elevation of 100–1,000 ft.	Mild winters with frost, hot humid summers Average annual temperature 60–75°F Growing season 200–300 days Average annual precipitation 40–60 in.	Marsh and swamp Broadleaf deciduous forest Needleleaf evergreen trees	Ultisols, vertisols
Map Code/Province: 232–Outer Coastal Plain, Mixed-Forest Province (173,800 mi²)			
Flat and irregular Atlantic and Gulf Coastal plain 50% of area gently sloping Elevation <300 ft.	Average annual temperature 60–70°F Average annual precipitation 40–60 in. Long growing season	Temperate rainforest Laurel forest Temperate evergreen forest Gum and cypress along coast	Wet, acidic, and low in nutrients Heavy clay, sand, and gravel
Map Code/Province: 234–Lower Mississippi Riverine Forest Province (44,300 mi²)			
Flat and gently sloping flood plain Low terraces Elevation sea level to 660 ft.	Average winter temperature 50–60°F Average summer temperature 70–80°F Average annual precipitation 43 in.	Bottomland, deciduous forest	Mosaic of inceptisols and mollisols
Map Code/Province: 242–Pacific Lowland-Mixed Forest Province (14,900 mi²)			
Valley lies between the coast ranges and the Cascade Mountains Elevation of sea level to 1,500 ft. Isolated hills and low mountains	Generally mild Average annual temperature 48–55°F Average annual rainfall 15–60 in. with predominant area receiving 30–45 in.	Coniferous forest, big-leaf maple, cottonwood Prairies with oak and Douglas fir Swamp and bog	Alfisols, inceptisols, ultisols

TABLE 15.7 (Continued)

Land Surface Form	Climate	Vegetation	Soil
Map Code/Province: M242–Cascade Mixed Forest-Coniferous Forest-Alpine Meadow Province (53,400 mi^2)			
Steep, rugged mountains Narrow coastal plain Elevation 1,000–14,000 ft.	Generally mild Average annual temperature 35–50°F Heavy rainfall, average 35–150 in. with maximum in winter East slopes drier than west	Conifer forest, Douglas fir, numerous shrubs, hemlock, silver fir, redwood	Andisols, moist inceptisols, drier in the east
Map Code/Province: 251–Prairie Parkland Province (Temperate) (218,200 mi^2)			
Alternating prairie and deciduous forest Elevation 300–2,000 ft.	Hot summers, cold winters Average annual temperature 40°F in north, 60°F in south Growing season 120 days in north, 235 days in south Average annual precipitation 20–40 in.	Intermingled prairie and groves of deciduous trees Grasses are predominant on prairie—big bluestem, little bluestem, switchgrass, Indian grass, and legumes	Mollisols dominate, limestone hills in west
Map Code/Province: 252–Prairie Parkland Province (Subtropical) (80,000 mi^2)			
Gently rolling to flat plains Elevation from sea level to 1,300 ft.	Warmer than site 251 Winter average temperature 50–60°F, summer average temperature 70–80°F Average annual precipitation 35–55 in.	Prairies and savannas Grasses predominate with bluestem as the principal variety	Mollisols, alfisols, and vertisols
Map Code/Province: 262–California Dry Steppe Province (19,200 mi^2)			
Central valley of California, broad valleys, terraces, and linear foothills Elevation of sea level to 500 ft.	Hot summers, mild winters Annual average temperature 60–70°F Precipitation primarily in Dec., Jan., and Feb. Average annual precipitation 6–30 in.	Annual grasses including avens, brome, fescue, and barley Greasewood, saltgrass, and shadscale in alkaline flats	Entisols, alfisols
Map Code/Province: M261–Sierran Steppe-Mixed Forest-Coniferous Forest-Alpine Meadow Province (68,300 mi^2)			
Steeply sloping to extreme mountain conditions interlaced by valleys, highly glaciated Elevation 2,000–14,000 ft.	Average annual temperature 35–52°F (cooler as elevation increases) Annual precipitation 10–15 in. (west slope), 40–50 in. (mostly snow) in high country	2,000–4,000 ft. covered by coniferous and shrub associations –Higher slopes have digger pine and blue oak –Low hills with ever-green scrub or chaparral 4,000–7,000 ft. covered by a variety of pine, fir and conifers 6,500–9,500 ft. covered by mountain hemlock and pine	Ultisols at mountain elevation, dry alfisols at lower elevations, entisols in flood plain

(continued)

TABLE 15.7 (Continued)

Land Surface Form	Climate	Vegetation	Soil
	Map Code/Province: M262–California Coastal Range-Open Woodland-Shrub-Coniferous Forest-Meadow Province		
Gently to steeply sloping low mountains (north), precipitous mountains (south) Narrow valleys Elevation 2,000–12,000 ft.	Hot, dry summers and wet, mild winters Average annual temperature 53–65°F (coast range), 32–60°F (southern California) Average precipitation 12–40 in. (mostly rain)	Alternating association of schlerophyll forest and chaparral shrub Schlerophyll forest comprised of various oaks, laurel, Pacific madrone, and Pacific bayberry Chaparral comprised of 40 evergreen shrub species including chamise and manzanita	Complex pattern of alfisols, entisols, and mollisols

[a]Provinces M222, M231, 261, 263, excluded.
Source: USDA: Forest Service.

TABLE 15.8 A Description of the Provinces of Dry Domain[a]

Land Surface Form	Climate	Vegetation	Soil
	Map Code/Province: 311–Great Plains Steppe and Shrub Province (17,600 mi^2)		
Irregular plains Elevation 1,600–3,000 ft.	Semiarid, subtropical Average annual temperature 57–64°F Growing season 105–230 days Average annual precipitation 19–29 in.	Tallgrasses predominate Bluestem-grama prairie, generally Oak savanna (east and along river valleys) Sandsage-bluestem prairie (west)	Mollisols
	Map Code/Province: 313–Colorado Plateau Semidesert Province (75,300 mi^2)		
Tablelands Narrow stream valleys (water level fluctuates widely) Elevation of plateau tops 5,000–7,000 ft. with local relief ranging from 500–3,000 ft. in deep canyons	Cold winters, hot summers with cool nights Average annual temperature 40–55°F Average annual precipitation 20 in. Thunderstorms in summer with predominate rainfall in winters	Lowest areas are arid grasslands, xeric shrubs, sagebrush, cactus, and yucca Woodlands dominated by pinyon pine and juniper; sparse coverage of grama and various shrubs Higher elevations characterized by ponderosa pine (south) and lodgepole pine (north)	Entisols (flood plains), aridisols (plateau tops)

TABLE 15.8 (Continued)

Land Surface Form	Climate	Vegetation	Soil
	Map Code/Province: 315–Southwest Plateau and Plains Dry Steppe and Shrub Province (160,900 mi²)		
Flat to rolling plains and plateaus Elevation from sea level to 6,500 ft.	Semiarid with long hot summers and short, mild winters Average annual temperature 60–70°F Growing season 130–300 days Annual precipitation range 10–30 in. (evaporation is about double precipitation from May to Oct.)	Arid grasslands Blue grama and buffalo grass interspersed with mesquite in plains Oak, juniper, and needlegrass in the plateaus	Entisols, mollisols
	Map Code/Province: M313–Arizona-New Mexico-Mountain-Semidesert-Open Woodland-Coniferous Forest-Alpine Meadow Province (50,200 mi²)		
Steep foothills and mountains Elevation 4,500–12,600 ft.	Highly variable with altitude Average annual temperature 55°F (lower foothills), (upper 40°F mountains) Average annual precipitation 10–35 in. Moisture deficient in late spring	Foothills covered by mixed grasses, chaparral brush, oak-juniper woodland, and pinyon-juniper woodland 7,000–9,000 ft. characterized by pine forest, fir, and aspen 9,000–11,000 ft. contains Englemann spruce and corkbark fir	Entisols in Four Corners region, alfisols and inceptisols in mountain region, stone and rock outcrops in many areas
	Map Code/Province: 321–Chihuahuan Desert Province (85,200 mi²)		
Desert plains Rio Grande and Pecos Rivers are only continuous streams of note Elevation typically at 4,000 ft. with scattered mountains rising to 9,000 ft.	Distinctly arid, spring and early summer extremely dry Average annual temperature 50–65°F Average annual precipitation 8 in.	Thorny shrub species associated with grama grass Extensive arid grasslands Prickly pear, yucca, honey mesquite, and creosote bush are predominant	Aridisols (west and north), aridisols and entisols (south)
	Map Code/Province: 322–American Semidesert and Desert Province (87,700 mi²)		
Extensive desert plains with isolated buttes and mountains Elevation 280 ft. below sea level to 4,000 ft. in basins; mountains rising to 11,000 ft. Colorado river is only significant stream	Extremely arid Average annual temperature 60–75°F Subject to occasional frosts Average annual precipitation 2–10 in. (valleys), 25 in. (mountain slopes) Western region may go without rain for extended periods Very high summer evaporation rate	Sparsely vegetated Cholla, creosote bush, paloverde, ocotillo, sagnero, and bitterbrush predominate	Gravel and bare rock Aridisols Some Entisols

(continued)

TABLE 15.8 (Continued)

Land Surface Form	Climate	Vegetation	Soil
Map Code/Province: 331–Great Plains-Palouse Dry Steppe Province (290,700 mi²)			
Rolling plains and tablelands Elevation 2,500–5,500 ft. (Rockies) 1,200–6,000 ft. (Palouse-Washington and Oregon)	Semiarid Average annual temperature 45–60°F Growing season 100–200 days Annual precipitation 10–25 in.	Short-grass prairie Buffalo grass, sunflower, locoweed, grama, wheatgrass, and needlegrass are predominate Bluestem, wheatgrass, fescue, and bluegrass predominate in Palouse region	Mollisols Contain a large excess of calcium carbonate and are high in bases
Map Code/Province: 332–Great Plains Steppe Province (134,000 mi²)			
Flat and rolling plains Elevation 1,000–2,500 ft.	Average annual temperature 40°F (north), 65°F (south) Average annual precipitation 20–35 in. (east), 15–25 in. (west)	Mixed-grass steppe with numerous forbs Blue grama, hairy grama, and buffalo grass are typical short grasses Little bluestem and needle-and-thread grasses are representative taller grasses	Mollisols Entisols (northeast sand hills)
Map Code/Province: M331–Southern Rocky Mountain Steppe-Open Woodland-Coniferous Forest-Alpine Meadow Province (102,300 mi²)			
Rugged glaciated mountains High elevation plateaus Elevation 6,000–14,000 ft.	Semiarid steppe Average annual temperature 35–45°F Annual precipitation 10–40 in. depending on elevation	Alpine tundra at highest elevations As elevation falls below timberline Englemann spruce and subalpine fir transition to ponderosa pine and Douglas fir Grass and sagebrush predominate as well. At the lowest elevation scrub oak and pinyon-juniper associations provide coverage	Mollisols, alfisols aridisols, inceptisols
Map Code/Province: M332–Middle Rocky Mountain Steppe-Coniferous Forest-Alpine Meadow Province (81,800 mi²)			
Granite batholith Mountains with alluvial fans at the base and floodplains in valley floors Elevation 3,000–10,000 ft.	Semiarid with generally mild conditions Average growing season 120 days Annual precipitation 20–30 in.	Douglas fir and grand fir below subalpine zone Lodgepole pine and grasses in mid-elevation zone Sagebrush semidesert in the lower slopes and plains	Inceptisols, alfisols, mollisols

TABLE 15.8 (Continued)

Land Surface Form	Climate	Vegetation	Soil
Map Code/Province: M333–Northern Rocky Mountain Steppe-Coniferous Forest-Alpine Meadow Province (38,100 mi²)			
High, rugged mountains Glaciated Flat valleys Elevation 6,000–9,000 ft.	Severe winters Average temperature of the coldest month to below 32°F Average temperature of warmest month is below 72°F Annual precipitation 20–40 in.	Mixed evergreen-deciduous forest predominates Douglas fir forest is significant Lower elevations interspersed with sagebrush and grass	Inceptisols predominate
Map Code/Province: 341–Intermountain Semidesert and Desert Province (107,000 mi²)			
Variety of interior basins Mountain originating from semiarid, sagebrush plains Heavy accumulation of alkaline and saline salts in lower regions	Average annual temperature 40–55°F Average precipitation 5–20 in. (mostly in the form of snow)	Sagebrush dominates at lower elevations Alkali-tolerant shrubs As elevation rises pinyon pine and juniper are more frequent Ponderosa pine then transitions to subalpine fir	Aridisols dominate
Map Code/Province: 342–Intermountain Semidesert Province (159,100 mi²)			
Plateaus and tablelands of the Northwest lava fields Elevation 3,000–10,000 ft. Broad intermountain basins and isolated mountains	Plateaus –Semiarid –Average annual temperature 50°F –Average annual precipitation 10–20 in. Basins –average annual temperature 50–52°F –Growing season 100–140 days –Average annual precipitation 5–14 in.	Sagebrush or shadscale mixed with grasses Alkali-tolerant greasewood Bunch grass in higher rainfall areas to the west	Extensive alkali deposits Mollisols Aridisols Some entisols
Map Code/Province: M341–Nevada-Utah Mountains Semidesert-Coniferous Forest-Alpine Meadow Province (43,600 mi²)			
Semiarid plains giving rise to mountain ranges High elevation plateaus Elevation 5,000–13,000 ft.	High altitude temperate desert Definite dry season Average annual temperature 50°F (valleys), 38°F (mountain slopes) Average annual precipitation 5–8 in. (valleys), 25–35 in. (mountains)	Sagebrush at lower elevations with shadscale, four-wing, saltbush, and rabbitbush Juniper-pinyon on lower slopes Ponderosa pine transitioning to alpine fir and Englemann spruce	Aridisols Mollisols Alfisols

[a]Province M334 excluded.
Source: USDA: Forest Service.

TABLE 15.9 Description of Major Soil Orders of Importance to Grass and Forage Production

Soil Order	% of U.S. Total Soils	Distribution	Productivity	General Description
Entisols	8	River floodplains, rocky soils in mountain areas, beach sands, rangeland	Variable from highly productive to infertile	Slight soil development, range from sand to river-deposited clay to volcanic ash deposits
Inceptisols	18.2	Middle Atlantic and Pacific states	Highly variable	Weakly developed, quick horizon formation
Andisols	1.9	Hawaii and Pacific Northwest	Highly productive if well managed	Weakly to moderately developed, rapidly weathered, mostly formed from volcanic activity
Aridisols	11.6	Western mountain and Pacific states	Very productive when irrigated and fertilized	Dry soils, prone to accumulation of soluble salts
Mollisols	25.1	Great Plains	High fertility due to higher humus and nutrient content	Dark soils of grasslands and some hardwood forests, very diverse
Vertisols	1.0	Southeastern and central Texas, western Gulf Coast states	Not easily cultivated, generally fertile, but restriction of root penetration	Originate from limestone, marl or basalt, high swelling-clay content
Alfisols	13.5	North central and mountain states	Productive crop-land, irrigation is less critical	Moist mineral soils, often developed under deciduous forests
Spodisols	4.8	New England, mid-Atlantic and northern Great Lakes states	Requires significant fertilization	High sand content, well leached, strongly acidic
Ultisols	12.8	Southern Atlantic, eastern south central and Pacific states	Very productive with high management and fertilization	Humid-area soils, clay layers

Source: Adapted from Miller and Donahue (1995) and Brady (1990).

Soil classification occurs using a system of orders, suborders, great groups, subgroups, families, and series. For the purposes of this chapter, only the orders will be mentioned. There are 11 orders of soils of which five exist in diverse climates while the other six are more regionalized. A description of soil types, distribution, and productivity is provided in Table 15.9.

MAJOR GRAZING REGIONS OF THE UNITED STATES

The West

The western half of the United States is highly dependent on snow accumulation in the mountains to supply needed water for irrigation and to enhance spring plant growth. Thus, many cattle producers in the West watch for mountain snow pack reports as they prepare forage management plans for the upcoming spring and summer.

Most cattle grazing operations in the Pacific Northwest are cow-calf operations. Nearly three-fourths of the brood cow's diet is from on-site grazed forage with supplemental hay fed in the wintertime. Northern California, with its mixture of valleys and timbered mountain ranges, has the typical cow-calf operations of the Pacific Northwest states. In the California annual grass type, there is no one predominant type of cattle operation. The mountain grasslands that lie between the valleys and timbered mountains typically are grazed for 2 months each in the spring and fall. Productivity of mountain bunchgrass is high, yet nutritional value drops below adequate levels in late summer and fall as forage matures. During about one-third of the fall period, seasonal rains followed by several days of adequate growing temperatures permit fall regrowth, which has good protein content.

The California annual grasslands, prominent in much of the California foothills, are used during fall, winter, and spring periods. They are a good source of forage in connection with yearling or feeder operations. Coupled with the use of annual grasslands and supplemented rations for feeders or stockers are improved pastures and grazing of crop aftermath. Some cow-calf operators make use of the annual grass type until it matures, then move their herds to higher elevation timber types or to renovated or suitable mountain shrub areas for summer grazing.

Ranchers living within a reasonable distance of mountain-timbered summer range often combine grazing of sagebrush range with mountain summer grazing to achieve forage supplies for all but the winter feeding period. Ranch operations in the heart of sagebrush country try to make this range serve their needs for spring, summer, and fall. Ideally, however, sagebrush range should be used as spring and late fall forage with summer grazing occurring in the mountains. Many producers have small tracts of irrigated meadows interspersed with sagebrush, or they provide supplemental feeds in addition to their sagebrush range.

Throughout the Pacific West, forests provide considerable livestock grazing on a range basis. The most valuable of these for range grazing is ponderosa pine. In prime condition, ponderosa pine is open forest, with an abundance of green feed of good protein content during the hot season. This characteristic gives the ponderosa pine type a premium or extra incremental value over a lower-elevation range.

Coastal Douglas fir and interior mixed conifer forests occupy western Oregon, western Washington, and the northerly slopes of all interior mountains. The grazing productivity of the forest floor is poor (0–50 lbs of air-dry herbage per acre) prior to logging but may increase greatly following logging. The openness of a particular logging or harvest system regulates the amount of usable ground cover produced. Because these tree species are so shade tolerant, they tend to naturally provide a closed canopy in about 20 years. After tree seedlings are reestablished, the forage produced during this 20-year period is often abundant, but the nutritive quality is variable and uncertain. With better utilization of forest residues, treatment of slash, and reseeding of improved ground cover, possibilities exist for summer season herbage production of 1,200–1,800 lbs per acre.

Livestock operators in the Intermountain region, which consists of Idaho, Nevada, Utah, western Colorado, and Wyoming, make use of seasonally productive rangelands by moving animals from one geographical range to another. The desert ranges are used during the winter (November to April), the foothill or intermediate elevation ranges are used during the spring (April to July), and some are used in the fall (October to November). Mountain ranges are used during the summer and fall (July until about mid-October).

Of great importance is the comparative nutrient value of different forage plants during various seasons and the ability or inability of these forage species to meet the requirements for optimum livestock production. Animals in the Intermountain area do not need a supplement during the spring and summer range grazing seasons if the plants are growing. During fall and winter, supplements often are necessary because the forage nutrient content is marginal for animal needs, and inclement weather may seriously reduce daily intake (Fig. 15.8).

A scarcity of suitable spring range in the Intermountain area generally is a limiting factor for successful cattle production. During recent years, it has become common practice to seed depleted foothill range with cool-season species to provide more suitable forage for spring grazing. Grass species such as crested wheatgrass, intermediate wheatgrass, or Russian wildrye often are planted to help provide needed nutrients when native foothill ranges are mature and deficient in nutrients (Fig. 15.9). An alternative, when foothill ranges become dry and dormant, is to move animals to higher elevations where feed is still green and growing. This is sometimes difficult because it puts an additional grazing load on ranges that must also be grazed in the summer.

The vegetation of desert ranges in the Intermountain-Great Basin region is composed primarily of browse species with various quantities of grass. Generally, desert browse plants meet the protein requirements for livestock during gestation and are exceptionally high in carotene. They may be slightly deficient in phosphorus, however, and decidedly low in energy-furnishing constituents. Grasses, during the winter, are markedly deficient in protein, phosphorus, and carotene but are good sources of energy. Forbs are generally sparse on desert ranges and are unimportant in the diet during winter grazing. If the diet is largely grass, then phosphorus and digestible protein may be markedly deficient; if the diet is largely browse, energy may be deficient.

FIGURE 15.8 Cattle may need a protein supplement, shown here in tub form, when forage matures and decreases in protein content.

FIGURE 15.9 Cattle grazing a cleared area, reseeded to crested wheatgrass, in western United States.
Source: Colorado Rancher and Farmer.

Animals on many winter ranges normally require a supplement to meet their requirements when properly grazed; however, with increased grazing intensity, the quantity and even the type of supplement needed may change. Overgrazing may create a need for a greater quantity or even a more expensive supplement over a longer period of time.

In the desert areas of the Southwest, small herds are common, with steers purchased during wet periods to use abundantly produced annual grasses and forbs.

Desert ranges can be grazed year-long. Grasses like black grama cure well and maintain some nutritive value year-long. Desert grasslands are quite valuable for grazing despite their aridity because most of the grasses are highly preferred. During the growing season, grasses have a crude protein content of 8% or higher, generally at or above minimum levels for livestock. Protein drops off to 4–5% in winter, requiring some supplementation. Desert shrub areas have a more erratic type of forage production. Most of the forage is produced following rains (usually in late winter and spring) and consists of annuals such as alfilaria, six-weeks grama, and other annual forbs and grasses. Some shrubs, principally four-wing saltbush, provide nutritious, palatable feed. Four-wing saltbush, for example, retains a year-long protein content in excess of 11%.

The foothills-type rangeland, which includes pinyon-juniper and chaparral-mountain shrub ecosystems, provides mainly spring-fall grazing for cow herds. The pinyon-juniper is an abundant ecosystem but relatively unproductive of forage. It has historically been heavily grazed because of its proximity to ranch headquarters and the better parts have been plowed for crop production. The chaparral-mountain shrub ecosystems generally provide limited grazing because the dense brush and tree cover make grazing difficult. Their primary use is spring-fall, or spring, summer, and fall.

The mountains of the Southwest commonly support a forest cover of ponderosa pine with occasional open, mountain grassland areas. Part of the precipitation falls in the winter and spring from Pacific storms, and part during the summer from storms originating in the Gulf of Mexico. Grazing by cow herds occurs during the summer and fall.

Throughout the Southwest, brush has increased since domestic livestock grazing began there several hundred years ago. Palatable, perennial grasses have decreased, and in some areas, like the Desert Shrub, the perennial grasses have been largely replaced by annual grasses and forbs.

The Great Plains

The Great Plains, which occupies approximately one-third of all U.S. land, includes parts of ten states (Montana, North Dakota, South Dakota, Wyoming, Nebraska, Kansas, Colorado, Oklahoma, Texas, and New Mexico). This vast area of semiarid, high-plateaued grasslands once known as the Great American Desert varies in topography from extremely flat to quite rolling. In general, it is covered with short grasses and mid-grasses, but in the Black Hills of South Dakota, it is covered with trees. On the eastern edge and scattered throughout on deep, sandy soils are large areas of highly productive tallgrass prairie.

The northern Great Plains covers an estimated 300,000 square miles, which is approximately one-tenth of the total land area of the United States. The range in the northern Great Plains occurs in the western three-fourths of North Dakota, South Dakota, and Nebraska, in the eastern two-thirds of Montana, and in the eastern one-half and eastern one-fourth, respectively, of Wyoming and Colorado. Sometimes the Great Plains is identified in three parts, for example, northern, central, and southern (Fig. 15.4) rather than the northern and southern as described here.

Approximately one-fourth of the northern Great Plains area is under cultivation. Production of both cultivated crops and range forage varies greatly from year to year and from one area to another. Droughts are frequent; records indicate that precipitation in certain parts of this area has dropped below 75% of the normal on the average of once in every 5 to 8 years. Native vegetation can best withstand these successive drought periods.

While cultivated crops may be of importance, especially in local areas, grasslands are the principal resource base for the agricultural economy prevalent in the northern Great Plains.

Livestock in the northern Great Plains graze the native range species for about 8–9 months of the year, from April to December. During cold and stormy winter months, cattle are generally fed in holding pastures near farmsteads or on range areas where shelter is provided by breaks in the terrain or creek bottoms where trees may be present. Range areas in the northern Great Plains furnish one-half of the nutrients needed by cattle. An additional 30% comes from grassland hay, which is useful during the long winter periods. The rest of the annual feed requirements come from supplemental feeds and cultivated cropland, including crop aftermath.

Some livestock operations in the northern Great Plains are cow-calf enterprises or yearling steer operations, where most producers sell weaner calves and yearlings. In cases where topography is highly variable and vegetation composition is complex, both cattle and sheep can be managed to utilize more fully the range forage resource.

The major range types in the northern Great Plains consist of short-grass and mixed-grass prairies. The Sandhill ranges are also important vegetation types and occur mainly in west-central Nebraska with smaller areas in adjacent states. Precipitation varies from 15 in. to as high as 20–22 in. annually. Since moisture is rapidly absorbed into the soil, there is little runoff; therefore, effective precipitation is greater in the sand areas.

The sagebrush-saltbush grassland of the northern Great Plains is called the Red Desert and covers some 43,000 square miles. It is the most arid part of the area. The Red Desert is grazed mainly by sheep in the fall and winter when snow is present.

In the Black Hills of South Dakota and the Bighorn Mountains of Wyoming and other local areas, grasses occur in an open forest type and include ponderosa pine, Douglas fir, and some spruce at higher elevations. This covers almost 20,000 square miles in the northern

Great Plains and produces considerable amounts of forage used primarily by cattle during the summer.

The southern Great Plains is a leading cattle-producing and livestock farming area that includes over 130 million acres of southeastern Colorado, western Kansas, western Oklahoma and Texas, and parts of northeastern New Mexico. Like the Plains area to the north, the climate of the southern Great Plains is extremely variable from year to year. Rainfall is generally limited, humidity is low, and high winds cause high evaporation rates.

Year-long grazing of range is a common practice in the southern Great Plains, especially with breeding herds. Cultivated crops are often used to supplement winter range, and wheat pasture is an important winter and early spring forage. In the extreme northern part of this area, ranges may be used from April or May to October or November, as is the case in the northern Great Plains.

The vegetation and topography of the southern Great Plains are excellent for cattle. The environment is generally well suited to the production of young animals. Calves are sold primarily as stockers and feeders.

Some supplemental feeding is practiced in winter for increasing protein intake. The use of crop residues in parts of the southern Great Plains and of wheat pasture in fall and spring is common. Using sorghum pastures for late summer grazing is rapidly increasing throughout most of the southern Great Plains.

The tallgrass prairie in the eastern portion of the area is among the world's most productive grassland areas. The vegetation is comprised primarily of tall grasses, some of which exceed 6 ft in height. The forage of this area can support many cattle. Compared to more drier areas, the vegetation does not cure well. Protein supplements are needed throughout the winter months. Areas that once largely grazed steers in the spring and summer are now being converted to cow-calf operations.

The principal range forage types in the western portion of the southern Great Plains consist of short- and mid-grass vegetation that occurs on the heavier textured soils. On more sandy soils along the streams in this area, mid-grass and tallgrass vegetation occurs. The short-grass vegetation type is dominated by blue grama and buffalo grass (Figs. 15.10 and 15.11).

Grazing Fees

Rangelands in the seventeen western States (including the Great Plains) have both private and public ownership. Many ranches have a combination of privately deeded land and grazing permits for use of public lands. Grazing associations also exist where several ranchers cooperatively graze cattle on land privately owned by a grazing association, public grazing land, or a combination of the two.

FIGURE 15.10 A brush and grass range forage combination in the Great Plains area. Note the water tank, filled by a windmill, which is common to several separately fenced pastures. *Source: Colorado Rancher and Farmer.*

FIGURE 15.11 Cattle grazing native range in the Flint Hills, a very productive grazing area in Kansas.
Source: Kansas Agric. Expt. Sta.

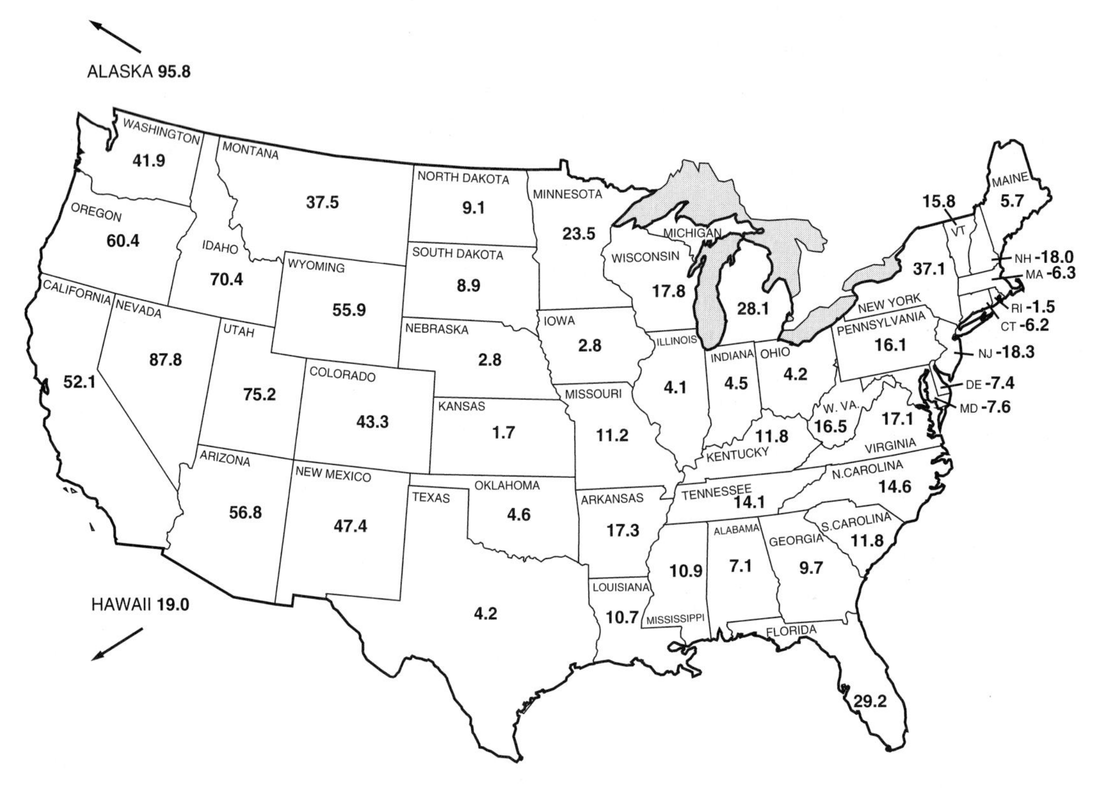

FIGURE 15.12 Percentage of federal and state government ownership of land.
Source: National Wilderness Institute.

Public lands are both state and federally owned (Fig. 15.12). State lands primarily involve millions of acres. These state lands are usually referred to as "school sections" since revenues from two sections, from each township (see Figs. A.4 and A.5 in the Appendix), were to help support public schools. Cattle producers often have lease agreements for this land with extended leases given to them if certain terms and conditions are met. Lessees own all improvements on the land. If the land is sold or leased to another individual, the original lessee is reimbursed for an appraised value of the improvements.

Federally owned grazing lands are typically managed by the U.S. Forest Service (USFS) or the Bureau of Land Management (BLM). Permits to graze Forest Service land are issued to qualifying permittees who have a privately owned livestock operation. Only cattle owned by permittees are allowed to graze the lands managed by the Forest Service. Forest supervisors control the pattern of range use. They may adjust the number of cattle permitted to graze a certain area and the length of time the cattle can be on the range. Grazing fees, however, are established by federal legislation. Therefore, grazing fees and stocking rates are established independently of each other.

Permits can be terminated if the Forest Service decides to use the land for other purposes. In recent years, there has been considerable discussion as to whether federally owned lands should have only a single use (recreation for the general public) or multiple uses, including cattle grazing.

The BLM manages the other public land in a manner similar to that of the Forest Service. Grazing permits are issued, and BLM personnel determine the manner in which the land will be grazed.

Annual grazing fees are assessed to beef producers who have grazing permits. Fees are determined on an animal unit month (AUM) basis. Total forage costs on national forests (BLM forage costs are similar) may include one or more of the following components: (1) the annual grazing fee, (2) nonfee costs (investment and maintenance of improvements, herding, transportation, and others), and (3) investment in the grazing permit. Grazing fees, which are governed by Congress, are often a source of controversy (Table 15.10).

One of the challenges in determining comparable value between private and public land grazing leases is accounting for nonfee costs and permit value. While the courts have ruled that permit values (economic value associated with holding a federal grazing permit) do not have to be accounted for in determination of federal grazing fees, they still have market value and, as such, remain a point of contention.

The 1995 report by the Grazing Fee Task Group found that on private leases, landlords provided all or part of property maintenance (51%), daily care of cattle (17%), water supply (53%), liability insurance (41%), utilities (34%), and provision for death losses (6%). On federal lands, neither the BLM or USFS provides these services. Therefore, the nonfee costs must be accounted for when comparing private and public lease rates (Table 15.11). The total cost analysis (Table 15.11) found that 34% of cattle producers on BLM lands and 62% of cattle producers on USFS lands paid more for grazing public lands as compared with private lands.

TABLE 15.10 Average Annual Grazing Fees for Federal Range, 1970–2001

Year	Forest Service/BLM—Cost per AUM ($)
1970	$0.60[a]
1975	1.11[b]
1980	2.36
1985	1.35
1990	1.81
1995	1.61
1998	1.35
2001	1.35

[a]Denotes USFS fee, BLM fee was $0.44 per AUM.
[b]Denotes USFS fee, BLM fee was $1.00 per AUM.

TABLE 15.11 Average Cattle Grazing Costs on Public and Private Leased Lands in Idaho, New Mexico, and Wyoming, 1992

	BLM ($)	USFS ($)	Private ($)
Lost animals	3.09	4.49	2.10
Association fees	0.20	1.07	0.00
Veterinary	0.08	0.12	0.12
Moving livestock[a]	2.83	4.94	2.08
Herding	3.63	5.00	2.94
Miscellaneous labor/mileage	0.61	0.77	0.18
Salt and feed	1.41	1.12	1.80
Water	0.47	0.24	0.11
Improvements maintenance	2.86	3.41	1.84
Other costs	0.69	0.96	0.26
Private land lease rate	—	—	7.71
Total Cost	15.41	21.89	19.04

[a]Includes cost of horse.
Source: The Value of Public Land Forage and the Implications for Grazing Fee Policy, 1995.

The Grazing Fee Task Group (1995) made several key recommendations in regards to the grazing fee issue:

1. The grazing fee should be established within the range of $3 to $5 per AUM. This recommendation assumes no allowance for the investments graziers have in grazing permits. Recognition of grazing permit value remains a key issue. The current grazing fee, or an even lower fee, would be appropriate if grazing permit investment were considered.
2. Any base grazing fee should be applied throughout the West.
3. Any base grazing value should be updated annually with the forage value index from the previous year.
4. The BLM and USFS should investigate the potential of implementing a competitive bid system to create a market for public land grazing.

Wheat Pasture

Cattle graze several types of small grain pastures throughout the United States. The most noted of these are the winter wheat pasture areas in several southern Great Plains states where millions of stocker cattle are grazed each year. Wheat pasture is also important in some adjoining states (Fig. 15.13), but to a much lesser extent than in Kansas, Oklahoma, and Texas.

Wheat is planted in late summer (typically the first week of September); depending on moisture and other conditions, the fall and winter growth is available for cattle grazing. Typically, cattle are removed from wheat pastures in early spring, and a crop of wheat for grain is then produced. The usual grazing period is from November to the middle of March. Grazing can continue to the end of May if a decision is made not to harvest a crop of grain.

Daily gains on wheat pasture are usually 1.5–2.5 lbs per day. There can be serious problems with bloat; average annual death losses of less than 1% can be expected, although losses may reach 3%. Bloat is most frequent during a 2–4-week period in the early spring when the wheat breaks dormancy. As the wheat matures, bloat is reduced. Some producers feed low-quality roughages (wheat straw or sorghum-sudan hay) in an attempt to improve the utilization of wheat pasture and reduce bloat. Research shows that *ad libitum* consumption of these roughages is low, with little effect on the incidence of bloat.

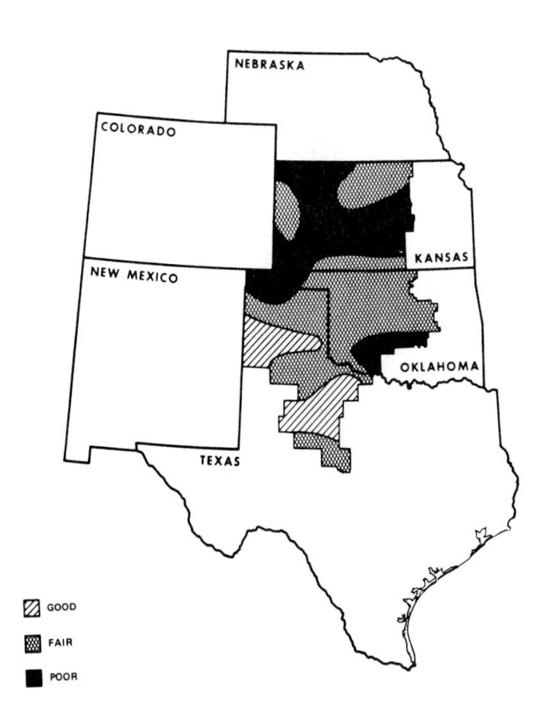

FIGURE 15.13 Major winter wheat pasture area in the United States. Amount of top growth available for grazing at one point in time is depicted.
Source: Western Livestock Round-Up.

Stocking rates vary with the amount of top growth available, and more numbers can typically be put on irrigated wheat than dry-land wheat. Stocking rates are more critical after the jointing of the wheat as it starts to mature and accelerate growth. If stocking rates are not high enough at this time, the wheat will mature ahead of the cattle's ability to graze it. Thus the cattle will spot graze and avoid the more mature wheat plants.

Cattle may have to be moved from wheat pasture during periods of heavy rainfall or irrigation to prevent trampling damage and a significant reduction to the wheat crop.

North-Central Area

A large concentration of highly fertile soil and favorable growing conditions is located in the Corn Belt states (Iowa, Missouri, Illinois, Indiana, Ohio, and parts of some adjoining states). In most cases, pasture production cannot compete economically on the most productive land where corn and soybeans can be grown. There are thousands of acres of productive land in the Midwest, where row-crop farming is not feasible on all these acres because of the slope of the land. This land has major erosion problems if it is plowed annually. Conservation compliance may result in more of this land being converted into permanent pasture. Pasture production on this type of land is excellent, and beef cow numbers are high in several of the states in the Midwest. For example, Iowa, with its 1 million head of beef cows, has more beef cows than any one of the 11 western states and ranks high in beef cow numbers in the United States. Beef cows are concentrated in the rolling hills area of southern Iowa, where it is estimated that more than 650,000 cows are located within a 75-mile radius of Ottumwa.

Smooth bromegrass, orchardgrass, fescue, and Kentucky bluegrass are among the most common grasses in the Midwest. They are grown in pure stand pastures or interseeded with common legumes, such as alfalfa, clovers, birdsfoot trefoil, and lespedeza. These cool-season pastures are highly productive during spring and early summer. Their productivity, however, is low in July and August. Thus, cattle performance declines during the midsummer months. If adequate amounts of rainfall are received during late summer and early fall, there will be a regrowth of highly nutritious cool-season forage that can be utilized. Freezing temperatures occurring in the fall terminate new growth, and beef producers rely on the grazing of matured forages and crop aftermath. Harvested forage is then provided when available forage for grazing is depleted, or when snow prevents an adequate amount of dry matter to be grazed.

There is increased utilization of warm-season grasses to complement cool-season forage. For example, in past years the tallgrass prairie extended into the eastern Corn Belt. Switchgrass, Indiangrass, and big bluestem are being utilized in Iowa and Missouri to provide available green forage during July and August. Preference is given to seeding these in pure stands because they are easier to manage. Also, warm-season grasses do not perform well when mixed with cool-season grasses or legumes because of differences in growth patterns and competition.

Warm-season grasses have to be managed differently than the same grasses grown farther west because higher rainfall in the Midwest and competition from weeds is a greater problem. Switchgrass is often the first choice among farmers trying a warm-season grass for the first time because it is easier to establish. Switchgrass, however, is considered a lower-quality grass compared with big bluestem or Indiangrass.

Grazing management of cool-season species is similar to warm-season grasses with one noticeable exception: Warm-season grasses store their energy reserve for regrowth above ground in the lower 8–10 inches of the plant, whereas cool-season grasses and legumes place their food reserves in the root systems and crowns. This is the main reason warm-season grasses do not tolerate close grazing.

Figures 15.14, 15.15, and 15.16 demonstrate an excellent planning system for providing a yearly forage supply under Midwest conditions. Figure 15.14 is a worksheet for estimating the forage demand while Figure 15.15 helps estimate forage availability. Figure 15.16 shows seasonal forage production for a variety of sites.

The South

Until the 1940s, cattle production in the South was incidental to the principal uses of the cleared eastern forestland for raising cotton and tobacco. Prior to this time, the relatively infertile soils were becoming depleted and cash returns were dropping. Ways were sought to rehabilitate seriously eroded cotton and tobacco lands and to provide alternate sources of cash income.

Many eroded croplands in the South were planted in trees; however, this did not solve the cash income problem since 15–20 years were required for the planted trees to reach pulpwood size. Interest in cattle raising as an income source increased, particularly when it was realized that forest soils could produce high yields of forage when soil fertility needs were met. In addition, high-yielding species or strains of grasses and legumes were developed, among them the pioneer Coastal bermudagrass.

As a result of improved pastures in the Southeast, beef cattle numbers increased dramatically. In twelve southeastern states (Florida, Mississippi, Louisiana, Alabama, Georgia, Arkansas, Kentucky, Tennessee, North Carolina, South Carolina, Virginia, and West Virginia), beef cow numbers increased from 2.8 million in 1950 to more than 7.9 million in 2001 (or approximately 25% of the nation's cow herd).

There is great diversity in the pasture operations of the Southeast (Fig. 15.17), including many highly developed and intensively managed pastures for purebred, commercial cow-calf, or yearling operations. Kentucky bluegrass is used extensively in the northeastern part of the region; tall fescue is also important. In the southern part of the region, bermudagrass, bahiagrass, and other warm-season grasses are common.

Kind/Class of Livestock	Animal Value Unit	# AUs	Month Jan	Feb	Mar	Apr	May	Jun	Jul	Aug	Sep	Oct	Nov	Dec
Mature cow	0.9	#	412	412	412							412	412	412
(nonlactating)		AUs	370.8	370.8	370.8							370.8	370.8	370.8
Mature cow	1.0	#				500	500	500	500	500	500			
(lactating)		AUs			88 head	500	500	500	500	500	500			
Replacement bred heifers	0.08	#	100	100								100	100	100
(18–24 months)		AUs	80	80								80	80	80
Replacement yearling heifers	0.7	#				125	125	125	125	125	125			
(12–17 months)		AUs				87.5	87.5	87.5	87.5	87.5	87.5			
Replacement heifer calves	0.5	#	125	125	125							125	125	125
(6–12 months)		AUs	62.5	62.5	62.5							62.5	62.5	62.5
Calves (3–4 months	0.3	#							480	480	480	81 hd terminal		
through weaning)		AUs							144	144	144			
Weaned steer-heifer calves	0.5	#	274	274	274							274	274	274
(6–12 months)		AUs	137	137	137							137	137	137
Yearling steers/heifers	0.7	#				226	226	226	226	226	226	sold		
(12–17 months)		AUs				113	113	113	113	113	113			
Young bulls	1.2	#												
(12–24 months)		AUs												
Bulls	1.5	#	15	15	15	15	26	26	26	26	15	15	15	15
(mature, 2–5 years)		AUs	20.25	20.25	20.25	20.25	35.1	35.1	35.1	35.1	20.25	20.25	20.25	20.25
Total		#												
		AUs	670.55	670.55	675.95	720.75	735.60	735.60	879.60	879.60	864.75	670.55	670.55	670.55

FIGURE 15.14 Worksheet for estimating forage demand (from Waller et al., 1986). Refer to Table 15.4 for assigning the correct AU.

Kind of Forage	AUM/ Acre	Acres	AUMs	Month											
				Jan	Feb	Mar	Apr	May	Jun	Jul	Aug	Sep	Oct	Nov	Dec
Summer Winter Grass	.429	9,820	4,214* ↓ 5,057						900	900	900	900		729	729
Alfalfa Residue	2	290	580										580		
Alfalfa	15.6	290	4,524	800	800	800	800	800					175	175	175
Total			10,161												
Total AUMs Allocated				800.00	800.00	800.00	800.00	800.00	900.00	900.00	900.00	900.00	755.00	904.00	904.00
Total AUMs Required				670.6	670.6	676.0	720.8	735.6	735.6	879.6	879.6	864.8	670.6	670.6	670.6
Total AUMs Excess/Deficiency				153.45	153.45	148.05	79.25	64.40	164.40	20.40	20.40	35.25	108.45	257.45	257.45

*Assumed additional AUMs can be harvested with 60% use (including deferred winter use pasture).

FIGURE 15.15 Worksheet for estimating forage availability (from Waller et al., 1986).

Forage	Jan	Feb	Mar	Apr	May	Jun	Jul	Aug	Sep	Oct	Nov	Dec
	Percent Produced by Months											
RANGELAND												
Tallgrass prairie	—	—	—	5	30	33	18	10	4	—	—	—
Mixed grass prairie	—	—	—	5	25	30	25	10	5	—	—	—
Sand prairie	—	—	—	5	25	30	25	10	5	—	—	—
Shortgrass plains	—	—	—	—	10	35	35	10	10	—	—	—
TAME PASTURE-Cool Season												
Kentucky bluegrass	—	—	5	10	30	30	—	—	29	10	—	—
Smooth bromegrass	—	—	5	15	30	20	—	—	8	17	5	—
Smooth bromegrass + N	—	—	5	15	50	10	—	—	5	10	5	—
Tall fescue	—	—	5	5	35	30	—	—	17	8	—	—
Tall fescue + N	—	—	5	15	35	15	—	—	8	17	5	—
TAME PASTURE-Warm Season												
Bermudagrass + N	—	—	—	—	15	30	20	15	20	—	—	—
TAME PASTURE-Irrigated												
Cool-season mixture	—	—	—	5	10	25	15	8	10	10	7	—
Interm. Wheatgrass	—	—	—	8	33	33	16	8	2	—	—	—
LEGUME PASTURES												
Alfalfa	—	—	—	—	10	30	30	20	10	—	—	—
Alfalfa-Brome	—	—	—	—	30	30	15	10	15	—	—	—
ANNUAL PASTURES												
Hybrid pearl millet	—	—	—	—	—	5	40	40	15	—	—	—
Rye	—	—	15	25	20	—	—	—	—	15	15	10
Spring oats	—	—	—	—	33	67	—	—	—	—	—	—
Summer annuals	—	—	—	—	—	10	45	35	10	—	—	—
Wheat (grain)	—	—	40	20	—	—	—	—	—	15	20	5
Wheat (graze out)	—	—	10	25	35	—	—	—	—	5	15	10
CROP RESIDUE												
Corn	10	—	—	—	—	—	—	—	—	20	50	20
Grain sorghum	10	—	—	—	—	—	—	—	—	20	40	30
Wheat	—	—	—	—	—	10	70	20	—	—	—	—

FIGURE 15.16 Forage resources and production distribution for grazing in Kansas. Adapted from data by Barnett, Murphy, Posler, and Owensby (Kansas State), Wedin (Iowa State), Moline et al. (Nebraska), Murphy et al. (Missouri), and McMurphy (Oklahoma State). Compiled by G. L. Posler and P. D. Ohlenbusch, Department of Agronomy, KSU.

FIGURE 15.17 Brahman cattle grazing Bahiagrass pasture in Louisiana.
Source: Dr. L. I. Smart, Louisiana State University Agricultural Center.

In some areas of the Southeast, cattle may graze entirely on farm woodland areas and other pastures (Fig. 15.18). Many depend largely on industrial forestlands and national forests. It is not unusual for cattle to graze year-long in the forest with no provision for rotation other than that associated with range burning. Systems for grazing pine forest ecosystems have been developed that are compatible with tree growing. Cattle also graze hardwood forests, but this is often highly damaging to tree growth.

An example of year-round grazing in the Southeast is shown in Figures 15.19 and 15.20. Cattle that can be grazed on green forage most of the year will have excellent performance. Small grains pasture or small grains and ryegrass overseeded into bermudagrass can provide excellent winter forage (Fig. 15.21).

The Northeast

A high proportion of the farms in the Northeast are dairy farms situated on a combination of hill and valley land, which requires the use of forage grown in rotation to maintain soil productivity. Nearly half of the region's farmland is occupied by hay and pasture crops because most of the land is poorly suited for intensive cultivation. The Northeast has traditionally imported much of its feed grains from the Midwest.

FIGURE 15.18 Cattle grazing mixed improved and native forage in a pine stand in Georgia.

FIGURE 15.19 Forage management fields for year-round forage supply in Kentucky. *Source:* Adapted from Absher, Murdock, and Kaiser (1979).

FIELD 1 (25-30 ACRES)
WINTER PASTURE
FESCUE & NITROGEN

FIELD 3 (25-30 ACRES)
SUMMER PASTURE
PERENNIAL GRASS & CLOVERS
(Red, common white and ladino clover must dominate the cool season perennial grass used in this pasture. Tall fescue should only be used if a dominant clover stand can be maintained. Could use bluegrass, orchardgrass or Bermudagrass).

FIELD 2 (40-50 ACRES)
SPRING & FALL PASTURE
PERENNIAL GRASS & CLOVERS
(Most likely tall fescue, red, white and ladino clovers. Other grasses could be orchardgrass or bluegrass).

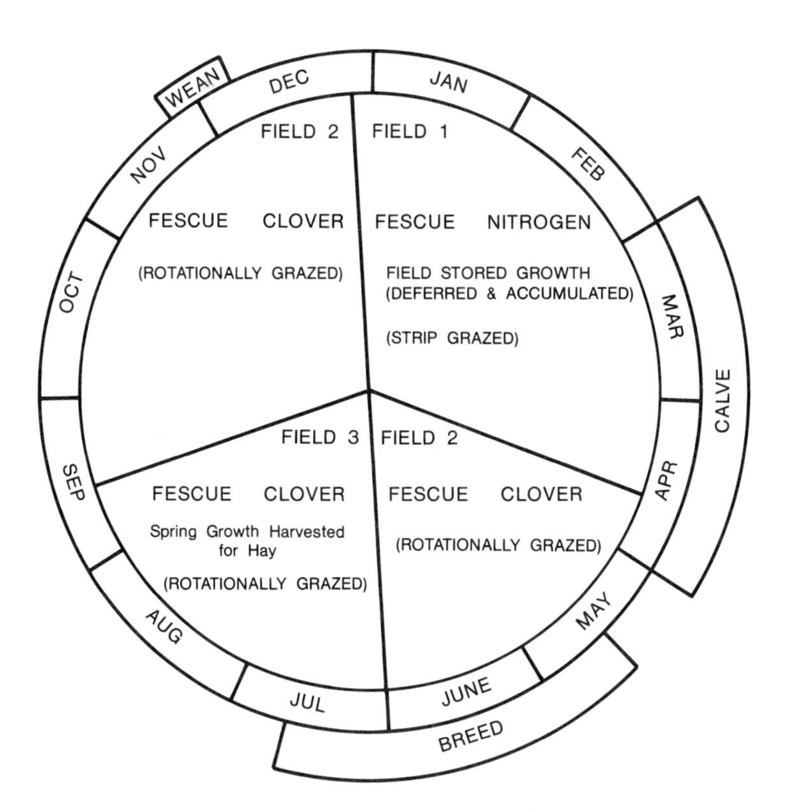

FIGURE 15.20 Cow forage management calendar-clock in Kentucky.
Source: Adapted from Absher, Murdock, and Kaiser (1979).

FIGURE 15.21 Steer grazing on rye and ryegrass pasture in northern Georgia.
Source: M. K. Cook, Univ. of Georgia.

Perennial grasses dominate most haylands and pastures in the Northeast. Timothy, sorghum-sudangrass, smooth bromegrass, and orchardgrass are the most commonly used forage grasses. Except for alfalfa, most perennial seedings include both grasses and legumes. Alfalfa is the most productive legume where it is adapted. It is used as hay, silage, or greenchop, or in rotation pastures. Red clover, ladino clover, and birdsfoot trefoil are other legumes successfully grown in the area.

In the southern area of the northeastern United States, cool-season annual grasses are an important source of forage. Winter cereals can be grazed by beef cattle during late fall and early spring. Winter rye can be sod-seeded where bermudagrass is grown to extend the grazing season into the cooler months. Warm-season annual grasses are used throughout the region as a source of supplemental summer feed.

HAYS

Hays are harvested from legumes, grasses, and cereal crops. Alfalfa hay comprises nearly 60% of the more than 60 million acres of hayland in the United States.

The primary objective in haymaking is to reduce the moisture content of green forage, which typically ranges from 65–85% but can be as low as 20%. Hay that is stored with excessive moisture can be destroyed by fire caused by spontaneous combustion, or mold can develop from heating and thus reduce the feeding value. Hay that is harvested too dry is likely to lose many leaves, which are much higher than the stems in protein, carotene, and other nutrients. In addition, hay that is bleached by the sun has a reduced carotene content. Bleaching of hay can be extensive when it is left too long in the windrow or when it is not properly stored outside.

Maturity at harvest has one of the most pronounced effects on the nutrient composition of hay accounting for approximately 70% of quality differences. The crude protein content of alfalfa, for example, shows the following values on a dry-matter basis: immature, 21.5%; prebloom, 19.4%; early bloom, 18.4%; midbloom, 17.1%; full bloom, 15.9%; and mature, 13.6%.

Hay exposed to moisture during the haymaking process is subject to extensive leaching of nutrients from the plant material. Studies have shown that alfalfa hay in the swath when exposed to 0.8 in. of rainfall can have its dry-matter yield reduced by 10% and its cellular proteins and carbohydrates reduced by 10–20%. The leaching losses from this amount of moisture are not as large as those that occur when the maturity of plants is allowed to advance seven days. Rainfall during the curing of hay can maintain moisture levels favorable to microbial decomposition, which causes the shattering of fragile leaves.

Hay production costs are often a significant factor influencing the cow-calf enterprise. The costs associated with hay production are outlined in Table 15.12.

It is not uncommon for hay to lose 30–50% of its dry matter from harvesting, storing, and feeding. Small bales of hay can be stored in well-constructed stacks in order to reduce losses in nutrient storage (see the Appendix). Losses from large round hay bales can be much greater than losses from smaller, square bales unless managed properly. A study conducted in a 19-in. rainfall area showed the following losses in storing large round bales: pyramid, 10.3%; individually separated, 4%; and end-to-end arrangement, 0.8%. If baled hay is to be stored outside, it is important to select a well-drained, gently sloping site with southern exposure. A barrier between the ground and the bale can significantly reduce spoilage and loss. The use of individual bale wrappers or a tarp covering for the stack is recommended especially when bales are in a multilayer formation.

If bale spoilage occurs, losses can be significant. For example, a round bale that is 4 ft wide and 5 ft in diameter will have 13% of the forage in the bale affected by spoilage if the weather penetration is 2 in. By contrast, if the weather penetration is 6 in. deep, 36% of the forage is

TABLE 15.12 1996 Costs Associated with Haymaking in Various Regions of Colorado

		Range/Most Common			
Activity	**Basis**	**SE**	**Western**	**NE**	**San Luis Valley**
Hay, mowing	$/acre	7–9.5/7	10–17.5/10	NA	NA
Raking	$/acre	2–6/4	5–17.5/7	4–7/5	1–3.50/1.50
Swathing	$/acre	5–10/9	10–17.5/12	4–12/7.5	9–12/10
Baling (small square)	$/bale	8.75–13/10	10–13/12	8.25–10/10	12–18/13
Baling (big square)	$/ton	5–15/12	10–22/15	2–15/12	10–14/13
Baling (big round)	$/bale	6–20/12	7–25/12	6–10/8	N/A
Load and stack	$/bale	.23–.40/.25	.30–.50/.40	.18–.35/.32	.20–.40/.25
Haul bales (big square)	$/ton	3–15/9	8–20/10	9–10/9	3–6/4
Stacking loose hay	$/ton	10–15/10	N/A/N/A	10–20/15	N/A/N/A
Tub grinding	$/ton	10–12/10	6–12/10	11–12/11	N/A/N/A

Source: Colorado State University.

FIGURE 15.22 Large, round hay bales can be satisfactorily fed to reduce the amount of hay wasted.

affected. It is extremely important to use feeding systems that reduce the amount of hay that is trampled and wasted (Fig. 15.22).

Because forage quality is maximized prior to attainment of maximum yields, forage to be utilized as feed for young, growing stock should be harvested early to take advantage of feed value while forage destined for use by mature cattle should be harvested later to gain increased tonnage. Furthermore, stockpiling forage, cool-season species in particular, may be an appropriate option. Stockpiled pastures are saved for late-season grazing. Stockpiled legume pastures should not be grazed until after a hard frost occurs.

FIGURE 15.23 Corn stalk fields are available to cattle after the grain has been harvested.
Source: Holly Foster.

CROP RESIDUES

Several crop residues provide a substantial amount of roughage and forage for beef cattle. The most common crop aftermath feeds that can be harvested or grazed are corn stalks, straw, or sorghum stubble and soybean residue. Most of these crop residues will contain what is left behind after machine harvesting. It is estimated that the 75 million acres of harvested corn and 35 million acres of harvested grain sorghum can produce approximately 200 million tons of crop residue each year, supplying the winter energy needs for 35–40 million cows. An additional 75 million tons of cereal grain residue (primarily straw) is available in the United States.

Crop residues, because of their relatively low availability of metabolizable energy, are most effectively utilized in maintenance rations for gestating cows when the energy requirement is less than during lactation. Typically, crop residues are low in protein and phosphorus, marginal in calcium, and high in fiber and lignin.

Grazing crop aftermath is usually the most economical method for utilizing feed; however, harvested material can be used when supplemented properly (Fig. 15.23). Snow and mud interfere with the effective utilization of grazed crop residues. During the past several years, more crop residues have been used for cattle feed due to the combined efforts of producers, researchers, and equipment manufacturers.

In semiarid regions, residues from feed grains and cereal crops function in water harvest (via snow collection) and erosion protection. Thus, their value as feed is less important than their value as ground cover in minimum tillage systems.

DROUGHT MANAGEMENT

Drought is defined as a prolonged period of several months or more of below-normal soil moisture. A severe drought occurs when precipitation is 25% or more below normal. While drought is usually identified as years of low rainfall, drought can occur when the seasonal distribution of rainfall is not favorable for plant growth or when temperatures are high.

Below-normal rainfall throughout the United States and reduced snow pack in the mountains of the arid West can result in extremely low forage production. These drought conditions can occur during one year or can extend through several years. If stocking rate is not reduced during drought, there is a reduction in the pounds of beef produced and usually costs increase. Drought can lead to lowered pregnancy rates, reduced rangeland productivity, and changes in plant community composition.

Drought management cannot always be precisely implemented. However, a critical analysis of weather trends and the development of a plan based on "what if" questioning can help producers mitigate the severity of impact from drought conditions. The first step is to assure that the land resource is continually managed with the goals of establishing organic matter and plant cover appropriate to a healthy rangeland. If soils and plants have already been stressed by poor management practices in the average to above-average rainfall years, then a drought year leads to catastrophic conditions for the resource.

Managers should have sufficient monitoring data to be able to establish a general forage grow curve for the ranch or farm. Based on these trends, producers should be able to determine a "date of no return" at which time forage growth is likely to be insufficient even if rainfall occurs.

The adjustment of stocking rate is critical in drought conditions, and changes must be made in a timely fashion to avoid losing flexibility altogether. Drought slows the ability of plants to grow and, therefore, resting pastures become of increasing importance. Movement of cattle more frequently is a likely strategy under these conditions.

Producers should have a destocking strategy in place based on their "what if" planning process. Reacting under stressful conditions is not desirable. An analysis of market conditions and the financial position of the enterprise can help determine whether to buy feed, transport cattle to nondrought areas, or to sell cattle. If selling cattle is the best option, marketing the least productive animals is a good place to start.

In some drought areas, producers will wean calves early so cows can maintain or increase body condition, thus preventing low reproductive performance the following year. Creep feeding calves prior to early weaning can be done to reduce stress following weaning. The early-weaned calves make an easier transition to the feed after weaning.

Income taxes should be evaluated as forced liquidation may dramatically increase the tax liability for that year. However, if a federal drought disaster is declared, special tax considerations may change what are the best management options.

MARKET CATTLE PRODUCTION ON GRAZED FORAGE

During the 1970s, the production of market cattle by using primarily forages was a topic of much discussion and research. Some cattle are fed grain while grazing forage from pastures (Figs. 15.24 and 15.25). Market cattle production from forage occurs when feed grain prices are very high and large losses are experienced by commercial feedlots. Feed grain prices are

FIGURE 15.24 Yearling heifers on a grain-on-grain system in Kentucky. Cattle graze pastures, then fed grain for 90 to 150 days to the select-low Choice grade. *Source:* Kentucky Agric. Expt. Station.

FIGURE 15.25 Calves can be self-fed grain while obtaining part of their nutrient supply from pasture. *Source:* Elanco Products Company.

variable depending on yearly production, exports, and government programs. Currently, most market cattle are "fed" cattle, which means they have been fed concentrates in a feedlot. A major challenge in producing forage-fed cattle in extensive forage areas such as the Southeast is the lack of accessible packing plants.

Some opponents of grain feeding argue that the grains should be reserved for human consumption. Feed grain prices are the primary determining factor in how much beef is produced from grain feeding. Some beef producers expect grass fattening of beef to increase in future years due to the rising costs of cultural energy and reductions in underground water supplies. The latter may reduce significantly the amount of grain grown in several important cattle feeding states in the Great Plains.

The large renewable forage base for cattle grazing assures people in the United States of a supply of beef even under adverse economic conditions. Most of the nation's beef supply comes from the commercial cattle feeding industry. Without a feeding industry, cows and yearlings would compete for the forage supply. Total annual tonnage of beef would likely be reduced and the supply would be more seasonal. In addition, the palatability characteristics of grass-finished beef would be less uniform than those experienced with grain-fed beef.

Economics Associated with Grazing

The most expensive asset purchased by an agricultural enterprise is land. Real estate market trends have a significant impact on the ability of existing enterprises to expand or for new producers to initiate a farm or ranch. Since 1992, the value of farm real estate, including land and buildings, has increased by nearly 30% in the United States (Table 15.13). Utah, Oregon, Maryland, and Alabama have all experienced increases in value of greater than 30%. Minneosta, South Dakota, Florida, Oklahoma, and California have less than 12% increases in farm real estate over the same time period. Factors influencing farm real estate values include pressure from alternative uses, proximity to population centers, availability of water, aesthetic value of the property, governmental regulations, and availability of credit. Given the rather dramatic increased farm real estate valuations in the 1990s, agricultural producers must carefully assess the costs and benefits associated with land purchases. Because of the

TABLE 15.13 Value of Farm Real Estate[a]

	Average Value per Acre as of January 1			
Region and State	1992 (dollars)	1996 (dollars)	2000 (dollars)	1992–2000 Change (percent)
Northeast	1,977	2,485	2,470	19.9
Maine	1,033	1,291	1,210	14.6
New Hampshire	2,103	2,578	2,300	8.6
Vermont	1,223	1,534	1,640	25.4
Massachusetts	4,340	5,597	5,900	26.4
Rhode Island	5,627	7,204	6,500	13.4
Connecticut	5,241	6,810	6,600	20.6
New York	1,139	1,333	1,410	19.2
New Jersey	6,710	8,172	7,100	5.5
Pennsylvania	2,073	2,505	2,620	20.9
Delaware	2,042	2,907	2,850	28.3
Maryland	2,530	3,826	3,500	27.2
Lake States	920	1,126	1,490	38.2
Michigan	1,106	1,470	2,000	47.3
Wisconsin	865	1,175	1,500	42.3
Minnesota	884	976	1,270	30.3
Corn Belt	1,190	1,578	1,840	35.3
Ohio	1,396	1,989	2,250	37.9
Indiana	1,325	1,801	2,210	40.0
Illinois	1,536	2,064	2,200	30.2
Iowa	1,153	1,442	1,750	34.1
Missouri	734	948	1,190	38.3
Northern Plains	400	478	526	23.9
North Dakota	318	383	415	23.4
South Dakota	286	319	380	24.7
Nebraska	517	632	695	25.6
Kansas	460	553	590	22.0
Appalachia	1,223	1,597	1,940	36.9
Virginia	1,643	1,925	2,130	22.9
West Virginia	843	965	1,060	20.5
North Carolina	1,455	1,970	2,400	39.4
Kentucky	988	1,377	1,590	37.8
Tennessee	1,130	1,526	2,100	46.2
Southeast	1,301	1,631	1,920	32.2
South Carolina	1,152	1,363	1,600	28.0
Georgia	1,025	1,358	1,800	43.0
Florida	2,033	2,306	2,400	15.3
Alabama	936	1,387	1,680	44.3
Delta States	820	1,009	1,230	33.3
Mississippi	754	917	1,180	36.1
Arkansas	815	989	1,250	34.8
Louisiana	926	1,176	1,250	25.9
Southern Plains	487	562	631	22.8
Oklahoma	482	547	634	23.9
Texas	488	566	630	22.5
Mountain	283	379	440	35.7
Montana	219	289	300	27.0
Idaho	680	905	1,170	41.9
Wyoming	145	206	235	38.3
Colorado	400	558	640	37.5
New Mexico	212	258	215	1.4

(continued)

TABLE 15.13 (Continued)

	Average Value per Acre as of January 1			
Region and State	**1992 (dollars)**	**1996 (dollars)**	**2000 (dollars)**	**1992–2000 Change (percent)**
Arizona	311	399	1,140	72.7
Utah	445	697	900	50.5
Nevada	262	332	440	40.4
Pacific	1,410	1,675	1,890	25.4
Washington	880	1,117	1,150	23.5
Oregon	607	928	1,020	40.5
California	2,157	2,404	2,850	24.3
48 States	713	890	1,050	32.1

[a]Including land and buildings.
Source: USDA, NASS, Agricultural Land Values.

TABLE 15.14 Cash Rent per Acre (1996–2000)

Region and State	1996 (dollars)	1998 (dollars)	2000 (dollars)
Northeast	NA	31.00	24.00
Lake States	NA	25.30	28.00
Minnesota	16.0	16.00	17.50
Corn Belt	NA	25.30	24.50
Iowa	28.90	34.00	29.00
Missouri	20.00	18.00	20.00
Northern Plains	NA	10.50	11.20
Kansas	11.90	13.00	12.80
Nebraska	10.00	10.70	11.30
North Dakota	8.50	9.20	9.50
South Dakota	9.10	9.70	11.00
Appalachian	NA	17.50	18.00
Tennessee	13.50	18.00	18.00
Virginia	15.00	15.50	16.00
Southeast	NA	15.90	17.20
Florida	17.40	14.00	15.00
Georgia	23.20	20.60	22.00
Delta States	NA	15.00	13.80
Louisiana	12.60	16.00	14.00
Mississippi	15.60	14.50	14.00
Southern Plains	NA	6.90	6.30
Oklahoma	8.00	7.80	7.80
Texas	5.40	6.60	6.00
Mountain	NA	4.00	3.80
Colorado	NA	5.00	5.20
Montana	5.00	4.50	4.80
New Mexico	NA	1.50	2.00
Utah	NA	11.00	11.00
Pacific	NA	13.00	11.00
California	NA	12.00	10.00
48 states	NA	8.80	8.50

Source: Agricultural Cash Rents, USDA, NASS.

generally high cost of purchasing land, many cow-calf producers and stocker cattle operators prefer to lease grazing. Grazing leases on private property are typically evaluated on a monthly cost for an animal unit or for a cow-calf pair. Stocker rates are also tabulated as a monthly fee per animal unit or per head. Table 15.14 illustrates private lands grazing fees in the 17 western states for 1996 and 2000 computed on an animal unit, cow-calf pair, or per head basis.

HEALTH PROBLEMS ASSOCIATED WITH GRAZING PLANTS

Poisonous Plants

Poisonous plants represent a major economical loss to the cattle industry, affecting 3–5% of cattle each year. The estimated $300 million losses in the 17 western states result from death, reduced gains, cattle treatment costs, abortions, photosensitizations, and birth defects. In attempting to reduce the number of cattle exposed to poisonous plants, additional economic losses include the costs of spraying, fencing, altering grazing programs, and preventing forage loss.

Hundreds of plants are poisonous to cattle throughout the United States. Some are toxic at all times; others are toxic only under certain conditions. Some poisonous plants are palatable to cattle, whereas others will be consumed when they are the only plants available in quantity. Cattle that are not watered regularly or that are allowed to become hungry are more likely to eat poisonous plants in lethal amounts.

There are very few known treatments for cattle poisoned by plants. Many times cattle are poisoned in remote range or pasture areas, and the effect of the poison is so rapid that the animal is seriously affected before treatment can be applied. Therefore, preventing the loss of cattle from poisonous plants is generally a matter of range, pasture, and cattle management. Producers should be aware of the poisonous plants in their areas knowing the conditions under which the plants can be toxic. In addition, producers should know the animal symptoms associated with plant poisoning, which can vary depending on the amount of poisonous plant ingested over a given period of time and variation in individual animal response. Other preventive measures include developing grazing plans to prevent poisonous plant ingestion, avoiding overgrazing, keeping cattle free from undue stress (including hunger and thirst), providing adequate salt and other necessary minerals, and controlling poisonous plants when physically and economically feasible. If cattle show signs of poisoning, a local veterinarian should be consulted for proper diagnosis, treatment, and prevention.

Table 15.15 lists some common poisonous plants and provides information on their distribution and control as well as the symptoms and treatment of poisoned cattle. However, producers will need more extensive information when plant poisoning problems are encountered. A good informational source is the USDA Poison Plant Research Laboratory (see the Appendix).

Fescue Toxicity

Fescue toxicity is a serious problem with cattle grazing tall fescue, especially in the tall fescue-growing areas of the eastern United States, where the grass is the principal diet of cattle. Fescue toxicity is often incorrectly referred to as "summer slump" because of the extreme symptoms it can cause during warm weather—poor animal gains, reduced conception rates, intolerance to heat, failure to shed winter hair coat, elevated body temperature, and nervousness characterize this syndrome. The economic impact of the tall fescue toxicity problem is at least $609 million annually.

TABLE 15.15 Selected Poisonous Plants[a] That Cause the Largest Economic Loss in Cattle

Plant	Distribution	Cause of Poisoning	Prevention or Control[b]	Sign of Poisoning	Effective Treatment
Arrowgrass	Widely distributed in marshy, alkaline pastures and native hay areas throughout the United States	Hydrocyanic acid	Prevent consumption—especially of plants retarded by drought or frost	Distress, cyanosis, rapid respiration, salivation, convulsions, rapid death	Limited; immediate treatment of solution (10 cc of 10% sodium 10 cc of 10% thiosulfate and sodium nitrate)
Astragulus (nitro-containing)	Western Canada, western United States, and northern Mexico	3-nitropropionic acid or 3-nitro-1-propanol	Use grazing programs that favor more desirable forages; application of herbicides; prevent grazing	*Acute:* Respiratory distress and muscular weakness, primarily in the pelvic limbs *Chronic:* Labored and rapid respiration, wheezing or roaring sound, weakness in pelvic limbs, knuckling at fetlocks, goose stepping, and knocking together of hind feet when walking	None
Bracken fern	Throughout the United States		Prevent overgrazing	High fever; loss of appetite, depression, labored breathing, salivation, nasal and rectal bleeding	None
Grass tetany	(See section in latter part of chapter)				
Fescue toxicity	(See section in latter part of chapter)				
Larkspur (two groups–tall and low, based on height at maturity)	Primarily in the eleven western states	Alkaloids	Grazing of infested areas later in grazing season. Apply herbicides if economical	Nervousness, staggering gait, salivation, bloat, sudden death	Limited; immediate injections of physostigmine salicylate can be helpful
Lupines	Foothills and mountain ranges in sagebrush and aspen areas (most western states)	Alkaloids	Birth defects prevented if consumption not permitted between 40 and 70 days of gestation. Apply herbicides if economical	Crooked-legged calves and other birth defects (Fig. 15.26) Nervousness, loss of muscular control, convulsions, coma, frothing at mouth	None
Milkvetch (several species)	Throughout most of the United States	Miserotoxin or nitro compounds	Prevent grazing as plants are palatable. Use herbicides on some species if economical	*Acute:* Tendency to stand on toes, weakness, irregular gait, posterior paralysis *Chronic:* Nervousness, muscular weakness, respiratory distress	None

TABLE 15.15 (Continued)

Plant	Distribution	Cause of Poisoning	Prevention or Control[b]	Sign of Poisoning	Effective Treatment
Milkweed (several species)	Most of 17 western states	Glucosidic substances (called cardenolides)	Adequate forage, as plant is unpalatable. Herbicides available but usually not economical	Difficult breathing (grunting), loss of muscular control, bloat, spasms	None
Nitrate poisoning	(See section in latter part of chapter)				
Ponderosa pine needles	Most 11 western states		Needles are usually grazed only when cattle are stressed (due to feed changes, weather, and hunger)	Abortion during late trimester, retained placenta, calves may be weak if near term	None
Selenium-accumulating plants (sometimes called *akalli disease*)	Primarily in North and South Dakota, Montana, Wyoming, Colorado, and Utah	More than 5 parts per million of selenium in plants *Acute:* Selenium accumulation greater than about 100 ppm *Chronic:* Grasses and grains containing 5–40 ppm selenium	Prevent consumption	*Acute:* Loss of appetite, depression, coma, death *Chronic:* Dullness, rough hair coat, emaciation, lameness, possible hoof deformities	Remove from range containing plants with high selenium concentrations
Sweet clover poisoning	Throughout most of the United States	Coumarin in clover is converted to dicoumarin, which prevents the synthesis and metabolism of vitamin K	Avoid feeding moldy sweet clover	Stiffness, lameness, and swellings (blood clots) beneath the skin	Blood transfusions; intramuscular injections of vitamin K
Threadleaf snakeweed (broomweed)	Drier range areas, primarily in the southwestern United States		Avoid grazing. Plants susceptible to 2, 4-D	Abortion, sometimes death; weight loss and diarrhea	None
Waterhemlock	Throughout most of the United States	Cicutoxin (a highly poisonous unsaturated alcohol that has a strong carrot-like odor)	Provide more palatable forage; prevent cattle from consuming roots brought to the surface by plowing or cleaning ditches; keep animals away (3 weeks) from plants sprayed with herbicide as palatability is increased; apply herbicide if economical	Cattle die suddenly with few observed signs of poisoning	None

[a]Other plants occasionally poisonous to cattle include bitter rubberweed, chokecherry, cocklebur, copperweed, desert baileya, death camas, drymary (inkweed), greasewood, groundsel (riddel and threadleaf) halogeton, hemp dogbane, horsetails, oak, pinque, poison hemlock, rayless goldenrod, St. John's wort, sneezeweed, spring parsley, tansy mustard, tansy ragwort, and western false hellebore.

[b]The use of some herbicides listed may be banned in certain states. Directions should be followed carefully.

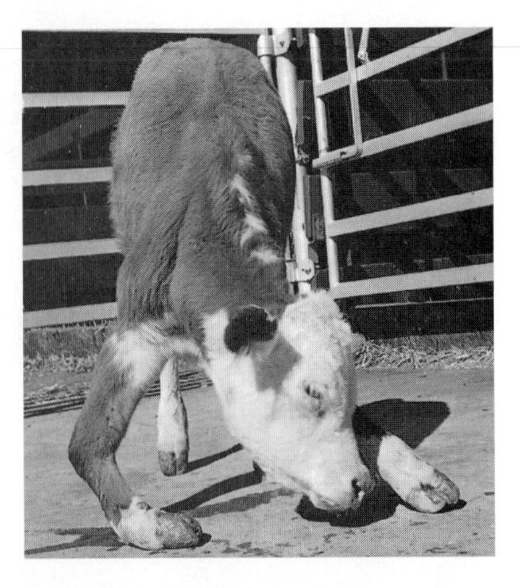

FIGURE 15.26 Crooked legs and other congenital deformities occur in newborn calves if cows graze toxic species of the lupine plant between 40 and 70 days after breeding. Cleft palate may also occur in this "crooked calf disease or syndrome."
Source: USDA Poisonous Plant Research Laboratory, Logan, Utah.

FIGURE 15.27 Yearling heifers grazing on low endophyte tall fescue in Kentucky. The fungus-free fescue shows great promise of reducing the protein of poor animal performance in hot weather.
Source: Kentucky Agric. Expt. Sta.

Fescue toxicity is associated with a fungus called an *endophyte.* The fungus lives within the leaves, stems, and seed of the tall fescue plant and is not visible externally. Tall fescue pasture, hay, and seed contain the fungus, although the fungus dies within about a year in stored seed. It is likely that the fungus causes the grass to produce a toxic compound (not yet identified) that is harmful to animals eating the forage. Surveys conducted in several southeastern states show that over 95% of tall fescue pastures are heavily infected with the fungus.

Average daily gain of steers on tall fescue pastures has been poor, usually about 1 lb per day or less for the grazing season. The introduction of endophyte-free fescue has increased gains dramatically (Fig. 15.27). A four-year grazing study showed significantly higher gains (1.82 lbs per day versus 1.09 lbs per day) for steers grazed on endophyte-free fescue. Another study with cows and calves showed average daily calf gains of 2.65 lbs for endophyte-free fescue and 1.92 lbs gains for calves on endophyte-infected grass. The average daily weight changes of cows were 1 lb on endophyte-free fescue and −0.40 lb on endophyte-infected fescue. The average daily gain of steers fed endophyte-free fescue hay was 1.45 lbs but only 0.63 lb for steers fed endophyte-infected hay.

The effects of fescue toxicity can be reduced by one or more of the following methods:

1. Growing a legume with tall fescue (e.g., planting ladino or red clover in fescue sod is inexpensive and practical).

2. Feeding hay other than tall fescue in winter when cattle are grazing toxic fescue pasture. Bermudagrass or sericea lespedeza hay are good choices.
3. Avoiding heavy applications of broiler litter or nitrogen fertilizer on tall fescue when it is the sole source of pasture and hay.
4. Replanting infected fescue pastures with stess-tolerant tall fescue with nontoxic endophytes. Old seed, where the fungus has died, can be planted, but the seed rate must be increased to make up for reduced germination. An infected fescue pasture should not be allowed to make seed during the year it is to be destroyed and replanted. The endophyte-free cultivars are less tolerant of drought and overgrazing so excellent management is required.
5. Planting seed of the new improved winter-productive Triumph variety.
6. Performing an economic analysis of the productivity and longevity of endophyte-free fescue (which have been shown to decrease in humid environments) as compared with endophyte-infected pastures.

Grass Tetany

Grass tetany, also known as winter tetany, grass staggers, wheat poisoning, magnesium tetany, and hypomagnesemia, occurs throughout the United States. Although it is usually observed in the spring, it can occur in the fall and winter. Most often it is observed when cattle are grazing cool-season forages that have a very rapid, lush growth. Cows nursing calves under two months of age are most frequently affected by grass tetany.

Many times the clinical symptoms of grass tetany are not observed, and dead animals are the only history of the problem. Affected animals may become excitable, expressing a wild stare with erect ears, and appear to be blind. They may appear uncoordinated, tending to lean backward or to stumble. An affected animal often has trembling muscles and grinding teeth followed by violent convulsions, deep coma, and death.

Positive diagnosis of grass tetany is difficult because the symptoms can easily be confused with other diseases. A blood test can be helpful as the serum of magnesium is typically low (the normal level is 2.25 mg per 100 ml of serum, whereas affected animals are usually below 1 mg).

Prevention of grass tetany may involve a combination of several factors. Magnesium mineral (magnesium oxide) should be provided. Legume or legume-grass pasture should be utilized, since tetany seldom occurs where legumes are grazed. Less susceptible animals (heifers, stockers, dry cows, or cows with calves over four months) can be grazed on high-risk pastures. Some producers feed a corn silage supplement while cows and calves are on wheat pasture, a practice that greatly reduces the incidence of grass tetany. If soil magnesium is low, a program can be implemented to increase its level in the soil. However, fertilization with magnesium is less effective than supplementing cattle with magnesium.

Early treatment of animals affected by grass tetany is important. Cattle that have collapsed and have been down more than 12–24 hours only seldom recover. Animals should be handled gently to prevent excitement. Two hundred cubic centimeters (cc) of a sterile solution of magnesium sulfate (epsom salts) injected under the skin in at least four different sites will give a high level of magnesium in the blood within 15 minutes.

Nitrate Poisoning

Nitrate poisoning occurs throughout the United States because any forage plant can accumulate nitrate (NO_3) under certain conditions such as drought. Nitrate is lethal to cattle when plants contain more than 0.9% nitrate (NO_3) and sublethal when plants contain 0.5–1.5% nitrate. The nitrate level of plants should be checked if nitrate poisoning is suspected or anticipated.

Nitrates are converted to nitrites, which produce methemoglobin, a type of hemoglobin that cannot carry oxygen. The acute signs of nitrate poisoning in cattle are a blue coloration of mucous membranes, a staggering gait, shortness of breath, then death. Some chronic signs of poisoning are watering eyes, unthriftiness, reduced milk flow, and reduced gains.

Administration of methylene blue to affected animals will convert methemoglobin back to hemoglobin. In chronic cases, the feed should be changed or mixed to reduce the nitrate intake of cattle. The use of careful testing of feeds for nitrate levels, the frequent observation of cattle, and the utilization of appropriate harvesting techniques can minimize the potential problem.

MANAGEMENT SYSTEMS HIGHLIGHTS

1. The ability of cattle to utilize roughage in the creation of protein is the basis for the beef industry.
2. Because of the extreme diversity of rangelands and pastures, management strategies must be site specific to account for variation in topography, precipitation, soil type, plant community composition, and forage quality.
3. Range improvements should be made based on an evaluation of geographical, ecological, economic, regulatory, and management expertise considerations.
4. Utilization of profitable forage management strategies provides opportunities for cow-calf producers to enhance their competitive position. However, expansion of the cow-calf enterprise on purchased versus leased land should be carefully evaluated on a cost-to-benefit basis.
5. Grazed forage is typically more cost-effective than purchased feeds or mechanically harvested forages.

REFERENCES

Publications

Absher, C. W., Murdock, L. W., and Kaiser, C. J. 1979. *A Beef Forage System.* Lexington, KY: Kentucky Agricultural Extension Service ID-5.

Bailey, R. G. 1995. *Description of the Ecoregions of the United States.* USDA Forest Service Publ. 1391.

Ball, D. M., Hoveland, C. S., and Lacefield, G. D. 1991. *Southern Forages.* Norcross, GA: Potash and Phosphate Institute.

Barnes, R. F., Miller, D. A., and Nelson, C. J., (eds.). 1995. *Forages: An Introduction to Grassland Agriculture.* 5th ed. Ames, IA: Iowa State Univ. Press.

Barnes, R. F., Miller, D. A., and Nelson, C. J., (eds.). 1995. *Forages: The Science of Grassland Agriculture.* Ames, IA: Iowa State Univ. Press.

Blasi, D. A., and Corah, L. R., 1993. *Assessment of Forage Resources for Determining the Ideal Cow.* Proceedings of the Cow-Calf Conference III, Salina, KS.

Brady, N. C. 1990. *The Nature and Properties of Soils.* 10th ed. New York: Macmillan.

Cook, C. W., et al. 1983. *Alternate Grass and Grain Feeding Systems for Beef Production.* Fort Collins, CO: Colorado State University Experiment Station Bull. 579S (rev.).

Dagget, D., and Dusard, J. 1995. *Beyond the Rangeland Conflict: Toward a West That Works.* Flagstaff, AZ: The Grand Canyon Trust.

Forage Facts Notebook. 1998. Kansas State University Agricultural Experiment Station and Cooperative Extension Service, Manhattan, KS.

Grazing Fee Task Group. 1995. *The Value of Public Land Forage and the Implications for Grazing Fee Policy.* New Mexico State Univ., AES Bull. 767.

Hart, R. H. 1987. *Economic Analysis of Stocking Rates and Grazing Systems.* Proceedings of the Range Beef Cow Symposium X, Cheyenne, WY.
Hart, R. H., et al. 1993. Grazing Systems, Pasture Size, and Cattle Grazing Behavior, Distribution and Gains. *J. Range Mgt.* 46: 81.
Heitschmidt, R., and Walker, J. 1983. Short Duration Grazing and the Savory Grazing Method in Perspective. *Rangelands* 5: 147.
Hermel, S. 1996. Calculating Grass Needs. *BEEF:* Spring Cow-Calf Special.
Holechek, J. L., Pieper, R. D., and Herbel, C. H. 1989. *Range Management: Principles and Practices.* Englewood Cliffs, NJ: Prentice Hall.
Holland, C., and Kezar, W. (eds.). 1995. *Pioneer Forage Manual: A Nutritional Guide.* Pioneer Hi-Bred International.
Hoveland, C. S. 2000. Achievements in Management and Utilization of Southern Grasslands. *J. Range Mgt.* 53: 17.
James, L. F., Keeler, R. F., Johnson, A. E., Williams, M. C., Cronin, E. H., and Olsen, J. D. 1980. *Plants Poisonous to Livestock in the Western States.* Washington, DC: USDA. Agric. Info. Bull. 415.
Launchbaugh, J. L., and Owensby, C. E. 1978. *Kansas Rangelands: Their Management Based on a Half-Century of Research.* Manhattan, KS: Kansas Agricultural Experiment Station Bull. 622.
Miller, R. W., and Donahue, Roy C. 1995. *Soils in Our Environment.* 7th ed. Englewood Cliffs, NJ: Prentice Hall.
Morrow, R., and Hermel, S. 1996. Calculating Grass Needs. *BEEF:* Spring Cow-Calf Special.
Neilsen, N. B., and James, L. F. 1992. *The Economic Impact of Livestock Poisoning by Plants.* In: Poisonous Plants. Proceedings of the Third International Symposium, Ames, IA: Iowa State Univ. Press.
Opportunities to Increase Red Meat Production from Ranges of the United States. Phase I—Non-Research. 1974. Washington, DC: USDA.
Parsons, S., and Pratt, D. 1992. *Ranching for Profit (Self Study Course).* Ranch Management Consultants, 7719 Rio Grande Blvd. N.W., Albuquerque, NM 87107.
Pastures for Profit. 1999. Minnesota Extension Bull. A3529. University of Minnesota.
Reece, P. E. 1991. *Evaluation and Practical Use of Research Results for Developing Grazing Strategies.* Proceedings of the Range Beef Cow Symposium XII, Fort Collins, CO.
Reid, R. L., and Klopfenstein, T. J. 1983. Forages and Crop Residues: Quality Evaluation and Systems of Utilization. *J. Anim. Sci.* 57: 534 (Suppl. 2).
Roath, R. 1997. *Applications of Monitoring for Producers.* Proceedings of the Range Beef Cow Symposium XV, Rapid City, SD.
Savory, A. 1999. *Holistic Management.* 2nd ed. Washington, DC: Island Press.
Smith, B., Leung, P., and Lore, G. 1986. *Intensive Grazing Management: Forage, Animals, Men, Profits.* The Graziers Hui, Kamuela, HI.
Stevens, R., Aljoe, H., Forst, T. S., Motal, F., and Shankles, K. 1997. How Much Does It Cost to Burn? *Rangelands* 19: 2.
Tranel, J. E., Sharp, R. C., and Kaan, D. A. 2001. *Custom Rates for Colorado Farms and Ranches in 2000.* Agriculture and Business Management Notes. Fort Collins, CO: Colorado State University.
Vallentine, J. F. 1989. *Range Development and Improvement.* San Diego: Academic Press.
Vallentine, J. F. 1990. *Grazing Management.* San Diego: Academic Press.
Waller, S. S., Moser, L. E., and Anderson, B. 1986. *A Guide For Planning and Analyzing a Year-Round Forage Program.* N.E. Coop. Ext. 5086-113.
White, R. S., and Short, R. E. (eds.). 1988. *Achieving Efficient Use of Rangeland Resources.* Miles City, MT: Montana State Agricultural Experiment Station.

Visuals

"Along the Waters Edge" (20 min.). Fisheries and Oceans Canada. Society for Range Management, 1839 York St., Denver, CO 80206.
"Buffalo Lessons" (22 min.). Agribase, Inc., 7509 Tiffany Springs Parkway, Kansas City, MO 64190.
"Cattlemen Care about the Environment" (7 min.). National Cattlemen's Beef Association, P.O. Box 3469, Englewood, CO 80155.

"Exploring Rangeland Ecosystems" (60 min.). Society for Range Management, 1839 York St., Denver, CO 80206.

"Grass Identification, I and II" (29–34 min.); "Forbs Identification" (25 min.); and "Legume and Woody Plant Identification"(25 min.). CEV, P.O. Box 65265, Lubbock, TX 79464-5262.

"Grass Tetany—Clinical Signs in a Natural Environment" (16 mm film or videotape). Southern Piedmont Conservation Research Center, USDA, ARS, P.O. Box 555, Watkinsville, GA 30677.

"Last Stand of the Tallgrass Prairie." (57 min.). Inland Sea Productions, Inc. Order from www.pbs.com.

"Livestock Grazing and Riparian Systems" (17 min.). Bureau of Land Management. Society for Range Management, 1839 York St., Denver, CO 80206.

Parsons, S. "Grazing for Profit" (1991; 2 hours). Ranch Management Consultants, 7719 Rio Grande Blvd., N.W., Albuquerque, NM 87107.

Computer Programs

The Grazing Manager. Contact Ray Hinnert, Texas A&M University, College Station, TX. Fax: 409-845-6430. Cost: $199.00.

Websites

Agriculture and Business Management Resource Manual—Colorado State University: http://dare.agsci.colostate.edu/index_extension.html

Range Plant Identification: www.incolor.com/gibbens/plant/main.htm

Selected North Dakota and Minnesota Range Plants: www.ext.nodak.edu/extpubs/ansci/range/eb69-1.htm

Noble Foundation Plant Image Gallery: www.noble.org/imagegallery/

Forage Identification NRCS Slide Series: www.caf.wvu.edu

Forage Id: www.agrg.purdue.edu/ext/forages/forageid.htm

Forage Identification CD-ROM: http://forages.orst.edu/resources/medial/idcdrom/default.html

Herd Health

Effective health programs involve both prevention and treatment of disease problems affecting beef cattle. The old adage that "an ounce of prevention is worth a pound of cure" applies to herd health management. Preventative measures are typically more cost-effective to implement than attempting to deal with a disease outbreak. This chapter focuses on helping producers understand the major health problems affecting beef cattle and the importance of planning economically effective prevention and treatment programs.

ESTABLISHING A HERD HEALTH PROGRAM

Beef producers should be knowledgeable of the disease agents, contributory risk factors, and other health-related conditions that may be unique to their area. This knowledge should be used in developing a specific health program for each operation. The importance of strict sanitation and biosecurity practices cannot be overemphasized. The most successful herd health management programs involve cooperative planning by producers and their veterinarians. Veterinarians need to understand the management programs of individual beef cattle operations, especially the major factors affecting profitability. Producers should be aware of their limitations in evaluating, treating, and caring for sick animals and of the proper timing for seeking a veterinarian's assistance. Veterinarians can also serve as liaisons in utilizing the services of other specialists and well-equipped diagnostic laboratories. It has been said that a "good herd health program does not come in a bottle." A sound herd health program is based on the following components:

1. Sound nutritional regime.
2. Continuous training of personnel.
3. Known source of livestock.
4. Sound sanitation management and biosecurity practices.
5. Excellent recordkeeping system accompanied by a sound monitoring and evaluation system.
6. Functional, well-maintained facilities.
7. Excellent relationship with a professional herd veterinarian.
8. A sound preventative vaccination system.

Implementation of an immunization protocol requires that producers and their herd health consultants identify and understand the following:

1. Herd health management history (which diseases are of significant risk).
2. Marketing goals for the herd and expectations of the customer.
3. Timing of vaccination to maximize immune response.
4. Avoidance of weaning, transport, and vaccination at the same time, if possible.
5. Correct vaccine administration in terms of temperature, date of expiration, route, and site.
6. Maintenance of biosecurity via strict control over the source of new cattle and good sanitation procedures in trucks, trailers, feedbunks, waterers, and receiving pens.

Early recognition of and response to emerging health problems is a key to avoiding the financial losses associated with isolated or widespread disease in a herd. Some of the early "signals" that an animal may be diseased are diarrhea, weight and condition loss, lameness, stiff movement, loss of appetite, general depression, low head carriage, droopy ears, nasal or ocular discharge, irregular respiration, rough and/or dull hair coat, and skin lesions.

STRESS AND HEALTH

Cattle have passive and active defense mechanisms that can counteract most disease organisms. Exposure to severe stress conditions, however, limits the ability of these defense mechanisms to overcome disease and can lead to sickness or death in animals.

Stress factors include fatigue, hunger, thirst, dust, weaning, castration, dehorning, shipping, mixing with other cattle, unnecessary or abusive handling, adverse weather, parasites, poor sanitation, ammonia buildup, and anxiety. Excessive stress can be avoided by vaccinating and processing calves 30 days prior to weaning, by avoiding inclement weather during cattle handling/processing/transport, and by maintaining a functional working facility where cattle can be handled quietly and gently.

Preconditioning is a complete health management program for reducing sickness and death rate and improving weight gains. Preconditioning can prepare calves to better withstand the stress of movement from their preweaning production site into and through the various production and marketing channels. A complete preconditioning program for calves is defined as follows:

Weaning—All calves should typically be weaned at least 30 days prior to their entry into the next production phase and (or) marketing channels.

Nutrition—Calves should be adjusted to trough and bunk and accustomed to a beginning feedlot ration. They should also be trained to drink from a nonstream water source.

Vaccination—Calves should be vaccinated against the *Clostridial* group, Infectious Bovine Rhinotracheitis, Para-Influenza$_3$, *Pasteurella,* and Bovine Respiratory Syncytial Virus at least 2 weeks prior to weaning.

Deworming—All calves should be dewormed with an injectable, drench, or feed additive anthelmentic when internal parasites are a problem in the area.

External parasite control—All calves should be treated with a recommended, appropriate, and approved external parasiticide at least 2 weeks prior to weaning.

Castration and dehorning—Calves should be castrated and dehorned at least 30 days prior to movement, preferably at 1–2 months of age.

Identification and certification—All preconditioned calves should be identified with an ear tag and should be accompanied by a properly signed certificate that states the specifics relative to the preconditioning process (timing and type of vaccine, etc.).

The concept of preconditioning is often poorly interpreted by producers, buyers, and veterinarians. Many buyers prefer to buy replacement cattle at the cheapest price and in thin condition. They frequently overlook the immediate health status and prior immunization of the animals they are purchasing, hoping to compensate for health losses through compensatory gains. As a result, the economic success of preconditioning varies considerably. However, preconditioning offers significant economic merit when it is implemented correctly.

Preconditioning programs should be assessed from a cost-to-benefit and risk-analysis standpoint. The specific preconditioning program that is best for a particular enterprise will depend on age of the cattle at time of marketing or movement into the next production phase, marketing channel utilized, the degree of discount/premium associated with preconditioning, and the costs of labor, vaccines, and feed. Consider the example in Table 16.1. Calves managed under Program 1 would have to yield $4.56 per head in additional revenue or productivity to break even on the additional costs as compared with Program 2. The effects of sick versus healthy calves in feedlot performance and profitability are illustrated in Table 16.2.

TABLE 16.1 Comparison of Costs of Preconditioning Programs

	Dollars per Head	
Input	Program 1 (intensive)	Program 2 (less intensive)
Vaccines (preweaning)		
7-way, Blackleg	$ 0.35	$ —
IBR, BVD, PI3, BRSV	1.15	1.15
Pasteurella	2.05	—
Dewormer	5.00	1.45
Implant	1.00	—
Vaccine booster at weaning		
IBR, BVD, PI3, BRSV	1.15	1.15
Subtotal	10.70	3.75
Labor (processing)		
4 employees × $6.50/hr[1]	2.08	2.08
Creep feed—30 days @ 5 lbs/head/day; $215/ton	16.13	—
Labor (feeding)—30 min/day @ $6.50/hr × 30 days	0.97	
Total cost	29.88	5.83
Addition to breakeven cost[1]	$ 5.73	$1.17

[1]Assumes 100 steers processed per group with sale weights of 521 lbs for Program 1 and 500 lbs for Program 2.
Source: Gutierrez and Hogan (1997).

TABLE 16.2 Comparison of Performance and Profitability of Sick Versus Healthy Calves

	Sick	Healthy
Number of head	218	1,000
Death loss (%)	5.5	0.7
Average daily gain (lbs/day)	2.65	3.08
Total cost of gain ($/cwt)	62.32	49.03
Medicine cost ($/hd)	26.78	0.00
Percent grading		
Choice (%)	37	53
Select (%)	53	43
Standard (%)	10	3
Net return ($/hd)	+22.31	+146.17
Difference in net return ($/hd)		+123.86

Source: Texas A&M (2000).

DISEASES AND HEALTH PROBLEMS

Table 16.3 identifies some common disease and health problems encountered by beef cattle. Accurate disease diagnosis is an essential element in a health management program. A thorough examination by a veterinarian plus diagnostic laboratory verification may be necessary to establish a specific disease diagnosis. Once a diagnosis has been made the veterinarian should prescribe treatment. Specific products used in prevention and treatment should be cleared through a veterinarian so producers can meet FDA regulations in product use and withdrawal times. Prevention and treatment guidelines are given in Table 16.3 for general information only. Specific treatments and preventive measures should be determined by establishing a valid veterinarian-client-patient relationship (VCPR). The components of a valid VCPR include assumption of responsibility by the veterinarian for making medical judgments regarding the health of the animal(s), the need for treatment, and the agreement of the client to follow the veterinarian's directives. Furthermore, the veterinarian must have sufficient knowledge of the case to make a diagnosis by virtue of an exam of the animal(s) and the premises where the animal(s) are kept. Finally, the practicing veterinarian must be readily available for follow-up evaluation.

TABLE 16.3 Common Diseases and Other Health Problems in Cattle

Disease and Cause(s)	Clinical Signs	Prevention	Treatment
Abomasitis Ulceration, sand colic abomasal bloat in calves, *Clostridium perfringens A*	Adults: anorexia, distended right flank, weak, dehydrated—Calves: unthrifty, distended abdomen, soft-discolored feces	Avoid trash (rags tarps, etc.) in pastures, supplement protein when ration is high in poor-quality roughage	Prognosis generally considered unfavorable; abomasotomy possible in calves
Acidosis, Rumenitis, Liver Abscess Complex Overloading on concentrate after a period of reduced feed consumption; increasing concentrate in diet too rapidly *Fusobacterium necrophorus; Actinomyces pyogenes; Bacteroides*	Initially, animals off feed clinical signs of "feed intoxication"; acidosis observed following slaughter	Reduction of concentrates (less than 75%), more roughage; Chlortetracycline, 70 mg/head/day; Tylosin, 60–90 mg head/day; use of buffers; use of ionophores	Antacids; antifermentatives; gastric lavage; fluid therapy; reduce concentrate levels in feed
Anaplasmosis *Anaplasma marginale*	Anemia, fever, icterus, weakness, and emaciation; use only laboratory diagnosis	Control of insects; care in spreading disease by veterinary instruments; vaccine is available—Chlortetracycline: Cattle up to 700 lbs—350 mg/head/day; cattle 700–1,000 lbs—500 mg/head/day; cattle 1,500 lbs and over—0.5 mg/head/day Chlortetracycline (for carrier stage of anaplasmosis): 5.0 mg/lb body weight/day for 60 days in the feed	Acute cases—blood transfusions; chlortetracycline; oxytetracycline

TABLE 16.3 (Continued)

Disease and Cause(s)	Clinical Signs	Prevention	Treatment
Anthrax *Anthrax bacillus*	Sudden death (1–2 hours after infection); "sawhorse on its side" appearance; failure of blood to clot; delayed rigor mortis; can cause disease in humans	Vaccination (recommended only in areas where disease occurs)	Antibiotics and antiserums (success relatively poor); contact state veterinarian; do not move or transport carcasses
Asthma (cow) Occurs in the fall when cows are moved from dry, sparse forage to lush, green pasture; may be caused by an improper amino acid (tryptophan) metabolism	Clinical signs vary from slight to severe; respiratory distress, with abnormal increase in depth and rate of respiration; a grunt usually accompanies respiration	Moderate the feed change; have cows filled with dry feed before moving to lush, green pasture; ionophores for 5d before change in feed to 7d after change in feed type	None very effective
Blackleg *Clostridium chauvoei* (bacteria)	Muscular depression, gaseous swelling in muscles; lameness	Vaccination of calves at branding and weaning and/or dam precalving	Penicillin
Bloat Variety of causes	Excessive accumulation of gas in the rumen and reticulum causes distension of left side; right side is also distended in more severe cases; labored breathing may occur	Poloxalene drench; if possible, feed stemmy roughage; avoid feeding alfalfa hay with high barley rations	Poloxalene oral drench; stomach tube; emergency rumenotomy
Bluetongue Virus spread by blood-sucking insects (primarily the genus *Culicoides*)	Varies—high fever; depression; profuse slobbering; crusty muzzle; ulcers of dental pad and lips; poor reproduction; coronary band lesions	Maintenance of closed herd and control of *Culicoides;* vaccine available, but recommended only in infected herd; vaccination of pregnant animals may cause brain damage to fetus	None; disease runs its course in several weeks; death may occur due to secondary infection of pneumonia
BVD (bovine viral diarrhea) Virus. Laboratory diagnosis imperative for accuracy	Feeder cattle: ulcerations throughout digestive tract; diarrhea (often contains mucus or blood)	Vaccination prior to exposure; avoid contact with infected animals	Symptomatic treatment; antibiotics; sulfonamides; force feed
	Breeding cattle: abortions, repeat breeding	Annually vaccinate cows 30 days prior to breeding	None; symptomatic treatment
Brisket Disease Congestive right heart failure due to stress of high altitude; greater predisposition if cattle graze on locoweed at high altitudes	Edema in brisket (Fig. 16.1); jugular vein distension; high neonatal calf mortality	Keep cattle below 5,000-ft altitude; practice genetic selection (primarily bulls) using pulmonary arterial pressure test	Take affected cattle to elevations below 5,000 ft
BRSV (bovine respiratory syncytial virus)	Labored breathing; pneumonia	Vaccination	Antihistamines Corticosterioids

(continued)

TABLE 16.3 (Continued)

Disease and Cause(s)	Clinical Signs	Prevention	Treatment
Brucellosis *Brucella abortus* (bacteria)	Abortions	Calfhood vaccination (in some states) at the age of 4–12 months	Test and slaughter; report reactors to state veterinarian
Campylobacteriosis (formerly vibriosis) *Campylobacter fetus*	Repeat breeding; abortions (1–2%)	Annual vaccination of females and bulls prior to breeding; use of artificial insemination; virgin bulls on virgin heifers; avoid sexual contact with infected animals; cull open in cows infected herds	None (consult herd veterinarian)
Cancer Eye Genetic predisposition; environmental factors (dust, wind and ultraviolet light) enhance development	Cancerous cell growth on the eye, eyelid, third eyelid, or conjunctivitis	Genetic selection; reduction of other predisposing factors	Surgery; immunotherapy; electrothermal; cryotherapy
Coccidiosis *Eimeria zurnii; Eimeria bovis* (Protozoa)	Fluid feces, bloody feces, and straining; occasionally central nervous system signs	Avoid crowding, wet pens, wet pastures, filth; remove affected animals from lots in early stages; Amprolium: 5mg/kg/day for 21 days; Decoquinate: 22.7 mg/100 lbs/ head/day for 28 days; Ionophores	Sulfadimethoxine, 25 mg/lb for first day followed by 12.5 mg/lb for 4 days; Amprolium
Diphtheria *Fusobacterium necrophorus* (bacteria); tissue damage to laryngeal area	Difficult breathing; painful coughing, hoarseness, rattling noise when breathing	Control other respiratory diseases	Penicillin Oxytetracycline
Enteric colibaccilosis or salmonellosis	Severe gastrointestinal inflammation, diarrhea, dehydration, high death loss of affected animals	Improve absorption of colostral immunoglobulins	Aggressive antibiotic and anti-inflammatory treatment plus supportive fluid therapy
Enterotoxemia "Overeating." Toxins of *Clostridium perfringens* Type D (bacteria)	Sudden death "downers"; animals usually fed on high-concentrate diets; diarrhea, though many animals die before clinical signs appear	Increase concentrate in diet at slow rate; vaccination	Reduce concentrate; sudden death usually precludes treatment; anti-serum
Fescue Toxicity (Chapter 15)			
Foot rot (infectious pododermatitis) *Fusobacterium necrophorus; Bacteroides* sp. (bacteria)	Lameness, foot swelling	Cattle up to 700 lbs: Chlortetracycline, 70 mg/ head/day; Ethylenediamine dihydroiodide, 10 mg/ head/day; pens and lots free of objects that can injure feet	Sulfonamides Oxytetracycline Chlortetracycline Erythromycin Tylosin (injectable)

TABLE 16.3 (Continued)

Disease and Cause(s)	Clinical Signs	Prevention	Treatment
Founder (laminitis) Similar to an allergic reaction associated with acidosis complex	Lameness, inflammation between bony part of the foot and the hoof wall, toes may grow long and turn up; painful straight-legged gait in extreme cases	Keep cattle from over-consuming grain; high level of management of high-concentrate rations; gradual adaptation to high-energy feeds	Trim long toes; sell animal for slaughter; antihistamines and steroid therapy may be useful but not economical
Grass Tetany (Chapter 15)			
Haemophilus Somnus (Infectious thromboembolic meningoencephalitis) *Haemophilus somnus* infections; difficult to diagnose; needs laboratory confirmation	Fever, incoordination, head-pressing; central nervous system signs	Vaccine (two injections) Chlortetracycline, 350 mg/head/day for 28 days; Oxytetracycline, 3–5 mg/lb daily for 4 days	Sulfonamides Oxytetracycline
Hardware Disease Ingestion of sharp objects that perforate the reticulum and cause severe damage to the abdominal cavity, heart sac or lungs	Loss of appetite, reduced milk production, abdominal pain, and sometimes labored breathing; bloat and diarrhea in chronic cases	Keep wire, nails, and other sharp objects from being eaten by cattle; magnets on feed equipment; intrarumenal magnets	Mild cases may heal if animal is stalled and feed intake is reduced; antibiotics; surgery may be performed in severe cases
IBR (Infectious Bovine Rhinotracheitis) Virus. Laboratory diagnosis imperative for accuracy	Pneumonia; fever; vaginitis; infertility and abortion in females; preputial infections in males	Vaccinate cows 40 days prior to breeding; vaccinate feeder cattle prior to exposure; semen from reputable bulls	Oxytetracycline, Penicillin to minimize bacterial infections
Johne's Disease, Paratuberculosis *Mycobacterium paratuberculosis* Laboratory confirmation necessary	Chronic diarrhea	Vaccine requires approval of state veterinarian; does not remove reactors; prevent infection; test: hygenic program for calf raising, separate from adult herd	Consult herd veterinarian
Leptospirosis *Leptospira* sp.	Fever, off feed, abortions, icterus, discolored urine	Vaccination at least annually (in high-risk areas, more frequently); proper water management; control rodents; avoid contact with wildlife and other infected animals	Dihydrostreptomycin Penicillin
Leukosis (Bovine; BLV) Virus. Laboratory diagnosis necessary	Weight loss, lymph node enlargement	Prohibit sale of animals for breeding purposes; prevention of infected animals going into noninfected herds	None; cull affected animals

(continued)

TABLE 16.3 (Continued)

Disease and Cause(s)	Clinical Signs	Prevention	Treatment
Listeriosis *Listeria monocytogenes*	Fever; circling; one eye or one ear paralyzed; sudden death	No vaccine; organism frequently found growing in moldy silage	Sulfmethazine Oxytetracycline
Lump Jaw (Actinomycosis, Actinobacillosis) *Actinomyces bovis*	Lumps on bony tissues of head	Removal of objects causing head punctures (nails, splinters, grass awns, stemmy feed) Ethylenediamine dihydroiodide, 10 mg/head/day in feed or salt	Ethylenediamine dihydroiodide, 10 mg/head/day Sodium iodide
Malignant Catarrhal Fever Virus. Frequently confused with other viral infections	High fever; inflammation of respiratory, digestive, and urinary systems	No vaccine; requires positive diagnosis to differentiate from other diseases; prevent contact with sheep that harbor organism	Treatment generally ineffective
Malignant Edema *Clostridium septicum* (bacteria)	History of wounds; fever and swelling around wound	Vaccination	Penicillin
Mastitis *Staphylococci* spp. *Streptococcus* spp. Others	Inflammation of mammary gland; udder is hard, hot, painful to the touch, and commonly discolored; milk is yellow, thick, and stringy	Reduce injury from rough handling and other types of trauma	Systemic antibiotics; frequent milking, anti-inflammatories
Navel Ill Neonatal septicemia and bacteremia; poor sanitation at calving	Umbilical abscesses and hernias; swollen joints in young calves	Iodine on or within umbilical cord immediately after birth	Penicillin-dihydro streptomycin combination (1–2 gm streptomycin every 12 hours for 3 days)
Neonatal septicemia or calf septicemia *septicemic colibacillosis, septicemia salmonellosis, pasteurellosis*	Occurs when infecting organism spreads via bloodstream to multiple organs, very high mortality rates, diarrhea plus increased heart and respiratory rates, weakness and recumbency	Improve colostral absorption of antibodies	Early and aggressive therapy is required; treatment includes fluid therapy, antibiotics, IV plasma, intranasal O_2
Pine Needle Abortion Cows eat needles or buds from Ponderosa pine	Abortion	Prevent consumption of needles	None
Pinkeye *Moraxella bovis* (bacteria, spread by insects, primarily the face fly)	Watery eyes, swelling, corneal opacity, and ulceration	Control of flies; isolate infected animals; select breeding cattle with eyelid pigmentation; vaccination as an adjunct to sound management	Oxytetracycline; patch over eye
Pneumonia *Pasteurella* spp. *Haemophilus somnus* (bacteria)	Pneumonia	350 mg/head/day of Sulfamethazine or Chlortetracycline	Broad spectrum antibiotics

TABLE 16.3 (Continued)

Disease and Cause(s)	Clinical Signs	Prevention	Treatment
Poisoning (plant) (see Chapter 15)			
Polioencephalomalacia H_2S toxicity	Sudden death; blindness; incoordination; "downers"	Gradual adaptation to high-energy feeds; thiamine supplementation; decrease sulfur intake in water or feed	Vitamin B complex (thiamine) may be helpful
Prolapse (See uterine and vaginal prolapse)			
Pulmonary emphysema (see asthma)			
Rabies Virus. Usually bites from infected mammals	Clinical signs vary from furious to dumb type; can be transmitted to humans; never insert hand in suspect animal's mouth	Vaccination of susceptible animals, especially dogs and cats that act as intermediary between humans and infected mammals when necessary (epidemic)	None; human health hazard
Red Water Disease, Bacillary Hemoglobinuria *Clostridium novyiType D*	Sudden deaths; red urine; bloody diarrhea; not to be confused with leptospirosis	Vaccination: *Clostridium hemolyticum*	Penicillin
Ringworm Fungus	Round, crusty, and thickened circles, with hair denuded inside circle; more common in winter months	Avoid contact with infected animals and areas	Local application of fungicide (equal parts of tincture of iodine and glycerin); avoid contact of lesions with hands, as the disease is contagious to humans
Salmonellosis *Salmonella typhimurium and S. dublin* (bacteria)	Diarrhea (bloody); high temperature; often confused with coccidia infections; highly fatal	Contaminated lots and feed must be eliminated; vaccinate in heavily infected feedlots	Neomycin, 140 mg per ton of complete feed for calves; Chlortetracycline 70 mg/head/day
Scours *E. coli K99* (bacteria), corona virus, rotavirus (viral), and cryptosporidia (protozoa)	Diarrhea, weakness, dehydration; and severe acidosis	Covered later in this chapter	Covered later in this chapter
Tetanus *Clostridium tetani* (bacteria)	Spasms, contractions of voluntary muscles, high mortality rate	Avoid contamination of open wounds; vaccination in high-risk areas	Three phases: (1) antibiotics; (2) tranquilizers or chloralhydrate to relax muscle; (3) high doses of tetanus antitoxin (up to 300,000 units every 12 hrs); keep cattle in quiet, dark area

(continued)

TABLE 16.3 (Continued)

Disease and Cause(s)	Clinical Signs	Prevention	Treatment
Trichomoniasis *Trichomonas foetus* (protozoa) Bulls are asymptomatic carriers	Infertility and abortion (2–4 months) pyometra	Maintain closed herd or introduce only virgin replacement heifers and bulls; cull open cows in infected herds	Cull carrier animals; report to state veterinarian
Tuberculosis	Usually none	Periodic testing; slaughter reactors	Test and slaughter reactors; report to state veterinarian
Ulcers, Gastric Sudden changes of feed ration too high in concentrates; some association with trace mineral deficiency (see acidosis complex)	"Tarry" stools; bloat; sudden death	Change roughage concentrate ratio; reduce stresses; proper balance of trace minerals	None
Urinary calculi, "Water Belly" Change in ration, mineral imbalance	Straining to urinate; dribbling of urine	Salt (NaCl) up to 4–5% of ration; ammonium chloride 0.74–1.25 oz per head per day or at 0.5% of ration; keep calcium levels higher than phosphorus in diet	Surgery
Uterine and Vaginal Prolapse Excessive straining at calving or from coughing or diarrhea	Vagina, cervix, uterus, or all three protrude through vulva	Not known	Disinfect prolapsed organ; reposition organs and suture around vulva
Vesicular stomatitis (virus)	Inflammation or blistering of tongue, lips, dental pad, nose, teats, and feet	Avoid contact with infected animals	None; report to state veterinarian
Warts, Viral Papillomatosis (virus)	Warts on all parts of body	Sanitation in vaccination; tattooing to prevent transmission of virus from animal to animal; vaccination against warts may be considered	Surgical removal
White Muscle Disease Vitamin E and selenium deficiency	Calves are stiff with arched back; diarrhea can occur; death from starvation	Selenium/vitamin E supplementation in some areas	Sodium selenite-vitamin E in aqueous solution (0.03 mg selenium per lb of body weight)

FIGURE 16.1 Fluid accumulation (edema) in the dewlap, brisket, and underline are characteristic signs of brisket disease. Congestive heart failure at high altitudes (over 7,000 feet), rather than an organism, causes the physiological problem.
Source: J. L. Shupe, Utah State University.

The need to pursue a specific diagnosis cannot be overemphasized. Cost-effective health program recommendations require an accurate diagnosis, an excellent recordkeeping system, and superior communication between producer and veterinarian. In regard to the responsible use of therapeutic and prophylactic drugs, it is important for beef producers to identify accurate and professional sources of information.

Producers can access regulatory disease information for each state by contacting the USDA State Regulation Retrieval System (1-800-545-8732). This voice response system allows 24-hour access to information about emergency notices, state regulations, or access to a state veterinarian.

Disease is defined as any abnormal structural or functional change in the tissues of the body. Cattle producers need to recognize abnormal changes in the condition, appearance, and behavior of animals that may be related to disease. Usually the first clinical signs of disease are often slight and may go unnoticed. Early detection of disease is usually associated with both success of treatment and lower health costs.

It is important to understand normal vital body signs since diseases typically affect them. Temperature, pulse, and respiration rate are the important vital body signs. An abnormal *temperature* is one of the first objective signs of a health problem. When an animal's temperature is above normal, it is considered a fever; when it is below normal, it is called hypothermia. Fever is more common than hypothermia. The absence of fever does not mean the animal is not sick in all instances. Infectious diseases may cause fever. Metabolic diseases are usually not associated with fever.

The normal temperature range for cattle is from 100.4°F to 103.1°F; factors such as time of day, amount of physical activity and others cause normal fluctuations in temperature. When body temperature goes one degree above the normal upper limit, the animal is considered to have a fever. Even with an extremely high fever, the temperature seldom goes above 107°F unless the animal experiences heat stroke, at which time the temperature may exceed 110°F.

Pulse is the rhythmic periodic thrust felt over an artery in time that is associated with the heartbeat. The rate of pulse can be obtained by putting the fingers over the superficial arteries and pressing the arteries against a hard or bony structure (the common location in cattle is the lower part of the jaw or under the tail 2–6 inches from its base). The pulse rate can vary depending on age, size, sex, atmospheric condition, time of day, excitement, and other factors. The normal range for pulse rate (heartbeats per minute) is 40 to 70 in mature cattle and it is higher in calves.

Respiration is the act of breathing—taking in oxygen and expiring carbon dioxide. The respiratory system is affected by many primary and secondary diseases. Respiration rate is the number of inspirations (expansion of chest or thorax area) per minute. The normal range is 10 to 30 per minute. Some of the same factors identified for temperature and pulse can cause variations in the rate of respiration.

Several other aspects of disease need to be recognized. Some diseases in beef cattle are *acute* in nature as they are typically severe but short term. These diseases, such as bloat or enterotoxemia, must be recognized early or prevented altogether, to avoid death losses. *Chronic* diseases are lingering dysfunctions that lead to the continual reduction of productivity and health status. Johnne's disease is an excellent example of this disease classification.

Some diseases are overt and thus can be clinically diagnosed as the symptoms are relatively apparent. Subclinical conditions are harder to detect as they may involve slow losses in animal productivity and generalized health status. These diseases require a greater attention to monitoring if they are to be detected early and prior to them having significant negative consequences.

Immunity

Infectious disease occurrence is the result of interactions between disease-causing agents (pathogens), the host animal, and the surrounding environment. Disease occurs when the pathogens overwhelm the animal's ability to eliminate them. Most frequently, the number of pathogens are low enough that their effects can be neutralized by the host. Environmental factors such as temperature and precipitation conditions can dramatically affect disease occurrence. For this reason, disease incidence may often follow a seasonal pattern.

There are a series of protective measures utilized by cattle to avoid development of disease. Innate disease protection is provided by the skin, mucous membranes, and other intact tissue barriers. Normal, healthy animals are also able to defend themselves from disease via an immune response.

Immunity is developed when a foreign substance (antigen) penetrates the innate defense mechanisms. The body recognizes the invader and then acts to rid itself of the particular antigen. Even though immunity is an enhanced defense against disease, it is not absolute in effectiveness. Immunity is acquired via an active response to an agent mediated by the lymphocytes. The body produces specific proteins against antigens called antibodies. Memory cells are produced after the first encounter with an antigen that boosts the immune response in subsequent encounters. This "memory system" explains why some vaccines are most effective when administered as an initial dose followed by a booster(s) to achieve the desired level of immunization. Vaccines are most effective when administered at a time when the animal's stress is low and before the specific disease challenge has occurred.

Immunity can also be acquired passively such as the ingestion of colostrum or first milk by newborn calves. The dam develops antibodies to the pathogens in her environment and concentrates them in her colostrum. This form of immune protection is somewhat short-term in that only in the first few hours of the calf's life is it able to effectively absorb antibodies from the colostrum through its intestinal wall. Furthermore, milk produced later in the lactation lacks the antibody concentration of colostrum. It should also be noted that a cow needs time to build colostral antibodies to assure passive immunity. A veterinarian should be consulted to make recommendations as to appropriate timing of vaccination.

Types of Vaccines

Biologicals are derived from living pathogenic organisms or their metabolic products and are used to enhance an animal's disease resistance. Vaccines, bacterins, and antitoxins are all examples of biologicals. The three common forms of vaccine utilized in the cattle industry are:

1. Killed organism vaccines—dead organisms or their parts are utilized. The chemical antigen remains intact to stimulate an immune response.
2. Inactivated bacterial toxins—toxicity of a pathogen is neutralized while the antigen's structure remains in place.
3. Modified-live organism vaccines—organisms are grown in unique mediums or mutated to eliminate disease-causing capability while retaining antigenic structure.

A sound vaccination program is designed to help keep an animal's disease resistance well above the level of disease challenge. Disease occurs when challenge exceeds resistance. Resistance fluctuates over time as a result of environmental stress, poor nutrition, and a variety of other factors that can lower an animal's immunity. Disease challenge is also variable. For example, the disease challenge on the ranch can be considerably different than at a commercial feedlot where cattle from a variety of sources are co-mingled. In some cases, vaccines need to be administered as a series and as such revaccination must be timely to insure adequate protection.

Administering Health Products

Administering health products is a vital part of disease prevention and treatment. Method of administration is important because it affects the speed with which the product enters the animal's system. Health products should always be measured accurately and administered according to the manufacturer's recommendations. Aseptic conditions should be utilized during vaccination since other diseases and infections can be easily transmitted.

Some vaccinations do not result in the animal becoming immunized. This occasionally happens because vaccines are not appropriately selected or handled properly, improper vaccination methods are used, or the individual animals have a unique physiological response. Killed organism and toxoid products require two doses. Failure to administer the second dose would result in immunization failure.

The proper use of animal health products is a requirement to assure superior efficacy and assurance of food safety to consumers. Federally licensed products should always be selected and used in accordance with the label instructions. Specific label information to be considered includes dosage, timing and route of administration, withdrawal periods and other warnings, storage conditions, and expiration dates. The use of any product in a manner not specified by the label requires a veterinarian's prescription.

Vaccines should never be combined as this will likely destroy product effectiveness. For reconstituted products, a sanitary transfer needle should be utilized. Reconstituted vaccines tend to lose their effectiveness within an hour after mixing. Direct sunlight and improper storage temperatures during processing often result in poor product performance.

Always mark and separate different syringes used with different products to prevent bacterins or killed-type products from damaging modified live vaccines. Syringes used to administer modified live vaccines should only be cleaned with hot water. Cleaning with disinfectants may destroy the efficacy of vaccines. Select the correct needle size (subcutaneous use 16- or 18-gauge needle, 1/2 to 3/4" in length; intramuscular use 16- or 18-gauge needle, 1 to 1 1/2" in length), assure that the air is out of the syringe prior to injection, and use appropriate injection techniques, route, and site. If using an implant, assure proper placement and good sanitation.

Following are several methods of administering vaccines and therapeutic agents:

Intramuscular (I.M.)—the preferred injection site is a well-muscled area of the neck. Absorption is rapid due to good blood supply. A 1 1/2 in. needle is commonly used, allowing complete penetration of the skin and partial penetration of the muscle. The volume of the injection in any one site should be moderate (<10 cc in adult cattle) to prevent an abscess from forming. Injection into the hip, lower leg, and shoulder are to be avoided so as not to create an injection blemish.

Intranasal—a method used to create local resistance to disease in the respiratory tract. The vaccine is administrated via inspired air.

Intravenous (I.V.)—used commonly to administer therapeutic agents to young calves. The drug is rapidly available to the animal's system in larger volume and tissue irritation is avoided. The best site is the jugular vein, located between the neck muscles and throat on the side of the neck (the ideal site is approximately one-third of the distance between the jaw and chest). The site should be cleansed and wetted with alcohol, and a 1.5–2-in. needle should be inserted at a 30-degree angle pointing toward the body. Slight suction on the syringe should allow aspiration of blood into the syringe to verify accurate placement and the injection should be slow and steady.

Oral—common in light weight calves, this method is used to administer therapeutic drugs to cattle. Administration is simple in smaller calves. The greatest danger is failure to get the drug far enough into the throat to ensure swallowing; aspiration of the drug can lead to pneumonia or, in large enough quantities, to drowning.

Subcutaneous (Sub.Q.)—an injection under the skin that results in slow but sustained absorption due to a relatively small blood supply. The injection site is where the skin is loose, usually in the neck. Utilize the "tented" technique whereby loose skin is gathered and pulled away from the body with one hand and the needle is inserted into the skin fold with the other. This assures that the product is administered under the skin but not into the muscle. The base of the ear is an alternative subcutaneous injection site.

Quality Assurance Programs

Beef Quality Assurance (BQA) is a program designed to ensure that cattle are managed in a way that results in the production of safe and wholesome beef products for consumers. The effects of a good BQA program include, but are not limited to, residue prevention, avoidance of pathogen contamination, and elimination of injection site blemishes. BQA efforts incorporate Deming's Total Quality Management (TQM) principles, NASA's Hazard Analysis Critical Control Point (HACCP) protocols, and biosecurity principles.

A multitude of cattlemen's organizations offer BQA programs on the state level and many individual cattle enterprises have developed on-farm quality plans (Fig. 16.2). *Biosecurity* measures are those procedures that can be implemented to minimize the risk of introducing and/or spreading disease. Biosecurity received considerable attention in 2001 due to the out-

FIGURE 16.2 The development of quality assurance protocols is becoming increasingly important to assure market access for beef cattle. These operating procedures can be obtained from a state or national program. They can also be written specifically to the needs of a particular enterprise.

break of hoof and mouth disease in Great Britain. Biosecurity measures go far beyond vaccination programs to include development of select supplier arrangements, improved sampling and monitoring procedures, and improved employee training. A set of good management practices developed by Colorado State University, the Colorado Livestock Association, and the Colorado State Veterinarians Office include the following:

1. Limit visitors entering with livestock or feed, and be watchful of unusual visitors or activities.
2. Consultants, veterinarians, and buyers should be required to wash and disinfect footwear before entering livestock feed storage areas. The choice of disinfectant will depend upon the infectious agents which are being targeted.
3. Isolate new livestock from resident animals for at least four weeks. Observe these new animals closely for signs of disease during this period, use separate tools to handle feed, bedding, and feces of these animals, and these animals should be handled last.
4. Minimize livestock handling and processing stress.
5. Regularly monitor livestock for signs of illness. Remove sick animals from unaffected animals until diagnoses is accomplished and appropriate treatment has been instituted.
6. Necropsy all deaths that are unexplainable.
7. Control birds and vermin.
8. Train employees not to walk in feed bunks and feed stuffs and avoid defecating around these areas and livestock.
9. Store feed in areas that will not contact drainage from livestock areas and manure piles.
10. Water tanks should be regularly cleaned.
11. Require commercial livestock trucks to be cleaned prior to loading livestock, especially if the truck has hauled another producer's animals previously.

In the case of those situations where highly contagious diseases are involved additional precautions may be required. These include restricting access to the farm premises, having employees change clothes and disinfect when both entering and leaving the farm, and disinfecting equipment, tack, and vehicles.

The use of BQA programs is becoming recognized as an advantage in the market place. For example, ConAgra Beef Company™ will only purchase cattle from feeders who are willing to sign a document stating: "To the best of my knowledge, all animals presented for slaughter have been handled in a manner to prevent a pharmaceutical or agricultural chemical residue violation. Label dosages, route of administration and withdrawal times have been followed, and only approved FDA pharmaceutical compounds have been used for treatment. Any exception to the previous statement has been administered under a recognized veterinarian-client-patient relationship and proper precautions taken for off-label use. Agricultural chemicals used for herd health management and production of feedstuffs have also been used in compliance with label directions and withdrawals."

The 2000 NCBA Beef Quality Audit found that producers lose an average of $18 per head to management-related quality problems such as injection site blemishes, hide damage, bruises, and liver condemnations. These losses to the industry could be considered small compared to the loss of reduced beef consumption if consumers lose confidence in the safety and wholesomeness of beef.

An example of a quality-related defect is related to intramuscular injections. The use of intramuscular injections can lead to associated lesions that result from tissue irritation by the injected substance (Table 16.4).

Better than 90 percent of the lesions identified in the various audits conducted by Colorado State University researchers were classified as "clear scars" or "woody calluses" which would have originated from injections given by cow-calf producers, stockers, or very early in the feeding period. While injection-site lesions are not a food safety concern, they do impact palatability and consumer perception. In fact, injections given by cow-calf producers to calves (Table 16.5) can have dramatic negative consequences on the palatability of steaks/roasts from those same animals. The shear force required to tear a steak core taken as far as 3 inches from the center of an injection-site lesion was 5.80 kg (acceptable shear forces are ≤ 3.86 kg for restaurant quality; ≤4.45 kg for retail trade) according to George, et al. (1997). Any intramuscular injection results in tissue irritation and scaring—these scars should be considered a life-long effect.

TABLE 16.4 Incidence of Injection-site Lesions in Top Sirloin Butts and Round

Year	Top Sirloin Butts (%)	Rounds (%)	Round Trim Loss (oz.)
1990	21.6	NA	—
1991	19.3	NA	—
1992	12.5	NA	—
1993	11.5	NA	—
1994	13.6	10.1	6.8
1995	10.5	11.6	7.3
1996	9.2	7.1	8.2
1997	6.2	4.4	15.7
1998	5.1	7.5	14.9
1999	4.0	5.1	10.0
2000	2.5	11.3	12.5

Source: Roeber, et al. (2000).

TABLE 16.5 Incidence of Injection-Site Lesions Associated With Product Administration at Branding and Weaning

Intramuscular Injection—Type of Product	Incidence of Associated Lesions at Slaughter[a] (%)
Branding (376 days prior to slaughter)	
Clostridial bacterin (2 ml)	72.5
Clostridial bacterin (5 ml)	92.7
Vitamins A and D^3	5.3
Long-lasting oxytetracycline	51.2
Weaning (225 days prior to slaughter)	
Clostridial bacterin (2 ml)	46.3
Clostridial bacterin (5 ml)	79.5
Vitamins A and D^3	10.0
Long-lasting oxytetracycline	92.3

[a]Ages at slaughter from 12 to 24 months.
Source: Colorado State University (1997).

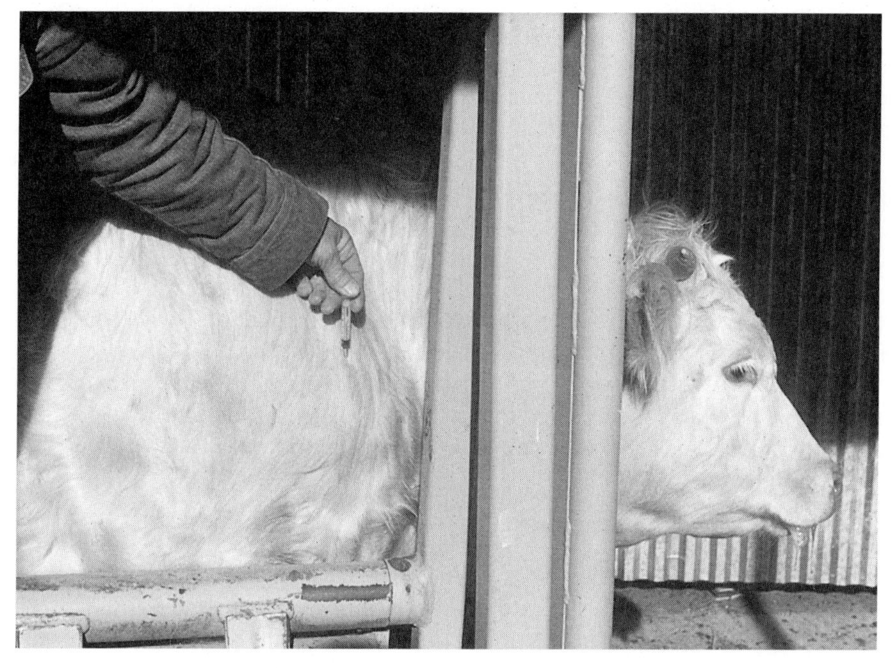

FIGURE 16.3 Selection of injection site is an important management decision in preventing blemishes to valuable cuts of meat.

Minimizing injection-site lesions can be accomplished by adopting the following management strategies:

1. Administer clostridial bacterins subcutaneously in the neck using the tented techniques (Fig. 16.3) or at the base of the ear.
2. Avoid unnecessary repeat or multiple injections of clostridial bacterins.
3. Avoid intramuscular injections whenever other routes of administration are listed in the label recommendations.
4. Inject no more than 10 ml per injection site (less in light calves).

TABLE 16.6 Average Percent of Intramuscular Injections Given by Cow-Calf Producers and Veterinarians by Site

	1993		1996	
	Producers	Veterinarians	Producers	Veterinarians
Neck	19.8	27.0	35.2	49.8
Shoulder	3.4	6.5	17.1	12.9
Hip	64.7	56.6	42.8	34.8
Lower rear leg	9.6	8.7	4.6	2.1
Other	2.5	1.2	0.3	0.4

Source: NAHMS.

5. Change needles at least every 15 injections (more often if cattle are dirty).
6. Appropriately discard bent and/or used needles—never straighten and reuse.
7. Choose products that have low-volume doses whenever possible.

As a result of educational efforts from quality assurance programs, both veterinarians and producers have altered the site of intramuscular injections (Table 16.6).

Hide damage can be reduced by placing brands low on the hip or between the hooks and pins high on the hip instead of on the sides, using a small brand, and using proper techniques during the branding process. Bruises can be prevented by using appropriate equipment and handling cattle carefully and quietly.

PARASITES

Parasites cause millions of dollars of loss to the cattle industry annually. These losses take the forms of reduced weight gains, increased amount of feed per pound of gain, lower milk production, reduced hide value, additional trim on carcasses, and death. The losses are not always apparent and may be excessive before cattle producers recognize there is a serious problem. There are two basic types of parasites: external and internal.

External Parasites

The two primary groups of *external parasites* are (1) insects (which include flies, lice, and mosquitos) and (2) arachnids (which include ticks and mites). Figures 16.4–16.7 identify some common flies, lice, mites, and ticks affecting cattle. External parasites live off the flesh and/or blood of cattle. They can mechanically transmit the organisms that cause pinkeye, mastitis, anaplasmosis, blue tongue, and other infectious diseases to cattle. Buzzing and biting of flies annoys cattle, decreasing gains and increasing feed costs. Lice and mites cause cattle to rub, which occasionally results in large areas of bare skin. Fence maintenance costs increase when posts and wire are used to satisfy the itching caused by certain lice and mites.

Lice are flat, wingless insects with three pairs of legs. The body of the louse is divided into three segments, with the body of most lice being less than 3/16-inch long. Lice are classified as biting lice or sucking lice. The blood-sucking lice are bluish-slate in color. Lice spend their entire lives on the host and are especially abundant during the winter when the host's hair is longer and thicker.

Mites have four pairs of legs and undivided bodies. They are microscopic in size and are whitish in color. They live off the skin tissue of the host and cause cattle diseases such as scab, mange, itch, or scabies. Mites are spread by contact with infected animals or from objects that

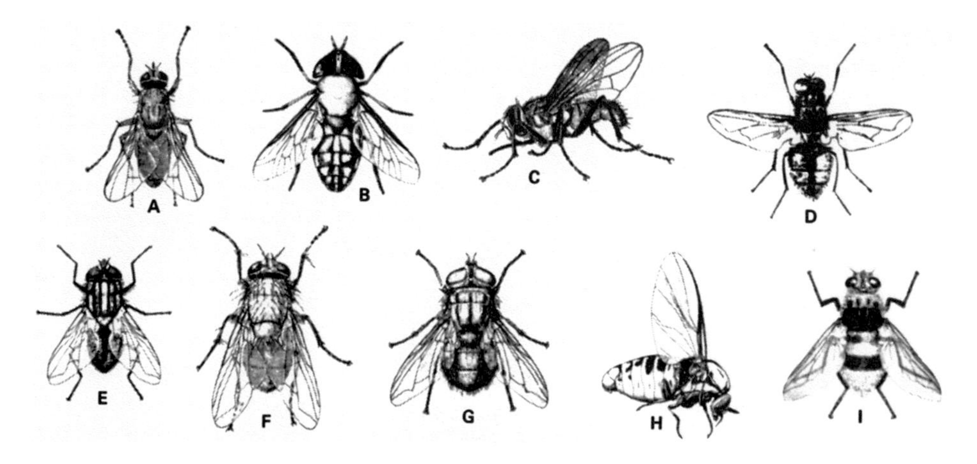

FIGURE 16.4 Flies. A Horn fly is the most common fly around cattle. It is about half as long as the housefly and similar in appearance. It spends most of the life cycle on cattle, typically resting around the base of the horn, except to lay eggs. B. Horsefly (also called deer fly, ear fly, or gadfly) ranges up to an inch in size. Specimen shown is Greenhead Horsefly, one of the many species. C. Stablefly, similar to common housefly, has a stinging bite. It doesn't remain on cattle, usually resting in the shade. D. Face fly (male). It is slightly larger than common housefly and breeds in fresh manure. Pinkeye is carried by this fly when it feeds on the mucous membranes around the eyes. E. Housefly has four black stripes on its back and feeds on secretion around eyes, nose, and other orifices. It can carry disease and eggs of other smaller parasites. F. Blowfly is usually metallic or yellow color and approximately 1/2 in. in length. The blowfly will lay eggs in an open wound where the blowfly maggot or larva feeds off the flesh, which increases the size of the wound and prevents healing. G. Screwworm fly (adult) is probably the deadliest fly. It is bluish-green, with a yellowish-red face and three blue stripes on its back between the wings. It resides only in the southern parts of the United States and Central America where the screwworm (maggots or larvae of the screwworm fly) feeds on the flesh of open wounds. H. Buffalo gnat (adult female) is a small black fly less than 1/6 in. long which sucks blood from the ears, eyes, and nose of cattle. I. Heel fly (adult female) has a 1/2 in. long black and yellow-striped, hairy body, with orange hairs at the posterior end of the abdomen. The eggs are laid on legs and lower parts of bodies of cattle during the spring. The eggs hatch, burrowing through the skin and migrating to the back of the animal forming bumps under the hide. The larvae (called grubs) winter in the back of cattle, emerging through the hide, then hatching and becoming an adult fly in the spring.
Source: USDA and Bull-O-Gram, Rainbow's End Ranch, Rt. 1, Box 43, Elfrida, AZ 85610.

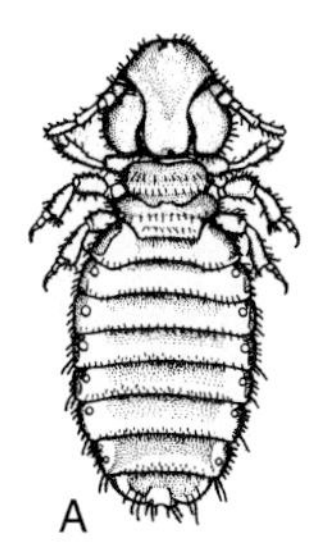

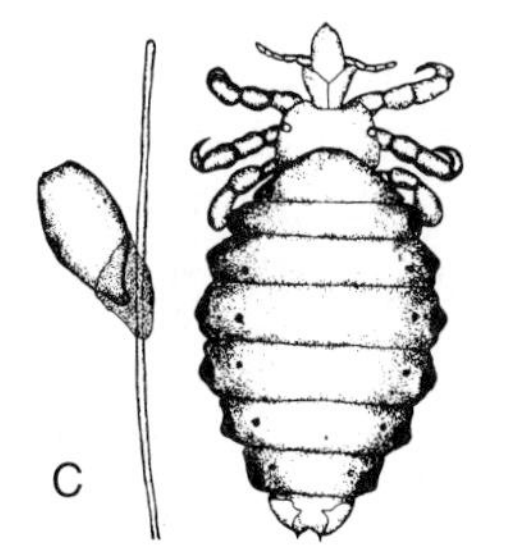

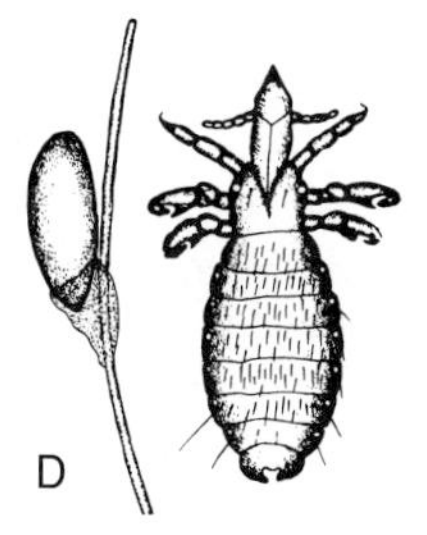

FIGURE 16.5 Lice. A. Chewing cattle louse (a biting louse) has a yellow body with eight dark crossbars. B. Hairy cattle louse (blood-sucking). C. Short-nosed cattle louse (blood-sucking), with a louse egg attached to an animal hair. D. Long-nosed cattle louse (blood-sucking), with a louse egg attached to an animal hair.
Source: USDA and Bull-O-Gram, Rainbow's End Ranch, Rt. 1, Box 43, Elfrida, AZ 85610.

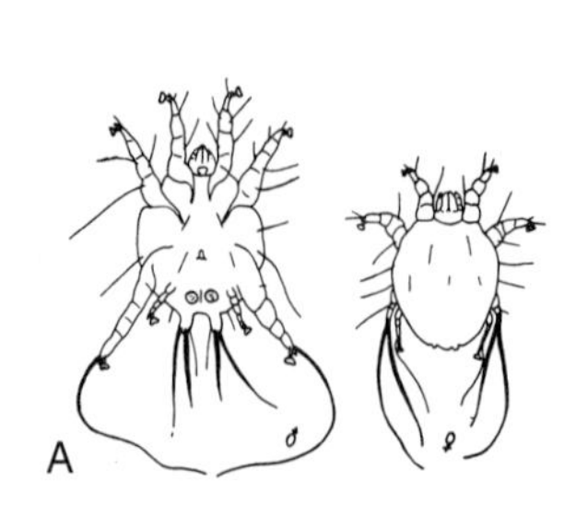

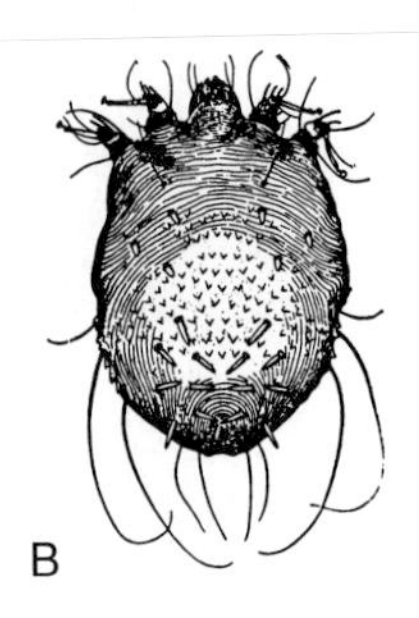

FIGURE 16.6 Mites. A. Chorioptic mite is responsible for tail mange which is the most common type of mange in cattle. It develops chiefly in the tail region and spreads to other areas of the body. B. Sarcoptic mite causes scabies which is the most severe form of mange. Lesions, characterized by a crusted appearance, and actual thickening of folds of skin, appear around the head and neck and spread to other parts of the body.
Source: USDA and Bull-O-Gram, Rainbow's End Ranch, Rt. 1, Box 43, Elfrida, AZ 85610.

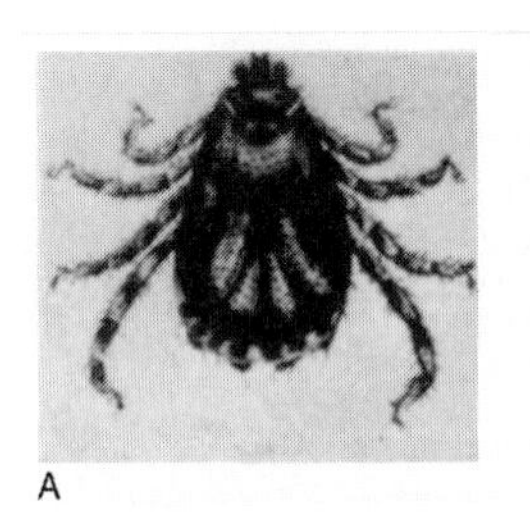

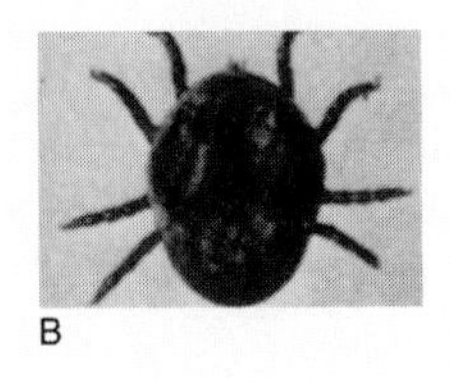

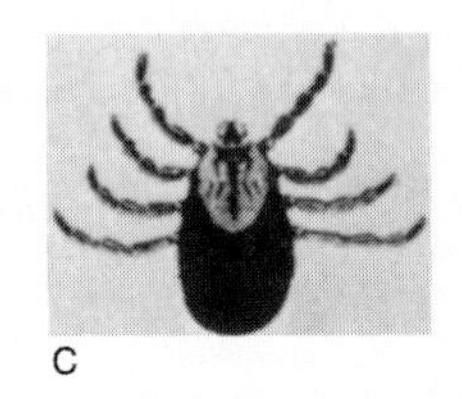

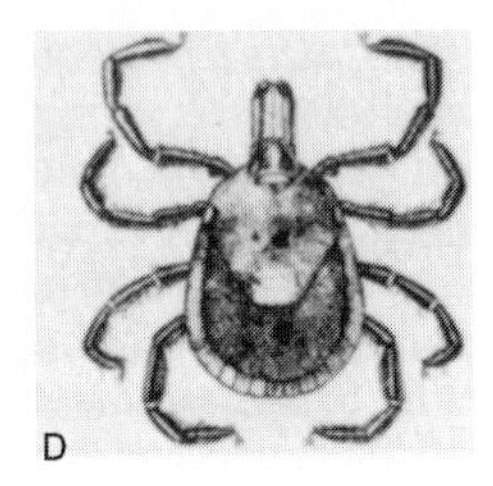

FIGURE 16.7 Ticks. A. Cattle fever tick (mature male) was previously a serious danger as a vector for the protozoa which caused Babesiasis. It has been eradicated in the United States and is no longer such a serious threat. B. Spinose ear tick nymph (fully engorged) is a soft tick whose larvae infest the ears of livestock. They are common in the arid region of the Southwest, especially in cattle kept in confinement. C. Winter tick is different from other hard ticks in that it is a single-host tick and attacks cattle in the winter and early spring. D. Lone star tick (female) is a hard tick with silvery markings on its shield. The male has a pattern of markings while the female has only one dot. The tick is a vector for Rocky Mountain spotted fever.
Source: USDA and Bull-O-Gram, Rainbow's End Ranch, Rt. 1, Box 43, Elfrida, AZ 85610.

have been in contact with the infected animals. This contagious aspect is heightened by increasing levels of confinement. However, seasonal-environmental effects are also realized. The psoroptic mite (not shown in the figures) causes common scab. If cattle scab is diagnosed, a report should be made to the state veterinarian. The mites feed on the skin of the host, which becomes covered with scabs that form over the mites and the hair comes out in patches. Scabs usually appear first on the neck and around the base of the horns.

Grubs infest cattle via the attachment of heel fly eggs to the hair of the animal. Larvae migrate down the hair and penetrate the skin. The larvae pass through the warble stage beneath the skin of the back. They make breathing holes through the hide which provide a migratory route to the outside several weeks later. Larvae first appear on cattle in the southern regions in September and on cattle in the northern regions in January or later. Grubs emerge from the animal two months later.

Controlling External Parasites.

Table 16.7 lists several commonly used insecticides for controlling external parasites. The most popular method of application is the use of systemic insecticides. In this situation, the material is applied to only one area of the body (usually the back) and the insecticide is absorbed, spreading throughout the entire system (Fig. 16.8). Insecticides can also be applied by

TABLE 16.7 Prevention and Control of External Parasites in Cattle

Parasite	Prevention	Treatment	Comments
Grubs	Learn life cycle of grubs and treat animals accordingly	Co-Ral, Ivomec, Spotton, Tiguron, Warbex, Dectomax	Follow labels carefully. Some products have side reactions. Should be used only on certain ages or types of cattle
Lice	Sanitation; use of backrubbers or periodic spraying with insecticide	Pour-ons: Ivomec, Lysoff, Neguvo, Permectin, Tiguron, Warbex, Ivomec Eprinix Injection: Ivomec Sprays: Taktic, Prozap	Watch withdrawal periods for all products used to control parasites
Mange	Sanitation; avoid contact with infected animals	Check with local veterinarians	Mange is a reportable disease for which animals are quarantined
Flies (house, stable)	Sanitation; primarily by removing manure and cleaning other areas where flies breed and multiply	Supplemental use of insecticides or biological controls can be effective	Do not contaminate feed or water
Flies (face, horn)	Apply ear tags or tape for pastured cattle	Cyanobenzeneacetate (Ectrin), Cyfluthrin (Cutter Gold), Cyhalothrin (Saber Extra), Cypermethrin + Chlor-Pyrifos (Max-Con), Diazinon (Terminator), Fenthion (Cutter Blue), Permethrin, Atroban Extra, Deckem, Ectiban, Expar Extra	In some areas flies have developed resistance to certain products. Producers should alternate products and control measures. The latter could include sprays, dust bags, backrubbers, pour-ons, or feed additives. Remove ear tags before slaughter

Source: Various sources.

spraying, fogging, or using back rubbers, ear tags, injections, or dust bags. They can also be applied by preparing a large quantity of solution and placing it in a below ground vat. The cattle are then either herded into the vat one by one or lowered into the vat using a hydraulic cage. This is the most thorough method of applying insecticides; however, it is usually the most expensive and least practical method for most cattle operations. The most commonly reported fly control methods are topical products (dustbags, dips, sprays, backrubs) and treated ear tags, with 60.6 and 30.8% of operations utilizing these control methods, respectively, according to the 1997 NAHMS survey.

Insecticides are used to break the life cycle of the parasite. Figure 16.9 shows the life cycle of the cattle grub and Figure 16.10 indicates when treatment should be given to break the life cycle. Several methods may have to be implemented to control external parasites in any one operation. Continuous inspection for parasite infestation is the best indicator of effective control. It is important to remember that insecticides are toxins and, as such, should be handled only by trained individuals following proper procedures and utilizing the correct equipment. Failure to do so may result in serious human health consequences.

FIGURE 16.8 A systemic insecticide being administered topically to the back of a yearling steer.

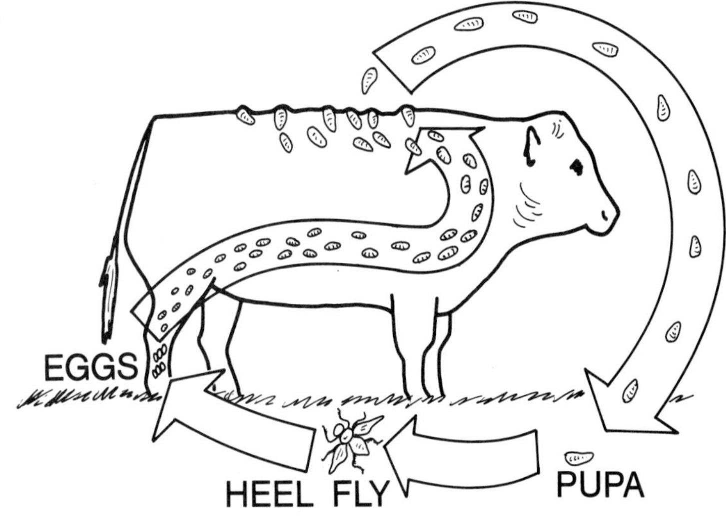

FIGURE 16.9 The life cycle of the cattle grub.
Source: Colorado State University.

Internal Parasites

While the effects of external parasites are often obvious to cattle producers, the impact of internal parasites can easily escape notice. Internal parasites are present inside the animal but the parasites and their eggs are microscopic in size. The economic loss to cattle producers resulting from internal parasites is often a slow, draining, continuous process that goes unnoticed because of the difficulty in measuring it.

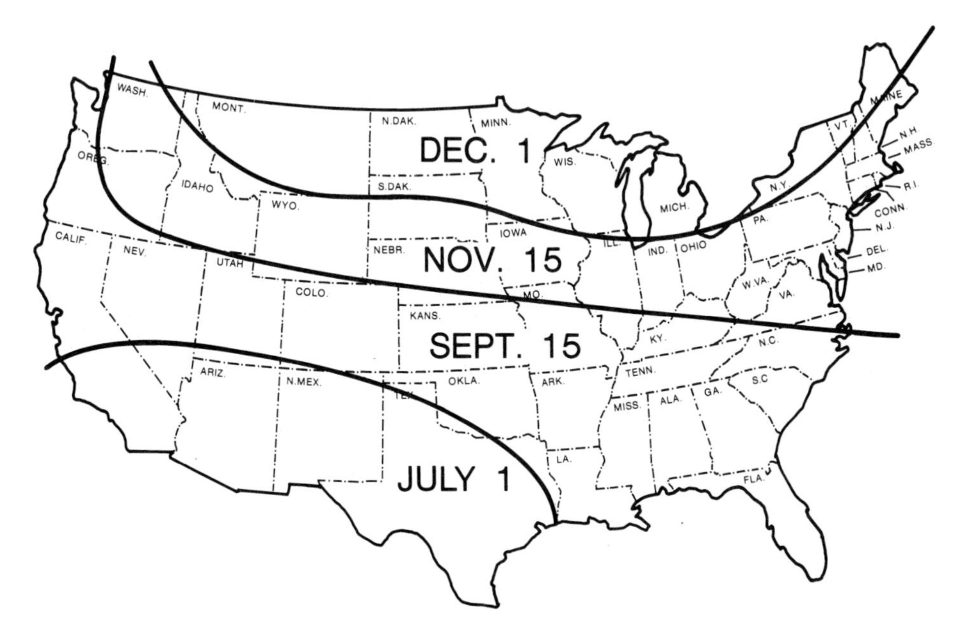

FIGURE 16.10 Cattle located in the areas outlined should be treated prior to the date specified for maximum grub control during heel fly activity. Dates given are subject to seasonal variation and apply to the common grub. The northern species appear later.
Source: Colorado State University.

The most common class of internal parasites is the roundworm (Nematodes) with the other two basic classes being tapeworms (Cestodes) and flukes (Trematodes). Typical internal parasites are identified in Table 16.8.

Controlling Internal Parasites

In most areas, internal parasites cannot be totally eliminated. Cattle producers, however, should work to keep the level of infection below the level of economic loss. Nearly 73% of all cow-calf operators dewormed at least part of their herd in 1996 according to the National Animal Health Monitoring System (NAHMS) Survey.

Control of internal parasites involves interrupting their life cycle. This can be accomplished in several ways: (1) the presence of unfavorable climatic conditions; (2) development of resistance in cattle; (3) management of cattle to prevent their ingestion of infective organisms; (4) destruction of intermediate hosts and environmental features hospitable to the parasite outside the cattle's bodies; and (5) therapeutic chemical treatment of cattle.

Wet and warm weather favor proliferation of certain internal parasites. Areas that are colder and more arid (such as the western states) and northern Great Plains do not have the same internal parasite problems as those found in the southern states. While cattle producers cannot control the climate, they can use weather to assist in controlling parasites. In areas where internal parasite control may not be economically feasible, producers should conduct fecal egg counts to determine if a treatment program is needed.

Some cattle are more resistant to internal parasites than are other cattle. This resistance may be limited to an immunity developed due to earlier infections. Younger cattle and cattle stressed with poor nutrition and diseased conditions are less resistant to internal parasites.

TABLE 16.8 Common Internal Parasites

Common Name	Scientific Name	Length and Shape	Location	Effect
Large stomach worm (barberpole or wire worm)	*Haemonchus placei*	15–40 mm	Abomasum	Sucks blood; anemia; diarrhea
Medium stomach worm (brown stomach worm)	*Ostertagia ostertagi*	6–9 mm	Abomasum	Reduction of nutrient absorption; profuse, watery diarrhea
Small stomach worm (bankrupt worm)	*Trichostrongylus axei, Trichostrongylus colubriformis*	5 mm	Abomasum	Reduction of nutrient absorption; profuse, watery diarrhea
Cooperids	*Cooperia* (various species)	5–8 mm	Small intestine (first 10–20 feet)	Profuse diarrhea; weight loss
Hookworm	*Bunostomum phlebotomum*	9–18 mm	Small intestine (first few feet)	Sucks blood; anemia; weight loss
Intestinal threadworm	*Strongyloides papillosus*	3.5–6 mm	Small intestine	Intestinal hemorrhages
Thread-necked intestinal worm	*Nematodirus (helvetianus)*	12–25 mm	Small intestine	Diarrhea; loss of appetite; weakness
Nodular worm	*Oesophagostonum radiatum*	12–15 mm	Colon	Loss of appetite and weight; diarrhea; weakness
Lungworm	*Dictyocaulus viviparus*	80 mm	Trachea and bronchioles	Coughing; rapid, shallow breathing; secondary bacterial infections
Tapeworm	Most common is *Moniezia expansa* and *Moniezia benedeni*	Several feet	Small intestine	Reduced nutrient absorption; occasional diarrhea
Common liver fluke	*Fasciola hepatica*	30 mm (leaflike)	Liver (bile ducts)	Hemorrhage of liver; hyperplastic and enlarged bile ducts
Giant liver fluke	*Fascioloides magna*	75 mm (oval)	Liver	Similar to common liver fluke
Lancet fluke (lesser liver fluke)	*Dicrocoelium dendriticum*	12 mm (long and slender)	Liver	Similar to common liver fluke
Stomach flukes (conical flukes or amphistomes)	*Paramphistomes* (several species)	15 mm	Rumen and reticulum	Diarrhea; loss of appetite and weight

Close confinement and overgrazing are both conducive to spreading parasitic infections. Under these conditions, cattle are more likely to eat infected manure or graze the lower parts of plants where parasites are more numerous. Pasture rotation can help reduce the number of parasites by causing infective organisms to die before they are ingested.

In some cases, the intermediate host can be destroyed, thereby reducing the number of parasites. An example is the snail, which is the intermediate host for the fluke. Scattering manure, renovating pastures, plowing, and draining marshes and swamps all upset the hatching process and reduce the concentration of infective organisms.

TABLE 16.9 Selected Anthelmintic Drugs Used to Control Internal Parasites

Chemical	Trade Name	Form	Effective Against
Thiabendazole	TBZ E-Z-Ex	Paste Pellet/Block	*Haemonchus, Ostertagia, Trichostrongylus* (adult and immature forms), *Cooperia, Nematodirus, Oesophagostomum radiatum*
Levamisole	Levasole Tramisol	Injectable Bolus Drench Soluble powder Feed additive Oral gel	*Haemonchus, Ostertagia, Trichostrongylus* (adult and immature forms), *Cooperia, Nematodirus, Oesophagostomum, Bunostomum, Chabertia, Dictyocaulus*
Morantel tartrate	Rumatel	Feed additive	*Haemonchus, Ostertagia, Trichostrongylus, Cooperia, Nematodirus, Desophayostomum*
Doramectin	Dectomax	Injectable	*Ostertagia* + 35 stages of internal and external parasites
Ivermectin	Ivomec-F	Injectable	*Haemonchus, Ostertagia, Trichostrongylus, Cooperia, Oesophagostomum, Bunostomum, Nematodirus* (adults only), *Dictyocaulus, Fascuola hepatica*

Anthelmintic drugs are used to kill parasites inside cattle. However, these chemical dewormers have varying degrees of effectiveness on different parasites and different forms of the same parasite (Table 16.9). Dewormers come in drench, feed additive, injection, or bolus form.

The best time to deworm depends on the type of operation and its location. In general, the best time to deworm a cow herd is just before calving. Calves may be dewormed when they are worked prior to weaning. In areas where internal parasites have not been a serious problem, a periodic fecal check for parasite eggs and a comparison of the performance of some treated and nontreated animals are useful in determining whether a total treatment program is necessary. Since feedlot cattle originate from many sources, a routine deworming of incoming cattle from geographic problem areas should be considered.

Avermectins are a class of parasiticides effective against both internal and external parasites. However, this class of products and all other products should be evaluated in terms of their biological and economic effectiveness.

COW-CALF HEALTH MANAGEMENT PROGRAMS

Respiratory, reproductive, and digestive diseases continue to be the major cause of disease loss in beef cattle. Costs of respiratory diseases to cattle producers, in terms of treatment, weight loss, death loss, and culling in weaning calves, are at a third of a billion dollars annually.

Bovine respiratory disease (BRD) is seldom the result of a single factor. BRD usually is caused by a combination of stress, virus infection, and invasion of the lungs by pathogenic bacteria such as *Pasteurella* and *Haemophilus.* Stress undermines the natural defenses built into the lining of the trachea and bronchi, and respiratory viruses (such as IBR, PI_3, BVD, and BRSV) further damage these natural defenses. Ultimately, pathogenic bacteria find a wide-open road into the lungs where they localize, multiply, and cause the severe damage called BRD, pneumonia, or shipping fever.

Immunity against IBR, BVD, BRSV, and PI_3 can be enhanced by the administration of virus vaccines to cattle. Modified live and inactivated virus vaccines are available in single and

TABLE 16.10 Percent of Death Losses in 1996 Attributed to Various Causes

Disease/Cause	Unweaned Calves	Breeding Cattle
Digestive problems	14.4	6.1
Respiratory problems	16.3	6.0
Weather	20.2	18.0
Calving problems	13.9	17.0
Poisoning	1.3	3.7
Predators	6.4	1.1
Theft	0.8	0.8
Other known	9.2	27.0
Unknown	17.5	20.3

Source: NAHMS (1997).

combination forms. The routes of administration of these vaccines are intramuscular (IBR, PI_3, BRSV, BVD) or intranasal (IBR, PI_3 only). Both intramuscular and intranasal vaccines provide adequate immunity. Best protection is obtained by administering vaccines to healthy, non-stressed animals.

Controlling Major Cattle Diseases

The general causes of death in unweaned calves and breeding cattle are categorized in Table 16.10. The primary causes of death in unweaned calves are digestive, respiratory, weather-related, dystocia, and unknown problems. Breeding cattle mortality rates are highlighted as being caused by dystocia and weather-related problems.

Management guidelines for controlling BRD and other diseases are put forth by the Animal Health Committee of the NCBA and the American Association of Bovine Practitioners. The management outline attempts to present the ideal immunization program. However, there are regional and area differences in disease incidence and the general health of cow herds. In addition, management needs vary from operation to operation. Therefore, individual producers should work closely with their local veterinarian in developing specific health programs for their operations. Several different management options are presented for each individual cattle operator's consideration.

Program A—Young Calves (1–3 months of age)

1. Respiratory diseases:
   a. IBR-PI_3 (use killed vaccine intramuscularly or modified live vaccine intranasally), also BRSV.
   b. *Pasteurella.*
   c. *Haemophilus somnus.*
2. Other diseases:
   a. Leptospirosis (single or multiple strains available).
   b. Clostridial diseases (includes *Cl. chauvei, Cl. septicum, Cl. novyi Type B, Cl. sordellii, Cl. perfringens* C and D, and *Cl. novyi Type D*).
3. Other procedures that should be followed at this time include the following:
   a. Implant cattle with approved growth stimulants (except the animals that will be kept for breeding).
   b. Castrate and dehorn. Male calves castrated at the initiation of a stocker program as compared with calves steered early in life experience higher morbidity and poorer average daily gains, which can significantly impact profitability.

c. External and internal parasite control. (Some products are not recommended for use with very young calves.)

Program B—Older Calves (3–4 weeks prior to weaning)

1. Respiratory disease:
   a. IBR-PI_3: If killed vaccine was administered at 1–3 months of age, it should be readministered as a booster, or a modified live vaccine administered intranasally or I.M. may be used instead of the killed product. If this is the initial vaccination and not a booster, however, use the modified live vaccine intranasally or I.M.; also BRSV.
   b. BVD: BVD vaccine is available singly or in combination with modified live IBR virus, PI_3 virus, or BRSV *Pasteurella.* Since calves are not apt to be heavily stressed when vaccinated 3–4 weeks before weaning, a "combination" BVD vaccine could be used at this time.

Note: Some cattle operators and veterinarians prefer not to administer BVD simultaneously with IBR-PI_3 modified live vaccine intramuscularly because of the added stress on the animal. This is true in very young or heavily stressed animals, which can be vaccinated against BVD 2–3 weeks later; in this case, at weaning time. It must be emphasized that calves stressed in any form (e.g., castrated, dehorned, branded, shipped, and so forth) should not be vaccinated against BVD; otherwise, adverse reactions could occur.

   c. *Pasteurella* and *Haemophilus somnus:* Two injections, 2–4 weeks apart, are needed whenever *Pasteurella* and *Haemophilus* bacterins are used. If the first injection is given at this time, follow it with a booster injection at weaning time.
2. Other diseases (not generally part of the BRD complex) that may be considered at this time include the following:
   a. *Clostridial* diseases booster.
   b. *Leptospirosis.*
   c. *Vibriosis* (heifer and bull calves kept for breeding).
   d. *Brucellosis* (for heifer calves). Requirements vary among different states. Consult a veterinarian regarding specific state regulations. It is a one-time vaccination and must be administered by a licensed, accredited veterinarian or another designated animal health official. All *brucellosis* vaccinations must be officially reported to the appropriate state agency.
3. Other procedures:
   a. Vitamin A (injectable preferred).
   b. Implant with approved growth stimulants (except for animals that will be kept for breeding).

Note: If the immunization program is started early enough, both the initial and booster shots may be administered prior to weaning. This will reduce stress at weaning and reduce shrink and amount of time required to regain weaning weight. This regimen is subject to local practices, individual herd health needs, and various combinations of vaccines. The above recommendations, early immunization, processing of cattle, and other management practices are designed to reduce stress at weaning.

   c. Treat for internal and external parasites.

Calves—Weaning Time

1. If the original and booster shots have both been given, no additional immunization is needed at this time. Treat for internal and external parasites, if not already done 3–4 weeks prior to weaning.
2. If the first vaccination series as listed in Program A has been given, give boosters as listed under Program B and treat for internal and external parasites.

3. For calves that have not previously been vaccinated, several alternatives are feasible, depending on type of confinement, feed, and equipment; availability of labor; and final disposition or destination of the calves. Calves must be handled twice for optimum results; Programs C and D are outlined for consideration.

Program C—At Weaning

1. Respiratory disease:
   a. IBR-PI_3 intranasally or I.M. (killed vaccine), also BRSV.
   b. *Pasteurella* bacterin.
   c. *Haemophilus* bacterin.
2. Other diseases:
   a. *Clostridial* bacterin.
3. Other procedures:
   a. Internal parasites.
   b. Implant.
   c. Vitamin A.

Fourteen to twenty-one days later (providing calves are consuming 2–3% of their body weight of feed, or are on pasture and eating well):

1. Respiratory disease:
   a. IBR-PI_3 booster if killed vaccine was used at weaning, also BRSV.
   b. BVD vaccine.
   c. *Pasteurella* booster.
   d. *Haemophilus* booster.
2. Other diseases (not generally considered part of BRD) that may be prevented at this time include the following:
   a. *Clostridial* booster.
   b. *Leptospirosis* bacterin.
   c. *Vibriosis* bacterin.
   d. *Brucellosis* (heifers).
3. Other procedures:
   a. Treat for external parasites

Program D—At Weaning

1. Respiratory disease:
   a. *Pasteurella* bacterin.
   b. *Haemophilus* bacterin.
   c. BVD vaccine.
2. Other diseases:
   a. *Clostridial* bacterin.
3. Other procedures:
   a. Treat for internal parasites.
   b. Implant.
   c. Vitamin A.

Fourteen to twenty-one days later (providing calves are consuming feed equivalent to 2–3% of their body weight or are on pasture and eating well):

1. Respiratory disease:
   a. IBR-PI_3 intranasally or I.M., also BRSV
   b. *Pasteurella* booster.
   c. *Haemophilus* booster.

2. Other diseases (not generally associated with BRD) that may be considered at this time include the following:
   a. *Clostridial* booster.
   b. *Leptospirosis* bacterin.
   c. *Vibrio* vaccine (heifers and bulls).
   d. *Brucellosis* (heifers).
3. Other procedures:
   a. Treat for external parasites.

Other Health Management Tips

High levels of antibiotics may be fed during the weaning period on the advice of a veterinarian. This practice should be utilized judiciously and not as an absolute. Do not attempt to feed antibiotics unless the feed consumption of calves is 2–3% of body weight. Prior to processing cattle, consult with a veterinarian concerning the use of epinephrine to treat shock or hypersensitivity reactions in cattle. It is a potent drug. Producers should know the clinical signs of shock and hypersensitivity, indications for epinephrine, and its dosage. Always observe cattle that have been treated.

Management of Calves at Weaning

1. Calves should be eating some dry feed 2–4 weeks prior to weaning.
2. Vaccination procedures should be reviewed and changed as necessary, depending on the health conditions of specific lots of cattle, environmental conditions, and the prevalence of various diseases in the immediate area.
3. An adequate, fresh water supply is essential, preferably from a source that cattle can see or hear running.
4. Vitamin A prior to or at weaning is generally recommended.
5. Check feed and water consumption—both should increase during the weaning period.
6. Provide good, high-quality hay. Calves should be consuming 2–3% of body weight of feed before either feed or water is medicated.
7. Check calves two or three times daily.
8. Seek professional help from a veterinarian when needed and before a major problem arises.

Marketing and Transportation

1. Minimize stress factors as follows:
   a. It is crucial to get calves moving through marketing channels quickly.
   b. Avoid crowding and bruising.
   c. Avoid conditions of extreme temperature variations, dust, or wetness.
   d. Feed and water calves before shipping.
2. Other factors to improve health:
   a. Upon arrival at the feedlot, cattle should first be fed hay prior to having access to water. Be sure to have adequate water facilities available. Begin a limited feeding of grain and protein supplement.
   b. Segregate sick animals.
   c. Tractor exhaust stacks must be tall enough for gases to clear the trailer well.
   d. Avoid ammonia buildup in trucks, yards, barns, and sheds from excess urine, manure, and moisture. Ammonia contributes to respiratory disease.
   e. Start adequate treatment promptly. Identify sick cattle and treat as recommended by a veterinarian. Maintain and utilize accurate records.

Cow Herd, Replacement Heifers, and Bulls

It must be emphasized that vaccination and adequate handling of calves are part of, but not a substitute for, a total herd health management program. It is essential that an adequate breeding herd vaccination program be implemented for maximum benefits to be expected from vaccinating or preconditioning calves or both.

A sound long-term herd health program depends on the systematic monitoring of all aspects of enterprise management that have an impact on animal well-being. This monitoring process includes observation of BCS changes, avoidance of nutrient or environmental conditions that create toxic or deficient states, awareness of actual growth rates relative to projected outcomes, and coordination of nutritional, genetic, reproductive, and biosecurity management.

The following recommendations for replacement heifers and bulls (at 10–15 months of age) are intended to ensure immunization of the breeding herd against diseases of recognized significance. The immunization outlined here should be boostered annually, no later than 30 days prior to breeding.

1. Replacement cattle with unknown history status. Heifers not pregnant.
   a. First processing:
      1. Immunization: (a) IBR-PI_3, (b) *Vibriosis,* (c) *Leptospirosis,* (d) *Clostridial* diseases, (e) *Haemophilus,* and (f) *Pasteurella.*
      2. Other treatment: (a) Treat for internal parasites and (b) Vitamin A.
   b. Second processing:
      1. Immunization: (a) BVD, (b) *Leptospirosis* booster, (c) *Vibriosis* booster, (d) *Clostridial* booster, (e) *Haemophilus* booster, and (f) *Pasteurella* booster.
      2. Other treatments: (a) Treat for external parasites, depending on grub development and season. (b) If heifers are of eligible age, vaccinate for *brucellosis.* If heifers are older than eligible vaccination age, they should be tested, possibly twice, for *brucellosis.*
2. Replacement cattle sufficiently immunized by calfhood and weaning programs. This is recommended as optimum management, starting after calves are weaned and assuming they have had at least minimal recommendations suggested for calves.
   a. Booster vaccinations: (1) IBR, (2) BVD, (3) *Leptospirosis,* (4) *Vibriosis,* and (5) *Clostridial.*
   b. Other treatments: (1) Internal and external parasites. (2) Vitamin A.

Mature Cows and Bulls

Assuming that the breeding herd of mature cows and bulls has been previously immunized, either as calves or as herd replacements, the following booster immunizations are recommended (Table 16.11):

TABLE 16.11 Booster Immunizations Recommended for Mature Breeding Cattle

Booster Immunizations	Not Less Than 30 Days Prior to Breeding		Last Trimester of Pregnancy
IBR	X	or	X-killed, chemically attenuated, or intranasal vaccine
BVD	X		
Clostridial	X	or	X
Leptospirosis	X		
Vibriosis	X		

Frequently, veterinarians recommend that IBR, BVD, and *Clostridial* immunizations be administered during the last trimester of pregnancy because it conveys a greater passive immunity to the calf. One should be cautioned, however, that if IBR is administered during the last trimester of pregnancy, that it be a killed product, chemically attenuated or an intranasal vaccine.

BVD vaccines are modified live virus vaccines and ordinarily should not be administered to pregnant animals. There are experimental studies and reports of practitioners having administered BVD vaccines during the last trimester of pregnancy to increase passive immunity in the calf without adverse effects. Killed BVD vaccines are also available and have different recommendations. Caution is emphasized that all biological products should be administered in accordance with the recommendations of the manufacturers and in consultation with a veterinarian. Annual boosters are frequently recommended. It is important, however, that cattle operators consult with a veterinarian as to the appropriate schedule of immunization for their herds.

Special Notice: *Brucellosis* Status

Cow-calf producers should pay special attention to *brucellosis* because the infection still exists in several herds in some states. Regulations regarding vaccination and testing are expected to change. If new cattle are to be brought into a herd, they should originate from a negative herd, be isolated from other animals in the herd of destination, and be retested before being commingled with the new herd. Vaccination should be used as recommended by a veterinarian. Vaccinate all eligible heifers for *brucellosis* and purchase only vaccinated heifers.

Calf Scours

Causes Since diarrhea in the calf is second only to reproductive diseases as a cause of losses to the cattle industry, it is discussed here in more detail than some of the other diseases. The four major causes of diarrhea are colibacillosis, enterotoxemia, salmonellosis, and coccidiosis. A summary of each of the latter three diseases is presented in Table 16.3.

Colibacillosis is by far the most important cause of diarrhea in calves. Although this type of scours can affect calves up to a month of age, it is most commonly observed in calves from birth up to two weeks of age. Scours, in general, may be caused by bacteria, viruses, or parasites (Table 16.12). Vaccination or immunity against one type has little effect against the others, and many different stress factors may predispose the calf to these organisms (Table 16.13). The variety of causative organisms and mitigating stress factors make the calf scours disease complex difficult to prevent or treat. According to a 14-year research study conducted at the Northern Agricultural Research Center, calves that scour have weaning weights 20 lbs lighter than healthy calves. This loss of performance equated to an economic loss of $8.50 per calf in the study population.

TABLE 16.12 Causes of Calf Scours

Pathogen	% of cases
Bacteria[a]	22
Viruses[b]	35
Parasites[c]	24
Other	19

[a]Primarily *E. coli K99* and *Clostridial* strains.
[b]Rotavirus and coronavirus.
[c]Cryptosporidia and coccidia.

TABLE 16.13 Potential Stress Factors That Can Contribute to the Incidence of Calf Scours

Management Area	Specific Issues
Sanitation	Overcrowding, poor ventilation, poor drainage
Climate	Wet weather, extreme temperatures
Nutrition	Poor quality colostrum, insufficient colostrum supply/intake (Table 16.14), parasite infestation, poor general nutrition, changes in ration, overeating, poorly balanced diet, appropriate pre- and postcalving dam nutrition
Dystocia	Stress on cow and calf (Table 16.14)

There are two clinical forms of calf scours. One is sudden death, often without signs of sickness and before diarrhea has had a chance to develop. It occurs because the organisms, or toxins produced by them, gain access to the blood, and death occurs within 4–24 hours. The other and more common clinical form of scours is diarrhea. Stools vary in consistency from watery to semisolid and are lighter in color than normal. This is in contrast to coccidiosis and enterotoxemia, in which the feces are darker than normal and may contain visible blood. In the early stages of the chronic form, the calf continues to eat but later becomes dehydrated and weak and then stops eating. The course of the disease generally is 2–3 days but may vary from one to seven days.

Prevention Scours occurs when a calf's exposure to the disease exceeds the calf's resistance to the disease. Prevention of calf scours occurs when exposure is decreased and resistance is increased.

Exposure to calf scours organisms can be decreased by calving and raising calves in dry, clean areas. If cows have to calve in a confined area, provide a dry, clean individual stall for the cow. The cow and calf should be moved to clean pasture as soon as possible after calving. Calving in corrals or keeping cows and calves in corrals increases the risk of exposing the animals to scour-causing organisms.

Bringing new cows or calves into the herd just prior to calving or during calving increases the risk of exposure to calf scours. Calves needed for grafting would best be selected within the herd rather than purchasing calves from outside the herd.

The major factor in increasing a calf's resistance to scours is to give the calf the first-milking colostrum, preferably within six hours after the calf is born. Colostrum contains immunoglobulins (antibodies) needed by the calf, since the calf is born with little natural immunity against disease organisms. Because of rapid changes in the digestive system, the calf's ability to absorb antibodies diminishes rapidly. Immunoglobulin absorption at calving is 100%, absorption rates decline to 50, 25, and 5% at 6, 12, and 24 hours postcalving, respectively. Furthermore, the acidosis experienced by calves that require assistance at birth inhibits antibody absorption (Table 16.14).

Cows can be vaccinated 30–40 days prior to calving, with a repeat injection two weeks following the first injection. The antibodies produced are for specific vaccine organisms only, and scours could still occur if caused by other organisms. Immunity, even for the specific organisms, is effective only when the colostrum is consumed by the calf and the antibodies are absorbed from the digestive tract.

A genetically engineered oral vaccine to prevent *E. coli K99-specific* infections can be given to calves shortly after they are born.

Treatment Treatment of calf scours is similar regardless of its cause. Diarrhea occurs because of alterations in intestinal function, resulting from increased secretion from the body or

TABLE 16.14 Effect of Dystocia on Calf Vigor and Immunoglobulin Concentration

	Calving Difficulty		
	Unassisted	Easy Pull	Hard Pull
Time from delivery to standing (min.)[a]	40	51	84
Calf serum IgG (mg/ml)[b]	2,401	2,191	1,918
Calf serum IgM (mg/ml)[b]	195	173	136

[a]A measure of calf vigor.
[b]Assisted females were milked out immediately after calving and calves were fed the colostrum.
Source: Colorado State University.

decreased absorption of water and electrolytes (sodium chloride, potassium, and bicarbonate) from the intestinal lumen. These important electrolytes are lost in the feces.

Loss of bicarbonate is a primary cause of acidosis. Loss of potassium causes lethargy and muscle weakness. Loss of water and sodium chloride results in dehydration. Treatment should be directed toward correction of the dehydration, acidosis, and electrolyte loss.

During diarrhea, cells that line the intestine are sloughed off, and it takes two or three days for new cells to develop. The major aspects of treatment, therefore, are to keep the calf alive while the affected intestinal cells are replaced and to prevent secondary diseases such as pneumonia.

Most dehydrated calves suffer from hypothermia (i.e., when the body temperature is lower than normal). Therefore, the calf should be provided with supplemental heat in a dry location.

Clinical signs of dehydration occur when fluid loss reaches approximately 6% of body weight. Fluid losses of 8% result in depression, sunken eyes, dry skin, and difficulty standing upright. With 10% fluid loss, the legs will be colder than the rest of the body. A 12% fluid loss usually results in death.

There are several commercial electrolyte powders for administering fluids to the calf. If they are not available, producers can economically prepare satisfactory mixtures as follows:

Formula 1
1 tablespoon baking soda
1 teaspoon salt
8 oz (250 cc) dextrose
(do not use table sugar)
Add warm water to make a gallon
Administer up to 1 quart orally
every 4–6 hours

Formula 2
1 package (1 oz) fruit pectin
1 teaspoon of Lite® salt
2 teaspoons baking soda
1 can beef consommé soup
Add warm water to make 2 quarts
Give 1 quart orally every 4–6 hours

Either of these homemade products administered at appropriate intervals with recommended volumes will provide an adequate nutrient supply over a period of 24–48 hours. Do not give milk or milk replacers to the calf at the same time as administering electrolytes (however, electrolytes do not replace the need for nutrition). Electrolytes interfere with the normal digestion of milk. Also, whole milk and milk replacers should not be altered as fluid sources because of altering digestive enzymes needs. Whole milk or milk replacers can be alternated at appropriate intervals with oral electrolytes as long as fluid needs are being met. In general, a calf requires 10–12% of body weight in oral fluids every 24 hours. Return the calf to the cow as soon as it is able to follow its mother.

Antibiotics (penicillin or tetracyclines) are given systemically during fluid therapy to prevent secondary infections. They should be given at proper treatment levels each day for at least three days or until the calf has recovered.

TABLE 16.15 Annual Veterinary, Drug, and Total Costs of Disease on a Per Cow Basis by Disease Class[a]

	Disease Class					
Type of Cost	Reproductive Tract	Sudden Death/ Clostridial	Misc.[b]	Enteric	Respiratory Tract	Total[c]
Veterinary service						
Average	$0.99	$0.02	$0.53	$0.28	$0.22	$2.04
Range	0–25.07	0–0.55	0–4.98	0–4.56	0–9.07	0–29.88
Drugs used in treatment						
Average	0.12	0.05	0.31	0.36	0.39	1.22
Range	0–1.77	0–2.91	0–3.54	0–5.44	0–5.80	0–9.62
Other costs						
Average	9.28	9.68	8.88	3.65	2.73	34.23
Range	0–58.86	0–108.85	0–76.07	0–22.71	0–37.33	0–135.33
Total costs						
Average	10.39	9.75	9.72	4.29	3.34	37.49
Range	0–70.29	0–108.85	0–76.07	0–27.37	0–39.00	2.12–148.90

[a]Data from 86 cow-calf operations in Colorado National Animal Health Monitoring System (NAHMS).
[b]Includes all diseases not covered in the other four categories—e.g., abcesses, cancer eye, foot rot, grass tetany, lameness, parasites, pink eye, and others.
[c]Producer's labor, death, loss of production, transportation, and others.
Source: Salman et al., 1991. *JAVMA* 198: 1739.

Disease Costs

There are few studies that have evaluated the total costs of disease in the cattle industry. Ballpark cost estimates are quoted for certain diseases, however, those estimates are not usually supported by a detailed cost analysis

A case study approach does allow for a degree of understanding about the economic impacts of disease occurrence versus disease prevention. The National Animal Health Monitoring System (NAHMS), evaluated disease rates and their associated costs over a three-year period (Table 16.15). The most costly diseases were reproduction (dystocia as the primary contributor), sudden death/clostridial disease (enterotoxemia), miscellaneous (Table 16.15 footnotes), enteric scours (unknown source), and infections of the respiratory tract.

Respiratory disease is of particular interest to feedyard managers. There is a relatively high risk of bovine respiratory disease (BRD) in lightweight, ranch fresh, highly stressed calves. The problem is magnified if weaning and transport to the feedyard occur in times of changing weather conditions and if the cattle originated from multiple sources. The cost of BRD is illustrated in Table 16.16.

The predominant message from the information in Table 16.16 is that healthy animals that are not overtreated with preventative measures are more profitable than those showing clinical BRD symptoms or those animals undergoing a costly perfect prevention program.

The challenge with BRD and many other disease classes is clearly identifying sick animals. Animals may be sick but asymptomatic. These undiscovered disease cases can slowly drain the profitability from individual cattle.

The design of a prevention-oriented health program that is cost-effective is a major component of running a profitable enterprise that yields value to customers via the production of safe and wholesome beef products.

TABLE 16.16 Effect of Bovine Respiratory Disease (BRD) on Production Costs and Profits

Production Factors	Healthy No BRD Impact	Treat Clinically Sick Cattle[a]	Perfect BRD Prevention[b]
BRD preventatives ($/hd)	0.00	2.00	20.00
Morbidity rate (%)	0.00	20.00	0.00
Loss due to chronics ($/hd)	0.00	2.59	0.00
Mortality rate (%)	0.30	1.30	0.30
Economic impact of sickness ($/hd)	0.00	3.30	0.00
Cost of treatment ($/hd)	0.00	10.23	0.00
Cost of death loss ($/hd)	1.85	6.77	1.85
Production cost ($/hd)	366.02	395.00	387.03
Breakeven ($/cwt)	68.00	70.42	69.75
Profit per animal ($/hd)	23.98	(5.00)	2.97
Percent return per $150 equity (%)	8.76	(1.95)	1.08

[a]7% decline in ADG, 50% lung lesions.
[b]Spending 5% of total production costs on prevention.
Source: Adapted from Griffin, Bryant, and Perino (2000).

MANAGEMENT SYSTEMS HIGHLIGHTS

1. A sound preventative health program is the best approach to minimizing the costs associated with diseases of beef cattle.
2. Minimizing intramuscular injections and giving required intramuscular injections only in the neck is an effective management strategy to enhance the value of beef.
3. Participation in quality assurance programs enhances the opportunity for producers to become select suppliers of cattle into a quality-driven marketplace.
4. Health programs should be tailored to the specific needs of a particular enterprise.
5. Herd health does *not* come in a bottle.

REFERENCES

Publications

Agri-Practice. The Journal of Medicine and Surgery for the Animal Practitioner (Monthly periodical). Veterinary Practice Publishing Co., P.O. Box 4457, Santa Barbara, CA 93140-9960.

Bailey, J. H. 1981. *Parasites: The Profit Robbers.* Proceedings of the Range Beef Cow Symposium VII (Dec. 7–9).

Byford, R. L., Craig, M. E., and Crosby, B. L. 1992. A Review of Ectoparasites and Their Effect on Cattle Production. *J. Anim. Sci.* 70: 597.

Compendium of Veterinary Products. 1995–1996. 3rd ed. Port Huron, MI: North American Compendiums.

Control of Bovine Respiratory Disease in the Cow-Calf Herd. 1981. NCA Animal Health Committee and American Association of Bovine Practitioners, Beef Business Bulletin, July 17.

Garry, F. 2000. *Newborn Calf Infectious Diseases: The Importance of Distinguishing Etiologies.* White paper. Department of Clinical Sciences, Colorado State University, Fort Collins, CO.

Garry, F. B., and Odde, K. G. 1991. *Calf Scours, Causes, Treatment and Prevention.* Proceedings of the Integrated Resource Management Program. Colorado State Univ. Ext. Service.

Griffin, D., Bryant, L., and Perino, L. 2000. Economic Impact Associated With Respiratory Disease in Beef Cattle. *CALF News* (Aug.).

Gutierrez, P., and Hogan, T. 1997. Pencil It Out. *Beef:* Spring Cow-Calf Mgt. Issue.

George, M. H., Tatum, J. D., Smith, G. C., and Cowman, G. L. 1997. Injection-Site Lesions in Beef Subprimals: Incidence, Palatability, Consequences, and Economic Impact. Compendium's *Food Animal Medicine and Management* 19: 2.
Herrick, J. B. 1991. Cattle Disease Guide. *Feedstuffs* (Reference Issue) 63: 102.
Improving the Quality, Consistency, Competitiveness and Market Share of Fed Beef. 2000. Final report of the Third National Quality Fed-Beef Audit. Englewood, CO: National Cattlemen's Beef Association.
Kimberling, C. V. 1981. *Brucellosis.* Proceedings of the Range Beef Cow Symposium VII (Dec. 7–9).
McCurnin, D. M. 1990. *Clinical Textbook for Veterinary Technicians.* Philadelphia: W. B. Saunders.
McNeill, J. 2000. Ranch to Rail Program Summary. College Station, TX: Texas A&M University.
National Animal Health Monitoring System. 1994–1998. *Beef Cow-Calf Survey Results.* USDA-Veterinary Service (see Websites).
Pfizer Animal Health. 1995. *Feedlot and Stocker Health Guide.*
Pfizer Animal Health. 1996. *Cow-Calf Health Reference Manual.*
Pierson, R. E., Kainer, R. A., and Teegarden, R. M. 1981. *Classifications of Pneumonias in Cattle.* Proceedings of the Range Beef Cow Symposium VII (Dec. 7–9).
Roeber, D. L., Speer, N. C., Gentry, J. G., Tatum, J. D., Smith, C. D., Whittier, J. C., Jones, G. F., Belk, K. E., and Smith, G. C. 2000. Feeder Cattle Health Management: Effects on Morbidity Rates, Feedlot Performance, Carcass Characteristics, and Beef Palatability. *Prof. Anim. Sci.* 17: 39.
Salman, M. D., King, M. E., Odde, K. G., and Mortimer, R. G. 1991. Cost of Vaccines/Drugs Used for Prevention and Treatment of Diseases in 86 Colorado Cow-Calf Operations Participating in the National Animal Health Monitoring System (1986–1988). *JAVMA* 198: 1739.
Schoenfelder, D. 1979. Herd Health: External and Internal Parasites. *Charolais Bull-O-Gram* (Aug.–Sept.).
Smith, B. P. 1990. *Large Animal Internal Medicine.* St. Louis: C. V. Mosby.
USDA. 1997. *Reference of 1997 Beef Cow-Calf Management Practices. Parts I and II.* National Animal Health Monitoring System. Fort Collins, CO: Centers for Epidemiology and Animal Health.
Ward, J. K., and Nielson, M. K. 1979. Pinkeye (Bovine Infectious *Keratoconjunctivitis*) in Beef Cattle. *J. Anim. Sci.* 38: 1179.
Williams, R. E., Westby, E. J., Hendrix, K. S., and Lemenager, R. P. 1981. Use of Insecticide-Impregnated Ear Tags for the Control of Face Flies and Horn Flies on Pastured Cattle. *J. Anim. Sci.* 53: 1159.

Visuals

"Beef Cattle Castration" (sound filmstrip). Vocational Education Productions, California Polytechnic State University, San Luis Obispo, CA 93407.
"Bovine Dehorning Procedure" (13 min.). Instructional Media Center, Michigan State University, East Lansing, MI 48824.
"Cattle Abnormalities" (26 min.); "Postmortem Abnormalities of Beef" (26 min.); and "Management Practices of Beef Cattle—II" (47 min.). CEV, P.O. Box 65265, Lubbock, TX 79464-5265.
"Dehorning Beef Cattle" (sound filmstrip). Vocational Education Productions, California Polytechnic State University, San Luis Obispo, CA 93407.
"Quality Begins with You" (Beef Quality Assurance Program Video and Brochure). SmithKline Beecham Animal Health, P.O. Box 80809, Lincoln, NE 68501-0809.

Websites

Beef Quality Assurance: www.bqa.org
Elanco Animal Health: www.elanco.com
Hi-Pro Animal Health: www.frionaind.com
Merck & Co, Inc.: www.merck.com
National Animal Health Monitoring System: www.aphis.usda.gov
Pfizer Animal Health: www.pfizerbeef.com
Schering-Plough Animal Health: www.sp-animalhealth.com
Texas A&M Ranch to Rail: www.animalscience-extension.tamu.edu/frameset.html

Growth, Development, and Beef Cattle Type

Many aspects of the growth, development, and beef cattle type are covered in other chapters of this book because they are closely related to nutrition, genetics, animal breeding, and meat science. However, it is helpful to discuss some specific relationships separately—those that allow producers and others to understand more fully the beef animal and its production and management.

GROWTH AND DEVELOPMENT

Beef type, or *beef conformation* (shape and form), of the live animal or carcass results from numerous factors affecting the process of growth and development. During growth and development, the beef animal changes in form and composition. *Growth* is defined in different ways, with no one definition being acceptable to all concerned. Generally, *growth* is an increase in weight until mature size is reached. More specifically, it implies the production of new biochemical units brought about by cell division, cell enlargement, or incorporation of materials from the environment. Finally, *development* is defined as the directive coordination of all diverse processes until maturity is reached. It involves growth, cellular differentiation, and changes in body shape and form. For the purpose of this discussion, growth and development are integrated into a single process.

Prenatal Growth

Prenatal growth, or *embryological development,* is fascinating to observe as a spherical mass of cells organizes itself via morphogenesis into different shapes of recognizable organs (Figs. 17.1 and 17.2). The endoderm eventually forms into the digestive tract, lungs, and bladder; the mesoderm forms the skeleton, skeletal muscle, and connective tissue; and the ectoderm

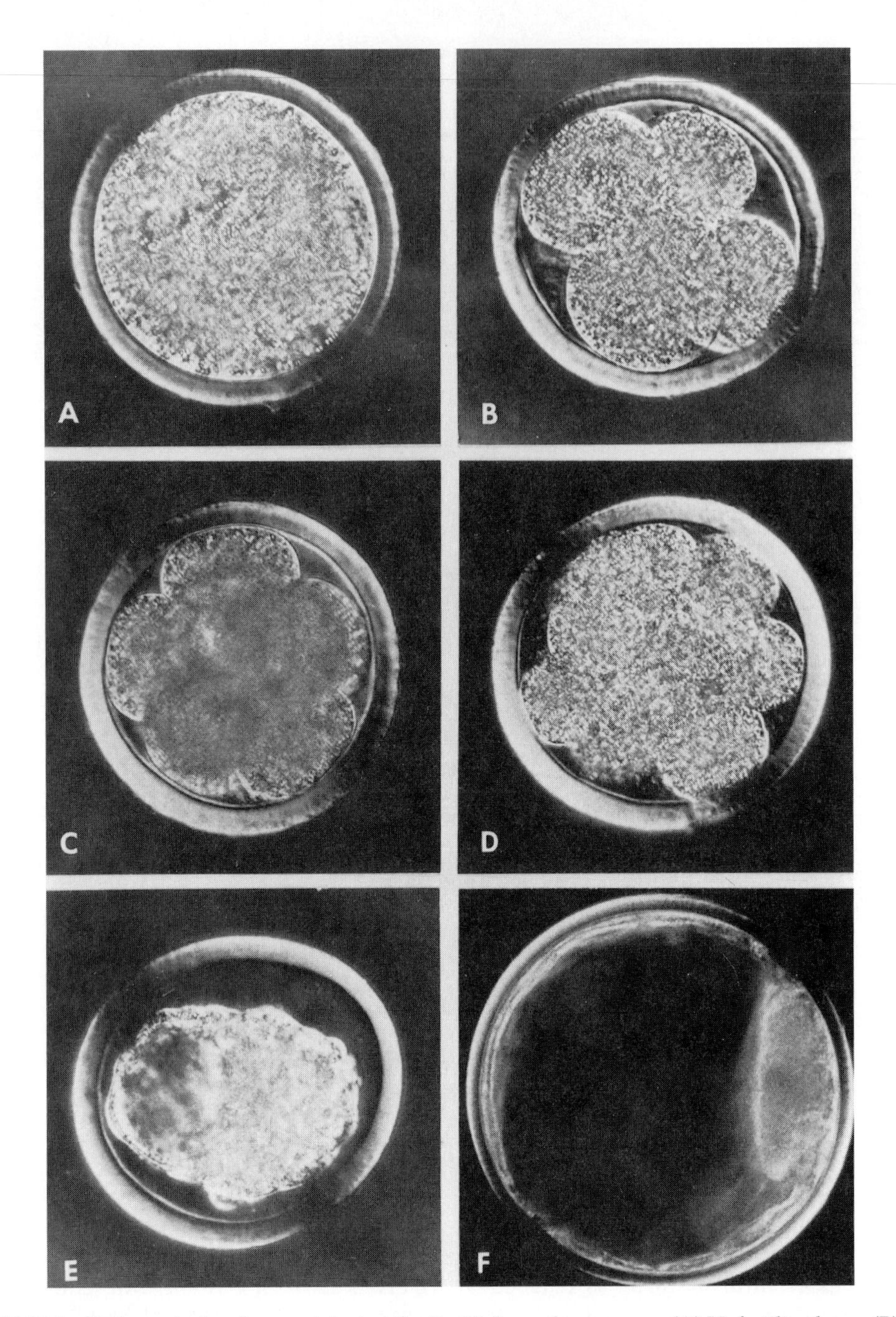

FIGURE 17.1 Embryonic development during the first 8 days of pregnancy. (A) Unfertilized egg; (B) 4-cell embryo on day 2 of pregnancy (estrus = day 0); (C) 8-cell embryo on day 3 of pregnancy; (D) an 8- to 16-cell embryo on day 4 of pregnancy; (E) very early blastocyst stage (approx. 60 cells); (F) expanded blastocyst (> 100 cells) recovered from the uterus on day 8 of pregnancy. Magnification approximately × 300. Courtesy of Seidel, G. E., Jr., 1981. Superovulation and embryo transfer in cattle. *Science* 211:351. Copyright 1981 by the American Association for the Advancement of Science.

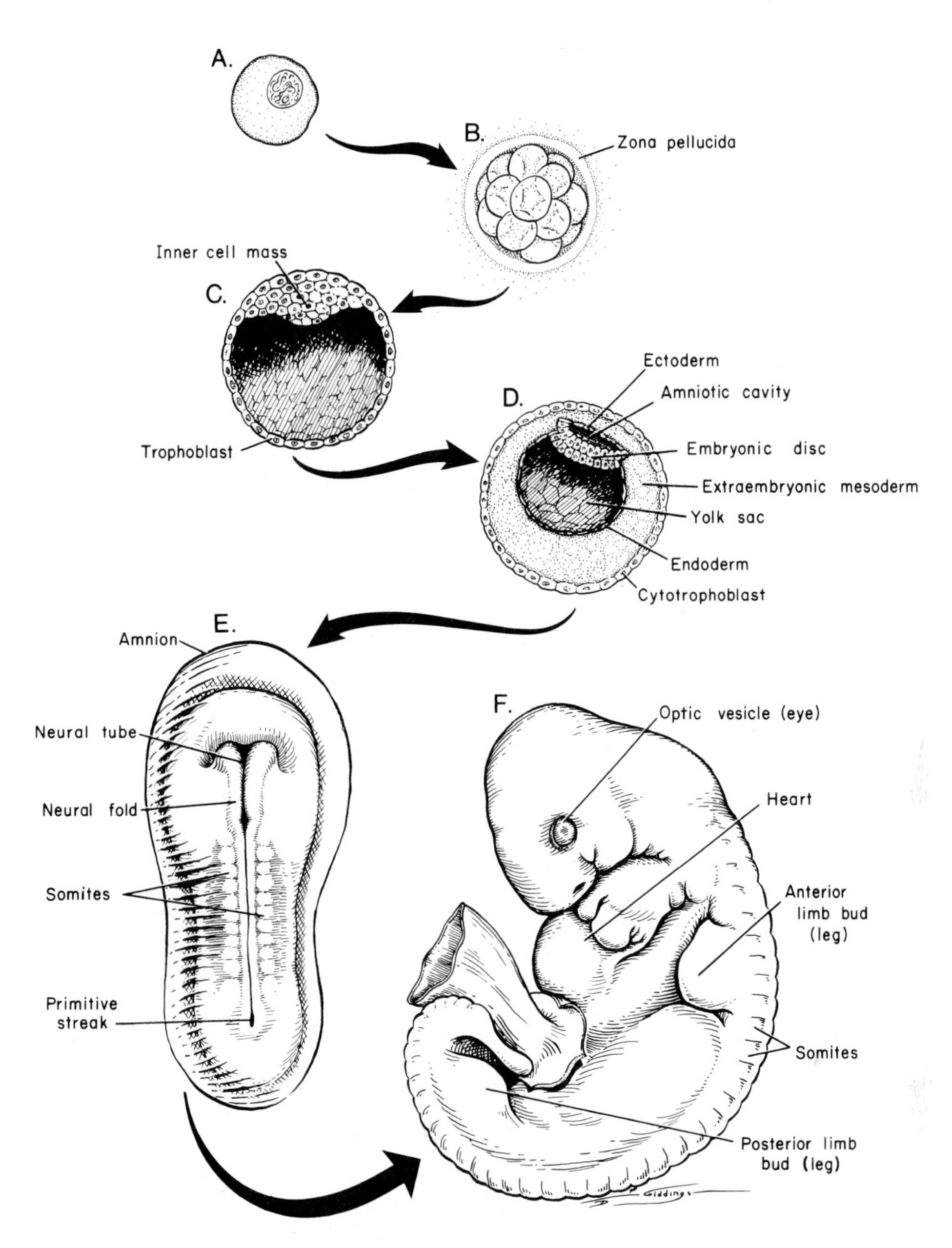

FIGURE 17.2 The morphogenesis of a single egg cell (A) into a morula; (B) then to a blastocyst; (C) (D) shows the stage at which the two cavities have formed in the inner cell mass; an upper (amniotic) cavity and a lower cavity yolk sac. The embryonic disc containing the ectoderm and endoderm germ layers is located between cavities. (E) is a cattle embryo showing the neural tube and the stomites. (F) illustrates the development of a 14-day cattle embryo.
Source: Colorado State University.

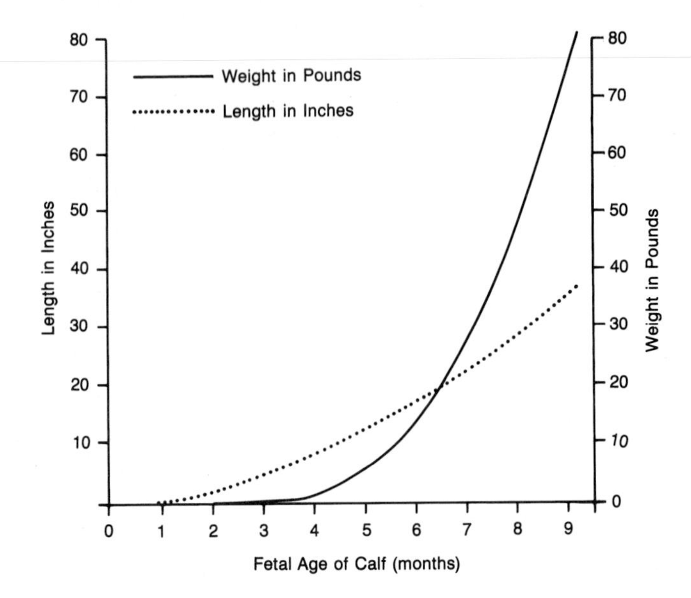

FIGURE 17.3 Growth of the fetus.
Source: Colorado State University.

forms into the skin, hair, brain, and spinal cord. This process of growth, development, and differentiation, involving primarily protein synthesis, is directed by DNA chains of chromosomes in the cells.

The three phases of prenatal life—the zygote, the embryo, and the fetus—are discussed in Chapter 11. The relative size of the fetus changes during gestation, with the largest increase in weight occurring during the last three months of pregnancy (Fig. 17.3). The fetus undergoes marked changes in shape and form during prenatal growth and development. Early in the prenatal period, the head is much larger than the body. Later, the body and limbs grow more rapidly than the other body parts. The order of tissue growth follows a sequential trend determined by physiological importance, starting with the central nervous system and progressing to bones, tendons, muscles, intermuscular fat, and subcutaneous fat.

During the first two-thirds of the prenatal period, most of the increase in muscle weight is due to hyperplasia (an increase in the number of muscle fibers). During the last three months of pregnancy, hypertrophy (an increase in the size of muscle fibers) represents most of the muscle growth. Individual muscles differ in their rate of growth; the larger muscles of the legs and back have the greatest rate of postnatal growth.

Birth

After birth the number of muscle fibers does not increase significantly; therefore, postnatal muscle growth is primarily by hypertrophy. All muscle fibers appear to be red or fast-twitch type at birth, but shortly thereafter some of them differentiate into white and intermediate muscle types.

At birth the various body parts have considerably different proportions when compared with mature body size and shape: the head is relatively large, the legs are long, and the body is small. In the mature animal, in contrast, the head is relatively small, the legs are relatively short, and the body is relatively large. Birth weight represents approximately 6–8% of the mature weight, while leg length at birth is approximately 60% and height at withers approximately 50% of those same measurements at maturity. Hip width and chest width at birth are approximately one-third of the same measurements at maturity. This shows that the distal parts (leg length and wither height) are developed earlier than the proximal parts (hips and chest).

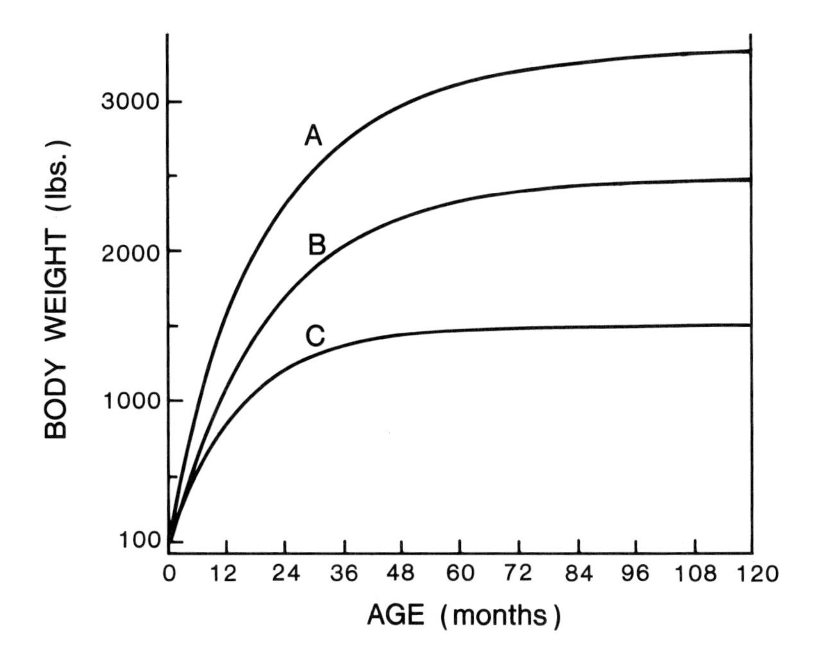

FIGURE 17.4 Growth curves representative of: (A) bulls of very large frame sizes, (B) bulls of intermediate frame sizes, and (C) bulls of small frame sizes. Growth curves for steers and heifers of the three frame sizes would be markedly less than their counterpart bulls.
Source: Colorado State University.

Liveweight Gain

Beef cattle growth is typically observed or measured by evaluating body weight changes per unit of time or by plotting body weight against age. Growth, expressed as average daily gain, weight per day of age, or pounds of lean (muscle) per day of age is particularly useful as a management tool. It is also useful to describe growth curves, which are similar in shape for various breeds and types of cattle but contain noticeable differences when mature size is reached relative to age (Fig. 17.4). Growth curve analysis receives increased emphasis as altering the shape of curves or achieving the best combination of rapid early growth in slaughter cattle and reducing mature size in breeding cattle become more important.

Carcass Composition

A superior carcass is characterized by a low proportion of bone, a high proportion of muscle, and an optimum amount of fat. During the past two to three decades, carcass composition has been improved in beef cattle primarily by reducing the amount of fat in the carcass.

Figure 17.5 shows expected changes in fat, muscle, and bone as cattle increase in liveweight during the linear phase of growth. Fat growth begins rather slowly, then increases exponentially as the animal enters the fattening phase. Bone has a smaller relative growth rate (2 lb of muscle to 1 lb of bone) than either fat or muscle. The muscle-to-bone ratio in the young calf is approximately 2:1; the ratio at a slaughter weight of 1,100–1,200 lbs for the typical slaughter steer is 3.5 to 4:1.

Sex affects tissue growth primarily in terms of the fat component (Fig. 17.6). Heifers fatten at an earlier weight than steers, which is why they are slaughtered at a weight 100–200 lbs less than steers of a similar frame size. Bulls have less fat than both steers and heifers at the same slaughter weight while also being superior in muscle mass.

Genetic differences in carcass composition are shown in Figure 17.7. These different maturity types can be described as breeds or biological types. However, it is not entirely clear how much of the breed differences can be accounted for by frame size alone. There are mature size differences within a breed that are reflected in large part by frame size differences.

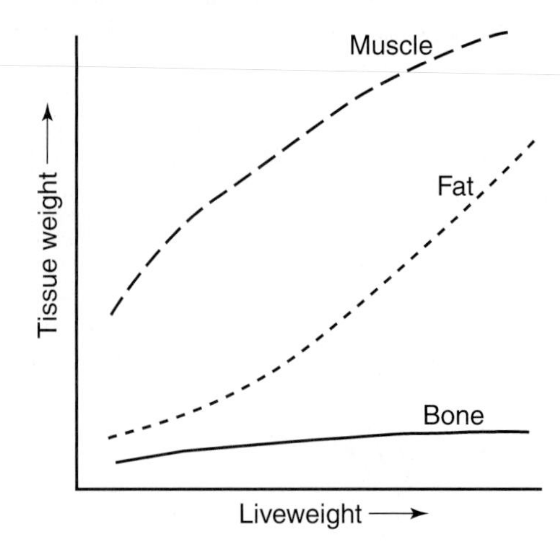

FIGURE 17.5 Tissue growth relative to liveweight.
Source: Adapted from American Society of Animal Science, Berg and Walters (1983).

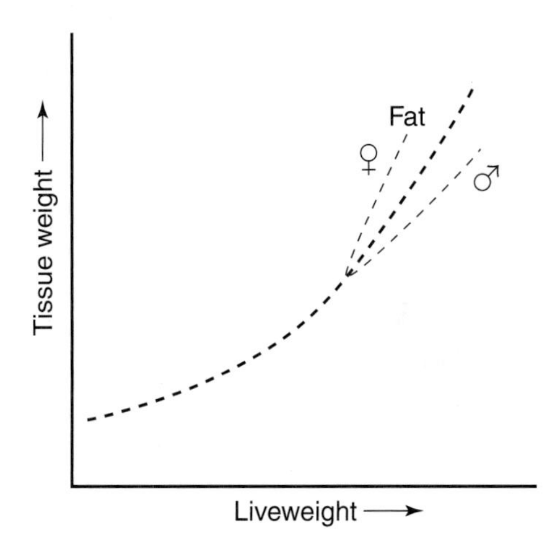

FIGURE 17.6 Sex effects on fat deposition.
Source: Adapted from American Society of Animal Science, Berg and Walters (1983).

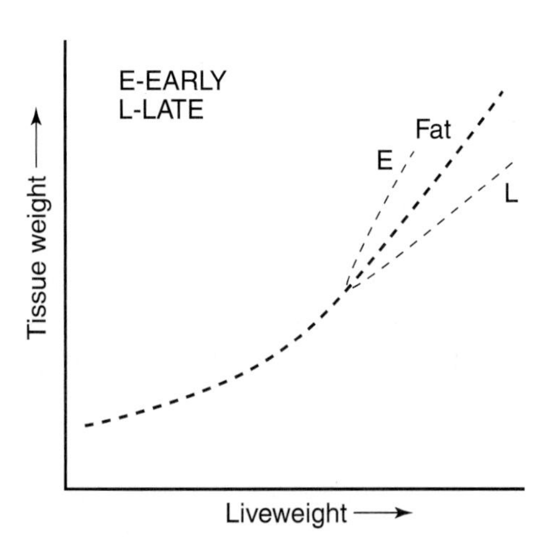

FIGURE 17.7 Effects of different maturity types on carcass fat deposition.
Source: Adapted from American Society of Animal Science, Berg and Walters (1983).

Fat Growth and Development

Plane of nutrition or level of energy in the ration changes the relative amount of fat that is deposited. Cattle fed high-energy diets deposit fat earlier than cattle on lower-energy diets, even when the cattle are evaluated at the same liveweights.

Where fat is deposited, as liveweight increases, is of high economic importance to the beef industry. Current market preferences favor an optimum combination of intramuscular fat (marbling) and other fat depositions (subcutaneous, cavity, and seam) that allow a carcass to be a Yield Grade 3.5 or lower. The use of both within-breed variation and between-breed differences can be employed to enable the consistent production of Choice, Yield Grade 2 carcasses.

Fat partitioning among the various fat depots has both genetic and environmental influences. Both genetic and environmental tendencies to earlier fattening result in increased proportion of subcutaneous fat to intermuscular fat, while delayed or decreased fattening has the opposite effect. Kidney and pelvic fat generally grows at the same relative rate as total fat, although there are some breed, sex, and plane of nutrition effects. Some evidence indicates that during the fattening phase, the amount of kidney and pelvic fat increases at a faster rate than intermuscular fat. Dairy breeds deposit a higher proportion of internal fat and a lower proportion subcutaneously than traditional beef breeds. Limited data indicate that at equal fatness the fat partitioning in steers, heifers, and bulls is quite similar.

Cattle deposit fat at relatively high levels in the flank and brisket as well as along the loin and rib from the tailhead if the diet is of sufficient energy level. Fat depositions are minimal at the forearm and stifle. Thus, these are excellent anatomical reference points to evaluate muscle, independent of fat differences.

Muscle Growth and Development

The proportion of muscle in the carcass varies indirectly with fat, and a higher proportion of fat is associated with a lower proportion of muscle and vice versa. Muscle has a much faster relative growth rate than bone (Fig. 17.5). Muscle weight relative to liveweight or muscle-to-bone ratio can be used as valuable measurements of muscle yield.

Relatively large genetic differences exist for muscling in cattle. These muscling differences are shown for lighter-muscled cattle in Figure 17.8, for heavy-muscled cattle in Figure 17.9, and for double-muscled cattle in Figure 17.10.

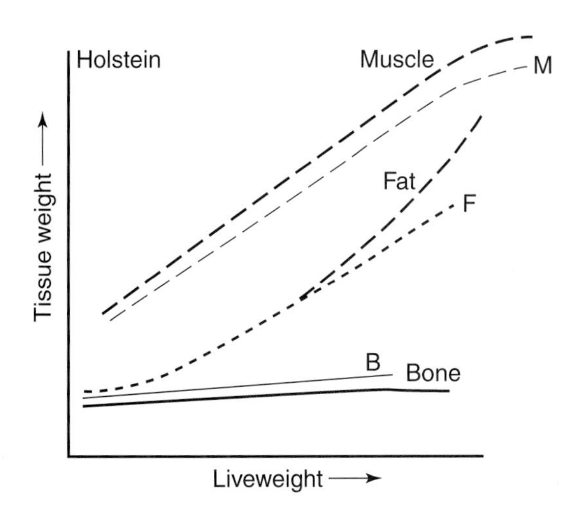

FIGURE 17.8 Lighter-muscled cattle (lighter lines) compared with average cattle.
Source: Adapted from American Society of Animal Science, Berg and Walters (1983).

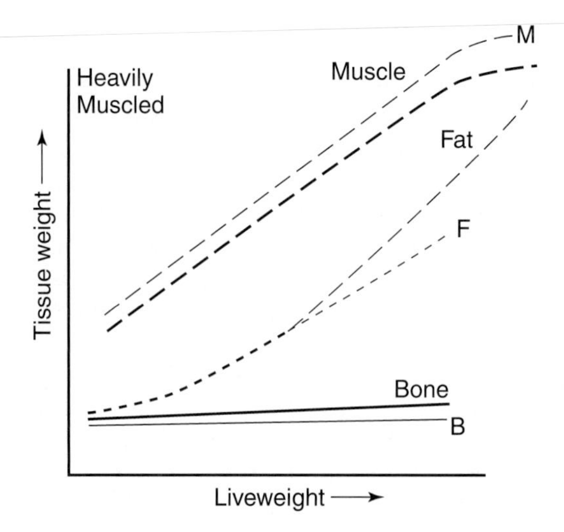

FIGURE 17.9 Heavy-muscled cattle (lighter lines) compared with average cattle. *Source:* Adapted from American Society of Animal Science, Berg and Walters (1983).

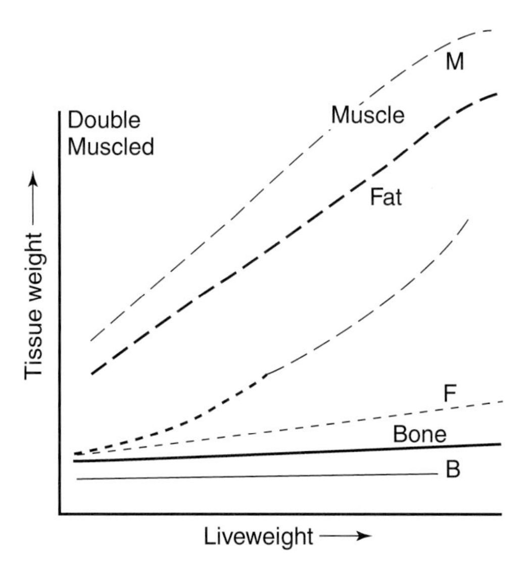

FIGURE 17.10 Double-muscled cattle (lighter lines) compared with average cattle. *Source:* Adapted from American Society of Animal Science, Berg and Walters (1983).

The current population of cattle in the United States has sufficient muscling variation to allow changes in muscle via selection or crossbreeding. Muscle yield relative to liveweight of more than 33–50% (muscle-to-bone ratio ranges of 3.5:1 to 5.0:1) has been observed in steers and heifers of slaughter weight without experiencing the extremes of double-muscling. These differences are economically important, particularly if more boneless cuts are merchandised.

Extreme muscling, as observed in double-muscling types, is associated with several problems of fitness, such as increased susceptibility to stress and reduced reproductive efficiency in the female. There may be some latitude in increasing the muscle-to-bone ratio in slaughter cattle without encountering negative side effects.

Individual muscles have different relative growth rates. The newborn calf has relatively well-developed muscles in the hind legs. As maturity continues, there is a progressive rise in relative growth rate from the rump to the neck, with the shoulder girdle and neck muscles having the highest growth rates of all muscles. Also, the front limb muscles develop more rapidly than the hind limb muscles as the weight of the animal shifts progressively forward. Muscle growth rate is relatively high in the abdominal and thoracic areas.

Muscle Distribution

Individual muscle dissection has shown a similarity of proportion of muscles to the total body muscle weight. Thus, increasing muscle weight in areas of high-priced cuts, at the expense of muscle weight in lower-priced cuts, is not commercially feasible. Some small, yet statistically significant differences in muscle distribution have been shown, but the differences do not seem to be economically significant for the industry. Double-muscled animals show the greatest evidence of a higher proportion of muscle weight in higher-priced muscle cuts.

Bone Growth and Distribution

Bone growth has the slowest growth rate when compared with muscle and fat (Fig. 17.5). There are only minor differences in the relative growth rates of different bones, although the growth rate for limb bones is somewhat slower.

Although breed differences in bone distribution have been reported, they have little commercial value.

Age and Teeth Relationship

For many years, producers have observed teeth of cattle to estimate their age or their ability to graze effectively with advancing age. Cattle are sometimes "mouthed" to classify them into appropriate age groups for specific showring classes. The latter approach is often inexact because of variations in teeth condition and the ability of some exhibitors to manipulate the condition of the teeth.

Table 17.1 gives some of the guidelines used in estimating the age of cattle by observing their teeth. The incisors (front cutting teeth) are the teeth used. There are no upper incisors and the molars (back teeth) are not commonly used to determine age. Figure 17.11 shows the incisor teeth of cattle of different ages. When the grazing area is sandy and the grass is short, the wearing of the teeth progresses faster than that previously described. Also, some cows may be "broken-mouthed," which means that they have lost some of their permanent incisor teeth.

BEEF TYPE

The term *beef type* is defined as the ideal or standard of perfection that combines all characteristics contributing to an animal's usefulness for a specific purpose. *Conformation,* used interchangeably with *type,* broadly refers to the form and shape of the animal. Both type and

TABLE 17.1 Estimating the Age of Cattle by Observing Their Teeth

Age	Description of Teeth
Birth	Usually only one pair of middle incisors
1 month	All eight, temporary incisors
1½–2 years	First pair of permanent (middle) incisors
2½–3 years	Second pair of permanent incisors
3¼–4 years	Third pair of permanent incisors
4–4½ years	Fourth (corner) pair of permanent incisors
5–6 years	Middle pair of incisors begins to level off from wear; corner teeth may also show some wear
7–8 years	Both middle and second pairs of incisors show wear
8–9 years	Middle, second, and third pairs of incisors show wear
10 years +	All eight incisors show wear

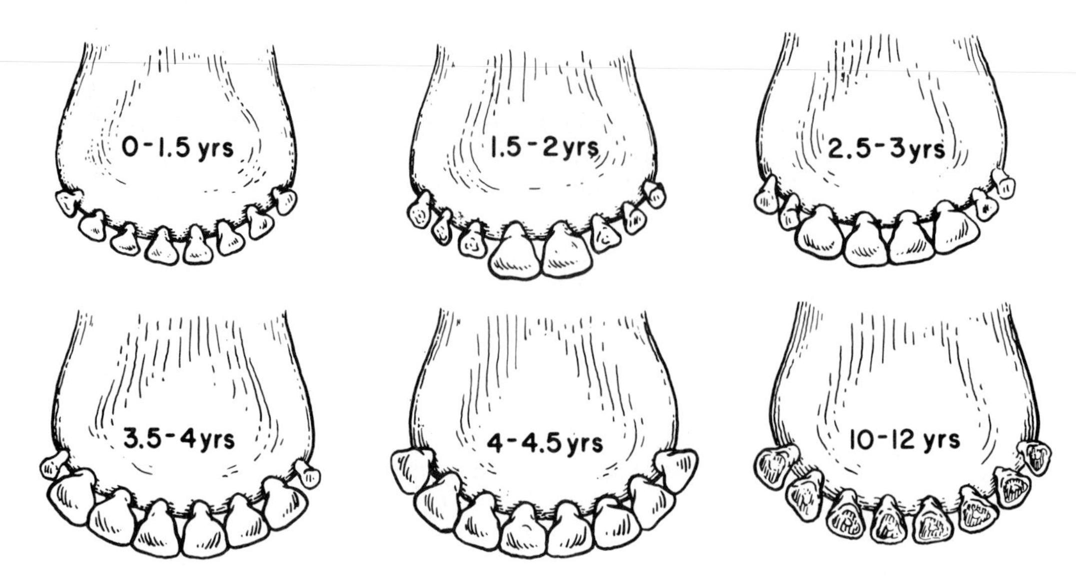

FIGURE 17.11 Incisor teeth of cattle of different ages.
Source: Colorado State University.

conformation involve external characteristics of the animal that can be evaluated visually or measured objectively. Most general descriptive terms involve length, height, thickness, and general symmetry. More specifically, body parts are given and are described both individually and how they blend together.

One of the first-known writings on livestock judging is Bourgelat of Lyon, *Traite de la Conformation Exterieure du Cheval* in 1768. Since that time, there has been much interest in the visual appraisal of cattle. Conformation (or form and shape) has been widely used in evaluating beef cattle and carcasses.

However, there are wide differences of opinion about the relationship of form and function in beef cattle. The roles of visual appraisal and measurements of body form continue to be debated as to their value in breeding programs, market grades, and defining "ideal" types of cattle. It is important to separate true relationships from opinion, particularly in the controversial area of beef form and function.

Obvious differences in cattle form and function come from a broad evaluation of the various uses of cattle. Development of cattle, over time, has brought several divergent types and forms into existence. Where cattle have been used as draft animals, the form generally has been one of large, massive size. Cattle selected primarily for milk production have developed into more angular animals. Cattle developed for maximum meat production express more thickness and total volume of muscle. Dual-purpose cattle, where the emphasis has been placed on meat, milk, and possibly draft, typically reflect some combinations of body form unique to each of those functions being combined. A study by Callow that evaluated beef, dual-purpose, and milk-type Shorthorn cattle found that those animals selected for beef were associated with higher levels of subcutaneous fat and lower amounts of internal fat. Milking Shorthorn were associated with less subcutaneous fat and more internal fat while dual-purpose types were intermediate.

Human beings have directed the change in cattle form under certain environmental conditions and are thus responsible for most differences in cattle types observed today (Fig. 17.12). A study conducted at Colorado State University for the American Hereford Association evaluated the type changes in the Hereford breed from the 1950s to the 1990s. Hereford

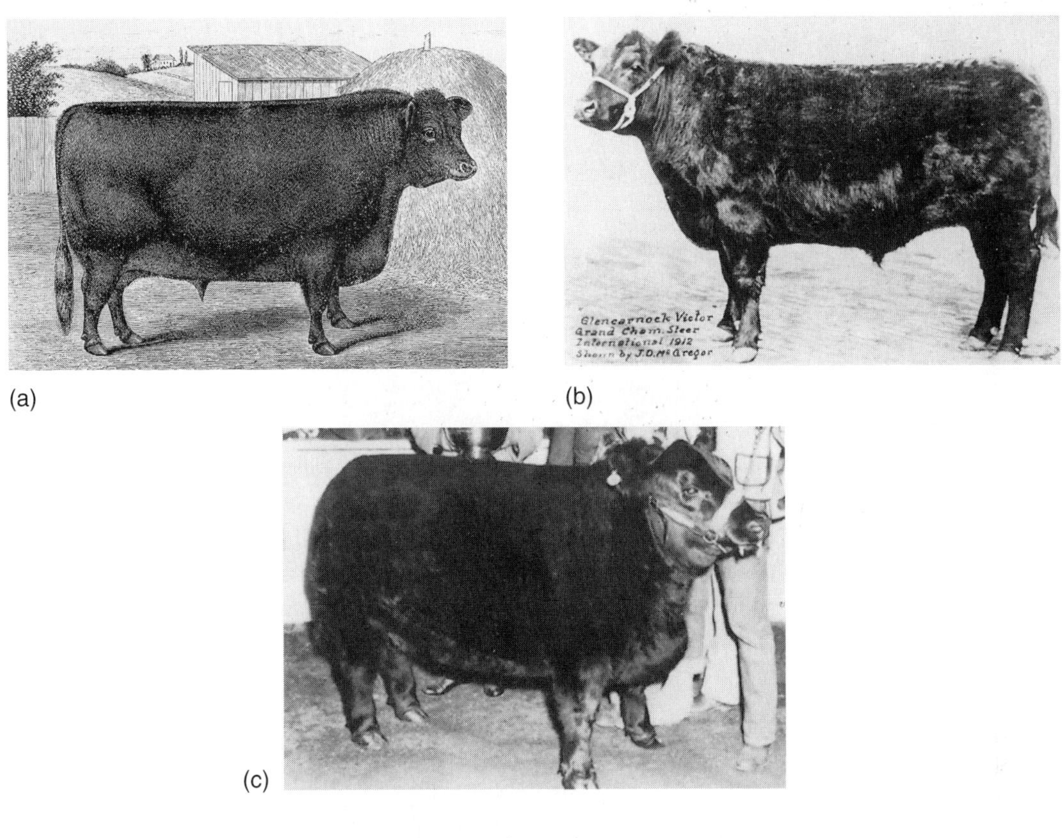

(a) (b) (c)

(d)

FIGURE 17.12 Body type changes, reflecting breeders' preferences, have been dramatic over the past 100 years—even within the same breed. (A) Angus steer Black Prince, imported from Scotland to United States was the champion steer, weighing approximately 2,400 lbs at the 1883 Chicago Fat Stock Show. (B) Champion steer at 1912 Chicago International (1,630 lbs). Also champion at the 1913 Chicago International weighing approximately 150 lbs less than 1912. (C) Champion steer at 1958 Chicago International, weighing less than 900 lbs. (D) Preferred type of Angus steer for the 1980s and 1990s, weighing approximately 1,200 lbs. Note the type resemblance of steers A and C; also steers B and D. *Source:* American Angus Association.

FIGURE 17.13 Left to right: 1970s sired steer, 1950s sired steer, and a 1990s sired steer.

TABLE 17.2 Average Growth and Carcass Performance of Three Generations of Hereford Cattle[a]

Trait	1950s	1970s	1990s
Birth weight (lbs)	82.5	85.9	91.4
Live weight @.45 in. external fat (days)			
Steers (lbs)	1,083	1,214	1,275
Heifers (lbs)	970	1,013	1,138
Frame size	3.7	4.9	5.5
Days on feed @.45 in. external fat (days)	109	111	111
Average daily gain (lbs)	3.81	4.01	4.25
Carcass weight			
< 600 lbs (%)	50	50	0
600–800 lbs (%)	15	85	0
> 800 lbs (%)	0	74	26
Ribeye area (in.)	11.5	11.8	12.3
Quality Grade			
Choice (%)	33	50	52
Select (%)	67	47	48
Standard (%)	0	3	0

[a]Reflects only sire-effect differences.
Source: Colorado State University.

sire from the 1950s, 1970s, and 1990s were randomly mated to commercial Hereford cows similar in age, pedigree, and frame size from a single Nebraska ranch. A sample of the progeny is shown in Figure 17.13. Their feedlot and carcass performance records are highlighted in Table 17.2. Producers attempt to appraise form and its relationship to function (growth, carcass, reproductive ability, and longevity). While some of these relationships have been observed and validated, the issue is far from resolved. For example, some cattle producers argued strongly that certain animal characteristics were more attractive or "more pleasing to the eye" even

TABLE 17.3 Expected versus Actual Differences in Growth Traits as Projected by EPDs

	1950–1970		1970–1990		1950–1990	
Trait	Actual	EPD	Actual	EPD	Actual	EPD
Birth weight (lbs)	+4	+6	+5	+4	+9	+10
Weaning weight (lbs)	+35	+32	+26	+25	+61	+57
Yearling weight (lbs)	+30	+54	+46	+50	+76	+104

Source: Colorado State University.

though they had little or no relationship to productivity. Those who could verbalize with great conviction, and particularly those who became recognized authorities in beef evaluation, greatly influenced the opinions of many others in the industry. As a result, the perpetuation of both valid and invalid opinions about the value of beef type and conformation continues.

Part of the beef production goal should be to put a highly productive animal into a merchandisable package that will command the highest market price. Productivity of breeding and slaughter cattle is best identified by using meaningful performance records and effective visual appraisal. It is well demonstrated that performance records are much more effective than visual appraisal in improving beef cattle productivity. In fact, objective measures of genetic merit in the form of expected progeny differences accurately predicted the differences between the three generations of Hereford cattle in the Colorado State University study (Table 17.3). These results demonstrate the effectiveness of expected progeny differences. The actual yearling performance was likely depressed by the limited energy backgrounding ration provided to the cattle from weaning to one year of age. Visual appraisal, however, is important in evaluating productivity, primarily in identifying reproductive and skeletal soundness and health status. Visual appraisal is also helpful in evaluating productive differences in carcass composition. Both performance records (such as ultrasound readings for fatness) and visual appraisal are important in evaluating slaughter animals. Visual appraisal of the slaughter cattle can be used effectively to predict carcass composition (fat, lean, and bone) when relatively large differences exist. Most slaughter cattle are purchased at market time on the basis of a visual appraisal of their carcass merit.

PARTS OF THE BEEF ANIMAL

Figure 17.14 shows the parts of the beef animal. Figures 17.15 and 17.16 show other descriptions of conformation that are used to identify the body dimensions of the beef animal. A knowledge of beef animal parts and their description is important for effective communication in any phase of the beef industry involved with live animal and carcass evaluation.

CARCASS CONFORMATION

When the USDA beef carcass quality grades were established in 1927, conformation was included as a factor in grade standards. In 1975, however, conformation was eliminated from quality grade standards because research studies had shown that it has no positive effect on beef palatability.

Fat thickness, carcass weight, and area of the ribeye are the carcass characteristics related to carcass conformation that are presently included in yield grade standards. Research reports disagree about the importance of carcass conformation, as described in the USDA quality grade standards prior to 1975, and yield of retail cuts. Reports range from a slightly positive relationship between conformation and retail cut yield to no relationship.

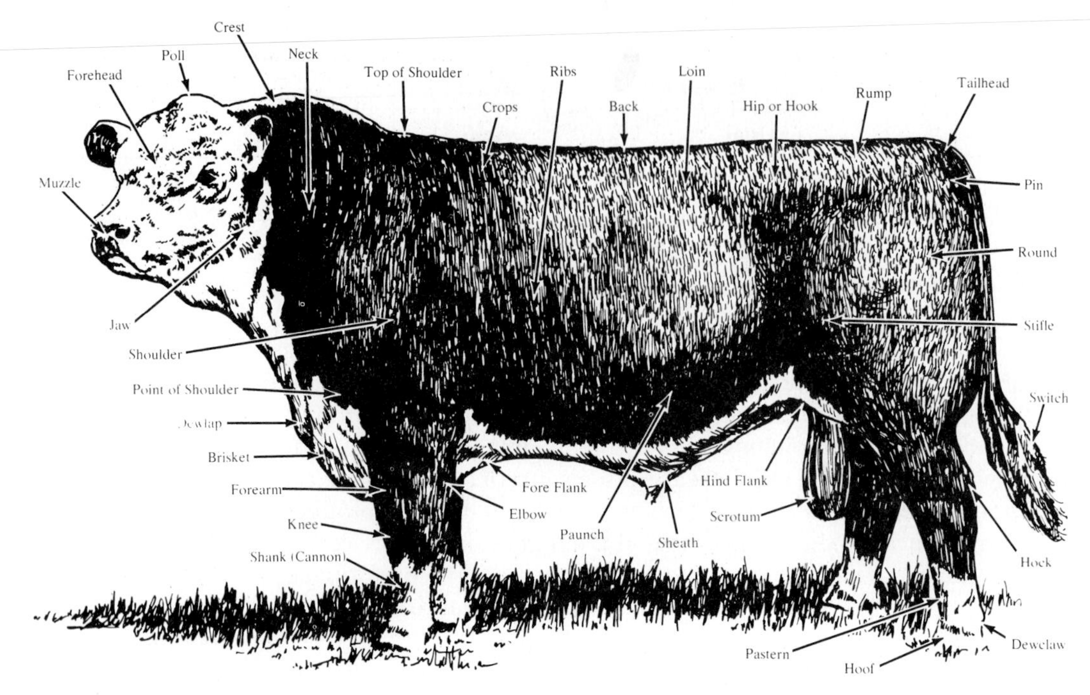

FIGURE 17.14 Parts of the beef animal.
Source: Courtesy of the American Polled Hereford Association.

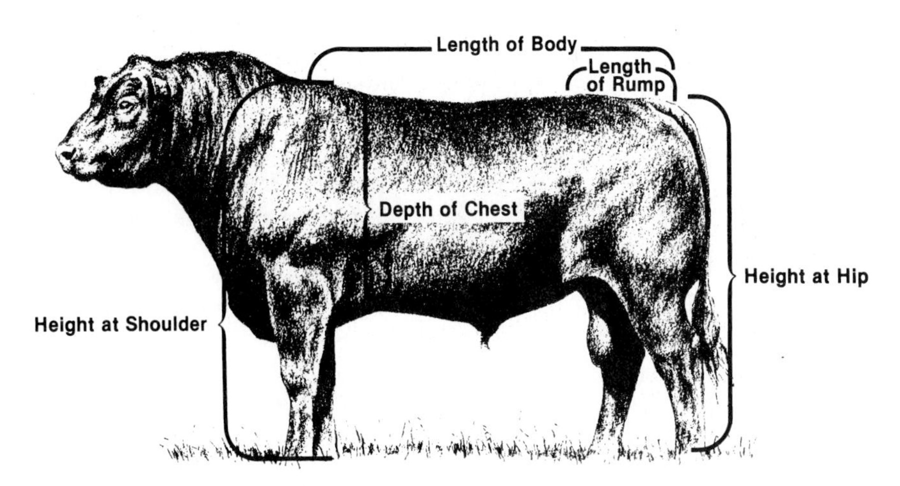

FIGURE 17.15 Parts of the beef animal used to identify body dimensions (side view).
Source: American Angus Association.

Fat greatly influences the form and shape of the carcass. Subcutaneous fat affects conformation the most; however, intermuscular fat also causes muscles to have an appearance of greater fullness and depth. There are large differences in the muscle-to-bone ratios in the carcass. These widely different ratios can be observed visually by noting the muscling areas of the carcass that are not greatly influenced by subcutaneous and intermuscular fat. Some packers and retailers still consider carcass conformation important from a marketing standpoint. For example, several branded beef programs discriminate against dairy carcass types. Preference is given to a thicker carcass with a thick, plump round.

FIGURE 17.16 Parts of the beef animal used to identify body dimensions (front and rear views).
Source: American Angus Association.

CONFORMATION OF SLAUGHTER CATTLE

Visual appraisal appears to be more accurate for assessing differences in carcass composition than for other economically important traits. The accuracy of individual animal estimates for quality grade is low; however, live animal estimates of carcass yield grades are more accurate. This accuracy is dependent on the experience of the appraisers and the variability in yield grades of cattle.

After becoming familiar with the external parts of the animal, effective visual appraisal involves understanding where major meat cuts are located in live animals. A carcass is evaluated after an animal has been slaughtered, eviscerated, and split into two halves. When the carcass is evaluated, it is hanging by the hind leg from the rail in the packing plant. This makes it difficult to perceive how the carcass would appear as part of the live animal standing on all four legs. Figure 17.17, which shows the location of wholesale cuts on the slaughter steer, helps to understand the major carcass component parts of the live animal.

The carcass is composed of fat, lean (red meat), and bone. The beef industry goal is to produce large amounts of highly palatable lean and minimal amounts of fat and bone. Effective visual appraisal of carcass composition requires knowing the body areas of the live animal where fat deposits and muscle growth occur. Color Plates F and G show the fat and lean composition at several carcass cross sections of a Yield Grade 2 and a Yield Grade 5 steer. Ribbon on the frozen carcasses shows location of the cross sections. Cross section 3 removes the bulge to the round, cross section 4 is in front of the hip bone down through the flank, cross section 5 is at the twelfth and thirteenth rib, and cross section 6 is at the point of the shoulder down through the brisket. Note the contrasts in fat (18% versus 44%) and lean (66% versus 43%) percentages. Both steers graded Choice and were slaughtered at approximately 1,100 lbs.

Color Plates E and H show another perspective of the fat, lean, and bone composition of slaughter steers. Removing fat and muscle, a layer at a time, gives a clear perspective of what is under the hide of slaughter steers. The conformation characteristics of the two steers (Color Plates F and G), which are primarily influenced by fat deposits, and some muscling differences are contrasted in Figures 17.18 and 17.19. Conformation of each steer is also contrasted with the conformation of a slaughter steer that is underfinished and thinly muscled (Fig. 17.20). All body parts referred to in these figures can be identified in Figure 17.14, with the exception of the twist. The twist is the distance from the top of the tail to where the hind legs separate as observed from a rear view (note no. 4 in Color Plate H).

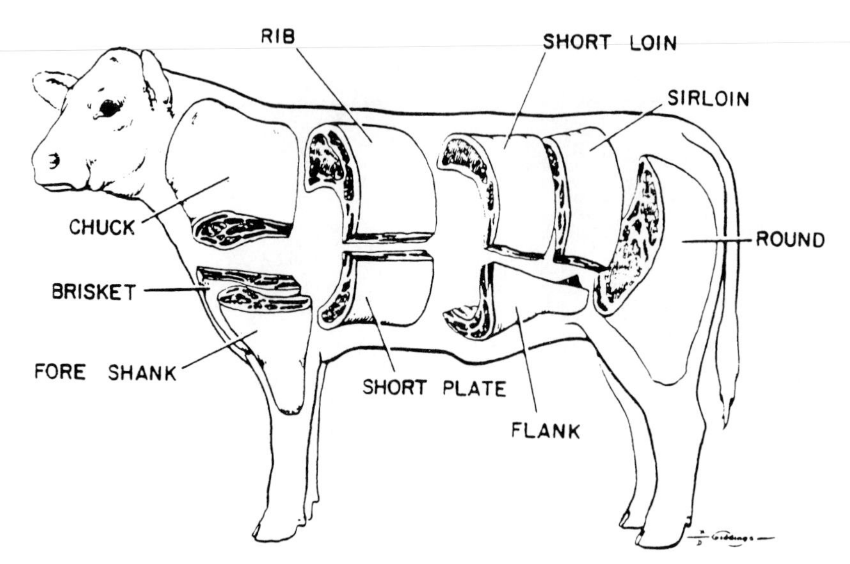

FIGURE 17.17 Location of the wholesale cuts on the live steer.
Source: Colorado State University.

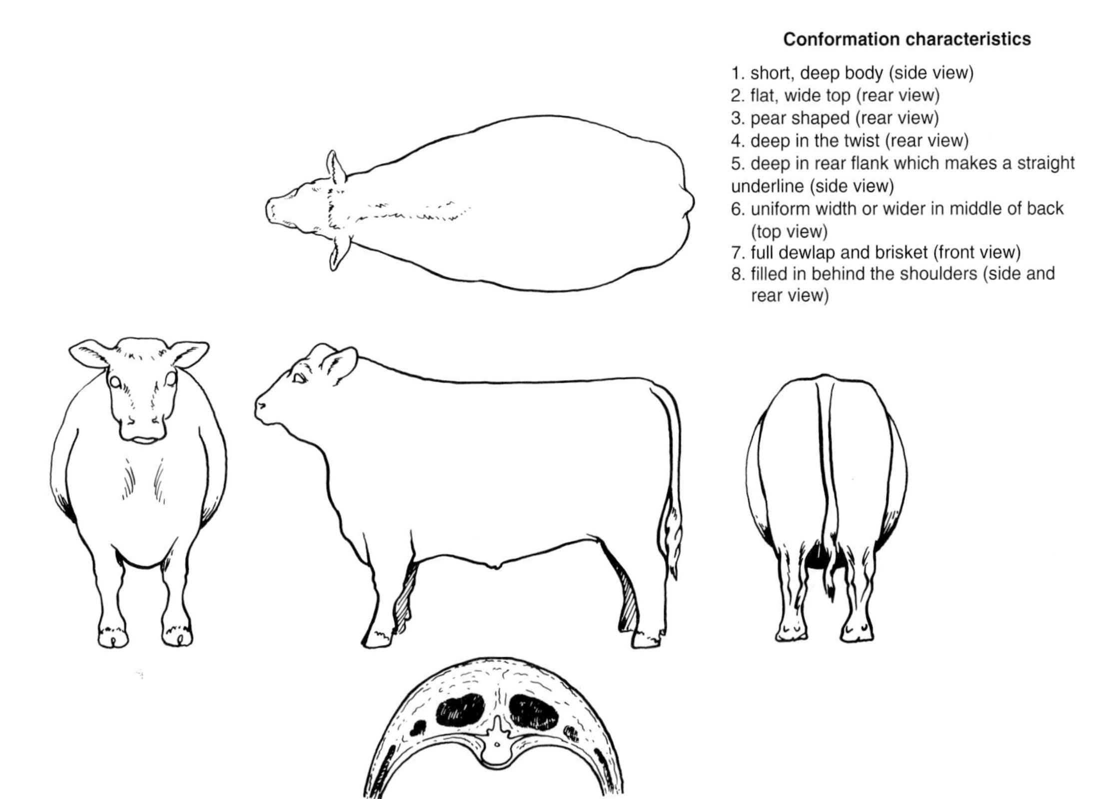

FIGURE 17.18 Slaughter steer typical of Yield Grades 4 or 5.
Source: Drawn by Dennis Giddings.

FIGURE 17.19 Slaughter steer typical of Yield Grade 2.

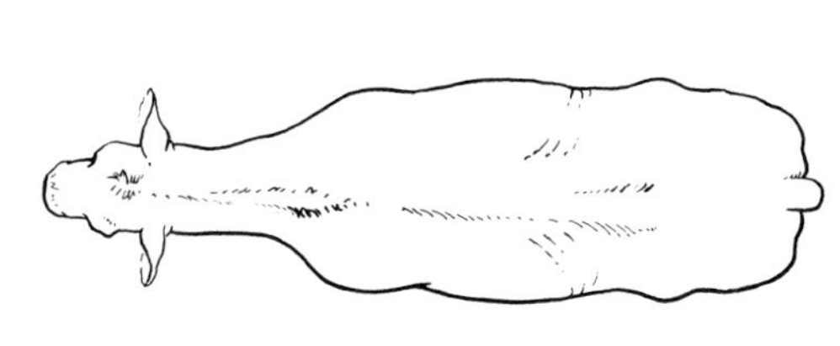

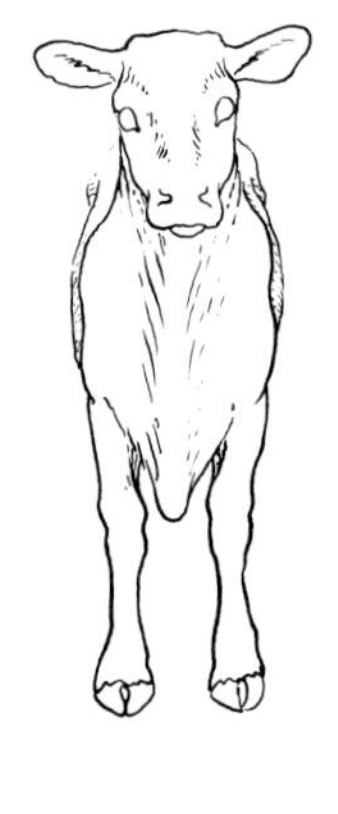
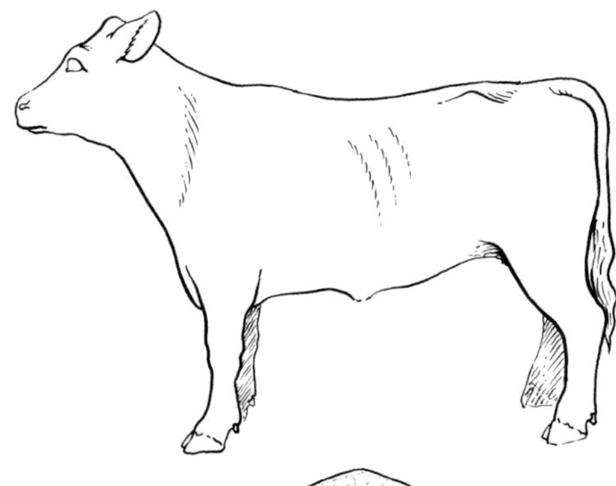
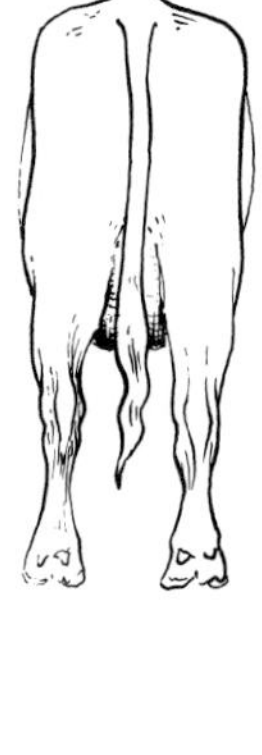
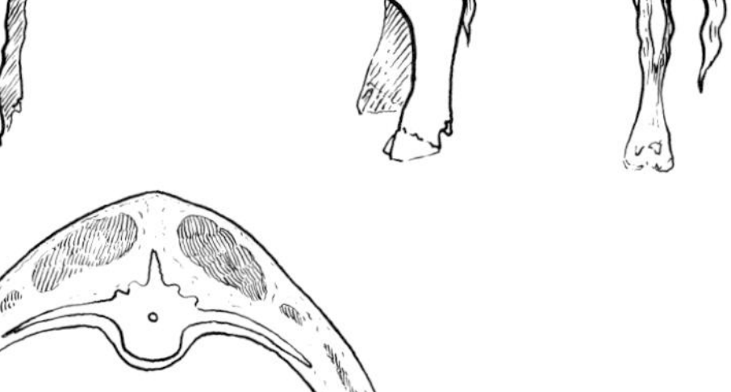

FIGURE 17.20 Slaughter steer that is underfinished and thinly muscled.
Source: Drawn by Dennis Giddings.

Accuracy in visually appraising slaughter cattle is obtained by making visual estimates of yield grades, then comparing the visual estimates with carcass measurements. Accurate visual appraisal can be used as one tool in producing slaughter cattle with a more desirable carcass composition of lean to fat.

The rear views in Color Plates F and G show thickness or muscling differences that can be appraised visually, even though there are disagreements about differences in muscling in slaughter cattle. Kauffman et al. (1973) reported that muscular-shaped steers contain 2% more fat-free muscle when compared with nonmuscular-shaped steers at similar weight and fat composition. In this study, muscular steers had a muscle-to-bone ratio of 3.8 while nonmuscular steers had a ratio of 3.0.

Muscling is best described by muscle-to-bone ratios, where observed differences typically range from 2.5:1 to 4.5:1 (and even higher for double-muscled steers). Muscling differences of 2.5 to 3.0:1 versus 4.0 to 4.5:1 can be visually appraised in slaughter cattle with a reasonable degree of accuracy. Plate E shows these muscling differences in carcasses that vary considerably in the thickness and fullness of the round. Although it is commonly assumed that large differences exist in muscle distribution in cattle (e.g., that a higher percentage of carcass weight is in the hind limb muscle than in the forelimb muscle or vice versa), there is no evidence to indicate any significant differences in muscle distribution in slaughter steers (Kauffman et al., 1973; Berg and Butterfield, 1976; and MARC data). Under today's marketing stipulations, then, market steers should have the following conformation: (1) a 1.5 to 3.0 Yield Grade carcass, (2) a 3.5 to 4.5:1 muscle-to-bone ratio, and (3) a 600–800-lbs carcass with a minimum of 0.2 in. of fat over the twelfth rib.

CONFORMATION OF FEEDER CATTLE

The feedlot performance of most feeder cattle is estimated by visual appraisal. The majority of commercial feedlot cattle buyers use visual criteria—skeletal size, finish, weight relative to age, and health status—to appraise cattle. Preference is given to cattle that show evidence of compensatory gain: large-framed, thin, and old for their weight.

Frame size has become quantified by measuring skeletal height at the shoulder or hip. Reference is occasionally made today to the Madison Type Conferences (Brungardt, 1972), where Angus, Charolais cross, and Hereford calves of various frame size (skeletal height and length) were evaluated for feedlot and carcass characteristics. A summary of the results of this 3-year study is shown in Table 17.4. The calves were born in April and May, fed for approximately 150 days after weaning (Charolais crosses were fed for 190 days) and slaughtered at approximately 14–15 months of age. Cattle of larger body types had to be fed to heavier weights to make similar carcass quality grades. At the respective slaughter weights, each body type graded between low Choice and average Choice. When fed to the same quality grade, there was little difference in feed efficiency. At similar weights, cattle that gained faster were more efficient in feed conversion. Profitability generally favored larger-framed feeder calves (primarily up through body type 5) when evaluated through the feedlot and the carcass.

Evidence shows that visual estimates of trimness, finish, length, and height in feeder cattle can give fair estimates of cutability at slaughter. In the same study, where feeder calves of widely different biological types were evaluated, no combination of feeder calf traits could make a meaningful estimate of carcass quality grades.

Feedlot gains are usually higher for longer, taller cattle when fed to a constant time on feed or weight-constant basis. When cattle are fed to a similar compositional endpoint, efficiency differences between frame sizes are minimized. However, large-framed cattle are typically faster growing than small cattle. This implies rather large differences in final carcass weights, which could have economic significance depending on carcass weight specifications.

TABLE 17.4 Feedlot and Carcass Characteristics of Various Body Types of Angus (A), Charolais Cross (CX), and Hereford (H) Calves (100 head per breed group)

Visual Body Types as Feeders	Ht at Shoulder at Slaughter (in.)			Slaughter Wt (lbs)			Avg. Daily Gain (lbs)			Fat Cover (in.)		
	A	CX	H	A	CX	H	A	CX	H	A	CX	H
1	42.1	—	42.4	877	—	889	2.7	—	2.7	0.57	—	0.53
2	43.8	—	43.4	984	—	958	2.8	—	2.8	0.65	—	0.50
3	45.5	45.5	45.3	1,024	1,029	1,055	2.9	2.9	3.0	0.73	0.35	0.66
4	45.9	46.8	47.0	1,035	1,093	1,107	2.9	3.0	3.0	0.74	0.34	0.56
5	47.1	47.9	48.4	1,106	1,133	1,175	2.9	3.1	3.2	0.80	0.30	0.60
6	—	49.2	—	—	1,137	—	—	3.0	—	—	0.33	—
7	—	49.5	—	—	1,211	—	—	3.2	—	—	0.34	—

Source: V. H. Brungardt, *Efficiency and Profit Differences of Angus, Charolais, and Hereford Cattle Varying in Size and Growth* (Madison, WI: University of Wisconsin, Research Reports R2397–R2401, 1972).

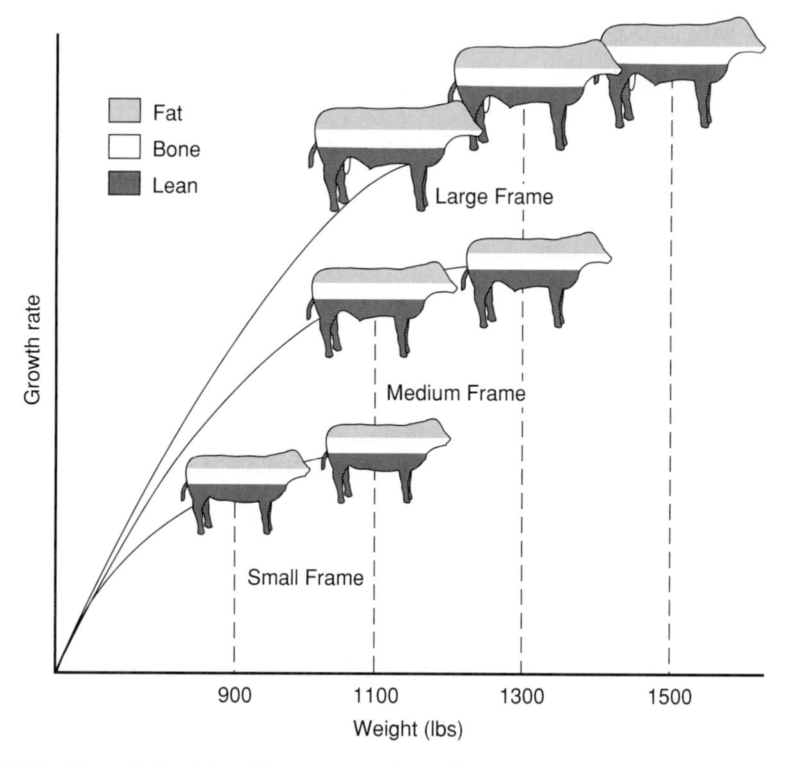

FIGURE 17.21 The relationship of frame size and weight to carcass composition in beef steers. *Source:* Colorado State University.

Large differences in rate and efficiency of gain do exist in cattle of similar frame size and environmental background. Visual appraisal is not able to detect these differences with any high degree of accuracy.

Frame size, whether appraised visually or measured as hip height, can be used as a management tool in determining the logical slaughter weight of cattle in the feedlot. Figure 17.21 shows compositional differences of different frame sizes of feeder cattle fed to different slaughter weights. No one frame size is superior to the others. Feed resources, market weight

a. Large frame

b. Medium frame

c. Small frame

FIGURE 17.22 The three frame sizes of the USDA feeder grades for cattle.
Source: USDA.

No. 1

No. 2

No. 3

FIGURE 17.23 The three thickness standards of the USDA feeder grades for cattle.
Source: USDA.

preferences, and other production costs determine the frame size preference for different feeders. A visual evaluation of frame size and condition can be used to sort feeder cattle into separate feeding groups. Days on feed can be projected for each group to reach similar yield grade and quality grade endpoints (Daley, Tatum, and Taylor, 1983b).

The updated USDA feeder grades are based on visual appraisal of frame size, thickness, and thriftiness (Figs. 17.22 and 17.23). There are three frame size specifications: (1) large (steers and heifers would have Choice carcasses at approximately 0.50 in. of fat at 1,250+ lbs and 1,150+ lbs, respectively); (2) medium (Choice carcasses for steers at 1,100–1,250 lbs liveweight and heifers at 1,000–1,150 lbs); and (3) small (steers grading Choice at less than 1,100 lbs and heifers at less than 1,000 lbs).

The frame sizes shown in Table 17.4 would relate to the feeder grade frame size descriptions as follows: small frame (frame score 1, 2, and lower half of 3); medium frame (upper half of 3 and 4 and lower two-thirds of 5); large frame (upper third of 5, 6, and 7). No. 1 thickness would be predominately for cattle with a high proportion of beef breeding with at least moderate thickness. No. 2 thickness grade includes cattle that are predominately beef type and slightly thick, while the No. 3 thickness grade includes cattle that are narrow-made and thinly muscled. No. 4 thickness cattle are extremely light muscled. The last grade of the proposed feeder grades is inferior. It includes unthrifty cattle and *double-muscled* cattle regardless of their frame size or thickness.

CONFORMATION OF BREEDING CATTLE

Carcass Composition

Cattle breeders should study carefully the value of conformation in carcasses, slaughter, and feeder cattle before making decisions on how conformation characteristics can be used in a breeding program. Packers and retailers must accept the conformation resulting from previous decisions made by cattle breeders and feeders.

In addition to some of the relationships mentioned earlier in the chapter, the breeder should recognize how sex differences affect conformation, particularly body composition. Figure 17.24 shows the compositional differences of heifers, steers, and bulls at different weights. This information tells producers why heifers are typically marketed 100–150 lbs lighter than their counterpart steers. The breeder needs to understand these relationships to be more effective in sire selection. When the breeder focuses on a particular steer endpoint (primarily weight and carcass composition), this dictates that the bull must be much more extreme in composition at the same weight; in other words, the bull must be approximately 150 lbs heavier than the steer to be similar in carcass composition.

Ultrasound or a mechanical backfat probe or visual evaluation can be used for accurately estimating subcutaneous fat. Recent studies show these methods to be similar in accuracy, accounting for approximately two-thirds of the differences in carcass subcutaneous fat thickness (Daley, Tatum, and Taylor, 1983a). Visual appraisal can have added value by evaluating predisposition to fat, primarily in the brisket and flank areas. The validity of such an appraisal is based on the relationship that fat cover over the ribeye does not always give an accurate prediction of how fat trim affects yield grade. The carcass yield grader makes adjustments to the carcass fat measurement by visually appraising fat accumulation in the brisket, plate, and flank areas.

The accuracy of ultrasound, fat probes, or visual estimates on live cattle is dependent on the experience and technique of the evaluator. Ultrasound or the fat probe, at the twelfth to thirteenth rib, is most useful on yearling bulls that have been fed relatively high-energy rations and are probed at weights similar to preferred slaughter steer weights (1,050–1,200 lbs). At these similar weights, approximately 0.20 in. should be added to the backfat probe of a bull to project to a slaughter steer basis.

Linear measurements of skeletal size appear to be most useful with breeding cattle, particularly with young bulls 12–15 months of age. Hip height is commonly used to measure size of skeleton. Body length is a questionable measurement because it is quite proportional

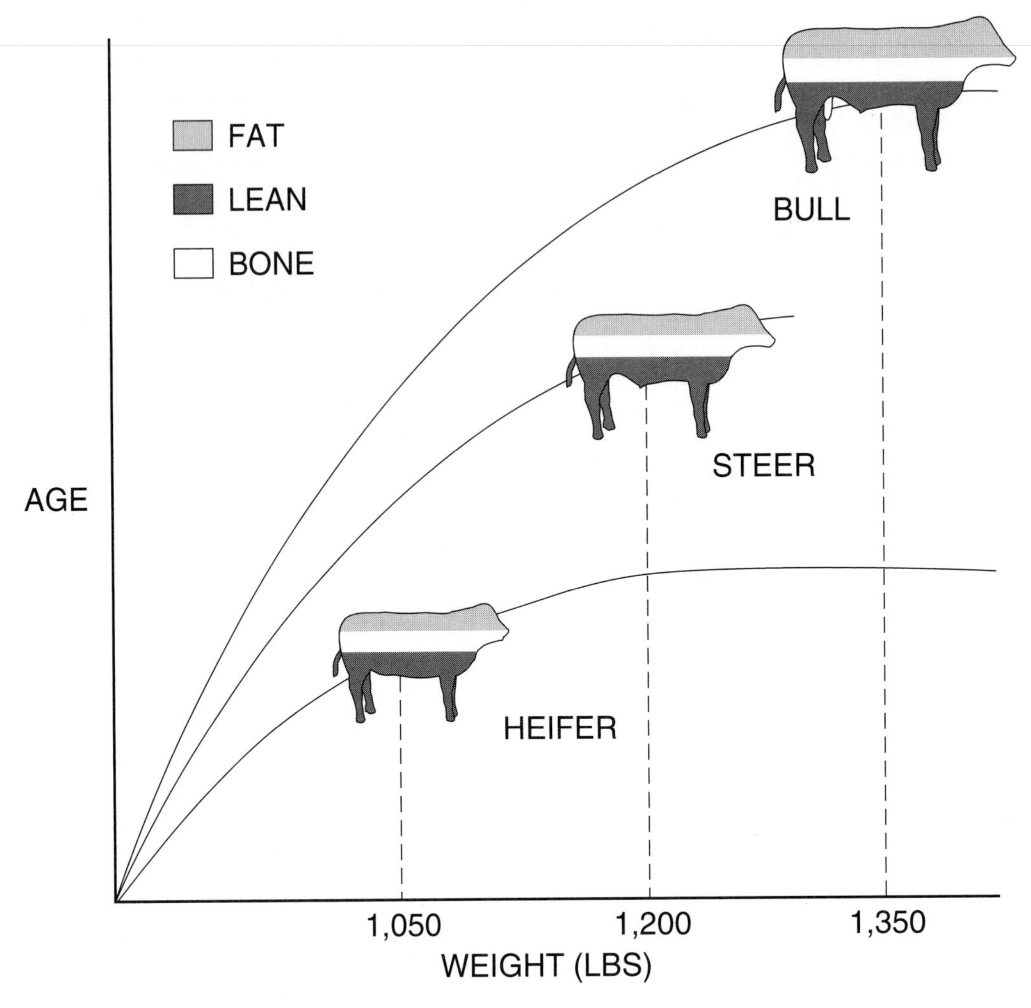

FIGURE 17.24 The influence of sex on body (carcass) composition at various weights.
Source: Colorado State University.

to skeletal height and adds little additional information. There is also more difficulty in obtaining accurate body length measurements. Tables 17.5 and 17.6 give the hip height measurements used to identify various frame sizes in bulls and heifers.

Weaning height for bulls and heifers can be adjusted for age of calf and age of dam. To adjust heights to 205 days for sex, (1) multiply the number of days *under 205* by 0.033 for bulls or 0.025 for heifers and add that figure to the actual height *or* (2) multiply the number of days *over 205* by 0.033 for bulls or 0.025 for heifers and subtract the result from the actual height. To adjust for age of dam, multiply the adjusted height for sex by the age-of-dam factor (Table 17.7).

Yearling height measurement can be adjusted for age of bull or heifer by using the same age-adjustment factors as for 205-day adjustments (e.g., for bulls, 0.033 in./day; for heifers, 0.025 in./day).

Sex differences affecting hip height are noted by comparing Table 17.5 with Table 17.6. Steers are intermediate between bulls and heifers when compared in the 5–21-month

TABLE 17.5 Bull Hip Height (inches) and Frame Scores

Age in Months	Frame Score[a]								
	1	2	3	4	5	6	7	8	9
5	33.5	35.5	37.5	39.5	41.6	43.6	45.6	47.7	49.7
6	34.8	36.8	38.8	40.8	42.9	44.9	46.9	48.9	51.0
7	36.0	38.0	40.0	42.1	44.1	46.1	48.1	50.1	52.2
8	37.2	39.2	41.2	43.2	45.2	47.2	49.3	51.3	53.3
9	38.2	40.2	42.3	44.3	46.3	48.3	50.3	52.3	54.3
10	39.2	41.2	43.3	45.3	47.3	49.3	51.3	53.3	55.3
11	40.2	42.2	44.2	46.2	48.2	50.2	52.2	54.2	56.2
12	41.0	43.0	45.0	47.0	49.0	51.0	53.0	55.0	57.0
13	41.8	43.8	45.8	47.8	49.8	51.8	53.8	55.8	57.7
14	42.5	44.5	46.5	48.5	50.4	52.4	54.4	56.4	58.4
15	43.1	45.1	47.1	49.1	51.1	53.0	55.0	57.0	59.0
16	43.6	45.6	47.6	49.6	51.6	53.6	55.6	57.5	59.5
17	44.1	46.1	48.1	50.1	52.0	54.0	56.0	58.0	60.0
18	44.5	46.5	48.5	50.5	52.4	54.4	56.4	58.4	60.3
19	44.9	46.8	48.8	50.8	52.7	54.7	56.7	58.7	60.6
20	45.1	47.1	49.1	51.0	53.0	55.0	56.9	58.9	60.9
21	45.3	47.3	49.2	51.2	53.2	55.1	57.1	59.1	61.0

[a]Frame Score = −11.548 + 0.4878 (ht) − 0.0289 (days of age) + 0.00001947 (days of age)^2 + 0.0000334 (ht) (days of age).
Source: Guidelines for Uniform Beef Improvement Programs, 1990. Stillwater, OK: Beef Improvement Federation.

TABLE 17.6 Heifer Hip Height (inches) and Frame Scores

Age in Months	Frame Score[a]								
	1	2	3	4	5	6	7	8	9
5	33.1	35.1	37.2	39.3	41.3	43.4	45.5	47.5	49.6
6	34.1	36.2	38.2	40.3	42.3	44.4	46.5	48.5	50.6
7	35.1	37.1	39.2	41.2	43.3	45.3	47.4	49.4	51.5
8	36.0	38.0	40.1	42.1	44.1	46.2	48.2	50.2	52.3
9	36.8	38.9	40.9	42.9	44.9	47.0	49.0	51.0	53.0
10	37.6	39.6	41.6	43.7	45.7	47.7	49.7	51.7	53.8
11	38.3	40.3	42.3	44.3	46.4	48.4	50.4	52.4	54.4
12	39.0	41.0	43.0	45.0	47.0	49.0	51.0	53.0	55.0
13	39.6	41.6	43.6	45.5	47.5	49.5	51.5	53.5	55.5
14	40.1	42.1	44.1	46.1	48.0	50.0	52.0	54.0	56.0
15	40.6	42.6	44.5	46.5	48.5	50.5	52.4	54.4	56.4
16	41.0	43.0	44.9	46.9	48.9	50.8	52.8	54.8	56.7
17	41.4	43.3	45.3	47.2	49.2	51.1	53.1	55.1	57.0
18	41.7	43.6	45.6	47.5	49.5	51.4	53.4	55.3	57.3
19	41.9	43.9	45.8	47.7	49.7	51.6	53.6	55.5	57.4
20	42.1	44.1	46.0	47.9	49.8	51.8	53.7	55.6	57.6
21	42.3	44.2	46.1	48.0	50.0	51.9	53.8	55.7	57.7

[a]Frame Score = −11.7086 + 0.4723 (ht) − 0.0239 (days of age) + 0.0000146 (days of age)^2 + 0.0000759 (ht) (days of age).
Source: Guidelines for Uniform Beef Improvement Programs, 1990. Stillwater, OK: Beef Improvement Federation.

TABLE 17.7 Age-of-Dam Adjustment Factors for Height at Weaning

Age of Dam (years)	Bulls (weaning ht)	Heifers (weaning ht)
2 and 13 or older	1.02	1.02
3 and 12	1.015	1.015
4 and 11	1.01	1.01
5 through 10	(no adjustment)	

Source: Guidelines for Uniform Beef Improvement Programs, 1990. Stillwater, OK: Beef Improvement Federation.

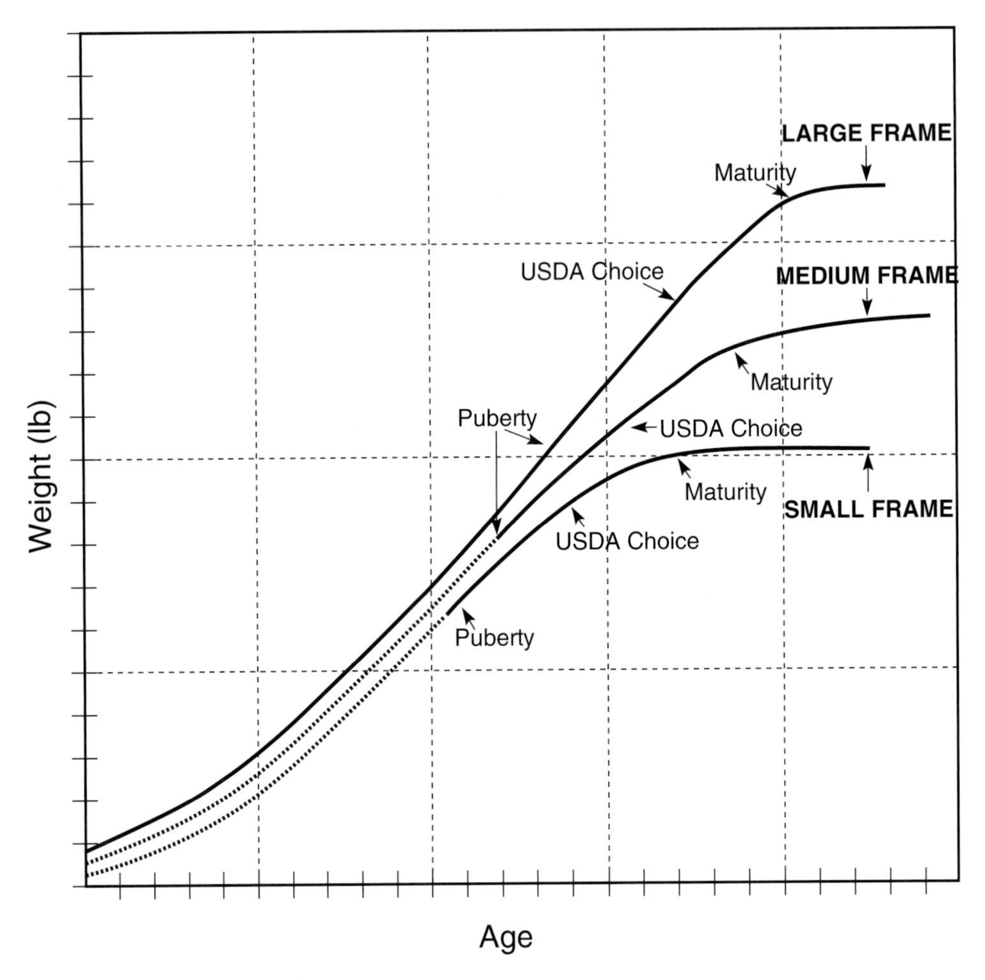

FIGURE 17.25 Frame size, weight, and age influence on maturity and carcass traits.
Source: Adapted from *BEEF*, Spring, 1987.

age range. Steers have 1 in. more than heifers and 1 in. less than bulls, when compared at the same age.

Hip height (frame score) is related to sexual maturity (puberty), carcass composition, and mature weight (Fig. 17.25). Table 17.8 shows how frame score relates to the approximate weight for steers to reach the USDA Choice grade.

TABLE 17.8 Relationship of Frame Score to Slaughter and Carcass Weights of Steers Grading USDA Choice (0.5 in. of fat)

Frame Score	Slaughter Weight (lbs)	Carcass Weight (lbs)
2	850–950	525–600
3	950–1,050	600–675
4	1,050–1,150	675–750
5	1,150–1,250	750–800
6	1,250–1,350	800–875
7	>1,350	>875

Source: Colorado State University.

Using Standards to Meet Market Targets

A case in point for the development of both live animal and carcass specifications to meet a specific market demand is Certified Angus Beef™. The live specifications include at least 51% black-hided and having Angus-type characteristics. The carcass specifications are: Modest or higher degree of marbling, medium or fine marbling texture, Yield Grade 3.9 or better, A maturity, at least moderately thick muscling, no evidence of Brahman influence (minimal hump—< 2 in.), no blood spots, and no dark cutters.

Approximately 8% of all U.S. cattle qualify for the Certified Angus Beef™ label (CAB®). Less than 20% of cattle meeting the live specification also qualify based on carcass performance. The primary reason that more than 75,000 cattle, studied by Iowa State University scientists, failed to meet the specifications was insufficient marbling (84%). Failure to meet Yield Grade standards accounted for an additional 14% of the failure rate. While the genetic correlation between subcutaneous fat and marbling deposition is nearly zero, there is a tendency for CAB® acceptance rates to increase with declining cutability (i.e., 11% acceptance for Y.G. 2.5 and 33% for Y.G. 3.5).

This analysis points to the need for increased selection to hit these carcass specifications as well as the need to better manage individual cattle to meet market targets. For example, it is more profitable to harvest cattle that meet a marbling target at desirable levels of composition rather than allowing them to become excessively fat prior to harvest.

The dilemma is finding ways to accurately assess both marbling and compositional differences in live cattle. In an attempt to overcome this challenge, individual cattle management (ICM) systems have been developed. ICM systems will also better enable cattle producers to meet demand for improved red meat yields as a result of an increase in case-ready processing. The net effect of case-ready beef production is that trim losses will exclusively be borne at the packing level. As such, those animals with less trim and superior red meat yield are likely to receive significant premiums in the marketplace.

ICM involves the application of a variety of technologies to diverse populations of cattle with a goal of sorting these animals into outcome groups. These groups can be appropriately managed by matching nutrition, implant protocols, and harvest timing to the growth potential of the animal. These systems typically utilize ultrasound measures of fat and muscle, video image analysis of body dimension, weight measures, and hip heights to sort cattle. These systems are typically tied to electronic identification approaches and to integrated information management databases. There is evidence that such approaches can add to profitability by sorting cattle into more uniform outcome groups.

Skeletal Soundness

Visual appraisal is used extensively by beef producers to evaluate skeletal soundness as it relates to longevity and productivity. Figure 17.26 shows feet and leg descriptions commonly used to define skeletal soundness. Showring standards have long dictated that each of the four legs should be straight and set on the corners, that the toes should point straight ahead, and that the animal should move with a long, easy stride. However, exactly what constitutes skeletal soundness is not well defined. Opinions vary and the research is sparse. The following points about skeletal soundness should be considered:

1. Cattle with hock angles greater than 150 degrees appear to be more predisposed to becoming "stifled." Cattle that are post-legged have more serious problems than those with sickle-hocked condition (Woodward, 1968).
2. Excessive hoof growth can result in lameness and economic concern if foot trimming is necessary. Hoof growth appears to be highly heritable (Brinks et al., 1979).
3. "Founder," where lameness is typically involved, can be a relatively serious condition under feedlot conditions. This structural soundness condition, usually observed when cattle are on high-energy rations, appears to have a genetic basis.
4. Many so-called points of abnormal conformation are more of aesthetic value than having practical importance. Some respected cattle judges have stated that show cattle, which are straight on their legs, are that way primarily because of numerous trips to the hoof-trimming table. They also state that productive cattle are naturally a little splayed in front (toed out) and slightly cow-hocked in back with some set in their hock.
5. At this time, there appears to be little value in selecting for bone size and so-called "bone quality." Bone size appears to be independent of performing ability, with the exception that a smaller bone, larger muscle combination contributes to higher muscle-to-bone ratios. Also, larger bone may contribute to increase calving difficulty.
6. Further evidence that selection for skeletal soundness may have been overemphasized in the past is the observation of cattle in closed, inbred herds over the past 20–30 years. Examples are the cattle at the U.S. Range Station in Miles City, Montana, and the San Juan Basin Research Center in Hesperus, Colorado. There have been marked improvements in productivity where the primary selection criteria have been weaning and yearling weights in linebreeding and inbreeding programs with minor emphasis on selection for skeletal soundness.

More research needs to be conducted to determine the relationship between skeletal soundness and productivity, particularly longevity. Currently, not enough is known about whether differences in bone structure and joints (Fig. 17.26) are more important from an aesthetic standpoint or as a measure of productivity. Beef cattle breeders are better served by focusing their attention to the selection of cattle with superior muscularity (Fig. 17.27).

Reproduction

Bonsma (1983) has demonstrated that cattle can be selected visually for reproductive performance, although Wilson et al. (1981) found no conclusive evidence in relating feminine features in cows to fertility. Most of Bonsma's work attempts to predict the number of calves produced by each cow. His appraisal appears to reflect the expression of hormone levels or hormone imbalances as they affect specific target locations of the neck, shoulders, hair, tail-head, and reproductive organs. This type of appraisal has not proven effective in heifer selection, where hormone expression has been limited. In addition, the use of good performance

FIGURE 17.26 Foot and leg structure as presently associated with structural soundness. *Source:* R. A. Long, Texas Tech. University.

Correct skeletal structure

Sickle-hocked and back at the knees

Post-legged and buck-kneed

(continued)

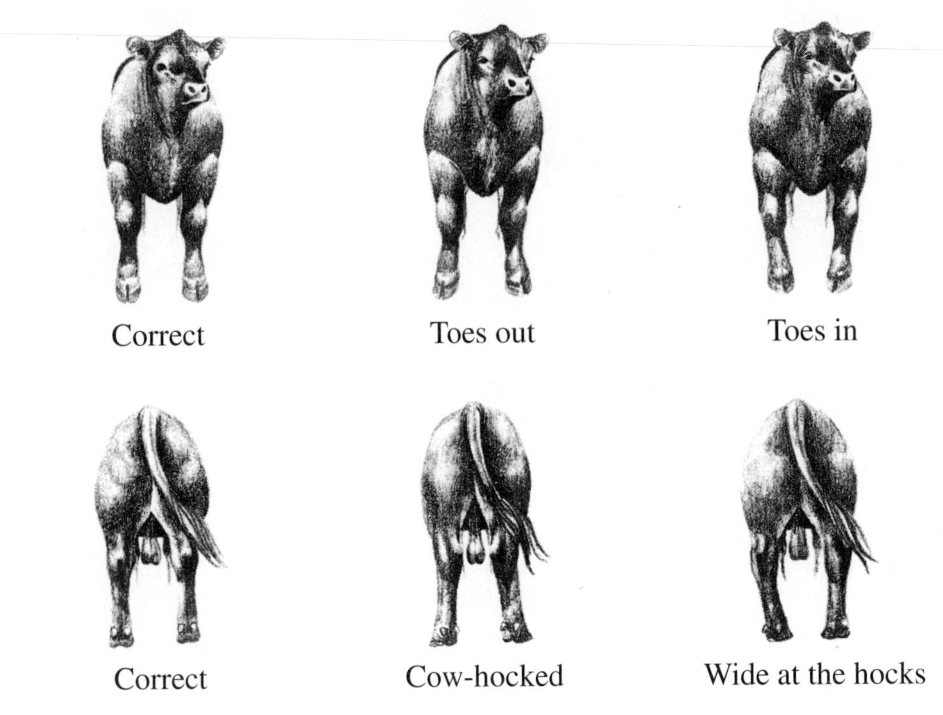

FIGURE 17.26 (Continued)

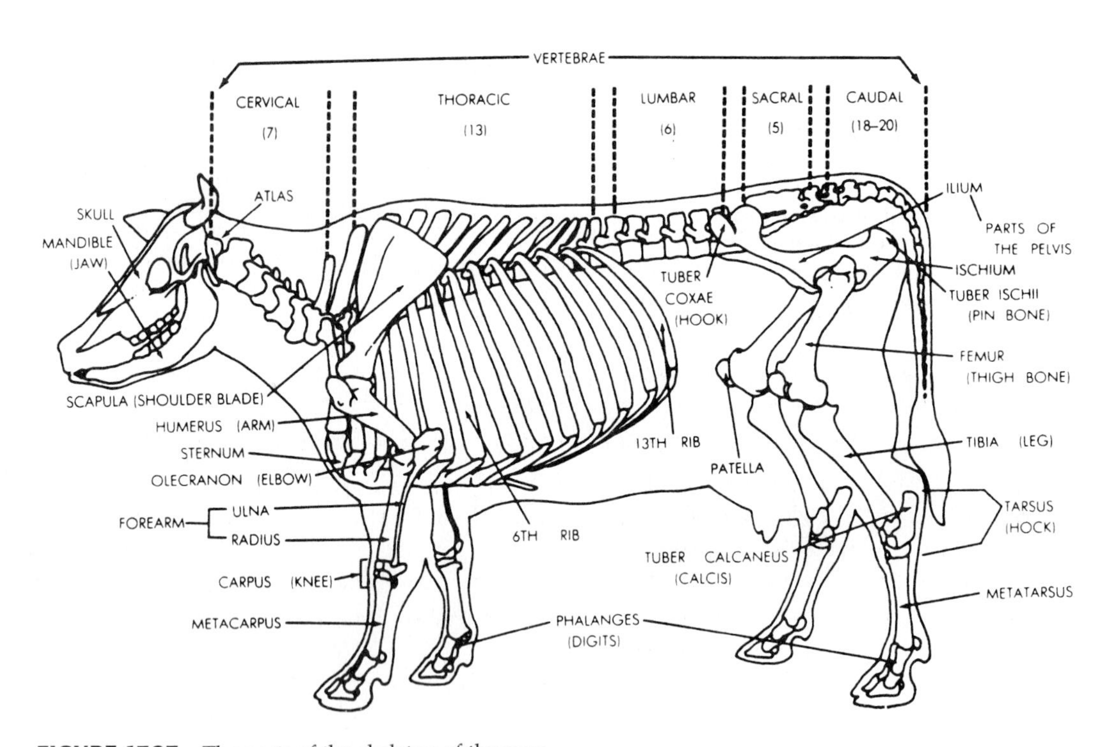

FIGURE 17.27 The parts of the skeleton of the cow.
Source: Battaglia, R. A., and Mayrose, V. B., 1981. *Handbook of Livestock Management Techniques.* Minneapolis: Burgess Publishing Co., copyright ©.

records in mature cows is far more economically feasible than trying to gain the experience needed to duplicate Bonsma's visual appraisal method.

Visual appraisal of body condition in breeding females is useful in assessing postpartum interval. Visual body scores have been shown to be highly related to carcass fat and carcass energy content (r = 0.80) and more useful than weight-to-height ratios.

Merchandising

Visual appraisal can be used to put productive animals into attractive packages. Body form and appearance are generally independent of productivity; however, it may be economically feasible for producers to combine highly productive animals with visual traits that are pleasing to the eye (Fig. 17.28). This is part of the merchandising effort whether the evaluation is focused on a retail cut, carcass, feeder calf, slaughter steer, or breeding animal. However, caution must be exercised in maintaining productivity as the priority, with only a secondary emphasis on eye appeal.

THE LIVESTOCK SHOW

The competitiveness of the beef cattle show has challenged the beef producer for centuries. Breeding and management programs have been designed on the basis of showring placings. Heated discussions have focused on the value of the showring to beef producers and the beef industry. As emotion began to give way to genetic research in the 1940s and 1950s, it became evident that major differences in showring cattle were created by environmental factors such as how the cattle were fed and fitted. Showring placings and high sale prices in many cases proved to be negatively related to productivity as measured by the commercial cattle industry.

FIGURE 17.28 Eye appeal or attractiveness may be useful in merchandising. However, visual appraisal should play a secondary role to the use of performance records when selecting cattle.

Today, most beef producers recognize the showring for what it really is—a tool for promotion and merchandising. It is not a place to measure genetic differences, determine breeding programs, or establish standards of productivity.

Cattle Judging

Numerous cattle fairs and shows require judges to rank cattle visually in groups or in individual placings. Usually one judge assumes this task, although occasionally an associate judge is also involved. Several major national beef shows use a three-judge system where the judges work independently of each other and the final placing of cattle is the result of the combined placings of either two or three judges. For each class, the three judges are randomly chosen to be judge A, judge B, and referee. The referee judge enters into the placing decision only when judges A and B disagree. The three-judge system is used to eliminate the bias that can occur with a single judge. Usually the three-judge system is more costly than other systems.

Most beef judges have developed their judging ability through participation in 4-H, FFA, junior college, and university judging contests that are usually associated with fairs and shows. Most land grant universities have courses in beef cattle and livestock judging. These courses improve students' abilities to make decisions and communicate more effectively rather than make marked improvements in their abilities to identify more productive cattle. Shanteau (1978) presents evidence that livestock judges have unusual abilities to make complex judgments, and the courses and training programs used to develop these judges have a clear impact on their abilities. Many successful cattle operators and other businesspeople attribute their decision-making success to participation in livestock judging programs.

Youth Projects

Youth programs (such as 4-H and FFA) typically use county, state, and national shows as the completion date for a beef project. Steer shows still have considerable participant and audience appeal. The Grand Champion steer receives major publicity because of its high sale price.

In recent years, several 4-H beef projects have become more consistent with commercial beef production. Carcass and rate-of-gain evaluations have been added to the live placing. In this situation, steers with the best combination of live placing, gain, and carcass are designated as the "Supreme Champions." Great strides in meaningful youth education have been achieved where this type of program receives primary emphasis.

The dimensions of realistic commercial beef marketing are still seriously lacking in many youth beef programs, particularly in the national shows. It is not uncommon for feeder steers to be purchased by the youth for several times the current commercial price. The motivation is still the live champion steer that sells for several times the commercial market value. One cannot argue the economics for the winner, but low economic return is evident for many steers placing lower than champion. Many of these unrealistic dimensions of the youth programs are cause for concern after reading the educational objectives of the 4-H program.

Steer shows have received considerable public attention and media appeal. The beef industry is fortunate to have this public interest and can continue to capitalize on this consumer focus on beef. Junior heifer shows, sponsored by several breed associations, have provided a meaningful stimulus for young people to initiate breeding herds of their own. These have merit in encouraging youth to become involved in the beef industry (Fig. 17.29).

In recent years, there have been some excellent accomplishments in combining the competition and excitement of the show with the valid performance comparison of animals. Most

FIGURE 17.29 Beef cattle projects give youth experience in assuming responsibility, understanding more about the beef industry, and becoming familiar with some of the factors affecting beef cattle productivity.

of these have centered around steers on a competitive feedlot and carcass test or bulls on a common 100–140-day-gain test, then exhibited in competition and sold. In other shows, EPDs and other performance data are provided to the judge(s). This type of show or exhibit, plus other new innovations, will continue to have appeal to beef producers and breeders in years to come. The cattle show or exhibit is not likely to disappear as some have predicted. It will be molded to fit the needs of the day. If for no other reason, the gathering of those with common interests and a desire to share friendship and ideas will perpetuate the beef show.

MANAGEMENT SYSTEMS HIGHLIGHTS

1. Understanding growth patterns of various biological types of cattle is critical in terms of matching cattle to forage conditions, marketing cattle at appropriate degrees of composition, and developing rations that meet animal requirements.
2. Humans can significantly influence the mature size, growth rate, and rate of maturity in a genetic selection. Producers are advised to study history prior to implementing selection programs that dramatically affect growth rates in any direction.
3. Objective measures of selection are typically most effective. However, trained evaluators can accurately assess compositional differences between animals.
4. Cattlemen are advised to recognize that while livestock shows provide excellent opportunities for merchandizing, promotion, and fellowship, they often do not provide the foundation for effective, low-cost selection programs.

REFERENCES

Publications

Berg, R. T. 1979. *Growth and Development of Carcass Components in Cattle.* Fayetteville, AR: Arkansas Agricultural Experiment Station Special Report 12.

Berg, R. T., and Butterfield, R. M. 1976. *New Concepts of Cattle Growth.* New York: Wiley.

Berg, R. T., and Walters, L. E. 1983. The Meat Animal: Changes and Challenges. *J. Anim. Sci.* 57: 133.

Boggs, D. L., and Merkel, R. A. 1993. *Live Animal, Carcass Evaluation, and Selection Manual.* 4th ed. Dubuque, IA: Kendall/Hunt.

Bonsma, J. C. 1983. *Man Must Measure: Livestock Production.* Cody, WY: Agi Books.

Briggs, H. M. 1981. *Uses and Abuses of the ShowRing.* Fayetteville, AR: Arkansas Experiment Station Special Report 93.

Brinks, J. S., Davis, M. E., Mangus, W. L., and Denham, A. H. 1979. *Genetic Aspects of Hoof Growth in Cattle.* Fort Collins, CO: CSU Gen. Series 982.

Brown, C. J., Brown, A. H., Jr., and Johnson, Z. 1983. *Studies of Body Dimensions of Beef Cattle.* Arkansas Agricultural Experiment Station Bull. 863.

Brungardt, V. H. 1972. *Efficiency and Profit Differences of Angus, Charolais, and Hereford Cattle Varying in Size and Growth.* Research Reports R2397–R2401. Madison, WI: University of Wisconsin.

Butterfield, R. 1974. *Exposing the Myths of Show Judging.* California Livestock Symposium.

Buttery, P. J., Hayes, N. B., and Lindsay, D. B. 1986. *Control and Manipulation of Animal Growth.* Stoneham, MA: Butterworth.

Callow, E. H. 1961. Comparative Studies of Meat: VII. A Comparison between Hereford, Shorthorn, and Friesian Steers on Four Levels of Nutrition. *J. Agr. Sci.* 56: 265.

Crickenberger, R. G., and Black, J. R. 1976. *Influence of Frame Size on Performance and Economic Considerations of Feedlot Cattle.* East Lansing, MI: Michigan State Agricultural Experiment Station Research Report 318.

Crouse, J. D., Dikeman, M. E., and Allen, D. M. 1974. Prediction of Beef Carcass Composition by Live-Animal Traits. *J. Anim. Sci.* 38: 264.

Currie, W. B. 1992. *Structure and Function of Domestic Animals.* Boca Raton, FL: CRC Press.

Daley, D. A., Tatum, J. D., and Taylor, R. E. 1983a. Accuracy of Subjective and Objective Preslaughter Estimates of Beef Carcass Fat Thickness. *J. Anim. Sci.* 56: 1.

Daley, D. A., Tatum, J. D., and Taylor, R. E. 1983b. Practical Methodology of Feeder Cattle Sorting. *J. Anim. Sci.* 57: 390 (Suppl. 1).

Hammond, J., Jr., Robinson, T., and Bowman, J. 1983. *Hammond's Farm Animals.* Baltimore: Edward Arnold University Park Press.

Hedrick, H. B. 1983. Methods of Estimating Live Animal and Carcass Composition. *J. Anim. Sci.* 57: 1316.

Kauffman, R. G., Grummer, R. H., Smith, R. E., Long, R. A., and Shook, G. 1973. Does Live-Animal and Carcass Shape Influence Gross Composition? *J. Anim. Sci.* 37: 1112.

Prior, R. L., and Lasater, D. B. 1979. Development of the Bovine Fetus. *J. Anim. Sci.* 48: 1546.

Shanteau, J. 1978. *Psychological Abilities of Livestock Judges.* Kansas Agricultural Experiment Station Bull. 620.

Swatland, H. J. 1984. *Structure and Development of Meat Animals.* Englewood Cliffs, NJ: Prentice-Hall.

Tatum, J. D., and Field, T. G. 1996. *Type Changes in the Hereford Breed—1950s to 1990s.* Research Report to the American Hereford Association. Fort Collins, CO: Colorado State University.

Trenkle, A. H., and Marple, D. N. 1983. Growth and Development of Meat Animals. *J. Anim. Sci.* 57: 273 (Suppl. 2).

Wilson, G. R., Reef, J. E., Turner, T. B., and Wilson, G. W. 1981. *Relating Visual Feminine Characteristics to Reproductive Performance in a Herd of Angus Cows.* Columbus, OH: Ohio Agricultural Research and Development Center Animal Science Series 81-1: 25.

Woodward, R., 1968. The Importance of Soundness in the Selection of Breeding Animals. *Charolais Banner* (July).

Visuals

"Beef Conformation" (carcass cross-section slides). Iowa State University, Ames, IA 50010.
"Breeding Cattle Evaluation" (1991; video, 42 min.); "Market Cattle Evaluation" (1991; video, 47 min.); and "Feeder Cattle Evaluation" (video, 28 min.). CEV, P.O. Box 65265, Lubbock, TX 79464-5265.
"Fitting and Showing Beef Cattle" (filmstrip and tape). Vocational Education Productions, California Polytechnic State University, San Luis Obispo, CA 93407.
"Shaping Carcass Cutability" (16mm film or video). Elanco Products Co., P.O. Box 1705, Indianapolis, IN 46206 or R. E. Taylor, Dept. of Animal Sciences, Colorado State University, Fort Collins, CO 80523.

Websites

Official livestock grading standards for USDA: www.ams.usda.gov
Microbeef Individual Management System: www.microbeef.com

Cattle Behavior, Facilities, and Equipment

Astute cattle producers know that cattle behavior has a direct relationship to cattle productivity. Disposition, especially when cattle are being handled, and social behavior during feeding are among the behavior areas that have economic importance.

Adequate facilities and equipment are necessary to handle, feed, and market cattle for optimum performance. Weight gain, reproduction, and bruises and injuries affect productivity and net returns. Poor facilities, rough handling, and wild dispositions of cattle can yield lower weight gains, poorer meat quality, increased injury to humans and cattle, and reduced overall performance. An understanding of cattle behavior is most useful in handling cattle, designing facilities, and planning equipment uses.

BEHAVIOR

It is important for producers to observe and understand cattle behavior, for they can influence cattle through their care, training, and selection and the social and physical environments provided. Producers can then make management decisions that will improve cattle productivity and profitability.

Disposition

Cattle producers consider the disposition or temperament of cattle to range from docile to wild or high-strung. This evaluation is usually made at calving time and when cattle are handled in corrals, pens, chutes, and other working facilities. Typically, cattle with a poor disposition are very fearful of people or they exhibit aggressive behavior.

There is evidence that cattle develop a good or poor disposition either from inheritance or from the way they are handled. The few heritability estimates for disposition are in the medium-to-high categories, indicating that the trait has a genetic basis and would respond to selection. Some producers cull cattle with poor dispositions because of economic losses (such as broken fences and other facility repairs, risk of injury, and reducing the excitability of other animals).

A 1997 survey of beef producers found that 8.3% of herd managers reported an increase of temperament/disposition-related problems over the previous five years. Managers in the Southeast (11.9%) and North Central (10.5%) regions were most likely to report an increase in disposition-related problems. Managers in the South Central region (4.2%) and the West (4.5%) were least likely to report a rise in temperament problems. The use of a simple temperament rating system based on observing cattle during handling for vaccinations or other procedures can be an effective culling tool. A four-point scale is used to evaluate cattle behavior during chute restraint: 1—stand quietly, 2—restless, 3—struggles constantly, and 4—frenzied activity. Cattle that are scored a 4 should probably be culled.

Communication

Communication exists when information is exchanged between individuals. This may occur with the transfer of information through any of the senses. A distress call, involving a distinct type of sound, occurs from either the cow or her calf when they become separated (Fig. 18.1). Although the cow and calf may recognize each other's vocal sounds, the most effective way for the cow to recognize her calf is by smell. Cows will more easily adopt other calves through the transfer of the odor of one calf to another. To reduce stress during weaning, calves and

FIGURE 18.1 A calf, separated from its dam, has a distinct bawl that communicates distress or dissatisfaction. *Source:* American Hereford Association.

cows should be separated by a single fence. It is preferred to accomplish weaning in a familiar pasture. Calves should not be weaned in a dirt lot to avoid respiratory distress. Calves that remain in visual contact with their mothers will bawl less than calves that are moved far away from their dams.

Cattle learn to respond to vocal calls, whistles, or other sounds made by the producer at feeding time. Cattle soon learn that the stimulus of the sound is related to being fed.

The bull vocally communicates his aggressive challenge to other bulls and intruders to his area through a deep bellow. This behavior is controlled by testosterone. The castrated male seldom exhibits similar behavior. The bull also issues vocal calls to cows and heifers, especially when he is separated from them but they are within his sight. A bull that has aggressive intent will perform a broadside threat display toward another bull or a person. He will stand sideways and hunch his shoulders to make himself appear large.

Cattle, which have 310- to 360-degree vision, are especially perceptive in their sight. This affects their behavior in many ways, such as when they are approached from different angles or when they are handled through various types of facilities.

Social Behavior

Some cows withdraw from the group to find a secluded spot just prior to calving. Cattle sometimes withdraw from the group if they are sick. Early and continuous association of calves is associated with greater social tolerance, delayed onset of aggressive behavior, and relatively slow formation of social hierarchies.

Status and social rank typically exist in a herd of cows with certain individuals dominating other, more submissive ones. The presence or absence of horns is important in determining social rank, particularly when strange cows are mixed together. In addition, horned cows usually outrank polled or dehorned cows in situations involving close contact, such as at feeding time.

Large differences in age, size, strength, genetic background, and previous experience have powerful effects in determining social rank. Once the rank is established in the cow herd, it tends to be consistent from one year to the next. There is evidence that genetic differences exist for social rank, both within and between breeds.

Research also indicates that animals fed together consume more feed than when fed individually. The competitive environment evidently stimulates the increased feed consumption. Dairy calves separated from their dams at birth appear to gain equally well whether fed milk in a group or kept separate. There is, however, evidence that they learn to eat grain earlier when group fed compared with being individually fed. Cattle individually fed in metabolism stalls consume only 50–60% the amount of feed they would eat if group fed.

When fed in a group of older cows, 2-year-old heifers have difficulty getting their share of supplemental feed (Table 18.1). However, the evidence indicates that 2-year-old heifers and 3-year-old cows can be fed together without affecting their competition for supplemental

TABLE 18.1 Weight Changes in 2-Year-Old Heifers Fed Separately or Together with Older Cows

Treatment	Weight Change
Pastured and fed with older cows	25 lbs loss
Pastured and fed separately	46 lbs gain

Source: K. A. Wagnon, *Social Dominance in Range Cows and Its Effects on Supplemental Feeding* (Davis, CA: California Agricultural Experiment Station Bull. 819, 1965).

feed (Wagnon, 1965). These behavior differences no doubt explain some of the nutrition, weight gains, and postpartum interrelationships that are age related when cows of all ages compete for the same supplemental feed.

Dominant cows raised in confinement usually consume more feed and wean heavier calves than submissive cows, whereas submissive cows wean lighter calves (25%) and become pregnant less often than aggressive cows. Figure 18.2 shows a 1-hour feeding pattern for two cows of different social rankings. A highly dominant cow may prevent other animals from drinking or eating. A dominant bull may prevent other bulls from mating in a multiple-sire pasture.

A dry, mature cow placed with newly weaned calves helps to offset the stress associated with weaning. Calves weaned in the presence of a mature female experience less morbidity than calves weaned without exposure to a mature cow.

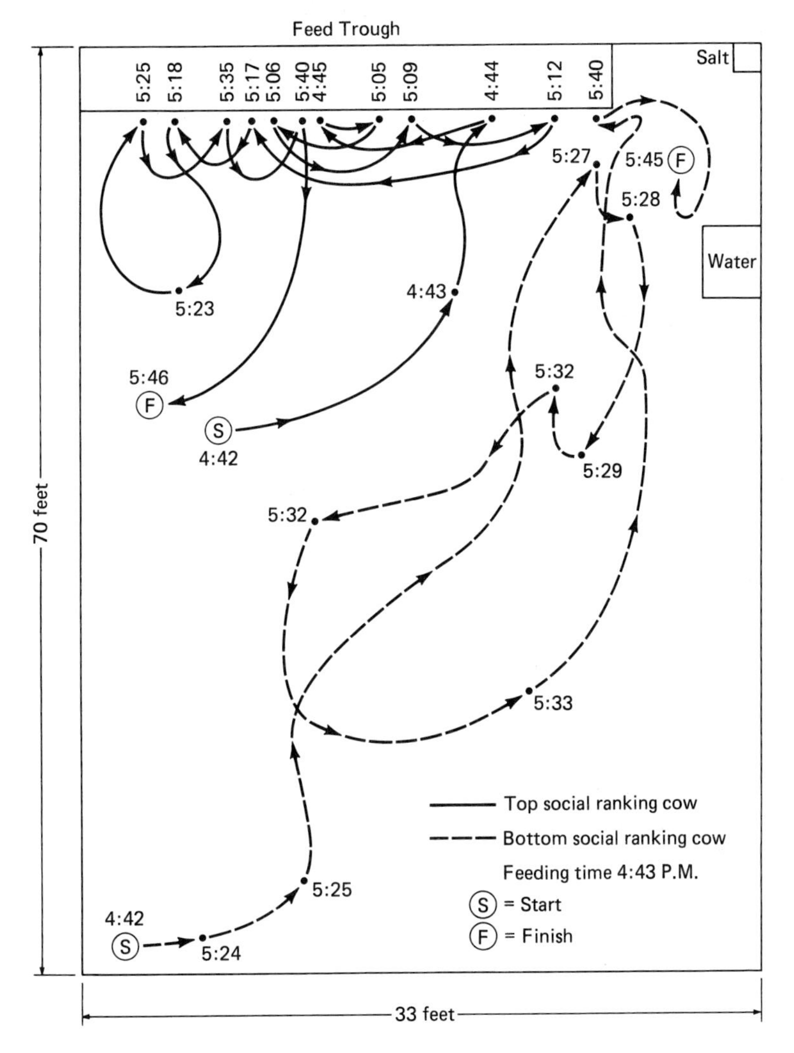

FIGURE 18.2 One-hour feeding pattern for two cows of different social rankings.
Source: Schake and Riggs (1972).

FIGURE 18.3 A bull exhibiting the flehmen response.

Reproductive Behavior

Some profound behavior patterns are associated with the sex or sex condition of cattle. This verifies the importance of the hormonal-directed expression of behavior. Bulls exhibit more aggressive behavior, whereas steers are more docile after losing their source of testosterone following castration.

The bull frequently curls his nose and upper lip (known as an olfactory reflex or *Flehmen*) after sniffing the urine or genitals of females (Fig. 18.3). This behavior is associated with inhalation through the upper respiratory tract. The bull, through the use of the vomeronasal organ, is identifying pheromones, which are chemical substances secreted in the urine of the proestrus or estrus female. In a sexually active group of cows, the bull is attracted to a cow in heat most often by visual means (observing cow-to-cow mounting) rather than by olfactory clues.

When females are sexually receptive, they usually seek out a bull if mating has not previously occurred. Primarily through hormonal influence, the female in estrus will stand for the bull when he mounts her. Cows in heat demonstrate male behavior by mounting and being mounted by other cows. A thrusting action is common in the cows as they terminate the mounting action.

The bull may guard a female that is approaching estrus. His success in guarding the female or actually mating with her is dependent on his rank of dominance in a multiple-sire herd. During the mating process, the bull will usually playfully nudge the female, with his tongue lick her side and back prior to mounting, and rest his chin on her rump. After mating, most bulls temporarily lose interest in the cow. Some bulls leave the cow and do not make repeated matings, while other bulls may breed the same cow numerous times even in the presence of other cows in estrus. These different behavior patterns may explain the wide variations in the success of individual bulls with varying bull-to-cow ratios.

Recent studies show that many individual bulls have sufficient sex drive and mating ability to fertilize more females than are commonly allotted to them. Some producers use an excessive number of males in multiple-sire herds to offset the few bulls that are poor breeders and to cover for the social dominance that exists among several bulls running with the same herd of cows. If low fertility exists in the dominant bull or bulls, then calf crop percentage will be seriously affected even in multiple-sire herds.

Tests have been developed to measure libido and mating ability differences in young bulls. While behavioral differences are evident between different bulls, tests, based on pregnancy rates, have not proven accurate for use by the beef industry.

Some cows eat or chew the afterbirth (placental membranes), and most cows will begin licking the calf after it is born. Exceptions to the latter occur when the cow does not feel well after a difficult birth or due to some illness. The licking of the calf provides an olfactory sense for identifying the calf in the future. The calf usually nurses with its rear end toward the cow's head. This allows the cow to smell the calf and decide to accept or reject it. A rejected calf is usually bunted with the head of the cow and kicked with the rear legs when it attempts to nurse.

Cows accept foster calves by smearing the foster calves with amniotic fluid previously collected from the second "water bag" or by putting the skin of the dead calf on the foster calf. Where foster calves have been raised on a cow, it has been shown that foster calves suckled less frequently and for shorter periods and weighed less than the cow's own calf.

Certain cows become very aggressive in protecting their calves shortly after calving. Serious injury can occur to producers who do not use caution with these cows.

Bulls being raised with other bulls commonly mount one another, have a penal erection, and occasionally ejaculate. Individual bulls can be observed arching their back, thrusting their penis toward their front legs, and ejaculating. Bulls can be easily trained to mount objects that provide the stimulus for them to ejaculate. AI studs commonly use restrained steers for collection of semen. Bulls soon respond to the artificial vagina, when mounting steers, which provides them with a sensual reward. Bull calves that are raised in a social group with other cattle are less likely to attack people. Bull calves reared in isolation are more likely to become dangerous.

Mating behavior has an apparent genetic base, as there is evidence of more frequent mountings in hybrid or crossbred animals.

Research indicates that more cows calve during periods of darkness than during daylight hours. The calving pattern, however, can be changed by altering when cows are fed. Cows fed during late evening will have a higher percentage of their calves during daylight hours.

Suckling Behavior

Cattle have certain instinctive behavior patterns, such as the newborn calf's searching movement of head and neck, sucking of protruding objects, and swallowing of fluids. Obviously, the function of these behaviors is to locate the mammary gland and suckle the teats for milk.

The number and duration of suckling incidences have been observed in range beef calves (Odde, 1983). In this study, the average time spent suckling during each of several 24-hour periods was approximately 45 minutes, with an average of five suckles per calf. Peaks in suckling activity occurred from 5:00 A.M. to 7:00 A.M., 10:00 A.M. to 1:00 P.M., and 5:00 P.M. to 9:00 P.M. The most suckles in a single hour occurred between 5:00 A.M. and 6:00 A.M. and the fewest between 10:00 P.M. and 11:00 P.M. Cows that gave more milk nursed less frequently, and heavier calves sucked less frequently. Age, breed (Polled Hereford and Simmental), and sex of calf did not influence the number or duration of suckles.

The calf is usually born away from the herd in a secluded spot if one is available. The calf remains secluded while the cow grazes or feeds elsewhere, usually with the herd. The cow will return to the calf several times each day to allow the calf to nurse. In a few days, the cow will lead the calf to the herd where both will remain.

Cross-nursing is more frequent for cows forced to accept a second calf than for cows raising only a single calf. In the latter situation, some cross-nursing of calves may be observed and occasionally a cow will nurse another cow. These behaviors can affect the validity of weaning weight comparisons.

Grazing Behavior

Cattle have or develop palatability preferences for certain plants and may have difficulty changing from one type of plant to another. Also, cattle prefer to graze lower areas, especially when they are near water. These grazing behaviors tend to cause overgrazing in certain pasture areas, which results in reduced weight gains. Also, young calves learn to graze the plant types that their mothers consume.

Age of cow and weather affect the behavior of cows grazing native range during the winter. At the Range Research Station at Miles City, Montana, cows grazed less as temperatures dropped below 20°F, and 3-year-olds grazed approximately 2 hours less than 6-year-olds. The colder the temperature, the longer cows waited before starting to graze in the morning. At 30°F, cows started grazing between 6:30 and 7:00 in the morning; at −30°F, they waited until about 10 A.M. to begin grazing.

In northern Utah, observations were made on cows grazing open, treeless winter range (crested wheatgrass) from mid-November to mid-January. Bitter cold temperatures, strong winds, and snow accumulations of up to 10 inches are common in this area. The average time cows spent in various activities is shown in Table 18.2. On the average, cows walked 3.6 miles per day in the course of eating and moving to and from water. When temperatures lowered and wind speeds were in excess of 2 mph, travel was restricted to as little as 1.5 miles per day, and the cows spent less time grazing.

Feedlot Behavior

Cattle kept in barns or sheds have been observed to be awake approximately 20 hours per day, and asleep the remaining 4 hours, with 40% of the day spent standing. Cattle tend to sleep less when on pasture than when in familiar buildings.

The "buller-steer" problem with steers in a feedlot is a unique behavior problem with serious economic significance to feedlot operators. Certain steers are singled out, probably by an olfactory identification through the vomeronasal organ, and ridden continually by other

TABLE 18.2 Activities of Cows Grazing on Winter Range

Activity	Hours
Grazing	9.4
Ruminating	
Standing	0.6
Lying	8.3
Idle	
Standing	1.1
Lying	4.0
Traveling	0.6
Total	24.0

Source: Utah State University.

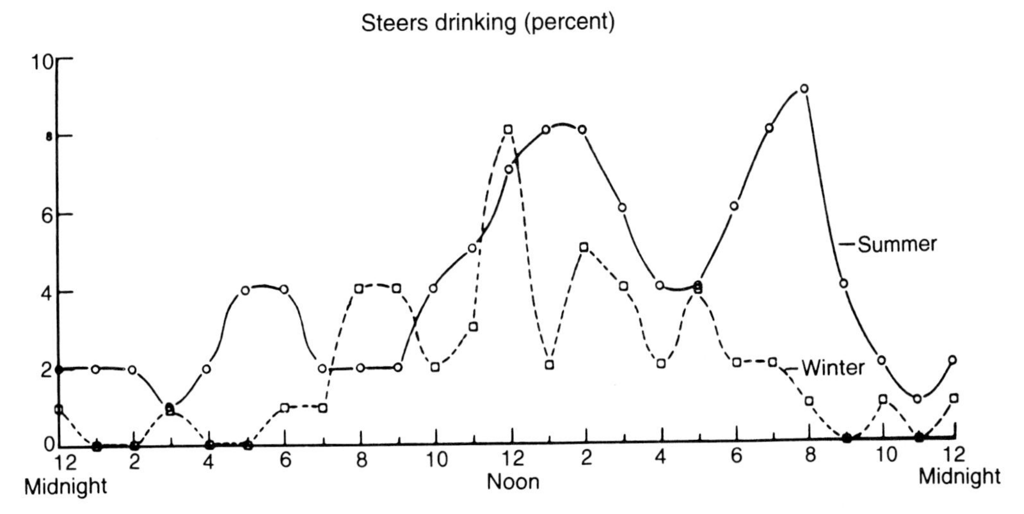

FIGURE 18.4 Seasonal influence on diurnal activity of yearling steers in the feedlot.
Source: Ray and Roubicek (1971).

steers. While the exact cause of the buller-steer syndrome is not known, there is evidence that the use of hormonal implants (especially estrogenic ones) have increased the incidence. Other factors affecting the syndrome appear to be keeping large numbers in a group and introducing new cattle into a pen. Providing adequate watering and bunk space may help prevent bullers. The behavior pattern reduces feedlot gains and usually causes the buller-steer to experience serious health problems, including death.

Bulls can become very aggressive toward one another in riding and fighting. A group of bulls can become extremely aggressive toward another bull that may have been separated from the group for a few days and then reintroduced into the group.

Feedlot steers have different patterns of water consumption throughout the day. Some of these patterns are significantly affected by season of the year (Fig. 18.4). The feeding behavior of steers in confined areas with concrete floors appears to be slightly different from steers in drylot, dirt pens. A higher percentage of steers on concrete eat during the night than during the middle of the day (10 A.M. to 3 P.M.). When the weather becomes extremely hot, cattle will have dramatically increased water consumption.

Behavior During Handling and Restraint

Most cattle are handled and restrained several times during their lifetime. Ease of handling depends largely on their temperament, size, and previous experience as well as the design of the handling facilities. Producers knowledgeable about animal behavior can prevent injury, undue stress, and physical exertion for both cattle and producers themselves. For example, an astute producer knows how to approach cattle so they will respond in the desired way (Fig. 18.5). Cattle have a "flight zone." When a person is outside the flight zone, cattle exhibit an inquisitive behavior and face the handler. When a person moves inside the flight zone, the animal usually moves away.

Blood odor appears to be offensive to cattle because it is novel and strange. Therefore, the reduction or elimination of such odors may well encourage cattle to move through handling facilities with greater ease. Cattle are easily disturbed by loud or unusual noises, such as those made by motors, pumps, and compressed air.

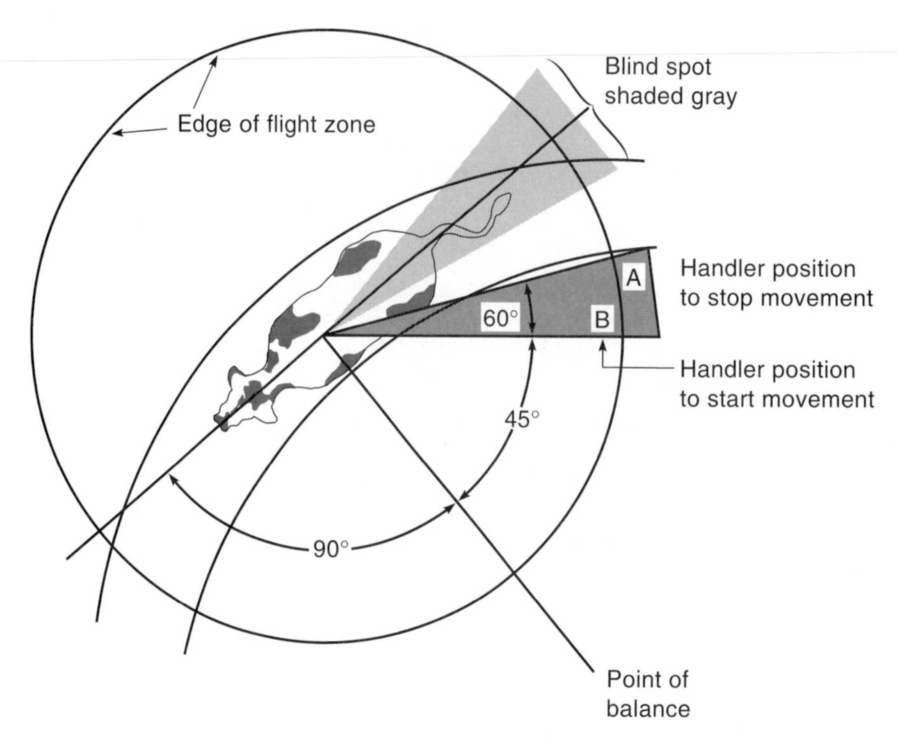

FIGURE 18.5 Flight zone of cattle with positions of handler to influence movement of the cattle.
Source: Temple Grandin.

Cattle can be handled most effectively by the producer who understands their behavior patterns and who provides handling facilities that complement those behaviors in positive ways. Calm cattle are easier to move and sort than excited cattle. Once agitated, it takes cattle about one-half hour to calm down. Following are some important facts about the behavior of cattle:

1. Because cattle have 310–360-degree vision, they are sensitive to shadows and unusual movements observed at the end of or outside of the chute (Fig. 18.6). Thus, cattle usually move with greater ease through chutes that are curved and that have solid sides (Fig. 18.7).
2. Cattle tend to move toward light (except blinding direct sunlight). Loading cattle at night is best accomplished by providing a nonglaring light near the truck or trailer gate. Cattle move easier into a dark area when in a single-file chute rather than as a group. Loading chutes and squeeze chutes should face north and south because cattle do not move easily into direct sunlight. White translucent skylights will facilitate cattle movement inside buildings. If the interior of a building is too dark, cattle will be hesitant to enter it.
3. Shadows across chutes and scales should be prevented because cattle are fearful of shadows.
4. Poor dispositions are developed by cows and calves handled in an abusive manner. Cattle remember painful and adverse experiences. A small flag on the end of a stick is useful for moving and sorting cattle. Most cattle can be easily moved without the use of a whip or electric prod.
5. Cattle usually respond negatively to abuse, loud noises, and other confusing situations that can excite them. Thus, noisy equipment should be kept as far as possible from cattle.

FIGURE 18.6 Shadows that fall across a chute can disrupt the handling of animals. The lead animal often balks and refuses to cross the shadow. Other distractions that will make cattle balk and refuse to move are: moving chain ends, reflections in puddles, seeing people in front of them, fan blades turning in the wind, changes in flooring type, or an object on a fence (such as a coat) that is flapping.
Source: Temple Grandin.

FIGURE 18.7 Animals move more easily through curved chutes with solid sides.
Source: Temple Grandin.

Metal chutes and alleys should be constructed so that loud clanging and banging noises are eliminated. Yelling at cattle is aversive and increases the stress levels of both cattle and handler.

6. Cattle are creatures of habit; an established, daily routine will result in ease of handling.
7. Handle animals in groups. A single animal often resists going into a chute or pen by itself. It may also become excited and injure itself or the handler or damage the facilities.
8. Curved chutes are preferred over straight chutes because cattle cannot see what is at the end of a curved chute until they are almost there (Figs. 18.6 and 18.7). A curved chute also utilizes the natural tendency of cattle to circle around the handler and to go back where they came from. The catwalk should be inside the curved chute so that the handler is positioned at the best angle for working with the cattle.
9. The cattle handler's movements should be slow and deliberate; any sudden movements will frighten cattle and make them difficult to handle.
10. By understanding the flight zone of cattle (Fig. 18.5), the handler can effectively work cattle in the corral or pasture (Figs. 18.8, 18.9, and 18.10). Cattle can be moved more easily if the handler works on the edge of the flight zone. The handler penetrates the flight zone to start cattle movement and retreats outside the flight zone to stop cattle movement. When the handler is positioned behind the point of balance at the shoulder, the animal will move forward. It will move backward when the handler is in front of the point-of-balance. Figure 18.11 shows how to use the point-of-balance to move an animal into a squeeze chute. The animal moves forward when the handler walks quickly past the point-of-balance at the shoulder.
11. If cattle balk and refuse to move through a chute, distractions such as a moving piece of chain or a coat hung on a fence should be removed.

Round pens (i.e., those without square corners) enhance cattle movement and prevent injury to more excitable cattle. Head catches are a necessity for immobilizing cattle of large sizes. If the head needs to be restrained, a halter is preferred to nose tongs.

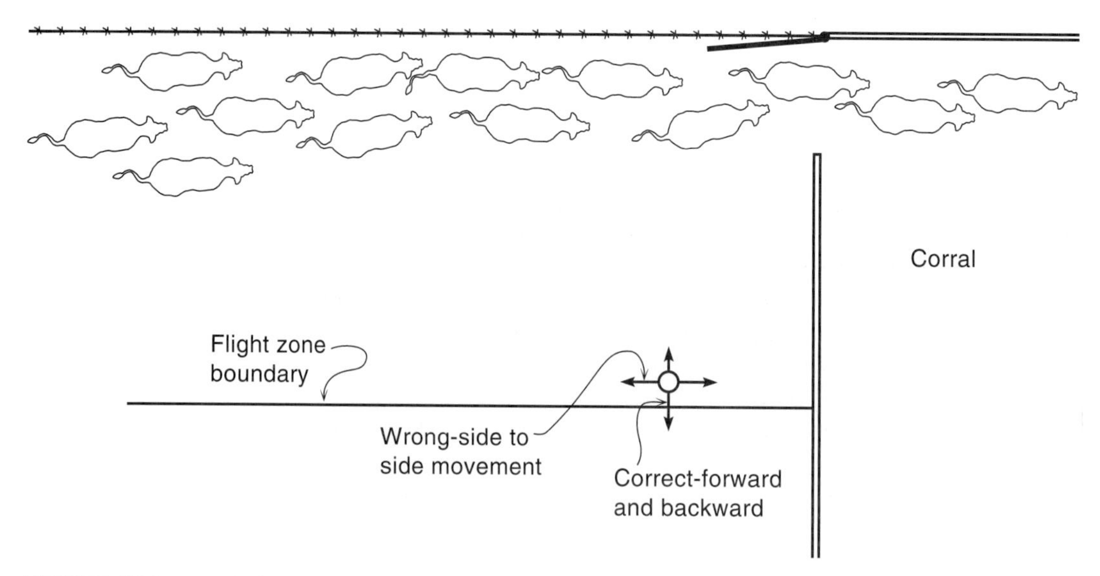

FIGURE 18.8 Leader handler position for filling corral.
Source: Temple Grandin.

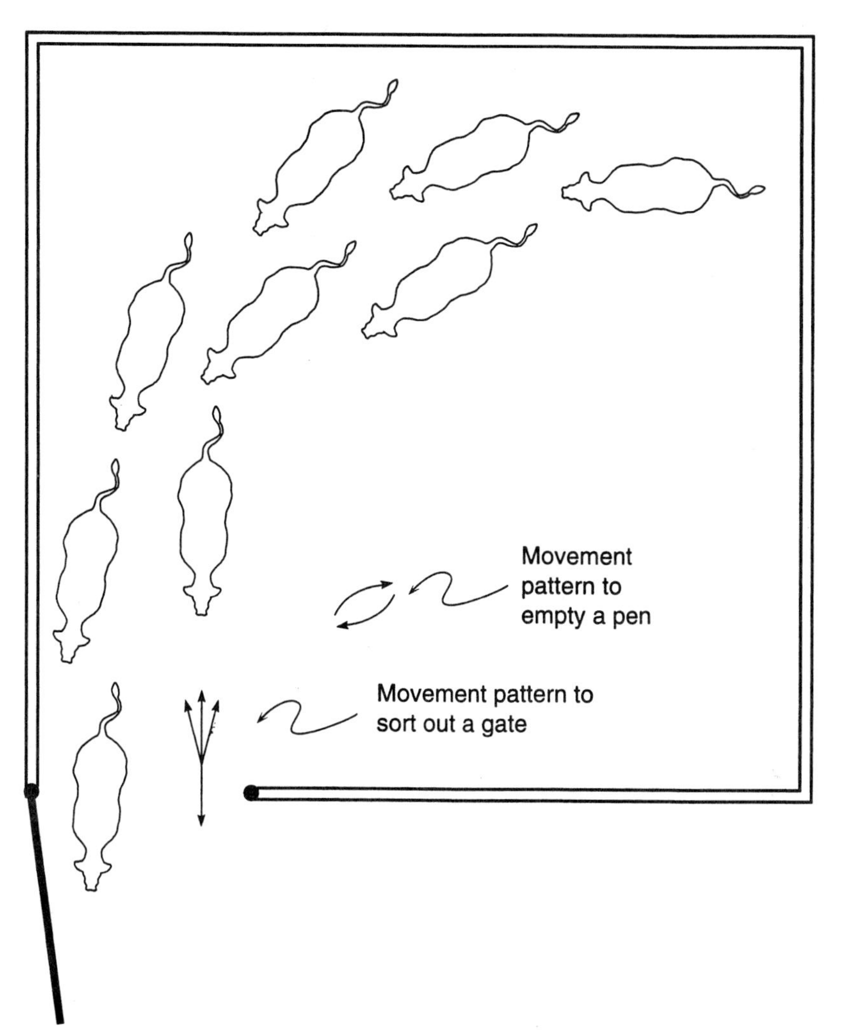

FIGURE 18.9 Handler positions for emptying a pen and sorting at a gate. The handler should control the animal's movement out of the pen. This is especially important when moving from one pasture to another so as to prevent the cows from leaving their young calves behind.
Source: Temple Grandin.

Gathering Cattle on Pasture and Training Cattle

Cows that are accustomed to people can be easily trained to come in when a horn on a vehicle is blown. Cattle that are not accustomed to people can be gathered by using the movement pattern shown in Figure 18.12. This movement pattern will trigger the animal's instinct to bunch together. The handler walks on the edge of the herd's collective flight zone. The handler must *not* circle around the cattle or chase stragglers. The handler must walk back and forth until the instinct to bunch is triggered. The principle is to trigger the bunching instinct before any attempt is made to move the cattle forward. Cattle are moved forward by increasing pressure on the collective flight zone. When the animals start moving the handler must practice the principle of pressure and release. After the herd starts moving, the handler should back off and when the herd slows down the handler can apply more pressure to the flight zone. Practicing pressure and release will help prevent running and breaking fences.

Cattle will be easier to handle when they are moved from a ranch to a feedlot if they have been habituated and trained to different methods of handling. They should become accustomed to vehicles, people on foot, and to riders on horseback. First experiences with new things

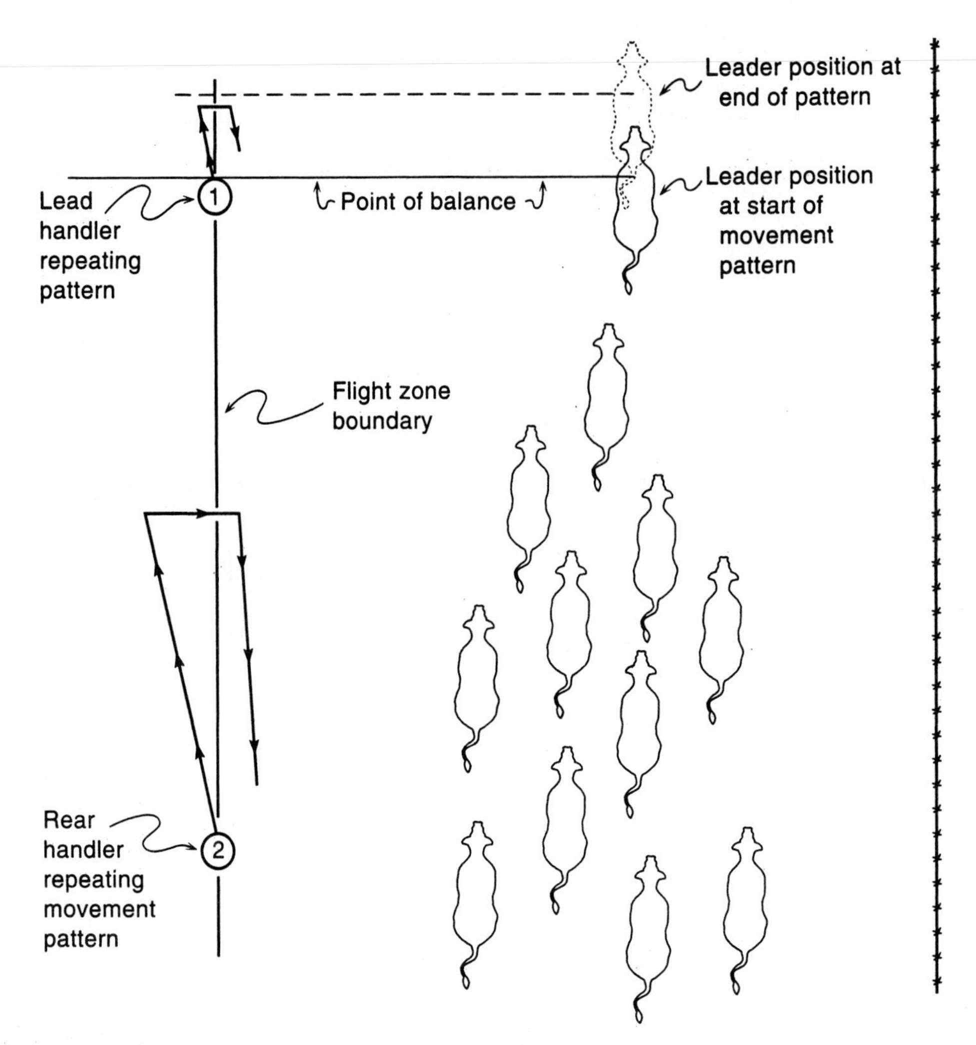

FIGURE 18.10 Handler positions to move groups of cattle on pasture.
Source: Temple Grandin.

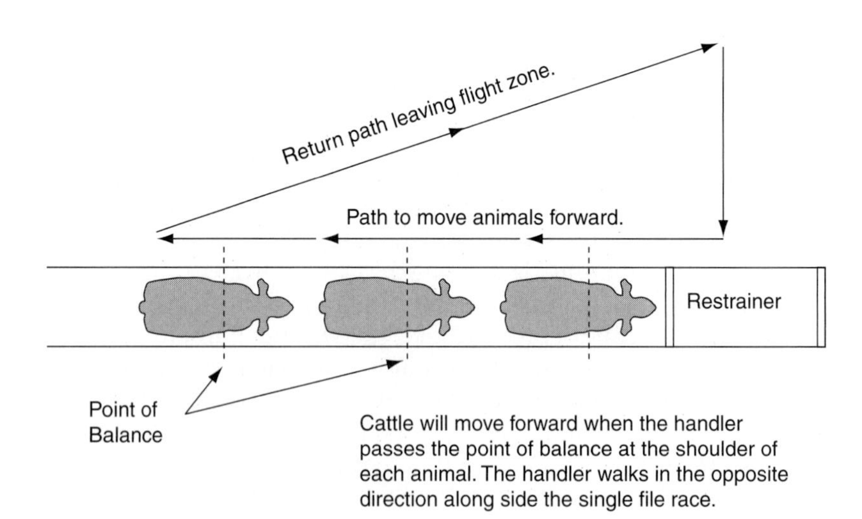

FIGURE 18.11 Handler movement pattern to keep cattle moving into a squeeze chute or restrainer. Cattle will move forward when the handler passes the point of balance at the shoulder of each animal. The handler walks in the opposite direction alongside the single-file race.
Source: Temple Grandin.

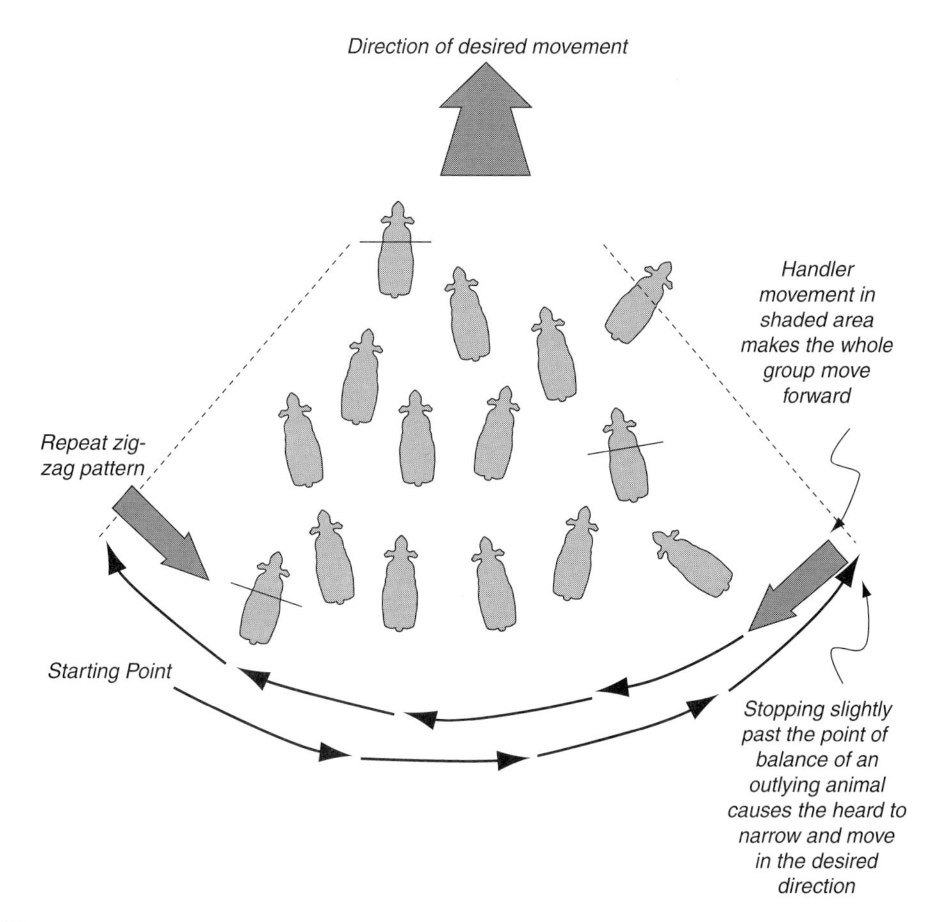

FIGURE 18.12 Handler zigzag movement pattern for use in open pastures—starting movement in the desired direction. The handler must zigzag back and forth to keep the herd going straight. Imagine that the leaders are the pivot point of a windshield wiper and the handler is out on the end of the blade sweeping back and forth. As the herd narrows and gets good forward movement, the width of the handler's zigzag narrows. *Source:* Temple Grandin.

make a big impression on cattle and they never forget. When they are first introduced to a new corral or a horse and rider, the first experience should be positive. The rider can ride quietly through the herd or the cattle could be fed the first time they are brought to a new corral.

CATTLE TRANSPORT

One of the most critical control points in the prevention of bruising and injury is in the handling of cattle during transport as well as the design of cattle trailers. The three primary sources of bruising during transport are rough handling, jostling and shifting in trucks or trailers, and catching hips or back on truck/trailer doors and gates.

Two strategies should be utilized to minimize bruising and injury. First, feedlot and transport personnel must be adequately trained in proper cattle handling procedures. Second, the design of transport vehicles should be carefully evaluated to remove sources of bruising. A study conducted in Ontario, Canada, utilized a system of cushions inside of trailers and in loading chutes to minimize bruising. These padded trailers reduced bruising by about 10%. This equates to an estimated economic savings of one million Canadian dollars annually if it were implemented on a national scale.

Cattle producers are advised to contract with only those trucking companies who properly train drivers, maintain cleanliness and upkeep of equipment, and ensure that footing in trucks and trailers is always solid. A producer can do everything correctly only to incur serious financial losses due to poor transport protocol.

FACILITIES AND EQUIPMENT

General Facility Guidelines

Facilities that are properly constructed and maintained in good working order will enhance the producer's time management and safety. Cattle disposition will be calm and docile if facilities are constructed to control or influence their behavior in a positive manner. Tables 18.3 and 18.4 give the space requirements and general facility guidelines needed by beef cattle.

Methods of Cattle Restraint

Restraining and controlling cattle are necessary for many production and management practices, such as health care, identification, showing, marketing, weighing, breeding, pregnancy checking, and implanting growth stimulants. Cattle restraining facilities should be constructed to provide safety for those working the cattle and to prevent unnecessary excitement

TABLE 18.3 Cattle and Facility Space Requirements

Corral and Pen Dimensions			
Facility Description	To 600 lbs	600–1,200 lbs	1,200 lbs or Cow-Calf
Holding area (sq ft per head)	14	17	20
Crowding pen (sq ft per head)	6	10	12
Working chute with vertical sides			
Width (in.)	18	22	26
Minimum length (ft)	20	20	20
Working chute with sloping sides			
Width at bottom inside clear (in.)	15	15	16
Width at top inside clear (in.)	20	24	26
Minimum length (ft)	18	18	18
Working chute fence			
Height solid wall (in.)	45	50	50
Depth of posts in ground (in.)	36	36	36
Overall height (7 ft minimum clearance below crossties to walk under)			
Top rail, medium-sized docile cattle (in.)	55	60	60
Top rail, large-sized or wild cattle (in.)	60	66	72
Corral fence			
Recommended height (in.)	60	60	60
Depth of posts in ground (in.)	30	30	30
Loading chute			
Width (in.)	26	26	26–30
Length (in.)	12	12	12
Rise (in. per ft)	3	3	3
Ramp height for gooseneck trailer (in.)	15	—	—
Pickup truck (in.)	28	—	—
Van-type truck (in.)	40	—	—
Tractor Trailer (in.)	48	—	—
Double-deck (in.)	100	—	—

Source: Adapted from multiple sources.

TABLE 18.4 Space Requirements and General Facility Guidelines for Feedlot Cattle

	Feedlot (sq ft per head)
20 sq ft in barn and 30 sq ft in lot	Lot surfaced; cattle have free access to shelter
50 sq ft	Lot surfaced; no shelter
150–800 sq ft	Lot unsurfaced except around waterers, along bunks, and open-front building and a connecting strip between them
20–25 sq ft	Sunshade
	Buildings with Feedlots (sq ft per head)
20–25 sq ft	600 lbs to market
15–20 sq ft	Calves to 600 lbs
1/2 ton per head	Bedding
	Cold Confinement Buildings (sq ft per head)
30 sq ft	Solid floor, bedded
17–18 sq ft	Solid floor, flushing flume
17–18 sq ft	Totally or partly slotted
100 sq ft	Calving pen
1 pen for 12 cows	Calving space
	Feeders (in. per head along feeder)
All animals eat at once:	
18–22 in.	Calves to 600 lbs
22–26 in.	600 lbs to market
26–30 in.	Mature cows
14–18 in.	Calves
Feed always available:	
4–6 in.	Hay or silage
3–4 in.	Grain or supplement
6 in.	Grain or silage
1 spacer per 5 calves	Creep or supplement
	Water Requirements
Gal per 1,000 lbs of cattle	29 gal per day at 50°F
	18 gal per day at 90°F
Automatic waterers for cattle:	1/2 gpm minimum
20–40 head per waterer	2 gpm preferred
	Waterers
40 head per available water space in drylot	
	Isolation and Sick Pens
40–50 sq ft per head	
Pens for 2–5% of herd	
	Mounds
25 sq ft per head	Minimum
50 sq ft per head	If windbreak on top of mound, 25 sq ft per head each side
	Slopes
Floors, pavements	
1/2–1/4 in. per ft	Bunk aprons with step
1 in. per ft	Bunk aprons without step
1/4–1 in. per ft	Solid floors toward slats, flumes
1/2% or more	Longitudinal bottoms of gutters, flumes
1%	Gravity pipe to lagoon
Earth	
4:1 to 5:1	Mound sideslopes
5% maximum	Mound longitudinal
4–6%	Lots
	Daily Manure Production Per 100 lbs Liveweight
60 lbs, 1 sq ft	Feces and urine
6.91 lbs, 0.12 sq ft	Solids

Source: Adapted from multiple sources.

or stress on the cattle. An understanding of cattle behavior is needed to construct or purchase working facilities and equipment that conform to cattle's natural instincts. Forcing cattle to do something against their will usually causes cattle to develop poor dispositions, increases risk of injury, and usually results in reduced productivity.

Working Chutes *Working chutes* can be used effectively to restrain cattle during such management practices as vaccinations, spraying, and applying pour-on insecticides while the animals are crowded head-to-tail in the chute. For more rigid confinement, a squeeze chute can be placed at the end of the working chute.

Many commercial and homemade working chute designs are available. The best working chutes have the following characteristics:

1. Shaped sides or an adjustable side (18–30 in.) so that large or small cattle can be handled.
2. The V-shaped sides should measure 14–17 in. wide at the bottom and remain so for the first 2 ft from the bottom, at which point they should flare out to approximately 24–30 in. at hip and head height.
3. Straight-sided, permanent chutes should measure 26–28 in. in width and the sides should be solid so cattle cannot see through them and become distracted.
4. The crowding pen and working chute should be curved with solid sides.
5. The working chute exit and squeeze chute entry gate are usually made of bars to give cattle the illusion of being able to escape and thereby encouraging them to enter.

When constructing a working chute, an access gate should be built adjacent to the squeeze chute end so that the working chute can be entered to perform palpations and castrations.

The step-by-step procedure for moving cattle into and through the working chute is as follows:

1. Move cattle to the working alley and crowding pen that funnel into the working chute. Avoid overfilling the crowding pen; cattle need room to turn. Only fill the crowding pen half full.
2. Start cattle into the working chute and let them follow the leader. It will be more efficient to get one animal started down the chute instead of trying at random to drive all of them at once. If the cattle will not cooperate, look to remove any distractions and ensure that people are in the right place. If it becomes necessary to enter the crowding pen and encourage them with a prodding stick or tail-twist, handler restraint must be exercised. When tail-twisting is used, the handler should stop twisting the tail when the animal cooperates by moving. The animal learns quickly that cooperation will be rewarded by a release of pressure on the tail. Cattle that get turned around in the crowd pen can be easily turned by shaking plastic streamers next to their heads. A plastic garbage bag taped to a broom handle works well. The animal will turn away from the plastic strips because their vision is blocked on one side.

   Furthermore, when several animals are in a relatively small area, the producer is at risk of being squeezed, stepped on, or kicked. Avoid using the electric prod or abusive hitting; this will not only make cattle nervous and less cooperative but will also increase the producer's risk of injury both now and later on. If an animal rears up in the single-file chute, the handler should back up and move away. The animal rears up in an attempt to get away from the handler who is deep in its flight zone.
3. Keep cattle moving down the working chute by walking along the outside and talking to them, slapping them on the rump, and tail-twisting as necessary. Walking along the chute toward the crowd pen will cause each animal to move forward as the handler moves past the point of balance at each animal's shoulder.
4. Place bars completely across the working chute behind the last and first animals in the chute. The bars, which should be placed just above the hock, will prevent the animal from

backing up. An alternative is to install a one-way entry in the working chute. Cattle can walk through the one-way passage, but it prevents them from backing up.
5. Open the gate to the squeeze chute and allow the first animal in the working chute to enter (Battaglia and Mayrose, 2001).

Squeeze Chute and Headgate The combination of *squeeze chute* and *headgate* can be used to advantage on any type of cattle farm or ranch to facilitate several management techniques, including dehorning, castrating, branding, implanting, ear tagging, stomach tubing, artificial insemination, and blood testing. A squeeze chute and headgate are made even more valuable when they are positioned at the terminal end of a working chute.

There are many commercial and homemade designs available for the squeeze chute and headgate. A workable combination should consist of the following: a squeeze mechanism, a headgate with head and nose bars, a tailgate, removable solid side panels measuring approximately 24 in. from the ground, and removable side bars for easy access to the animal's side (Fig. 18.13).

Three basic headgate designs are used in beef operations: *straight-bar* headgate, *positive-type* headgate, and *curved-bar* headgate. The straight-bar headgate generally is designed to catch an animal automatically as it walks through the chute. The greatest advantage of the straight-bar headgate is its protection against choking the animal. It is also the recommended type if a body squeeze chute is not available. Its main disadvantage is that the animal can move its head up and down easily, which can create problems with techniques that require head immobilization. The positive-type headgate, in contrast, operates somewhat like a guillotine. Its main advantage is almost complete head control, both sideways and up and down. The headgate almost completely immobilizes the head without the necessity of a head and nose bar. However, this type of headgate can choke an animal so it is recommended only for wild cattle. Finally, the curved-bar headgate has vertical neck bars that are curved, providing more head control than the straight-bar headgate but with a greater risk of choking. Curved-bar stanchion headgates are used on most hydraulic squeeze chutes in large feedlots. This headgate is a good compromise for most cow-calf producers and should be used in conjunction with body restraint to prevent the animal from lying down and choking.

It is absolutely essential that the headgate be adjusted to the correct height of the animal being worked in order to prevent choking. In addition, it is mandatory that the animal's head

FIGURE 18.13 A squeeze chute and headgate. This piece of equipment is manually operated.

be released from the headgate before the body squeeze is released. This helps prevent the animal from lunging forward and possibly injuring itself. If this is not done, the animal can lunge forward and break its neck or choke to death. Regardless of the headgate design used, it is important that the headgate be adjusted to the size of the animals being worked so to prevent choking or escape. If an animal shows signs of choking, it should be released immediately. Calm handling of cattle prior to their entry into the chute will help prevent most problems at the head catch from occurring. Calves can be worked through a conventional chute, but additional labor is typically required to prevent the animals from turning around (Fig. 18.14).

Vertical Tilt Tables In many cow-calf operations, *vertical tilt calf tables* are used for restraining the calf during the branding, vaccination, castration, and dehorning process. The calves are relatively small because they are only a few weeks to a few months old.

Some cow-calf operations, however, continue to restrain calves by roping their hind legs from a horse while another person "tails" or "flanks" the calf to its side and sits on it as others dehorn, castrate, vaccinate, and brand the calf.

Another type of vertical tilt equipment is used for hoof trimming (Fig. 18.15). Hooves can also be trimmed in squeeze chutes and in upright nontilting chutes constructed for hoof trimming.

Rope Halters *Rope halters* are considered a necessity for restraining cattle on many farms and ranches. Halters are used to teach cattle to lead and are used when the head needs to be

FIGURE 18.14 Dehorning a calf in a calf table.

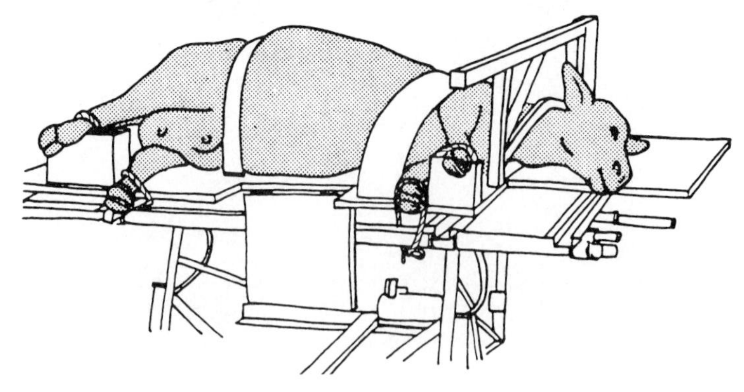

FIGURE 18.15 Animal restrained on tilting hoof-trimming table. Note feet are tied to minimize movement during the trimming process.
Source: Battaglia, R. A., and Mayrose, V. B., 2001. *Handbook of Livestock Management.* Upper Saddle River: Prentice Hall. ©.

tightly restrained. Figure 18.16 shows the construction of a low-cost, adjustable rope halter. The following procedure is used to make a rope halter:

1. Select 12–15 ft of half-inch, three-strand rope. Almost any rope type—from cotton to nylon—will work depending on its strength, durability, and cost.
2. Rope ends that are left unfinished will fray and deteriorate. Finish one end of the rope by whipping, clamping, dipping, or heat-treating (the method selected depends on personal

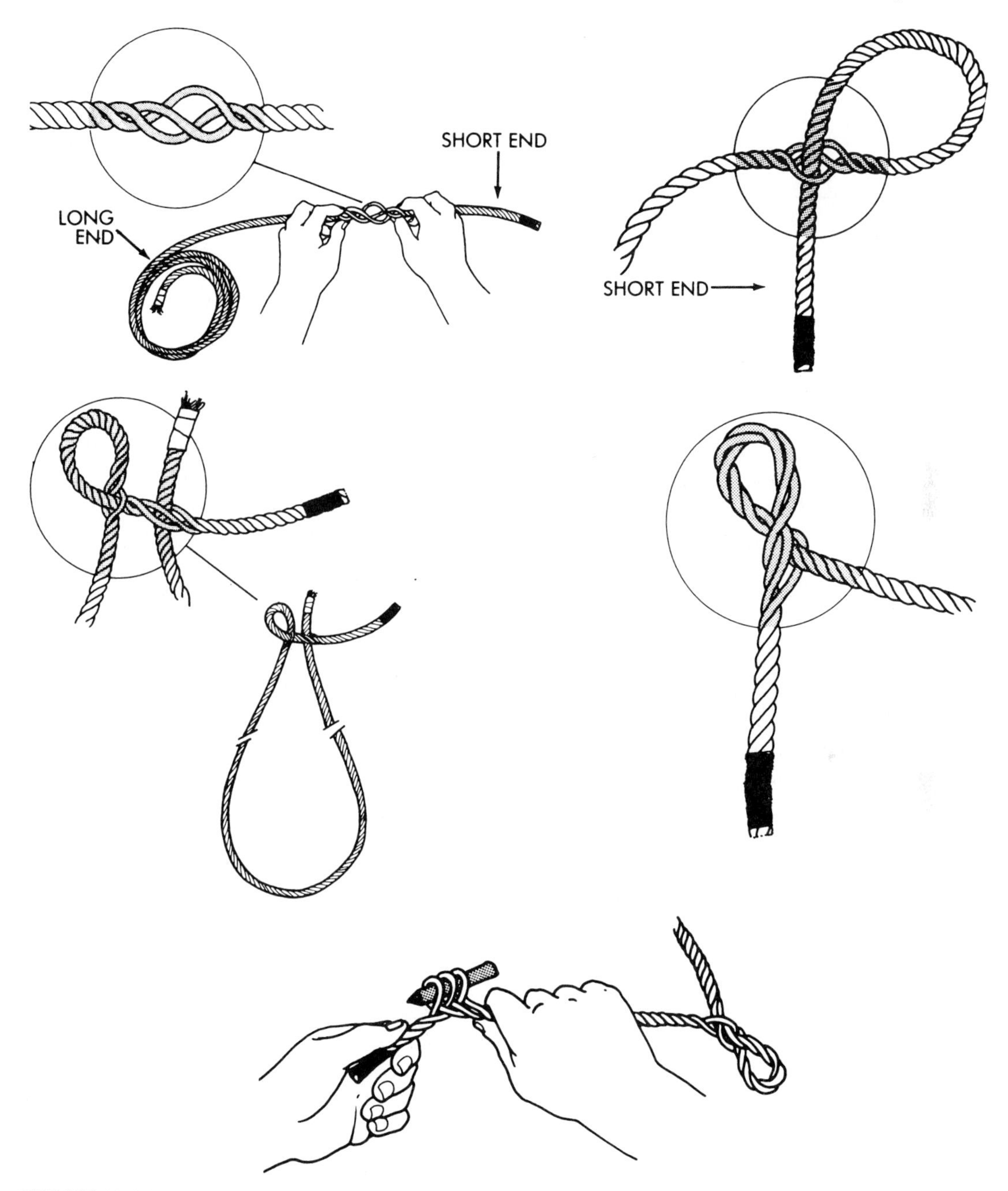

FIGURE 18.16 Diagrams and description for making an adjustable rope halter.
Source: Battaglia, R. A., and Mayrose, V. B., 2001. *Handbook of Livestock Management*. Upper Saddle River: Prentice Hall, ©.

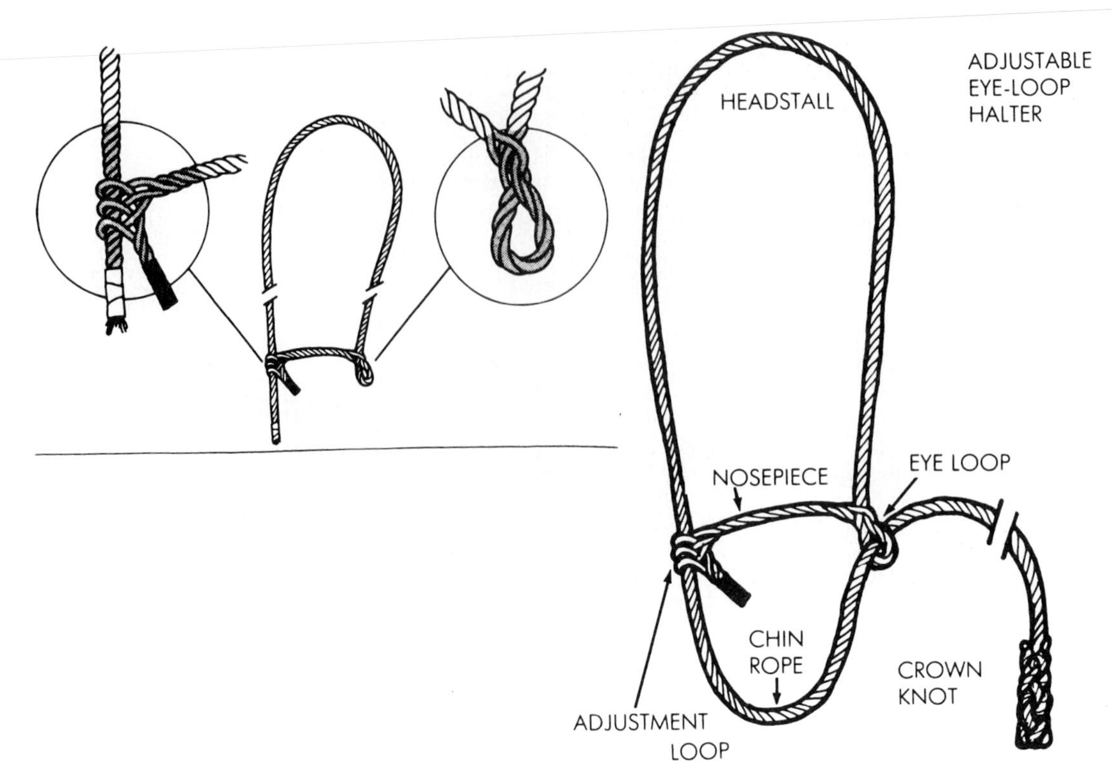

FIGURE 18.16 (Continued)

preference and the type of rope selected). Temporarily finish the other end of the rope with tape or string. A crown knot may be formed into this end after the halter is constructed.

3. Mark a point about 12–15 in. from the whipped end of the rope. Refer to this 12–15 in. length as the *short end* of the rope and to the remaining length as the *long end.*
4. Place the short end of the rope to your right and the long end to your left. Grasp the rope at the 12–15-in. mark between the thumb and first two fingers of both hands. Separate your right and left hands by about 2 in. Rotate the rope clockwise with your right hand and counterclockwise with your left hand in order to open the strands of the rope between your hands.
5. Isolate any one of the opened strands with the thumb and index fingers of your left hand. Use your right hand to insert the whipped end of the short end of the rope under this strand opening until the loop formed has an eye opening of about double the rope's diameter.
6. Now position the rope so that the eye loop is in your left hand, the short end points toward 3 o'clock, and the long end exits toward 6 o'clock. Grasp the eye loop and the single strand running across the short end of the rope between your left thumb and index finger. With your right thumb and index finger, grasp the short end of the rope at a point near the eye loop. Twist the eye loop and short end of the rope with your hands until you have isolated two strands between your right thumb and index finger.
7. Use your left hand to insert the long end of the rope from bottom to top under and through these two strands. Pull it completely through until all the slack is gone. If done properly, one side of the loop will show three strands lying smoothly side by side. This is important because they will be positioned against the animal's face.
8. With the eye loop to your right, grasp the short end of the rope between your left thumb and index finger about 2 in. from the whipped end. Two inches farther from the whipped end, grasp it in the same manner with your right hand. Open the strands by twisting

FIGURE 18.17 The use of a grooming chute allows for the humane restraint of the animal while still allowing the herdsman access to the animal for the purposes of fitting and grooming prior to a show.

clockwise with your right hand and counterclockwise with your left hand. When the strands are opened wide, push your hands together. This will cause the strands to "buckle" and fold over, forming three loops.

9. Line up these three loops and work into them a sharpened stick of a diameter equal to that of the rope. Use your right hand to feed the long end of the rope into the loops, starting at the one closest to the eye loop. Remove the stick from one loop at a time as you run the long end of the halter through them.
10. Run the long end of the rope into and through the eye of the loop to complete the halter.
11. Permanently finish the long end of the rope in the chosen manner. Consider crowning the end because a crown knot and back splice create a convenient handle.
12. Always place the halter on the animal so that the eye loop is on the left side. Lead from the left as well.

Grooming Chute The *grooming chute* is used only with cattle that are gentle and trained to lead. It is not meant to replace the squeeze chute or foot-trimming chute for restraint. A grooming chute is used to clip and prepare an animal for show or sale.

Many commercial and homemade designs are available for use. The best grooming chutes have a metal frame made of pipe or square tubing, a rough wood floor, a simple headgate mechanism, and swingable or removable side bars that allow easy access to the side of the animal (Fig. 18.17).

Cow-Calf Handling Facilities

Figure 18.18 shows some simple handling facilities for doctoring, marketing, weighing, and performing other tasks for small cow herds. A corral design for large cow herds is shown in Figure 18.19. As noted earlier in the chapter, cattle move easier through facilities with a

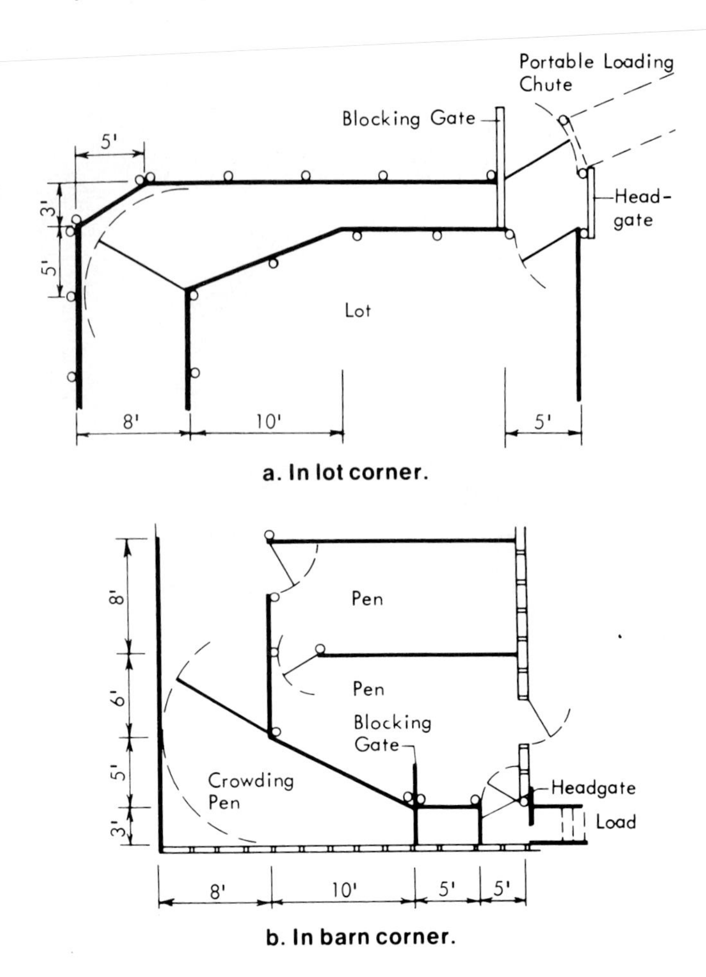

FIGURE 18.18 Simple layout of handling facilities for small herds. *Source: Beef Housing and Equipment Handbook.* MWPS-6, 4th ed., 1986 ©. Midwest Plan Service, Ames, Iowa 50011.

crowding area and chutes that are curved with solid sides. Figure 18.20 shows a basic curved handling facility for a feedlot or cow-calf operation.

AI facilities should permit cows to be handled quietly and carefully. Facilities where cows have previously felt pain should not be used for artificial insemination. Pregnancy rates are likely to be higher if stress is minimized by handling cattle calmly. AI facilities can be adapted to existing facilities (fences, pens, corrals, and gates) to make them more cost-effective. A breeding chute with a headgate can be used, although some producers prefer using a breeding box, which can be constructed separately or placed in front of existing chutes. As its name implies, the breeding box is a dark, quiet, solid-sided box that usually has a chain restrainer at its rear. A portable breeding box can be constructed to fit existing working facilities. It can be made from 2 × 6-ft uprights and half-inch plywood sides with a 3/4-in. plywood front, door, and top. If not slipped into a chute, the box can be held in place by four metal posts tied across the top with wire.

Dark box chutes can be used either singly or in a herringbone configuration when large numbers of cows are to be bred (Fig. 18.21). In some breeding facilities, a "pacifier" cow is put in a breeding box to keep the other cows calm. The cow to be inseminated rests her head on the pacifier cow's rump. Afterwards, the inseminated cow is released through a side gate and the pacifier cow remains in the box.

Cow-calf producers need useful equipment and facilities at calving time, particularly to assist first-calf heifers with difficult births. Some of this equipment is discussed in Chapter 11.

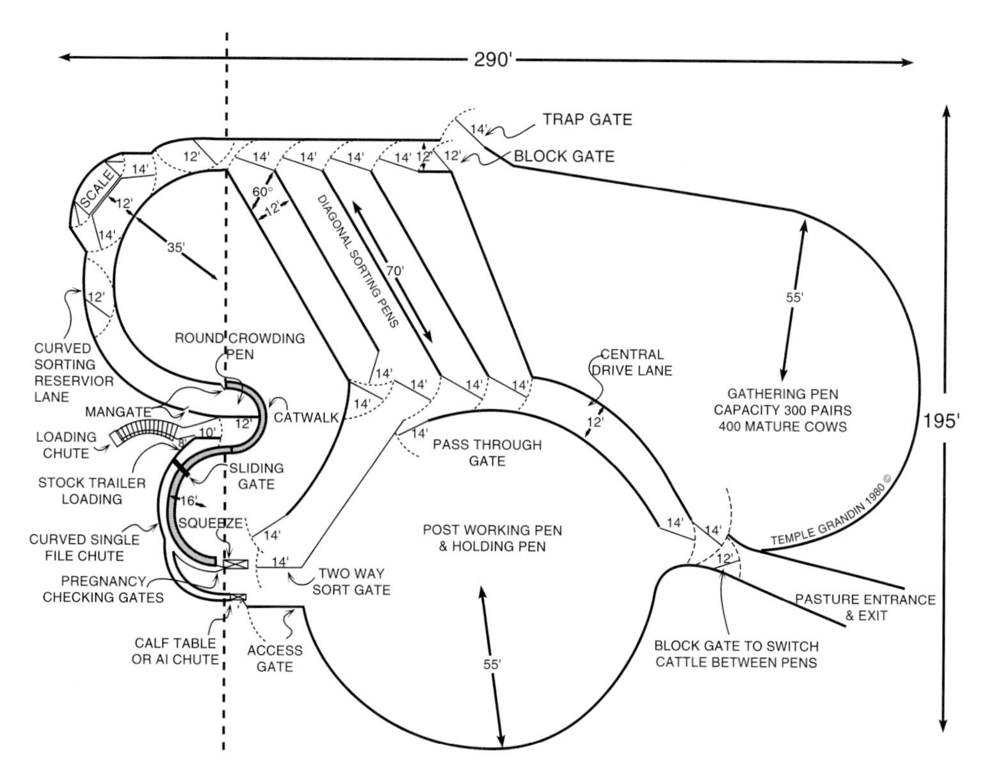

FIGURE 18.19 Corral system for a large ranch.
Source: Temple Grandin.

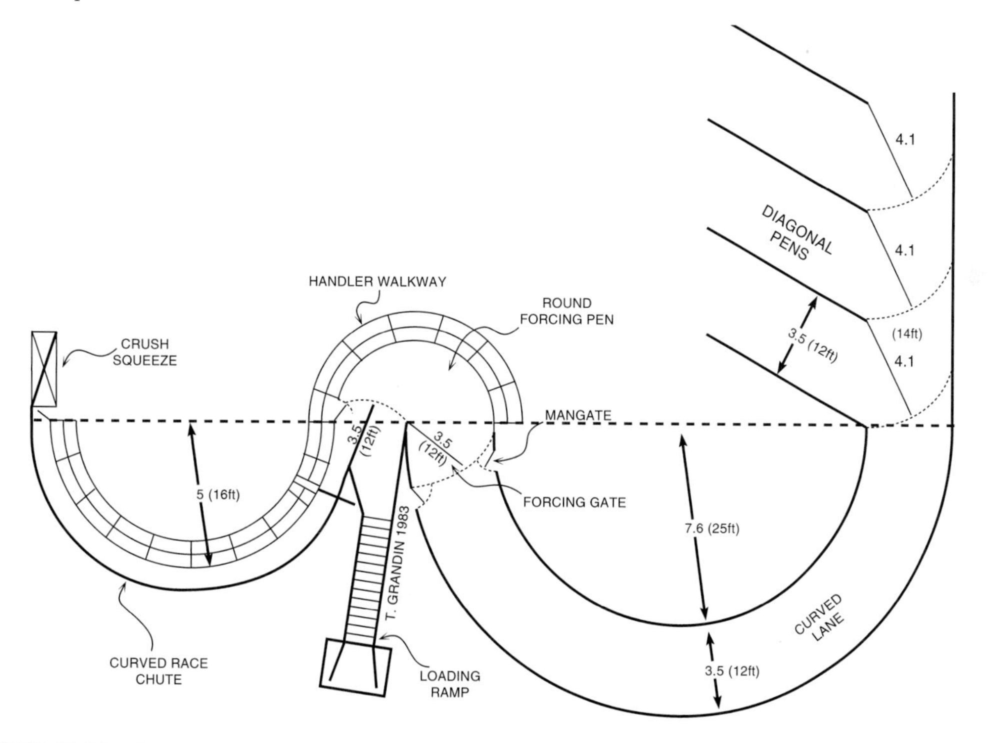

FIGURE 18.20 Basic curved handling facility.
Source: Temple Grandin.

FIGURE 18.21 Herringbone configuration of breeding boxes to allow several cows to be ready for breeding or palpation at one time.

FIGURE 18.22 A straight-sided head catch with swinging gates is excellent equipment to assist with difficult births at calving time. Note that the sides of the chute can be swung out to allow more freedom of movement for both the cow and the handler during an assisted birth.

Figure 18.22 shows an excellent head catch with gates that help restrain the female—yet swing out of the way when calving assistance is being given.

Feedlot Facilities

Figure 18.23 shows several items that could be considered in the construction of facilities for a commercial feedlot operation. Mounds to keep cattle dry and out of the mud, location of feed bunks and waterers, sick pens, feed alleys, and working alleys for cattle are all important

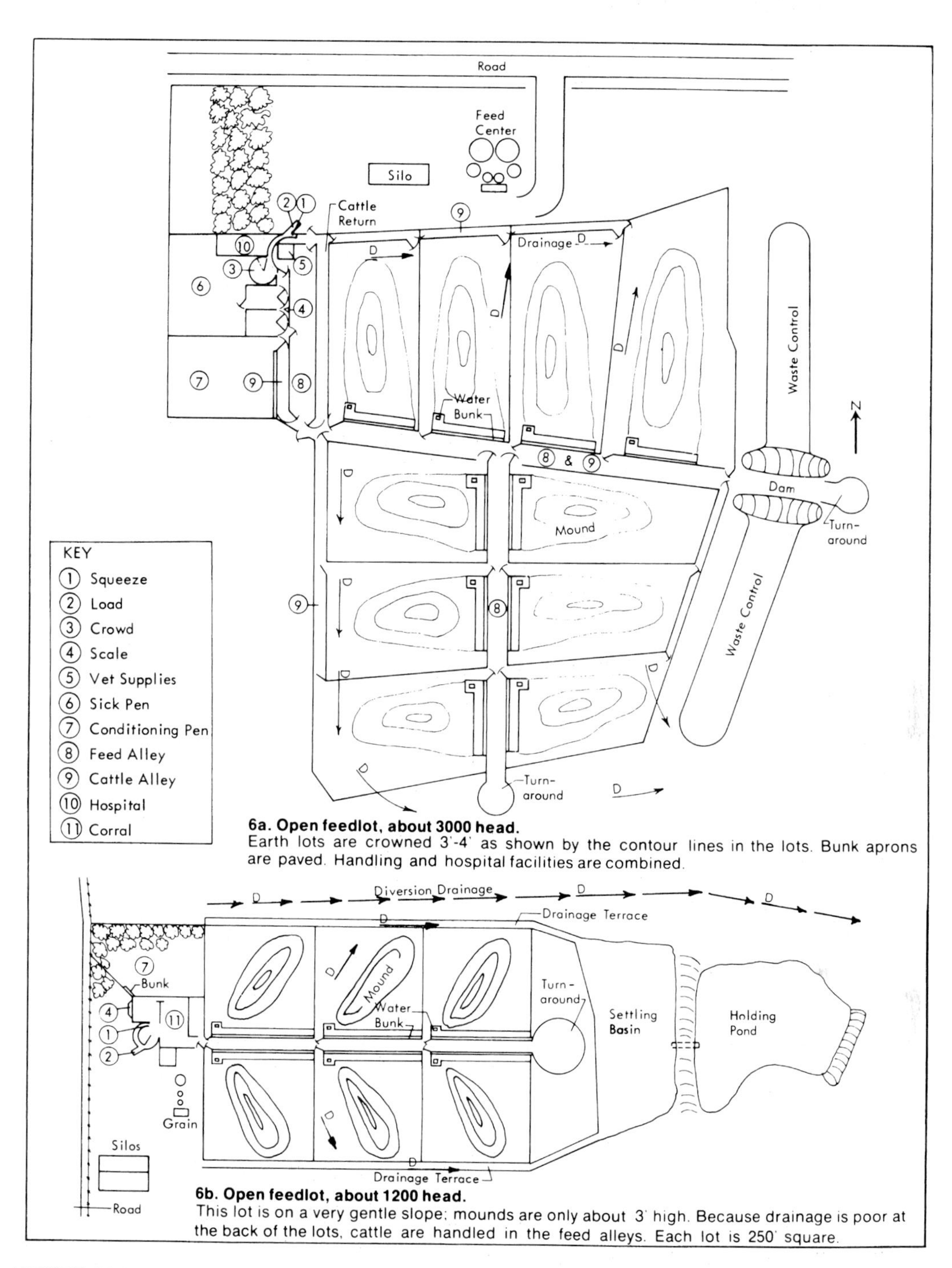

FIGURE 18.23 Open feedlot layout for a 3,000-head capacity feedlot. For good drainage, feedlot pens should have a 2–3% slope away from the feedbunk to help keep the pens dry.
Source: Beef Housing and Equipment Handbook. MWPS-6, 4th ed. 1986. Midwest Plan Service, Ames, IA 50011.

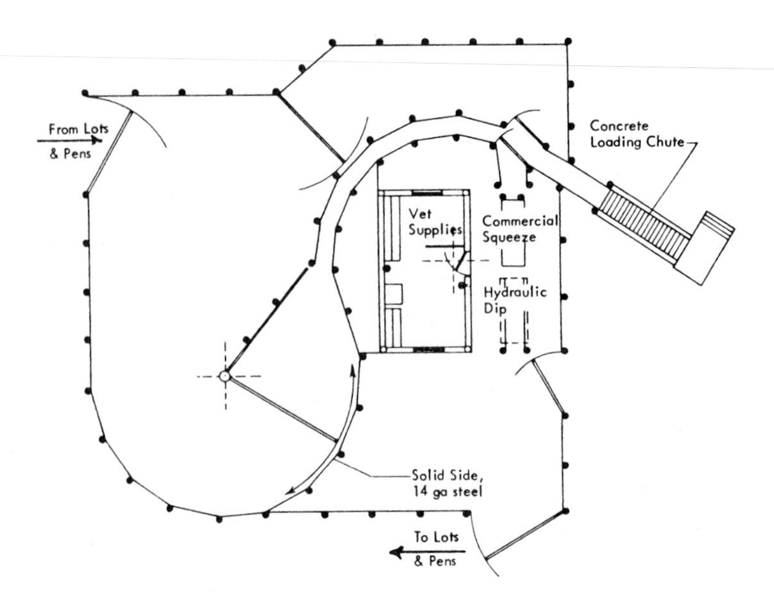

FIGURE 18.24 Working area for a feedlot.
Source: Beef Housing and Equipment Handbook, MWPS-6, 4th ed. 1986 ©. Midwest Plan Service, Ames, Iowa 50011.

considerations. Well-drained pens with waste control are necessary for feedlots to meet government pollution regulations and to prevent disease in cattle. Feed storage and processing, cattle loading and unloading facilities, and cattle processing facilities are all necessary for a cattle feeding operation. It is important that many of these processes operate simultaneously in order to use labor and equipment efficiently. A selected working area design for a feedlot is shown in Figure 18.24.

MANAGEMENT SYSTEMS HIGHLIGHTS

1. Understanding the behavior of beef cattle can assist producers in making better management decisions.
2. Effective use of the flight zone and point-of-balance assists managers in handling cattle with less stress and fewer injuries to animals or people.
3. Designing cattle-handling facilities suited to the behavioral tendencies of the animal helps to assure more humane and efficient restraint and processing.

REFERENCES

Publications

Battaglia, R. A., and Mayrose, V. B. 2001. *Handbook of Livestock Management.* 3rd ed. Upper Saddle River, NJ: Prentice Hall.
Beef Housing and Equipment Handbook. 1986. Ames, IA: Midwest Plan Service, Iowa State University.
Blockey, M. A. de B. 1976. Sexual Behavior of Bulls at Pasture: A Review. *Theriogenology* 6: 387.
Craig, J. V. 1981. *Domestic Animal Behavior.* Englewood Cliffs, NJ: Prentice-Hall.
Curtis, S. E., and Houpt, K. A. 1983. Animal Ethology: Its Emergence in Animal Science. *J. Anim. Sci.* 57: 234 (Suppl. 2).
Ewing, S. A., Lay, D. C., and Von Borrell, E. 1999. *Farm Animal Well-Being, Stress Physiology, Animal Behavior and Environmental Design.* Upper Saddle River, NJ: Prentice Hall.
Grandin, T. 1989. Behavioral Principles of Livestock Handling. *Prof. Anim. Sci.* 5: 1.
Grandin, T. 1997. Assessment of Stress During Handling and Transport. *J. Anim. Sci.* 75: 249.

Grandin, T. 2000. *Livestock Handling and Transport.* 2nd ed. CAB International, Wallingford, United Kingdom.

Hernsworth, P. H., and Coleman, G. J. 1998. *Human Livestock Interactions.* CAB International, Wallingford, United Kingdom.

Kidwell, J. F., Bohman, V. R., and Hunter, J. E. 1954. Individual and Group Feeding of Experimental Cattle as Influenced by Hay Maturity. *J. Anim. Sci.* 13: 543.

Kilgour, R., and Dalton, C. 1983. *Livestock Behavior: A Practical Guide.* Boulder, CO: Westview Press.

Odde, K. G. 1983. *The Postpartum Cow: Effect of Early Weaning and Limited Suckling on Cow and Calf Performance, and Suckling Behavior in Range Calves.* Ph.D. Thesis, Kansas State University.

Pierson, R. E., Jensen, R., Brady, P. M., Horton, D. A., and Christie, R. M. 1976. Bulling Among Yearling Feedlot Steers. *J. Amer. Vet. Med. Assoc.* 169: 521.

Ray, D. E., and Roubicek, C. B. 1971. Behavior of Feedlot Cattle During Two Seasons. *J. Anim. Sci.* 33: 72.

Rile, R. W., MacNeil, M. D., Jenkins, T. G., and Koong, L. J. 1982. *A Simulation Model of Grazing Behavior in Cattle.* Proceedings of the Western Section American Society of Animal Science.

Rupp, G. P., Ball, L., Shoop, M. C., and Chenoweth, P. J. 1977. Reproductive Efficiency of Bulls in Natural Service: Effects of Male to Female Ratio and Single vs. Multiple-Sire Breeding Groups. *J. Amer. Vet. Med. Assoc.* 171: 639.

Schake, L. M., and Riggs, J. K. 1970. Activities of Beef Calves Reared in Confinement. *J. Anim. Sci.* 31: 414.

Schake, L. M., and Riggs, J. K. 1972. *Behavior of Beef Cattle in Confinement.* College Station, TX: Texas Agricultural Experiment Station Technical Report 27.

Sherry, C. J., Klemm, W. R., Sis, R. F., and Schake, L. M. 1982. *Reproductive and Feedlot Behavior: The Role of the Vomeronasal Organ.* College Station, TX: Beef Cattle Research in Texas, PR 3923.

Voisinet, B. D., Grandin, T., Tatum, J. D., O'Connor, S. F., and Struthers, J. J. 1997. *Bos indicus*-cross Feedlot Cattle with Excitable Temperaments Have Tougher Meat and a Higher Incidence of Borderline Dark Cutters. *Meat Science* 46: 367–377.

Voisinet, B. D., Grandin, T., Tatum, J. D., O'Connor, S. F., and Struthers, J. J. 1997. Feedlot Cattle with Calm Temperaments Have Higher Average Daily Gains Than Cattle with Excitable Temperaments. *J. Anim. Sci.* 75: 892–896.

Wagnon, K. A. 1965. *Social Dominance in Range Cows and Its Effects on Supplemental Feeding.* Davis, CA: California Agricultural Experiment Station Bull. 819.

Wagnon, K. A., Loy, R. G., Rollins, W. C., and Carroll, F. D. 1966. Social Dominance in a Herd of Angus, Hereford, and Shorthorn Cows. *Anim. Behavior* 14: 474.

Warnick, V. C., Arave, C. W., and Mickelsen, C. H. 1977. Effects of Group, Individual, and Isolation Rearing of Calves in Weight Gain and Behavior. *J. Dairy Sci.* 60: 947.

Wyatt, R. D., Gould, M. B., and Totusek, R. 1977. Effects of Single vs. Simulated Twin Rearing on Cow and Calf Performance. *J. Anim. Sci.* 45: 1409.

Visuals

"Bovine Restraint" (video 13 min.). Instructional Media Center, Michigan State University, East Lansing, MI 48824.

"Cattle Handling Principles to Reduce Stress" (video 58 min.). Temple Grandin, Fort Collins, CO.

"Management Practices for Beef Cattle—I" (video 25 min.). CEV, P.O. Box 65265, Lubbock, TX 79464-5265.

Websites

Livestock Behavior, Design of Facilities and Humane Slaughter : www.grandin.com

Improved Livestock Trailer Design: www.gov.on.ca:80/OMAFRA/English/livestock/beef/cushion/livestockcushion.htm

Managing Information Resources

Valid information is needed to make effective management decisions. The primary sources of information are (1) information generated from within the operation (presented primarily in Chapter 3) and (2) information that comes from outside the operation. This latter source of information is covered in this chapter. If useful information is obtained and implemented, changes can be effectively managed. The most critical tests for determining the usefulness of information for the current beef industry are (1) if the information will help lower production costs (e.g., breakevens) and (2) if the information will help increase consumer market share (e.g., the percent of total meat expenditures spent for beef).

New information usually provides the stimulus for change. It has been noted that cattle producers may be engulfed with change or engulfed because they do not change. Change does not come without frustration; as one producer puts it, "I'm all for progress; it's just those changes I don't like." Another remarked, "The best way to tell if change is present is to determine how uncomfortable you feel." However, it is comforting to know that change can be effectively managed.

Not all change is good. Undirected or mismanaged change may be more disastrous than no change at all. The balance needed is summarized in the statement, "Tradition for tradition's sake is no worse than change for change's sake." One needs to reflect occasionally on the purpose and source of change. People are motivated to seek the unknown so they can understand and influence the world around them. For example, the application of the biological principles of nutrition, health, and genetics has resulted in more productive cattle and beef products that are lower in cost, more highly palatable, safe, and nutritious. The application of financial and economic principles has reduced production costs and increased continuing net profit.

Fluctuating economic conditions dictate a considerable amount of change in the beef industry. Economic conditions can change slowly or rapidly. The latter may create the need for immediate change.

Intelligent producers recognize that change is inevitable and can be managed by a continual learning process. Those who handle change best are often involved in directing change. They are the leading problem solvers and the innovators.

How much more knowledge (true relationships) is waiting to be discovered? No doubt Sir Isaac Newton, who discovered many great truths, had vision and understanding when he said, "I am but a child playing on the sands by the sea. I have found a few pretty pebbles washed by the spray, but beyond me, unfathomed and unexplored, lies the great ocean of whose depth and mysteries I know nothing." A considerable amount of human effort is spent identifying, teaching, and learning true principles and relationships.

Claims have been made that there have been more scientific achievements and advancements in technology during the past 150 years than at any other recorded time in world history. This explosion of knowledge continues at a rapid pace in a society that is increasing in specialization. Therefore, producers and others in the beef industry find themselves surrounded with voluminous amounts of information. Some of this information can or will have a direct effect on their specific operations by changing levels of productivity or changing the economic environment in the area, country, or world.

The extensive amount of available information and limitations of time force an individual to set priorities as to how much time and money can be spent with the many sources of information. Table 19.1 gives guidelines in managing the information resource most effectively. The management plan that is developed and continually assessed for a specific operation should help set these priorities.

The importance of having useful information is best summarized by the following statement: "No decision is better than the information on which it is based."

It is important to recognize that not all written or verbalized information is valid or useful. Some research reports contain gross errors while some popular publications capitalize on the emotional and sensational. Occasionally, information is highly selective. The decision maker must move from the "opinion type" or "selected type" of information to the "verified type" of information. Information must be critically evaluated in terms of its validity and reliability in providing useful information before applying it to a specific operation. Some primary sources of information for the decision maker are shown in Figure 19.1.

Research uses the scientific method to discover new knowledge (see Fig. 19.2). Concepts and ideas that have cause-and-effect relationships are added to the pool of known information. The educational process makes these principles known so that people can understand and effectively apply them. The purpose of the scientific method in beef production is to analyze well-designed experiments to better understand the biological relationships in cattle. Producers need to know, for example, how the treatment of cattle affects the predictability of their responses.

TABLE 19.1 Components of Effective Information Management

- Ask the right questions (focus on mission statement and goals)
- Identify true principles
- Must be cost and time effective
- Information from inside the business
  - — written records and reports
  - — observations of/from team members
- Information from outside the business
  - — publications
  - — people

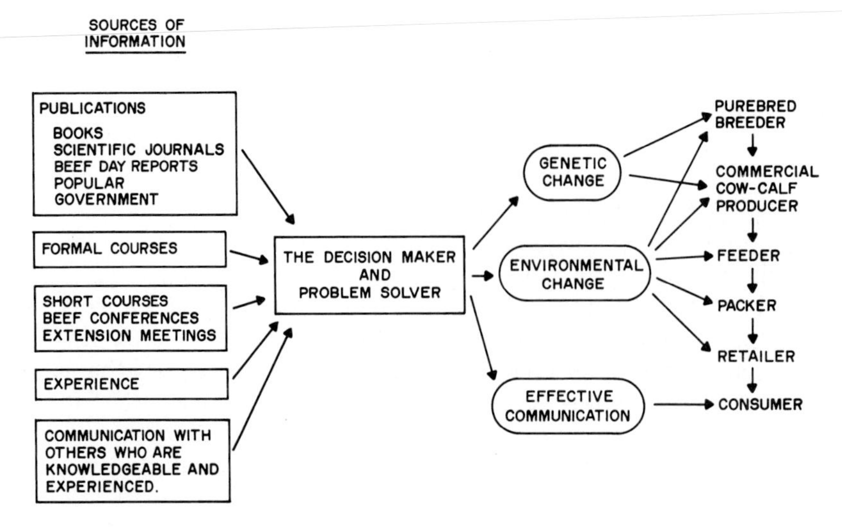

FIGURE 19.1 Primary sources of information which implement change in the various beef industry segments.
Source: Colorado State University.

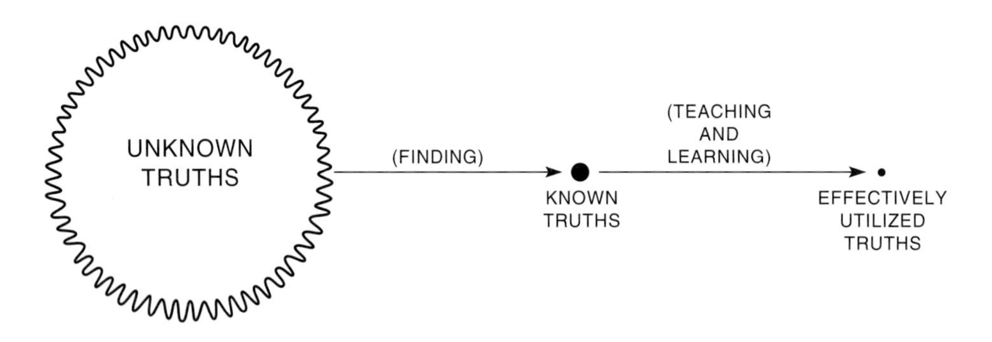

FIGURE 19.2 Human process of finding and utilizing true relationships. Size of each area approximated for demonstrational purposes. The magnitude of the unknown truths cannot be quantified at this time.
Source: Colorado State University.

Evaluating statistical information in research reports is described in the 1996 Kansas State University Cattlemen's Day Report. It is quoted here for reference:

> The variability among individual animals in an experiment creates problems in interpreting the results. Although the cattle on treatment X may have had a larger average daily gain than those on treatment Y, variability within treatments may mean that the difference was not the result of the treatment alone. You can never be totally sure that the difference you observe is due to treatment, but statistical analysis lets researchers calculate the probability that such differences are from chance rather than treatment.
>
> In some of the articles that follow you will see the notation ($P < .05$). That means the probability of the differences resulting from chance is less than 5%. If two averages are said to be "significantly different," the probability is less than 5% that the difference is from chance—the probability exceeds 95% that the difference results from the treatment.

Some papers report correlations measures of the relationship between traits. The relationship may be positive (both traits tend to get bigger or smaller together) or negative (as one gets bigger, the other gets smaller). The perfect correlation is either +1 or −1. If there is no relationship at all, the correlation is zero.

You may see an average given as 2.50 ± .10. The 2.50 is the average; .10 is the "standard error." That means there is a 68% probability that the "true" mean (based on an unlimited number of animals) will be between 2.4 and 2.6. "Standard deviation" is a measure of variability in a set of data. One standard deviation on each side of the mean is expected to contain 68% of the observations.

Many animals per treatment, replicating treatments several times with pens of animals, and using uniform animals increase the probability of showing the real differences when they actually exist. Statistical analysis allows more valid interpretation of the results regardless of the number of animals in the experiment. In the research reported herein, statistical analyses are included to increase the confidence you can place in the results.

SOURCES OF INFORMATION ON BEEF CATTLE

The primary sources of information are people, hard-copy publications, and the electronic media. Some of these sources identify principles that do not change, while other sources convey beef cattle management practices that may frequently change.

People

People are excellent sources of information based on what they have read and analyzed, their communication with other people, and their experiences in leadership and management. Outstanding managers have a network of people with whom they periodically synergize in sharing useful information. These people usually include other managers, industry leaders, marketing specialists, and university research and extension specialists. Development of a credible and effective human network is one of the most important activities for managers and leaders in the beef industry.

Publications

Books provide broad coverage of beef production and the beef industry. They cover the basic principles of nutrition, breeding, reproduction, meats, and others in addition to information on other resources and beef industry segments. Refer to the References that appear at the end of each chapter in this text as they provide additional sources of information on beef cattle and related resources.

Books are excellent sources of material for an overview of basic principles. The information provided by books, however, is often not as current compared with information contained in periodicals, particularly when the current information on the beef industry is needed.

Scientific periodicals (Fig. 19.3) typically contain research articles that are published monthly or several times a year by universities and industry. These publications usually are peer reviewed for validity of the research—experimental design, statistical analysis, and interpretation. Many of these articles are most useful for contribution of basic knowledge with limited application to the beef production system. The most applicable articles are those that review the previous research on a specific subject.

Popular periodicals (Figs. 19.4–19.6) are usually weekly or monthly publications that are extensively utilized by beef producers, processors, and other industry personnel. Articles can

FIGURE 19.3 These journals are examples of scientific publications in which beef cattle research results are published.
Source: Colorado State University.

FIGURE 19.4 Selected monthly publications that help beef producers access information resources.
Source: Colorado State University.

FIGURE 19.5 Selected weekly publications that emphasize current articles, sales, and market reports.
Source: Colorado State University.

FIGURE 19.6 Selected breed publications containing information primarily for purebred breeders.
Source: Colorado State University.

FIGURE 19.7 Publications of the National Cattlemen's Beef Association.
Source: National Cattlemen's Beef Association.

cover several different resource areas and industry trends. Some articles are written with the interpretation by an author associated with the publication or by a researcher who is highlighting some recent research. Caution should be exercised with some articles as they may represent "linear thinking" with selected information that may be more sensational than factual. The National Cattlemen's Beef Association publishes several magazines and newsletters (Fig. 19.7) designed to inform producers about issues, industry trends, political and regulatory decisions, and market information.

Proceedings from conferences, short courses, seminars, and symposia (Fig. 19.8) are usually published and available free of charge or at a minimal cost. Several organizations hold meetings where speakers address current beef industry topics, which may be narrowly focused or cover a wide array of beef-related resources. See examples in the section on Cattle Resources later in the chapter.

Additional *university publications* include experiment station bulletins, extension service bulletins, newsletters, annual beef research reports (Fig. 19.9), handbooks (Fig. 19.10), and fact sheets. Many of these are semi-technical research reports and extension publications. They usually do not meet the same criteria as scientific publications, where a detailed statistical analysis and peer review has been done prior to publication. Most universities have Web sites or catalogs that list the research, extension bulletins, and other publications that are available.

Information about university publications on beef cattle can be obtained from extension beef specialists located at land-grant universities. Addresses of land-grant universities are included to obtain publications and to contact beef extension and other specialists in agricultural economics, forage (agronomy and range science), and veterinary medicine. Extension beef specialists can be contacted at the animal science departments of the universities that are

FIGURE 19.8 Proceedings from conferences and short courses provide useful reference material for many beef producers.
Source: Colorado State University.

FIGURE 19.9 Selected beef research reports published annually by land-grant universities (departments of animal science) in states where beef cattle are of primary importance.
Source: Colorado State University.

FIGURE 19.10 Beef cattle handbooks are available from beef cattle extension specialists in several states. *Source:* Colorado State University.

listed. An Internet link to cooperative extension services in all states can be utilized through www.reeusda.gov.

Auburn University
Auburn, AL 36830

University of Arizona
Tucson, AZ 85721

University of Arkansas
Fayetteville, AR 72701

University of California
Davis, CA 95616

Colorado State University
Fort Collins, CO 80523

University of Connecticut
Storrs, CT 06268

University of Delaware
Newark, DE 19711

University of Florida
Gainesville, FL 32611

University of Georgia
Athens, GA 30602

University of Hawaii
Honolulu, HI 96822

University of Kentucky
Lexington, KY 40506

Louisiana State University
Baton Rouge, LA 70893

University of Maine
Orono, ME 04473

University of Maryland
College Park, MD 20742

University of Massachusetts
Amherst, MA 01002

Michigan State University
East Lansing, MI 48824

University of Minnesota
St. Paul, MN 55108

Mississippi State University
Mississippi State, MS 39762

University of Missouri
Columbia, MO 65201

Montana State University
Bozeman, MT 59717

University of Illinois
Urbana, IL 61801

Purdue University
West Lafayette, IN 47907

Iowa State University
Ames, IA 50011

Kansas State University
Manhattan, KS 66506

New Mexico State University
Las Cruces, NM 88003

Cornell University
Ithaca, NY 14853

North Carolina State University
Raleigh, NC 27607

North Dakota State University
Fargo, ND 58102

Ohio State University
Columbus, OH 43210

Oklahoma State University
Stillwater, OK 74074

Oregon State University
Corvallis, OR 97331

Pennsylvania State University
University Park, PA 16802

University of Rhode Island
Kingston, RI 02881

Clemson University
Clemson, SC 29631

University of Nebraska
Lincoln, NE 68503

University of Nevada
Reno, NV 89507

University of New Hampshire
Durham, NH 03824

State University of New Jersey
New Brunswick, NJ 08903

South Dakota State University
Brookings, SD 57007

University of Tennessee
Knoxville, TN 37901

Texas A&M University
College Station, TX 77843

Utah State University
Logan, UT 84322

University of Vermont
Burlington, VT 05401

Virginia Polytechnic Institute
Blacksburg, VA 24061

Washington State University
Pullman, WA 99164

West Virginia University
Morgantown, WV 26506

University of Wisconsin
Madison, WI 53706

University of Wyoming
Laramie, WY 82071

The Agriculture Research Service (ARS) of the USDA has several beef cattle research centers and laboratories located throughout the United States. Their research is published periodically in cooperation with several universities or in such publications as the following:

Beef Research Program Progress Reports, No. 1 (1982), No. 2 (1985), No. 3 (1988), No. 4 (1993), and No. 16 (1997). U. S. Meat Animal Research Center, P.O. Box 166, Clay Center, NE 68933.

Research for Rangeland Based Beef Production, 1990. Fort Keough Livestock and Range Research Laboratory and Montana State University, Rt. 1, Box 2021, Miles City, MT 59301.

THE ELECTRONIC MEDIA

Electronic media includes the Internet, fax, video/television, CD-ROM, and satellite broadcasting. Several of these can be used in combination. Some of the useful videos, CD-ROM programs, and Web sites are listed at the end of chapters throughout the book.

The Internet

The Internet is referred to as the Information Superhighway. It allows communication throughout the world and can be used to access vast amounts of information including text, graphics, sound, and video. Someone has said that the Information Superhighway is not yet perfectly paved so one can expect some bumps, detours, and roadblocks.

The World Wide Web may be the most complete realization of the Internet to date. It is a graphical environment that can be navigated through hyperlinks. From one site, a person can click on hyperlinks to go to any number of related sites. The address site on the Web usually begins with http://www. The end of the Web address and other Internet addresses is called the domain. Examples of domain names are com (a commercial organization, business, or company), edu (an educational institution), int (an international organization), gov (a nonmilitary government entity), mil (a military organization), net (a network administration), and org (other organizations).

U. S. Government Information

The U. S. Department of Agriculture (USDA) has a large database on beef cattle and related agricultural products. The Economic Research Service (ERS), National Agricultural Statistics Service (NASS), and Foreign Agricultural Service (FAS) of the USDA each have home pages on the World Wide Web that provide easy access to the wide range of information and data produced. Following is a list of these Internet sites:

ERS http://www.ers.usda.gov
NASS http://www.usda.gov/nass
FAS http://www.fas.usda.gov

Through a cooperative project with Cornell University, the Albert R. Mann Library distributes ERS, NASS, and World Agricultural Outlook Board (WAOB) periodicals and data. Internet users can access more than 400 reports annually and more than 7,500 data files free of charge. All NASS reports, ERS situation and outlook reports (or their summaries), and WAOB's World Agricultural Supply and Demand Estimates are available electronically within three hours of release. The USDA Economics and Statistics System at the Mann Library can be accessed at http://usda.mannlib.cornell.edu. There is also the option of receiving reports via e-mail.

INFORMATION FOR MANAGEMENT SYSTEMS

Books on Integrated and Holistic Systems

Deming, W. E. 1994. *The New Economics* (Chap. 3, Introduction to a System). Cambridge, MA: Massachusetts Institute of Technology.
Savory, A. 1999. *Holistic Management: A New Framework for Decision Making.* Washington, DC: Island Press.
Senge, P. M. 1990. *The Fifth Discipline.* New York: Doubleday. Also available as a cassette tape series.
Taylor, R. E., and Field, T. G. 1999. *Beef Production and Management Decisions.* Upper Saddle River, NJ: Prentice-Hall.

BEEF INDUSTRY OVERVIEW

Periodicals

Beef Business Bulletin. Published weekly for NCBA members. National Cattlemen's Beef Association (NCBA), Box 3469, Englewood, CO 80155. Internet: http://www.beef.org; click on NCBA; click on Members Only; click and enter user name and password (membership number at top of mailing label).

National Cattlemen. P.O. Box 3469, Englewood, CO 80155. Monthly publication of the National Cattlemen's Beef Association. Internet: http://www.beef.org; click on NCBA; click on Members Only; click and enter user name and password (membership number at top of mailing label).

State Livestock and Crop Statistics. Most states have a Crop and Livestock Reporting Service within the State Department of Agriculture that cooperates with the USDA. An annual agricultural statistics book that lists cattle numbers, prices, and so on is usually published by states. The addresses of departments of agriculture for specific states can be obtained by contacting the National Association of State Departments of Agriculture, 1616 H St., N.W., Washington, DC 20006.

Internet Links

http://www.cowtown.org/
http://www.beef.org/ (click on library; click on hot links)

CONSUMER/BEEF PRODUCTS

Books

Anderson, B. A., and Hoice, I. M. 1990. *Composition of Foods: Beef Products.* Agriculture Handbook 8-13. Washington, DC: USDA.

Hedrick, H., Aberle, E., Forrest, J., and Judge, M. 1994. *Principles of Meat Science.* Dubuque, IA: Kendall/Hunt.

Romans, J. R., Costello, W. J., Carlson, C. W., Greaser, M. L., and Jones, K. W. 2001. *The Meat We Eat.* Danville, IL: Interstate Publishers.

Popular Periodicals

Food Review. Washington, DC, USDA: http://www.foodreview.com
Meat and Poultry. P.O. Box 1959, Mill Valley, CA 94942.
Supermarket News. 7 East 12th St., New York, NY 10003.

Scientific Periodicals

Journal of Animal Science. American Society of Animal Science, 1111 North Dunlap Ave., Savoy, IL 61874.
Journal of Food Science. Institute of Food Technologists, Suite 2120, 221 N. LaSalle St., Chicago, IL 60601.
Meat Science. Elsevier Applied Science Publishers, Ripple Road, Barting, Essex, England.

Internet

American Meat Institute: www.meatami.com
The Food Marketing Institute: www.fmi.org

HUMAN RESOURCES

Books

Covey, S. R. 1989. *The 7 Habits of Highly Effective People.* New York: Simon and Schuster.
Covey, S. R. 1991. *Principle-Centered Leadership.* New York: Simon and Schuster.
Covey, S. R. 1994. *First Things First.* New York: Simon and Schuster.
Gersick, K. E., Davis, J. A., Hampton, M. M., and Lansberg, I. 1997. *Generation to Generation: Life Cycle of the Family Business.* Boston: Harvard Business School Press.

Kotter, J. P. 1996. *Leading Change.* Boston: Harvard Business School Press.
O'Toole, J. 1995. *Leading Change.* San Francisco: Jossey-Bass Publishers.
Peters, T. J. 1997. *The Circle of Innovation.* New York: Alfred A. Knopf. www.tompeters.com.
Peters, T. J., and Waterman, R. H., Jr. 1982. *In Search of Excellence.* New York: Harper and Row.

FINANCIAL/ECONOMIC RESOURCES

Books

Boehlje, M. D., and Erdman, V. R. 1984. *Farm Management.* New York: John Wiley & Sons.
Deming, W. E. 1994. *The New Economics.* Cambridge, MA: Massachusetts Institute of Technology.
Kay, R. D., and Edwards, W. M. 1998. *Farm Management.* New York: McGraw-Hill.
Lee, W. F., Boehlje, M. D., Nelson, A. G., and Murray, W. C. 1998. *Agricultural Finance.* Ames, IA: Iowa State University Press.

Popular Periodicals

Choices. The magazine of food, farm, and resource issues. Published quarterly by the American Agricultural Economics Assoc., Iowa State University, 80 Healy Hall, Ames, IA 50011.

Internet

Ag Econ-Agribusiness: www.agecon.ksu.edu
Agriculture and Business Management Resource Manual:
www.colostate.edu/Depts/CoopExt/NWR/Abm/abmndx.htm
Farm Business Management and Marketing: www.ext.vt.edu/resources/
Farm Financial Management Resources: www.agecon.okstate.edu/ffmr.htm
Western Rural Development Center: http://extension.usu.edu/wrdc/

FORAGE RESOURCES

Books

Barnes, B. F., Miller, D. A., and Nelson, C. J. (eds.). 1995. *Forages. An Introduction to Grassland Agriculture.* Vols. I and II. Ames, IA: Iowa State University Press.
Holechek, J. L., et al. 1989. *Range Management Principles and Practices.* Englewood Cliffs, NJ: Prentice-Hall.
Smith, B., Leung, P., and Love, G. 1986. *Intensive Grazing Management: Forage, Animals, Men, Profits.* Kingsbery Communications, P.O. Box 1988, Woodinville, WA 98072.
Valentine, J. F. 1989. *Range Developments and Improvements.* New York: Academic Press.
Valentine, J. F. 1990. *Grazing Management.* New York: Academic Press.

Popular Periodicals

Hay and Forage Grower. P.O. Box 12951, Overland Park, KS 66282.
Rangelands. Society for Range Management, 445 Union Blvd., Suite 230, Lakewood, CO 80228-1259.
Stockman Grass Farmer. P.O. Box 2300, Ridgeland, MS 39157.

Scientific Periodicals

Journal of Range Management. Society of Range Management, 445 Union Blvd., Suite 230, Lakewood, CO 80228-1259.

Internet

Eco Results: Working Together to Heal the West: www.ecoresults.org
Morgans Forage Site contains information about forage, hay, silage, and pasture: www.forage.com
Natural Resources Conservation Service: www.nres.usda.gov
Samuel Roberts Noble Foundation: www.noble.org
The Allen Savory Center for Holistic Management: www.holisticmanagement.org

CATTLE RESOURCES

Books

Albin, R. C., and Thompson, 1990. *Cattle Feeding: A Guide to Management.* Amarillo, TX: Trafton Printing.
Bearden, H. J., and Fuquay, J. W. 1997. *Applied Animal Reproduction.* Upper Saddle River, NJ: Prentice-Hall.
Boggs, D. L., and Merkel, R. A. 1993. *Live Animal and Carcass Evaluation.* Dubuque, IA: Kendall/Hunt.
Bourdon, R. M. 1997. *Understanding Animal Breeding.* Upper Saddle River, NJ: Prentice-Hall.
Cattlemen's Library. 1997. University of Idaho, Ag Science Room 10, Moscow, ID 83844.
Ensminger, M. E., and Perry, R. C. 1997. *Beef Cattle Science.* Danville, IL: Interstate Publishers.
Field, M. J., and Sand, R. S. (eds.). 1994. *Factors Affecting Calf Crop.* Boca Raton, FL: CRC Press.
Guidelines for Uniform Beef Improvement Programs. 1996. Beef Improvement Federation, Ronnie Silcox, Animal and Dairy Science Complex, University of Gerogia, Athens, GA 30602-2771.
Jurgens, M. H. 1997. *Animal Feeding and Nutrition.* Dubuque, IA: Kendall/Hunt.
Nutrient Requirements of Beef Cattle. 1996. Washington, DC: National Research Council.
Price, D. P. 1992. *Real World Answers to Cattle Management Problems.* Las Cruces, NM: SWI Publishing.
Taylor, R. E., and Field, T. G. 1999. *Beef Cattle Production and Management Decisions.* Upper Saddle, NJ: Prentice-Hall.

Popular Periodicals

BEEF. 7900 International Drive, Suite 300, Minneapolis, MN 55425. http://www.beef-mag.com
Beef Today. Published by Farm Journal, 230 West Washington Square, Philadelphia, PA 19105. www.beeftoday.com
Breed publications. Most breed associations publish a breed journal or magazine. Contact specific breed associations (see addresses in Appendix) to identify the specific periodical.
Drovers Journal. 10901 W. 84th Terr., Suite 200, Lenora, KS 66214. www.drovers.com
Feed Lot Magazine. P.O. Box 850, Dighton, KS 67839. www.feedlotmagazine.com
Feedstuffs. 12400 Whitewater Dr., Minnetonka, MN 55343. www.feedstuffs.com
The Record Stockman. 4800 Wadsworth Blvd., Wheat Ridge, CO 80034. www.recordstockman.com
Southern Livestock Review. P.O. Box 423, Somerville, TN 38068. www.southernlivestockrev.com
Western Livestock Journal. P.O. Drawer 17F, Denver, CO 80521. www.wlj.net
Western Livestock Reporter. 18th and Minnesota Ave., Billings, MT 59101.

Scientific Periodicals

American Journal of Veterinary Research. American Veterinary Medical Association, 930 W. Meacham Road, Schaumburg, IL 60196. www.avma.org
Canadian Journal of Animal Science. Agricultural Institute of Canada, Suite 907, 1512 Slater St., Ottawa, Canada KIP 5H4.
Journal of Animal Science. American Society of Animal Science, 1111 N. Dunlap Ave., Savoy, IL 61874. Paper copy and electronic version available. www.asas.org

Professional Animal Scientist. American Registry of Professional Animal Scientists, 1111 N. Dunlap Ave., Savoy, IL 61874. www.arpas.org

Proceedings

Beef Improvement Federation. Ronnie Silcox, Animal and Dairy Science Complex, University of Georgia, Athens, GA 30602-2771.

Cattlemen's College. National Cattlemen's Beef Association (NCBA), Box 3469, Englewood, CO 80155.

Cornbelt Cow-Calf Conference. Ottumwa, IA. Contact Byran Leu, 422 McCarroll Dr., Ottumwa, IA 52501.

Range Beef Cow Symposium is held on odd-numbered years. Location alternates among Nebraska, Wyoming, South Dakota, and Colorado. Contact the beef extension specialist in one of these states.

Various land-grant university-sponsored conferences and workshops. Contact the State Beef Cattle Extension Specialist.

Internet

Veterinary Services, APHIS:USDA. Animal Health Monitoring and Programs: www.aphis.usda.gov/vs/cattle.htm

MARKETING

Books

Stasko, G. F. 1997. *Marketing Grain and Livestock.* Ames, IA: Iowa State University Press.

Internet

Ag Infolink: www.aginfolink.com
Agriculture Marketing Service of USDA: www.ams.usda.gov
Ag Span: www.agspan.com
Cattleinfo Net: www.cattleinfonet.com
Chicago Mercantile Exchange: www.cme.com
Livestock Marketing Information Center: www.lmic1.co.ncrs.usda.gov/
The Market Advisor: www.ag.ndsu.nodak.edu/cow/
Vantage Point Network™: www.vantagepoint.com

Market Services

Cattle-Fax. Box 3947, Englewood, CO 80155. A marketing report and analyst service provided to members. Marketing information provided by telephone and a weekly publication (UPDATE). www.cattlefax.org

INTERNATIONAL

Books

Allen, D. 1990. *Beef Production and Marketing.* London: BSP Professional Books.

Baker, M. J. (ed.). 1993. *Grasslands for Our World.* Wellington, New Zealand: SIR Publishing.

Cooper, M. M., and Willis, M. B. 1989. *Profitable Beef Production.* Great Britain: Redwood Buen Ltd.

FAO Production Yearbook. Published annually by the Food and Agriculture Organization of the United Nations, 00100 Rome, Italy. It contains information on land area, population numbers, beef numbers, beef production, beef and food consumption, and additional agricultural statistics. www.fao.org

Hodgson, J., and Illius, A. W. 1996. *The Ecology and Management of Grazing Systems.* Wallingford, Oxon, UK: CAB International.
Hoffman, D., Nari, J., and Petheram, R. J. (eds.). 1996. *Draught Animals in Rural Development.* Canberra: Australian Centre for International Agricultural Research.
Payne, W. J. A., and Hodges, J. 1997. *Tropical Cattle, Origins, Breeds, and Breeding Policies.* Ames, IA: Iowa State University Press.
Preston, T. R., and Leng, R. A. 1987. *Matching Ruminant Production Systems With Available Resources in the Tropics and Sub-Tropics.* Penabul Books, P.O. Box 512, Armidale, N.S.W., 2350, Australia.
Schmidt, P. J., and Yeates, N. T. M. 1985. *Beef Cattle Production (Australia).* Stoneham, MA: Butterworth Publishers.
Seré, C., and Steinfeld, H. 1996. *World Livestock Production Systems.* Rome: Food and Agriculture Organization (FAO) of the United Nations.
World Animal Science. Amsterdam, the Netherlands: Elsevier Science Publishers, R.V. The following books are selected from 34 available volumes of *World Animal Science.*
Hickman, C. G. (ed.). 1991. *Cattle Genetic Resources.*
Morley, F. H. W. (ed.). 1981. *Grazing Animals.*
Nestle, B. (ed.). 1984. *Development of Animal Production Systems.*
Tullock, N. M., and Holmes, J. H. G. (eds.). 1992. *Buffalo Production.*

Popular Periodicals

U.S. Meat Export Analysis and Trade News. Published monthly by the Meat Export Research Center, Iowa State University, 215F Meat Lab, Ames, IA 50011-1070. http://www.ag.iastate.edu/centers/merc/news.
World Development Form. Published by the Hunger Project, 1300 19th Street, N.W., Suite 407, Washington, DC 20036.

Scientific Periodicals

Animal Feed Science and Technology. Elsevier Science Publishers, P.O. Box 211, 1000 AE Amsterdam, the Netherlands.
Animal Production. Journal of the British Society of Animal Production. Scottish Academic Press, 33 Montgomery St., Edinburgh, Scotland EH7 5JX.
Animal Reproduction Science. Elsevier Science Publishers, P.O. Box 211, 1000 AE Amsterdam, the Netherlands.
Australian Journal of Agricultural Research. CSIRO, 314 Albert St., East Melbourne, Victoria, Australia 3002.
Journal of Animal Breeding and Genetics. Paul Parey Scientific Publishers, 35–37 W. 38th St., #3W, New York, NY 10018.
Journal of Animal Physiology and Animal Nutrition. Paul Parey Scientific Publishers, 35–37 W. 38th St., #3W, New York, NY 10018.
Livestock Production Science (official journal of the European Association for Animal Production). Elsevier Science Publishers, Box 211, 1000 AE Amsterdam, the Netherlands.
New Zealand Society of Animal Production (Proceedings). Hamilton, New Zealand: New Zealand Society of Animal Production.
Queensland Agricultural Journal. GPO Box 46, Brisbane 4001, Australia.
Reproduction in Domestic Animals. Paul Parey Scientific Publishers, 35–37 W. 38th St., #3W, New York, NY 10018.
South African Journal of Animal Science. P.O. Box 1758, Pretoria 0001, Republic of South Africa.
Theriogenology. An International Journal of Animal Reproduction. Geron-X, Inc., Box 1108, Los Altos, CA 94022.
Tropical Agriculture. Butterworth Scientific Limited, 88 Kingsway, London, WC2 6AB, UK.
Tropical Animal Health and Production. Longman Group Limited. Fourth Avenue, Harlow, Essex, England CM19 5AA.

World Animal Review. Food and Agriculture Organization of the United Nations, Va delle Terne di Caracalla, 00100 Rome, Italy.

World Review of Animal Production. International Publishing Enterprises, via di Tor Vergata, 8t/87 Rome, Italy.

Internet

Australian Lot Feeders Association: www.information.com.au/alfa/default.asp
Beef New Zealand: www.beef.org.nz
Canadian Cattlemen's Association: www.cattle.ca
Foreign Agricultural Service: www.ffas.usda.org
Meat and Livestock Australia: www.mla.com.au
United States Meat Export Federation: www.usmef.org
Winrock International: www.winrock.org
World Trade Organization: www.wto.org

MANAGEMENT SYSTEMS HIGHLIGHTS

1. Information is power—the power to lead and manage change in management systems.
2. How people access and utilize information will classify them into one of three groups: (1) those who make things happen, (2) those who watch things happen, and (3) those who wonder what is happening.
3. Information that helps obtain and validate principles is most useful. New information will not result in time-tested principles such as the law of gravity or honesty and integrity to be replaced. These and other principles should be retained while continuing to identify other highly valued principles to complement those already known.
4. The biggest challenge is to integrate the important principles into management systems where mission statements and goals can be accomplished.

REFERENCES

Publications

Gum, R., and Tronstar, R. 1996. The World Wide Web: Applications to Agricultural Economics. *Choices* (First Quarter).

Holin, F. 1997. Foraging through the Internet. *Hay and Forage Grower* (Feb.).

James, H. 1996. *The Farmer's Guide to the Internet.* Lexington, KY: TVA Rural Studies, University of Kentucky.

Northcutt, S. L., Burditt, L., and Buchanan, D. S. 1996. *World Wide Wisdom.*

Owens, F. N. 1997. Sources of Animal Science Research Information: Usefulness and Reliability. *J. Anim. Sci.* 75: 331.

Products and Services for ERS-NASS. USDA's Economic Agencies. Annual Issue 1997.

Tobin, D. R. 1998. *The Knowledge-Enabled Organization.* New York: AMACOM (A division of American Management Association).

The Beef Industry's Past and Future

The beef cattle industry's exciting past is captured in many books and films. The future of the beef industry will likely be equally as exciting and challenging. Considerable insight into the present and future cattle industry can be gained by studying its past:

> *The farther back you look, the farther forward you are likely to see.*
>
> —Winston Churchill
>
> *Those who don't remember the past are compelled to relive it.*
>
> —George Santayana

The latter quote is especially meaningful to those who have observed or studied the repeated beef cattle cycles (see Chapter 9) or who have observed several significant changes in beef cattle type (see Chapter 17).

SCIENTIFIC AND HISTORICAL EVIDENCE OF EARLY CATTLE

Historical events are presented in chronological order that highlight the early history of cattle and the significant events in the development of the cattle industry throughout the world. It is interesting to relate these events to the current U.S. beef industry not only to understand its heritage, but also to realize how the current industry has been shaped by the past.

Three to Four Million Years Ago

During the Miocene era extensive grasslands were present that allowed the proliferation of grazing animals in both numbers and species—the forebearers of modern cattle among them.

Fossil remains of cattle have been found in northwestern India dating cattle back to this time period. One species is believed to be a direct ancestor of the ox. The head was relatively long and narrow. Bulls were horned and females were hornless. The horn spread was between 6 and 7 ft. They were approximately 5 1/2 ft high at the shoulders.

Fossil remains were also found in west central Italy (after those found in India). These had more slender limbs than those found in Asia. They were given the name *Leptobos etruscus.* Other specimens were found in southern France.

One Million Years Ago

Fossils have been identified that provide recognition of three different forms of ox in Europe as well as a bison and a water buffalo. The most noted of these was the Auroch (the great wild ox), classified as *Bos primigenius.* They were extremely large, standing some 6 ft high at the shoulder. Measurements of 10–12 ft between the tips of the horns were common.

One-Half Million Years Ago

Competition for feed from other forage-loving animals occurred. Over the next several hundred thousand years, four glaciation periods posed climatic challenges to cattle. It is believed that cattle were protected by their heavy layers of thick, shaggy hair, similar to the Scotch Highland cattle today. During this time, cattle dealt with a new predator—people who came in search of food and skins. The ability of humans to control fire provided a springboard to move beyond life as a scavenger and gatherer.

40,000–50,000 B.C.

Fossilized bones reveal that prehistoric people cooked meat and cracked open the bones for the marrow inside.

30,000–15,000 B.C.

Paintings of bison were found in caves in northern Spain in 1875. More spectacular paintings (varying from 3 in. to 17 ft in length) of Aurochs and *Bos longifrons* were found in the Lascaux Caves in France in 1940. There is evidence that some of these paintings found in the same caves were created thousands of years apart. It is likely that cattle were incorporated into primitive religion as totems or icons.

10,000–8000 B.C.

Neanderthal humans, prevalent in Europe, chased the Auroch for food. They were likely assisted in the hunt by the first domestic animal—the dog. Cro-Magnon humans improved their hunting skills by driving the cattle over steep cliffs, later practiced by Native Americans on the Plains when hunting bison. Evidence exists that prehistoric humans stored meat (Mastadon) by anchoring it to the bottom of cool lakes in Michigan.

8000–4000 B.C.

The domestication of cattle occurred during this time period, although insufficient information exists to identify the exact time of domestication. Fossil remains give evidence that Denmark, Switzerland, Iraq, Palestine, Egypt, and Iran were the early locations of domestic cattle, where the remains of Aurochs (*Bos primigenius*), Celtic Shorthorns (*Bos longifrons*), and cattle of Zebu descent have been identified.

6500 B.C.
Some evidence indicates that the earliest domestication date for cattle was in Greece at this approximate date.

4500 B.C.
The oldest archaeological record for Zebu cattle is from a figurine found in Mesopotamia that dates back to this time.

3000 B.C.
The fossil remains of humped cattle and artwork showing Auroch-type cattle in Crete date back to 3000 B.C. Pre-Roman colonizers, migrating from Greece, had cattle. Humped cattle also appeared in Egypt.

Bull worship was present in Egypt in 3000 B.C. and possibly occurred even earlier. Certain sacred bulls were considered the earthly incarnations of a god. Records from the First Dynasty note that a festival was held regularly to celebrate the appearance of a sacred bull. Strabo, a Greek geographer, recorded in 60 B.C. a visit to an underground burial chamber for the sacred bulls of Egypt.

Auguste Mariette, a noted Egyptologist and founder of the Egyptian Museum in Cairo, rediscovered the underground burial chambers of the sacred bulls in 1851. The chambers contained 28 sarcophagi (stone coffins), each weighing approximately 80 tons and holding a bull that was mummified similar to the Pharaohs. Only one sarcophagus not previously robbed contained a mummified bull that is on display in the Agricultural Museum in Cairo.

Oxen pulled the scratch plow signifying human transition to cultivated agriculture.

2500 B.C.
The tombs in Egypt show the following drawings as evidence of early cattle production:

- The tomb of Achti-Hotep shows hornless cattle being led by a man.
- The tombs of Deir show a calf being delivered. Also shown are cattle being thrown for slaughter or branding.
- There is a drawing of a group of six oxen pulling a plow.
- The tomb of Huy depicts cattle being branded.
- The tomb of Auta contains a picture of a bull branded with the number "113."
- Oxen and donkeys were yoked to two-wheel carts.
- There are drawings of cattle being fattened and of being stunned for slaughter, skinned, and their meat being boned.
- There are models of slaughterhouses, killing floors, smokehouses, and meat shops.
- There are drawings of water buffalo domesticated to provide meat, milk, and draft power primarily to people in tropical conditions.

2000–1500 B.C.
The year 2000 B.C. marks the estimated beginning of the patriarchs of the Old Testament (Abraham, Jacob, and others), in which many references are made to cattle. The treasured Ark and Covenant was carried on a cart pulled by a select yoke of oxen (1 Sam. 6:7).

A "golden" calf figurine, estimated at 3,500 years of age, was excavated in 1990 from an ancient Canaanite temple in Israel. This and biblical records suggest the Canaanites worshiped the golden calves; for example, ". . . there are with you golden calves, which Jeroboam made you for gods" (2 Chron. 13:8). These ancient people believed the gods rode calves, thus the animals represented the gods and the worshipers made sacrificial offerings to statues of the calves.

500 B.C.

Romans were probably the first people to discover the principles of good farming and stock raising.

65 B.C.

Julius Caesar provided the first historical record of wild cattle (Uri or Aurochs) when he wrote:

> There is a third kind of these animals that are called Uri. In size these are but little inferior to elephants, although in appearance, color, and form they are bulls. Their strength and speed are great. They spare neither men or beasts when they see them. . . . Those who kill most receive great praise when they exhibit their horns as trophies of their success. These Uri, however, even when young cannot be tamed. In the expanse of their horns, as well as in form and appearance, they differ much from our domesticated oxen.

Auroch cattle existed in the wild state for several centuries. The last known survivor died in 1627 in a Polish park.

30 B.C.

Roman cattle producers received advice from Virgil (a farmer's son, poet, and small-time husbandman) in his four poems called *Georgics.* Following is an excerpt from one of the poems:

> Next, when calving is o'er, men's whole thought goes to the offspring; and they stamp them anon with brands distinguishing each one as preference dictates: these of maintaining a true breed, these of sacred office, and these as laboring oxen upturning the rugged loamclods and straining across them while others at grass go forth as an army to pasture.

The Greeks also wrote on husbandry and the Romans made extensive use of this information.

A.D. 100

Trajan built his forum in Rome at this time, and some of its decorative bull sculptures showed an advanced stage of breeding for meat production.

A.D. 300

This was the age of the Barbarians, who killed cattle, scattered the herds, and killed many knowledgeable cattle producers. It was a setback to sound cattle breeding and management.

A.D. 493

There was a revival of fairs in Italy at this time, held to stimulate trade with other countries. There were impressive sales of work oxen and cows at the fairs. Other significant cattle fairs were held in France and Russia.

HISTORICAL LANDMARKS IN THE AMERICAN CATTLE INDUSTRY

A.D. 981

Eric the Red, the famous Norseman, brought cattle to Greenland. The cattle were supposedly of Spanish (Andalusian) origin. During the fifteenth century, several thousand people succumbed to illness and died along with their livestock.

1131

The oldest fair for the exhibition of cattle was established at Smithfield, in London. The fair, considered the granddaddy of all fat stock shows in England and America, had a very slow start. It was not until 1798 that the Smithfield Show emerged as an important influence on the cattle industry.

1200

The Castile area in north central Spain became one of the first-known areas of running domesticated cattle on the open range. The Spanish probably adopted ranching from the Moors who ruled Spain from A.D. 711 to almost 1492. The practice spread southward into Andalusia (extreme southern Spain). Cattle ranching was practiced here, and its influence was carried into North America and South America when the cattle of Andalusia were transported to the West Indies. When Columbus left Spain for the New World, cattle ranching in Andalusia was two and a half centuries old.

1273

The Mesta of Shepards is formed in Spain. The Mesta was the first livestock protection organization. This would be comparable to the NCBA.

1493

On his second voyage, Columbus brought cattle to Santo Domingo (the West Indies), considered by many to be the first significant introduction of cattle to the New World. These first cattle, in addition to later shipments of cattle from Spain to the West Indies, developed into a thriving cattle industry over the next 20 years. Also during this time period, the number of cattle reached surplus levels and some cattle ran wild. The several hundred head of cattle shipped from Spain to the West Indies would become the progenitors of millions of head of cattle that would later range from Canada to Argentina. No other cattle types would mix with these Spanish cattle until after the beginning of the nineteenth century.

1513

Ponce de Leon brought cattle into Florida on his first expedition (in 1513) and on his second trip some eight years later. However, there is no evidence that these cattle or the settlers survived. Several other cattle importations into Florida during the next 40 years failed to be successful because of resistance by Native Americans. Eventually successful importation of cattle into the region was achieved and these became the progenitors of the breed known as Florida Crackers.

1519

Cortez was the first individual to bring livestock from Spain to the American mainland via Cuba (although there is some question about whether cattle were included in all of his expeditions). Cortez not only conquered Mexico but also started the cattle ranching business there. He branded some of his Aztec prisoners on their cheek. Thus, it is likely that the first cowboys on the American continent were branded before the first cows. His brand was three crosses—symbolic of the Holy Trinity.

1521

de Villalobos shipped cattle from Santo Domingo into Mexico. The cattle were likely descendants of the first cattle introduced by Columbus in 1493.

1528
de Vaca was the first Spaniard to see the American bison while crossing Texas. The French, who were among the first white explorers in North America, called the large, humpbacked animals they encountered on the Plains *les boeufs,* meaning "oxen" or "beeves." Through a sequence of word changes, *boeuf* became the English *buff,* then *buffle* or *buffler,* and finally *buffalo.* Scientists use the Latin terminology of *Bison bison* (genus and species) for this animal, pointing out that its correct name is "Bison." The true buffalo has no hump, thus the name buffalo belongs only to the water buffalo in Asia and to the African buffalo (these two buffalo belong to the genus *Bubalus*). The Spanish unsuccessfully attempted to domesticate buffalo.

1529
A brand book was established in Mexico City, where all cattle owners were required to register their brands. The Mexican Mesta was formed and served as an early form of the modern livestock producers' association.

1540
Coronado was the first of Spanish origin to enter the southwestern United States with cattle on foot. He entered into what is now known as Arizona with 500 cattle that were part of 6,500 head of livestock associated with Coronado's traveling army. Some exhausted cattle left behind became ancestors of thousands of wild cattle in the territory.

1565
Cattle were taken from the West Indies into Florida. These Spanish cattle were the first to persist in a breeding herd in the territory, later to be introduced to the United States. St. Augustine has been considered the first permanent settlement of Europeans in the United States.

1598
Juan de Onate founded the first self-supporting colony of San Juan at the junction of the Chama and Rio Grande Rivers in New Mexico. Stockbreeding began in this area, where 7,000 head of livestock were originally involved.

1607
Cattle were brought to Virginia from England when the first English colony was founded at Jamestown. In 1620, Virginia had 500 cattle. The first cattle were eaten as the colony struggled to become established. However, by the 1640s, cattle numbers had increased to nearly 20,000 head.

1624
Captain Edward Winslow brought the first cattle, three heifers, and a bull from England to the Plymouth Colony in Massachusetts.

1625
The Dutch, who settled in New York in 1614, brought from Holland the Flemish-type bulls and cows.

1630
The Massachusetts Bay Colony imported thirty cows. During the next 3 years, cattle arrived on almost every ship. No doubt many choice cattle of various breeds found their way from

England to the colonies, including the so-called Rubies or the bright Red Devon cattle. During this same time period, New Hampshire founder John Mason started a Danish colony and imported several groups of the large, yellow Danish cattle for draft purposes. Few cattle were imported from England after 1650 as a result of the English civil war. The early pioneers, carving out new homes in a wilderness, had little time for selective cattle breeding. Crosses of cattle from the Red Devons, the yellow Danish cattle, and the red-and-white and black-and-white Dutch cattle were evident. While these cattle provided the foundation for what were termed "native cattle," this population was eventually replaced in total by the breed introductions that would come later.

1636

William Pynchon and his son, John, established the first meatpacking plant in Springfield, Massachusetts. Beef and pork were slaughtered and packed in barrels; thus, the terms *beef packer* and *packing plant* emerged.

1650

American colonies begin exporting cattle by ship. The likely destination was the sugar plantations of the West Indies.

1660s

The cattle industry began to flourish in the Carolinas. Also at this time, some European-bred cattle began to encounter numbers of wild Spanish cattle. The latter cattle were descendants of those brought by the Spaniards from the West Indies to Florida. Early cattle producers followed the explorers and trappers, but basically let their cattle run wild. A reporter traveling in Maryland and Virginia said cattle were seldom fed and were killed "fat out of the woods."

1691

While it was the Spanish conquistadors that trailed the first cattle into the Southwest, it was the Spanish padres who ensured the survival of cattle ranching in this area. One pioneer missionary who had a great influence on the development in the Southwest was Eusebio Francisco Kino, who founded and managed a livestock breeding establishment at his first mission, Delores, in Old Mexico. From that location he sent hundreds of cattle into 19 other missions located in Arizona, Senora, and southern California. Kino personally supervised many of these livestock operations but never profited from them. He gave cattle freely to the Native Americans. During this same time period, the first cattle entered Texas, near the Louisiana line, also a result of the extension of Spanish missions. The first Vaqueros were Indians working at the missions.

1756

Brighton Market, near Boston, became the first recorded public auction market. It also became the slaughter center for the Northeast, even though most slaughtering was still done on farms.

The Late 1700s

The farmer and cattle raiser began to push westward from the original colonies, following earlier hunters, trappers, and traders. The Allegheny Mountains presented a physical barrier, but soon the emigrants brought hundreds of cattle into the eastern Mississippi Valley by way of the Allegheny and Ohio Rivers. Daniel Boone's wilderness road through the Cumberland

Gap, a southern trail around the Appalachians into Tennessee, and more importantly, trails across Pennsylvania, provided access routes for the westward movement of surplus cattle from the eastern United States. In fact, three major railroads would eventually follow the original cattle trails through the mountains.

1769

The first cattle (of Spanish origin) arrived in California and were part of the mission founded at San Diego.

1779

First cattle trail in North America from San Antonio, Texas, to the Louisiana Territory.

1783

Shorthorn cattle were exported from England to Virginia. The Shorthorn had the first herd book record of any breed, established by George Coates in 1822.

1788

Cattle, probably of Longhorn origin, were shipped by boat from California to Oregon territory during this year.

1817

Grain feeding of cattle occurred in Ohio in the early 1800s. New York City received its first shipment of Ohio grain-fed steers in June 1817. They were driven for nearly 1,000 miles at a rate of approximately seven miles per day. Cattle on these drives to the eastern United States weighed 900–1,000 lbs at the start of the trip and usually lost 100–150 lbs during the drive period. During the next few decades, Illinois became the center of surplus cattle production. Many cattle produced in Illinois were driven to Ohio, where they were fattened and sold as Ohio cattle.

The westward movement of cattle was based primarily on the economics of fed-beef production. Ohio enjoyed a near monopoly of beef production until a few years later when central Indiana took over. By the mid-1850s Illinois assumed that leadership, but then in the late 1880s Iowa became a more efficient producer of beef. In the early 1900s, the westward shift of efficient beef production had moved into eastern Nebraska.

The famous importation of Shorthorn cattle (ancestors of the American Shorthorns or Durhams) was made by Lewis Sanders of Kentucky when he brought them from England in 1817. Henry Clay also imported Herefords from England to Kentucky in this same year.

1828

Urea was first synthesized, but was not used in cattle feeding until World War II, when protein costs were very high.

1834

During this year, the 21 existing Spanish missions reported total cattle numbers of 396,400. By 1842, cattle numbers in the missions had decreased to approximately 29,000, in part because of the drought in 1840 but primarily because Mexican Republic authorities enforced secularization. This policy was to emancipate Native Americans and divide mission properties among them. The padres retaliated by slaughtering cattle, primarily in 1834. Since the primary value of cattle was in their hides and tallow, they were slaughtered for that purpose. Most of these products were shipped by ox-carts and mules to San Pedro (now Los Angeles). Boats then took the products primarily to the Boston area. In 1834, the year's shipment from San Pedro was 100,000 hides and 2,500 centals (100 lbs to the cental) of tallow and several cargoes of locally made soap. The tallow wrapped in hides were called *botas*.

1836

The first public cattle auction sale was held in Ohio in 1836.

1830–1870

The development of the western United States was initiated by missionaries striving to convert the Native Americans, the gold seekers, those seeking freedom from religious persecution (Mormons), and those who hungered for the acquisition of land. The pioneers usually had three types of cattle—oxen, milk cows, and cattle raised primarily for meat. Oxen were primarily responsible for the movement of thousands of emigrants and their tons of supplies (see Fig. 20.1). Most covered wagons were drawn by six- or eight-yoke teams of oxen. Their hoofprints dug deep into the major trails (Oregon, California-Overland, and Santa Fe) and their many side trails. An estimated 250,000 emigrants traveled the California-Overland trail between 1859 and 1869. On August 14, 1850, during the Gold Rush, the register at Fort Laramie recorded the presence of more than 39,000 men, 3,000 women and children, 36,000 oxen, 7,300 cows, 9,000 wagons, and other livestock.

1840

Corn shellers and hammer mills were utilized in cattle feeding operations.

1842

Cattle from Texas moved northward into Missouri. A few years later, herds of 1,000+ head moved from Texas into Ohio and Illinois.

FIGURE 20.1 United States postage stamps commemorating the pioneers. Cattle (oxen) were important in the transportation process.

1849

The first importation of Brahman cattle to America occurred when James Bolton Davis brought two cows and a bull to his South Carolina plantation. It is believed that these Brahmans were killed during the Civil War. Later, in 1854, two Brahman bulls were brought to Louisiana and used in crossbreeding. Chicago was identified as the meatpacking center of the United States in 1849.

1850s

Cattle from Texas followed the Gold Rush into California. Cattle prices in Texas of $14 per head and in California of $100 per head greatly stimulated the movement of thousands of cattle. Invention of the corn planter also took place at this time.

1853

Joseph Mallory of Piatt County, Illinois, created a cattle sensation when he moved a herd from Native American territory to his farm, where he fattened the cattle and later shipped them into New York. Soon after, Illinois became an active feeding center for Texas and Native American territory cattle. The cities of Independence, Westport, and Kansas City became the sites of large cattle markets.

1860s

The Civil War changed the U.S. cattle scene, first in Texas and then in the North. The Union Army blocked the gulf ports and the Confederate Army demanded the services of all able-bodied residents. During the war, cattle that were not cared for ran wild and they multiplied at a very rapid rate. By 1865, cattle numbers had reached 6 million head in Texas. At the war's end, cattle prices had decreased to $5 per head in Texas and $50 per head in the North. The inequity in cattle numbers and prices set the scene for the historic cattle drives and their well-known trails. During the next 30 years, some 10 million head of cattle would be driven over such famous trails as the Shawnee, Chisholm, Western, and Goodnight-Loving. The trails ended at the Kansas railheads, terminated at Dakota–Wyoming–Montana pastures for fattening, or connected with feeder trails leading into Colorado, Kansas, or Missouri (see Fig. 20.2).

Also during this period and for the next several decades, manure from cattle feeding was a highly valued product since commercial fertilizers were not available. Manure was valued at $1.20–$1.50 per ton, while corn was selling for 60 cents a bushel and slaughter cattle for less than $50 per head.

1861

Joseph McCoy, a promoter and entrepreneur, rode to the railhead town of Abilene, Kansas, and purchased it for $5 an acre. He then advertised throughout Texas that he would double the price of a Texas steer delivered to Abilene. The first herd arrived three months later. Before leaving Chicago, McCoy had bragged that he could bring 200,000 head to Chicago within 10 years. He actually brought 4 million head in four years. This feat inspired the saying "It's the real McCoy."

1862

Land-grant colleges were endowed under the Morrill Act, or Land Grant Act, of 1862. Congress granted each state 30,000 acres of land for each senator and representative it had in Congress. The lands were to be sold, the proceeds invested, and the income used to create and

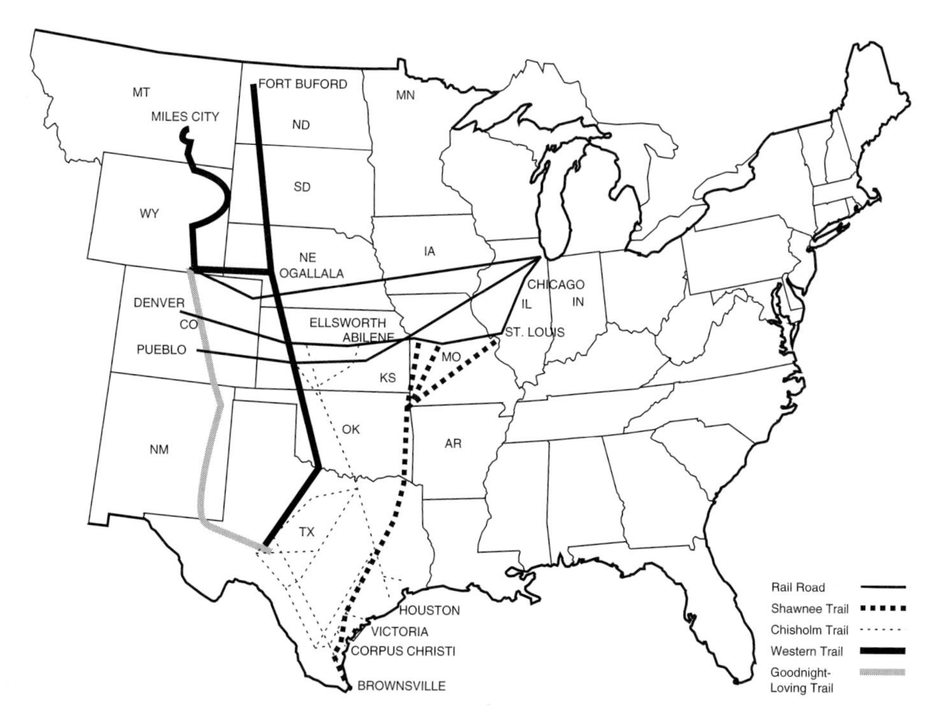

FIGURE 20.2 Major trails for the cattle drives to the railroads.
Source: Courtesy of USDA.

maintain a college for agriculture and mechanic arts. The U.S. Department of Agriculture (USDA) was established as a separate government agency in 1862. The Homestead Act encouraged land settlement and family farms, which pushed cattle grazing farther west.

1865
The Union Stockyards and Transit Company of Chicago opened for business and became the hub of the livestock industry.

1866
Gregor Mendel published a paper in 1866 on the genetic basis of inheritance, but it was overlooked. The merit of his work would be recognized several decades later.

1869
The first transcontinental railroad was completed.

1870 to 1890s
Railroads stimulated the development of new meatpacking centers, as stockyards with affiliated meat plants were built in midwestern cities. Yards were founded in Kansas City in 1871, in St. Louis in 1873, in Omaha in 1884, in Denver and St. Paul in 1886, in Sioux City and St. Joseph in 1887, and in Oklahoma City in 1910.

FIGURE 20.3 During the late 1800s and early 1900s, the English breeds (primarily Angus, Hereford, and Shorthorn) increased in numbers, which resulted in the Longhorn range cattle losing their identity. The American Angus Association used this oil painting in 1973 to help commemorate the 100th anniversary of importation of Angus cattle into the United States. The painting represents the first Angus bulls brought to Victoria, Kansas, by George Grant to cross on his Longhorn cows.
Source: Frank C. Murphy and the American Angus Association.

1871
A peak year for cattle drives, most of which involved Longhorns.

1873
Barbed wire was invented. Angus were imported (see Fig. 20.3).

1875
Silos and refrigerator cars came into existence. Chilled beef shipped from New York to Europe in refrigerated ships.

1870s–1880s
Silage was introduced in cattle feeding but its use grew slowly until the turn of the century.

1880
Many cattle died in the Great Plains and western United States during the severe winter of 1880. Cattle producers then realized the need to store and preserve winter feed.

1881
The *Breeders' Gazette* originated and the American Hereford Association was established.

1882
The American Shorthorn Association was organized.

1883
The American Angus Association was established. National Cattle Growers' Association organized in Chicago.

1884
Federal disease eradication programs were initiated. Animal and Plant Health Inspection Service (APHIS) is established.

1885
Texas breeders imported two Nellore-type Brahmans from Calcutta, India. These importations and the grading-up breeding process increased Brahman numbers, particularly in the southeastern United States.

1886
Severe blizzards, drought, and overgrazing plagued cattle producers in the northern Great Plains. Coupled with westward expansion, these severe weather conditions brought the unregulated open range livestock industry to a close.

1887
The Hatch Act, passed by Congress in 1887, created agricultural research funds for college experiment stations. Interstate Commerce Act passed.

1888
Crested wheatgrass was introduced to the United States from Russia.

1889
The commercial feed industry began in Chicago. The Oklahoma Land Rush occurred.

1890
By 1890, it was estimated that 10 million cattle had been trailed from Texas to the north and northwest. They were driven in herds of approximately 3,000 head by 11 men with 6 horses per man. Each man received $30 per month with the boss getting $100 per month. The 3,000 head traveled about 3,000 miles with a total cost of $3,000.

This year witnessed the development of the first side-delivery rake and hay loaders, the first Meat Inspection Act passed, and the Sherman Anti-Trust Act became law.

1892
A researcher at the USDA determined that Texas Fever was carried by ticks. This breakthrough provided the initial credibility for the agency.

1895
Sears, Roebuck Company started catalog business.

1896
Simmentals were introduced but they had little influence on the cattle industry at this time. Wilson was appointed Secretary of Agriculture by President McKinley. During his 16 years of leadership at the USDA, research was elevated to a top priority.

1897
The forerunner of the American National Cattlemen's Association, the National Livestock Association was established in 1897 with headquarters in Denver. Also, the Amalgamated Meat Cutters and Butchers Workmen of North America (AFL-CIO) was founded.

1898

The first edition of *Feeds and Feeding* written by W. A. Henry of the University of Wisconsin was published. Later editions by Frank B. Morrison became known as the "Nutrition Bible," one of the most widely used reference books in the world.

1900

The International Livestock Exposition in Chicago first opened in 1900. It was held each year shortly after Thanksgiving so that it could be a culminating show of the year after all other state and regional shows had been held. Also, the show created a great source of Christmas beef, which was a traditional dish in Chicago and surrounding areas.

The Polled Hereford Association was organized. In addition, Mendel's research on inheritance was rediscovered and validated independently in Germany, Holland, Austria, and the United States.

1901

The Forest Service was established as the Bureau of Forests, then renamed the Forest Service in 1905 and transferred to the Department of Agriculture. It allowed grazing in the National Forests under its own set of regulations. Theodore Roosevelt became president.

1902

Free mail delivery to rural areas became permanent.

1903

The emergence of the five major meatpacking companies—Swift, Armour, Cudahy, Wilson, and Morris—occurred in 1903. Also, the first gasoline-powered tractors were developed. The first airplane flight was made. The Saddle and Sirloin Club was formed in conjunction with the International Livestock Exposition in Chicago. Each year since, the portraits of notable livestock industry leaders were hung in the club. The tradition continued after the International Livestock Exposition was moved to Louisville, Kentucky.

1904

The value of protein supplementation to grains was first shown at this time. The first livestock auction was held in Union, Iowa. The Caterpillar truck was invented.

1905

Dairy Herd Improvement Association (DHIA) records became a database for selection. The dairy performance records system later became a model for beef performance records. *The Jungle,* written by Upton Sinclair, reported on the poor sanitation conditions in packing plants. This eventually led to the passage of the Meat Inspection Act and the formation of the FDA.

1906

This year witnessed the introduction of Ford's famous Model T and the establishment of the Beefmaster breed. In addition, the Meat Inspection Act was passed; it provided for the inspection (for wholesomeness, freedom from disease, and adulteration) of all meat intended for interstate and foreign commerce. The Food and Drug Administration (FDA) was established to ensure food safety and labeling of foods. The National Live Stock Association and American Stock Growers' Association merge.

1908
The American Society of Animal Science was organized. Hybrid vigor in corn was discovered but not applied commercially until Henry A. Wallace started Pioneer Hybrid Seed Company in 1926.

1909
William Howard Taft was elected president.

1912
The term *vitamin* was coined as a result of biochemical research at the University of Wisconsin, and the first successful AI of cattle occurred (in Russia).

1914
The Smith-Lever Act, passed by Congress in 1914, brought the Cooperative Extension Service into existence at the federal level.

1915
Transcontinental phone service was established in the United States.

1916
The Stock Raisers Homestead Act of 1916 permitted the homesteading of 640 acres for stock-raising purposes.

1917
The United States entered World War I.

1918
The Santa Gertrudis breed was founded at the King Ranch in south Texas.

1919
Institute of American Meat Packers, now American Meat Institute (AMI), was organized.

1920
The use of tractors expanded in 1920; it had a tremendous effect on the cattle industry by increasing forage and grain production. Mechanization increased the feed supply, and the reduction of 24 million head of horses and mules (1920–1960) made additional feed available to the cattle industry. In addition, the first self-service meat counters opened in retail stores.

1920s
The largest importations of Zebu cattle came into the United States from Brazil through Mexico.

1921
The Packer and Stockyards Act, passed by Congress in 1921, provided marketing regulations. Warren G. Harding was elected 29th president.

1922
The National Live Stock and Meat Board, founded in 1922, was the first livestock industry agency to enhance the public image of meat and build consumer dollar demand.

1923
Calvin Coolidge became president.

1924
The American Brahman Breeders Association was organized.

1926
The USDA Federal Meat Grading Service, set up in 1926, officially graded beef carcasses on a voluntary basis only.

1928
The Future Farmers of America organization was founded.

1929
Herbert Hoover was elected 31st president. Stock market prices crash on October 29 (Black Tuesday).

1930
More than 1,300 banks close. The U.S. Range Livestock Research Station was established at Miles City, Montana, where performance records for beef cattle were first researched.

1931
Bank failures reached 2,294 as unemployment reached nine million. Drought and depression plagued the cattle industry.

1932
Drought in the Great Plains led to the Dust Bowl. The Brangus breed was established.

1933
Franklin Delano Roosevelt was elected 32nd president.

1934
The Taylor Grazing Act was passed by Congress. It came into existence to control the unregulated grazing of livestock on the millions of acres of nonhomesteaded land in the West.

1935
Social Security Act reserved funds for pensions and unemployment.

1936
Charolais were imported from Mexico and artificial insemination began in dairy cattle.

1940
The self-tying hay baler was developed.

1941
Japan bombed Pearl Harbor. The first performance bull-testing station (in Texas) was established.

1942
Big commercial feedlots emerged. Large numbers of semitrailer trucks hauled both grain and cattle.

1943
Rationing of sugar, gasoline, meat, and other items began. Cattle tick fever eradicated.

1945
The National Association of Animal Breeders was organized to provide self-regulation for the AI industry. Also, anhydrous ammonia was first used as a fertilizer. Harry S. Truman became 33rd president. World War II ended.

1946
The Grazing Service (originally created to manage the Taylor Grazing Act of 1934) and the General Land Office were combined to form the Bureau of Land Management (BLM).

1948
The Society for Range Management was organized.

1950
The beef carcass grade standards were amended in 1950—the grades of Prime and Choice were combined into Prime and the Good grade was changed to Choice. Also, antibiotics were shown to increase growth in cattle; over the next two decades, antibiotics became widely used in feedlot rations.

1951
The American National Livestock Association changed its name to American National Cattlemen's Association. Snorter dwarfism was first reported. The first calf from bovine embryo transfer was born.

1952
The first successful breeding using frozen semen occurred at this time as a result of research at Cambridge in the United Kingdom. Also, the American National Cowbelles, the women's auxiliary of the American National Cattlemen's Association, was created to promote beef and the beef industry. Dwight D. Eisenhower was elected 34th president.

1953
Stilbestrol was discovered and the defined level to enhance feedlot performance was patented. Crick and Watson determined the structure of DNA. From 1953 to 1975, per capita

beef consumption more than doubled as a result of offering grain-fed versus grass-fed beef to United States' consumers. This change in beef's image was critical to the emergence of the modern industry.

1954
The Red Angus Association of America was formed.

1955
The first Beef Cattle Improvement Association was organized in Virginia and other states soon followed. McDonald's chain restaurants began. The consumption of ground beef grew rapidly as a result of growth in the fast-food restaurant business.

1956
The Performance Registry International (PRI), established in Texas in 1956, recorded performance records of beef cattle. The USDA carcass grades were amended—the Commercial grade was divided into two grades, Standard and Commercial. The Soil Bank was created, which paid farmers to remove cropland from production.

1958
The Humane Slaughter Act, passed by Congress in 1958, required packing plants to slaughter animals humanely if they desired to sell products to federal agencies. In addition, the hay baler for large round balers was developed.

1959
The Angus Herd Improvement Records (AHIR) was established.

1960
AI studs began emphasizing beef AI in 1960. Also, transportation of cattle shifted from rail to truck as new slaughter plants were built near feedlots rather than in large metropolitan areas. Over the next several years, the large packing plants at the terminal markets were closed. Retail chain stores processed beef centrally. Boxed beef processing began to influence the industry.

1961
John F. Kennedy was inaugurated as 35th president.

1963
Lyndon B. Johnson became 36th president.

1964
Chicago Mercantile Exchange (CME) offered first live cattle contract. USDA feeder cattle grades were officially accepted. Also, the American Hereford Association created the Total Performance Records (TPR) program and the American Polled Hereford Association established its Guidelines Program (for performance records). The Vietnam War began.

1965
Carcass yield grades were officially adopted. Computer technology started to revolutionize livestock data analysis.

1966
Screwworms were eradicated in the United States. Simmentals were reintroduced (*see* 1896), and Limousins were first introduced by way of Canadian importations. Beef boycotts begin in Denver.

1967
The Wholesome Meat Act, passed by Congress in 1967, extended meat inspection to products in interstate commerce.

1968
The Beef Improvement Federation was formed to coordinate the performance record programs of state and breed associations. Also established were the American Simmental Association and the North American Limousin Foundation. Cattle-Fax was created by ANCA.

1969
Surplus grassland was produced by government cropland diversion programs, as 56 million acres were diverted from crop production in 1955–1969. Richard M. Nixon was elected 37th president.

1970
Consumer activism surfaced, focusing on such issues as high beef prices, nutrition, and health concerns. During the 1970s, the Great Plains became the leading U.S. cattle feeding area. The Environmental Protection Agency (EPA) was created.

1971
Union Stockyards in Chicago closed after 106 years of operation. The American Angus Association provided selection worksheets with estimated breeding values. The American Simmental Association published the first National Sire Summary. Guidelines for national sire evaluation policies were published by BIF.

1972
The American Angus Association allowed registration of calves sired by non-owned sires via AI. Other breeds soon followed.

1973
First peacetime price freeze imposed by President Nixon leads to "the wreck" for cattlemen. Fed cattle prices dropped from $80 to $38 per hundredweight in 1973. Cost of gain increased from 30 cents to 50 cents a pound because of higher grain prices. During the next several years, some cattle feeders lost as much as $200 per head, with bankruptcy common among feedlots.

1974

Gerald Ford became 38th president. Monensin (trade name Rumensin®), a feed additive discovered in 1974, improved feed efficiency. The first Angus, Hereford, and Polled Hereford sire summaries using reference sires and designed tests were published.

1975

An all-time cattle inventory record of 131.8 million head was reached in 1975. The high-voltage electric fence was developed in New Zealand.

1976

The USDA carcass grade standards were revised. The major changes included (1) that all federally graded beef had to be yield-graded as well as graded for quality, (2) conformation was eliminated as a quality grade factor, (3) the marbling requirement for Prime, Choice, and Standard was reduced by one degree of marbling, and (4) the range of marbling for the Good grade was narrowed.

1977

Jimmy Carter was inaugurated as 39th president. The National Cattlemen's Association (NCA) was organized in 1977 through a merger of the American National Cattlemen's Association and the National Livestock Feeders' Association. Also, a controversial report ("Dietary Goals for the United States") published by a U.S. Senate committee in 1977 recommended less meat in the American diet due to health concerns.

1978

The first U.S. calf from a frozen embryo was born. The Humane Methods of Slaughter Act (1978) amended the Meat Inspection Act of 1906, requiring that all meat to be inspected and approved come only from livestock slaughtered in accordance with humane methods.

1979

Revision of the USDA feeder cattle grades, where new grades were based on frame size and thickness. DES (diethylstilbestrol) was banned from use in cattle feeding and the first prostaglandin (Lutalyse) was approved for use in estrous synchronization.

1980

Prime interest rates peaked at 21.5%, highest since Civil War. At this time, many large forage-producing areas were converted to other crops because, since the early 1970s, the grain export market had increased from $7.7 billion to nearly $40 billion (1972–1980). Also, in 1980, Integrated Reproductive Management (IRM) projects were first funded. The IRM concept started 2–3 years earlier through the efforts of the National Extension Industry Beef Resource Committee. Later in the 1980s, the "R" in IRM was changed to "Resource" to broaden the scope of management that would include all resources, not just reproduction. The first sire summary using field data from the AHIR database was published by the American Angus Association.

1981

Ronald Reagan was inaugurated as 40th president. The first test-tube calf in the world was born in the United States (PA). Fertilization of this Holstein calf occurred in a laboratory dish about the size of a dollar.

1982

The first identical twin calves were born in the United States (CO) in 1982 resulting from microsurgery (splitting the genetic material of a single 5–6-day embryo). The first identical twin calves from this same process were born in Great Britain in 1981. Lasalocid (trade name Bovatec®) was approved for use in feedlot rations; the antibiotic improves both rate and efficiency of gain.

1983

Dairy PIC (Payment in Cash) bailed out dairymen, left cattlemen holding the bag.

Mid-1980s

Land values decreased by 30–60% between the early and mid-1980s. Many cow-calf producers could not service a heavy debt load because of recent land purchases or because they had borrowed heavily on appreciating land values. Loan agencies only provided capital if the operations had a positive cash flow. Many producers exited the business because of these financial pressures.

1985

Beef Promotion and Research Act became law. The University of Georgia computed National Cattle Evaluation genetics estimates whereby all animals, not just sires, received genetic predictions for several beef breeds.

1986

"War on Fat" was launched. The first genetically engineered calf was born in 1986 in Canada. The calf had a human gene that produced interferon to improve disease resistance. The results of the National Consumer Retail Beef Study were announced and the national beef checkoff program of "one dollar per head marketed" was started. Its objective was to improve profitability for beef producers.

1987

The quality grade Good was officially changed to Select. EPDs gained industrywide acceptance as a selection tool. The Stock market crash on October 19 precipitated a greater percentage of value lost than in the crash of 1929. American Heart Association endorsed beef as a healthy food. Cattle first sold by satellite in "video auction."

1988

The National Beef Market Basket Study showed 27% less trimmable fat in all retail beef cuts, while more than 42% of the retail cuts had no external fat. External fat for all retail cuts was 1/8 in., which was a marked change from the 1/4 in. in 1986 and the 1/2 in. prior to 1986. The USDA agreed to use these findings in its beef nutrition handbook.

Beef producers supported the beef checkoff program, with nearly 80% in favor of a national referendum. The Surgeon General's Report on Nutrition and Health presented evidence that lean beef was important in a healthy diet.

1989

The first genetically engineered calves were produced in 1989 in the United States. The calves had genes from several species, including the human species. The introduced genes were intended to increase growth and lean production. George Bush was inaugurated as 41st president. European Economic Community (EEC) imposed a hormone ban on United States beef. The Berlin Wall opened to the West after 28 years.

The USDA uncoupled the Quality and Yield grades. "Uncoupling" allowed packers to Quality grade or Yield grade or do both to facilitate marketing. The NCA established the Beef Safety/Quality Assurance Task Force to deal with present and potential problems in food safety.

1990

The NCA's value-based marketing task force declared a "war on fat" in 1990. The National Beef Tenderness Survey showed that 20% of the loin and rib cuts and 60% of the retail cuts from the round were not satisfactory for tenderness. The Spotted owl was listed as an endangered species.

1991

The National Beef Quality Audit (NBQA-91) gave benchmarks never previously available to the beef industry. Japan officially liberalized its beef import policy. The Soviet Union was divided as the Iron Curtain fell.

1992

The 1991 National Beef Quality Audit reported that the beef industry was losing $280 per fed beef animal, with $219 of the total due to waste fat. Also in 1992, Standardized Performance Analysis (SPA)—a cow-calf enterprise production and financial performance analysis system—was introduced by the national IRM committee of the NCA. Value of U.S. beef exports exceeded imports for the first time.

1993

William Clinton was elected 42nd president. *E. coli* outbreak stepped up food safety efforts. Fed-cattle prices reached a record $85.50/cwt. GATT approved by Congress. NAFTA went into effect. The Strategic Alliance Field Study, supported by several beef industry organizations, demonstrated that production costs can be reduced and product quality improved through the cooperative efforts of several beef industry segments.

1994

The Bureau of Land Management and the Forest Service proposed regulations known as Rangeland Reform '94. The proposed changes affected the federal grazing fee and grazing program administration. This proposal yields community-based action that strengthens coalitions between cattlemen's groups and local environmental entities.

1995

The National Beef Quality Audit (NBQA-95) showed additional beef industry benchmarks when compared with the NBQA-91. The first successful use of sexed semen was used for artificial insemination.

1996

The National Cattlemen's Association (NCA) and the Beef Industry Council (BIC) of the National Live Stock and Meat Board merged, creating the National Cattlemen's Beef Association (NCBA).

1997
Major recalls of ground beef occurred when *E. coli* was detected. The FDA approved irradiation to beef to control pathogens. Federal tax legislation gave cattlemen relief on estate taxes and capital gains taxes. The World Trade Organization issued a report saying the European Union (EU) ban on meat produced with growth promotants was not based on science.

1998
NCBA initiated Brand-Like Initiative report encouraging the use of value-added steps in production and processing.

1999
Consolidation at the retail grocery level saw the "Top 25" become the "Top 10." Trend for beef consumption became positive again following a 20-year decline. NCBA puts up $250,000 prize for the best beef convenience product.

2000
IBP introduced its branded product line named Thomas E. Wilson. For the first time in history, more than $50 billion was spent on beef by American consumers with a tally of $54 billion.

2001
George W. Bush became the nation's 43rd president. Tyson purchased IBP.

THE INFLUENCE OF CATTLE ON HUMAN CULTURE

The Early Era

Prehistoric people hunted cattle to meet some of their life-sustaining needs. The flesh of cattle provided food, the hides supplied various forms of clothing, and the bones were used to make weapons and tools. Cattle's first contribution to early humans, then, was to assist them in survival. Later, cattle dung was found useful first as a fuel in wood-sparse areas, then as a building material, and eventually as a fertilizer.

From a technological standpoint, the most important contributions of cattle were as beasts of burden and later of traction. The development of wheeled transport was probably associated with the domestication of cattle.

A Measure of Wealth

Long before the monetary system was created around 700 B.C., trade was based on barter and exchange. Because cattle were highly valued, they became the basis for value comparisons in the barter system. In Homer's *The Iliad*, for instance, prizes for wrestlers were based on cattle value:

> A massy tripod for the victor lies,
> Of twice six oxen its reputed price;
> And next, the loser's spirits to restore,
> A female captive, value but at four.

The noun *cattle* has the same origin as *chattel,* which means "possession." The Romans took the Latin word *decus* (cattle) to give origin to the word *pecunia* (money). The present word *pecuniary* refers to or relates to money.

The first Roman coins, made in about 350 B.C. were crude, shapeless lumps of bronze. Around 300 B.C., the bronze was cast into metal bars whose shape resembled flat, thin bricks. Called *aes signatum,* the bars were meant to represent the value of cattle in that both sides displayed the figure of a bull. Although the bars were inconvenient to handle, they represented the first step toward eliminating cattle as a medium of exchange.

Religion and Mythology

Evidence indicates that early humans held a sacred, mysterious image of cattle. Some historians interpret a deep human reverence for cattle in the drawings within the Lascaux Caves of France, which date back some 15,000 to 30,000 years B.C. It is argued that early humans believed that, by capturing the image of cattle in such drawings and other rites the actual chase or hunt would be made easier and more productive.

Cattle worship was probably most popular in ancient Egypt. The evidence suggests that Egyptians began worshipping cattle prior to 3000 B.C. and continued to do so for centuries. They believed that the Apis bull originated from a cow impregnated by lightning from heaven and thus regarded it as the earthly incarnation of Osiris (god and king of the dead). Persons who harmed these bulls were tortured and hanged. Egyptians' worship of the bull was based on their belief in its strength, fierceness, and enviable reproductive powers.

The Israelites at the base of Mount Sinai, influenced by Egyptian bull worshipping for centuries, worshiped the "Golden Calf." Even the prophet Moses, attempting to convey to them a new God, could not sway the Israelites from some of their traditional beliefs. Centuries later, the worshipping of calf idols was still being practiced: ". . . have made molten images of their silver . . . let the men that sacrifice kiss their calves" (Hosea 13:2).

Not only Egyptians worshiped and revered cattle, however. Many other early peoples believed that cattle had supernatural powers. It was claimed, for example, that the assassination of Caesar and the destruction of Pompeii were both foretold by an ox.

Today, the sacred reverence of cattle still exists in a number of countries. Cattle are used as ceremonial beasts in some African countries, where their slaughter is limited or prohibited. Cattle are also sacred to the Hindus of India, which has the largest cattle population of any country in the world. Many of India's nearly 200 million head of cattle roam the streets and countryside while some of its people starve—a vivid display of the deep reverence that Hindus have for cattle.

Astrologers who attempt to predict the fortunes of people by studying the stars, believe the zodiac (a band of stars that appears to encircle the earth) influences people's lives. They draw up charts called horoscopes to represent the various zodiac signs. Developed by astrologers more than 2,000 years ago, the signs of the zodiac still have meaning for many people today. Taurus, the second sign of the zodiac, is represented by a symbol of a bull. People born under the sign of Taurus (from April 20 to May 20) are believed to be affectionate, conservative, loyal, patient, trustworthy, and practical. Other characteristics associated with the Taurus sign include an even-tempered disposition that turns fierce when angered.

The Alphabet and Vocabulary

The first letter of the alphabet took its shape from an Egyptian hieroglyphic (a picture symbol) that depicted the head of an ox. The Semites, who lived in Syria and Palestine, changed the symbol slightly and called it *aleph,* meaning "ox." The ancient Greeks used a similar symbol and called it *alpha* (see Fig. 20.4).

FIGURE 20.4 The letter *A* evolved from an ox head.
Source: Colorado State University.

Reference was made earlier in the chapter to the origin of the word *cattle* and to cattle's influence on the Roman alphabet. The early ancestors of cattle were given many different names. From Latin and Greek, we get *bos,* with the genetive *bovis,* from which are derived the words *bovine, beeves,* and *beef.* From the Teutonic, we get the German *ochs,* the Danish *oxe,* and the English *ox.* Combining this with the old sanskrit word *ur,* meaning "forest" or "stony place," we get *urochs* or *aurochs,* which signifies wild cattle. The word *ur* was likely connected with cattle and was used for centuries; eventually it was Latinized into *urus* or the plural *uri.* The latter word was used by Caesar (65 B.C.) to describe the big, wild cattle hunted at that time.

Bos became the name of the genus in the zoological classification. Cattle producers using such jargon as "bossy," "so-o-o bos," and "come, bos" did not realize they were speaking in such scientific terms.

The word *Brahman,* which is a Zebu breed in this country, is probably a misnomer. Brahman is the name of the highest, strictly vegetarian caste in the Hindu hierarchy.

The British talked and wrote about the word *neat.* They borrowed the Icelandic word *naut,* meaning "ox," changed it to *neat* and associated it with cattle only. Colonists at Plymouth Rock referred to a shipment of neats from England. Even today, reference is made to neat's-foot oil and neat's leather.

Bulls and bears are the popular names for two points of view among those who invest in stocks or commodities. When more people want to buy than sell, prices of stocks or commodities rise. This is called a bull market. When more people want to sell than buy, then prices fall. This is a bear market. The expressions are believed to come from the way the two animals attack. The bull attacks by tossing his horns up in the air and the bear attacks by sweeping his paws downward. The term "stock market" comes from markets used to sell livestock. Wall Street originates from the use of a wall built in colonial New York City to control wandering livestock.

Entertainment

Humans have pitted themselves against the fighting bull for centuries. It appears that humankind's desire for blood and courage, still popular today, is as old as civilization itself. As early as 2500–1500 B.C. the sports arena existed on the Island of Crete, where "rodeos" or "bull grappling" involved women and men. Scenes from some of the world's first "rodeos" were cast in imperishable bronze, capturing for us a clear vision of what they were like. The bull grappler would first attempt to grab the charging bull by its horns; once this was accomplished, the grappler would hold onto the horns as the bull flipped its head, causing the grappler to turn a somersault and land on the bull's back. The grappler would then vault to the ground or into the arms of an accompanying performer.

Similar kinds of "rodeos" soon appeared in neighboring Greece and Italy. In Rome, wild cattle and lions were turned loose in an arena with unarmed Christians resulting in the deaths of large numbers of people. Also during this time, spectators watched as wild cattle battled gladiators armed with swords and spears.

The history of bullfighting in Spain, and eventually in Mexico, is older than the Spaniards themselves, since the natives were known as Iberians. The exact origin of this sport in Spain is not known, but records indicate that people were competing against bulls as early as the eleventh century. During the 1700s, bullfighting became a commercial enterprise in Spain as professionals were paid to engage in the sport. Breeding establishments also came into existence at this time; they bred bulls for superiority in fighting instinct. Today, thousands of 4- and 5-year-old bulls are used for bullfights in Spain, Mexico, and South America.

In the United States today, cattle are an integral part of many sporting events and provide entertainment for thousands of Americans. The rodeo would be far from complete without steer wrestling, team roping, calf roping, and bull riding. The latter is considered one of the most dangerous events of the rodeo. Many activities affiliated with horse shows such as cutting have their origin in the cattle industry (Fig. 20.5). County, state, and national fairs are a source of pleasure and entertainment to many people. The vast majority of the U.S. population in urban centers enjoys the opportunity to see their beef on-the-hoof.

Modern America

Cattle have had a tremendous influence on the development of the United States. As the original colonies grew in population, people looked westward for more land, more grass, and more opportunities. Cattle were second only to gold in helping stimulate the development of the West. Some of the towns named after cattle include Beefhide, Kentucky; Bulltown, West Virginia; Greybull, Wyoming; Cowtown, New Jersey; Cowpens, South Carolina; Holstein, Iowa; and Hereford, Texas.

Today, the glands, organs, and other body parts of cattle yield products that help relieve serious health problems in humans. Beef is a highly preferred food in American society. Although some consider beef a luxury item in the American diet, others consider it a necessity for a satisfying meal.

FIGURE 20.5 The sport of cutting evolved from the skills required to successfully handle and herd cattle from horseback on the ranches of the West and Southwest.
Source: Fred Field.

THE FUTURE OF THE BEEF INDUSTRY

Predicting the future is challenging, uncertain, and risky. Yet, in the beef industry, there are events from the past that continue to repeat themselves (e.g., beef cattle cycles) and current trends allow some events in the near future to be forecast with reasonable accuracy. Individuals must also be prepared to deal with the unpredictable surprises that occur from time to time.

In years past, there have been numerous times the future of the industry looked bleak because of the uncertainties at the time. In 1957, a noted cattle producer who had survived the Depression years told cattle producers at a conference that at no previous time could he recall the industry being so concerned about its future. Yet, out of that uncertainty would rise the feedlot industry and the emergence of demand for grain-fed beef.

During the 1970s and 1980s, the beef industry experienced numerous volatile conditions that raised serious questions about the future. These crises periodically centered on such issues as consumer reaction to high beef prices, consumer health concerns, market prices fluctuating widely (sometimes from week to week) causing significant changes in profitability, severe cost-price squeezes, decreased land values resulting in increased producer bankruptcies, and increased competition from other meats.

The mid-to-late 1990s had several issues raised that were projected to impact the future of the beef industry. Several beef producers expressed strong feelings that packer concentration and captive supplies were depressing cattle prices and would do the same in future years. Several studies showed that it was primarily large cattle numbers, not packer concentration, that were lowering cattle prices. These same studies projected that cattle prices would be higher in a few years if the cattle cycle reflected fewer cattle numbers.

During the mid-to-late 1990s, two trends were noted by industry leaders that could very possibly impact the beef industry during the next decade and beyond. These two trends are (1) higher beef production costs compared with the production costs of competitive meats—poultry and pork, and (2) the loss of consumer market share to poultry. These two trends are noted in the following summary statements by beef industry leaders:

> *Producers survival/consumer market share.* Beef industry has 5 years to position itself or it will become as the sheep industry—essentially extinct. (Harlan Ritchie, *BEEF,* 9/95, p. 20)
>
> *Low-cost producers* (in all segments of the production chain) *will survive* in this system of competitive markets. Others (*high-cost* producers) will eventually be unable to compete and *will exit* the business. (NCA Task Force Report)
>
> *The cattle industry doesn't have many options left.* They've got to produce premium beef . . . consistent quality and tenderness . . . and as quickly as possible. Take a lesson from the lamb industry . . . lamb producers missed the quality boat . . . annual consumption dropped to 1 lb per person. (W. D. Farr, Aug. 1995)
>
> *Loss of market share* is by far our greatest challenge; cow/calf producers can make a difference in beef's competitive position by 1) vigorously attacking the quality and consistency problems, and 2) increasing production efficiencies to narrow the price spread. (Maddux, 1995. *Range Beef Cow Symp.* XIV, p. 3)
>
> *Beef industry is faced with a daunting challenge:* Accomplish sweeping transformation in production, distribution, and marketing—or face continuing erosion of market share, per capita consumption, and ultimately, industry viability itself. (*Beef 2000 and Beyond Project Report,* NCA and The Beef Board, Oct. 1995)
>
> *The beef industry is segmented; the beef industry must change its infrastructure* . . . to provide the consumer with a tasty, tender, convenient beef product or beef will become a specialty product. (Topper Thorpe, Cattle-Fax, *National Beef Cattlemen,* July 1997; *W. Livestock J.,* 6/17/96)
>
> *We have serious choices to make about the future.* We have a few years to make the tough decisions that face this industry and have the guts to carry them out or get out of the way. Otherwise, we're going to be run over by competitors who are breathing down our necks. This is make or break time. (Bill Mies, *Adapting Total Quality Management to Feedyard Management.* NCBA Circular, 1993)

In the face of these challenges, the leadership of the National Cattlemen's Beef Association and the beef industry created a long-term strategic plan to correct these challenges. The vision statement was "A dynamic and profitable beef industry, which concentrates resources around a unified plan, consistently meets global consumer needs and increases demand." The two primary goals to help the industry achieve the vision were:

1. Stabilize beef demand by 2001.
2. Increase the opportunity for industry profitability.

Furthermore, the following areas of emphasis received the greatest investment of both time and dollars:

1. An emphasis on improving beef's nutritional image with a significant focus on food safety.
2. Adding value throughout the production, processing, and marketing chain.
3. Improve delivery of convenient, consistent, and highly palatable beef products via new product development.
4. Keep issues management in the forefront as a way to assure a legislative, regulatory, and judicial climate conducive to profitability.

As part of the process drivers of demand (Table 20.1) and drivers of profitability (Table 20.2) were identified.

TABLE 20.1 Demand Drivers for Beef

FOOD SAFETY

Assumptions

1. Food safety concerns affect the demand for beef.
2. Some government regulations will not be scientifically based.
3. Unified industry efforts will improve food safety.
4. Consumer groups will continue to challenge industry food safety practices and credibility.
5. Media will continue to sensationalize beef food safety issues.
6. If the industry does not address accountability, government will.
7. Food safety will continually be redefined by the public, government, and industry.
8. Food handling practices by all segments of the industry, including the customer, affects food safety.
9. Technology to identify new organisms or substances in beef will out-pace the scientific world's ability to evaluate/understand what is harmful vs. benign.

Outcomes

1. Consumer confidence in beef safety will increase.
2. Government regulations relative to food safety will increasingly be used by all segments of the industry.
3. Good management practices to improve beef safety will increasingly be used by all segments of the industry.
4. Legislation will be enacted removing liability for naturally occurring pathogens in pre-harvest beef operations.

PALATABILITY

Assumptions

1. The opportunity to increase tenderness exists both pre- and post-harvest and through preparation.
2. Wide variations in tenderness create consumer dissatisfaction.
3. Permanent individual animal identification will facilitate information exchange and add value throughout the production chain.

TABLE 20.1 (Continued)

4. Customer satisfaction will be enhanced by reconfiguring single or like-muscle fabrication.
5. Customer satisfaction will be enhanced by preparation information and precooked product.
6. Genetic markers and carcass EPDs for palatability attributes will gradually be utilized.
7. Processing technology will play a greater role in developing new products and improving existing products for consumers.
8. Taste will remain the number one consumer preference for repeat sales.

Outcomes

1. Predictable palatability will improve product value and customer satisfaction.
2. Objective measures utilizing instrument systems to predict palatability will be developed.

HEALTH & NUTRITION

Assumptions

1. Positive nutrition messages about beef are available.
2. Misinformation will continue and beef's opponents will continue to try to reduce beef in the diet.
3. The industry must be prepared to substantiate or challenge nutritional claims.
4. Nutrition perceptions influence consumer behavior.
5. Consumers' weight concerns influence diet.
6. Media attention on human nutrition will accelerate.
7. Concerns about dietary fat intake continue to influence beef demand.
8. Poultry has a perceived advantage over beef.
9. Merchandising of closely trimmed product is increasing.
10. New products will respond to health/nutrition concerns.
11. The beef industry will have to manage technology issues.

Outcomes

1. Consumers and influencers will have an increasingly positive attitude toward beef as healthy and nutritious relative to other nutrient sources.
2. Research will be done that will demonstrate the positive attributes of beef and these results will be effectively communicated.
3. A consistent nutrition message is delivered from all segments of the industry.
4. Governmental nutrition policy will increasingly recognize beef's positive role in the diet.

CONSUMER FRIENDLY PRODUCTS

Assumptions

1. Minimal time will be available for consumer shopping, preparation, and cleanup of meals.
2. Consumer demand for value will continue to be increased and redefined.
3. Currently, beef lags two-to-one behind competing meats in offering further processed, convenient products and will fall further behind unless new product development is accelerated.
4. Innovative partnerships within the beef system will deliver an increasing number of consumer friendly products.
5. Seamless communication must exist in the industry to meet consumer expectations.
6. New products will be developed to meet ethnic and demographic preferences.
7. Discipline to tight brand specifications will impact repeat sales and growth in beef demand.

Outcomes

1. The pace of new beef product development will be accelerated to meet consumers' time constraints, lifestyle expectations, and taste preferences.
2. Packaging will meet consumers' expectations and communicate the industry's messages.
3. Proprietary branded beef will supplement USDA grades as a standard by which beef is purchased at retail.
4. The variety of qualities and prices available to the consumer will increase through greater product differentiation.

Source: NCBA.

TABLE 20.2 PROFITABILITY DRIVERS FOR BEEF

COST EFFICIENCIES

Assumptions

1. Inefficiencies can be reduced at all levels of the industry.
2. The beef industry will continue to be capital-intensive and land-based.
3. Industry cooperation could enhance value by improving performance.
4. Change is necessary for increased profitability and survival.
5. Campaigns against the subtherapeutic use of antibiotics, if successful, will lower efficiency and raise costs for all food animal production but have the largest impact on total confinement systems, such an integrated broilers and swine operations.
6. Beef cannot compete with pork and poultry as the low-cost producer.
7. The three branches of government can have a negative impact on industry profitability.
8. Environmental philosophies will impact beef production costs.
9. A significant portion of the producers are not profit oriented, resulting in slower responses to economic signals.
10. Implementation of current and future technology will continue to be a challenge.
11. Consolidation within segments and between segments will continue.

Outcomes

1. The industry will continue to focus on programs to improve production efficiency.
2. The widespread use of benchmarking systems will allow producers to compare their production costs and performance efficiencies with other similar operations.
3. Cooperation and increased information transfer within industry will increase profit opportunities.
4. The regulatory climate for the beef industry will improve.

VALUE ENHANCEMENT

Assumptions

1. Value enhancement opportunities exist to develop economic and production practices that make cattle and beef products more convenient and/or cost effective to the next customer in the beef industry chain.
2. Value-adding opportunities exist at every stage of production and processing.
3. The customer defines value in every segment.
4. Distrust exists between segments of the industry resulting in missed opportunities.
5. Building trust between segments will accelerate information flow and enhance value.

Outcomes

1. All segments of the industry have increased opportunity for profitability.
2. Wider use of instrument technologies to measure animal and carcass values will allow more accurate pricing signals.
3. The sharing of information up and down the supply chain will cause barriers to fall and will result in greater cooperation.
4. Employment of production technologies, such as EPDs and gene markers, will allow the system to add value to the product.

Source: NCBA.

Seven trends will likely help shape the future of the beef industry in the United States. These trends are as follows:

1. International beef producers will become more competitive. However, demand for beef will increase on a global scale as more people gain more income. As the information in Chapter 18 indicated, the beef market is becoming more global in nature. U.S. producers will have to adapt to the demands of various market niches while proactively meeting the challenges of other nations competing for a share of those markets.
2. The competition at the meat case will intensify. Market segmentation will be met with an increasing number of value-added and/or branded products. Furthermore, the mindset of the future will be "product line" not "a product." As defined in Chapter 2, the beef indus-

try responded, albeit slowly, to the challenge put forth by the poultry industry's effective use of value-added, ready to heat and eat, and further processed product lines.

Perhaps the most important innovations by the beef industry were the efforts to find better protocols to fabricate muscles from the chuck and round that would be utilized to add value to what had previously been lower-priced wholesale cuts. Innovations in packaging, processing, and merchandising must be sustained if the beef industry is to continue to capture market share from its competitors.

3. Natural resource management and conservation of agricultural land are keys to the successful continuation of the beef industry. At the same time, cow-calf production will likely be shifted onto more marginal lands. The future of the cattle industry is inextricably tied to the management of forage resources (Chapter 15). Producers will benefit from the creation of natural resource planning and monitoring procedures designed to assure the long-term viability of the resources.

   One of the most daunting challenges facing agricultural production and rural communities is the continued loss of agricultural land to residential, commercial, and recreational development. As was pointed out in Chapter 15, it has been clearly documented that large blocks of agricultural land provide not only a source of food and fiber, but also preservation of open space, wildlife habitat, and rural culture. It will be increasingly important that producers find ways to reap economic benefits from the full spectrum of goods and services provided.

4. The focus on improvement of the palatability characteristics of beef must be balanced with the production of cattle that have acceptable levels of cutability, feedlot performance, and productivity at the cow-calf level. Single-trait selection has never served the industry well, nor will it in the future.

   The inherent need for improving knowledge and understanding about the systems approach to beef production will drive the need for new tools to assist producers with selection and management decisions. Selection indices and computerized decision-aid models will become the standard.

5. Commercial cow-calf enterprises and industry alliances focused on cost efficiencies and meeting diverse consumer demands will dictate the selection criteria of seedstock producers. Staying close to the customer will not be an option but rather a prerequisite. Seedstock producers will have to undertake considerable efforts to align their businesses with the needs of the beef marketing and production chain if they are to remain a viable segment of the industry.

   Furthermore, the successful seedstock producer of the future will offer a spectacular array of information-based services that go far beyond current standards. Genetic suppliers will be differentiated in the market by their ability to provide not only the best genotypes but also supporting tools for measuring their performance. Helping customers gain access to markets, improving management of specific genetic types, and enhancing their knowledge level through a variety of educational programs are other services that will be offered in the future.

6. The value of information in the future will increase dramatically. In short, those with documented information on their cattle will have the opportunity to participate in the capture of value-added returns. Those without information will be resigned to trying to survive in an increasingly limited commodity market.

   Breaking out of the commodity mentality will require that management systems conform to tighter market specifications while improving quality by eliminating mistakes and shortfalls. Just as the information age has changed virtually every other industry in the economic landscape, the need to collect, summarize, and utilize information will become a fact of life for beef producers.

7. The beef industry must find ways to engage a new generation of people who have the "right stuff" to ensure a successful future. The foundation of the industry is the creativity of people who choose to make a contribution to society by the production of food.

   Attracting the "best and the brightest" is of critical importance. The average age of U.S. beef producers continues to increase while each successive generation has fewer people choose to remain involved in agriculture.

The opportunities in the beef industry have never been greater. Demand has risen in response to the production, processing, and marketing innovations implemented during the past several years. Access to global markets creates additional opportunity as does increased levels of knowledge about reproduction, genetics, nutrition, and other management topics.

Motivated people who possess superior decision-making and problem-solving skills can find not only profitability but also personal fulfillment by pursuing careers in the beef cattle industry.

MANAGEMENT SYSTEMS HIGHLIGHTS

1. President John F. Kennedy stated that change is the law of life and those who look only in the past and the present are certain to miss the future. Other astute observers have said that we should learn from the past but not live in it, and where we have been and where we are today will largely dictate what the future will look like.
2. The future belongs to those who prepare for it—those who recognize that not only will changes occur, but they can direct the changes needed to accomplish their vision and goals (see Chapter 3).
3. While some changes cannot be controlled or significantly influenced (e.g., weather), it will be the decision of people (primarily beef producers and consumers) that will determine the future of the beef industry.
4. Management systems should change and implement the needed changes as the beef industry moves from the past and present into the future.

REFERENCES

Publications

Baker, F. H. 1967. History and Development of Beef and Dairy Performance Programs in the United States. *J. Anim. Sci.* 26: 1261.

Ball, C. E. 1998. *Building the Beef Industry. A Century of Commitment.* Englewood, CO: The National Cattlemen's Foundation.

Ball, C. E., and Cornett, S. 1990. Cattle Feeding in the United States: The First 300 years. *Cattle Feeding: A Guide to Management.* Amarillo, TX: Trafton Printing.

Betteridge, K. J. 1981. A Historical Look at Embryo Transfer. *J. Reprod. Fertil.* 62: 1.

Bourdon, R. 1992. Seedstock Selection in the Future. *Gelbvieh World* (April).

Brief History of the U.S. Meat and Livestock Industry. 1980. Meat Industry Magazine and National Live Stock and Meat Board.

Casteret, N. 1948. Lascaux Cave, Cradle of World Art. *National Geographic* 94: 771.

Clemen, R. A. 1923. *The American Livestock and Meat Industry.* New York: Ronald Press Co.

Dale, E. E. 1960. *The Range Cattle Industry: Ranching on the Great Plains from 1865 to 1924.* Norman, OK: University of Oklahoma Press.

Dobie, J. F. 1941. *The Longhorns.* Boston: Little, Brown.

Koch, R. M., and Alego, J. W. 1983. The Beef Cattle Industry: Changes and Challenges. *J. Anim. Sci.* 57: 28.

Maddux, J. 1995. *The Future of the Beef Industry: How Can We Stay Competitive.* Proceedings of the Range Beef Cow Symposium XIV, Gehring, NE.
Meat for Multitudes. 1981. *National Provisioner* (July 4): Vols. I, II.
National Cattlemen (annual July issue). Englewood, CO: National Cattlemen's Beef Association.
Purcell, W. D. 1997. *Future of the Beef Industry.* Proceedings of the Range Beef Cow Symposium, Rapid City, SD.
Qeuner, F. E. 1963. *A History of Domesticated Animals.* London: Hutchison and Co., Ltd.
Rouse, J. E. 1977. *The Criollo: Spanish Cattle in the Americas.* Norman, OK: University of Oklahoma Press.
Sanders, J. O. 1980. History and Development of the Zebu Cattle in the United States. *J. Anim. Sci.* 50: 1188.
Simoons, F. J. 1968. *A Ceremonial Ox of India.* Madison, WI: University of Wisconsin Press.
Skaggs, J. M. 1988. *Prime Cut: Livestock Raising and Meatpacking in the United States, 1607–1983.* College Station, TX: Texas A&M University Press.
Towne, C. W., and Wentworth, E. N. 1955. *Cattle and Men.* Norman, OK: University of Oklahoma Press.
Wade, L. C. 1987. *Chicago's Pride: The Stockyards, Packingtown, and Environs in the Nineteenth Century.* Chicago: University of Illinois Press.
Willham, R. L. 1982. Genetic Improvement of Beef Cattle in the United States: Cattle, People, and Their Interaction. *J. Anim. Sci.* 54: 659.
Willham, R. L. 1986. From Husbandry to Science: A Highly Significant Facet of Our Livestock Heritage. *J. Anim. Sci.* 62: 1742.

Visuals

"Buffalo Lessons" (22 min. VHS). An AgriBase, Inc. production.
"Taking Stock" (23 min. VHS). The Upjohn Company, Sigma Group, and AgriBase, Inc.

Glossary

abomasum Fourth stomach compartment of cattle that corresponds to the true stomach of monogastric animals.
abortion Delivery of fetus between time of conception and normal parturition.
abscess Localized collection of pus in a cavity formed by disintegration of tissues.
accrual method of accounting Accounting method whereby revenue and expenses are recorded when they are earned or incurred regardless of when the cash is received or paid.
accuracy or ACC (of selection) Confidence that can be placed in the EPD (expected progeny difference); for example, high (0.70 and above), medium (0.40–0.69), and low (below 0.40).
acetonemia *See* ketosis.
acidosis A high-acid condition in the rumen (pH 5.3–5.7) caused by rapid consumption or overconsumption of readily fermentable feed; may cause digestive disturbance and/or death.
ADG *See* average daily gain.
adjusted weaning weight Weaning weights of calves are adjusted to a standard age (205 days) and age of dam (5–9 years of age).
adjusted yearling weight Yearling weights of calves are adjusted to a standard age (365 days) by adding (160 times average daily postweaning gain) to the adjusted 205-day weight.
ad lib *See ad libitum.*
ad libitum (ad lib) Free choice; allowing cattle to eat all they want.
afterbirth Fetal membranes that are expelled after parturition. *See also* placenta.
AHIR *See* Angus herd improvement records.
AI *See* artificial insemination.
AI certificates Certificates issued by some breed associations that must be submitted before AI calves can be registered.
alliance An organization in the beef industry (horizontal or vertical) designed to improve profitability by improving coordination of beef production, processing, and merchandizing.
American Meat Institute (AMI) Association of meatpacking and processing companies.
American National Cattlewomen (ANCW) Organization of women involved in the promotion of beef through education and consumer relations programs.
AMI *See* American Meat Institute.
amnion Fluid-filled membrane located next to the fetus.
ANCW *See* American National Cattlewomen.
anestrous Period of time when the female is not in estrus; the nonbreeding season.
Angus Herd Improvement Records (AHIR) Performance records program administered by the American Angus Association.
animal unit (AU) A generalized unit for describing stocking density, stocking rate, and carrying capacity. Usually accepted to be a 1,000-lb cow with calf or 1.4 yearling cattle.
annual cow cost Cost (dollars) to keep a cow for a year.

animal unit month (AUM) Amount of feed or forage required to maintain one animal unit (e.g., a 1,000-lb cow and calf) for one month.
ante mortem Before death.
anthelmintic Drug or chemical agent used to kill or remove internal parasites.
antibiotic Product produced by living organisms such as yeast that destroys or inhibits the growth of other organisms, especially bacteria.
antibody Specific protein molecule that is produced in response to a foreign protein (antigen) that has been introduced into the body.
antigen Foreign substance that, when introduced into the blood or tissues, causes the formation of antibodies. Antigens may be toxins or native proteins.
appreciation Increase in the value of a capital asset (e.g., land) due to external influences such as inflation.
arteriosclerosis Disease resulting in the thickening and hardening of the artery walls.
artificial insemination (AI) Placing semen into the female reproductive tract (usually the cervix or uterus) by means other than natural service.
artificial vagina Device used to collect semen from a male while he mounts in a normal manner to copulate. The bull ejaculates into this device, which simulates the vagina of the female in pressure, temperature, and sensation to the penis.
as fed Feeding of feeds that contain their normal amount of moisture.
assets Items of value owned by a beef business or producer.
atherosclerosis Form of arteriosclerosis involving fatty deposits in the inner walls of the arteries. *See also* arteriosclerosis.
atrophy Shrinking or wasting away of tissue.
auction Market for cattle through which an auctioneer sells cattle to the highest bidder.
AUM *See* animal unit month.
autopsy Postmortem examination in which the body is dissected to determine cause of death.
average daily gain Pounds of liveweight gained per day.

backcross Mating of a two-breed crossbred offspring back to one of its parental breeds.
backfat Amount of fat over the animal's back, usually measured at the twelfth to thirteenth rib.
backgrounding Growing program (grazing or fed harvested feed) for feeder cattle from time calves are weaned until they are on a finishing ration in the feedlot.
balance sheet Financial statement that summarizes assets, liabilities, and net worth at a specific point in time. Also called a *net worth statement.*
balling gun Instrument inserted into the animal's throat to discharge pills.
Bang's disease *See* brucellosis.
barren Incapable of producing offspring.
basis Difference between the cash market price and the futures market price.
BCS *See* body condition score.
BCTRC Boneless, closely trimmed retail cuts from round, loin, rib, and chuck.
beef Meat from cattle (bovine species) other than calves. Meat from calves is called *veal.*
beef belt Area of the United States where commercial beef production, slaughtering, and processing are concentrated.
Beef Breeds Council National organization of beef breed associations.
beef checkoff program Beef Promotion and Research Act established in October 1986. Each time cattle are marketed, $1 per head is paid by the seller to the Beef Industry Council (BIC). Money is used in promotion, research, and education. Generates approximately $70 million per year.

Beef Improvement Federation (BIF) A federation of organizations, businesses, and individuals interested or involved in performance evaluation of beef cattle.
Beef promotion and research program *See* beef checkoff program.
Beef Quality Assurance (BQA) Program(s) designed to help beef producers assure that their production methods are not causing defects in beef products.
BIF *See* Beef Improvement Federation.
bioeconomic trait Any biological trait of economic importance.
biological efficiency Ratio of physical input to physical output (e.g., pounds of feed per 100 lbs of gain).
biological type Usually refers to size of cattle (large, medium, or small), growth rate, milk production (high, medium, or low), and age at puberty.
biotechnology Use of microorganisms, plant cells, and animal cells or parts of cells (such as enzymes) to produce industrially important products or processes.
birth weight (BW or B.Wt.) Weight of the calf taken within 24 hours after birth.
birth weight EPD The expected average increase or decrease in birth weight (lb) of a bull's calves when compared with other bulls in the same sire summary. A plus figure indicates an increase in birth weight, while a negative value is a decrease. The value is a measure of calving ease. *See also* expected progeny difference (EPD).
birth weight ratio Compares the individual birth weight of a calf to the herd average. Usually calculated within sex.
bloat Abnormal condition characterized by a distention of the rumen, usually seen on the left side, due to an accumulation of gases.
"bloom" Haircoat usually has a luster (shine) that gives the appearance of a healthy animal.
BLUP Best linear unbiased prediction method for estimating the breeding values of breeding animals.
body condition score (BCS) A visual score (usually 1 = thin; 9 = very fat) for body fatness, which is related to postpartum interval in beef females.
bolus (1) Regurgitated food. (2) Large pill for treating cattle.
Bos indicus Zebu (humped) cattle, including the Brahman breed in the United States.
Bos taurus Includes most cattle found in the United States and their European ancestors.
bovine Refers to a general family grouping of cattle.
bovine spongiform encephalopathy A degenerative disease that affects the central nervous system of cattle.
bovine viral diarrhea (BVD) Viral disease in cattle that can cause diarrhea, lesions of the digestive tract, repeat breeding, abortion, mummification, and congenital defects.
boxed beef Cuts of beef put in boxes for shipping from packing plant to retailers. These primal (rounds, loins, ribs, and chucks) and subprimal cuts are intermediate cuts between the carcass and retail cuts.
BQA *See* Beef Quality Assurance.
brand (1) Permanent identification of cattle, usually made on the hide with hot-iron or freeze branding. (2) Process of branding.
branded beef product A specifically labeled product that is differentiated from commodity items by its brand name. Certified Angus Beef, Laura's Lean, or Cattlemen's Collection are examples.
breakeven price Volume of output required for revenue to equal the total of fixed and variable expenses.
breaking Cutting carcasses into primal and subprimal cuts.
bred Female has been mated to a bull, usually assumed to be pregnant.
breech Buttocks. A breech presentation at birth occurs when the rear portion of the fetus is presented first.

breed Cattle of common origin and having characteristics that distinguish them from other groups within the same species.
breed complimentarity Combining breeds to take advantage of breed superiority for specific traits.
breeder In most beef breed associations, the owner of the dam of a calf at the time she was mated or bred to produce that calf.
breeding soundness examination (BSE) Evaluation of the reproductive potential of the bull giving the reproductive trait; a physical examination, measuring involves scrotal circumference, and evaluating a semen sample for motility and morphology.
breeding value Value of an animal as a parent. The working definition is twice the difference in performance between a very large number of progeny and the population average when individuals are mated at random within the population and all progeny are managed alike. The difference is doubled because only a sample half (one gene of each pair) is transmitted from a parent to each progeny.
brisket disease Noninfectious disease of cattle characterized by congestive right heart failure. It affects animals residing at high altitudes (usually above 7,000 ft). Sometimes referred to as "high mountain disease" or "high altitude disease."
British breeds Breeds of cattle, such as Angus, Hereford, and Shorthorn, originating in Great Britain.
brockle-faced White-faced with other colors splotched on the face and head.
broken mouth Some teeth are missing or broken.
broker Individual or firm that buys and sells options, futures contracts, and stocks and bonds for a commission fee.
browse Woody or brushy plants (e.g., sagebrush, shadscale, and other shrubs and bushes). Cattle feed on the tender shoots or twigs.
brucellosis Contagious bacterial disease that results in abortions; also called *Bang's disease.*
BSE *See* breeding soundness examination or bovine spongiform encephalopathy.
budget Financial form used to examine alternative plans for a beef operation and to estimate the profitability of each alternative.
bull Bovine male. The term usually denotes animals of breeding age.
buller-steer syndrome Behavior problem in which a steer is sexually attracted to other steers in the pen. The steer is ridden by the other steers, resulting in poor performance and injury.
bulling Term describing a cow or heifer in estrus.
bullock Young bull, typically less than 20 months of age.
butt-branded Hides from cattle that are hot-iron branded on the hip.
buttons May refer to cartilage on dorsal processes of the thoracic vertebrae. *See also* cotyledon.
BVD *See* bovine viral diarrhea.
bypass protein Feed protein that escapes microbial degradation in the rumen and is digested in the small intestine.
by-product Product of considerably less value than the major product. For example, the hide and offal are by-products while beef is the major product.

C-section *See* caesarean section.
Caesarean section Delivery of fetus through an incision in the abdominal and uterine walls.
calf Young male or female bovine animal under 1 year of age.
calf crop *See* percent calf crop.
calorie Amount of heat required to raise the temperature of 1 g of water from 15°C to 16°C.
calve Giving birth to a calf. Same as parturition.

calving difficulty (dystocia) Abnormal or difficult labor, causing difficulty in delivering the calf.
calving interval Time (days or months) between the birth of a calf and the birth of a subsequent calf, both from the same cow.
calving season Season(s) of the year when calves are born.
cancer eye Cancerous growth on the eyeball or eyelid.
carcass evaluation Techniques for measuring components of meat quality and quantity in carcasses.
carcass merit Value of a carcass for consumption.
carotene Orange pigment found in leafy plants (e.g., alfalfa), yellow corn, and other feeds that can be broken down to form two molecules of vitamin A.
carrier Heterozygous individual having one recessive gene and one dominant gene for a given pair of genes (alleles).
carrying capacity The maximum stocking rate that will achieve a target level of animal performance on a particular grazing unit under a specified grazing method. Or, the potential number of animals or liveweight that may be supported on a unit area for a grazing season based on forage potential.
case-ready Beef cuts received by the retailer that do not require further processing before they are put in the retail case for selling.
cash flow Cash receipts and cash expenses.
cash-flow budget Detailed estimate of the projected cash receipts and expenses over a future period of time used to evaluate the financial feasibility of a plan.
cash-flow statement Financial statement summarizing all cash receipts and disbursements over a period of time (usually monthly for a year).
cash market price Price that results when cattle go to market.
cash method of accounting An accounting method by which revenue and expenses are recorded when the cash is received or paid.
castrate (1) To remove the testicles. (2) An animal that has had its testicles removed.
cattalo Cross between domestic cattle and bison.
Cattle-Fax Nonprofit marketing organization governed by cattle producers. Market analysis and information are provided to members by a staff of market analysts.
Cattlemen's Beef Board (CBB) Responsible for the management of the beef checkoff program, oversees the collection of $1 per head on domestic cattle as well as the equivalent on imported beef, beef products, and cattle.
CBB *See* Cattlemen's Beef Board.
central test Location where animals are assembled from several herds to evaluate differences in certain performance traits under uniform management conditions. Usually involves breeding bulls, though some slaughter steer and heifer tests exist.
Certified Angus Beef Branded beef product supplied by Angus or Angus crossbred cattle that meets certain carcass specifications.
cervix Portion of the female reproductive tract between the vagina and the uterus. It is usually sealed by thick mucus except when the female is in estrus or delivering young.
checkoff *See* beef checkoff program.
chorion Outermost layer of fetal membranes.
chromosome Rodlike or stringlike body found in the nucleus of the cell that is darkly stained by chrome dyes. The chromosome contains the genes.
chronic Regular appearance of a symptom or situation.
chuck Wholesale cut (shoulder) of the beef carcass.
class Group of cattle determined primarily by sex and age (e.g., market class or showring class).

clitoris A highly sensitive organ in females located inside the ventral part of the vulva. It is homologous to the penis in the male.
clone Genetically identical organisms produced by nucleus substitution or embryo division.
closed herd Herd in which no outside breeding stock (cattle) are introduced.
cod Scrotal area of steer remaining after castration.
cold shortening Sarcomeres as part of the muscle fiber, shorten too rapidly during the chilling of the carcass, thus decreasing meat tenderness.
collagen Primary protein in connective tissue. Collagen envelops individual muscle fibers and attaches muscles to bones.
collateral relatives Relatives of an individual that are not its ancestors or descendants. Brothers and sisters are examples of collateral relatives.
colon Large intestine from the end of the ileum and beginning with the cecum to the anus.
Colorado branded Hides from cattle hot-iron branded on the ribs.
colostrum First milk given by a female following delivery of her calf. It is high in antibodies that protect the calf from invading microorganisms.
compensatory gain Faster-than-normal rate of gain following a period of restricted gain.
compensatory growth *See* compensatory gain.
complementarity Using breed differences to achieve a more optimum additive and nonadditive breed composition for production and carcass traits of economic value.
composite breed Breed that has been formed by crossing two or more breeds.
composition Usually refers to the carcass composition of fat, lean, and bone.
Compudose® Growth implant containing estradiol and progesterone.
computer Electronic machine which by means of stored instructions and information performs rapid, often complex, calculations or compiles, correlates, and selects data.
ConAgra One of the three largest beef-packing companies.
concentrate Feed that is high in energy, low in fiber content, and highly digestible.
conception Fertilization of the ovum (egg).
conditioning Treatment of cattle by vaccination and other means prior to putting them in the feedlot.
conformation Physical form of an animal; its shape and arrangement of parts.
congenital Acquired during prenatal life. Condition exists at birth. Often used in the context of congenital (birth) defects.
contemporaries Group of animals of the same sex and breed (or similar breeding) that have been raised under similar environmental conditions (same management group).
continental breed *See* European breed.
continuous grazing A method of grazing where animals have unrestricted access to an entire grazing unit throughout a large portion or all of a grazing season.
controlled grazing Grazing management designed to improve utilization of forage either by (1) allocating pasture in subunits with grazing periods typically less than 5 days or (2) varying stocking rate to match forage growth rate and availability (put-and-take stocking).
cooler A room in packing plant where carcasses are chilled after slaughter and prior to processing.
corpus luteum Yellowish body in the ovary. The cells that were follicular cells develop into the corpus luteum, which secretes progesterone. It becomes yellow in color from the yellow lipids that are in the cells.
correlation coefficient Measure of how two traits vary together. A correlation of +1.00 means that two traits will move in the same direction (either increase or decrease). A correlation of −1.00 means that as one trait increases the other decreases—a perfect negative, or inverse, relationship. A correlation of 0.00 means that as one trait increases, the other

may increase or decrease—no consistent relationship. Correlation coefficients may vary between +1.00 and −1.00.

cost of gain Total of all costs divided by the total pounds gained; usually expressed on a per pound basis.

cotyledon Area where the placenta and the uterine lining are in close association such that nutrients can pass to and wastes can pass from the circulation of the developing young. Sometimes referred to as *button*.

cow Sexually mature female bovine animal that has usually produced a calf.

cow-calf operation Management unit that maintains a breeding herd and produces weaned calves.

cow hocked Condition in which the hocks are close together but the feet stand apart.

creep Enclosure where calves can enter to obtain feed but cows cannot enter. This process is called creep feeding.

creep feeding *See* creep.

creep grazing The practice of allowing calves to graze areas that cows, with a lower nutritional requirement, cannot access.

crest Bulging, top part of the neck on a bull.

crossbred Animal produced by crossing two or more breeds.

crossbreeding Mating animals from different breeds. Utilized to take advantage of hybrid vigor (heterosis) and breed complementarity.

cryptorchidism Retention of one or both testicles in the abdominal cavity.

cud Bolus of feed that cattle regurgitate for further chewing.

cull To eliminate one or more animals from the breeding herd or flock.

currentness Marketing term indicating how feedlots market fed cattle. If current, then feedlots market cattle on schedule. If feedlots are not current, then a backlog of cattle usually results—these cattle typically have higher slaughter weights, poorer yield grades, and usually lower prices.

custom feeding Cattle feeders who provide facilities, labor, feed, and care as a service but they do not own the cattle.

cutability Fat, lean, and bone composition of the beef carcass. Used interchangeably with yield grade. *See also* yield grades.

cutting chute Narrow chute where cattle go through in single file, with gates such that selected animals can be diverted into pens alongside the chute; also referred to as a sorting chute.

cwt Abbreviation for hundredweight (100 lbs).

cycling Infers that nonpregnant females are having estrous cycle.

dam Female parent.

dark cutter Color of the lean (muscle) in the carcass has a dark appearance, usually caused by stress (excitement) to the animal prior to slaughter.

deflation General decrease in prices that increases the purchasing power of a dollar.

dehorn To remove the horns of an animal.

deoxyribonucleic acid (DNA) Molecule that comprises the genetic material of animals. Genes are units of DNA. *See also* gene.

depreciation Decrease in the value of an asset due to age, use, and obsolescence; the prorated expense of owning an asset.

dewclaws Hard, horny structures above the hoof on the rear surface of the legs of cattle.

dewlap Loose skin under the chin and neck of cattle.

digestibility Quality of being digestible. If a high percentage of a given food taken into the digestive tract is absorbed into the body, that food is said to have high digestibility.

direct selling Selling cattle from one ranch to another, from ranch to feedlot, or from feedlot to packer.

disease Any deviation from the normal state of health.

DM *See* dry matter.

DNA *See* deoxyribonucleic acid.

DNA markers Areas of the genome at which differences in the DNA sequence can be visually detected. A marker locus by itself may not have a direct effect on a phenotypic trait, but it may be located close to a gene that does directly affect a trait. Markers can serve as location reference points for gene mapping and marker-assisted selection.

DNA probe A method to determine an animal's genotype for a particular gene or marker.

dominance One allele masks the effect of another (recessive) allele.

double-entry accounting System of bookkeeping in which every transaction is recorded as a debit in one or more accounts and as a credit in one or more accounts such that the total of the debit entries equals the total of the credit entries.

double muscling A simple recessive trait evidenced by an enlargement of the muscles with large grooves between the muscle systems, especially noticeable in the hind leg.

drench To give fluid by mouth.

dressed beef Carcasses from cattle.

dressing percentage Percentage of the live animal weight that becomes the carcass weight at slaughter. It is determined by dividing the carcass weight by the liveweight then multiplying by 100. Also referred to as *yield.*

drop Body parts removed at slaughter, primarily the hide, head, shanks, and offal.

drop credit Value of the drop.

dropped Being born (e.g., "the calf is dropped").

dry (cow) Refers to a nonlactating female.

dry matter Feed after water (moisture) has been removed (100% dry).

dystocia Difficult birth. *See* calving difficulty.

earmark Method of permanent identification by which slits or notches are placed in the ear.

ear tag Method of identification by which a numbered, lettered, and/or colored tag is placed in the ear.

early maturity Early puberty as the animal begins to fatten early, sometimes before desired slaughter weight is obtained.

EBV *See* breeding value; expected progeny difference (EPD).

economic efficiency Ratio of output value to cost of input.

economic value The net return within a herd for making a pound or percentage change of the trait in question.

edema Abnormal fluid accumulation in the intercellular tissue spaces of the body.

efficiency Ratio of output to input. *See also* biological efficiency; economic efficiency.

80%–20% rule Basic rule of management. Too often managers expend 80% of their efforts on "the trivial many" problems that produce only 20% of the results. Effective managers recognize that spending time (20%) on problems or situations that count most will produce 80% of the desired results.

ejaculation Discharge of semen from the male.

emaciation Thinness; loss of flesh such that bony structures (hips, ribs, and vertebrae) become prominent.

embryo Fertilized egg in its early stages of development; after body parts can be distinguished, it is known as a fetus.

embryo splitting Dividing an embryo into two or more similar parts to produce several calves from a single embryo.

embryo transfer (ET) Transfer of fertilized egg(s) from a donor female to one or more recipient females.

Endangered Species Act (ESA) A regulatory statute intended to protect threatened and endangered species by preserving the ecosystems on which they depend.

endocrine gland Ductless gland that secretes a hormone into the bloodstream.

energy Force, or power, that is used to drive a wide variety of systems. It can be used as power of mobility in animals, but most of it is used as chemical energy to drive reactions necessary to convert feed into animal products and to keep the animals warm and functioning.

enterprise Segment of the cattle business or an associated business that is isolated by accounting procedures so that its revenue and expenses can be identified.

enterprise budget Detailed list of all estimated revenue and expenses associated with a specific enterprise.

environment Total of all external (nongenetic) conditions that affect the well-being and performance of an animal.

Environmental Protection Agency (EPA) Independent agency of the federal government established to protect the nation's environment from pollution.

enzyme Complex protein produced by living cells that causes changes in other substances in cells without being changed itself and without becoming a part of the product.

EPA *See* Environmental Protection Agency.

EPD *See* expected progeny difference.

epididymis Long, coiled tubule leading from the testis to the vas deferens.

epididymitis Inflammation of the epididymis.

epistatis Situation in which a gene or gene pair masks (or controls) the expression of another nonallelic pair of genes.

equity *See* net worth (equity).

eruction (or eructation) Elimination of gas by belching.

esophageal groove Groove in the reticulum between the esophagus and omasum. Directs milk in the nursing calf directly from the esophagus to the omasum.

estrogen Any hormone (including estradiol, estriol, and estrone) that causes the female to express heat and to be receptive to the male. Estrogens are produced by the follicle of the ovary and by the placenta and have additional body functions.

estrous Adjective meaning "heat" that modifies such words as *cycle.* The estrous cycle is the heat cycle, or the time from one heat to the next.

estrous synchronization Controlling the estrous cycle so that a high percentage of the females in the herd express estrus at approximately the same time.

estrus Period of mating activity in the heifer or cow. Same as *heat.*

ET *See* embryo transfer.

ethology Study of animal behavior.

EU *See* European Union.

European breed Breed originating in European countries other than England (these are called British breeds); a larger dual-purpose breed such as Charolais, Simmental, and Limousin; also called *continental* or *exotic breed* in the United States.

European Union Group of fifteen countries (Austria, Belgium, Denmark, Finland, France, Germany, Great Britain, Greece, Ireland, Italy, Luxembourg, Portugal, Spain, Sweden, and The Netherlands) whose major objective is to coordinate the development of economic activities. Previously called the European Economic Community, European Community, and Common Market.

eviserate Removal of the internal organs during the slaughtering process.

Excel One of the three largest U.S. beef-packing companies.

exotic breed *See* European breed.
expected progeny difference (EPD) One-half of the breeding value of a sire or dam; the difference in expected performance of future progeny of a sire, when compared with that expected from future progeny of bulls in the same sire summary.

F_1 Offspring resulting from the mating of a purebred (straightbred) bull to purebred (straightbred) females of another breed.
fabrication Breaking the carcass into primal, subprimal, or retail cuts. These cuts may be boned and trimmed of excess fat.
fat thickness Usually refers to the amount of fat that covers muscles; typically measured at the twelfth and thirteenth rib as inches of fat over the *longissimus dorsi* muscle (rib eye).
FDA *See* Food and Drug Administration (FDA).
feces Bodily wastes; excretion product from the intestinal tract.
fed cattle Steers and heifers that have been fed concentrates, usually for 90–120 days in a feedlot or until they reach a desired slaughter weight.
feed additive Ingredient such as an antibiotic or hormonelike substance that is added to a diet to perform a specific role.
feed bunk Trough or container used to feed cattle.
feed conversion *See* feed efficiency.
feed efficiency (1) Amount of feed required to produce a unit of weight gain or milk. (2) Amount of gain made per unit of feed.
feed markup Per ton feed cost charged to the customer by the feedyard for the cattlefeeding services it provides.
feeder (1) Cattle that need further feeding prior to slaughter. (2) Producer who feeds cattle.
feeder grades Grouping of feeder cattle to predict the slaughter weight endpoint to a desirable fat-to-lean composition. Frame size and thickness are the two criteria used to determine feeder grade.
feedlot Enterprise in which cattle are fed grain and other concentrates for usually 90–120 days. Feedlots range in size from less than 100-head capacity to many thousands.
feedyard Cattle-feeding facility.
femininity Well-developed secondary female sex characteristics, udder development, and refinement in head and neck.
fertility Capacity to initiate, sustain, and support reproduction.
fertilization Process by which a sperm unites with an egg to produce a zygote.
fetus Late stage of individual development within the uterus. Generally, the new individual is regarded as an embryo during the first half of pregnancy and as a fetus during the last half.
fill Contents of the digestive tract.
financing Acquiring control of assets by borrowing money.
finish (1) Degree of fatness of an animal. (2) Completion of the last feeding phase of slaughter cattle.
finished cattle Fed cattle whose time in the feedlot is completed and are now ready for slaughter.
finishing ration Feedlot ration, usually high in energy, that is fed during the latter part of the feeding period.
fitting Proper feeding, grooming, and handling of an animal, usually to prepare it for the showring.
fixed cost Costs incurred whether or not production occurs (e.g., interest, taxes).
flehmen Pattern of behavior expressed by the bull during sexual activity. The upper lip curls up and the bull initiates the smelling process in the vicinity of the vulva or urine.

flushing Placing females on a high level of nutrition before breeding to decrease postpartum interval and possibly stimulate an increased conception rate.

FMD *See* foot and mouth disease.

FMI *See* Food Marketing Institute.

FOB (or fob) Free on board; buyer pays freight after loading.

follicle Blisterlike, fluid-filled structure in the ovary that contains the egg.

follicle-stimulating hormone (FSH) Hormone produced and released by the anterior pituitary that stimulates the development of the follicle in the ovary.

Food and Drug Administration (FDA) U.S. government agency responsible for protecting the public against impure and unsafe foods, drugs, veterinary products, biologics, and other products.

Food Marketing Institute (FMI) National association of food retailers and wholesalers located in Washington, DC, that conducts programs of research, education, and public affairs for its members.

foot and mouth disease (FMD) Highly contagious disease affecting many species of livestock including cattle. This disease is of particular concern in that it can lead to loss of export markets.

footrot Disease of the foot in cattle.

forage Grazed or harvested herbaceous plants that are utilized by cattle.

forage production The total amount of dry matter produced per unit of area on an annual basis (e.g., lb/acre/year).

forb Weedy or broadleaf plants (unlike grasses) that serve as pasture for animals (e.g., clover, alfalfa).

forward contracting Future delivery of a specified type and amount of product at a specified price.

founder Nutritional ailment resulting from overeating. Lameness in front feet with excessive hoof growth usually occurs.

frame score Score based on visual evaluation of skeletal size or by measuring hip height (from ground to top of hips). This score is related to the slaughter weights at which cattle grade Choice or have comparable amounts of fat cover over the loin eye at the twelfth to thirteenth rib.

frame size Usually measured by frame score or estimated visually.

freemartin Female born twin to a bull (approximately 90% of such heifers will never conceive).

FSH *See* follicle-stimulating hormone (FSH).

full sibs Animals having the same sire and dam.

futures market Electronic market through which buyers and sellers trade contracts on commodities or raw materials. Futures contracts are available for a variety of delivery months. However, delivery of actual products seldom occurs. Futures markets are used as a risk management tool or as a speculative venture.

GATT (General Agreement on Tariffs and Trade) An agreement originally negotiated in Geneva, Switzerland, in 1947 among 23 countries, including the United States, to increase international trade by reducing tariffs and other trade barriers. The agreement provides a code of conduct for international commerce and a framework for periodic multilateral negotiations on trade liberalization and expansion.

gene Segment of DNA in the chromosome that codes for a trait and determines how a trait will develop.

gene map A blueprint of the chromosomes of a species indicating the relative order of location of genes and DNA markers.

generation interval Average age of the parents when the offspring are born.

generation turnover Length of time from one generation of animals to the next generation.

genetic correlation Correlation between two traits that arises because some of the same genes affect both traits. *See* correlation coefficient.

genetic engineering Changing the characteristics of an animal by altering or rearranging its DNA. It is an all-embracing term for several techniques: (1) manipulations at a cellular level (cloning); (2) manipulation of the DNA itself (gene manipulation); (3) changing the DNA sequence through the selection and mating of cattle.

genome Total number of genes in a species.

genotype Genetic constitution or makeup of an individual. For any pair of alleles, three genotypes (e.g., AA, Aa, and aa) are possible.

genotype–environmental interaction Variation in the relative performance of different genotypes from one environment to another. For example, the superior cattle (genotypes) for one environment may not be superior for another environment.

gestation Time from conception until the female gives birth, an average of 285 days in cattle.

goal Target or desired condition that motivates the decision maker.

gonad Testis of the male; ovary of the female.

gonadotrophin Hormone that stimulates the gonads.

grade augmentation Supplementation of traditional USDA visual carcass grading using objective instrumentation.

grade and yield Marketing transaction whereby payment is made on the basis of carcass weight and quality grade.

grading up Continued use of purebred sires of the same breed in a grade herd.

grass tetany Disease of cattle marked by staggering, convulsions, coma, and frequently death, which is caused by a mineral imbalance (magnesium) while grazing lush pasture.

grazier A person who manages grazing livestock.

grazing cell A parcel of land subdivided into paddocks and grazed rotationally.

grazing cycle The length or passage of time between two grazing periods in a particular paddock of a grazing unit. One grazing cycle includes one grazing period and one rest period.

gross margin Difference between the revenue and variable production cost for one unit (one acre or one animal) of an enterprise.

growing ration Usually a high-roughage ration whereby gains of 0.25–2 lb per day are anticipated.

growth Increase in protein over its loss in the animal body. Growth occurs by increases in cell numbers, cell size, or both.

grubs Larvae of the heel fly found on the backs of cattle under the hide.

half-sib Animals having one common parent.

hand mating Bringing a female to a male for breeding, after which she is removed from the area where the male is located (same as hand breeding).

hanging tenderloin Part of the diaphragm muscle, not to be confused with the tenderloin of the carcass.

"hard keeper" Term used when an animal does not perform well; it may have hardware, parasites, or show the effects of disease.

hardware disease Ingested sharp objects perforate the reticulum and cause infection of the heart sac, lungs, or abdominal cavity.

Hazard Analysis Critical Control Point (HACCP) A process used to identify those steps in production where mistakes may critically damage the final performance of the product and to establish a system of monitoring and intervention to avoid these mistakes.

heart girth Circumference of the animal's body, measured just behind the shoulders.
heat *See* estrus.
heat increment Increase in heat production following consumption of feed when an animal is in a thermoneutral environment. It includes additional heat generated in fermentation, digestion, and nutrient metabolism.
hedge Risk management strategy that allows a producer to lock in a price for a given commodity at a specified time.
heifer Young female bovine cow prior to the time that she has produced her first calf.
heiferette Heifer that has calved once and is then fed for slaughter; the calf has usually died or been weaned at an early age.
herd Group of cattle (usually cows) that are in a similar management program.
heredity The transmission of genetic or physical traits of parents to their offspring.
heritability Portion of the phenotypic differences between animals that is due to heredity.
hernia Protrusion of an intestine through an opening in the body wall (also commonly called rupture). Two types of hernias—umbilical and scrotal—occur in cattle.
heterosis Performance of offspring that is greater than the average of the parents. Usually referred to as the amount of superiority of the crossbred over the average of the parental breeds. Also called *hybrid vigor.*
heterozygous Designates an individual possessing unlike genes for a particular trait.
hides Skins from cattle.
high mountain disease *See* brisket disease.
hiplock Condition at calving in which the hips of the calf cannot pass through the pelvis of the cow.
homozygous Designates an individual whose genes for a particular trait are alike.
hormone Chemical substance secreted by a ductless gland. Usually carried by the bloodstream to other places in the body where it has its specific effect on another organ.
hot carcass weight Weight of carcass just prior to chilling.
"hot fat trimming" Removal of excess surface fat while the carcass is still "hot," prior to chilling the carcass.
HRI (hotel, restaurant, and institutional) Used in the context that some beef is supplied to the HRI trade.
hybrid vigor *See* heterosis.
hydrocephalus Condition characterized by an abnormal increase in the amount of cerebral fluid, accompanied by dilation of the cerebral ventricles.
hypothalamus Portion of the brain found in the floor of the third ventricle that regulates reproduction, hunger, and body temperature and performs other functions.

IBP One of the three largest beef-packing companies.
immunity Ability of an animal to resist or overcome infection.
implant To graft or insert material to intact tissues.
inbreeding Mating of individuals more closely related than the average individuals in a population. Inbreeding increases homozygosity in the cattle population but does not change gene frequency.
income Difference between revenue and expenses that is referred to as net income; gross income refers to total income.
income statement Financial statement that summarizes all revenues and expenses and used to determine the net income or net loss for a given period of time, usually a year.
independent culling level Selection method whereby minimum acceptable phenotypic levels are assigned to several traits.
index Overall merit rating of an animal.

inflation General increase in prices that decreases the purchasing power of a dollar.
insemination Deposition of semen in the female reproductive tract.
intake The amount of feed consumed by an animal per day. Intake is usually expressed as a percent of body weight or in pounds per day.
integrated resource management (IRM) Multidisciplinary approach to managing cattle more efficiently and profitably; management decisions are based on how all resources are affected.
integration Bringing together of two or more segments of beef production and processing under one centrally organized unit.
intensive grazing management (IGM) Grazing management where a grazing unit is subdivided into subunits (paddocks) with grazing periods typically less than five days.
intensive rotational grazing Synonymous with "intensive grazing management."
interest rate Charge or fee associated with borrowed money.
intermuscular fat Fat located between muscle systems. Also called *seam fat*.
intramuscular fat Fat within the muscle or marbling.
inter se mating Mating of animals within a defined population. Literally to mate among themselves.
intravenous Within the vein. An intravenous injection is made into a vein.
in vitro Outside the living body; in a test tube or artificial environment.
in vivo Within the living body.
involution Return of an organ to its normal size or condition after being enlarged (e.g., the uterus after parturition). A decline in size or activity of other tissues; the mammary gland tissues normally involute with advancing lactation.
ionophore Antibiotic that enhances feed efficiency by changing microbial fermentation in the rumen.
IRM *See* integrated resource management (IRM).

joint venture Any business arrangement whereby two or more parties contribute resources to and engage in a specific business undertaking.

kidney knob The kidney and the fat that surrounds it.
kidney, pelvic, and heart fat (KPH) The internal carcass fat associated with the kidney, pelvic cavity, and heart expressed as a percentage of chilled carcass weight. The kidney is included in the estimate of kidney fat. Used in the calculation of yield grade.
ketosis Condition characterized by a high concentration of ketone bodies in the body tissues and fluids. Also called acetonemia.
kosher meat Meat from ruminant animals (with split hooves) that have been slaughtered according to Jewish law.

labor (1) Parturition or the birth process. (2) Human resource that produces goods or provides services.
lactation Secretion and production of milk.
LEA *See* loin-eye area (LEA); rib-eye area (REA).
lethal gene A gene that causes the death of an individual at some stage of life.
legume Any plant type within the family Leguminosae, such as pea, bean, alfalfa, and clover.
leucocytes White blood cells.
LH *See* luteinizing hormone (LH).
liabilities Obligations or debts owed by a business or person to others.

libido Sex drive or the male's desire to mate.

lice Small, flat, wingless insects with sucking mouthparts that are parasitic on the skin of animals.

limited partnership Partnership consisting of at least one general partner who is responsible for the management and liabilities of the business, and at least one limited partner whose liability is limited to his or her investment.

linear programming Mathematical technique used to find profit-maximizing combinations of production activities or cost-minimizing combinations of ingredients subject to a number of linear relationships that constrain the activities or ingredients.

linebreeding Form of inbreeding whereby a bull's genes are concentrated in a herd. The average relationship of the individuals in the herd to this ancestor (outstanding individual or individuals) is increased by linebreeding.

linecrossing Crossing of inbred lines.

liquidate To convert to cash; to sell.

liver flukes Parasitic flatworm found in the liver.

load Pounds (number) of cattle that can be hauled on a large cattle truck. For example, pot load is 42,000–52,000 lbs (40–42 head of slaughter steers, 72 yearlings, or 100 calves).

locus Place on a chromosome where a gene is located.

loin-eye area (LEA) Area of the *longissimus dorsi* muscle, measured in square inches between the twelfth and thirteenth ribs. Usually referred to as rib-eye area (REA).

long yearling Animal between 18 months and 2 years of age.

longevity Life span of an animal; usually refers to the number of years a cow remains productive.

longissimus dorsi *See* rib-eye area (REA).

lousy Infested with lice.

luteinizing hormone (LH) Protein hormone produced and released by the anterior pituitary that stimulates the formation and retention of the corpus luteum. It also initiates ovulation.

maintenance Condition in which the body is maintained without an increase or decrease in body weight and with no production or work being done.

mammary gland Gland that secretes milk.

management Act, art, or manner of managing, handling, controlling, or directing a resource or integrating several resources.

management systems Methods of systematically organizing information from several resources to make effective management decisions. *See also* integrated resource management (IRM).

marbling Flecks of intramuscular fat distributed in muscle tissue. Marbling is usually evaluated in the rib eye between the twelfth and thirteenth ribs.

MARC *See* Meat Animal Research Center (MARC).

margin (1) "Earnest money" that serves as default protection in a futures transaction. (2) Difference between prices at different levels of the marketing system. (3) Difference between cost and sale price.

marker-assisted selection A method of genetic evaluation that takes into consideration the DNA marker genotype along with conventional selection procedures.

market class Cattle grouped according to their use, such as slaughter, feeder, or stocker.

market grade Cattle grouped within a market class according to their value.

market niche Segment of consumer demand targeted by a specialized production and marketing plant. Examples include the "white tablecloth" restaurant trade, health foods, and convenience foods.

masculinity Well-developed secondary sex characteristics in the neck, chest, and shoulders of the bull.
masticate To chew food.
mastitis Inflammation of the mammary gland.
maternal Pertaining to the female (cow or heifer).
maternal first-calf calving ease Ease with which a sire's daughters calve as first-calf heifers (under 33 months of age). Reported as a ratio or an EPD.
maternal heterosis Heterosis for those traits influenced by the cow genotype. For example, maternal heterosis of weaning weight refers to the increase in weaning weight from being raised on a crossbred cow rather than a straightbred cow.
maternal traits All the traits expressed by the cow. A limited definition implies milk and weaning weight production of the cow.
maternal weaning weight Weaning weight of a bull's daughter's calves. The EPD value predicts the difference in average 205-day weight of a bull's daughter's calves compared with daughters of all other bulls evaluated. It can be calculated by adding one-half of the bull's EPD for weaning weight to his milk EPD.
maturity An estimation of the chronological age of the animal or carcass.
maverick Unbranded animal, usually on the range.
M/B or M:B ratio *See* muscle-to-bone ratio.
mean (1) Statistical term for average. (2) Term used to describe cattle having bad behavior.
meat Tissues of the animal body that are used for food.
Meat Animal Research Center (MARC) Large U.S. government research center located in Clay Center, NE, that conducts numerous beef cattle research projects.
Meat Export Federation (MEF) *See* U.S. Meat Export Federation (MEF).
MEF *See* U.S. Meat Export Federation (MEF).
melengestrol acetate (MGA) Feed additive that suppresses estrus in heifers; used in estrus synchronization and feedlot heifers.
MERCOSUR (Common Market of the South) A customs union implemented in January 1995, and including Argentina, Brazil, Paraguay, and Uruguay. MERCOSUR represents the culmination of bilateral negotiations started by Argentina and Brazil in 1986.
metabolic body size Weight of the animal raised to the 3/4-power ($W^{0.75}$); a figure indicative of metabolic needs and of the feed required to maintain a certain body weight.
metabolism (1) Sum total of chemical changes in the body, including the "building up" and "breaking down" processes. (2) Transformation by which energy is made available for body uses.
metabolizable energy Gross energy in the feed minus the sum of energy in the feces, gaseous products of digestion, and energy in the urine. Energy that is made available for body uses.
metritis Inflammation (infection) of the uterus.
MGA *See* melengestrol acetate (MGA).
middle meats Rib and loin of a beef carcass. These primals generally yield the highest-priced beef cuts.
milk EPD Estimate of the milking ability of a bull's daughters compared with the average of the daughters of other bulls. Reported in pounds of weaning weight; positive values indicate above-average performance and negative numbers indicate below-average maternal ability. *See also* expected progeny difference (EPD).
mill feed Any feed that is subjected to the milling process.
minimum culling level Selection method in which an animal must meet minimum standards for each trait desired in order to qualify for being retained for breeding purposes.
mites Very small arachnids that can be parasites of cattle.

morbidity Measurement of illness; morbidity rate is the number of individuals in a group that become ill during a specified time period.
mortality rate Number of individuals that die from a disease during a specified time period, usually one year.
most probable producing ability (MPPA) Estimate of a cow's future productivity for a trait (such as progeny weaning weight ratio) based on her past productivity. For example, a cow's MPPA for weaning ratio is calculated from the cow's average progeny weight ratio, the number of her progeny weaning records, and the repeatability of weaning weight.
mouthed Examination of an animal's teeth.
MPPA *See* most probable producing ability (MPPA).
muley Term used to describe the polled (hornless) condition.
muscle-to-bone (M/B) ratio Pounds of muscle divided by pounds of bone. For example, 4:1 ratio means that there is 4 lbs of muscle to 1 lb of bone (usually on a carcass basis).
muscling Amount of lean meat in a slaughter animal or carcass. Estimated on the live animal by thickness of forearm muscle or stifle thickness. Ultimately, it is the ratio of muscle to bone or lean yield of the carcass after fat and bone are removed.
muzzle Nose of cattle.
myofibrils Primary component part of muscle fibers.

NAFTA (North American Free Trade Agreement) A trade agreement involving Canada, Mexico, and the United States, implemented on January 1, 1994, with a 15-year transition period.
National Cattlemen's Beef Association (NCBA) National organization for cattle breeders, producers, feeders, and affiliated organizations with offices in Englewood, CO, and Washington, DC. Previously known as the National Cattlemen's Association or NCA.
National Live Stock and Meat Board (NLSMB) Organization located in Chicago that provides nutrition, research, education, and promotional information on beef, pork, and lamb. *See also* Beef Industry Council (BIC).
national sire evaluation Programs of sire evaluation conducted by breed associations to compare sires on a progeny-test basis. Carefully conducted national reference sire evaluation programs give unbiased estimates of expected progeny differences. Sire evaluations based on field data rely on large numbers of progeny per sire to compensate for possible favoritism or bias for sires within herds.
native hides Hides from cattle that have not been hot-iron branded.
natural beef Refers to beef from cattle that have not been fed growth stimulants or antibiotics.
natural fleshing Lean meat or muscle.
navel Area where the umbilical cord was formerly attached to the body of the offspring.
NCBA *See* National Cattlemen's Beef Association.
necropsy To perform a postmortem examination.
NEg Net energy for gain.
NEl Net energy for lactation.
NEm Net energy for maintenance.
net energy Metabolizable energy minus heat increment, or the energy available to the animal for maintenance and production.
net income Total revenue earned minus expenses incurred for a given period of time.
net worth (equity) Represents the owner's claim on the assets of a business: net worth = assets − liabilities.
net worth statement *See* balance sheet.

nicking Way in which certain lines, strains, or breeds perform when mated together. When outstanding offspring result, the parents are said to have nicked well.
nipple *See* teat.
NPN (nonprotein nitrogen) Nitrogen in feeds from substances such as urea and amino acids, but not from preformed proteins.
nutrient (1) Substance that nourishes the metabolic processes of the body. (2) End product of digestion.
nutrient density Amount of essential nutrients relative to the number of calories in a given amount of food.

obesity Excessive accumulation of body fat.
offal All organs and tissues removed from inside the animal during the slaughtering process.
off feed Animal refuses to eat or consumes only small amounts of feed.
omasum One of the stomach components of cattle that has many folds.
on full feed Refers to cattle that are receiving all the feed they will consume. *See also ad libitum.*
open Refers to nonpregnant females.
operating expenses Expenses incurred in the usual production cycle, such as seed, fuel, feed, and hired labor costs.
opportunity cost Cost of using a resource based on what it could have earned using it in the next best alternative use.
optimize To make as effective as possible.
optimum Amount or degree of something that is most favorable to some end (e.g., the best combination of resources associated with cattle production yields the highest sustainable net return).
optimum level of performance Performance level of a trait or traits that maximizes net profit. Resources are managed (including a balance of traits) that sustain high levels of profitability.
outbreeding Process of continuously mating females of the herd to unrelated males of the same breed.
outcrossing Mating of an individual to another in the same breed that is not related to it. Outcrossing is a type of outbreeding.
ova Plural of *ovum,* meaning eggs. *See also* ovum.
ovary Female reproductive organ in which the eggs are formed and progesterone and estrogenic hormones are produced.
overhead Expenses incurred in the operation of the business that cannot conveniently be attributed to the production of specific commodities or services.
ovulation Shedding or release of the egg from the follicle of the ovary.
ovum Egg produced by a female.

packing plant Facility in which cattle are slaughtered and processed.
paddock A pasture subdivision within a grazing unit.
palatability Degree to which food (e.g., beef) is acceptable to the taste or sufficiently agreeable in flavor, juiciness, and tenderness to be eaten.
palpation Feeling or examining by hand (e.g., the reproductive tract is palpated for reproductive soundness or pregnancy diagnosis).
parasite Organism that lives a part of its life cycle in or on, and at the expense of, another organism. Parasites of farm animals live at the expense of the animals.
parity Number of different times a female has had offspring.

parrot mouth Upper jaw is longer than the lower jaw.

partial budget Budget that includes only those revenue and expense items that would change as a result of a proposed change in the business.

parturition Process of giving birth.

pasture rotation Rotation of animals from one pasture to another so that some pasture areas have no livestock grazing on them during certain periods of time.

patchy Uneven fat accumulations, usually lumps of exterior fat around the tailhead and pin bones.

paternal Refers to the sire or bull.

pathogen Biologic agent (e.g., bacteria, virus, protozoa, nematode) that may produce disease or illness.

paunch *See* rumen.

paunchy Heavy middled.

pay weight Actual weight for which payment is made. In many cases, it is the shrunk weight (actual weight − pencil shrink).

pedigree Records of the ancestry of an animal.

pelvic area Size of pelvic opening determined by measuring pelvic width and length and used to predict calving difficulty.

pen rider Person who rides through feedlot pens and checks cattle.

pencil shrink Deduction from an animal's weight, often expressed as a percentage of liveweight, to account for fill (usually 3% for off-pasture weights and 4% for fed-cattle weights).

pendulous Hanging loosely.

percent calf crop The percentage of calves produced within a herd in a given year relative to the number of cows and heifers exposed to breeding.

per capita Per person.

performance data Records on individual animal's reproduction, production, and possibly carcass merit. Traits included are birth, weaning, and yearling weights; calving ease; calving interval; milk production; and others.

performance pedigree Includes the performance records of ancestors, half- and full sibs, and progeny in addition to the usual ancestoral pedigree information. The performance information is systematically combined to list estimated breeding values on the pedigrees by some breed associations.

performance test Evaluation of an animal according to its performance.

pharmaceutical Medicinal drug.

phenotype Characteristics of an animal that can be seen and (or) measured (e.g., presence or absence of horns, color, or weight).

phenotypic correlations Correlations between two traits caused by both genetic and environmental influences. *See* correlation coefficient.

pheromones Chemical substances that attract the opposite sex.

photoperiod Time period when light is present.

pituitary Small endocrine gland located at the base of the brain.

placenta Membranes that form around the embryo and attach to the uterus. *See also* afterbirth.

Plains states Include Texas, Oklahoma, Kansas, Nebraska, South Dakota, and North Dakota and the eastern parts of New Mexico, Colorado, Wyoming, and Montana; often referred to as the "Beef Belt."

pluck Organs of the thoracic cavity (e.g., heart and lungs).

pneumonia Inflammation or infection of alveoli of the lungs caused by either bacteria or viruses.

polled Naturally or genetically hornless.
pons Accumulation of fat over pin bones.
portion-controlled beef products Retail cuts of beef that meet size and form specifications.
postnatal *See* postpartum.
postpartum After birth.
postpartum interval Days from calving until the cow returns to estrus, or days from calving until cow is pregnant again.
pounds of retail cuts per day of age A measure of cutability and growth combined; it is calculated as follows: cutability times carcass weight divided by age in days.
pounds of calf weaned per cow exposed Calculated by multiplying percent calf crop by the average weaning weight of calves.
preconditioning Preparation of feeder calves for marketing and shipment; may include vaccinations, castration, and training calves to eat and drink in pens.
prenatal Prior to being born; before birth.
prepotent Ability of a parent to transmit its characteristics to its offspring so that they resemble that parent, or each other, more than usual. Homozygous dominant individuals are prepotent. Also, inbred cattle tend to be more prepotent than outbred cattle.
price cycle Traditional or historic changes in prices (usually by months, seasons, or years).
price discovery Process that shows how the specific price for a given quantity and quality of beef is determined.
primal cuts Wholesale cuts—round, loin, flank, rib, chuck, brisket, plate, and shank.
production testing Evaluation of an animal based on its production record.
progeny Offspring of the parents.
progeny testing Evaluation of an animal based on the performance of its offspring.
progesterone Hormone produced by the corpus luteum that stimulates progestational proliferation in the uterus of the female.
prolapse Abnormal protrusion of part of an organ, such as the uterus or rectum.
prostaglandins Chemical mediators that control many physiological and biochemical functions in the body. One prostaglandin, $PGF_{2\alpha}$, can be used to synchronize estrus.
prostate Gland of the male reproductive tract located just behind the bladder, which secretes a fluid that becomes a part of semen at ejaculation.
protein Substance made up of amino acids that contains approximately 16% nitrogen (based on molecular weight).
protein supplement Any dietary component containing a high concentration (at least 25%) of protein.
puberty Age at which the reproductive organs become functionally operative.
purebred Animal eligible for registry with a recognized breed association.
purveyor Firm that purchases beef (usually from a packer), then performs some fabrication before selling the beef to another firm.

qualitative traits Those in which there is a sharp distinction between phenotypes (e.g., red or black color). Usually, only one or two gene pairs are involved.
quality Quality is something special about an object that makes it what it is; a characteristic, attribute, or excellence. Quality is the composite or attribute of an animal or product that has economic or aesthetic value to the user; meeting or exceeding each customer's expectations at a cost that represents value to the customer every time.
quality grades Grades such as Prime, Choice, and Select that group slaughter cattle and carcasses into value- and palatability-based categories. Grades are determined primarily by marbling and age of animal.

quantitative traits Those in which there is no sharp distinction between phenotypes, with a gradual variation from one phenotype to another (such as weaning weight). Usually, many gene pairs are involved, as well as environmental influences.

Ralgro® Growth implant containing zeranol (a fermentation product exhibiting estrogenic activity).
random mating System of mating whereby every female (cow and/or heifer) has an equal or random chance of being assigned to any bull used for breeding in a particular breeding season. Random mating is required for accurate progeny tests.
ration Feed fed to an animal during a 24-hour period.
REA *See* rib-eye area (REA).
reach *See* selection differential.
realizer Feedlot animal that is removed before the end of the feeding program. Only part of the animal's potential value is realized because of disease, injury, or the like.
recessive gene A gene that has its phenotypic expression masked by its dominant allele when the two genes are present together in an individual.
rectal prolapse Protrusion of part of the large intestine through the anus.
red meat Meat from cattle, sheep, swine, and goats. *See also* white meat.
reference sire Bull designated to be used as a benchmark in progeny testing other bulls (young sires). Progeny by reference sires in several herds enable comparisons to be made between bulls not producing progeny in the same herd(s).
registered Recorded in the herdbook of a breed.
regurgitate To cast up undigested food to the mouth as is done by ruminants.
replacement heifers Heifers, usually between 6 months and 16 months of age, that have been selected to replace cows in the breeding herd.
replacements Cattle that are going into feedlots or breeding herds to replace those being sold or that have died. *See also* replacement heifers.
reproductive tract score Numerical score based on palpation of the heifer's reproductive tract (1 = not cycling; 5 = heifer cycling).
resource Input or factor used in production, such as cattle, labor, or land.
retail cuts Cuts of beef in sizes that are purchased by the consumer.
retained ownership Usually refers to cow-calf producers maintaining ownership of their cattle through the feedlot.
retained placenta Fetal membranes (afterbirth) are not expelled through the reproductive tract in the normal length of time following calving.
reticulum One of the stomach components of cattle that is lined with small compartments giving a honeycomb appearance.
rib-eye area (REA) Area of the *longissimus dorsi* muscle, measured in square inches, between the twelfth and thirteenth ribs. Also referred to as the *loin-eye area (LEA).*
rib-eye area per cwt carcass wt Rib-eye area divided by carcass weight.
risk Possibility of suffering economic loss. Sources of risk include climate, disease, and changes in the marketplace.
risk management Managing risks in ways that allow a desired outcome to be achieved.
rotational crossbreeding Systems of crossing two or more breeds whereby the crossbred females are bred to bulls of the breed contributing the least genes to the females' genotype.
roughage Feed that is high in fiber, low in digestible nutrients, and low in energy (e.g., hay, straw, silage, and pasture).
rugged Big and strong in appearance; usually heavy boned.

rumen A compartment of the ruminant stomach that is similar to a large fermentation pouch where bacteria and protozoa break down fibrous plant material swallowed by the animal. Sometimes referred to as the paunch.
ruminant Mammal whose stomach has four parts—rumen, reticulum, omasum, and abomasum. Cattle, sheep, goats, deer, and elk are ruminants.
rumination Regurgitation of undigested food that is chewed and then swallowed again.

scale (1) Size of cattle. (2) Equipment on which an animal is weighed.
scours Diarrhea; profuse watery discharge from the intestines.
scrotal circumference Measure of testes size obtained by measuring the distance around the testicles in the scrotum with a circular tape. Related to the bull's semen-producing capacity and age at puberty of his daughters.
scrotum Pouch that contains the testicles. Also a thermoregulatory organ that contracts when cold and relaxes when warm, thus tending to keep the testes at a lower temperature than that of the body.
scurs Small growths of hornlike tissue attached to the skin of polled or dehorned animals.
seam fat *See* intermuscular fat.
seedstock Breeding animals. Sometimes used interchangeably with *purebred.*
seedstock breeders Producers of breeding stock for purebred and commercial breeders.
Select USDA carcass quality grade between Choice and Standard. It replaced the Good grade in 1988.
selection Differential reproduction (e.g., a bull or cow may leave several, one, or no offspring in a herd).
selection differential (reach) Difference between the average for a trait in selected animals and the average of the group from which they come. Also called *reach.*
selection index Formula that combines performance records from several traits or different measurements of the same trait into a single value for each animal. A selection index combines traits after balancing their relative net economic importance, their heritabilities, and the genetic association among the traits.
self-management Managing oneself as part of human resource management (e.g., time management, information management, self-motivation, honesty).
semen Fluid containing sperm that is ejaculated by the male. Secretions from the seminal vesicles, prostate gland, bulbourethral glands, and urethral glands provide most of the fluid.
seminal vesicles Accessory sex glands of the male that provide a portion of the fluid of semen.
served Female is bred but not guaranteed pregnant.
service To breed or mate.
settle To become pregnant.
shipping fever Widespread respiratory disease of cattle.
short yearling Animal is over one year of age but under 18 months of age.
"show list" or "show pens" Slaughter cattle that are ready for the cattle feeder to "show" the packer buyers.
shrink Loss of weight; commonly used in the loss of liveweight when animals are marketed.
sib Brother or sister.
sick pen Isolated pen in a feedlot where cattle are treated after they have been removed from a feedlot pen. Sometimes referred to as a hospital pen.
sickle hocked Hocks that have too much set, causing the hind feet to be too far forward and too far under the animal.

silage Forage, corn fodder, or sorghum preserved by fermentation that produces acids similar to the acids used to make pickled foods for people.
sire Male parent.
sire summary Published results of national sire evaluation programs that give EPDs and accuracies for several economically important traits. Several major breed associations publish their own sire summaries.
size Usually refers to weight, sometimes to height.
skins *See* hides.
skirt Diaphragm muscle in the beef carcass.
software Program instructions to make computer hardware function.
sonoray *See* ultrasound.
soundness Degree of freedom from injury or defect.
SPA *See* Standard Performance Analysis.
spay To remove the ovaries.
sperm A mature male germ cell.
specifications A detailed description, with numerical designations, of animal performance or product quantity.
spermatogenesis Process of spermatozoa formation.
splay footed *See* toeing out.
stag Castrated male that has reached sexual maturity prior to castration.
standard deviation For traits having a normal distribution characterized by a bell-shaped curve, 68% of the population = mean (average) $\pm$ 1 standard deviation, 95% = mean $\pm$ 2 standard deviations, and 99% = mean $\pm$ 3 standard deviations.
Standard Performance Analysis (SPA) Program to determine the unit cost of production for the cow-calf enterprise.
steer Bovine male castrated prior to puberty.
sterility Inability to produce offspring.
stifle Joint of the hind leg between the femur and tibia.
stifled Injury of the stifle joint.
stillborn Offspring born dead without previously breathing.
stocker Weaned cattle that are fed high-roughage diets (including grazing) before going into the feedlot.
stocking rate The number of animals, animal units, or total animal liveweight assigned to a grazing unit for an extended period of time. Stocking rates are usually expressed on a per acre basis.
stocking density The number of animals, animal units, or total animal liveweight present at a particular point in time on a defined area (paddock). Stocking density is usually defined on a per acre basis.
stockpiling The practice of allowing forage to accumulate for grazing at a later date. Most commonly done with late summer and fall forage growth for fall and (or) winter grazing.
strip grazing The practice of dividing a larger pasture into strips with movable fences to control grazing access.
straightbred Animal whose parentage has been from one breed.
stress Unusual or abnormal influence causing a change in an animal's function, structure, or behavior.
subcutaneous Situated beneath, or occurring beneath, the skin. A subcutaneous injection is an injection made under the skin.
subprimal cuts Smaller-than-primal cuts, such as when the primal round is split into top round, bottom round, eye round, and sirloin tip. Subprimal cuts are used in boxed beef programs.

success Progressive realization of predetermined, worthwhile goals that are based on true principles.
suckling gain Gain that a young animal makes from birth until it is weaned.
superovulation Hormonally induced ovulation in which a greater-than-normal number of eggs are typically produced.
sweetbread Edible by-product also known as the pancreas.
switch Tuft of long hair at the end of the tail.
syndactyly Union of two or more digits; for example, in cattle the two toes would be a solid hoof.
Synovex® Growth implant containing estradiol. Different types are used in calves (Synovex-C®), feeder steers (Synovex-S®), and feeder heifers (Synovex-H®).
synthetic breeds *See* composite breed.
systems analysis *See* management systems.

tariff A tax imposed on commodity imports by a government. A tariff may be either a fixed charge per unit of product imported (specific tariff) or a fixed percentage of value (ad valorem tariff).
tagging Usually refers to putting ear tags in the ear.
tandem selection Selection for one trait for a given period of time followed by selection for a second trait and continuing in this way until all important traits are selected.
TDN *See* total digestible nutrients (TDN)
teat Proturberance of the udder through which milk flows.
terminal crossbreeding *See* terminal sires.
terminal market Large livestock collection center where an independent organization serves as a selling agent for the livestock owner.
terminal sires Sires used in a crossbreeding system in which all their progeny, both male and female, are marketed. For example, crossbred dams could be bred to sires of a third breed and all calves marketed. Although this system allows maximum heterosis and complementarity of breeds, replacement females must come from other herds.
testicle Male sex gland that produces sperm and testosterone.
testosterone Male sex hormone that stimulates the accessory sex glands, causes the male sex drive, and results in the development of masculine characteristics.
tie Depression or dimple in the back of cattle caused by an adhesion of the hide to the backbone.
time management Manner in which time is utilized to achieve specific goals.
toeing in Toes of front feet turn in. Also called *pigeon toed.*
toeing out Toes of front feet turn out. Also called *splay footed.*
total digestive nutrients (TDN) Sum of digestible protein, nitrogen-free extract, fiber, and fat (multiplied by 2.25).
trait ratio Expression of an animal's performance for a particular trait relative to the herd or contemporary group average. It is usually calculated for most traits as:

$$\frac{\text{Individual record}}{\text{Average of animal in group}} \times 100$$

transgenic An organism or animal whose genome includes "foreign" genetic material. Foreign genetic material would be a DNA sequence or gene that does not normally occur in the species of the host organism or animal.
tray-ready beef Retail cuts that are cut and packaged at the packing plant for retail sales; also referred to as *case-ready.*
tripe Edible product from the walls of the ruminant stomach.

twist Vertical measurement from the top of the rump to the point where the hind legs separate.

type (1) Physical conformation of an animal. (2) All physical attributes that contribute to the value of an animal for a specific purpose.

udder Encased group of mammary glands of the female.

ultrasound Using high-frequency sound waves to show visual outlines of internal body structures (e.g., fat thickness, rib-eye area, and pregnancy can be predicted). The machine sends sound waves into the animal and records these waves as they bounce off the tissues. Different wavelengths are recorded for fat and lean.

umbilical cord Cord through which arteries and veins travel from the fetus to and from the placenta, respectively. This cord is broken when the young are born.

uncoupling Term used to consider separating quality grading and yield grading.

unsoundness Any defect or injury that interferes with the usefulness of an animal.

urinary calculi Disease that causes mineral deposits to crystallize in the urinary tract.

USDA *See* U.S. Department of Agriculture.

U.S. Department of Agriculture (USDA) An executive department of the U.S. government that helps farmers supply farm products for U.S. consumers and overseas markets. *See* Appendix for organizational structure.

U.S. Meat Export Federation (USMEF) Organization that works to increase consumer demand for red meats and by-products in overseas markets. Members include NCA, state cattle associations, beef councils, farm and commodity groups, packers, and agribusiness companies. Funds come from its members and the USDA.

uterus That portion of the female reproductive tract where the young develop during pregnancy.

vaccination The act of administering a vaccine or antigens.

vaccine Suspension of attenuated or killed microbes or toxins administered to induce active immunity.

vagina Copulatory portion of the female's reproductive tract. The vestibule portion of the vagina also serves for passage of urine during urination. The vagina also serves as a canal through which young pass when born.

value-based marketing Marketing system based on paying for individual animal differences rather than using average prices.

variable costs Costs that change with the amount produced. If the manager decides to cease production, these costs are avoidable.

variance Variance is a statistic that describes the variation we see in a trait.

variety meats Edible organ by-products (e.g., liver, heart, tongue, tripe).

vas deferens Ducts that carry sperm from the epididymis to the urethra.

veal Meat from very young cattle (under three months of age). Veal typically comes from dairy bull calves.

video image analysis (VIA) A video image is analyzed via sophisticated computer techniques to estimate factors associated with carcass value.

virus Ultramicroscopic bundle of genetic material capable of multiplying only in living cells. Viruses cause a wide range of diseases in plants, animals, and humans, such as rabies and measles.

viscera Internal organs and glands contained in the thoracic and abdominal cavities.

vitamin Organic catalyst, or component thereof, that facilitates specific and necessary functions.

volatile fatty acids (VFA) Group of fatty acids produced from microbial action in the rumen; examples are acetic, propionic, and butyric acids.
vulva External genitalia of a female mammal.

wasty Excessive accumulation of fat.
wattle Method of cattle identification in which 3–6-inch strips of skin are cut on the nose, jaw, throat, or brisket.
weaner Calf that has been weaned or is near weaning age.
weaning (wean) Separating young animals from their dams so that the offspring can no longer suckle.
weaning weight Weight of the calf at approximately 5–10 months of age when the calf is removed from the cow.
weaning weight EPD Estimate of the weaning weight (lb) potential of a sire's progeny. Positive numbers indicate above-average performance while negative values indicate below-average weights when compared with other bulls in the same sire summary. This estimate is for direct growth, as maternal effects are removed in the calculations. *See also* expected progeny difference (EPD).
weaning weight ratio The weaning weight of a calf divided by the herd average. Usually done within sex.
weight per day of age (WDA) Weight of an individual animal divided by days of age.
white meat Meat from poultry. *See also* red meat.
white muscle disease Muscular disease caused by a deficiency of selenium or vitamin E.
wholesalers Beef operations that buy and sell beef to other firms; considered the middlemen between the packer and consumer segments.
"window of acceptability" Identifies the acceptable minimum and maximum amounts of fat in meat on the basis of meat palatability and human health.
"with calf" Heifer or cow is pregnant.
withdrawal time Amount of time before slaughter during which a drug cannot be given to an animal.
"woody" Opposite of "bloom"—that is, the animal's haircoat appears dull, not shiny. Associated with unthrifty calves. *See also* "bloom."
World Trade Organization (WTO) Established on January 1, 1995 as a result of the Uruguay Round, the WTO replaces GATT as the legal and institutional foundation of the multilateral trading system of member countries.

yardage Per head daily fee charged by the feedlot to the customer owning the cattle. This fee is usually in addition to the cost of medicine and the feed markup.
yearling Animals that are approximately one year old (usually 12–24 months of age).
yearling weight Weight when approximately 365 days old.
yearling weight EPD Estimate of the yearling weight (lb) potential of a bull's progeny compared with progeny from other bulls in the same sire summary. Positive numbers indicate above-average performance while negative values indicate below-average performance. *See also* expected progeny difference (EPD).
yearling weight ratio Yearling weight of a calf divided by the herd average. Usually calculated within sex.
yield *See* dressing percentage.
yield grades USDA grades identifying differences in cutability—the boneless, fat-trimmed retail cuts from the round, loin, rib, and chuck.

Appendix

Contents

THE METRIC SYSTEM

The metric system has been accepted as the preferred system in nearly all countries of the world. The United States is the only major trading nation worldwide that does not presently use the metric system.

The National Bureau of Standards recommended in 1971 that the United States convert to the metric system. The president signed the Metric Conversion Act in 1975, establishing the national goal of converting, voluntarily, to predominantly metric measurements. Currently, it

TABLE A.1 Prefixes in the Metric System

Prefix	Symbol	Power and Meaning
tera	T	10^{12}
giga	G	10^{9}
mega	M	10^{6} 1,000,000 times base
kilo	k	10^{3} 1,000 times base
hecto	h	10^{2} 100 times base
deca	da	10 10 times base
deci	d	10^{-1} tenth
centi	c	10^{-2} hundredth
milli	m	10^{-3} thousandth
micro	μ	10^{-6} millionth
nano	n	10^{-9}
pico	p	10^{-12}
femto	f	10^{-15}
atto	a	10^{-18}

TABLE A.2 Weight–Unit Conversion Factors

Unit Given	Unit Wanted	For Conversion Multiply by
lb	g	453.6
lb	kg	0.4536
oz	g	28.35
kg	lb	2.2046
kg	mg	1,000,000
kg	g	1,000
g	mg	1,000
g	μg	1,000,000
mg	μg	1,000
mg/g	mg/lb	453.6
mg/kg	mg/lb	0.4536
μg/kg	g/lb	0.4536
Mcal	kcal	1,000
kcal/kg	kcal/lb	0.4536
kcal/lb	kcal/kg	2.2046
ppm	μg/g	1
ppm	mg/kg	1
ppm	mg/lb	0.4536
mg/kg	%	0.0001
ppm	%	0.0001
mg/g	%	0.1
g/kg	%	0.1

is necessary to understand both measurement systems and be able to convert one to the other. Weights and measurements with their conversions are shown in Tables A.1 and A.2.

Other Weights, Measures, and Sizes

Table A.3 gives the bushel weights of grains, while the capacities and volumes of silos are shown in Tables A.4, A.5, and A.6. Nail sizes and descriptions are identified in Table A.7.

VOLUMES AND WEIGHTS OF STACKED AND BALED HAY

Determining the volume and weight of hay is important whenever hay is sold or yields are determined. Volume of hay is expressed in cubic feet. Weight is expressed in pounds for individual bales, or tons for stacks and loads. Converting from volume, which is reasonably easy to measure, to weight requires that density (weight per cubic foot) of the hay be measured or estimated.

TABLE A.3 Average Bushel Weights of Selected Grains and Seeds

Grain or Seed	Average lb/Bushel
Alfalfa	60
Barley	48
Bluegrass	14–30
Clover	60
Corn	56
Oats	32
Orchardgrass	14
Sorghum	56
Soybeans	60
Wheat	60

TABLE A.4 Determining the Amount of Corn Silage in Trench Silo (Tons = average width × length × average depth of silage [in feet] × tons per cubic foot of depth)[a]

Depth of Settled Silage (ft)	Tons/ft^3	Depth of Settled Silage (ft)	Tons/ft^3
1	0.00925	9	0.01320
2	0.00985	10	0.01365
3	0.01040	11	0.01405
4	0.01090	12	0.01445
5	0.01140	13	0.01490
6	0.01190	14	0.01530
7	0.01235	15	0.01565
8	0.01280	16	0.01605

[a]*Example:* A trench silo measures 8 feet wide at the top and 12 feet wide at the bottom, and is 50 feet long. The silage averages 8 feet deep. The calculation for total tons in the trench silo is as follows.

Solution:

$$\frac{8 + 12}{2} \times 50 \times 8 \times 0.0128 = 51.2 \text{ tons}$$

TABLE A.5 Silo Volumes Per Foot and Pounds Per 2-Inch Layer

Silo Diameter	Volume/Ft of Depth (ft^3)	Lb Silage in 2-in. Layer Based on 40 lb/ft^3
12	113.1	755
14	153.9	1,025
16	201.1	1,340
18	254.5	1,696
20	314.2	2,094
22	380.1	2,534
24	452.4	3,015
26	530.9	3,539
28	615.8	4,105
30	706.9	4,712
34	908.0	6,053
38	1134.1	7,560
42	1385.4	9,235
50	1963.5	13,089
60	2827.4	18,847

TABLE A.6 Approximate Silo Capacity in Tons

Inside Diameter of Silo (ft)	Silo Height (ft)					
	20	30	40	50	60	70
10	33	56	77			
12	48	80	110			
14	66	109	150	193		
16	86	143	196	252		
18		180	248	320	392	
20		223	307	394	483	574
22		270	371	477	585	694
24		321	442	570	697	827
26		377	520	668	818	970
28			600	773	947	1125
30			690	886	1087	1290

Stacks

The formula commonly used for estimating the volume of loose hay in stacks is:

$$V = \frac{(O - 5.6W)}{2} \times W \times L$$

V = Stack volume in cubic feet
O = Average distance over stack in feet
W = Average stack width in feet
L = Stack length in feet

TABLE A.7 Nail Sizes and Descriptions

Size	Length (in.)	Approx. no./lb
Common nails		
2d	1	847
3d	1 ¼	543
4d	1 ½	294
5d	1 ¾	254
6d	2	167
7d	2 ¼	150
8d	2 ½	101
9d	2 ¾	92
10d	3	69
12d	3 ¼	63
16d	3 ½	49
20d	4	31
30d	4 ½	24
40d	5	18
50d	5 ½	14
60d	6	11
Spikes		
10d	3	32
12d	3 ¼	31
16d	3 ½	24
20d	4	19
30d	4 ½	14
40d	5	12
50d	5 ½	10
60d	6	9
5/16	7	6
3/8	8–12	5–3

The over measurement (O) can be obtained using a tape or string with attached weight that is thrown over the stack. The stack should be checked in about four places, then those measurements averaged.

Bales

The volume of a stack of baled hay can be determined by measuring length, width, and height, then multiplying these together. Another technique used with baled hay is to count the number of bales, then multiply by an estimated or determined weight per bale to determine total weight. The volume of the bales is occasionally needed for storage purposes. With round bales an estimate of total tonnage can be made by weighing a few bales and counting the number of bales involved.

Weight of Hay

The density, or pounds per cubic foot, of both stacked and baled hay varies greatly. The following table gives some guides. It is always better to calculate the density after weighing a few bales and determining the volume. It may also be possible to weigh a stack or portion of a stack. Remember that stacked hay settles over time.

	Weight of Loose Hay (lbs/ft^3)	Weight of Baled Hay (lbs/ft^3)
Alfalfa	4–5	8–14
Grass hay	3–5	6–10
Straw	2–3	4–6

Stacking Baled Hay

Figures A.1, A.2, and A.3 show how baled hay can be stacked to reduce nutrient loss.

ROUND GRAIN BIN VOLUMES

The following formula assumes that grain has fallen freely into the bin with the top forming a cone.

$$\text{Volume in bushels} = (0.974 \times d^2) \times \left(h + \frac{d}{20}\right)$$

d = diameter (in feet)
h = height (in feet)

Example: $d = 18 \quad h = 12 \quad V = (0.974 \times 18^2) \times \left(12 + \frac{18}{20}\right) \quad V = 4{,}071$ bushels

MEASURING IRRIGATION WATER FLOW

Water measurement is a necessary part of water management. Measurement is also used to verify water rights.

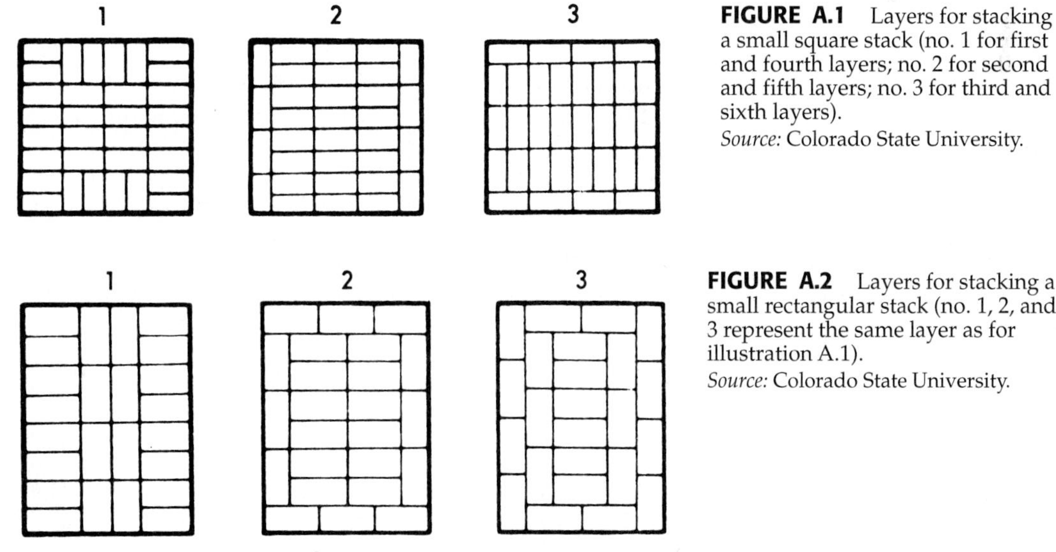

FIGURE A.1 Layers for stacking a small square stack (no. 1 for first and fourth layers; no. 2 for second and fifth layers; no. 3 for third and sixth layers).
Source: Colorado State University.

FIGURE A.2 Layers for stacking a small rectangular stack (no. 1, 2, and 3 represent the same layer as for illustration A.1).
Source: Colorado State University.

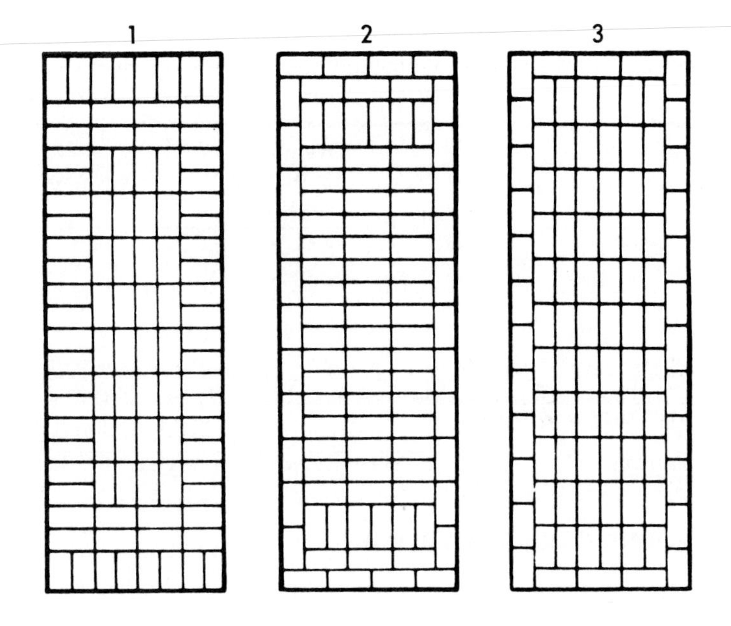

FIGURE A.3 Layers for a large rectangular stack (no. 1 or first, fourth, seventh, and tenth layers; no. 2 for second, fifth, and eighth layers; no. 3 for third, sixth, and ninth layers).
Source: Colorado State University.

Several devices can be installed for measuring water delivery in open-ditch irrigation. Common devices are weirs and flumes. For accurate water-flow measurements, weirs and flumes must be properly constructed, installed, and maintained. Details for constructing and using these measuring devices are given in the publication *Irrigation Water Measurement,* (Wyo. Agric. Ext. Serv. Bull. 583R, April 1978).

Inches of Water Applied with Hand Move and Wheel Line Sprinklers

Factors that affect the inches of water applied are pressure, nozzle size, sprinkler spacing, and time. Pressure is measured with a gauge. The size is stamped on the nozzle, but older sprinklers should be verified using a drill bit as a measure. Spacing is the distance between sprinklers and line positions. Use Tables A.8, A.9, and A.10 in sequence, to determine inches of water applied.

Examples:

1. Pressure of 50 psi and 13/64-inch nozzle from Table A.8 read the discharge as 8.71 gpm.
2. Sprinklers spaced 40 feet apart and lines 60 feet apart. From Table A.9 estimate the precipitation rate as 0.35 in. per hour.

TABLE A.8 Discharge in GPM for Various Nozzle Sizes and Pressures

	Nozzle Diameter (in.)					
PSI	9/64	5/32	11/64	3/16	13/64	7/32
30	3.26	4.01	4.83	5.75	6.80	7.85
40	3.74	4.61	5.54	6.64	7.80	9.02
50	4.18	5.15	6.19	7.41	8.71	10.10
60	4.50	5.65	6.80	8.12	9.56	11.05
70	4.96	6.10	7.34	8.78	10.32	11.95
80	5.29	6.52	7.84	9.39	11.02	12.74

TABLE A.9 Precipitation Rates in Inches Per Hour for Various Sprinkler Discharges and Spacings

	Sprinkler Discharge (gpm)						
Spacing Feet	3.0	4.0	5.0	6.0	8.0	10.0	12.0
30 × 50	0.19	0.25	0.32	0.38	0.51	0.64	0.76
30 × 60	0.16	0.21	0.27	0.32	0.43	0.53	0.64
40 × 50	0.14	0.19	0.24	0.29	0.38	0.48	0.58
40 × 60	0.12	0.16	0.20	0.24	0.32	0.40	0.48

TABLE A.10 Water Applied Per Set in Inches @ 75% Efficiency

	Precipitation Rate (in./hr)					
Hours	0.20	0.30	0.40	0.50	0.60	0.70
6.0	0.90	1.13	1.80	2.25	2.70	3.15
8.0	1.20	1.80	2.40	3.00	3.60	4.20
12.0	1.80	2.70	3.60	4.50	5.40	6.30
24.0	3.60	5.40	—	—	—	—

TABLE A.11 Letter Codes Used to Indicate Year of Birth in Identification Systems

X — 1988	J — 1999
Y — 1989	K — 2000
Z — 1990	L — 2001
A — 1991	M — 2002
B — 1992	N — 2003
C — 1993	P — 2004
D — 1994	R — 2005
E — 1995	S — 2006
F — 1996	T — 2007
G — 1997	U — 2008
H — 1998	W — 2009

3. Sprinklers are run for 12 hours. From Table A.10, estimate the amount of water entering the root zone as 3.15 inches.

LAND DESCRIPTION FOR LEGAL PURPOSES

The federal government has established land surveys that are used for the legal description of land. Townships, 6 miles square, are located north or south of standard parallels and east and west of prime meridians.

Each township is divided into thirty-six sections, each section being a mile square. The sections are numbered 1 through 36, starting with number 1 in the northeast corner of the township and ending with number 36 in the southeast corner (see Fig. A.4).

Sections are further divided into smaller units so that the location of every land parcel can be identified for legal purposes (see Fig. A.5). For example, the legal description of a certain 40 acres may read: The south half of the west half of the southwest quarter of section 6 in township 32, north of range 4 west.

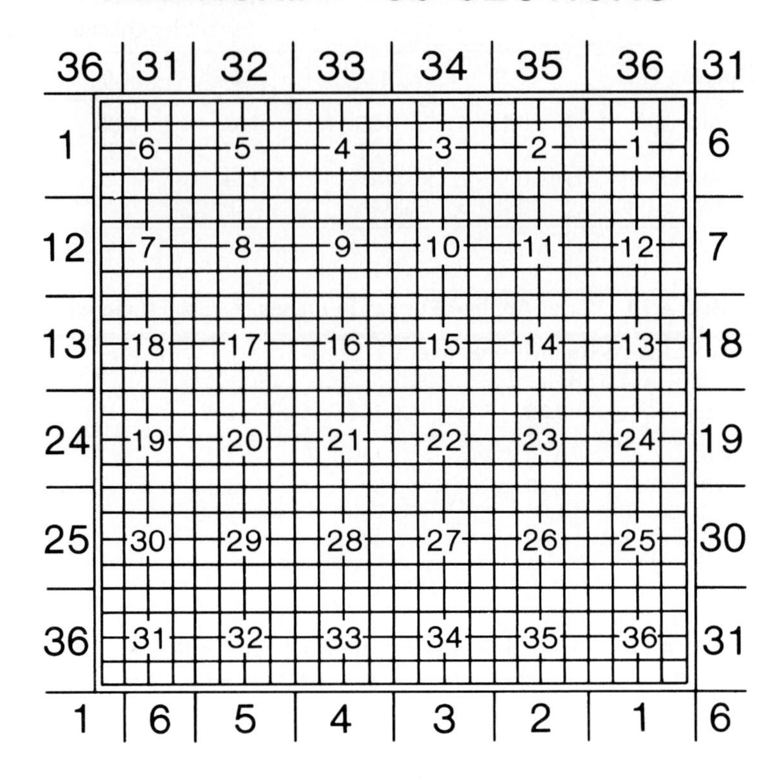

FIGURE A.4 A township divided into thirty-six sections. Adjoining sections are also shown.
Source: Colorado State University.

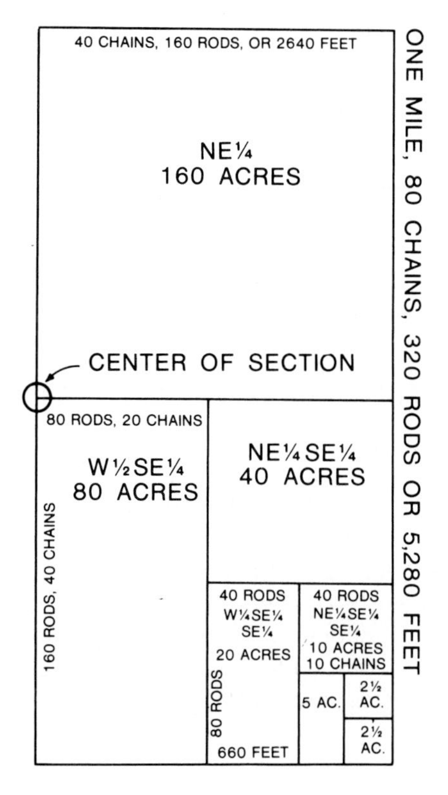

FIGURE A.5 A half-section divided into several component units.
Source: Colorado State University.

MAJOR ORGANIZATIONS WITHIN OR AFFECTING THE BEEF INDUSTRY

The names, addresses, and functions of the major organizations representing and influencing the beef industry are listed here for reference.

American Association of Bovine Practitioners
P.O. Box 1755
Rome, GA 30162-1755
www.aabp.org

A veterinary supplier and service group that promotes the interests and advancement of dairy and beef cattle veterinarians.

American Association of Meat Processors (AAMP)
224 East High St.
Elizabethtown, PA 17022
www.aamp.com

Retail operators of meat-processing plants, locker plants, frozen-food centers, and freezer-food suppliers form this association.

American Farm Bureau Federation
225 Touhy Ave.
Park Ridge, IL 60068
www.fb.com

The Farm Bureau is a general farm organization with a voluntary membership that represents producers of all agricultural commodities.

American Feed Industry Association (AFIA)
1501 Wilson Blvd., Suite 1100
Arlington, VA 22209
www.afia.org

AFI is the only national organization devoted exclusively to representing the regultory, business, and governmental interests of the animal feed and pet food industries and their suppliers.

American Forage and Grassland Council (AFGC)
P.O. Box 94
Georgetown, TX 78627
www.afgc.org

A spokesgroup for North America's forage-based agriculture. Members are representatives of agencies that provide research, educational, and public services.

American Institute of Food Distribution (AIFD)
28-12 Broadway
Fair Lawn, NJ 07410

Canners, packers, freezers, manufacturers, brokers, wholesalers, retailers-cooperatives, chains, independent retailers, growers, trade associations, banks, advertising agencies, government agencies, supply houses, and others working for or with the food trades comprise the AIFD. It serves as a central information service for the food trades.

American Meat Institute (AMI)
P.O. Box 3556
Washington, DC 20007
www.meatami.org
See Chapter 1.

American Meat Science Association
National Live Stock and Meat Board
444 North Michigan Ave.
Chicago, IL 60611
www.metscience.org

This organization is engaged in meat research, extension, and education in universities, industry, government, and other organizations. It encourages the exchange of ideas and information and seeks to foster education, research, and development in the field of meat science.

American Minor Breed Conservancy (AMBC)
Pittsboro, NC

The AMBC was organized in 1977 to conserve and preserve rare breeds and stocks of livestock and poultry. Some of its members raise the rare breeds. A semen bank has been established, in which Milking Devon, Dutch Belted, Florida Cracker Cattle, Dexter, Red Poll, and Lineback are among the first to be stored.

American National Cattlewomen (ANCW)
5420 South Quebec
P.O. Box 3881
Englewood, CO 80155
www.ancw.org
See Chapter 1

American Public Health Association
800 I Street, NW
Washington, DC 20001-3710
www.apha.org

Professional organization of physicians, nurses, educators, sanitary engineers, environmentalists, social workers, optometrists, podiatrists, community health specialists, pharmacists, dentists, industrial hygienists, and interested consumers. It seeks to protect and promote personal and environmental health.

American Registry of Professonal Animal Scientists (ARPAS)
1111 N. Dunlap Avenue
Savoy, IL 61874
www.arpas.org

The ARPAS certifies qualified animal scientists as credible experts in their fields based on intense scholarly preparation and practical experience.

American Society for the Prevention of Cruelty to Animals (ASPCA)
424 East 92nd Street
New York City, NY 10128
www.aspca.org

The ASPCA opposes raising or using any animal under cruel and inhumane conditions. Practices falling under this definition include veal calf farming, branding the faces of cattle, factory farming, and using animals in rodeos.

American Society of Animal Science (ASAS)
1111 N. Dunlap Ave.
Savoy, IL 61874
www.asas.org

ASAS works to increase knowledge and understanding of animals, especially farm animals, and to improve the care and productivity of animals both in commercial production and in research. Publishes the *Journal of Animal Science.*

American Veal Association
1500 Fulling Road
Middleton, PA 17057
www.vealfarm.com

Represents the more than 1,300 family veal farmers, producers, and associated industry members.

The Animal Transportation Association
P.O. Box 631206
Houston, TX 77263-1206
www.npscmgmt.com

It was organized in 1976 to assist in animal transporation by air. Promotes safe and humane transportation of animals by land, air, and sea.

Animal Industry Foundation (AIF)
1501 Wilson Blvd., Suite 1100
Arlington, VA 22209
www.aif.org

AIF attempts to redefine animal agriculture as high-quality animal care, correcting misinformation and violence aimed at farmers, ranchers, researchers, processors, auction markets, and related businesses. It is working to be the single, clear voice speaking to the American public on behalf of livestock and poultry producers.

Animal Liberation Front (ALF)

ALF is not legally recognized as an organization. The group has no official office and it operates underground. ALF aims to liberate animals used for entertainment, food, clothing, and research. Raids, break-ins, fires, vandalism, and theft are all characteristic of ALF.

Animal Rights Organizations
http://animalconcerns.netforchange.com

See American Society for the Prevention of Cruelty to Animals, Animal Liberation Front, Animal Welfare Institute, Earth First, Farm Animal Reform Movement, Farm Sanctuary, Humane Farming Association, Humane Society of the United States, and People for the Ethical Treatment of Animals.

Beef AI Organizations

AI organizations are the primary sources of commercial semen used in artificial insemination programs. These organizations publish sire directories, which list bulls and their available semen.

ABS Global, Inc.
1525 River Road
Deforest, WI 53532
www.absglobal.com

Accelerated Genetics
E10890 Penny Lane
Baraboo, WI 53913
www.accelgen.com

Alta Genetics, Inc.
R.R. 2, Balzae
Alberta, Canada TOM OEO
www.altagenetics.com

Elgin Breeding Service
P.O. Box 68
Elgin, TX 78621

Genetic Cooperative, Int'l.
Cooperative Resources
100 MBC Drive
Shawano, WI 54166
(Contact for affiliates and regional centers.)
www.crinet.com

Genetic Horizons
151 E. Drippings Springs Road
Columbia, MO 65202
www.semexusa.com

KABSU
1401 College Avenue
Manhattan, KS 66502

Select Sires, Inc.
11740 U.S. 42
Plain City, OH 43064
(Select Sires is a group of 10 co-ops located throughout the U.S. Contact Select Sires for their addresses.)
www.selectsires.com

There are additional AI organizations that do custom semen collection, while certain organizations sell semen for others. Some of these organizations are

Cottage Farm Genetics, 971 Old Bells Road, Jackson, TN 38305
Frontier Genetics (88), 80756 Hickey Lane, Hermiston, OR 97838
Great Lakes Sire Service, Inc. (179), Himebaugh Road, Bronson, MI 49028
Great Plains Breeders Service/Taurus (157), Box 468, N. Hwy. 83, Shamrock, TX 79079
Hawkeye Breeders Service (54), 3257 Old Portland Road. Adel, IA 50003, www.hawkeyebreaders.com
High Plains Genetics Research (75), HC 80, Box 835-10, Piedmont, SD 57769
Hoffman AI Breeders (99), 1950 S. Hwy. 89-91, Logan, UT 84321, www.hoffmanaibreeders.com
Interglobe Genetics (138), 14814 N. 1500 E. Road, Pontiac, IL 61764
JLG Enterprises (100), P.O. Box 1375, (11116 Sierra Road), Oakdale, CA 95361
Nebraska Bull Service, Inc. (106), P.O. Box 70, Wellfleet, NE 69170
Nichols Cyro-Genetics (137), 8827 N.E. 29th, Ankeny, IA 50021
Nokota Genetics(34), 6921 Hwy. 83 N., Minot, ND 58701-0241
North American Breeders (36), P.O. Box 228, Berryville, VA 22611
Reproduction Entp., Inc. (109) 908 N. Prairie Road, Stillwater, OK 74074, www.reprod-ent.com

Roberts Cattle Services (178), 6028 Victoria Lane, Billings, MT 59106
Rocky Mtn. Sire Service (59), 1616 Manila Road, Bennett, CO 80102
Sire Management Services LLC (41), 355 Hwy. 26 E., Elko, GA 31205, www.sales-synergy.com
Sire Technology (140), 5001 County Line Road, Springfield, OH 45502, www.siretech.com
Southeastern Semen Services (153), 16878 45th Street, Wellborn, FL 32094
Taurus Service, Inc. (76), P.O. Box 164, (Grist Flat Road), Mehoopany, PA 18629, www.Taurus-service.com
Ultimate Genetics/MVG (56), P.O. Box 535 (2097 CR 4516), Castroville, TX 78009, www.ultimategenetics.com
Vogler Semen Centre Lab, Inc., (129), 27104 Church Road, Ashland, NE 68003

Beef Improvement Federation (BIF)
Animal Science Dept.
University of Georgia
Athens, GA 30602
www.beefimprovement.org

BIF coordinates the performance testing programs of approximately 50 organizations from several states, including the national breed associations. Its primary objective is to establish accurate and uniform procedures for measuring and recording beef cattle performance data.

Breed and Breeder Associations

Members of the following associations are breeders or owners of specific breeds of cattle. Their purposes are to promote the breed, record the performance of the cattle, issue registrations, and keep the herd book. Dairy breeds are included because they are used in some crossbreeding programs, and all cattle eventually produce beef products.

American Breed
American Breed International Assoc.
306 S. Avenue A
Portales, NM 88130

Amerifax
Amerifax Cattle Association
P.O. Box 149
Hastings, NE 68901

Angus
American Angus Association
3201 Frederick Boulevard
St. Joseph, MO 64501
www.angus.org

Ankole-Watusi
Ankole-Watusi International Registry
22484 W. 239th Street
Spring Hill, KS 66083
www.AWIR.com

Aubrac Alliance
611 Sudbury Drive
Columbia, MO 65203

Ayrshire
Ayrshire Breeders Association
P.O. Box 1608
Brattleboro, VT 05302-1608

Barzona
Barzona Breeders Association
Box 631
Prescott, AZ 83602

Beefalo
American Beefalo World Registry
3772 12th Street
Allegan, MI 49010

Beef Friesian
Beef Friesian Society
25377 Weld County Road 17
Johnstown, CO 80534

Beefmaster
Beefmaster Breeders Universal
6800 Park Ten Blvd.
Suite 290 West
San Antonio, TX 78213
www.beefmasters.org

Belgian Blue
American Belgian Blue Breeders, Inc.
P.O. Box 35264
Tulsa, OK 74153
www.belgianblue.org

Belted Galloway
Belted Galloway Society, Inc.
5584 Shaver Mill Road
Linville, VA 22834

Bison and Buffalo
National Bison Association
100 Livestock Exchange Bldg.
401 Marion Street
Denver, CO 80216
www.nbabison.org

Blonde D'Aquitaine
American Blonde D'Aquitaine Association
P.O. Box 12341
North Kansas City, MO 64116

Braford
United Braford Breeders
422 E. Main, Suite 218
Nacogdoches, TX 75961
www.brafords.org

Brah-Maine
International Brah-Maine Society
RR 1, Box 233
Franklin, TX 77856

Brahman
American Brahman Breeders Association
3003 S. Lomp West, Suite 540
Houston, TX 77054
www.brahman.org

Brahmousin
American Brahmousin Council
Box 12363
1912 Clay Street
North Kansas City, MO 64116

Bralers
American Bralers Association
Box 75
Burton, TX 77835

Brangus
International Brangus Breeders Association
5750 Epsilon
San Antonio, TX 78249
www.int-brangus.org

Braunvieh
Braunvieh Association of America
P.O. Box 6396
Lincoln, NE 68506
www.braunvieh.org

British White
British White Cattle Association of America
P.O. Box 281
Bells, TX 75414
www.britishwhite.org

Brown Swiss
Brown Swiss Cattle Breeders Association
Box 1038
Beloit, WI 53511

Buelingo Cattle Society
6570 S. Hwy. 215
Charleston, AR 72933
www.buelingo.com

Charbray
American-International Charolais Association
Charbray Division
Box 20247
Kansas City, MO 64195
www.charolaisusa.com

Charolais
American-International Charolais Association
Box 20247
Kansas City, MO 64195
www.charolaisusa.com

Char-Swiss
Char-Swiss Breeders Association
407 Chambers Street
Marlin, TX 76661

Chiangus
American Chianina Association
Box 890
Platte City, MO 64079
www.chicattle.org

Chianina
American Chianina Association
Box 890
Platte City, MO 64079
www.chicattle.org

Chiford
American Chianina Association
Box 890
Platte City, MO 64079
www.chicattle.org

Chimaine
American Chianina Association
Box 890
Platte City, MO 64079
www.chicattle.org

Corriente
North American Corriente Association
P.O. Box 12359
N. Kansas City, MO 64116

Cracker Cattle[a]
Florida Cracker Cattle Breeders Association
Room 428, Mayo Bldg.
Tallahassee, FL 32399

Devon
Devon Cattle Association, Inc.
1082 Richie Road
Bunkie, LA 71322

Dexter
American Dexter Cattle Association
Rt. 1, Box 378
Concordia, MO 64020

Dutch Belted
Dutch Belted Cattle Association
Box 358-Highland County
Venus, FL 33960

[a]Also referred to as Florida Scrub or Piney Woods cattle. These cattle are descendants of the Spanish cattle brought to Florida from the islands of Hispaniola, Jamaica, and Cuba. They may also contain some breeding from mixed types of cattle developed in the eastern United States from cattle brought from Europe.

Galloway
American Galloway Breeders Association
310 West Spruce
Missoula, MT 59802
www.Galloway-world.org/agba/entrol.htm

Gelbray
Gelbray International, Inc.
Rt.1, Box 273C
Madell, OK 73446

Gelbvieh
American Gelbvieh Association
10900 Dover St.
Westminster, CO 80021
www.gelbvieh.org

Geltex
Geltex Breeders Association
Rt. 1, Box 114
Taylor, TX 76574

Guernsey
American Guernsey Organization
7614 Slate Ridge Blvd.
Reynoldsburg, OH 43068

Hereford
American Hereford Association
P.O. Box 014059
Kansas City, MO 64101
www.hereford.org

Herens
American Herens Association
122 N. Court St.
Lewisburg, WV 24901

Highland
American Highland Breeders Association
200 Livestock Exchange Bldg.
Denver, CO 80216
www.highlandcattle.org

Holstein-Friesian
Holstein-Friesian Association
1 Holstein Place
Battleboro, VT 05301
www.holsteinusa.com

Irish Blacks
(Registrations recorded in Beef Friesian Society Herdbook)

Jersey
The American Jersey Cattle Association
6486 East Main
Reynoldsburg, OH 43068
www.usjersey.com

Limousin
North American Limousin Foundation
7383 South Alton Way, Box 4467
Englewood, CO 80111
www.nalf.org

Lineback
Lineback Cattle Registry Association
c/o Paul Daniels
Daniels Farm Rd.
Irasburg, VT 05845

Longhorn
Texas Longhorn Breeders Association of America
2315 N. Main, Suite 402
Fort Worth, TX 76106
www.tlbaa.org

Maine-Anjou
American Maine-Anjou Association
760 Livestock Exchange Building
Kansas City, MO 64102
www.maine-anjou.org

Mandalong Special
Tri-State Breeders
E10890 Penny Lane
Baraboo, WI 53913

Marchigiana
American International Marchigiana Society
Box 198
Walton, KS 67151-0198
www.marchigiana.org

Milking Shorthorn
American Milking Shorthorn Society
P.O. Box 449
Beloit, WI 53512

Murray Grey
American Murray Grey Association
P.O. Box 188
Mayport, PA 16240
www.murraygreycattle.org

Normande
American Normande Association
11538 Spudville Rd.
Hibbing, MN 55746

Parthenais
Parthenais Cattle Breeders Association of America
Box 550
Bells, TX 75414
www.parthenaiscattle.org

Piedmontese
Piedmontese Association of the U.S.
108 Livestock Exchange Bldg.
4701 Marion St.
Denver, CO 80216

Pinzgauer
American Pinzgauer Association
P.O. Box 147
Bethany, MO 64424
www.afr.org/~greatcow/

Red Angus
Red Angus Association
4201 I-35 North
Denton, TX 76207
www.redangus1.com

Red Brangus
American Red Brangus Association
3995 E. Highway 290
Dripping Springs, TX 76207
www.brangusassc.com

Red Poll
American Red Poll Association
Box 157
Bethany, MO 64424
www.redpollusa.org

Romagnola
American Romagnola Association
2000 Flagstone Road
Reno, NV 89520
www.americanromagnola.com

RX_3
American RX_3 Cattle Registry
4568 T. Avenue
Oelwein, IA 50661

Salers
American Salers Association
7383 Alton Way, Suite 103
Englewood, CO 80112
www.salersusa.org

Salorn
International Salorn Association
Box 198
Granby, MO 64844
www.salorn.com

Santa Cruz
King Ranch
Box 1090, 201 E. Kleberg
Kingsville, TX 78364

Santa Gertrudis
Santa Gertrudis Breeders International
P.O. Box 1257
Kingsville, TX 78363
www.sgbi.org

Senepol
Senepol Cattle Breeders Association
Box 808
Statham, GA 30606
www.senapolecattle.com

Shorthorn/Polled Shorthorn
American Shorthorn Association
8288 Hascall St.
Omaha, NE 68124
www.beefshorthornusa.com

Simbrah
American Simmental Association
1 Simmental Way
Bozeman, MT 59715
www.simmgene.com

Simmental
American Simmental Association
1 Simmental Way
Bozeman, MT 59715
www.simmgene.com

South Devon
North American South Devon Association
2514 Avenue F
Sante Fe, TX 77520
www.southdevon.com

Tarentaise
American Tarentaise Association
P.O. Box 34705
N. Kansas City, MO 64116

Texon
International Texon Cattle Association
Rt. 1, Box 163R
Duncan, OK 73533

Tuli
North American Tuli Association
424 Tarrow
College Station, TX 77840

Wagyu
American Wagyu Association
Box 4071
Bryan, TX 77805
www.wagyu.com

Water Buffalo
American Water Buffalo Association
Box 13533
University of Florida
Gainesville, FL 32604

Watusi
World Watusi Association
P.O. Box 14
Crawford, NE 69339
www.watusicattle.com

Welsh Black
Welsh Black Cattle Association
208 N. Hymera East
Shelburn, IN 47879

White Park
American British White Park Cattle Association
2173 170th Street
West Point, IA 52656
www.amer-britishwhitepark.org

White Park Registry
HC 87, Box 2214
Big Timber, MT 59011

Zebu
International Zebu Breeders Association
1901 Miller Road
Rowlett, TX 75088

Bureau of Labor Statistics
Division of Information Services
2 Massachusetts Avenue, N.E., Room 2860
Washington, DC 20212
www.bls.gov

Cattle-Fax
Cattle Marketing Information Service, Inc.
Highland Place II—9110 East Nichols Avenue
Centennial, CO 80112
www.cattle-fax.com

A nonprofit corporation governed by cattle producers. Cattle-Fax has a staff of skilled, knowledgeable market analysts that provides local, state, regional, national, and international market information and analysis as well as analysis of factors affecting the market.

Centers for Disease Control and Prevention
Washington Office
200 Independence Avenue, S.W., Room 746G
Washington, DC 20212
www.cdc.gov

The federal agency responsible for protecting the public health of the nation by preventing unnecessary disease, disability, and premature death by promoting healthy lifestyles.

Chicago Mercantile Exchange (CME)
30 S. Wacker Dr.
Chicago, IL 60606
www.cme.com
See Chapter 17

Community Nutrition Institute (CNI)
910 17th Street, N.W., #143
Washington, DC 20006
www.cgc.apc.org

CNI is a nonprofit citizen's organization specializing in food and nutrition.

Consumer Federation of America (CFA)
1424 16th Street, N.W.
Washington, DC 20036
www.consumerfed.org

CFA is a federation of organizations that advances a proconsumer policy before Congress, the administration, regulatory agencies, and the courts. CFA is called one of the ten most influential lobbying groups in Washington.

Council for Agricultural Science and Technology (CAST)
4420 West Lincoln Way
Ames, IA 50014-3447
www.cast-science.org

CAST organizes task forces of food and agricultural scientists and technologists from the relevant disciplines to assemble and interpret factual information on food and agricultural issues of public concern. It disseminates this information in a usable and effective form to the public, news media, and government as appropriate.

Earth First
P.O. Box 3023
Tucson, AZ 85703
www.earthfirstjournal.com

A radical animal rights and environmental group that supports the removal of livestock from all public lands. Individual chapters plan their own agendas as there is no professional staff, formal leadership, or board of directors.

Ecoresults!
P.O. Box 23713
Flagstaff, AZ 86002
www.ecoresults.com

Ecoresults is an organization designed to promote restoration of the natural environment. This organization focuses on collaboration and teamwork to solve problems in environmental management.

Environmental Protection Agency
See U.S. Environmental Protection Agency (EPA).

Farm Animal Reform Movement (FARM)
P.O. Box 30654
Bethesda, MD 20824
www.farmusa.org

FARM is a vegetarian-oriented animal rights organization.

Farm Sanctuary
P.O. Box 150
Watkins Glen, NY 14891
www.farmsanctuary.org

An animal rights organization founded to end factory farming.

Federal Trade Commission (FTC)
600 Pennsylvania Avenue, N.W.
Washington, DC 20580
www.ftc.gov

The FTC works to keep the free enterprise system from becoming stifled by monopolies or restraints on trade or corrupted by unfair or deceptive trade practices.

Federation of Scientific Agricultural Societies (FSAS)

The FSAS represents scientific agricultural societies, particularly in identifying national goals and priorities related to education and research in food, agriculture, and natural resources, and in urging executive and legislative support of the priority areas.

Food and Drug Administration (FDA)
Department of Health, Education, and Welfare
5600 Fishers Lane
Rockville, MD 20857
www.fda.gov

Activities relate to protecting public health as it may be impaired by foods, drugs, biological products, cosmetics poisons, pesticides, and food additives. FDA ensures that foods are safe, pure, and wholesome.

Food and Drug Law Institute
1000 Vermont Avenue, N.W., Suite 200
Washington, DC 20036
www.fdli.org

Manufacturers and distributors of food, drugs, and cosmetics are the members of this organization, which promotes knowledge about the laws governing these and other products.

Food Marketing Institute (FMI)
Suite 700
1750 K Street, N.W.
Washington, DC 20006
www.fmi.org
See Chapter 1

Grazing Lands Conservation Institute (GLCI)
501 W. Felix Street, FWFC Bldg. 23
Fort Worth, TX 76115-3494
www.glci.org

GLCI provides high-quality technical assistance to private landowners and works to increase awareness of the importance of grazing resources.

Humane Farming Association (HFA)
P.O. Box 3577
San Rafael, CA 94912
www.hfa.org

An animal rights organization that opposes stall-raised veal and battery cage layers.

Humane Society of the United States (HSUS)
Farm Animals and Bioethics Division
2100 L Street, N.W.
Washington, DC 20037
www.hsus.org

Largest animal protection and rights organization in the United States. Activities include public education, investigation of reported abuse, and legislation. Supports the elimination of animal drug use, artificial supplementation of feeds, and auction markets.

Institute of Food Technologists (IFT)
221 North LaSalle, Suite 300
Chicago, IL 60601-1291
www.ift.org

IFT is a worldwide society of professional food technologists, scientists, engineers, executives, and educators in the field of food technology. Other individuals interested in food technology because of their work in closely related fields are also members.

Inter-American Confederation of Cattlemen
Avenue Justo Arosemena Y Calle 32, Segundo Piso, Edif. Vallarino
Panama City, Panama

Federation of associations of cattle operators in the Americas that works to promote the welfare of the cattle industry, exchange information, and organize forums.

International Embryo Transfer Society
www.iets.uiuc.edu

An organization of professional persons with an interest in animal embryo transfer.

Leather Industries of America
2501 M Street, N.W., Suite 350
Washington, DC 20037
www.leathernet.com

A trade association that represents tanners and allied industries.

National Institute for Animal Agriculture (NIAA)
1910 Lyda Avenue
Bowing Green, KY 42104
www.animalagriculture.org

NIAA serves as a clearinghouse for all sectors of the livestock and meat industry in sponsoring research and educational programs designed to eradicate diseases among livestock, to promote a safe and wholesome food supply, and to promote best practices in environmental stewardship, animal health, and well-being.

Livestock Marketing Association (LMA)
7509 Tiffany Springs Parkway
Kansas City, MO 64153
www.lmaweb.com

The LMA's Trade Group is composed of five interrelated organizations that promote the maintenance of a free, competitive marketing sector. The Trade Group's other major activities are to develop and provide commercial services for its subscribers.

Livestock Publications Council
910 Currie Street
Fort Worth, TX 76107
www.livestockpublications.com/index.htm

Among the many objectives of this organization are (1) to promote understanding and cooperation among publications serving the livestock industry; (2) to encourage and support research and activities designed to further the livestock industry; (3) to foster, through cooperative effort, relations among publishers, legislators, government administrators, and people in all segments of the livestock industry and allied enterprises.

National Academy of Sciences (NAS)
2101 Constitution Ave., N.W.
Washington, DC 20418
www.nas.edu

The NAS is a private honorary organization that serves as an independent adviser to the federal government on matters of science and technology. In each case, such counsel is provided as a formal report of the deliberations of a study project and its conclusions and recommendations; all reports are publicly available. The NAS created the National Research

Council (NRC) as part of its structure; it is the operating arm of the NAS and currently comprises some 900 committees and panels. Each study made by the NAS is conducted by one or more of the NRC committees or panels, the costs being met out of contracts among the NAS, the government, and private foundations.

National Agri-Marketing Association
11020 King Street, Suite 205
Overland Park, KS 66210
www.nama.org

This association is made up of persons engaged in agricultural advertising and marketing for manufacturers, advertising agencies, and the media. It promotes high standards of agri-marketing, provides for the exchange of ideas, and encourages study and better understanding of agricultural advertising, selling, and marketing.

National Agricultural Marketing Officials
c/o CA Department of Food and Agriculture
1200 "N" Street, Room A270
Sacramento, CA 95814
www.naamo.org/default.htm

State officials responsible for the administration of state agricultural marketing programs work through this organization to improve the marketing, handling, storage, processing, transportation, and distribution of North American agricultural products worldwide.

National Industrial Transportation League
1700 N. Moore Street, Suite 1900
Arlington, VA 22209-1904
www.nitl.org

Truckers, truck brokers, and others concerned with long-haul agricultural trucking are members of this organization.

National Association of Animal Breeders (NAAB)
P.O. Box 1033
Columbia, MO 65205
www.naab-css.org

Farmer cooperatives and private businesses interested in the improvement of farm livestock make up the NAAB, which works to stimulate and encourage research in artificial insemination and reproduction. It regularly establishes guidelines for AI, certified semen services (CSS), disease prevention, CSS health programs, and state and foreign health requirements.

National Auctioneers Association
8880 Ballentine
Overland Park, KS 66214
www.auctioneers.org

This association of professionals works to promote and advance the auction profession and the mutual interests of its members.

National Cattlemen's Beef Association (NCBA)
5420 S. Quebec Street, P.O. Box 3469
Greenwood Village, CO 80111
www.beef.org
See Chapter 1

National Consumer's League (NCL)
1701 K Street, N.W., Suite 1201
Washington, DC 20006
www.nclnet.org

NCL targets regulations on meat and poultry inspection, which questions the effectiveness of the U.S. meat and poultry inspection program to ensure a safe meat supply.

National Farm-City Council, Inc.
225 Touhy Avenue
Park Ridge, IL 60068
www.edu-ag.org/farmcity.html

Its purpose is to bring about better understanding between the rural and urban segments of society. Members of the council are organizations and individuals prominent in the agribusiness complex.

National Grain and Feed Association (NGFA)
1250 Eye Street, N.W.
Suite 1003
Washington, DC 20005-3922
www.ngfa.org

The NGFA is the national spokesgroup for grain, feed, and realted commercial industries.

National Food Brokers Association (NFBA)
1010 Massachusetts Avenue, N.W., 6th Floor
Washington, DC 20001

The NFBA represents qualifed firms and persons who sell food and nonfood products (at the wholesale-buying level) for a commission or brokerage fee.

National Grange
1616 H Street, N.W.
Washington, DC 20006
www.nationalgrange.org

A fraternal organization of rural families. It promotes general welfare and agriculture through legislative, social, education, community service, home economics, youth, aid for handicapped, cooperatives, insurance, and credit-union programs.

National Hot Dog and Sausage Council
P.O. Box 3556
Washington, DC 20007
www.hot-dog.org

This council conducts public-relations campaigns on behalf of hotdog and sausage manufacturers.

National Meat Association (NMA)
1400 16th Street, N.W., Suite 400
Washington, DC 20036
www.nmaonline.org

The NMA represents independent meatpackers and meat processors and suppliers to the industry. It has divisions in beef, pork, and processed meats.

National Meat Canners Association
P.O. Box 3556
Washington, DC 20007
www.meatami.org

Its members are 35 companies whose primary or secondary business is sterile processed meat products.

National Restaurant Association
1200 17th Street, N.W.
Washington, DC 20036
www.restaurant.org

This association represents restaurants, cafeterias, clubs, contract feeders, drive-ins, caterers, institutional food services, and other members of the food-service industry. It supports food-service education and research in several educational institutions, conducts traveling management courses and seminars for restaurant personnel, and distributes educational pamphlets, books, and films about the industry.

National Renderers Association, Inc.
801 N. Fairfax Street, Suite 207
Fairfax, VA 22314
www.rederers.org

Trade association of by-product rendering companies.

North American Meat Processors Association
1910 Association Drive
Reston, VA 20191
www.namp.com

A nonprofit organization made up of hotel, restaurant, and institution (HRI) supply houses that purvey meats and other food items to food service establishments and that have joined together cooperatively to enhance the interests of the industry.

Occupational Safety and Health Administration (OSHA)
U.S. Department of Labor
200 Constitution Avenue, N.W.
Washington, DC 20210
www.osha.gov

OSHA was established pursuant to the Occupational Safety and Health Act of 1970. It develops and promulgates occupational safety and health standards, develops and issues regulations, and conducts investigations and inspections to determine the status of compliance with safety and health regulations.

Public Citizen—Congress Watch
215 Pennsylvania Avenue, S.E.
Washington, DC 20003
www.citizen.org/congress

Nonpartisan political interest group that monitors consumer rights, governmental and corporate accountability, campaign finance reform, and environmental, safety, and labor issues.

Public Lands Council
425 13th Street, N.W., #1020
Washington, DC 20004

Established in 1969, the council represents the interests of ranchers who hold leases and permits to graze livestock on public lands in the western states. It is dedicated to the principle of sound management of federal lands for grazing and all other multiple-use purposes. Membership of the council includes both individual livestock producers and sheep and cattle organizations in the 13 western states.

Samuel Roberts Noble Foundation (NF)
P.O. Box 2180
Ardmore, OK 73402
www.noble.org

NF conducts agricultural, forage, biotechnical, and plant biology research. It also provides farmer-rancher consultation services.

Sierra Club
85 Second Street, 2nd Floor
San Francisco, CA 94105-3441
www.sierraclub.org

This organization has intensive lobbying efforts that influence government decisions on use of land and natural resources. The Sierra Club is involved with the major conservation issues.

Society for Range Management
445 Union Blvd., Suite 230
Lakewood, CO 80228
www.srm.org

This is an educational and research group that promotes the understanding of rangeland and its ecosystems and uses.

State Beef Council

Contact the National Cattlemen's Beef Association (www.beef.org) for addresses for specific states.

State Cattle and Feeders Association

Contact the National Cattlemen's Beef Association for addresses for specific states. Several states have their own state cattle association that includes both cow-calf producers and feeders; other states have separate organizations representing each group. Members discuss and take action on legislation, research, marketing, and other current issues influencing beef producers in their respective states and in the nation. Most state cattle organizations are affiliated with the National Cattlemen's Beef Association (NCBA) (www.beef.org).

The Allan Savory Center for Holistic Management
1010 Tigeras, N.W.
Albuquerque, NM 87102
www.holisticmanagement.org

The Savory Center is focused on restoration of landscapes and the lives of people dependent on them using practical, low-cost problem solving.

U.S. Beef Breeds Council
(No established office—correspondence is addressed to the current president's office—contact major breed association for current address.)

It functions to create a more unified effort by purebred breeders and to promote beef and the industry.

U.S. Hide, Skin, and Leather Association (USHSLA)
1700 N. Moore Street, Suite 1600
Arlington, VA 22209
www.meatami.com

USHSLA is the exclusive representative of the hide and skin industry. Its goals are to promote the marketing and processing of these primary by-products.

U.S. Department of Agriculture (USDA)
Washington, DC 20250
www.USDA.gov

The USDA works to improve and maintain farm income and develop and expand markets abroad for agricultural products. It also helps curb poverty, hunger, and malnutrition; and works to enhance the environment and maintain production capacity by helping landowners protect soil, water, forests, and other natural resources. Rural development, credit, and conservation programs are key resources for carrying out national growth policies. The USDA, through inspection and grading services, safeguards and assures standards of quality in the daily food supply. See the USDA organization chart in Figure A.6.

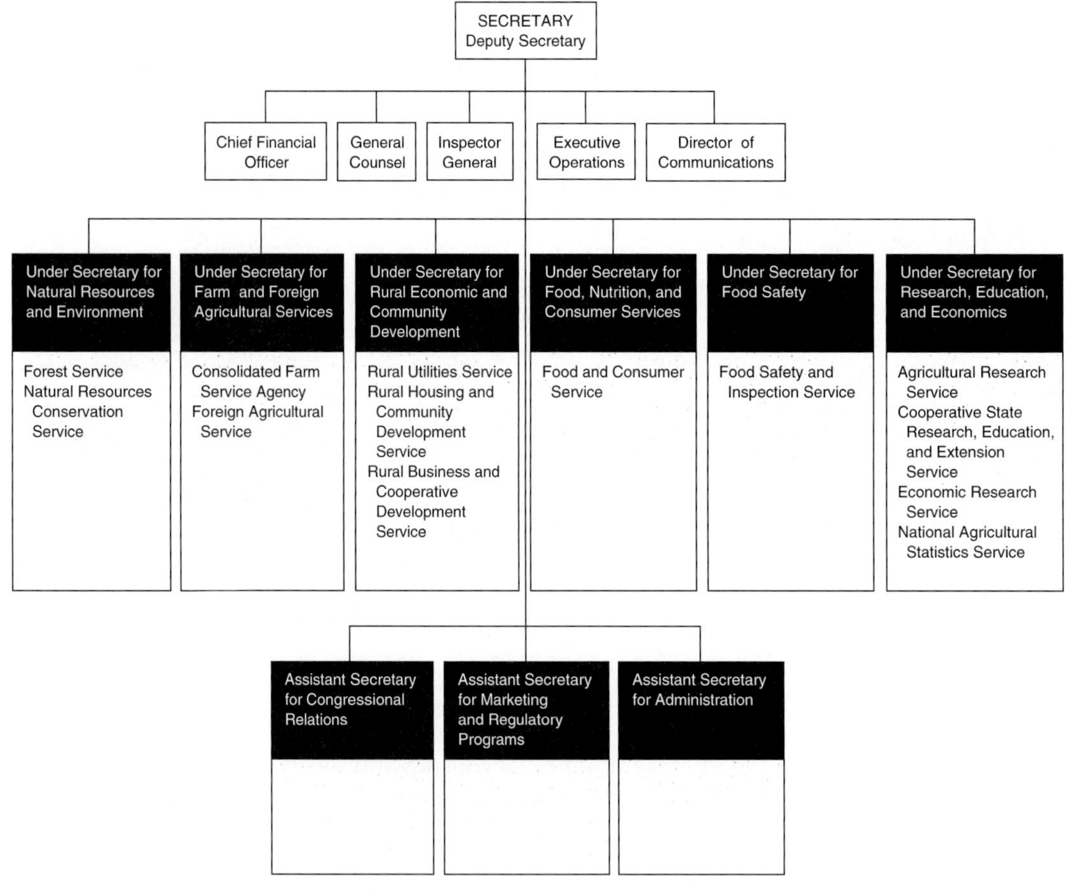

FIGURE A.6 United States Department of Agriculture organization chart.
Source: USDA.

USDA—Agricultural Marketing Services (AMS)
www.ams.usda.gov

AMS provides market news reports; develops quality grade standards for agricultural commodities; provides voluntary grading services for livestock, meat, poultry and more; administers marketing regulatory programs, marketing agreements and orders, and research and promotion programs; administers national organic standards activities; administers federal-state marketing improvement programs, wholesale facilities research programs, and food purchases; and is the coordinator for USDA's pesticide data program activities.

USDA—Agricultural Research Service (ARS)
www.ars.usda.gov

ARS conducts mission-oriented research to ensure adequate protection of food and agricultural products to meet nutritional and other needs of American consumers in animal and plant sciences including disease and pest controls, soil and water conservation, postharvest processing and storage of commodities, safety of food, human nutrition, and integration of agricultural systems.

USDA—Animal and Plant Health Inspection Service (APHIS)
www.aphis.usda.gov

APHIS administers regulatory programs to control or eradicate animal and plant pests and diseases; enforces domestic and port-of-entry agricultural quarantines; licenses and ensures safety and effectiveness of veterinary biological products; enforces the Animal Welfare and Horse Protection acts; administers programs providing protection to livestock and crops from depredation by rodents, birds, and predatory animals; and conducts cooperative programs to eradicate animal and plant pests and diseases in other countries.

USDA—Economic Research Service (ERS)
www.ers.usda.gov

ERS conducts research in domestic and foreign agricultural economics; analyzes factors affecting agriculture, farm productivity, financing, use of resources, and potentials of rural areas; evaluates marketing potential and development and marketing costs; studies U.S. trade in agricultural products and the role of agriculture in economic development of other countries; and reports agricultural situation and outlook, commodity projections, price spreads, and analysis of U.S. farm commodity programs.

USDA—Extension Service
www.reeusda.gov

The Extension Service serves as a partner with state governments through its land-grant universities and the county governments, forming the Cooperative Extension Service. These levels of government finance and conduct educational programs to help the public learn about and apply the latest technologies and management techniques developed through the research of the land-grant universities, the USDA, and other sources. Its major areas of assistance include agricultural production, marketing, natural resources, home economics, food and nutrition, 4-H youth development, and community and rural development.

USDA—Farm Service Agency (FSA)
www.fsa.usda.gov

The Farm Service Agency administers commodity production adjustment and support programs; conservation cost-sharing with farmers and ranchers; the conservation reserve program; natural disaster assistance to agricultural producers through payments and cost-sharing; and certain national emergency preparedness activities. The agency also provides services for the Commodity Credit Corporation and for state and county committees.

USDA—Food Safety and Inspection Service
www.usda.gov/agency/fsis

FSIS administers the federal meat and poultry inspection program to ensure safety, wholesomeness, and truthful labeling of meat and poultry products; and conducts food-safety consumer educational programs.

USDA—Foreign Agricultural Service (FAS)
www.ffas.usda.gov

FAS serves as a basic source of information to U.S. agriculture on world crops, policies, and markets; administers agricultural import regulations; assists in the export of U.S. farm products; represents U.S. agriculture in foreign trade matters; and administers USDA's responsibility for P.L. 480, CCC export credit programs, and the reporting of export sales.

USDA—Forest Service
www.fs.fed.us

The Forest Service has the federal responsibility for national leadership in forestry. Some of its goals are generation of forestry opportunity to accelerate rural community growth and encouragement of optimum forest land-ownership patterns.

USDA—Natural Resources Conservation Service (NRCS)
www.nres.usda.gov

The Natural Resources Conservation Service carries out a national soil and water conservation program with cooperation of landowners and operators in local soil and water conservation districts, and with other governmental agencies; administers USDA's Great Plains conservation program, and watershed protection and flood prevention program; leads national cooperative soil survey; and provides USDA leadership in assisting landowners and local groups in resource conservation and development projects.

USDA—Packers and Stockyards Administration (P&SA)
www.usda.gov/gipsa

P&SA enforces the Packers and Stockyards Act, an antitrust, fair-trade practice and payment protection law, designed to ensure free and open competition and prevent unfair and deceptive practices in the marketing of livestock, meat, and poultry.

USDA—Poison Plant Research Laboratory
1150 E. 14th N.,
Logan, UT 84321
www.pprl.usu.edu

This laboratory conducts research on major problems associated with poisonous plants.

USDA—U.S. Meat Animal Research Center (MARC)
www.marc.usda.gov

MARC develops, through research, new technology for meat animal production. Its goals are to improve carcass merit and reduce production costs of cattle, sheep, and hogs. The center is being developed by the federal government and is administered by the Agricultural Research Service of the USDA.

U.S. Environmental Protection Agency (EPA)
1200 Pennsylvania Avenue, N.W.
Washington, DC 20460
www.epa.gov

The EPA was created to permit coordinated and effective governmental action on behalf of the environment. It endeavors to abate and control pollution systematically by proper integration of a variety of research, monitoring, standard setting, and enforcement activities. The EPA is designed to serve as the public's advocate for a livable environment.

U.S. Meat Export Federation (USMEF)
Independence Plaza
1050 17th Street, Suite 2200
Denver, CO 80265
www.usmef.org
See Chapter 1

RECORDS SYSTEMS AND SOFTWARE

Beef Record System
American Angus Association
3201 Frederick Avenue
St. Joseph, MO 64506
www.angus.org

Cattle Pro 2000—Prime Plus, Prime, Choice Plus, Choice, Select
Bowman Farm Systems, Inc.
Rt. 5, Box 25
Cynthiana, KY 41031
www.cattlepro.com

Chaps III and Dataline
North Dakota State University
1089 State Avenue
Dickinson, ND 58601
www.CHAPS2000.com

Cow-Calf Commercial and Cow-Calf Production
Red Wing Business Systems
P.O. Box 19
Red Wing, MN 55066
www.redwingsoftware.com

Cow Calf 5
University of Nebraska
Great Plains Veterinary Educational Center
Box 187
Clay Center, NE 68933
www.gpvec.unl.edu

CowSense
Midwest MicroSystems, Inc.
208 East 6th Street
Ainsworth, NE 69210
www.miswestmicro.com

Farmstock
Farm Works Software
CTN Data Service Software, Inc.
P.O. Box 250
Hamilton, IN 46742-0250
www.farmworks.com

Gelbvieh Herd Track Software
American Gelbvieh Association
10900 Dover Street
Westminster, CO 80021
www.gelbvieh.org

Herd Handler 2000
American Simmental Association
1 Simmental Way
Bozeman, MT 59718
www.simmental.org

PC-Cowcard
University of Nebraska
P.O. Box 830918
Lincoln, NE 68583-0918
www.lanr.unl.edu

Standardized Performance Analysis (SPA)
Dept. of Agricultural Economics
Texas A&M University
2124 TAMU
College Station, TX 77843-2124
http://agecoext.tamu.edu/spa/

Index

TABLE C Quality Assurance Marketing Code of Ethics

I will only participate in marketing cattle that:

– Do not pose a known public health threat

– Have cleared proper withdrawal times

– Do not have a terminal condition (Including advanced lymphosarcoma, septicemia, etc.)

– Are not disabled

– Are not severely emaciated

– Do not have uterine/vaginal prolapses with visible fetal membrane

– Do not have advanced eye lesions

– Do not have advanced Lumpy Jaw

Furthermore, I will:

Do everything possible to humanely gather, handle, and transport cattle in accordance with accepted animal husbandry practices.

Finally, I will:

Humanely euthanize cattle when necessary to prevent suffering and to protect public health.